S. Braun, H.-O. Kalinowski, S.

150 and More Basic N

Related Titles from WILEY-VCH:

H. Friebolin
Basic One- and Two-Dimensional NMR Spectroscopy
3rd ed., 1998. XXIV, 386 pages with 170 figures and 49 tables.
Softcover. ISBN 3-527-29513-5

From the new Series *Spectroscopic Techniques: An Interactive Course*

E. Pretsch / J. T. Clerc
Spectra Interpretation of Organic Compounds
1997. XIII, 175 pages. Hardcover with CD-ROM.
ISBN 3-527-28826-0

P. Bigler
NMR-Spectroscopy: Processing Strategies
1997. XVII, 249 pages with 115 figures. Hardcover with CD-ROM.
ISBN 3-527-28812-0

U. Weber / H. Thiele/G. Hägele
NMR-Spectroscopy: Modern Spectral Analysis
1998. XVI, 393 pages with 198 figures. Hardcover with CD-ROM.
ISBN 3-527-28828-7

In preparation:

T. Jenny
NMR-Spectroscopy: Data Acquisition
1999. Approx. 250 pages. Hardcover with CD-ROM.
ISBN 3-527-28827-9

J. Fröhlich
NMR-Spectroscopy: Intelligent Data Management
1999. Approx. 250 pages. Hardcover with CD-ROM.
ISBN 3-527-28859-7

S. Braun, H.-O. Kalinowski, S. Berger

150 and More Basic NMR Experiments

A Practical Course – Second Expanded Edition

Weinheim · New York · Chichester · Brisbane · Singapore · Toronto

Siegmar Braun
Institut für Organische Chemie
Technische Hochschule Darmstadt
Petersenstraße 22
D-64287 Darmstadt

Hans-Otto Kalinowski
Institut für Organische Chemie
Justus-Liebig-Universität Gießen
Heinrich-Buff-Ring 58
D-35392 Gießen

Stefan Berger
Institut für Analytische Chemie
Universität Leipzig
Linnéstraße 3
D-04103 Leipzig

1st edition 1996 ("100 and More Basic NMR Experiments")
2nd expanded edition 1998

Library of Congress Card No. applied for.

A catalogue record for this book is available from the British Library.

Die Deutsche Bibliothek – CIP-Einheitsaufnahme
Braun, Siegmar:
150 and more basic NMR experiments : a practical course / S. Braun ; H.-O. Kalinowski ; S. Berger. – 2., expanded ed. – Weinheim ; New York ; Chichester ; Brisbane ; Singapore ; Toronto : Wiley-VCH, 1998
1. Aufl. u.d.T.: Braun, Siegmar: 100 and more basic NMR experiments
ISBN 3-527-29512-7

Printed on acid-free and chlorine-free paper.

Printing: betz-druck, D-64291 Darmstadt
Bookbinding: Großbuchbinderei Wilh. Osswald, D-67433 Neustadt/Wstr.
Printed in the Federal Republic of Germany

Preface

The 50th anniversary of NMR spectroscopy was celebrated a few years ago, and it is fascinating to observe how this method is still developing and forging further ahead.

In the field of structure elucidation of chemical, especially organic, compounds, progress occurs in such large strides and new experiments and new pulse sequences are presented in such rapid succession that a normal user or even an advanced practitioner may lose sight of the overall picture, and many a novice may give up in view of the overwhelming variety of methods.

It was therefore the aim of our first book in this field, which was published in early 1996 and entitled "100 and More Basic NMR Experiments", to give a compass, a kind of Ariadne's thread, to enable the reader to chart a straightforward path through the field of NMR experiments for structure determination. It was a work-book with 115 experiments described in detail to enable the reader to learn by doing.

Encouraged by the wide acceptance and the good resonance of that book, we decided to publish a new and up-to-date edition taking into account the newest developments, introducing a new chapter with an introduction to the field of solid-state NMR, and adding further established NMR experiments of practical importance, often taking up suggestions made by users of our first book. The inclusion of merely educational experiments has been found to be valuable and is kept up. As before, the material presented here originates from our various educational and advanced training activities.

The book on hand is now much enlarged and contains 162 experiments arranged in 14 chapters:

An introductory chapter on the FT NMR spectrometer and on practical aspects such as probe-head tuning, lock operation and shimming is followed by descriptions of the experiments, mostly on ^{1}H and ^{13}C, arranged in chapters according to common features, each with a short survey at the beginning. The content ranges from pulse-length determinations, routine spectra and test procedures, through decoupling techniques, variable temperature and LIS measurements, to 1D multipulse sequences and the observation of heteronuclides such as ^{6}Li, ^{15}N and ^{17}O. More demanding experiments include those using selective pulses, introducing the second and third dimension, applying field gradients and, as a new topic, observing solid samples. Because of the newly recognized importance of field gradients in high-resolution NMR, two separate chapters have now been devoted to this area.

Again all experiments have been specially performed for this book, exactly as described and depicted. Three compounds have been chosen as the main demonstration samples, namely chloroform, ethyl crotonate, and strychnine (the latter should be used as a sealed sample).

Since our standard scheme seems to be proven, it has been further applied for the detailed description of the experiments, organized in the following specific sections:

Purpose explains the idea and goal of an experiment and refers the user to related experiments.

Literature presents references to the original publication and to improvements and/or sections in monographs and reviews.

Pulse Scheme and Phase Cycle gives the pulse sequence in an instrument-independent graphic and self-explanatory form and the full phase cycles of the pulses and the receiver even if the particular experiment can be performed with only one transient.

Acquisition is the main section, with instructions on the sample to be used, the spectrometer configuration, the type of program, and finally the parameters necessary for the performance of the experiment.

Processing describes how to treat the time-domain data.

Results presents the spectrum obtained by following the procedure exactly, and includes some remarks concerning the interpretation.

Comments contains an explanation of the most important steps of the pulse sequence, sometimes with a description using the product operator formalism for which elementary rules are collected in the Appendix.

Own Observations may encourage the user to add his or her own remarks, corrections or hints important for the performance of an experiment on the particular spectrometer used.

It is obvious that the acquisition program of a specific experiment cannot be given explicitly, since the languages of the various manufacturers are different. Moreover, since even one and the same manufacturer may have different dialects in use, an acquisition program is not necessarily transferable to another spectrometer of the same company. Corresponding to the spectrometers used by the authors, the nomenclature of the experimental parameters follows the Bruker notation; a glossary of abbreviations and symbols of different dialects and languages of other equally well-known manufacturers is given in the Appendix for comparison. It should be noted that the "Time Requirement" given at the beginning of the section "Acquisition" only includes the measuring time and and should only be regarded as a rough indication; the smallest time unit is 5 min.

For some of the experiments, especially the newest ones, the manufacturer's software may not contain the specific acquisition program. In this case, ask an application chemist of your manufacturer for support or, if you need the program for the particular instrument referred to in the experiment described here (AM-, AC-, AMX-, ARX- or Advance-spectrometer), just send a fax or an e-mail.

In general, the conditions for the experiments have not been optimized. In all cases the results are presented exactly as they were obtained, without cosmetic retouching; sometimes the samples used even show impurities.

Some recommendations regarding how best to use the collection presented here will help the user to get maximum benefit. In principle, one could just jump into the chapter at the experiment one wants to perform, since each experiment is self-contained. However, the novice is recommended to first read Chap. 1 and the introductory remarks to Chap. 3 to perform the standard ^{1}H and ^{13}C experiments 3.1 and 3.2 (using the current settings of 90° pulse lengths etc. for the instrument being used). Then one should determine the pulses oneself, completing at least Experiments 2.1 to 2.3 before going on. In each case, whether a beginner or not, one should read the whole description first, including the "Comments". By doing so one will get information about the context, about essential prerequisites, and about possible problems. One will also find references to experiments which require less sophisticated equipment. When planning more advanced experiments one may start with an already known one near the level of the intended experiment, as a check and in order to become familiar with the notation used in the descriptions. It should be noted that no exhaustive theory is given; the references may serve as stimuli for further studies leading to a deeper insight and understanding.

The selection of the experiments is admittedly to some extent a matter of our subjective preferences. If the reader fails to find his or her favorite experiment, he or she should not hesitate to notify us. In general we encourage all users, in order that they gain maximum benefit from this learning medium, to send comments, suggestions for improvements, or hints on mistakes and inconsistencies. e-mail has proven to be a quick and informal means for communication between the users and the authors, leading to a kind of a living work-book. In the Internet, too, you will in the future find a list, hopefully short, of the more serious bugs.

Note that not only the content and therefore the title of the book have changed, but also the address of the contact-author:

Prof. Dr. Stefan Berger
Institut für Analytische Chemie
der Universität Leipzig
Linnéstr. 3
D-04103 Leipzig
e-mail: stberger@rz.uni-leipzig.de
Fax: + 49 341-9736115 or -9711833
Internet: http://www.uni-leipzig.de/~nmr/STB/stb.html

Finally, we would like to thank many colleagues and readers for suggestions and corrections, our graduate students for helpful criticism, and Dr. R. G. Leach for proof reading and finishing the new parts of the text.

(March 1998)

S. Braun
H.-O. Kalinowski
S. Berger

Longum iter est per praecepta,

breve et efficax per exempla.

L. A. Seneca, *Ad Lucilium Epistulae Morales, VI*

Contents

Chapter 1

1 The NMR Spectrometer

1.1 Principles of an NMR Spectrometer

1.1.1 The Magnet

In most current NMR spectrometers the magnetic field is generated by a superconducting magnet (Fig. 1.1). The first stage in reaching the very low temperature needed is an outer stainless steel or aluminum dewar which contains liquid nitrogen. Typically, this has to be refilled every ten days. In practice, it is advisable to do this refilling on a fixed day every week. An inner dewar contains the superconducting coil (4) immersed in liquid helium, which has to be refilled, depending on the construction, every two to four month. The *helium* refill should be carried out only by experienced people. A room-temperature bore is fitted with the shim coils (7), providing a room temperature homogeneity adjustment, and a spinner assembly (5), which contains the turbine system for spinning the NMR sample tube. The probe-head (8) is usually introduced into the magnet from the bottom. The probe-head is connected to at least three radio frequency (r.f.) cables providing the ^{2}H lock, ^{1}H frequency, and one X-nucleus frequency. Additional devices to control temperature (heater, thermoelement, air, sometimes water to insulate the probehead from the magnet) are needed.

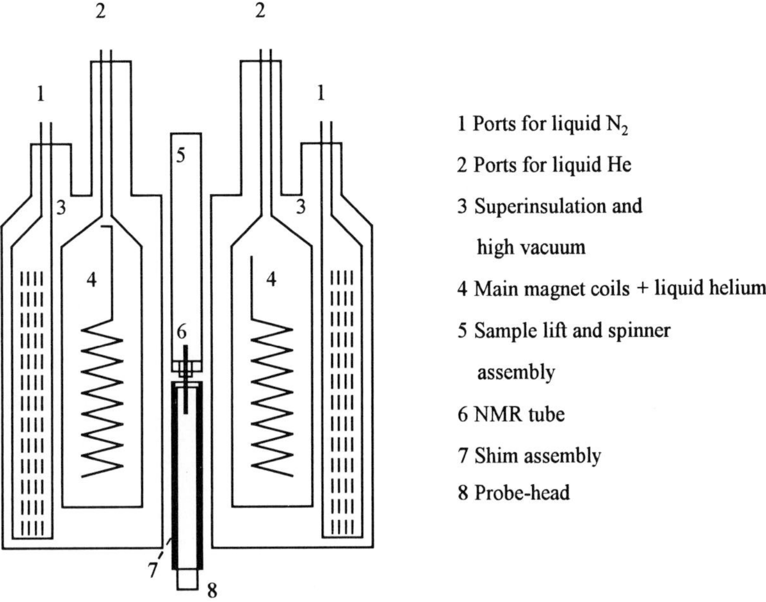

Fig. 1.1 Principles of a superconducting magnet

1.1.2 The Spectrometer Console

The spectrometer console provides at least three radiofrequency channels. Usually these frequencies are derived from digital frequency synthesizers which are phase-locked to a central quartz oscillator. These frequencies are controlled, amplified, pulsed, and transmitted to the probe-head. The various signals are preamplified, then mixed with the local oscillator frequency to yield the intermediate frequency (i.f.) The i.f. signal is further amplified, then in a second mixing stage the NMR audio signal is obtained after quadrature phase detection. The two signal components are digitized in the ADC and fed into the computer memory or, in the case of the lock signal, used for field/frequency regulation. Figures 1.2 and 1.3 show the principles of the system.

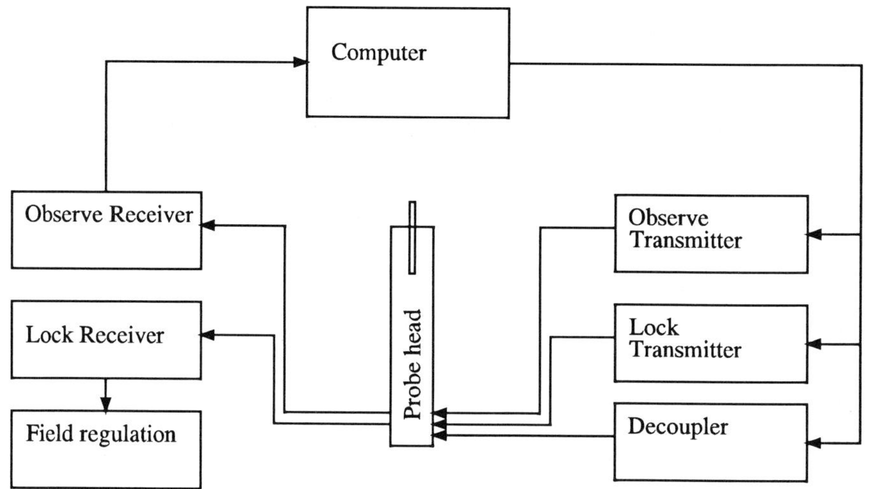

Fig. 1.2 Principles of an NMR spectrometer

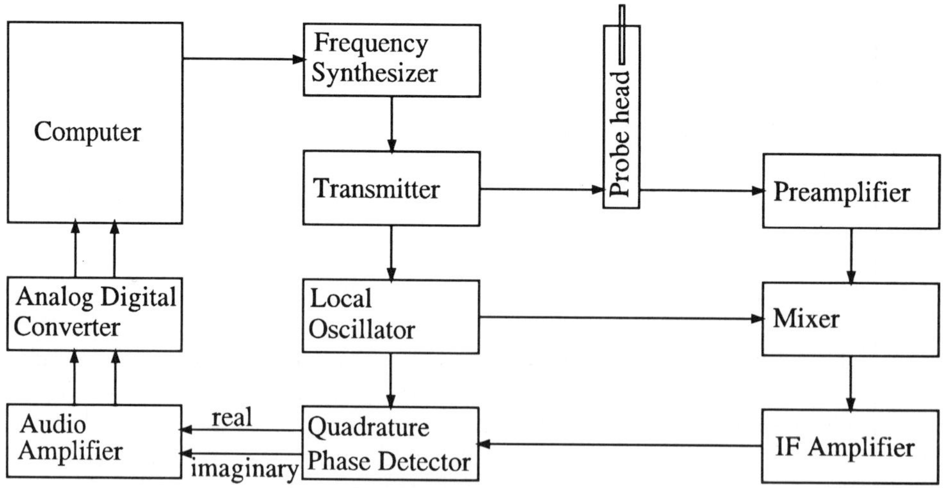

Fig. 1.3 Components of the observe channel

1.1.3 The Workstation

Modern NMR instruments are controlled by a workstation, commonly based on the UNIX operating system, however a change to Windows-NT can be foreseen. In addition, one finds a process controller integrated into the spectrometer console. The computing system has, in principle, two different tasks. First, the process controller must have on line control of many spectrometer functions such as lock, generation and timing of r.f. pulses, digitization and accumulation of the NMR signal (FID, free induction decay). Less time demanding are the other, mainly graphic, tasks in the processing of the NMR spectra. However, the massive amount of data in modern two- or multidimensional NMR spectroscopy techniques requires high storage capacities both on disk and in RAM, and a very high speed of computing.

1.1.4 Maintenance

Although most parts of modern NMR spectrometers are more or less maintenance-free, it is the experience of the authors that careful and regular checking of several components can save considerable money and time. Most important is the regular checking of the cryogens, which should be replenished on a strict schedule. Of course all magnet openings equipped with O-rings have to be carefully monitored. This is especially important for very low temperature work which can lead to icing of the O-rings. Regular checking of several hidden fans within the spectrometer console is advisable.

1.2 Tuning a Probe-Head

With a high field superconducting NMR spectrometer it is essential for obtaining a good signal-to-noise ratio, and for some advanced experiments to get any meaningful results at all, that the probe-head should be correctly tuned to the observe frequency with the particular sample of interest. There can be a huge difference depending on whether a compound is dissolved in water or in an organic solvent.

Although the construction of the resonant circuits of different probe-heads may vary considerably one has in general two capacitors to adjust, one which tunes the circuit to the desired resonance frequency (tuning) and one which performs the necessary impedance matching of the network (matching). However, these are mutually interactive and therefore they have to be adjusted in turn.

There are at least three common procedures for obtaining a global best, both on tuning and matching, and these are described below.

1.2.1 Tuning and Matching with a Reflection Meter

The simplest procedure works with a reflection meter introduced into the signal path. A tuning file on the computer should provide uninterrupted rapid pulsing in order to obtain a steady signal at the reflection meter. Turn the spinner off to avoid mechanical interference. Typical NMR parameters are p1 = 40 μs, td = 1 k, sw = 20 kHz, and ns = 64 k, which provide an acquisition time aq and hence a repetition time of 0.0256 s. Be sure that all filter boxes and other unnecessary devices are removed from the signal path in order to tune only the probe-head itself.

The goal is to minimize the strength of the reflected signal at the reflection meter. The procedure has to be done for all frequencies of the probe head starting from the lowest frequency. Thus, in the usual dual probe-heads (^{1}H and ^{13}C) adjust the ^{13}C coil first.

Typically, proceed as follows:

Obtain a first minimum on tuning and then on matching. Then alter the matching capacitor in one direction and check whether the tuning can be improved compared with the previous minimum. If yes, alter the matching further in the same direction as before and try to further improve the tuning. If not, alter the matching in the other direction and search for a better minimum, until a global minimum of both matching and tuning is obtained.

1.2.2 Tuning and Matching with an R.F. Bridge and an Oscilloscope

The probe-head is connected to an r.f. bridge and an oscilloscope as shown in Figure 1.4. The oscilloscope must be set to display the r.f . pulses. Tuning and matching is now adjusted as described above to obtain an oscilloscope pattern such that in the center of the pulse the r.f. amplitude is minimized, and only on the edges some signal is seen, as shown in Figure 1.5. The advantage of this method is that the tuning position can be found even in a grossly misadjusted probe-head. For experienced users this method is much more sensitive than using the reflection meter, but is more laborious.

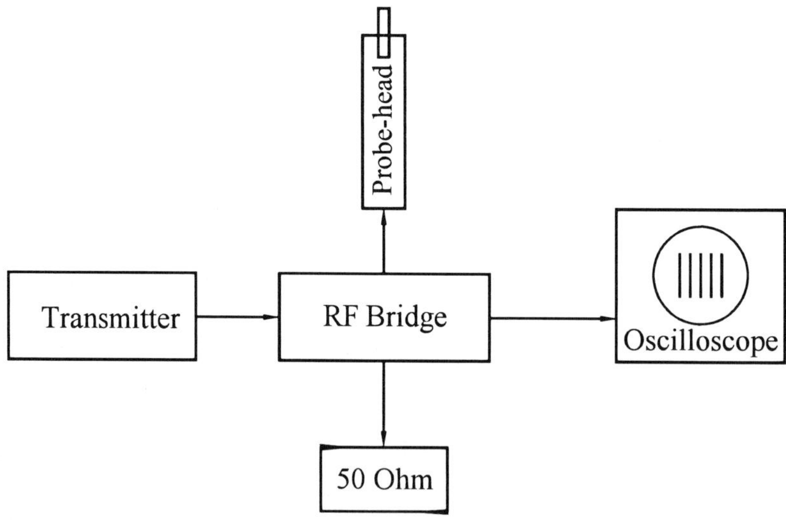

Fig. 1.4 R.f. bridge arrangement for probe-head tuning

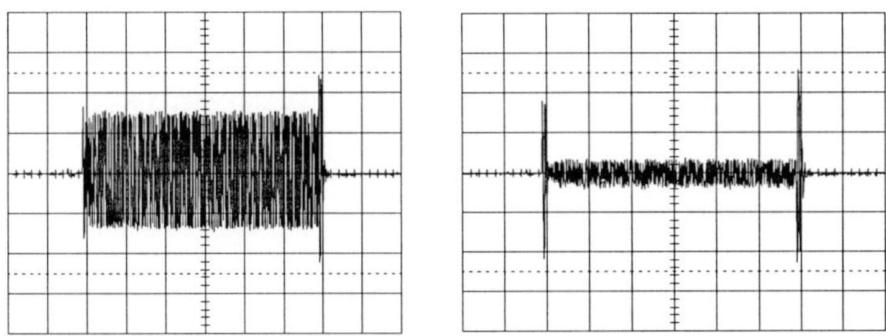

Fig. 1.5 Oscilloscope pattern before and after tuning

1.2.3 Tuning and Matching with a Wobble Generator

Professionals tune the probe head with a wobble generator, which, in addition, provides a symmetry information of the frequency dependence of the tuning. In recently built NMR spectrometers wobbling functions are programmed in the software, thus making tuning and matching a very easy process which can be followed on the computer screen. One simply has to obtain the lowest point of the wobbling curve. Such a curve is shown in Figure 1.6.

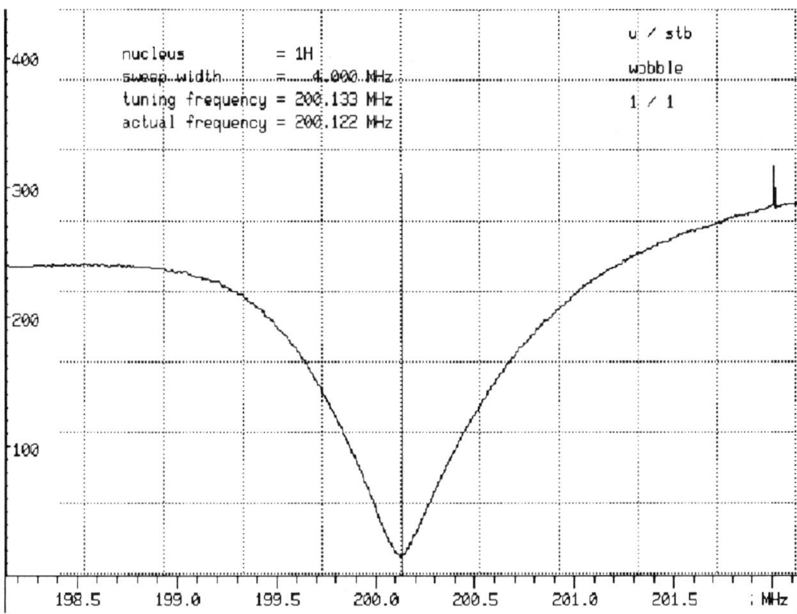

Fig. 1.6 Wobbling curve during probe-head tuning

1.3 The Lock Channel

As neither magnetic fields nor frequencies derived from synthesizers are stable enough for a long period of time, high resolution NMR measurements require a special field/frequency stabilization to allow accumulation of signals, which may be separated by less than one Hz. The basic idea of this stabilization device, called the "lock", is to hold the resonance condition by a separate NMR experiment, which runs parallel to the one in the observe channel. As long as the lock signal is held in resonance the field/frequency relationship is defined also for the observe channel. Figure 1.7 shows the principles of the lock channel.

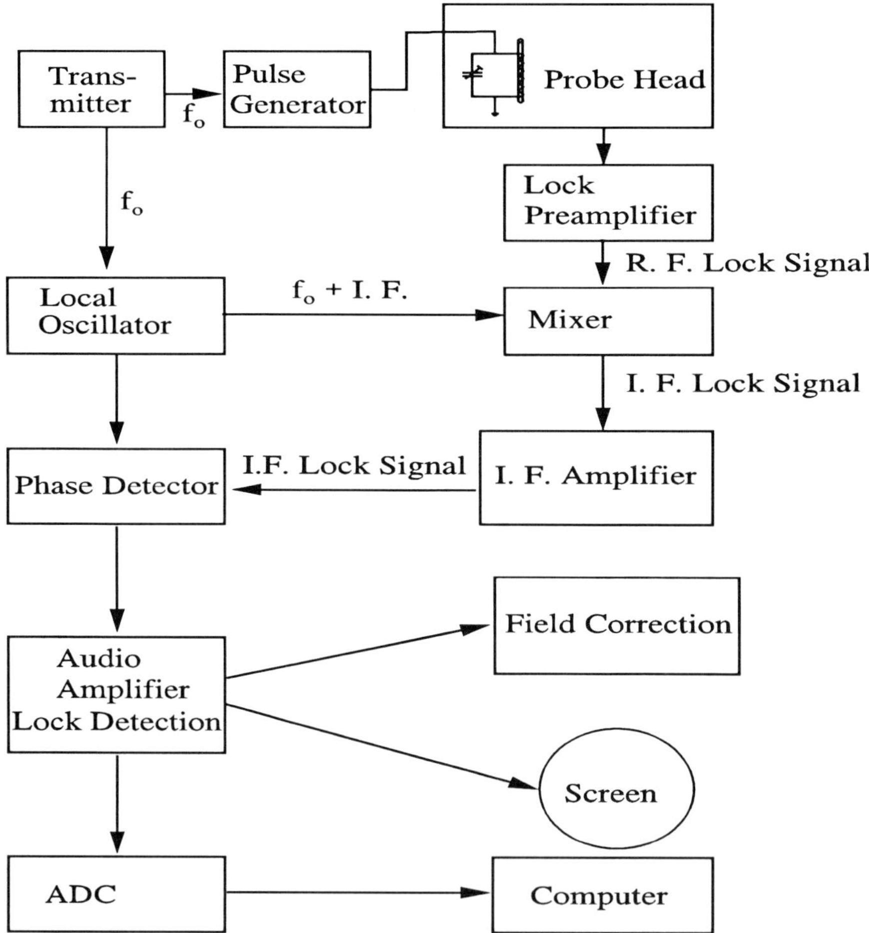

Figure 1.7 Schematic arrangement of the lock channel

Usually the ^2H resonance of the deuterated solvent is used to provide the NMR lock signal. Thus, an extra ^2H lock transmitter is needed, which transmits its frequency in pulsed form to the probe-head, in which the ^1H coil is often doubly tuned to both the ^1H and ^2H frequencies. The deuterium signal is preamplified and processed in the same way as the normal NMR signals in the observe channel. However, the final audio signal is used in dispersion mode to derive a negative or positive control voltage, which regulates the field position. Very recent developments employ a so-called digital lock, where the lock i.f. is fed directly into the ADC. The lock signal is displayed on the computer screen and provides a means of shimming the magnet. This is possible, because a narrower lock signal results in a higher d.c. voltage after rectification. Thus, by adjusting the various shim currents one aims for an optimum lock signal.

For special cases, e. g. for ^{2}H NMR spectroscopy, an ^{19}F lock is used instead of the ^{2}H resonance. The lock substance can be just the solvent, as described above, or may be provided within a capillary for chemical reasons. Special applications, such as probe-heads used as detectors for LC-NMR, use an external lock derived from an extra capillary within the probe-head.

On current NMR instruments, which are equipped with automatic sample changers, the lock capture and lock-in procedure is done by the instrument itself. However, any beginner in the NMR field should first learn how to do it manually. This is the basic start of any NMR experiment and a meaningful shimming procedure is only possible after having properly locked in.

There are several parameters which control the lock display on the computer screen. First, one needs a device to sweep the magnetic field (mostly forward and backward) over the lock resonance position, usually a sawtooth modulation. Its amplitude and sweep rate can be adjusted. Secondly, the position of the magnetic field must be adjusted to find the lock signal and to fix it at its lock-in position. The r.f. power of the lock transmitter, the gain of the lock receiver, and the phase and d.c. offset of the lock signal have to be correctly chosen. As for any NMR application, the lock transmitter power should not saturate the signal, and thus the lock transmitter must be sufficiently attenuated. The noise level of the receiver, however, should not be excessive. The lock-in procedure (i. e. pressing the "lock" button) automatically switches off the field sweep and holds the lock signal at its resonance position. After the lock-in procedure the operator should fine-adjust the magnetic field homogeneity by maximizing the lock signal level. Further attenuation adjustment of the transmitter power may be needed to ensure that the lock signal is not saturated.

If, for some reason, one wants to measure NMR spectra without the lock, one must turn off the field modulation manually; however, one has to be aware of the magnet field drift.

1.4 The Art of Shimming

The process of optimizing the magnetic field homogeneity for recording high resolution spectra is called "shimming" a magnet. Usually this is done by observing an NMR signal which has a natural line-width less than 0.1 Hz. This line-width corresponds to a homogeneity of the magnetic field better than 1 ppb for a 100 MHz spectrometer. Adjusting the homogeneity can be performed in different ways, by observing (i) a swept NMR signal (without lock), (ii) the lock level (with locking), or (iii) the FID or the area of the FID on the observe channel. The homogeneity is checked by the procedures described in Experiments 3.3, 3.4, and 3.6.

In the very first days of NMR spectroscopy shimming was performed mechanically (in the original meaning shims are small pieces of metal), but now an electronic device called the shim system is used for the shimming process. This device is essentially a set of coils causing very specific magnetic field contours. Because the homogeneity must be maintained over the total volume of NMR observation (probe coil), the shim system is installed in the room-temperature bore of the magnet and surrounds the probe-head and especially the sample region. The currents for the shim coils can create various gradients of any desired strength and shape and can be controlled separately by potentiometers from the spectrometer console. Table 1.1 shows the common room-temperature shims together with their specific functions and their interaction order. There is a second set of shims called cryoshims, which are adjusted during the installation of the magnet.

Table 1.1 Common room temperature shims with function and interaction order

Common Shim Name	Function	Gradient order	Inter-action order
$Z0$	1	0	0
$Z1$	z	1	0
$Z2$	$2z^2 - (x^2 + y^2)$	2	1
$Z3$	$z[2z^2 - 3(x^2 + y^2)]$	3	2
$Z4$	$8z^2[z^2 - 3(x^2 + y^2)] + 3(x^2 + y^2)^2$	4	2
$Z5$	$48z^3[z^2 - 5(x^2 + y^2)] + 90z(x^2 + y^2)^2$	5	2
X	x	1	0
Y	y	1	0
ZX	zx	2	2
ZY	zy	2	2
XY	xy	2	1
$X^2 - Y^2$	$x^2 - y^2$	2	1
Z^2X	$x[4z^2 - (x^2 + y^2)]$	3	2
Z^2Y	$y[4z^2 - (x^2 + y^2)]$	3	2
ZXY	zxy	3	2
$Z(X^2 - Y^2)$	$z(x^2 - y^2)$	3	2
X^3	$x(x^2 - 3y^2)$	3	1
Y^3	$y(3x^2 - y^2)$	3	1

1.4.1 The Shim Gradients

The different shims are also called shim gradients. One has to adjust the shim currents so that they cancel any gradients in the NMR sample as accurately as possible. There are two types of gradients: spinning ($Z0$–$Z5$) and non-spinning shims where z is the coordinate direction of the field B_0. Spinning the sample averages the field inhomogeneities along two axes but not along the axis about which the sample is spun. Therefore the shim procedure can be divided into two steps: shimming with and shimming without spinning the sample. Usually the sample spinning produces an amplitude modulation of the NMR signal, which gives rise to spinning sidebands on both sides of the signal. The spinning sidebands occur at integer multiples of the spinning frequency and become smaller as the homogeneity increases or the spinning rate is increased. Shimming is not a simple maximization process, because the shims have different gradient order and different interaction order (see Table 1.1). For the shim process you should use sample tubes with a filling height of 6 cm to avoid vortices. The following classification follows the gradient order. The total number of available shim gradients increases with the magnetic field strength of the magnet:

Zero order: The *Z0* shim is the only zero order shim. This is the field position in most instruments.

First order: The *Z1*, *X* and *Y* shims are first order shims. These gradients produce a linear variation of magnetic field strength and have shapes like the p atomic orbitals. They are optimized by a simple maximization process; this corresponds to an interaction order of 0.

Second order: There are five second order shim gradients (see Table 1.1; on older instruments *Z2* is called curvature), which have shapes like d atomic orbitals, e.g. *Z2* corresponds to the dz^2 orbital. These gradients cause quadratic variations in field strength. For three of them the interaction order is 1, for the other two the interaction order is 2. First order interaction means that the shims are adjusted by an successive iterative process. After the adjustment of the complete set of shims, you have to readjust the first shim of the set and you will find a different optimum. Successive iterations will lead to smaller and smaller changes on readjustment until no further change is observed. A typical example is *Z1* and *Z2*. After optimization of *Z1* followed by *Z2*, you will find a new optimum for *Z1* when readjusted. With an interaction order of 2 you have to change a given shim first and then adjust others before any improvement of the homogeneity can be observed. This means: change the shim a measured amount and optimize the other shims of the set. If this leads to a better response proceed to change the shim in the same direction another measured amount and repeat the process until the response (lock level or FID area) starts to decline. If the initial response is worse try the other direction.

Third order: The complete set of third order shims has seven different gradients corresponding to the shape of the seven f atomic orbitals. A complete set of these gradients is found on 600 and 750 MHz spectrometers. These gradients produce cubic variations of field strength. Usually there is only one 4th order and one 5th order shim gradient on high field instruments.

1.4.2 The Shimming Procedure

In the following shimming procedure, which is described very precisely by Conover [1], it is assumed that the sample is in the center of the shim-set. If this is not the case the center of the shim-set has to be located first. This is done by moving the sample with respect to the receiver coil. Usually the field centering has been performed by the manufacturer's engineer in the course of the installation of the magnet.

If the magnetic field is in a state of unknown homogeneity or is known to have poor homogeneity use the swept NMR signal, usually the deuterium lock signal, for the first steps in the shimming process. Otherwise proceed with the second round.

1. Spin the sample (20 to 30 Hz) and adjust the phase of the lock signal for absorption. The signal-to-noise ratio should be sufficient to allow height and the ring-down pattern (wiggles) to be observed. The ring down pattern can be used for the final adjustment. Adjust *Z1* and *Z2* interactively to produce the tallest swept signal (first order process).
2. Stop the spinner and adjust *X* and *Y* for the tallest swept signal response (first-order process).
3. Adjust *X* and *ZX* for the tallest swept signal (second-order process).
4. Adjust *Y* and *ZY* for the tallest swept signal (second-order process).
5. Adjust *XY* and *X2 – Y2* for the tallest swept signal (first-order process)
6. If any large shim changes were observed in the above process than repeat the process from 1. The NMR spectrometer should now be capable of operating with a field-frequency lock.

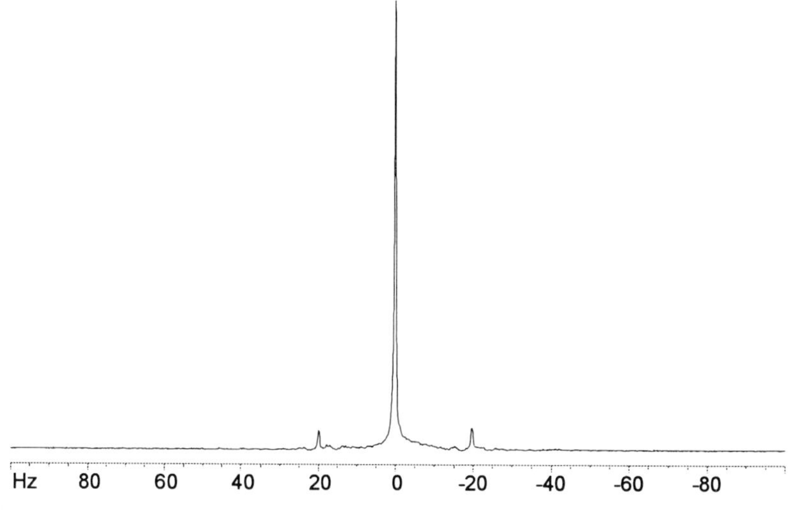

Fig. 1.8 Spinning sidebands obtained after incorrectly setting the X gradient

Second Round (Spinning Shims)

Spin the sample at 20 to 30 Hz, make sure that there is no vortex, especially if using a probe-head for 10 mm sample tubes. A vortex will lead to a false shim optimum, especially for *Z2*. If the lock signal is used for shimming, avoid saturation by using as low a lock power as possible. If the FID or FID area is used for shimming, use a pulse repetition time which is long enough for full relaxation; otherwise the NMR signal is saturated. The lock phase should be carefully adjusted and reexamined each time a large change is made to a shim with an even interaction order [*Z3*, *Z4*, *Z5*, *ZX*, *ZY*, Z^2X, Z^2Y, *ZXY*, $Z(X^2 - Y^2)$].

1. Optimize *Z1* and *Z2* (first order process).
2. Optimize *Z3* (second order process). Note the setting of *Z3* and the response. Change *Z3* to degrade the response by 20–30%. Repeat the process in step 1. If the new setting for *Z3* has yielded a better response then continue in the same direction. If the new response is less then try the other direction for *Z3*.
3. Optimize *Z4* (second order process). Note the position of *Z4* and the response. Change *Z4* to degrade the response by 30–40%. Repeat the process in step 1. Adjust *Z3* to provide the optimum response. If the *Z3* shim change is considerable, then repeat step 1 again and readjust *Z3* for maximum response. If, after optimizing *Z3*, *Z2*, and *Z1*, the new response is better than the previous one, continue in the same direction. If the response is worse then try the other direction.
4. The *Z5* shim normally needs to be adjusted only with wide-bore magnets and large-diameter sample tubes. Change *Z5* enough to degrade the response by 30–50%. Repeat step 1 and reoptimize *Z3*. Adjust *Z4* for maximum response. If either *Z3* or *Z4* changed by a considerable amount, repeat step 1 and reoptimize *Z3* and *Z4*. If the new response obtained after this procedure is better than before, continue in the same direction. If the response is worse, try the other direction with *Z5*.

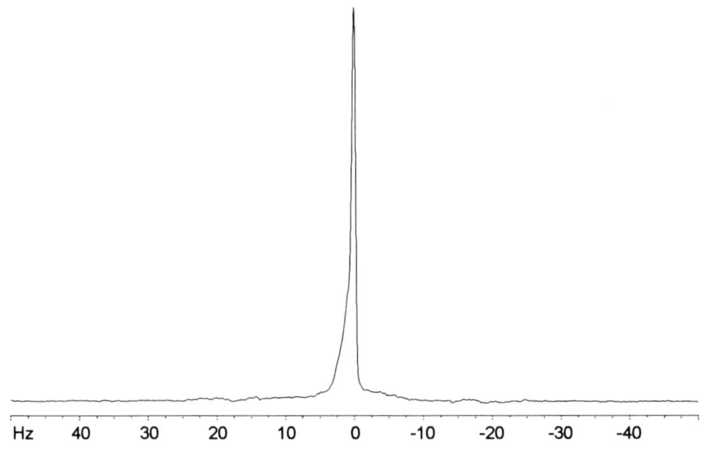

Fig. 1.9 Typical result obtained after incorrectly setting the Z4 gradient

Third Round (Non-spinning Shims)

This shim-set has to be adjusted while the sample is not spinning. Changing the shim gradients with *Z*-components causes changes in the spinning shim set. The spinning shim sequence should be repeated after completion of the non-spinning shim procedure, especially if one of the non-spinning shims changes significantly.

1. Turn the spinner off. Adjust X and Y for maximum response (first-order process).
2. Note the position of ZX and the response. Change ZX to degrade the response by 10% and adjust X for a maximum response. If the new response is better, continue in the same direction with ZX. If the response is less, try the opposite direction with ZX.
3. Repeat step 2 but using the Y and ZY shims.
4. Adjust XY and $X^2 - Y^2$ interactively (first-order process) for maximum response. If either XY or $X^2 - Y^2$ changed significantly then repeat steps 2 and 3.
5. Adjust Z^2X (second-order process). Note the position of Z^2X and the response. Change Z^2X enough to degrade the response by 30%. Maximize the response with ZX. Optimize the response with X. If the new response is larger than the initial response continue with Z^2X in the same direction. If the response is less then try the other direction.
6. Repeat step 5 but using Z^2Y, ZY and Y.
7. Adjust ZXY (second-order process). Note the position of ZXY and the response. Change ZXY enough to degrade the response by 20%. Maximize the response with XY. If the new response is larger than the initial one, continue with ZXY in the same direction. If the response is less, try the other direction.
8. Repeat step 7 but using $Z(X^2 - Y^2)$ and $X^2 - Y^2$.
9. Adjust X^3 and X interactively for maximum response (first-order process).
10. Adjust Y^3 and Y interactively for maximum response (first-order process).
11. If the non-spinning shim settings have significantly changed, then repeat the second round. If there are significant changes in the spinning shims, repeat the non-spinning shim procedure also.

Final Round

After all spinning and non-spinning shim gradients have been optimized the NMR instrument should be delivering less than 0.5 Hz line-width with a good line-shape (see Exp. 3.5) and minimal spinning side bands.

1.4.3 Gradient Shimming

Recent developments use a probe-head with x, y, and z pulsed field gradients. With such a device it is possible to record an image of the homogeneity. With this the computer calculates the required changes for good homogeneity and finds the optimum after a few iterations [3]. This procedure can also be performed with a z-only gradient probe-head providing that the shims containing x and y elements have been adjusted by hand.

In practice one starts by generating a field map which indicates how the probe-head in use reacts towards the settings of the shims. This is done with a sample giving a strong signal, usually water. Figure 1.10 shows a typical field map for a z gradient probe-head. On the x-axis of the plot the length of the r.f. coil is measured and the y-axis gives in relative units the signal response towards changes of the shim settings. The field map has, in principle, to be created only once for each probe head.

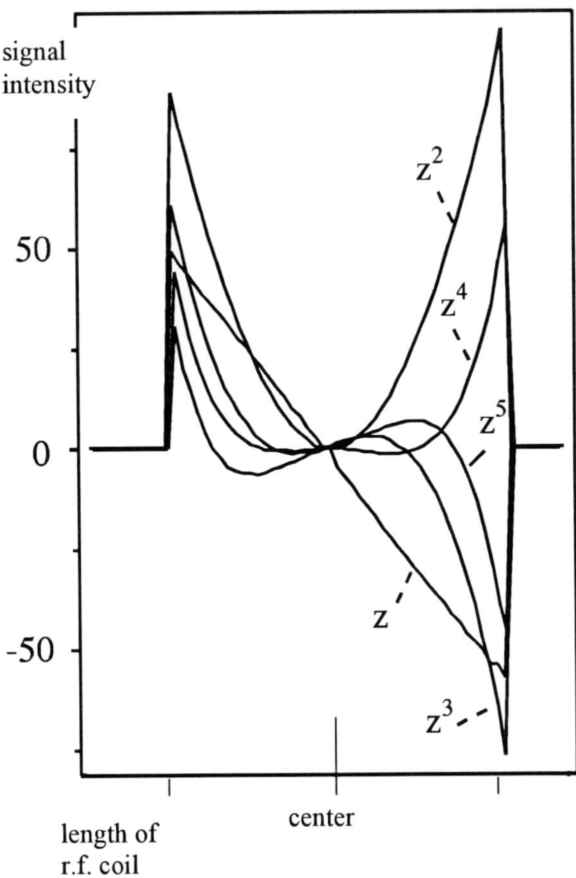

Figure 1.10 Field map obtained with a z gradient probe head

Using the values of the field map the actual shims are adjusted in several iteration steps; again the sample should contain just one strong signal. Since the x and y shim groups do not change too much in practice, the z gradient shimming method is a time-saving approach to obtain very good z shims, especially for biological samples dissolved in water. A typical result depicting the z homogeneity achieved across the sample is shown in Figure 1.11. Note the change in vertical scale compared to Figure 1.10.

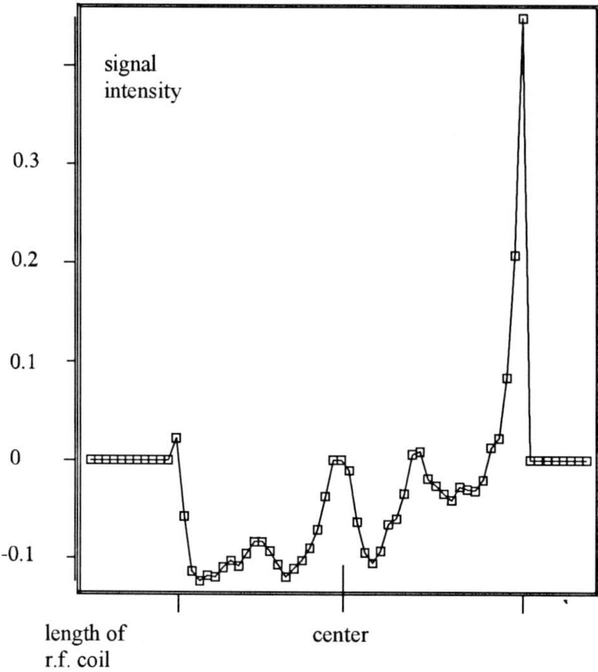

Figure 1.11 *z* Homogeneity obtained with a *z* gradient probe head after gradient shimming

Literature

[1] W. W. Conover, *Top. Carbon-13 NMR Spectrosc.* **1984**, *4*, 37–51.

[2] *SAM 1.0, Shimming Simulation Software package for IBM-PC compatible Computers*, ACORN NMR, 46560 Fremont Blvd., Fremont, CA 94538-6482.

[3] J. Hu, T. Javaid, F. Arias-Mendoza, Z. Liu, R. McNamara, T. R. Brown, *J. Magn. Reson. Ser. B* **1995**, *108*, 213–219.

Chapter 2

Determination of the Pulse-Duration

In pulsed Fourier transform NMR spectroscopy there is nothing more important than the use of correct radiofrequency pulses. This applies not only for advanced multipulse and multidimensional methods but for the most simple routine experiments as well. The use of a wrong excitation pulse can render all FT experiments insensitive and useless. Regular determinations of pulse-duration (also often called pulse-width or pulse-length) are also necessary for instruments working in an automatic mode with a sample changer. Aging or malfunctioning components can increase the pulse-width corresponding to the normal setting (e.g. 90°) and therefore degrade the performance of these spectrometers, if not corrected.

In this first experimental chapter we therefore provide five basic methods for measuring the pulse-width. First we describe the calibration of transmitter pulse-width both for ^{1}H and ^{13}C (Exps. 2.1–2.2) and of the ^{1}H decoupler pulse-width (Exp. 2.3). Corresponding experiments are then performed for the inverse mode of operation (Exps. 2.4–2.5). We demonstrate the use of composite pulses (Exp. 2.6) and the effect of radiation damping which renders the determination of the 90° pulse duration in normal water difficult (Exp. 2.7). At the end of the chapter we show the relationship between pulse and reciever phases (Exp. 2.8) and how the pulse-length is connected to radiofrequency power and the excitation bandwidth (Exp. 2.9). This provides important knowledge for the setting up more advanced experiments like those using spin-locks or selective pulses. The calibration of selective pulses is demonstrated in Chapter 7.

Literature

[1] A. E. Derome, *Modern NMR Techniques for Chemistry Research*, Pergamon, Oxford, **1987**, 67–77.

[2] E. Fukushima, S. B. W. Roeder, *Experimental Pulse NMR,* Addison Wesley, Reading, **1981**, 47–60.

Experiment 2.1

Determination of the 90° ^{1}H Transmitter Pulse-Duration

1. Purpose

One of the basic requirements of NMR spectrometer operation is the knowledge of the 90° pulse-length. The 90° or $\pi/2$ pulse or in general the flip angle θ is important not only for routine operation, but also for 1D multipulse and multidimensional NMR experiments. The flip angle depends on the r.f. magnetic field strength B_1, the pulse-length p, and the gyromagnetic ratio γ of the nucleus under observation, as expressed in radians and in degrees by Equations (1) and (2).

$$\theta\,[\text{rad}] = \gamma B_1 \cdot p \tag{1}$$

$$\alpha\,[°] = (360/2\pi)\,\gamma B_1 \cdot p \tag{2}$$

Usually the 90° or $\pi/2$ pulse-duration is determined by measuring the 180° (π) or 360° (2π) pulse-lengths since these pulses give a minimum signal. Here we present a description for the ^{1}H transmitter pulse calibration.

2. Literature

[1] J. K. M. Sanders, B. K. Hunter, *Modern NMR Spectroscopy,* 2nd Edition, Oxford University Press, Oxford,. **1993**, 33–34.

[2] D. Shaw, *Fourier Transform NMR Spectroscopy*, 2nd Edition, Elsevier, Amsterdam, New York, **1984**, Ch. 7.

[3] A. E. Derome, *Modern NMR Techniques for Chemistry Research*, Pergamon, Oxford, **1987**, 153–157.

3. Pulse Scheme and Phase Cycle

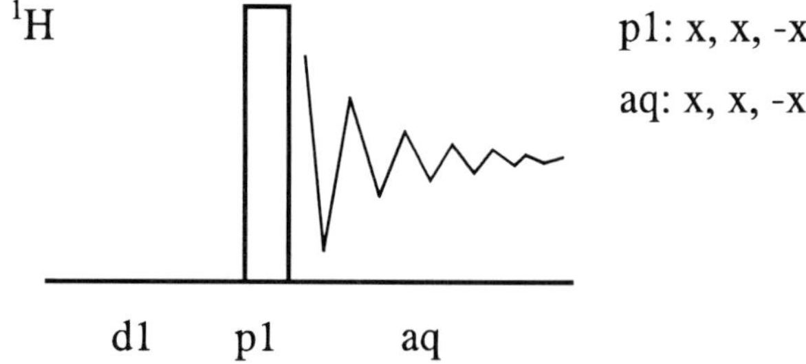

p1: x, x, -x, -x, y, y, -y, -y

aq: x, x, -x, -x, y, y, -y, -y

4. Acquisition

Time requirement: 30 min

Sample: 10% $CHCl_3$ in [D_6]acetone; *do not use a degassed and sealed sample*, since that would make the relaxation time of the $CHCl_3$ protons exceedingly long.

The probe-head should be tuned to the sample. Load standard ^{1}H parameters, record a normal ^{1}H NMR spectrum and note the offset of $CHCl_3$. On older instruments change to the absolute intensity mode. You have to set:

td: 4 k
sw: 500 Hz
o1: on resonance of $CHCl_3$ signal
p1: ^{1}H transmitter pulse, to be varied, 1 μs as initial value
d1: 30 s
rg: receiver gain for correct ADC input
transmitter attenuation (3 dB)
ns: 1

Increase p1 in 2 μs steps until the intensity of the processed signal begins to drop to nearly zero. Now use smaller steps in increasing p1, e.g. 0.1 μs, to find the minimum for the 180° pulse. With the determined pulse-length check the 90° pulse (maximum positive intensity), the 270° pulse (maximum negative intensity) and the 360° pulse (minimum intensity). If there are small deviations, calculate an average value for the 90° pulse-length. In case of large deviations repeat the procedure. If the deviations are still present, the probe-head may be arcing; increase the transmitter attenuation.

5. Processing

Use standard 1D processing (see Exp. 3.1) applying an exponential window with a line-broadening factor lb = 1 Hz. Adjust the phase of the first spectrum for pure absorption and always use the same phase correction.

6. Result

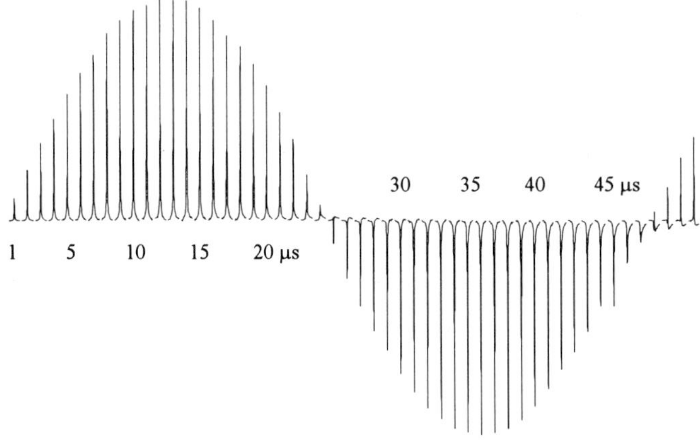

The figure on page 17 shows the full sinusoidal dependence of the signal intensity obtained on an AM-400 spectrometer with a 5 mm dual probe-head. The pulse-width was incremented in steps of 1 µs; the 90° pulse-duration was determined as 12.5 µs.

7. Comments

It is important to avoid using too short a pulse repetition time. The delay between successive measurements should be 5 times T_1. Even for protons in small molecules like $CHCl_3$ this can pose a problem, if the sample is degassed. For nuclei with long relaxation times it is more convenient to measure the 360° pulse-length.

In the rotating frame the r.f. pulse rotates the magnetization vector $M_0 = M_z$, creating an observable transverse magnetization M_{xy}. After a 180° pulse M_{xy} is theoretically zero and $M_0 = -M_z$, but in practice residual signals caused by inhomogeneities in B_1 can be seen.

The typical pulse-length for all probe-heads should be recorded in the logbook of the instrument.

8. Own Observations

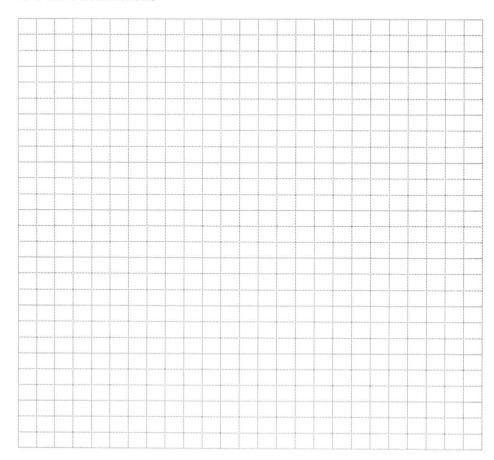

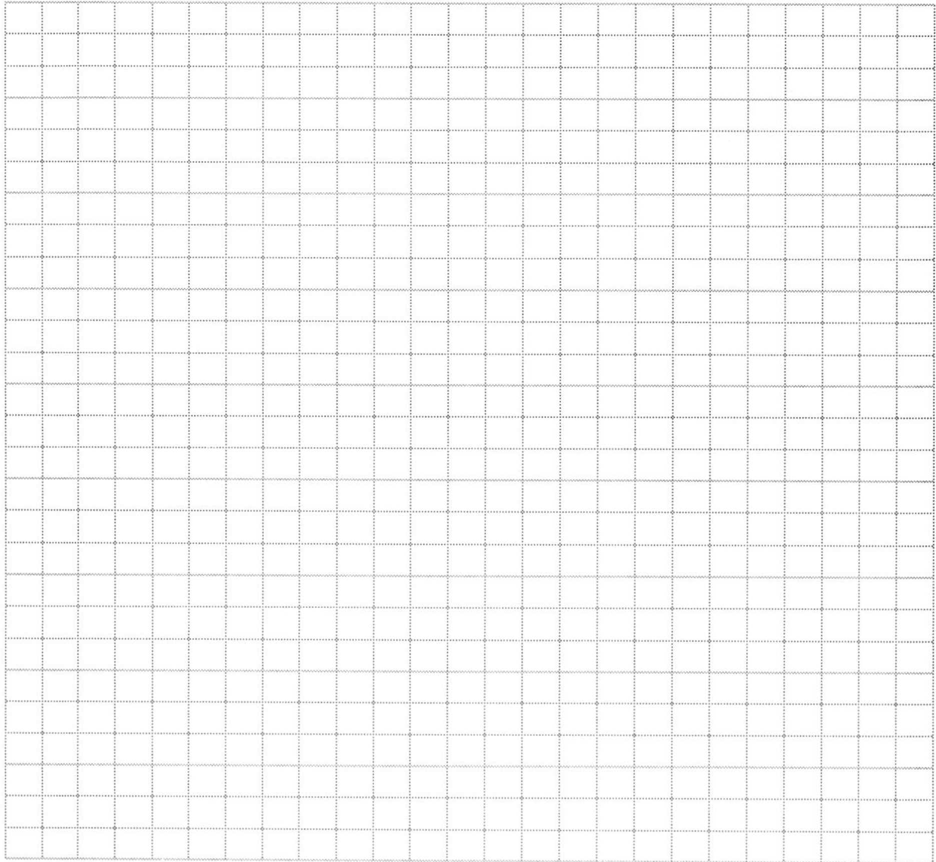

Experiment 2.2

Determination of the 90° ^{13}C Transmitter Pulse-Duration

1. Purpose

This experiment is very similar to Experiment 2.1. Here, however, the pulse-width determination is described for the ^{13}C nucleus and has to be performed with ^{1}H broadband decoupling. Although the experiment could be performed exactly like exp. 2.1 we describe here a procedure given by [4] which yields the 90° pulse by just two measurements. This method is helpful for very slowly relaxing nuclei.

2. Literature

[1] J. K. M. Sanders, B. K. Hunter, *Modern NMR Spectroscopy,* 2nd Edition, Oxford University Press, Oxford, **1993**, 33–34.
[2] D. Shaw, *Fourier Transform NMR Spectroscopy*, 2nd Edition, Elsevier, Amsterdam, New York, **1984**, Ch. 7.
[3] A. E. Derome, *Modern NMR Techniques for Chemistry Research*, Pergamon, Oxford, **1987**, 153–157.
[4] E. Haupt, *J. Magn. Reson.* **1982**, *49*, 358–364.

3. Pulse Scheme and Phase Cycle

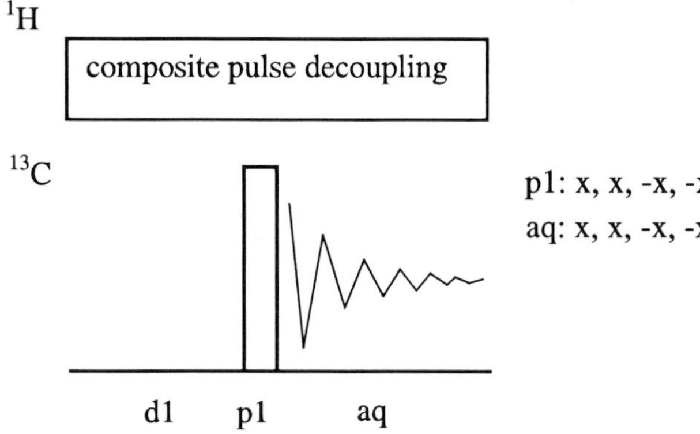

p1: x, x, -x, -x, y, y, -y, -y
aq: x, x, -x, -x, y, y, -y, -y

4. Acquisition

Time requirement: 15 min

Sample: 40% $CHCl_3$ in [D_6]acetone; *do not use a degassed and sealed sample*, since that would make the relaxation time of the $CHCl_3$ carbons exceedingly long.

The probe-head should be tuned to the sample. Load standard ^{13}C parameters with ^{1}H broad-band decoupling, record a normal ^{13}C NMR spectrum and note the offset of $CHCl_3$. On older instruments change to the absolute intensity mode. You have to set:

td: 4 k
sw: 500 Hz
o1: on resonance of ^{13}C signal
o2: middle of ^{1}H NMR spectrum
p1: ^{13}C transmitter pulse, 7 μs for experiment **a** and 14 μs for experiment **b**
d1: 60 s
rg: receiver gain for correct ADC input
transmitter attenuation (3dB)
decoupler attenuation and 90° pulse for CPD
ns: 1

Record spectrum **a** with p1 = 7 μs and spectrum **b** with p1 = 14 μs. Note the heights I_1 and I_2 of the $CHCl_3$ signal in these two spectra. If the pulse-width in the second experiment is double that in the first experiment, Equation (1) holds, from which the pulse angle α_1 (in degrees) of the first experiment can be calculated:

$$\alpha_1 = \arccos(0.5\, I_2/I_1) \tag{1}$$

Using Equation (2) the 90° pulse-length p_{90} can be calculated, where p1 is the pulse-length of the *first* experiment. p_{90} should be checked by applying the corresponding 180° pulse.

$$p_{90} = 90\ \mathrm{p1} / \alpha_1 \tag{2}$$

5. Processing

Use standard 1D processing (see Exp. 3.2) applying an exponential window with a line-broadening factor lb = 1 Hz. Adjust the phase of the first spectrum for absorption and use the same phase correction for the second experiment.

6. Result

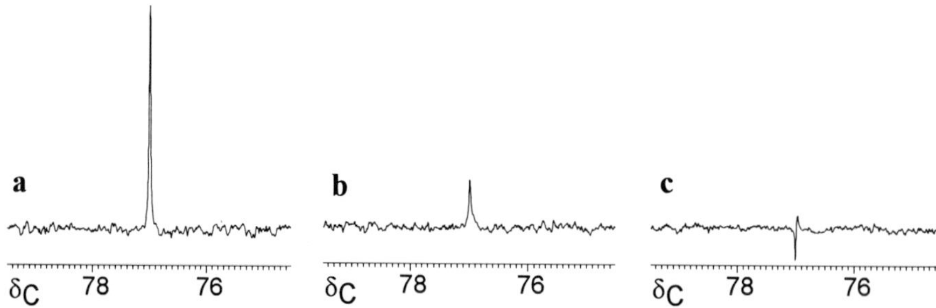

The figure on page 18 shows the result obtained on an AM-400 spectrometer with a 5 mm dual probe-head. Spectrum **a** (p1: 7 μs) yielded I_1 = 12.43, spectrum **b** (p1: 14 μs) yielded I_2 = 2.56 . According to Equation (1) α_1 is calculated to be 84°, and using Equation (2) gives p_{90} to be 7.5 μs. This was checked in **c** using p1 = 15 μs.

7. Comments

This procedure is very convenient if one does not know anything about the pulse-length, e.g. if one is using a probe-head for the very first time or studying a heteronuclide for the first time. The method works best if α_1 can be estimated to about 60°.

The pulse-width determinations in Experiments 2.1 and 2.2 use a narrow spectral width of only 500 Hz and the transmitter on resonance. Note that radiofrequency pulse-lengths are offset-dependent, the 180° pulse-length especially is quite different for signals with different offsets.

8. Own Observations

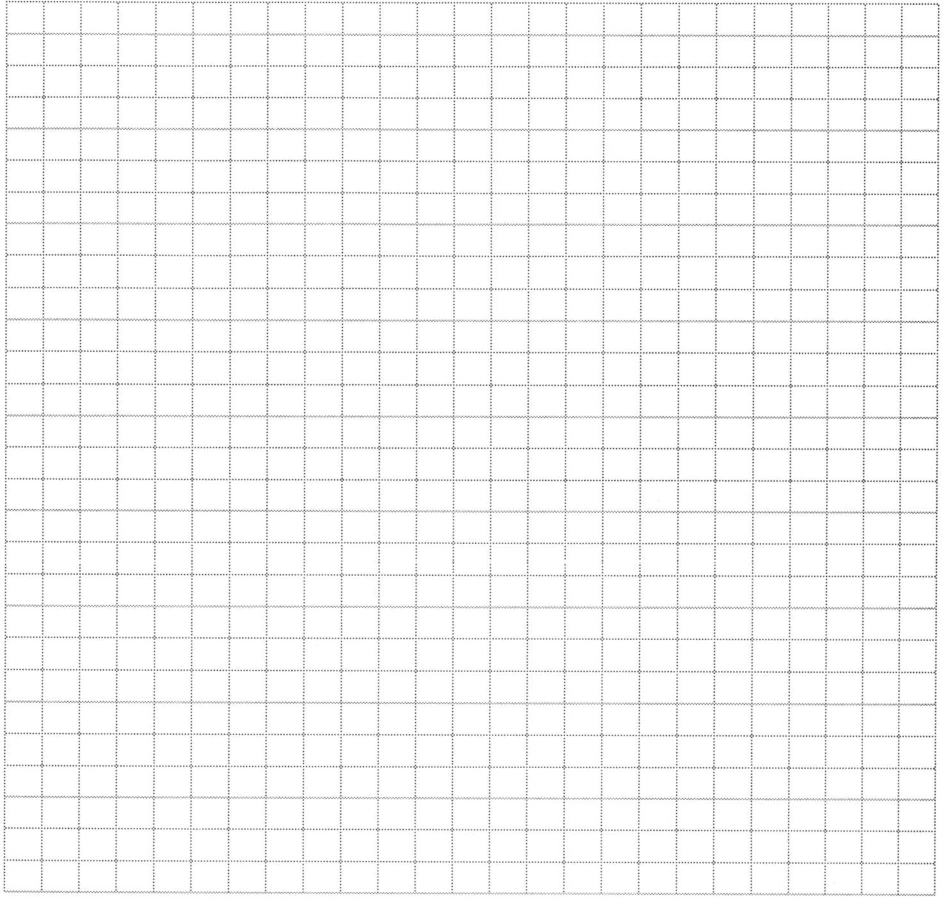

Experiment 2.3

Determination of the 90° ^{1}H Decoupler Pulse-Duration

1. Purpose

Many 1D and 2D multipulse sequences use defined ^{1}H decoupling pulses. Without knowledge of these pulses some important experiments such as DEPT (Exp. 6.9) or HETCOR (Exp. 10.10) cannot be performed. Furthermore, the common ^{1}H broadband decoupling technique for ^{13}C NMR spectroscopy, which uses composite pulses (CPD), fails if the decoupler pulse is wrong. This experiment has typically to be performed twice, once with low attenuation of the decoupler to calibrate the "hard" pulses which are used during a pulse sequence and once with high attenuation of the decoupler to define the pulses used for CPD.

2. Literature

[1] A. Bax, *J. Magn. Reson.* **1983**, *52*, 76–80.
[2] N. C. Nielsen, H. Bildsoe, J. Jakobsen, O. W. Sørensen, *J. Magn. Reson.* **1986**, *66*, 456–469.

3. Pulse Scheme and Phase Cycle

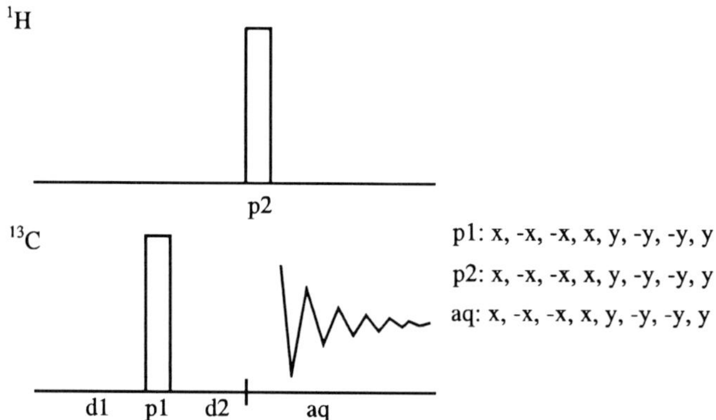

4. Practical Procedure

Time requirement: 10 min

Sample: 50% $CHCl_3$ in [D_6]acetone, *do not use a degassed and sealed sample* since that would make the relaxation times of carbons and protons exceedingly long.

Obtain both offsets for the ^{1}H and ^{13}C signals of the sample. Load the pulse program for the sequence shown above. You have to set:

td: 4 k
sw: 500 Hz
o1: on resonance of ^{13}C signal
o2: on resonance of ^{1}H signal
p1: 90° ^{13}C transmitter pulse (Exp. 2.2)
p2: ^{1}H decoupler pulse, 0 μs as starting value, to be varied
d1: 60 s
d2: $1/[2\,J(C,H)]$ = 2.36 ms, calculated from $^{1}J(C,H)$ = 212 Hz
rg: receiver gain for correct ADC input
decoupler attenuation [0 dB]
ns: 1

Record a first spectrum with p2 = 0 and adjust the doublet in antiphase. Then repeat the experiment with increasing pulse lengths p2 until the signals disappear, which corresponds to the 90° hard decoupler pulse.

In a second set of experiments use high decoupler attenuation for CPD [22 dB] and vary it so that p2 is in the region of 100 μs.

5. Processing

Use standard 1D processing (see Exp. 3.2) applying an exponential window with a line-broadening factor lb = 2 Hz.

6. Result

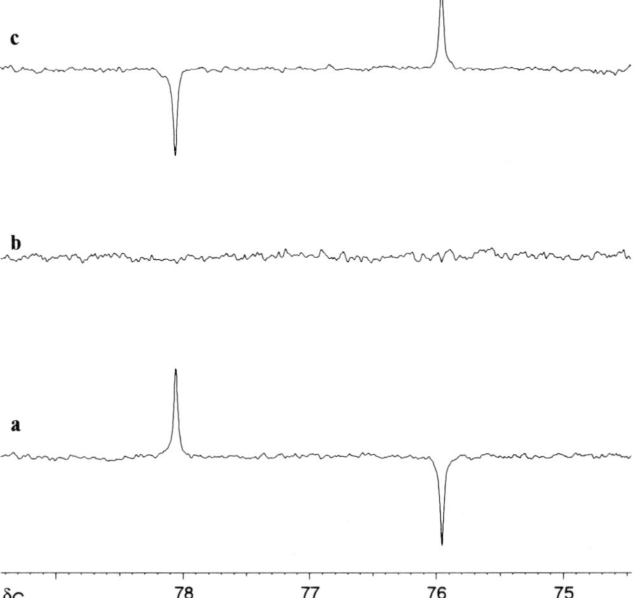

The figure on page 21 shows the result of a 0° decoupling pulse (**a**) and of a 90° decoupling pulse (**b**) obtained on an AM-400 spectrometer with a 5mm dual probe-head. Note the disappearance of the signals in **b.** The spectrum **c** was obtained with a 180° pulse.

7. Comments

We consider a ^{13}C,^{1}H spin pair. The equilibrium magnetization is converted by the 90° ^{13}C pulse into a transverse ^{13}C magnetization as described by Equation (1).

$$I_{Cz} + I_{Hz} \xrightarrow{90^\circ I_{Cx}} -I_{Cy} + I_{Hz} \qquad (1)$$

During the period τ = d2 = $1/2J$ spin–spin coupling between proton and ^{13}C evolves, as in Equation (2).

$$-I_{Cy} + I_{Hz} \xrightarrow{\pi J \tau 2 I_{Hz} I_{Cz}} 2I_{Cx}I_{Hz}\sin\pi J\tau - I_{Cy}\cos\pi J\tau + I_{Hz} \qquad (2)$$

Since τ was set to $1/2J$, (2) simplifies to (3).

$$2I_{Cx}I_{Hz}\sin\pi J\tau - I_{Cy}\cos\pi J\tau + I_{Hz} = 2I_{Cx}I_{Hz} + I_{Hz} \qquad (3)$$

A 90_x ^{1}H pulse converts $2I_{Cx}I_{Hz}$ into double quantum magnetization as in Equation (4), from which no observable ^{13}C signal can be generated. Thus, if the decoupling pulse is exactly 90°, the doublet disappears.

$$2I_{Cx}I_{Hz} + I_{Hz} \xrightarrow{90^\circ I_{Hx}} -2I_{Cx}I_{Hy} - I_{Hy} \qquad (4)$$

Note that with this method, in contrast to Experiments 2.1 and 2.2, the 90° pulse yields a minimum signal, whereas the 180° pulse inverts the initial phases of the doublet.

8. Own Observations

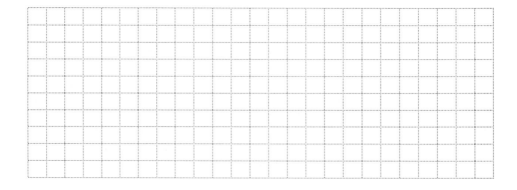

Experiment 2.4

The 90° ^{1}H Pulse with Inverse Spectrometer Configuration

1. Purpose

In "inverse" experiments one observes protons and decouples heteronuclei X (e.g. ^{13}C, ^{15}N). On NMR instruments built later than 1991 the pulse-lengths in the normal and the inverse mode are usually quite similar. Older instruments, however, use different r.f. sources and signal routings in these two configurations. Therefore, before starting an inverse experiment, such as an HMQC experiment (Exp. 10.13), the pulse-durations of protons and X nuclei have to be determined in this spectrometer configuration. This can be done exactly as described in Experiment 2.1. Here, however, we describe the 360° method.

2. Literature

[1] E. Fukushima, S. B. W. Roeder, *Experimental Pulse NMR* Addison Wesley, Reading, **1981**, 47–60.

3. Pulse Scheme and Phase Cycle

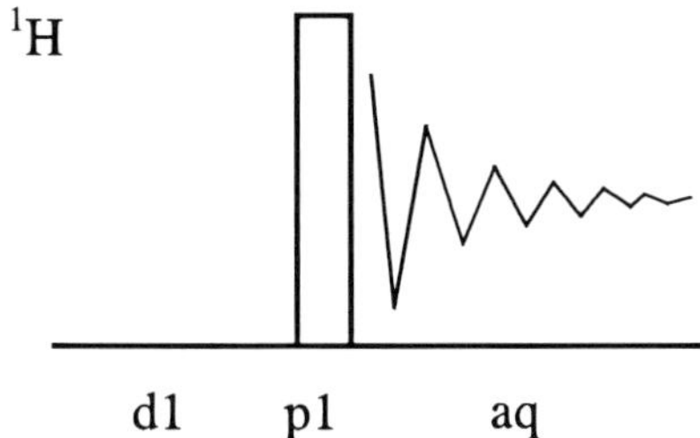

p1: x, x, -x, -x, y, y, -y, -y

aq: x, x, -x, -x, y, y, -y, -y

4. Acquisition

Time requirement: 10 min

Sample: 3% $CHCl_3$ in [D_6]acetone, *do not use a degassed and sealed sample* since that would make the relaxation times of the protons exceedingly long.

An inverse probe-head (inner coil ^{1}H, outer coil ^{13}C or other heteronuclei) is placed into the magnet, both coils are tuned to the sample. On older instruments change to the inverse set-up. Since the inverse set-up varies widely depending on the year of spectrometer's construction it cannot be discussed here. Load the correct pulse program to obtain a ^{1}H NMR spectrum in the inverse mode. You have to set:

td: 4 k
sw: 500 Hz
o1: 100 Hz towards higher frequency of $CHCl_3$ signal
p1: ^{1}H transmitter pulse, near 360° as starting value, to be varied
d1: 5 s
rg: receiver gain for correct ADC input
ns: 8

Set the pulse-width approximately to 360°, typically in the order of 40 μs, and record 8 transients. If the pulse-width is not exactly 360° a large FID signal will build up during the accumulation. Change the pulse-width until you observe a minimum FID signal. Divide the value by 4 and by 2 and check with the 90° and 180° pulses using one transient in each case.

5. Processing

No signal processing is required, since the FID is directly observed.

6. Result

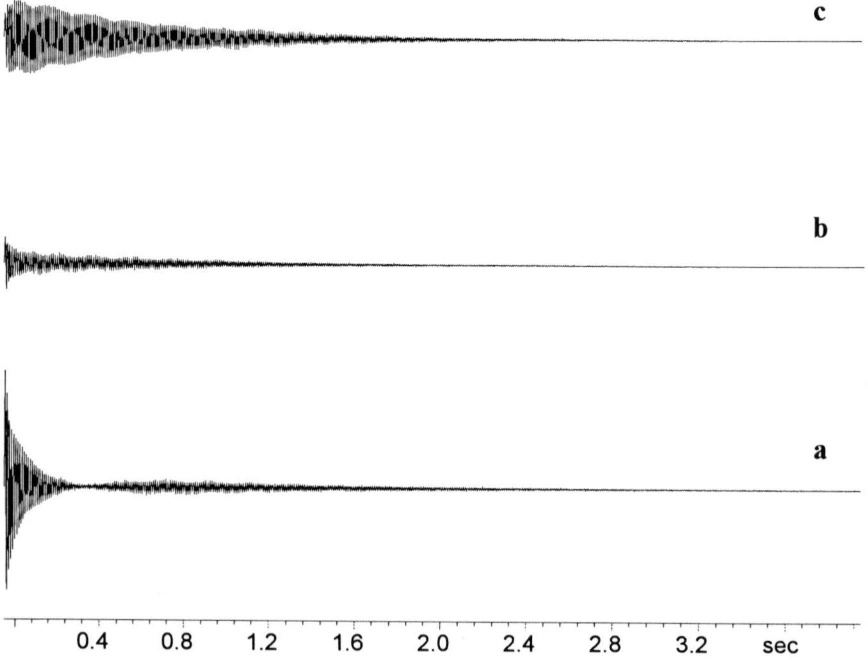

The figure shows the result obtained on an AM-400 spectrometer in the inverse mode. In **a** p1 was 39 μs, in **b** 40 μs and in **c** 41 μs. Note how the signal area changes if the pulse-width is increased only by 1 μs, which corresponds to a change of 0.25 μs for the 90° pulse.

7. Comments

The advantage of the method is that no long waiting times are needed to allow spin–lattice relaxation. R.f. pulses which are close to 360° will change the magnetization vector only slightly. The method is very sensitive, since already small deviations from 360° will cause a large FID signal.

Inverse experiments can be performed in normal probe-heads as well. As an exercise you may measure the pulse-length using a standard dual probe-head.

8. Own Observations

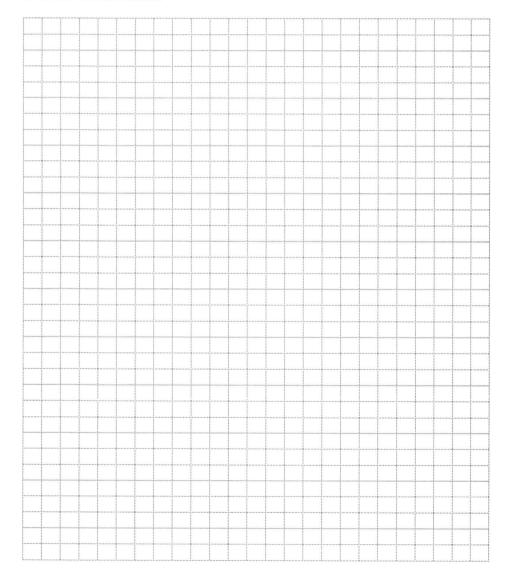

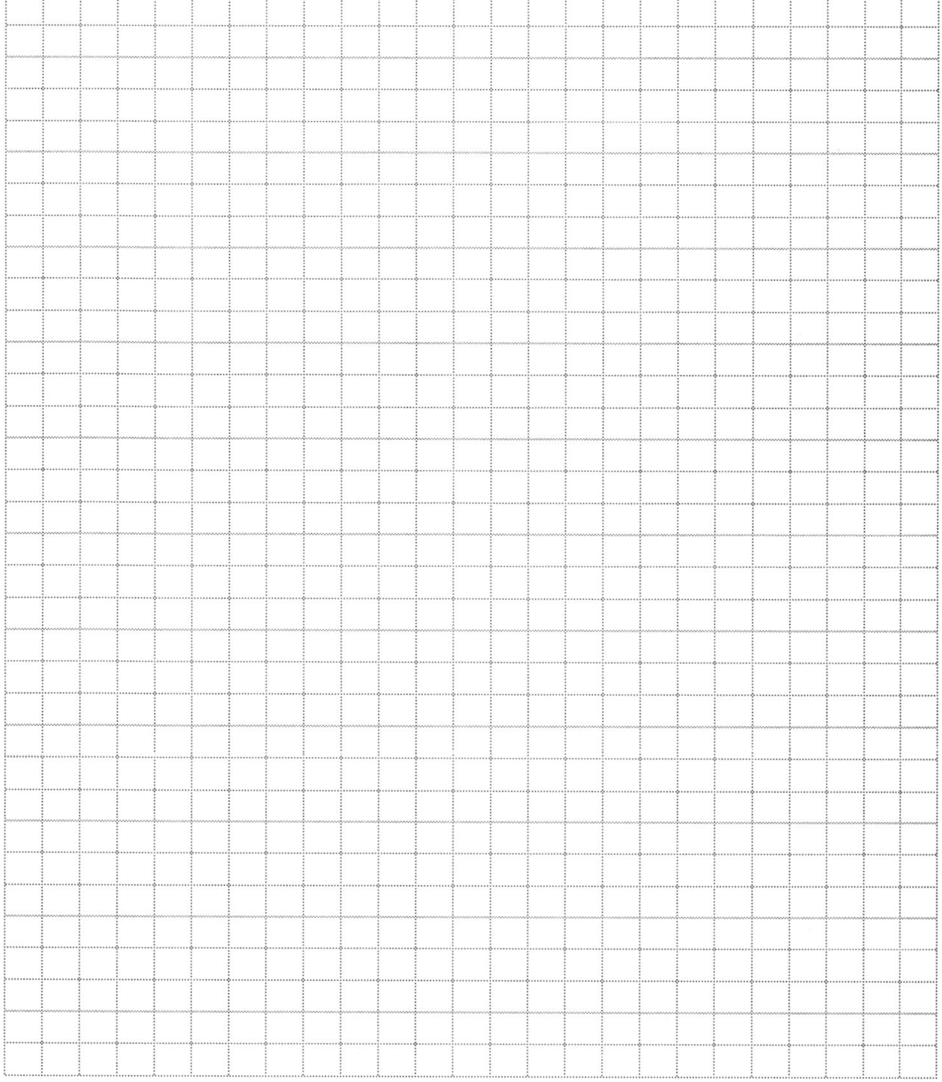

Experiment 2.5

The 90° ^{13}C Decoupler Pulse with Inverse Configuration

1. Purpose

In "inverse" experiments one observes protons and decouples heteronuclei X (e.g. ^{13}C, ^{15}N). On NMR instruments built later than 1991 the pulse-lengths in normal and inverse set-up are usually quite similar. Older instruments, however, use quite different signal routings and r.f. sources in these two configurations. Therefore, before starting an inverse experiment, such as an HMQC experiment (Exp. 10.13), the pulse-lengths have to be determined in this spectrometer configuration. Here the ^{13}C decoupling pulse is determined; prior to this experiment you have to perform Experiment 2.4. Depending on the attenuation of the decoupler the pulse-durations are quite different. Short ("hard") pulses with low decoupler attenuation are used for 90° or 180° pulses during a pulse sequence, whereas long ^{13}C pulses with high decoupler attenuation are needed for composite pulse decoupling such as GARP during acquisition. Thus, this experiment has typically to be performed twice, once with low and once with high attenuation of the decoupler.

2. Literature

[1] A. Bax, *J. Magn. Reson.* **1983**, *52*, 76–80.

3. Pulse Scheme and Phase Cycle

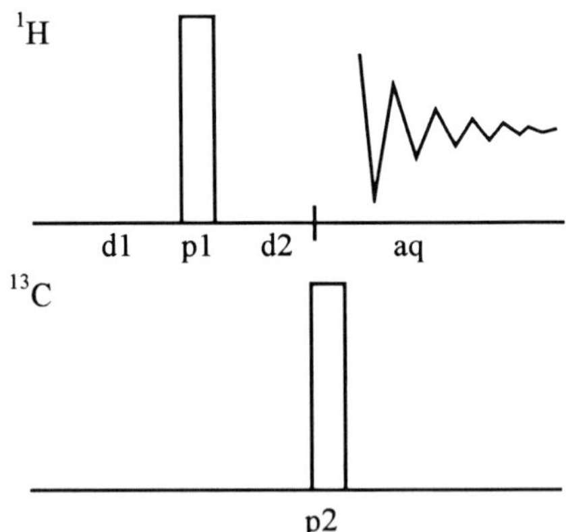

p1: x, -x, -x, x, y, -y, -y, y

p2: x, -x, -x, x, y, -y, -y, y

aq: x, -x, -x, x, y, -y, -y, y

4. Acquisition

Time requirement: 10 min

Sample: 10% $CHCl_3$ in [D_6]acetone, *do not use a degassed and sealed sample* since that would make the relaxation times of carbons and protons exceedingly long.

Obtain both offsets for the ^{1}H and ^{13}C signals of the sample with the spectrometer in the normal mode and tune both coils to the sample. Older spectrometers are then changed to inverse set-up, and an inverse probe-head (inner coil ^{1}H, outer coil ^{13}C) is placed into the magnet. Since the inverse set-up varies widely dependent on the year of spectrometer's construction it cannot be discussed here. Load the correct pulse program to obtain an inverse ^{1}H NMR spectrum with ^{13}C decoupling. You have to set:

td: 4 k
sw: 500 Hz
o1: on resonance of ^{1}H signal
o2: on resonance of ^{13}C signal
p1: 90° ^{1}H transmitter pulse in inverse configuration (Exp. 2.4)
p2: ^{13}C decoupler pulse in inverse configuration, 0 μs as starting value, to be varied
d1: 20 s
d2: $1/[2\,J(C,H)] = 2.33$ ms for $^{1}J(C,H) = 215$ Hz
rg: receiver gain for correct ADC input
^{13}C decoupler attenuation [0 dB]
ns: 1

Record a first spectrum with p2 = 0 and adjust the phase of the big signal stemming from the protons bound to ^{12}C in dispersion; look for a clean antiphase pattern of the ^{13}C satellites. Then repeat the experiment with increasing pulse-durations p2 until you get a zero for the satellites, which corresponds to the 90° decoupler pulse.

In a second set of experiments use a high decoupler attenuation for GARP [13 dB] and vary it so that p2 is in the region of 70 μs.

5. Processing

Use standard 1D processing (see Exp. 3.1) applying an exponential window with a line-broadening factor lb = 1 Hz.

6. Result

The figure on page 30 shows the results of a 0° decoupling pulse (**a**) and of a 90° decoupling pulse (**b**) obtained on an AMX-500 spectrometer with an inverse probe-head. Note the disappearance of the satellites in (**b**).

7. Comments

The product operator formalism is exactly the same as given in Experiment 2.3, only the notation for C and H spins has to be interchanged. If this method is used to determine ^{13}C decoupler pulses for GARP decoupling, be sure that the chosen ^{13}C offset is correct.

Inverse experiments can be performed in normal probe-heads as well. As an exercise you may measure the pulse-length using a standard dual probe-head.

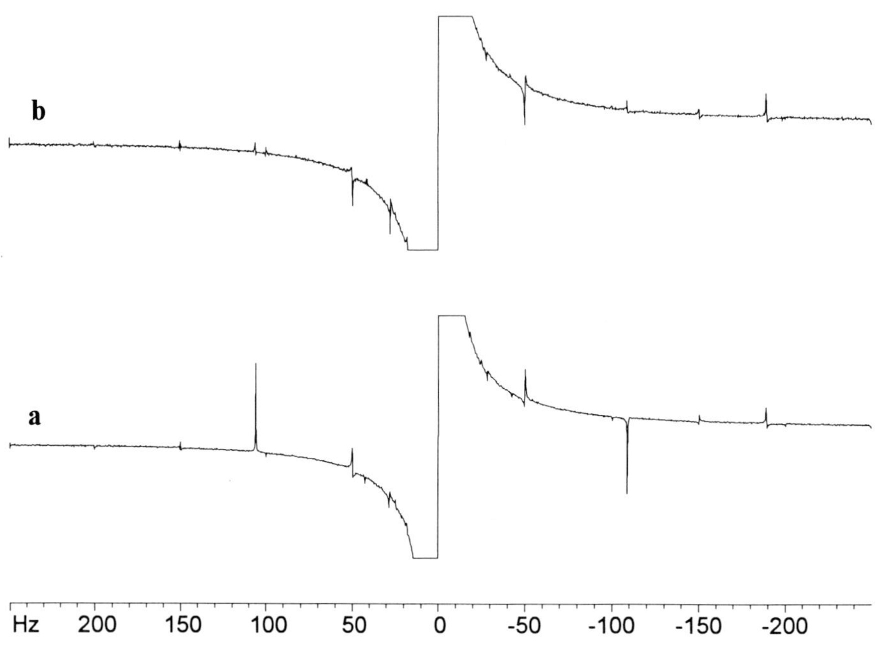

8. Own Observations

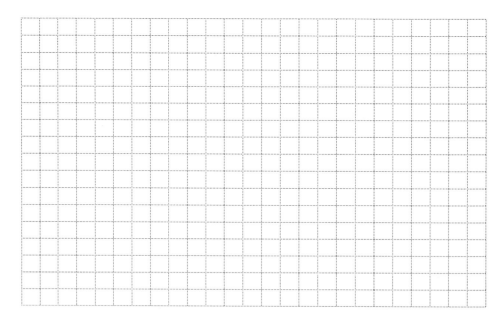

Experiment 2.6

Composite Pulses

1. Purpose

Ideally, the strong r.f. pulses used in NMR are rather short (~ 10 μs), have true rectangular shape, and excite the resonances present in the sample equally without a marked offset dependence. In reality these conditions are rarely met. Therefore composite pulses have been designed to compensate for pulse imperfections and offset dependent deviations. In addition, composite pulses are now widely used in all current broadband decoupling schemes and within spin-locks. This educational experiment demonstrates the inversion performance of a composite 180° pulse on chloroform at a large offset. For the composite pulse, the sequence $90°_y$, $180°_x$, $90°_y$ is chosen.

2. Literature

[1] M. H. Levitt, *Prog. NMR Spectrosc.* **1985**, *18*, 61–122.

3. Pulse Scheme and Phase Cycle

Experiment **a**

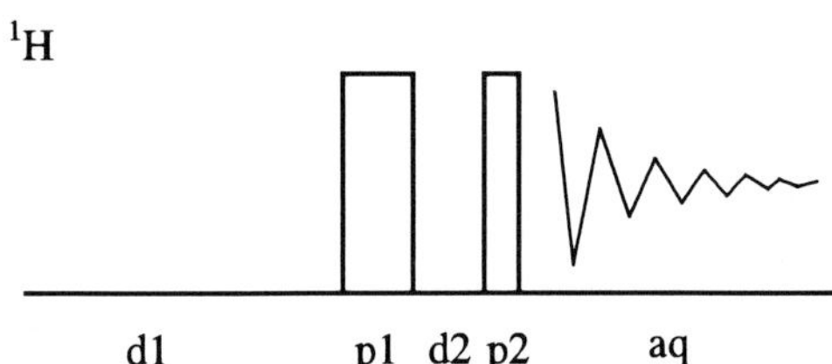

Experiment **b**

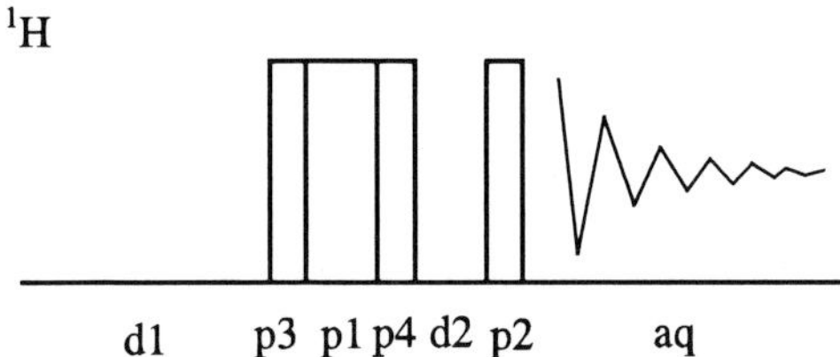

p1: x, -x

p2: x, x, -x, -x, y, y, -y, -y

p3, p4: y, -y

aq: x, x, -x, -x, y, y, -y, -y

4. Acquisition

Time requirement: 10 min

Sample: 10% $CHCl_3$ in [D_6]acetone; *do not use a degassed and sealed sample,* since that would make the relaxation time of the $CHCl_3$ protons exceedingly long.

The probe-head should be tuned to the sample. Load standard ^1H parameters, record a normal ^1H NMR spectrum and note the offset of $CHCl_3$. Determine exactly the 90° pulse-duration according to Experiment 2.1. The experiment compares the inversion property of a normal 180° pulse with that of a composite one. Therefore you have to perform two experiments **a** and **b** according to the two pulse schemes shown above. You have to set:

td: 64 k
sw: 80 ppm
o1: 10 kHz towards higher frequencies from the resonance of the $CHCl_3$ signal
p2, p3, p4: 90° ^1H transmitter pulse
p1: 180° ^1H transmitter pulse
d1: 30 s
d2: 10 ms
rg: receiver gain for correct ADC input
transmitter attenuation (3 dB)
ns: 2

5. Processing

Use standard 1D processing (see Exp. 3.1) applying an exponential window with a line-broadening factor lb = 1 Hz. Adjust the phase of the $CHCl_3$ signal to be negative.

6. Result

The figure shows the results obtained on an AMX-500 spectrometer in a 5 mm multinuclear probe-head. In **a** the signal after inversion by a normal 180° pulse is given, in **b** the same signal is shown, but inverted by the composite pulse, leading to nearly fourfold greater intensity.

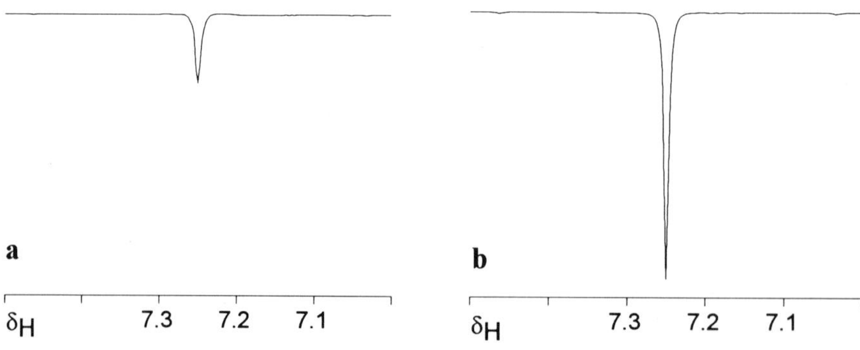

7. Comments

In both experiments **a** and **b** the magnetization is first inverted by a 180° pulse and after a short relaxation delay d2 read by a 90° pulse, similar to the T_1 determination as given in Experiment 6.1. A large offset is chosen to demonstrate the offset dependence of normal r.f. pulses as used in experiment **a**, which can be compensated by the use of composite pulses as shown in experiment **b**. Here the first $90°_y$ pulse of the composite pulse turns the magnetization towards the x-axis; however, because of pulse imperfections and the offset dependence, we assume that the magnetization ends somewhat above the x-axis. A perfect $180°_x$ pulse would now align the magnetization into the mirror position beneath the x-axis, and the subsequent imperfect $90°_y$ pulse would turn the magnetization exactly into $-z$. Since the deviation after the first $90°_y$ pulse is only small, even an imperfect 180° pulse will be able to correct the situation to the effect that the total performance of the composite 180° pulse is far better than that of a single 180° pulse, as borne out by the experiment.

There are many different varieties of composite pulses serving various purposes. Composite pulses are the standard building blocks of current decoupling techniques and should be used where large offsets are required, such as in INADEQUATE experiments. Many current spin-lock schemes (MLEV, DIPSI etc.) use composite pulses.

8. Own Observations

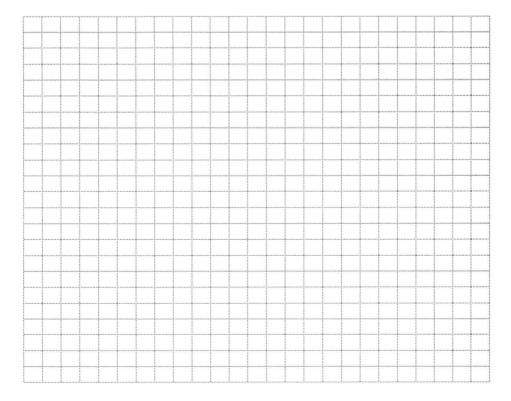

Experiment 2.7

Radiation Damping

1. Purpose

The pulse-length determination on samples dissolved in water often reveals a strong and special signal pattern at or near the 180° pulse which is caused by an effect called radiation damping. This effect, although already described in detail in the early days of NMR, becomes a problem especially at high magnetic fields using probe heads with a high Q-factor. Transverse magnetization created by an r.f. pulse induces a voltage in the nmr coil. This oscillating signal is amplified in the preamplifier and detected by the spectrometer. However, at the same time, the voltage in the nmr coil results in a current, producing in turn an r.f. field, which lags behind the transverse magnetization by 90°. Therefore it provides a torque to restore the magnetization towards the $+z$ axis, leading to a much faster relaxation than expected from natural T_1 and T_2 processes. This is the cause of the large line-width of the water signal after a 90° pulse. In the educational experiment described here, radiation damping is envisaged on a water sample.

2. Literature

[1] N. Bloembergen, R. V. Pound, *Phys. Rev.* **1954**, *95*, 8–12.

[2] A. Szöke, S. Meiboom, *Phys. Rev.* **1959**, *113*, 585–586.

[3] A. Abragam, *The Principles of Nuclear Magnetism*, Oxford Univ. Press **1962**, 73–74.

[4] R. Freeman, *A Handbook of Nuclear Magnetic Resonance*, Longman, Harlow, **1987**, 177–179.

[5] A. Sodickson, W. E. Maas, D. G. Cory, *J. Magn. Reson. Ser. B* **1996**, *110*, 298–303.

[6] X.-A. Mao, C.-H. Ye, *Concepts in Magn. Reson.* **1997**, *9*, 173–187.

3. Pulse Scheme and Phase Cycle

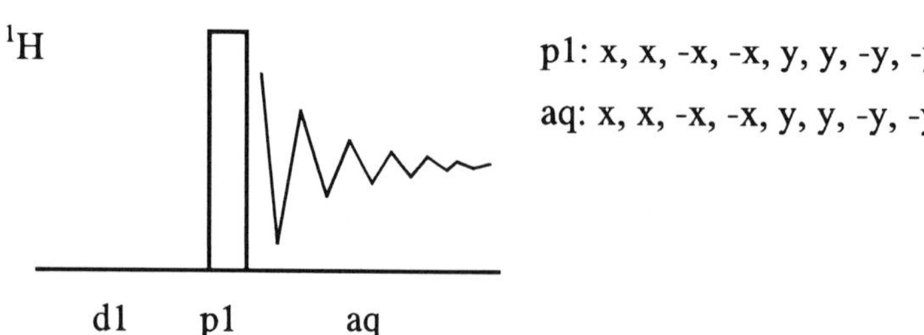

4. Acquisition

Time requirement: 10 min

Sample: 90% H_2O with 10% D_2O

The probe-head should be tuned to the sample. Load standard ^1H parameters, record a normal ^1H NMR spectrum and note the offset of H_2O. You have to set:

td: 4 k
sw: 500 Hz
o1: on resonance of H_2O signal
p1: **a**: 360° and **b**: 180° ^1H transmitter pulse
d1: 2s
rg: receiver gain for correct ADC input
transmitter attenuation (3 dB)
ns: 1

Determine the 360° pulse as described in Experiment 2.1 and record the spectrum **a**. In addition, record the spectrum **b** with a 180° pulse and compare the residual signal strength in both experiments.

5. Processing

Use standard 1D processing (see Exp. 3.1) applying an exponential window with a line-broadening factor lb = 0.3 Hz.

6. Result

The figure on page 36 shows the result obtained on an AMX-500 spectrometer with an inverse probe-head. In **a** the absence of radiation damping after a 360° pulse, in **b** the hyperbolic secant signal behavior produced by radiation damping after a 180° pulse is shown. Both traces are plotted on the same vertical scale!

7. Comments

The nutational behavior of the signals produced by radiation damping can be calculated from the Bloch equations [3,5]. Radiation damping can be observed even after a perfect 180° pulse, since transverse components can be created by thermal noise or r.f. leaking effects. Thus, if one applies a pulsed field gradient (see Chapter 11) after the 180° pulse to destroy any residual transverse components, the effect may still be observed [5]. In pulse sequences which establish $-z$-magnetization, radiation damping may cause loss of signal during the following delays, since the magnetization is driven back by the r.f. field in the coil. Radiation damping may further cause line-broadening and distortion in the relative intensities of multiplets. Possible remedies are detuning of the probe-head or, more recently, use of probe-heads with Q-switching or with active electronic feedback circuits.

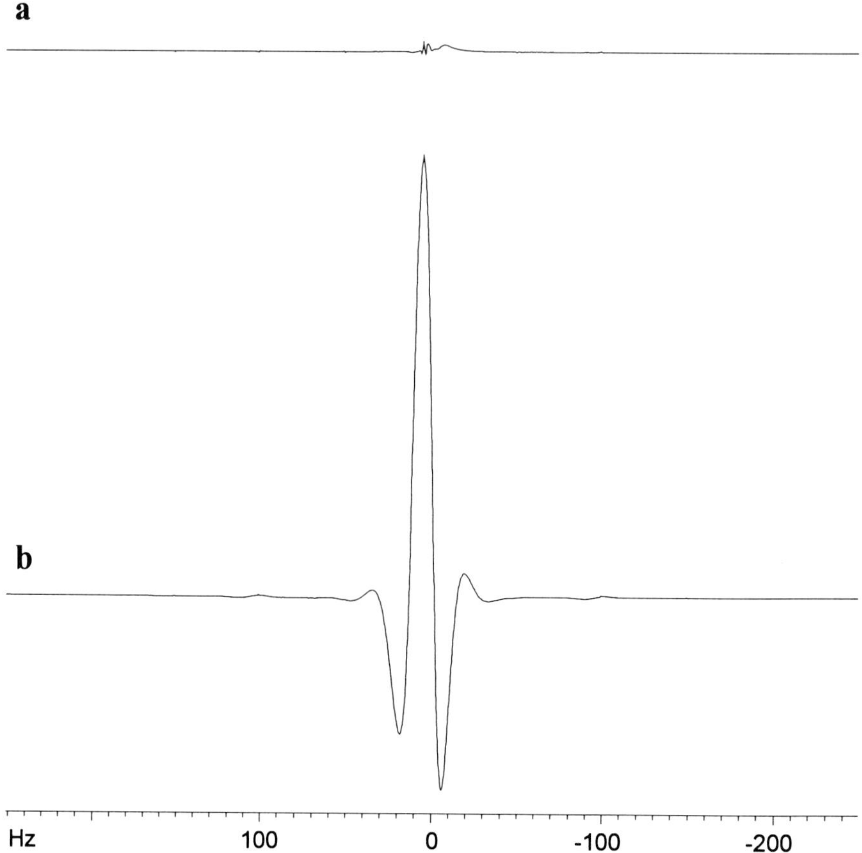

8. Own Observations

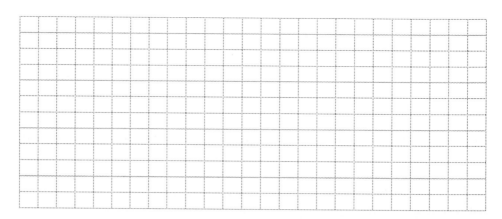

Experiment 2.8

Pulse and Receiver Phases

1. Purpose

The radio-frequency pulses of the transmitter and the NMR receiver have phases which are given in the pulse schemes throughout this book. By use of phase-cycling procedures (i.e. systematic variation of transmitter and receiver phases), several important features of NMR experiments can be realized, such as the suppression of artefacts, e.g. quadrature images, or the selection of desired coherences, e.g. double quantum signals, and in 2D or 3D spectroscopy the sign discrimination of the signals in the indirect dimension. Thus the understanding of the basic phase behavior of transmitter and receiver is of fundamental importance. In this educational experiment we describe a method of studying the phases generated by the NMR instrument using the single-line spectrum of chloroform.

2. Literature

[1] A. E. Derome, *Modern NMR Techniques for Chemistry Research*, Pergamon, Oxford, **1987**, 73–85.

[2] M. H. Levitt, *J. Magn. Reson.* **1997**, *126*, 164-182.

3. Pulse Scheme and Phase Cycle

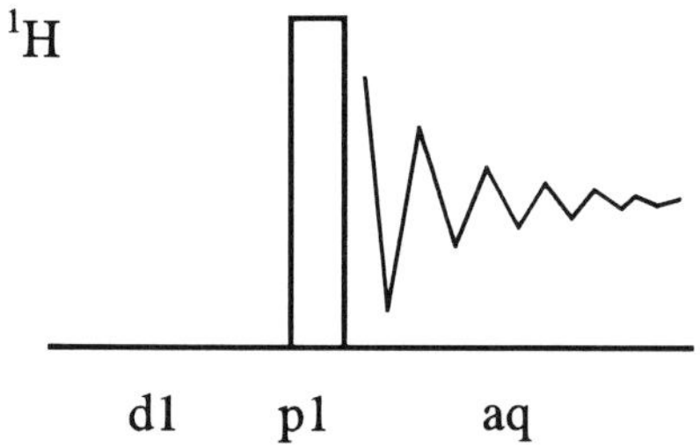

phase of p1 and aq to be varied

4. Acquisition

Time requirement: 20 min

Sample: 10% $CHCl_3$ in [D_6]acetone with added $Cr(acac)_3$

Load standard ^{1}H parameters, record a normal ^{1}H NMR spectrum and note the offset of $CHCl_3$. You have to set:

td: 4 k
sw: 500 Hz
o1: 50 Hz off resonance of $CHCl_3$ signal

p1: 90° ^{1}H transmitter pulse
d1: 1 s
rg: receiver gain for correct ADC input
transmitter attenuation (3 dB)
ns: 1

First set the offset on resonance of the chloroform signal, and set the instrument in the mode to display both quadrature channels of the receiver. Record an FID and, in your pulse program, change the transmitter phase (or on recent instruments the corresponding phase correction parameter) so that only the left quadrature channel receives a signal. Then set the offset 50 Hz off resonance and repeat the experiment. The left quadrature channel will display a cosine FID whereas the right channel will display a sine FID. Transform this FID and adjust the processing phases for absorption. Now change the transmitter phase in 90° steps and observe the changes on both FID channels and on the spectrum. Repeat this procedure, but leave the transmitter phase unchanged and cycle instead the receiver phase. In a final experiment use two transients and observe the adding of the FIDs. Then introduce a 180° transmitter or receiver phase shift for the second transient in your pulse program and observe the subtraction of the FIDs.

5. Processing

Use standard 1D processing (see Exp. 3.1) applying an exponential window with a line-broadening factor lb = 1 Hz. Adjust the phase of the first spectrum for pure absorption and always use the same phase correction.

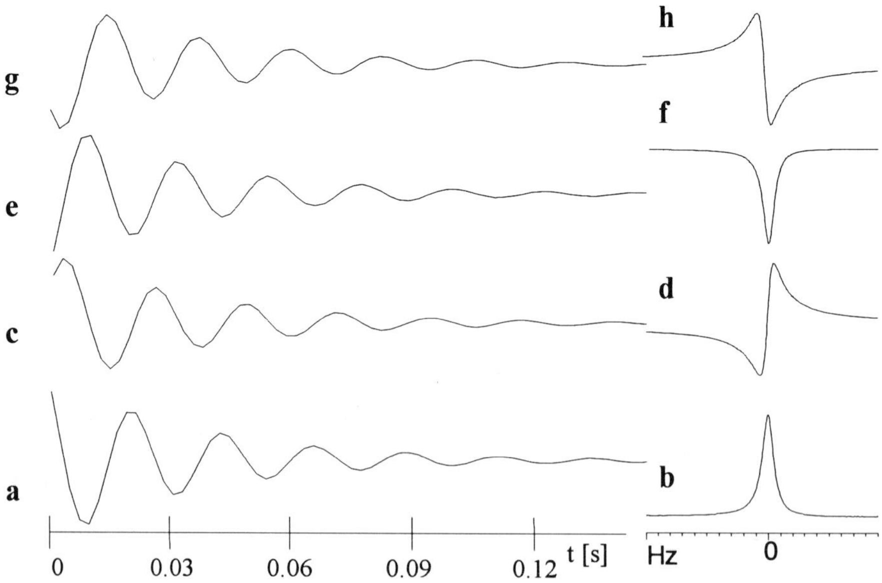

6. Result

The figure on page 38 shows the result obtained on an ARX-200 spectrometer in a dual probe head. An expansion of the FID in the left quadrature channel is plotted with the corresponding transformed spectrum. In **a** and **b** the initial FID and spectrum is shown, whereas in **c/d**, **e/f**, and **g/h** the transmitter phase was incremented by 90° steps.

7. Comments

If the transmitter is exactly on resonance, no chemical shift can develop after excitation. Therefore only one channel of the phase-sensitive detector will receive a signal. If, however, the transmitter offset is somewhat displaced with respect to the resonance frequency, chemical shift evolution will take place to form a sinusoidal FID as given by the product operator treatment in Eq. (2). The cosine and the sine components are detected separately by the quadrature receiver.

$$I_{Hz} \xrightarrow{90^\circ I_x} -I_{Hy} \tag{1}$$

$$-I_{Hy} \xrightarrow{\Omega I_z t} -I_{Hy}\cos\Omega t + I_{Hx}\sin\Omega t \tag{2}$$

As demonstrated in this experiment, the phase of a signal can be changed with either transmitter or receiver phase. This is the basis of all phase-cycling procedures used throughout this book.

8. Own Observations

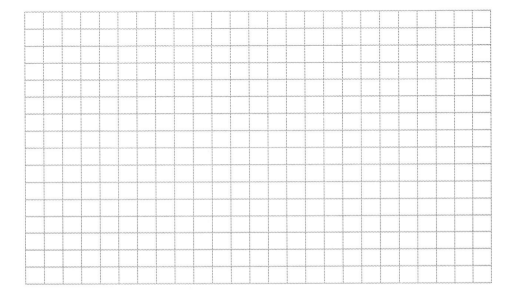

Experiment 2.9

Determination of Radiofrequency Power

1. Purpose

In many experiments the transmitter or decoupler power is attenuated to produce a radiofrequency field of a certain strength. This is needed e.g. for presaturation to suppress water signals (Exp. 6.17), for SPT investigations (Exp. 4.6), for TOCSY (Exp. 10.18), or for ROESY (Exp 10.20) experiments. The NMR literature uses a variety of measures to describe the power of a frequency source. In the experiment described here the fundamental parameters 90° pulse-width [μs], transmitter attenuation [dB], transmitter power [W], radiofrequency field strength [Hz] and peak to peak voltage [V] are measured and interrelated in a tabular form. The example is chosen from ^{1}H NMR spectroscopy, but the procedure can be used for any nucleus. In addition to a deeper understanding of the function of a frequency source, this experiment provides a check on the performance of the transmitter and the attenuators.

2. Literature

[1] M. L. Martin, J. J. Delpuech and G. J. Martin, *Practical NMR Spectroscopy*, Heyden, London, **1980**, 100–101.

3. Pulse Scheme and Phase Cycle

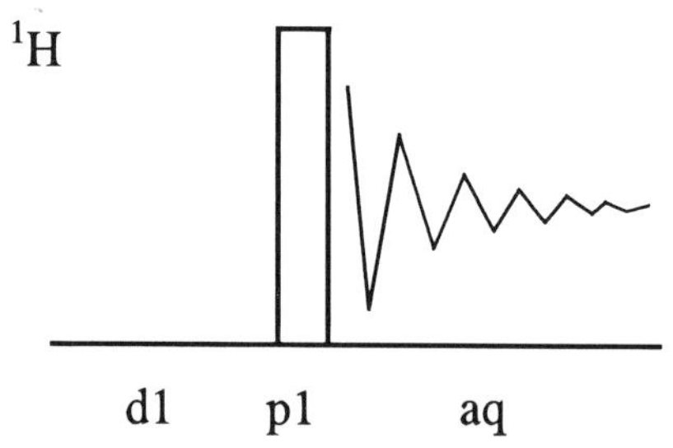

p1: x, x, -x, -x, y, y, -y, -y

aq: x, x, -x, -x, y, y, -y, -y

4. Acquisition

Time requirement: 2 h

Sample: 10% $CHCl_3$ in [D_6]acetone

Experimentally, this is a 90° pulse-width determination (actually performed by finding the 180° pulse), exactly as described in Experiment 2.1. However, in addition, we vary the transmitter attenuation and determine the pulse-width as a function of transmitter attenuation. On older instruments which do not allow variation of the transmitter power, the experiment can be performed in the inverse mode using the decoupler as the frequency source (Exp. 2.4). Load standard ^1H NMR parameters, record a normal ^1H NMR spectrum, and note the offset of $CHCl_3$. You have to set:

td: 4 k
sw: 500 Hz
o1: on resonance of ^1H signal
p1: 90° ^1H transmitter pulse, to be determined for each attenuation level
d1: 60 s
ns: 1
transmitter attenuation: 0 dB initial value, to be increased in 3 dB steps

Determine the 90° pulse-width at 0 dB attenuation. Then change the attenuation in 3 dB steps and redetermine the 90° pulse width for each attenuation level by varying p1. If available, measure with an oscilloscope the peak-to-peak voltage of your pulse (for this set ns to 64 k, d1 to 10 ms, p1 to 40 μs and td to 256 points in order to obtain rapid pulsing; connect the transmitter cable to an external attenuator and an oscilloscope with 50 Ω impedance).

5. Processing

Use standard 1D processing (see Exp. 3.1) applying an exponential window with a line-broadening factor lb = 1 Hz. From the 90° pulse-width calculate the radiofrequency field using Equation (3). Verify that the expected values correspond to the attenuation. Note the large power range, from 0 to 90 dB, required in different types of NMR experiments.

6. Result

The table lists results obtained with an ARX-200 spectrometer.

Attenuation [dB]	90°Pulse width [μs]	γB_1 [Hz]	U_{pp} [V]	P [W]
0	7.8	32051	107	28.6
3	10.3	24271	82	16.8
6	13	19230	59	8.7
9	19.5	12820	41	4.2
12	27.5	9090	29	2.1
15	39.5	6329	20	1.
18	55	4545	15	0.56
21	75	3333	11	0.30
24	105	2380	7.9	0.16
27	145	1724	5.9	0.087
30	200	1250	4.3	0.046
33	290	862	3.3	0.027
36	425	588	2.2	0.012

39	600	416	1.6	0.0064
42	825	303	1.3	0.0042
45	1100	227	1	0.0025
48	1600	156	0.74	0.0014
51	2400	104	0.46	0.00053
54	3500	71	0.33	0.00027
57	5000	50	0.23	0.00013
60	6800	37	0.17	0.000072
63	9000	28	0.13	0.000042
66	12500	20	0.099	0.000024
69	19000	13	0.073	0.000013
72	28500	9	0.054	0.0000073
75	41000	6	0.043	0.0000046
78	55600	4.5	*	
81	80000	3	*	
84	110000	2.3	*	
87	145000	1.7	*	
90	200000	1.25	*	

* oscilloscope used not sensitive enough

7. Comments

The magnetization vector precesses around the radiofrequency field B_1 according to Equation (1).

$$\omega = 2\pi\nu = \gamma B_1 \tag{1}$$

The angle of precession θ, measured in radians, is proportional to the pulse-width p and is given by Equation (2). For a 90° pulse ($\theta = \pi/2$) Equation (3) follows from (1) and (2).

$$\theta = \gamma B_1 \cdot p \tag{2}$$

$$\gamma B_1 = \frac{1}{4p} \tag{3}$$

Thus, knowing the pulse-width of a 90° pulse at a certain transmitter attenuation, we can estimate the radiofrequency field strength γB_1 measured in Hz from the simple relationship in Equation (3).

The dB unit used on NMR instruments is defined by Equation (4).

$$\mathrm{dB} \equiv 10 \log (P/P_0) \tag{4}$$

Thus, attenuation by 3 dB means, that the ratio P/P_0 decreases by a factor of 2. The power P of a transmitter measured in watt is given by Equation (5).

$$P = \frac{U_{\text{eff}}^2}{R} \tag{5}$$

U_{eff} is the effective voltage, which is equal to $\frac{U_{\text{pp}}}{2\sqrt{2}}$, and R is the load resistance. By measuring the peak to peak voltage U_{pp} on a 50 ohm load one can calculate the transmitter power in watt. The strength of the radiofrequency field is proportional to the square root of the transmitter power, as evident from Equation (6), and is therefore proportional to the peak to peak voltage measured with an oscilloscope. It can be calculated in tesla (SI magnetic field unit) from equation (6) if the quality factor Q, the frequency ν, and the volume V of the NMR coil are known.

$$B_1 = 3 \cdot 10^{-4} \sqrt{\frac{PQ}{\nu V}} \tag{6}$$

On recent NMR instruments one also finds the unit dBm. Whereas dB is a relative unit, dBm refers to a P_0 of 1 mW, as given in Equation (7).

$$\text{dBm} = 10 \log(P/\ 1\ \text{mW}) \tag{7}$$

Thus, the power of 1 mW corresponds to a dBm value of 0, 1 W corresponds to 30 dBm and 1 μW to –30 dBm.

8. Own Observations

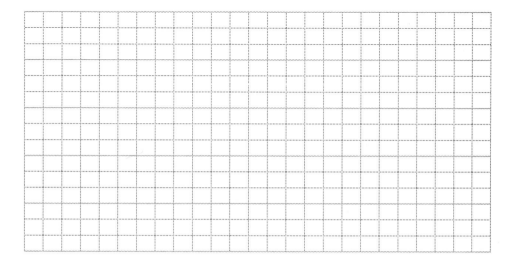

Chapter 3

Routine NMR Spectroscopy and Standard Tests

This chapter begins by describing how to record standard ^{1}H and ^{13}C NMR spectra. These descriptions are somewhat detailed, so that a beginner in this field should find sufficient advice. These experiments are followed by a description how to apply various window functions (Exp. 3.3) and how to use a PC for computer aided spectral analysis (Exp. 3.4). Furthermore, we describe in this chapter several important test procedures which are essential for the maintenance of instruments getting on in years. Performed regularly they give an early indication of developing problems.

Since a great variety of spectrometers and, much more important, of field strengths exists, one should have in mind all the possible consequences thereof when performing the experiments at a field strength different from that used here and comparing the results. These include the following:

1. The experimental time given depends on the acquisition time, since it is determined by the spectral width, entered in Hz and thus depending on field strength (and sometimes on the way the data are transferred to disk). It should be mentioned here that the time given for each experiment only roughly represents the time from the starting command just to the last acquisition step; preparations, setting-up, processing and output are not included; the smallest time unit is 5 min.
2. At different field strength the appearance of a spectrum may change due to differences in the extent of signal overlap and to higher-order effects in spectra of spin-coupled nuclei.
3. The field strength also affects the digital resolution (for a given spectral width and time-domain data file); the data file and/or the processing parameters may be changed accordingly.
4. At other field strengths the relaxation behaviour may be different, with consequences for the line-widths.
5. The signal-to-noise ratio depends on field strength (also, for example, on the probe-head used and on its tuning).

Literature

[1] M. L. Martin, J. J. Delpuech, G. J. Martin, *Practical NMR Spectroscopy*, Heyden, London, **1980**.

[2] A. E. Derome, *Modern NMR Techniques for Chemistry Research*, Pergamon, Oxford, **1987**.

[3] E. Fukushima, S. B. W. Roeder, *Experimental Pulse NMR*, Addison-Wesley, Reading, **1981**.

Experiment 3.1

The Standard ^{1}H NMR Experiment

1. Purpose

The aim of the standard ^{1}H NMR Experiment is to record a routine proton NMR spectrum in order to get structure-related information for the protons of the sample, i.e. chemical shifts, spin–spin couplings, and intensities. Here we apply this standard procedure to ethyl crotonate.

2. Literature

[1] A. E. Derome, *Modern NMR Techniques for Chemistry Research*, Pergamon, Oxford, **1987**.
[2] K. K. M. Sanders, B. K. Hunter, *Modern NMR Spectroscopy*, 2nd Edition, Oxford University Press, Oxford, **1993**.
[3] H. P. Friebolin, *Basic One- and Two-Dimensional NMR Spectroscopy*, 2nd Edition, VCH, Weinheim, **1993**.
[4] H. Günther, *NMR Spectroscopy*, 2nd Edition, Wiley, Chichester, **1995**.

3. Pulse Scheme and Phase Cycle

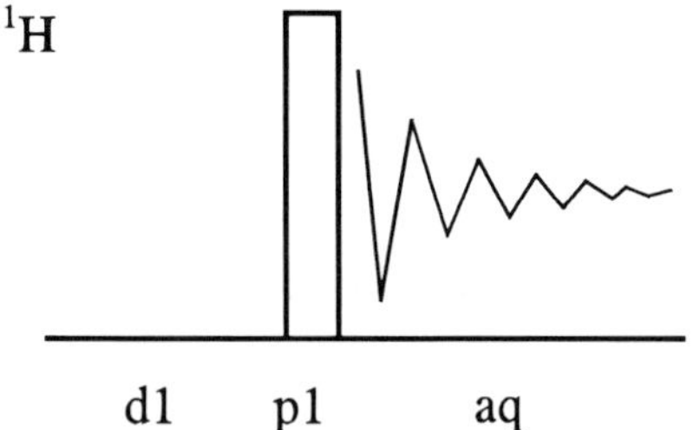

p1: x, x, -x, -x, y, y, -y, -y
aq: x, x, -x, -x, y, y, -y, -y

4. Acquisition

Time requirement: 5 min

Sample: 5% ethyl crotonate in $CDCl_3$ with TMS as standard

For sample preparation fill a clean and dry 5 mm sample tube with 0.7 ml $CDCl_3$ (5 cm filling height), 40 μl ethyl crotonate, and few drops of a solution of 3% TMS in $CDCl_3$ (for easier dosage). Clean the tube outside, close it with a cap, mark it (do not use a flag-like label which hinders rotation) and put it into the spinner turbine. Be sure that it is tight, but not too tight in the spinner, and adjust the depth of the tube using the depth gauge.

Put the tube into the magnet by means of the air lift, adjust the rotation frequency (20 Hz), and display the lock signal on the screen. Perform the lock procedure and optimize the field homogeneity (see Ch. 1, Section 1.4).

The spectrometer is to be adjusted for ^1H observation in quadrature detection mode. Load the ^1H acquisition program, which comprises the following basic commands: zero memory, set the relaxation delay, set the excitation pulse, perform acquisition, and write the data on file.

For acquisition the following parameters have to be set:

td: 32 k

td (time domain) is the number of points at which the free induction decay (FID) is sampled and the data stored. This parameter has to be chosen according to the desired digital resolution of the FID, which in turn depends on the spectral width sw. The settings of td and sw determine the acquisition time aq.

sw: 20 ppm

sw, the spectral width, has to be chosen so that all types of protons are within this spectral window; otherwise folding occurs. Folding of noise is prevented by appropriate filters which are set by the software corresponding to sw. The settings of sw and td determine the *acquisition time* aq, the time during which the FID data are acquired. The relationship between these three parameters, where aq is usually the dependent variable and sw has to be expressed in Hz, is described by the following fundamental equation (1):

$$\mathrm{aq} = \mathrm{td}/(2\ \mathrm{sw}) \tag{1}$$

Thus, with the above settings at 300 MHz, the acquisition time is 2.7 s.

o1: frequency (offset) *of the r.f. pulse at the center of the ^1H NMR spectrum:*

In quadrature detection mode the frequency of the exciting r.f. pulse (often called transmitter offset) is positioned in the center of sw, e.g. at about $\delta_H = 7$. On older instruments without digital lock, o1 depends on the field position of the deuterium lock and thus on the deuterated solvent used.

p1: 30°

p1, the ^1H transmitter pulse, creates an observable *x,y*-magnetization by tipping the magnetization vector towards the –y-axis through an angle α, determined by Equation (2):

$$\alpha = (360/2\pi)\ \gamma B_1 \cdot p \tag{2}$$

where γ is the proton gyromagnetic ratio, γB_1 the field strength of the radiofrequency field, and p the duration of the pulse. Since p1 is known for a flip angle $\alpha = 90°$ (see exp. 2.1), the duration of p1 corresponding to 30° is easily obtained.

Although a 90° pulse gives maximum signal intensity (Exp. 2.1), a shorter pulse-length of about 30° is used for routine work with data accumulation in order to avoid a long pulse repetition time of 5 T_1, which is necessary after a 90° pulse (see d1).

It should be mentioned here that between the irradiating pulse and the opening of the detector a *preacquisition delay* is introduced. It prevents the break-through of the transmitter pulse into the detector, is in the order of 10 to 100 μs, and is automatically calculated and set by the software.

d1: 0.1 s

d1, the relaxation delay, is introduced in order to establish thermal equilibrium of the spin system before the excitation pulse is applied. In routine ^{1}H NMR work with small flip angles and relatively long acquisition times, this delay may be short, especially for larger molecules with short T_1-values, since it is the *pulse repetition time*, the sum of aq and d1, which serves for spin–lattice relaxation.

rg:

rg, the receiver gain, has to be carefully adjusted so that the incoming FID does not exceed the ADC input limits. Otherwise signal distortions at the baseline will occur after Fourier transformation.

ns: 8

ns, the number of "scans" (individual FIDs), is chosen so that a reasonable signal-to-noise ratio is obtained in the final spectrum. Because of phase cycling a multiple of 8 and a minimum of 4 is advisable.

5. Processing

After data acquisition you have to process the data. Perform the following steps using the given settings:

size of the processed real data file: set **si = 16 k**. Normally si corresponds to td/2 (see acquisition), since, after Fourier transformation, half of the td data adresses contain the real part of the spectrum. This gives a digital resolution (Hz/data point) after Fourier transformation of 2 sw/td. (Note that on the older Bruker instrument series the parameter SI corresponds to 2 si of the newer ones.)
However, si may be set to a value greater than td/2, e.g. si = 32 k; this procedure is termed "zero filling" and leads, according to Equation (3) for the general case, to a higher digital resolution:

$$\text{digital resolution} = \text{sw/si} \quad (3)$$

With the above data, the digital resolution is 0.37 Hz/point, a reasonable value in comparison to the line-width usually obtained for a non-degassed sample. If a higher digital resolution is desired, one may optimize sw, increase td, or perform zero filling.

baseline correction: perform this correction on the *FID* in order to eliminate the d.c. offset between the two channels used in quadrature detection mode.

digital filtering: for standard applications use as window function an exponential window on the FID, characterized by the line-broadening factor lb; here set **lb = 0.1 Hz**. This type of digital filtering generally improves the signal-to-noise ratio, but at the cost of resolution. Here a very mild filtering is used since there is no signal-to-noise problem. On the other hand, by applying a Lorentz–Gauss multiplication, a resolution enhancement may be achieved; here one has to pay with a reduction of the signal-to-noise ratio.

Fourier transformation: use the correct type in accordance with the simultaneous or sequential quadrature acquisition mode.
phase correction: adjust to achieve pure absorption mode signals.

baseline correction: perform this correction on the *spectrum* in order to remove baseline rolling.

referencing: set the TMS signal to $\delta_H = 0$.

peak picking: choose the desired level.

integration: carefully generate the integrals.

plot: set parameters and plot spectrum, including integrals, peak picking, and the relevant acquisition and processing parameters.

6. Result

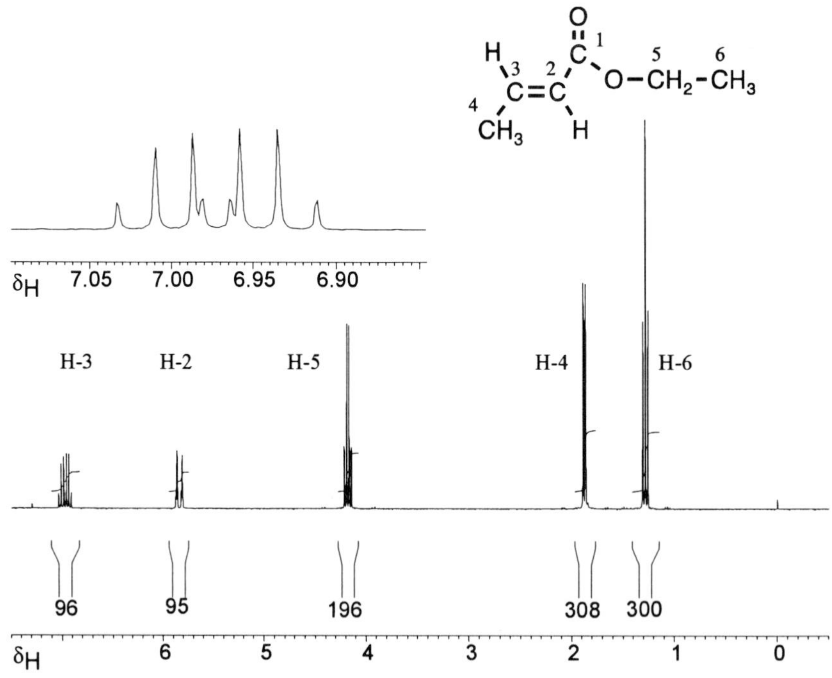

The figure shows the 300 MHz ^1H NMR spectrum of ethyl crotonate recorded on an ARX-300 spectrometer equipped with a dual probe-head; the region $\delta_H = -0.5$ to 7.5

including the integrals is displayed. The insert contains an expansion of the signal at ca. $\delta_H = 7$ (H-3) so that the coupling constants can be determined.

A closer inspection of the integrals reveals that those of H-2 and H-3 are too small as compared to the integrals of the two methyl groups. This may be due to the fact that the T_1 values of the protons of ethyl crotonate are relatively long *and* somewhat shorter for the two CH_3 groups than for H-2 and H-3. As an exercise you should perform the experiment with d1 = 3 s, corresponding to a pulse repetition time of nearly 6 s. (Concerning the determination of T_1 values see Exp. 6.1).

7. Comments

The excitation pulse p1 converts the equilibrium magnetization I_{Hz} of the ^{1}H nuclei into a transverse magnetization $-I_{Hy}$ as shown in Equation (1). During the acquisition time chemical shifts and spin–spin couplings develop in the *x,y* plane, as shown seperetely in Equations (2) and (3) and are detected by the receiver in the *x,y* plane in quadrature mode.

$$I_{Hz} \xrightarrow{90^\circ I_x} -I_{Hy} \tag{1}$$

$$-I_{Hy} \xrightarrow{\Omega I_z t} -I_{Hy}\cos\Omega t + I_{Hx}\sin\Omega t \tag{2}$$

$$-I_{Hy} \xrightarrow{\pi J\, 2I_{1z}I_{2z}t} -I_{Hy}\cos\pi J t + 2I_{Hx}I_{Hz}\sin\pi J t \tag{3}$$

8. Own Observations

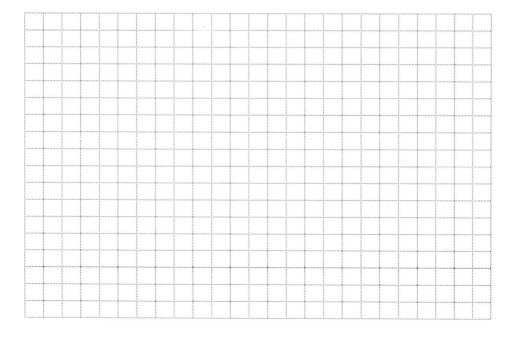

Experiment 3.2

The Standard ^{13}C NMR Experiment

1. Purpose

The aim of the standard ^{13}C NMR experiment is to record a ^{13}C NMR spectrum with proton broad-band decoupling and data accumulation in order to get chemical shift information for structure determination. Here we apply this standard procedure to ethyl crotonate.

2. Literature

[1] H.-O. Kalinowski, S. Berger, S. Braun, *Carbon-13 NMR Spectroscopy*, Wiley, Chichester, **1988.**
[2] A. E. Derome, *Modern NMR Techniques for Chemistry Research*, Pergamon, Oxford, **1987**.
[3] K. K. M. Sanders, B. K. Hunter, *Modern NMR Spectroscopy*, 2nd Edition, Oxford University Press, Oxford, **1993**.

3. Pulse Scheme and Phase Cycle

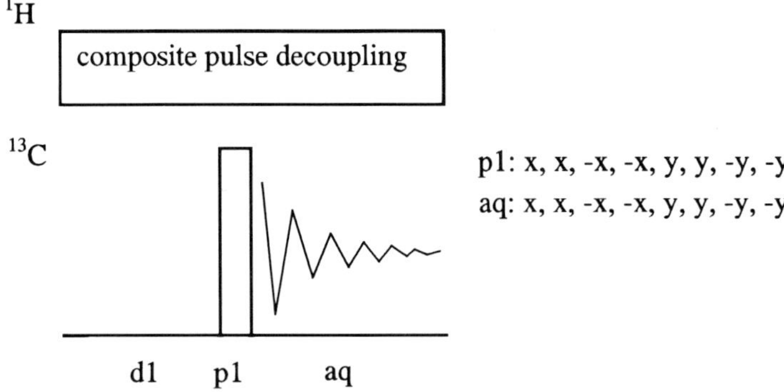

4. Acquisition

Time requirement: 5 min

Sample: 20% ethyl crotonate in $CDCl_3$ with TMS as standard.

For sample preparation fill a clean and dry 5 mm sample tube of a quality appropriate to the spectrometer frequency with 0.7 ml $CDCl_3$ (5 cm filling height), 150 µl ethyl crotonate, and one drop of TMS. Clean the tube outside, close it with a cap, mark it (do not use a flag-like label which hinders rotation) and put it into the spinner turbine. Be sure that it is tight, but not too tight in the spinner, and adjust the depth of the tube using the depth gauge.

Put the tube into the magnet by means of the air lift, adjust the rotation frequency (about 20 Hz), and display the lock signal on the screen. Perform the lock procedure and optimize the field homogeneity (see Ch. 1, Section 1.4).

The spectrometer is to be adjusted for ^{13}C observation in quadrature detection mode. Load the ^{13}C acquisition program which comprises the following basic commands: zero memory, set ^{1}H broad-band decoupling (CPD-mode), set the relaxation delay, set the excitation pulse, perform acquisition, and write the data on file.

For acquisition the following parameters have to be set:

td: 32 k

td (time domain) is the number of points at which the free induction decay (FID) is sampled and the data stored. This parameter has to be chosen according to the desired digital resolution, which in turn depends on the spectral width sw. The settings of td and sw determine the acquisition time aq.

sw: 250 ppm

sw, the spectral width, has to be chosen so that the resonance frequencies of all types of ^{13}C nuclei are within this spectral window; otherwise folding occurs. Folding of noise is prevented by appropriate filters which are set by the software corresponding to sw. The settings of sw and td determine the *acquisition time* aq, the time during which the FID data are acquired. The relationship between these three parameters, where aq is usually the dependent variable and sw has to be expressed in Hz, is described by the following fundamental equation (1):

$$aq = td/(2\ sw) \tag{1}$$

Thus, with the above settings and a ^{13}C resonance frequency of 75 MHz, the acquisition time is 0.9 s.

o1: frequency (offset) *of the r.f. pulse at the center of the ^{13}C NMR spectrum*:
In quadrature detection mode the frequency of the exciting r.f. pulse (often called transmitter offset) is positioned in the center of sw, e.g. at about $\delta_C = 120$. On older instruments without digital lock, o1 depends on the field position of the deuterium lock and thus on the deuterated solvent used.

o2: frequency (offset) *of the decoupler at the center of the ^{1}H NMR spectrum*:
The frequency of the decoupling r.f. pulse (often called decoupler offset) is positioned in the center of the ^{1}H NMR spectrum, e.g. about $\delta_H = 7$. On older instruments without digital lock, o2 depends on the field position of the deuterium lock and thus on the deuterated solvent used.

p1: 30°

p1, the ^{13}C transmitter pulse, creates an observable *x,y*-magnetization by tipping the magnetization vector towards the –y-axis through an angle α, determined by eq. (32:

$$\alpha = (360/2\pi)\ \gamma B_1 \cdot p \tag{2}$$

where γ is the gyromagnetioc ratio for ^{13}C, γB_1 the strength of the radiofrequency field, and p the duration of the pulse. Since p1 is known for $\alpha = 90°$ (see Exp. 2.2), the duration of p1 for a flip angle of 30° is easily obtained. Although a 90° pulse gives maximum signal intensity (Exp. 2.2), a shorter pulse-length of about 30° is used for routine work with data accumulation; thus the long *pulse repetition time* of $5T_1$, which is necessary after a 90° pulse, may be reduced.

It should be mentioned here that between the irradiating pulse and the opening of the detector a *preacquisition delay* (de) is introduced. It prevents the break through of the transmitter pulse into the detector, is in the order of 10 to 100 μs and is normally set by the computer.

d1: 0.4 s

d1, the relaxation delay, is introduced in order to establish thermal equilibrium of the spin system before the excitation pulse is applied. However, in routine ^{13}C NMR spectroscopy with long relaxation times, especially those of quaternary carbon nuclei, one accepts reduced intensities of their signals and uses a short d1 for time saving reasons. Often, especially for smaller molecules with longer T_1 values, thermal equilibrium is not reestablished; instead there is a steady state, reached initially by the introduction of a few dummy scans (ds) at the beginning of the experiment.

decoupler attenuation and 90° pulse-duration for CPD

In routine work proton broad-band decoupling is usually performed by CPD (composite pulse decoupling), for which the 90° decoupler pulse and the attenuation have to be known (see Exp. 2.3).

ds: 2

ds, dummy scans, are inserted before accumulation starts in order to establish a steady state.

rg:

rg, the receiver gain, has to be carefully adjusted so that the incoming FID does not exceed the ADC input limits. Otherwise signal distortions at the baseline will occur after Fourier transformation.

ns: 128

ns, the number of "scans" (individual FIDs), is chosen so that a reasonable signal-to-noise ratio is obtained in the final spectrum. Because of phase cycling a multiple of 8 is advisable.

5. Processing

After data acquisition you have to process the data. Perform the following steps using the given settings:

size of the processed real data file: set **si = 16 k**. Normally si corresponds to td/2 (see acquisition), since, after Fourier transformation, half of the td data adresses contain the real part of the spectrum. This gives a digital resolution (Hz/data point) after Fourier transformation of 2 sw/td. (Note that on the older Bruker instrument series the parameter SI corresponds to 2 si of the newer ones.)

However, si may be set to a value greater than td/2, e.g. si = 32 k; this procedure is termed zero-filling and leads, according to Equation (3) for the general case, to a higher digital resolution:

$$\text{digital resolution} = \text{sw/si} \qquad (3)$$

With the above data and at 75 MHz, the digital resolution is 1.1 Hz/point, a reasonable value in comparison to the line-width and peak separations normally encountered in ^{13}C NMR spectroscopy. If a higher digital resolution is desired, one may optimize sw, increase td, or perform zero-filling.

baseline correction: perform this correction on the *FID* in order to eliminate the dc offset between the two channels used in quadrature detection mode

digital filtering: for standard applications use as window function an exponential window function on the FID, characterized by the line broadening parameter lb; here set **lb = 2 Hz.**

This type of digital filtering generally improves the signal to-noise-ratio, but at the cost of resolution. Usually a value equal to the line-width obtained without application of an exponential multiplication is used.

Fourier transformation: use the correct type in accordance with the simultaneous or sequential quadrature acquisition mode.

phase correction: adjust to achieve pure absorption mode signals.

baseline correction: perform this correction on the *spectrum* in order to remove baseline rolling.

referencing: set the TMS signal to $\delta_C = 0$.

peak picking: choose the desired level.

plot: set parameters and plot spectrum, including integrals, peak picking, and the relevant acquisition and processing parameters.

6. Result

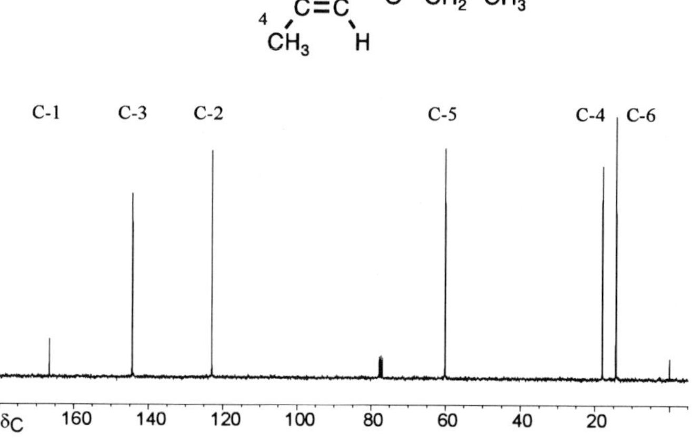

The figure on page 53 shows the ^1H broad-band decoupled ^{13}C NMR spectrum of ethyl crotonate as obtained on an ARX-300 spectrometer using a dual probe-head (region $\delta_C = -5$ to 180). Note that as usual no integration is performed, since under routine conditions the signal areas are not necessarily proportional to the number of ^{13}C nuclei giving rise to that signal (see especially the signal of the quaternary carbon C-1).

7. Comments

The excitation pulse p1 converts the equilibrium magnetization I_{Cz} of the ^{13}C nuclei to transverse magnetization $-I_{Cy}$ as shown in Equation (1). During the acquisition time the chemical shift develops in the x,y-plane as shown in Equation (2), and the resulting magnetization is detected by the receiver in the x,y-plane in quadrature mode.

$$I_{Cz} \xrightarrow{90^\circ\, I_x} -I_{Cy} \qquad (1)$$

$$-I_{Cy} \xrightarrow{\Omega I_z t} -I_{Cy}\cos\Omega t + I_{Cx}\sin\Omega t \qquad (2)$$

8. Own Observations

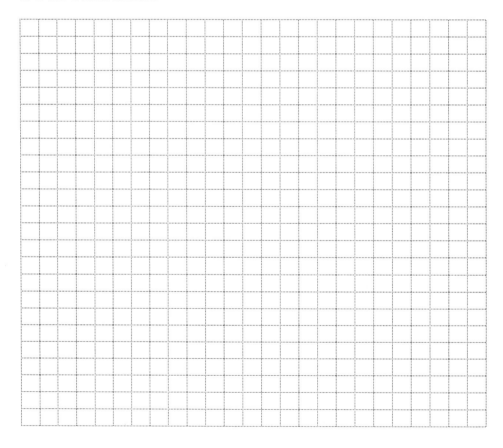

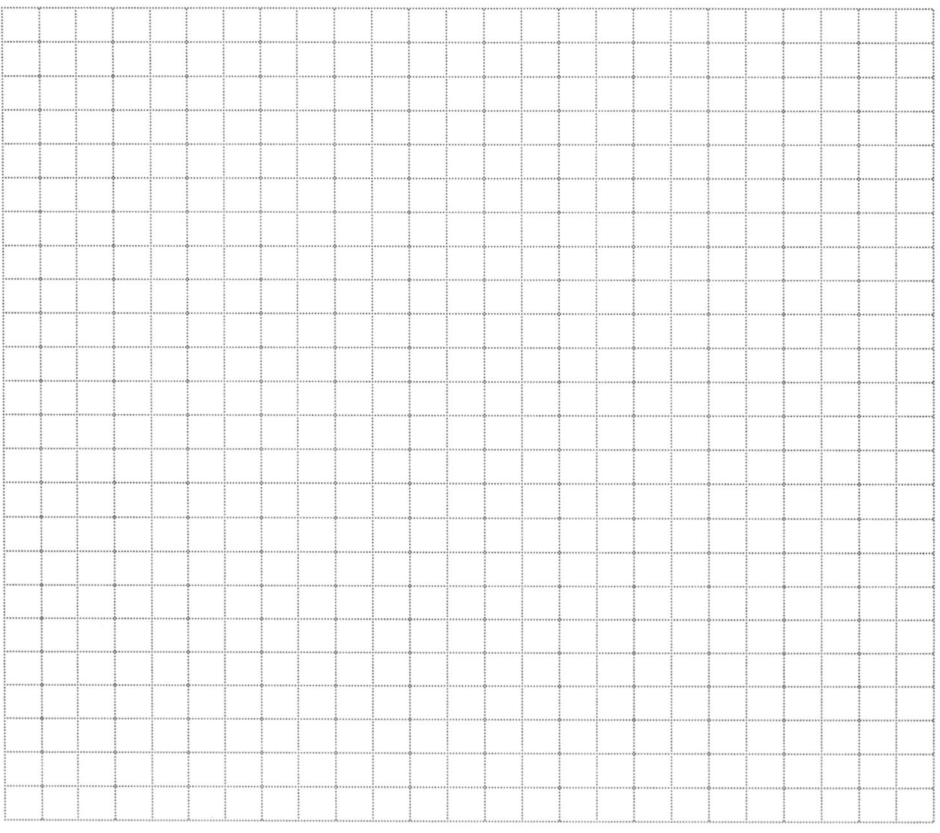

Experiment 3.3

The Application of Window Functions

1. Purpose

Once the FID is measured, it can be Fourier transformed into the frequency domain to yield the NMR spectrum. However, before this transformation it may be digitally filtered to enhance either the signal-to-noise ratio or the resolution and to remove apodization artefacts. This is done by multiplying the FID by a window function provided by the software and relies on the fact that in the time domain a good signal-to-noise ratio is present mainly at the beginning of the FID, whereas the resolution information develops at a later stage. Typical window functions supplied by the software are the exponential, Lorentz-Gaussian, sinusoidal, squared sinusoidal and trapezoidal. In this experiment we demonstrate the use of exponential weighting for sensitivity enhancement and the Lorentz-Gaussian window for resolution enhancement. These are the most important functions used for 1D NMR spectroscopy, whereas the sinusoidal functions are mainly used in multidimensional NMR.

2. Literature

[1] R. R. Ernst, *Adv. Magn. Reson.* **1966**, *2*, 1–135.
[2] E. D. Becker, J. A. Feretti, P. N. Gambhir, *Anal. Chem.* **1979**, *51*, 1413–1420.
[3] J. C. Lindon, A. G. Ferrige, *Prog. Nucl. Magn. Reson.* **1980**, *14*, 27–66.
[4] A. E. Derome, *Modern NMR Techniques for Chemistry Research*, Pergamon, Oxford, **1987**, 24–27.

3. Pulse Scheme and Phase Cycle

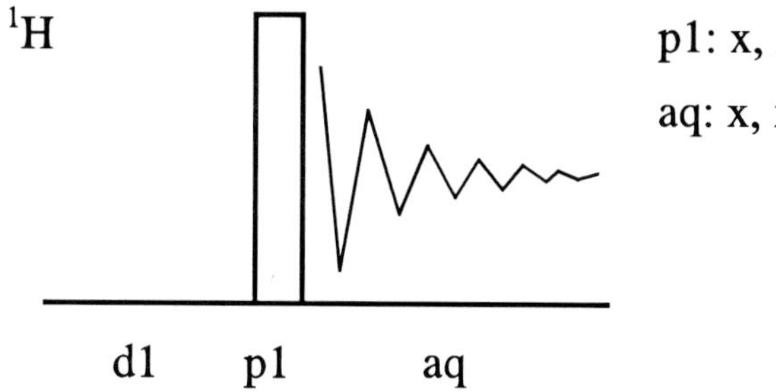

p1: x, x, -x, -x, y, y, -y, -y

aq: x, x, -x, -x, y, y, -y, -y

4. Acquisition

Time requirement: 5 min

Sample: 10% *ortho*-dichlorobenzene (ODCB) in [D$_6$]acetone, *degassed and sealed*

Experimentally this is identical with the resolution test as described in Exp. 3.6. You have to set:

td: 32 k
sw: 1 ppm
o1: center of ODCB multiplet
p1: 90° ^{1}H transmitter pulse
d1: 1 s
rg: receiver gain for correct ADC input
ns: 1

5. Processing

First transform the FID without any weighting function and observe the line-width of the eighth signal from the left, which should, after good shimming, be in the order of 0.1 Hz.

The theory of the "matched filter" for sensitivity enhancement requires multiplication of the original FID with a function having the same decay constant. This doubles the original line-width and yields the best signal-to-noise ratio without introducing too much distortion or reduction of fine structure. Thus, having measured a line-width of 0.07 Hz in this example, an exponential function with lb = 0.07 Hz was applied to the FID.

The application of Lorentz-Gaussian resolution enhancement first requires enough data points to make the improvement visible; thus, zero filling should be applied before. Adjust the data size si from 16 k to 32 k, which gives a zero filling of 16 k data points. The Lorentz-Gauss function has two adjustable parameters, gb and lb (Bruker software). The first determines where the maximum of the function is and is given as a fraction of the FID length. A gb of 0.33 thus puts the maximum of the window function at the end of the first third of the FID, reducing the initial fast-decaying components. The parameter lb is similar to the lb used in exponential weighting, but with several software packages applied with a negative sign for distinction. Especially for the Lorentz-Gauss function it is very advantageous to test the result *interactively*, which is allowed in recent software packages. Thus one can fine adjust, under observation of a certain multiplet, both parameters to yield optimum results.

6. Result

The figure on page 57 shows in **a** the left part of the AA'XX' spectrum and the FID obtained on an ARX-200 spectrometer in a dual probe-head without digital filtering, and in **b** after application of exponential weighting with lb = 0.07 Hz. Note the improvement in signal-to-noise as judged from the baseline, but also the loss of resolution as seen best for the lines 5 and 6 from the left. In **c** the result of Lorentz-Gaussian multiplication is given with gb = 0.25 and lb = –0.06. Note the decrease in signal-to-noise, but the improved resolution for lines 5 and 6. Often the Lorentz-Gaussian filtering is excessively applied, leading to negative overshoots at the feet of the signals.

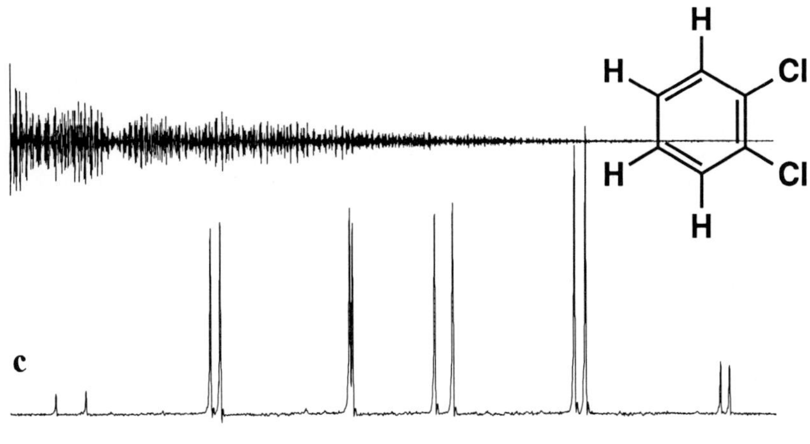
H
H
Cl
H
Cl
H
c

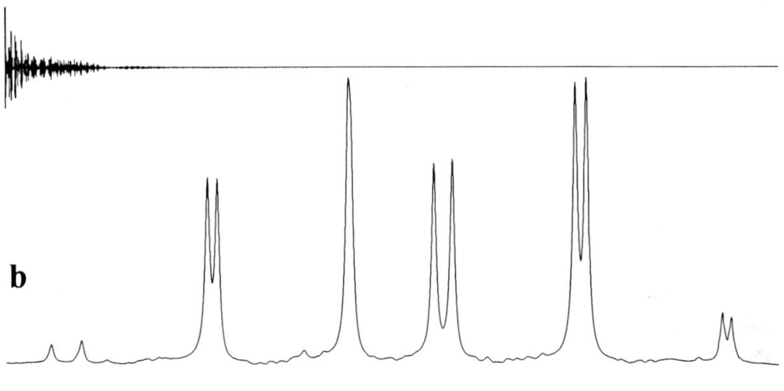
b

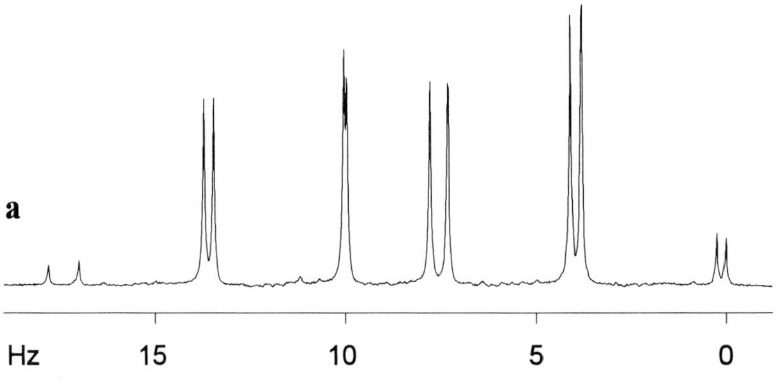
a
Hz
15
10
5
0

7. Comments

A procedure called convolution difference should be mentioned, which tries to separate sharp signals from broad background signals. The FID is multiplied by an exponential function corresponding to the line-width of the broad background signal. The result is subtracted from the original FID, leaving an FID with only the slowly decaying components.

In 2D NMR one has to cope with two problems. If the determination of the sign of the frequency in F_1 leads to addition of cosine and sine components within one FID (N or P type detection) one obtains a skewed line-shape. Furthermore, because of the limited number of data points in F_1 and F_2 these FIDs are often cut off without having decayed smoothly. Both problems are somewhat remedied by the application of sinusoidal window functions which significantly narrow the foot of a signal and decay to a true zero. With the software routines the sinusoidal functions can be shifted by a fraction of π.

8. Own Observations

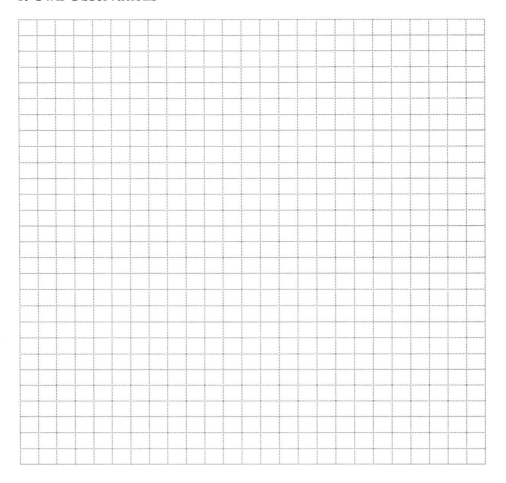

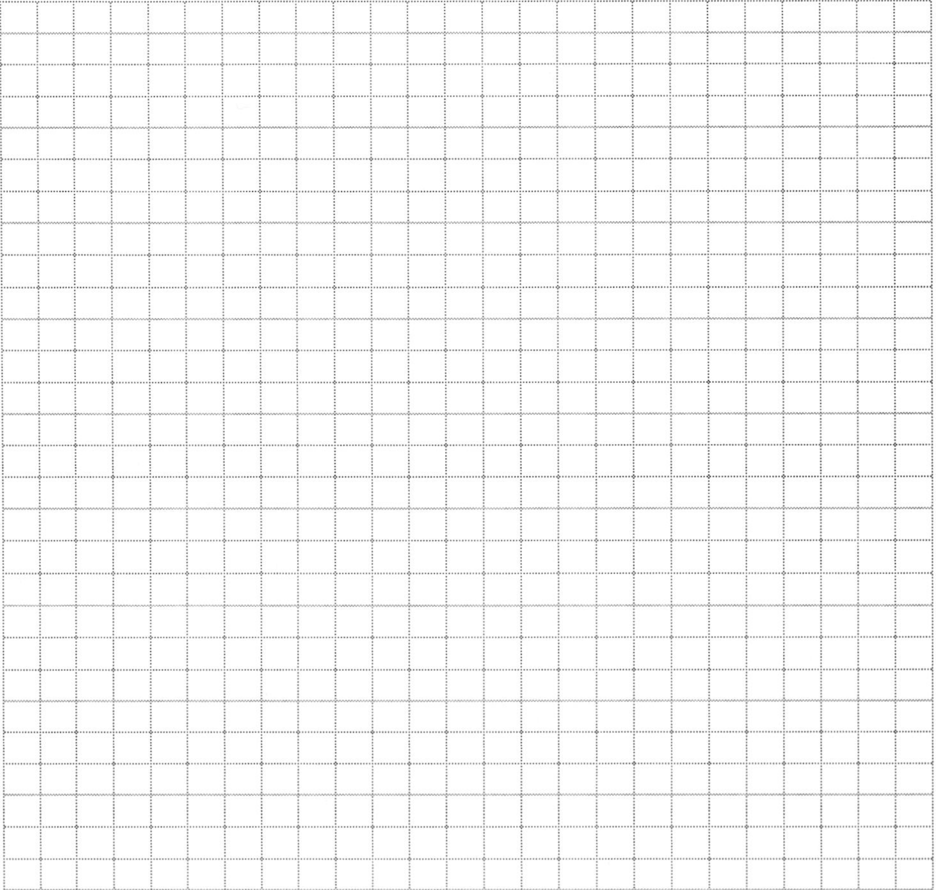

Experiment 3.4

Computer-aided Spectral Analysis

1. Purpose

Having acquired an ^{1}H NMR spectrum (see Exp. 3.1) or an ^{1}H coupled ^{13}C NMR spectrum (see Exp. 4.11) the chemist has to extract correct chemical shifts and spin coupling constants. This task may be difficult because of higher order effects of the spin systems. In such cases a computer-aided spin simulation with following iteration may be performed.

A spin simulation starts with a set of chemical shift values and spin coupling constants taken from the experimental spectrum and from chemical experience with similar compounds. Simple simulation programs available as public-domain software for PCs are able to generate spectra which can be compared with the experimental result. For a correct solution iterative programs are required. Only if both the theoretical and the experimental spectra are identical can the evaluated chemical shifts and spin coupling constants be considered to be correct

The LAOCOON type programs (**L**east squares **A**djustment **O**f **C**alculated **O**n **O**bserved **N**MR spectra [2]), which require input and appropriate conversion of the experimental spectra, can be obtained for a variety of platforms.[3,4] These take only transition frequencies into account, and the solution can only be considered to be correct if the intensities are also matched. Iterative programs [5,6] working on the full line-shape of the NMR spectrum provide the ultimate solution for the spectral evaluation and are usually available for workstations. In the experiment described here we demonstrate the performance of a PC-based LAOCOON-type program on *ortho*-dichlorobenzene.

2. Literature

[1] H. Günther, *NMR Spectroscopy*, 2nd Edition, Wiley, Chichester, **1995**.

[2] S. Castellano, A. A. Bothner-By, *J. Chem. Phys.* **1964**, *41*, 3863–3869.

[3] CALM, written by V. N. Khlopkov, E. V. Kirev, A. G. Shakhatani, A. O. Krasevin, ftp: //bloch.cchem.berkeley.edu/pub/nmr/ms-dos.

[4] K. Marat, *XSIM, The University of Manitoba NMR Spectral Simulation and Analysis Package*, **1996**.

[5] D. S. Stephenson, G. Binsch, *J. Magn. Reson.* **1980**, *37*, 395–408; 409–430.

[6] G. Hägele, M. Engelhardt, W. Boenigk, *Simulation und automatisierte Analyse von Kernresonanzspektren*, VCH, Weinheim, **1987**.

3. Pulse Scheme and Phase Cycle

see Experiment 3.6

4. Acquisition

perform Experiment 3.6

5. Processing

Transfer the spectrum obtained in Experiment 3.6 to a PC. The procedure shown here was performed with the CALM software [3], obtainable on the Internet. The next step is to convert the experimental spectrum into the format used by CALM with a program called CODER7 (part of the CALM package) and to extract only the desired part of the spectrum for the simulation.
Reference in the experimental spectrum the lowest frequency line of the left part of the AA'BB' pattern to $\delta_H = 0$ Hz, as seen in the Figure of Exp. 3.6. The centers of the two parts of the AA'BB' pattern will then be at 8 and −33 Hz respecetively (200 MHz Spectrometer). Calculate a trial spectrum using $\delta_1 = \delta_4 = 8$ Hz and $\delta_2 = \delta_3 = -33$ Hz, $J(1,2) = J(3,4) = 7.5$ Hz, $J(1,3) = J(2,4) = 1.5$ Hz, $J(2,3) = 8$ Hz and $J(1,4) = 0.3$ Hz. Display both the experimental and calculated spectra. For the iteration step they should look rather similar. Use the assignment mode and assign each theoretical transition to an experimental one. For the iteration allow the program to change all parameters; however be sure to use the inherent symmetry option, which means, that e.g. $J(1,2)$ is always equal to $J(3,4)$. The program tries to adjust the chemical shifts and spin coupling constants according to the least squares principle. The rms value obtained should be significantly less than 10%.

6. Result

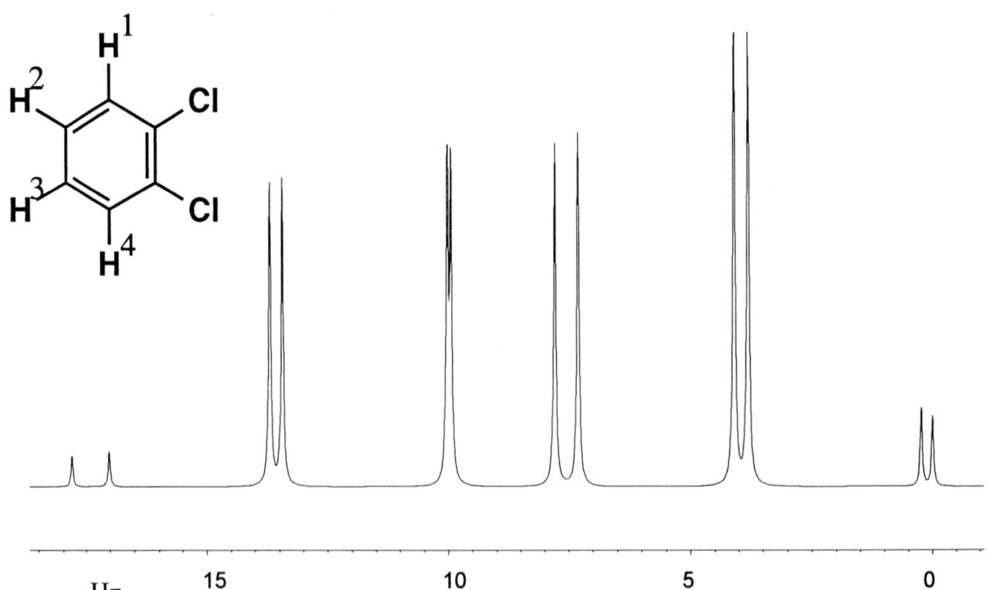

The Figure on page 60 shows the left part of the theoretical spectrum of *ortho*-dichlorobenzene obtained from the iteration. After assigning all 24 lines of the spectrum an rms value of 0.03 was obtained with the following parameters:

```
1 W(1)   =  -33.85972 W(4)   =  -33.85972  +/-  0.00098
2 W(2)   =    8.36557 W(3)   =    8.36557  +/-  0.00098

3 J(1,2) =   8.08109   J(3,4) =   8.08109  +/-  0.00193
4 J(1,3) =   1.52451   J(2,4) =   1.52451  +/-  0.00187
5 J(1,4) =   0.32835  +/-  0.00134
6 J(2,3) =   7.48863  +/-  0.00160
```

This output is directly reproduced from the computer program, so don't be disturbed by the high digital "accuracy". In your printout check on the differences between experimental and calculated line frequencies. For the calculation of the theoretical spectrum a line-width of 0.07 Hz was applied as taken from the experimental spectrum.

7. Comments

The first and basic requirement for any kind of spectral analysis is a NMR spectrum recorded and processed with high resolution. There is no point to iterate on badly resolved spectra. The choice of the type of spin simulation software used is very dependent on the purpose and availability. Simple simulators very often give a quick and sufficient answer, whether the principal understanding of the spin system and the signs or magnitudes of the coupling constants are correct. The LAOCOON type iteration programs require, that the starting spectrum is already rather close to the experimental spectrum. The line shape analysis programs are less demanding on a good starting model, but are costly and require fast workstations for reasonable performance. In these programs it is of coarse necessary to mark spectral regions, where impurities or solvent signals are present, in order not to iterate on these signals.

Another important distinction between the available programs is the number of independent spins they are able to handle and whether they allow for symmetry groups which may be very important especially for spin systems looked at in inorganic chemistry. Here the DAISY package [5] and XSIM [6] seem currently to be the most advanced systems.

8. Own Observations

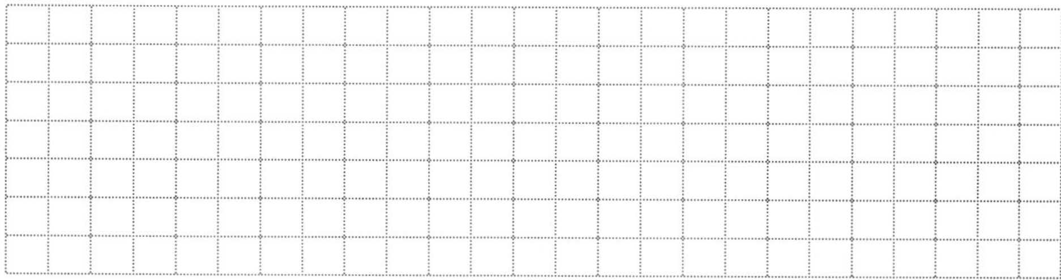

Experiment 3.5

Line-Shape Test for ^{1}H NMR Spectroscopy

1. Purpose

A good lineshape, as well as high resolution and high sensitivity, are the most important features for the performance of an NMR spectrometer (Exps. 3.5–3.10). In the line-shape test, also often called the hump test, the ^{1}H NMR signal of $CHCl_3$ is tested with regard to its line-width by measuring not only the width at half height (50%), but also at the height of the ^{13}C satellites (0.55%) and at 1/5 of this height (0.11%). NMR signals should have a Lorentzian line-shape. Therefore the widths at the latter two heights should be 13.5 and 30 times the half-height line-width $\Delta\nu_{1/2}$ (e.g. for $\Delta\nu_{1/2}$ = 0.2 Hz these "hump" values are calculated to be 2.7 and 6 Hz). Deviations from these ratios indicate a non-Lorentzian line-shape; such a situation should be avoided, even if the measured values are smaller than the calculated ones.

2. Literature

[1] A. E. Derome, *Modern NMR Techniques for Chemistry Research*, Pergamon, Oxford, **1987**, 38–42.

[2] *SAM 1.0, Shimming Simulation Software package for IBM-PC compatible Computers*, ACORN NMR, 46560 Fremont Blvd., Fremont, CA 94538-6482.

[3] C. Anklin, D. v. Ow, H. Ruegger, *Standard Test Procedures for Supercon Systems*, Spectrospin AG, Fällanden, **1988**.

3. Pulse Scheme and Phase Cycle

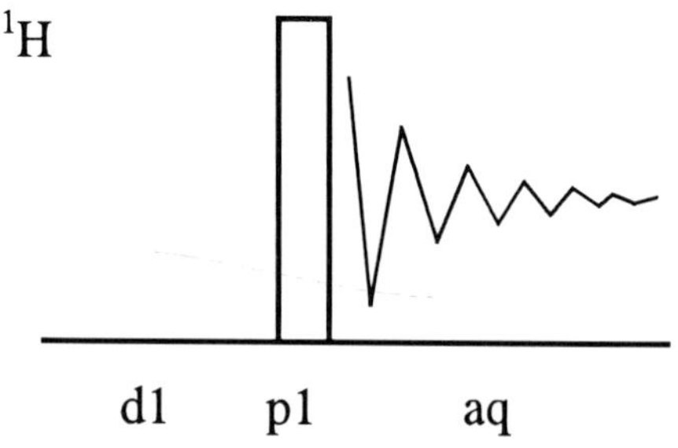

p1: x, x, -x, -x, y, y, -y, -y

aq: x, x, -x, -x, y, y, -y, -y

4. Acquisition

Time requirement: approx. 1 h, very dependent on the skill of the operator and the state of the system

Sample: 10% $CHCl_3$ in [D_6] acetone, *degassed and sealed*

The magnet is shimmed as well as possible (see Ch. 1, Section 1.4). Record a normal ^1H NMR spectrum (see Exp. 3.1) and note the offset of the $CHCl_3$ signal. You have to set:

spinning rate: 20 Hz
td: 32 k
sw: 500 Hz
o1: on resonance of ^1H signal
p1: 90° ^1H transmitter pulse
d1: 60 s
rg: receiver gain for correct ADC input
ns: 1

5. Processing

Use standard 1D processing (see Exp. 3.1) with zero-filling to 32 k and no window multiplication. Set the intensity of the main signal to 1000 and check whether the satellites have a height of 5.5. Measure the line-width at half height, at the height of the satellites, and at 1/5 of their height.

6. Result

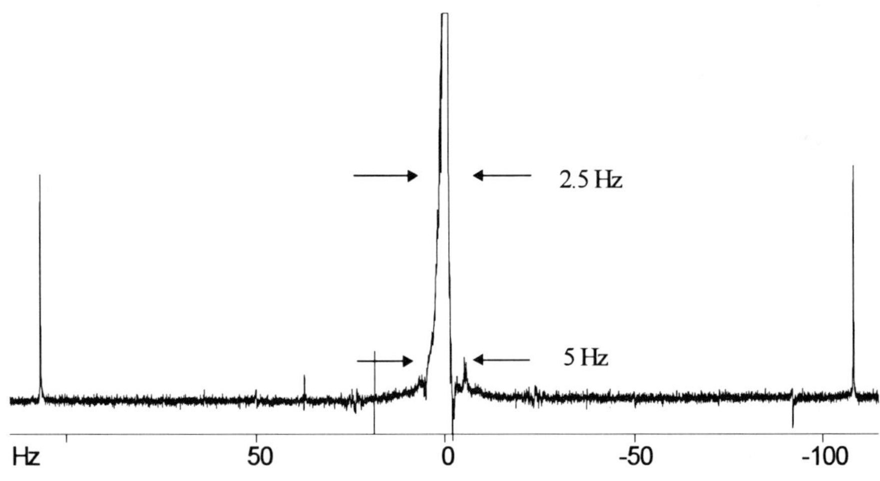

The figure shows the result obtained with a 5 mm dual probe-head on an ARX-200 spectrometer. Note that no spinning sidebands can be seen. The line-shape test is also a test for spinning sidebands, which must not exceed the height of the ^{13}C satellites.

7. Comments

The protons bound to ^{13}C have a shorter relaxation time than those producing the main signal. Therefore a "good" but false hump test result may be obtained if too short a relaxation delay d1 is used. A bad hump results in a severe loss of sensitivity, since the main part of the signal intensity lies in the foot of the signal. The hump test should be performed regularly for all probe-heads available and recorded in the log-book of the instrument.

8. Own Observations

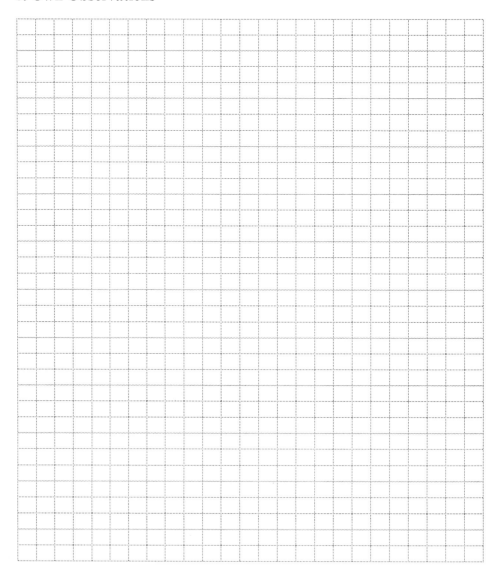

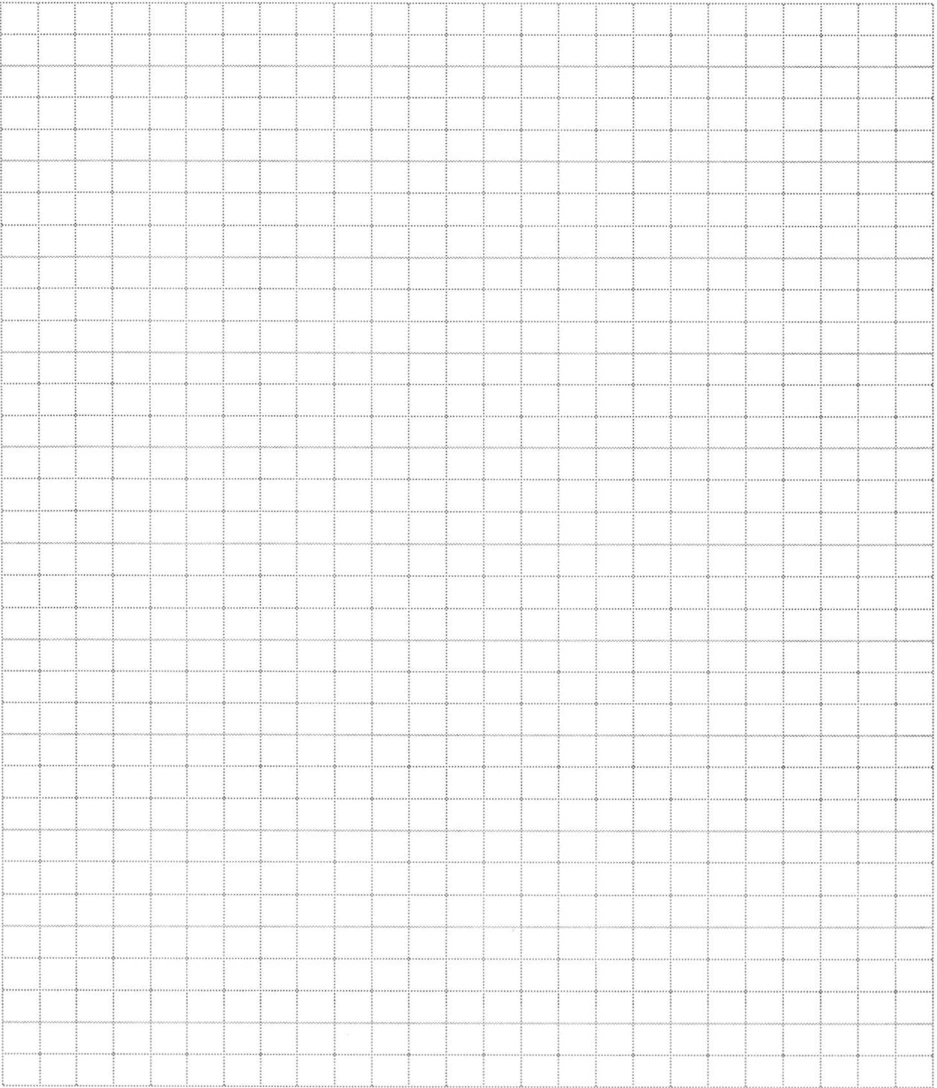

Experiment 3.6

Resolution Test for ^{1}H NMR Spectroscopy

1. Purpose

Resolution is the ability of an NMR spectrometer to observe resonance lines which are very close together as separate lines. This ability is measured by the line-width at half height, $\Delta\nu_{1/2}$, which is usually greater than the natural line-width. The experimental line-width $\Delta\nu_{1/2}$ is determined by the homogeneity of the magnetic field. For the standard ^{1}H NMR resolution test *ortho*-dichlorobenzene (ODCB) which gives an AA'BB' pattern, is used. Usually the eighth line from the left is used for resolution measurement.

2. Literature

[1] A. E. Derome, *Modern NMR Techniques for Chemistry Research*, Pergamon, Oxford, **1987**, 38–42.

[2] *SAM 1.0, Shimming Simulation Software package for IBM-PC compatible Computers*, ACORN NMR, 46560 Fremont Blvd., Fremont, CA 94538-6482.

[3] C. Anklin, D. v. Ow, H. Ruegger, *Standard Test Procedures for Supercon Systems*, Spectrospin AG, Fällanden, **1988**.

3. Pulse Scheme and Phase Cycle

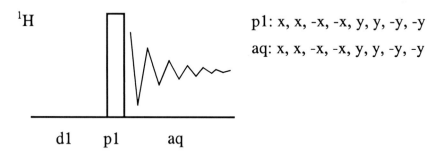

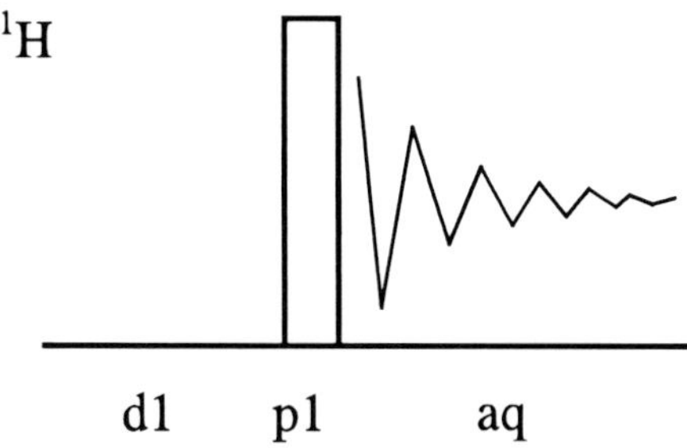

p1: x, x, -x, -x, y, y, -y, -y

aq: x, x, -x, -x, y, y, -y, -y

4. Acquisition

Time requirement: approx. 1 h, very dependent on the skill of the operator and the state of the system

Sample: 10% *ortho*-dichlorobenzene in [D$_6$]acetone, *degassed and sealed*

The magnet is shimmed as well as possible (see Ch. 1, Section 1.4). Record a normal ^{1}H NMR spectrum (see Exp. 3.1) and note the offset of the center of the ODCB multiplet. You have to set:

spinning rate: 20 Hz
td: 32 k
sw: 1 ppm
o1: center of ODCB multiplet
p1: 90° ^{1}H transmitter pulse
d1: 1 s
rg: receiver gain for correct ADC input
ns: 1

5. Processing

Use standard 1D processing as described in Experiment 3.1 with zero-filling to 32 k and no window multiplication.

6. Result

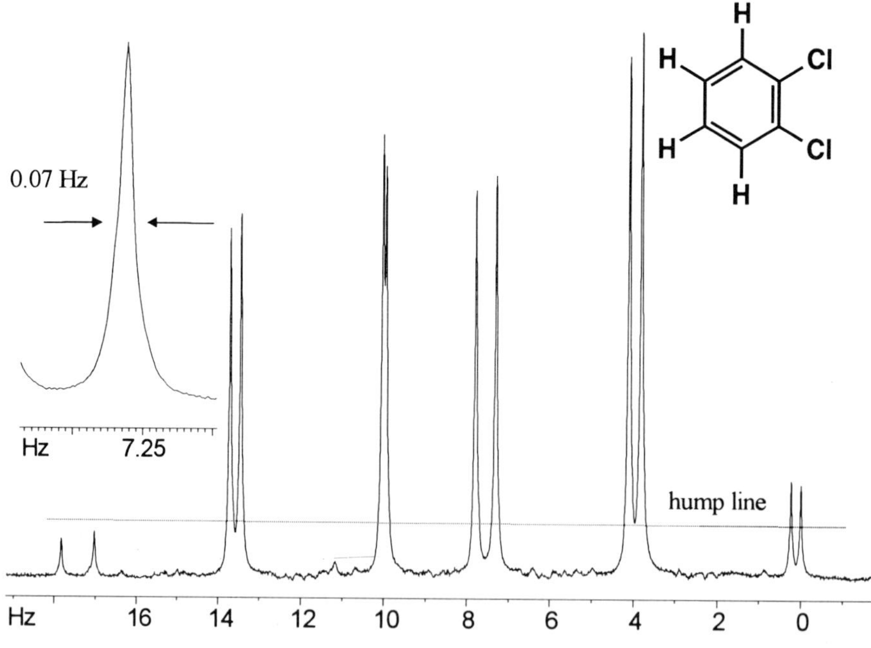

The figure shows the result (only the left half of the full AA'BB' pattern) obtained with a 5 mm probe-head on an ARX-200 spectrometer. The line-width at half-height measured on the eighth signal from the left was 0.07 Hz.

7. Comments

Note that the AA'BB' pattern is field dependent and slightly different splittings can be observed at higher field strength. The hump should not exceed 50% of the small inner signals (at 0 Hz above), indicated by the dotted line in the figure on page 66. The ODCB test should be performed regularly and the results should be recorded in the log-book of the instrument. Compared with the line-shape test in Experiment 3.3 the resolution test can be more rapidly achieved in practice.

8. Own Observations

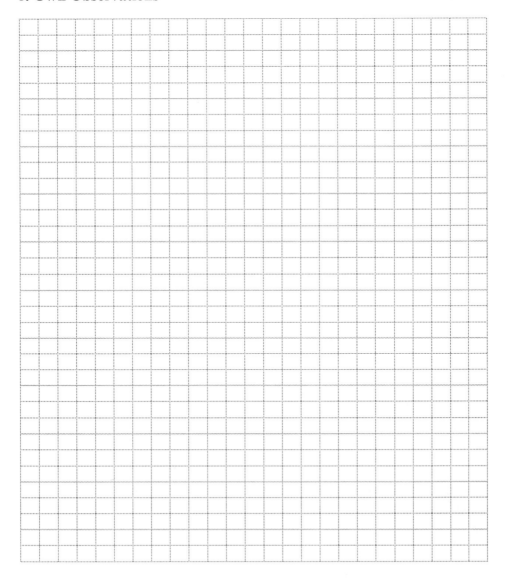

Experiment 3.7

Sensitivity Test for ^{1}H NMR Spectroscopy

1. Purpose

Sensitivity is one of the most debated points between manufacturers and customers buying new NMR instruments. Furthermore, sensitivity plays a central role concerning the performance of an NMR instrument in its everyday use. Therefore standardized tests have been developed which must be critically and honestly performed to yield meaningful results. In this experiment we describe the standard ^{1}H sensitivity test using ethyl benzene.

2. Literature

[1] A. E. Derome, *Modern NMR Techniques for Chemistry Research*, Pergamon, Oxford, **1987**, 50–56.

[2] C. Anklin, D. v. Ow, H. Ruegger, *Standard Test Procedures for Supercon Systems*, Spectrospin AG, Fällanden, **1988**.

3. Pulse Scheme and Phase Cycle

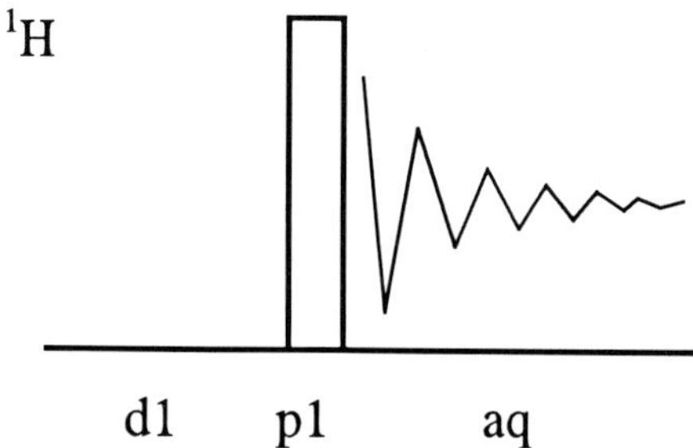

p1: x, x, -x, -x, y, y, -y, -y

aq: x, x, -x, -x, y, y, -y, -y

4. Acquisition

Time requirement: 10 min

Sample: 0.1% ethyl benzene in $CDCl_3$, *degassed and sealed*

The magnet is shimmed as well as possible (see Ch. 1, Section 1.4). For a meaningful sensitivity test, the line-shape and resolution tests (Exps. 3.5 and 3.6) should be satisfactory. Load standard ^{1}H NMR parameters. You have to set:

td: 32 k
sw: 10 ppm
o1: middle of ^{1}H NMR spectrum

p1: 90° ^{1}H transmitter pulse
d1: 60 s
rg: receiver gain for correct ADC input
ns: 1

5. Processing

Apply standard 1D processing (see Exp. 3.1); zero filling to 32 k and an exponential window with lb = 1 Hz should be used. The full spectrum should be plotted and the noise between δ_H = 3 and 5 enlarged to allow a correct peak to peak noise measurement.

6. Result

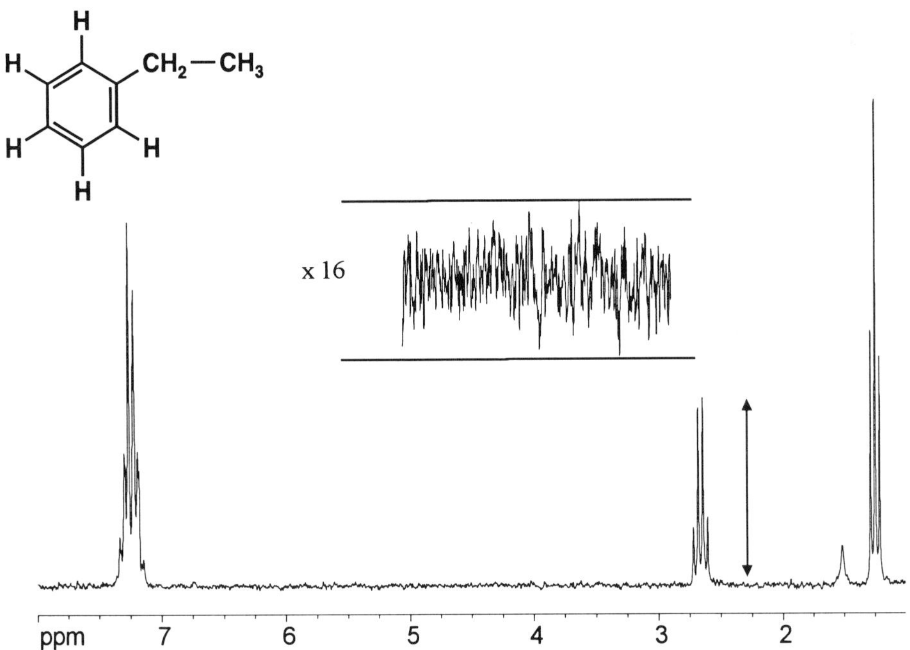

The figure shows the result obtained on an ARX-200 spectrometer with a 5 mm dual probe-head. The signal height of the CH_2 group (largest quartet line) was measured to be 29 mm, and the 16 times enlarged peak-to-peak noise was 25 mm. From these numbers a signal to rms noise ratio of 46:1 is calculated.

7. Comments

One usually divides the peak-to-peak noise by a factor of 2.5 yielding the rms noise, thus the signal-to-noise ratio is given by equation (1), where S_H is the signal height and N_{pp} the peak-to-peak noise amplitude.

$$S/N = 2.5\ S_H / N_{pp} \tag{1}$$

There are many "dirty tricks" to increase *S/N* ratios during instrument demonstrations; however, the only meaningful results are those which you can reproduce readily in your laboratory. Although current software allows calculation of the *S/N* ratio this parameter is still traditionally evaluated on paper with a ruler using the "spectroscopic eye". If the sensitivity of the instrument falls one can take as a rule of thumb that a factor of 2 may be due to bad resolution, whereas larger factors indicate hardware failures. Sensitivity tests should be performed regularly and the results should be recorded in the log-book of the instrument.

8. Own Observations

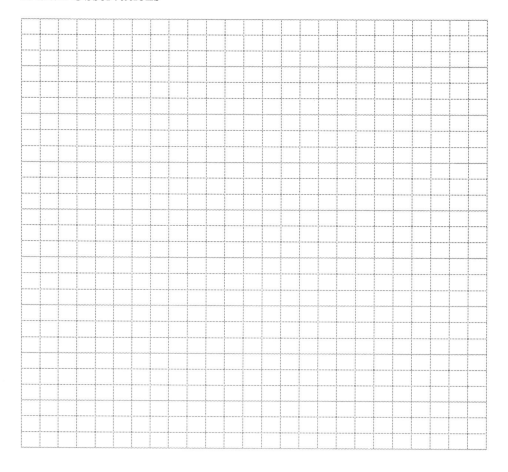

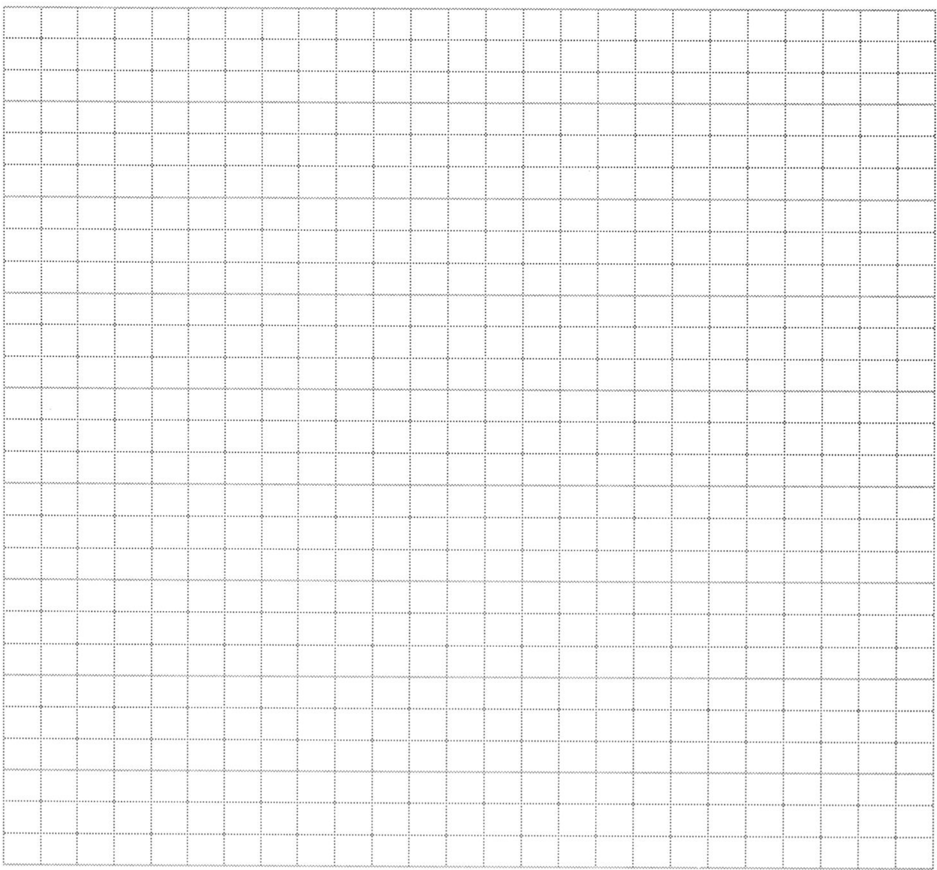

Experiment 3.8

Line-Shape Test for ^{13}C NMR Spectroscopy

1. Purpose

In this line-shape test, also often called the hump test, the ^{13}C NMR signal of benzene is tested with regard to its line-width by measuring not only the width at half height (50%), but also at the height of 0.55% and 0.11%. NMR signals should have a Lorentzian line-shape. Therefore the widths at the latter two heights should be 13.5 and 30 times the half-height line-width $\Delta\nu_{1/2}$. Deviations from these ratios indicate a non-Lorentzian line-shape: such a situation should be avoided. The ^{13}C line-shape test not only detects bad shimming or a defective probe-head but also checks for sufficient ^{1}H decoupling power.

2. Literature

[1] A. E. Derome, *Modern NMR Techniques for Chemistry Research*, Pergamon, Oxford, **1987**, 38–42.
[2] C. Anklin, D. v. Ow, H. Ruegger, *Standard Test Procedures for Supercon Systems*, Spectrospin AG, Fällanden, **1988**.

3. Pulse Scheme and Phase Cycle

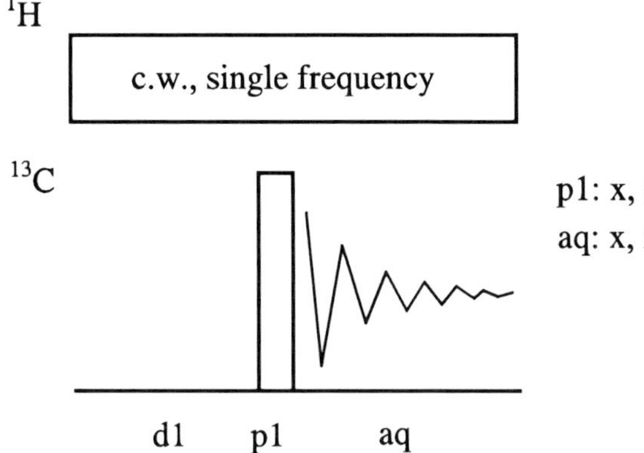

p1: x, x, -x, -x, y, y, -y, -y
aq: x, x, -x, -x, y, y, -y, -y

4. Acquisition

Time requirement: approx. 1 h, very dependent on the skill of the operator and the state of the system

Sample: 80% benzene in [D_6]acetone, *degassed and sealed*

The magnet is shimmed as well as possible (see Ch. 1, Section 1.4). Record normal ^{1}H and ^{13}C NMR spectra (see Exp. 3.1 and 3.2) and note the offsets of the benzene signals. You have to set:

spinning rate: 20 Hz
td: 16 k
sw: 200 Hz
o1: on resonance of ^{13}C signal
o2: on resonance of ^{1}H signal
p1: 90° ^{13}C transmitter pulse
d1: 1 s
decoupler attenuation for continous wave decoupling (see Exp. 4.2)
rg: receiver gain for correct ADC input
ns: 1

5. Processing

Use standard 1D processing with zero-filling to 16 k and no window multiplication. Set the intensity of the main signal to 1000 and check the line-width at heights 500, 5.5, and 1.1.

6. Result

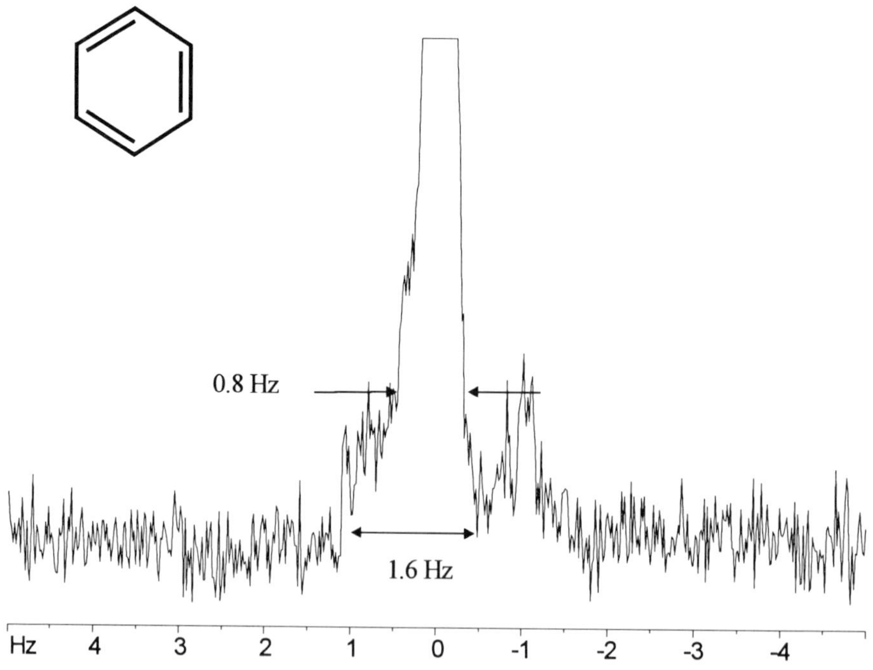

The figure on page 72 shows the result obtained on an ARX-200 spectrometer with a 5 mm dual probe-head. The small signals towards lower frequency from the main signal arise from benzene isotopomers which contain two ^{13}C nuclei.

7. Comments

A bad hump results in a severe loss of sensitivity, since the main part of the signal intensity lies in the foot of the signal. For this test on-resonance c.w. decoupling rather than the usual broad-band CPD decoupling technique is used, since the former is superior if only one signal has to be decoupled. The hump test should be performed regularly and recorded in the log-book of the instrument.

8. Own Observations

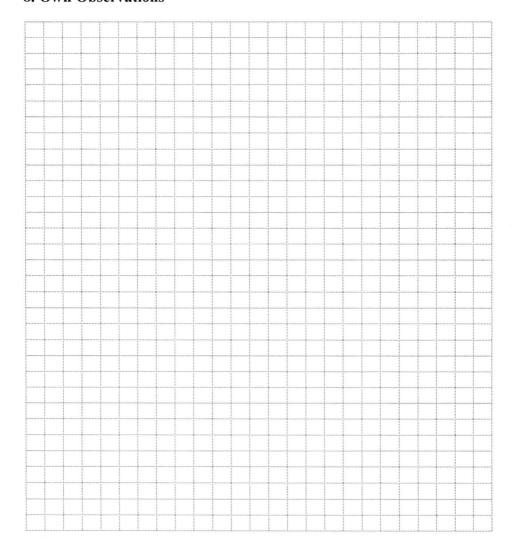

Experiment 3.9

ASTM Sensitivity Test for ^{13}C NMR Spectroscopy

1. Purpose

Good ^{13}C sensitivity is one of the most important points concerning the performance of any routine NMR service instrument in its everyday use. Therefore standardized tests have been developed which must be very critically and honestly performed to yield meaningful results. For ^{13}C two different tests are in common use. The ASTM (**A**merican Society for **T**esting and **M**aterials) procedure using [D_6]benzene in dioxane described here checks only the ^{13}C performance on the transmitter channel, whereas the sensitivity test with ethyl benzene (Exp. 3.8) also tests the ^{1}H decoupling efficiency at the same time.

2. Literature

[1] *Standard Practice for Data Presentation Relating to High Resolution NMR Spectroscopy*, American Society for Testing and Materials, Designation E 386-90, Annual Book of ASTM Standards, Philadelphia, **1990**; reprinted in: H. Günther, *NMR Spectroscopy*, 2nd Edition,Wiley, Chichester, **1995**.

[2] A. E. Derome, *Modern NMR Techniques for Chemistry Research*, Pergamon, Oxford, **1987**, 50–56.

[3] C. Anklin, D. v. Ow, H. Ruegger, *Standard Test Procedures for Supercon Systems*, Spectrospin AG, Fällanden, **1988**.

3. Pulse Scheme and Phase Cycle

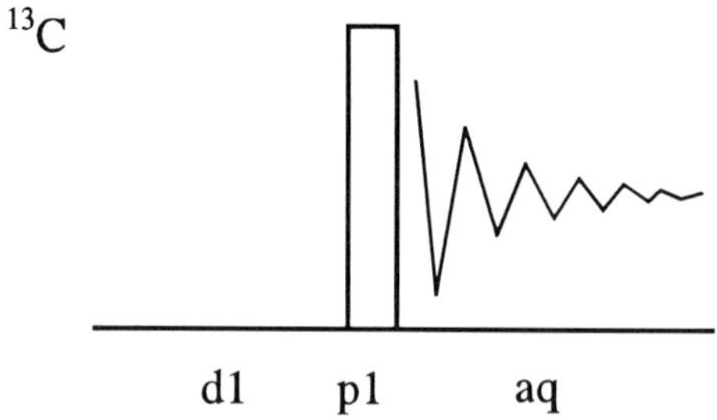

p1: x, x, -x, -x, y, y, -y, -y
aq: x, x, -x, -x, y, y, -y, -y

4. Acquisition

Time requirement: 10 min

Sample: 60% [D_6]benzene in 1,4-dioxane, *degassed and sealed*

The magnet is shimmed as well as possible (see Ch. 1, Section 1.4). For a meaningful sensitivity test, the ^{13}C line-shape test (Exp. 3.8) should be satisfactory. Load standard ^{13}C NMR parameters. You have to set:

td: 32 k
sw: 200 ppm
o1: middle of ^{13}C NMR spectrum
p1: 90° ^{13}C transmitter pulse
d1: 300 s
rg: receiver gain for correct ADC input
decoupler off
ns: 1

5. Processing

Apply standard 1D processing (see Exp. 3.2); zero filling to 64 k and an exponential window with lb = 3.5 Hz should be used. The full spectrum should be plotted and the noise between δ_C = 120 and 80 enlarged to allow a correct peak-to-peak noise measurement.

6. Result

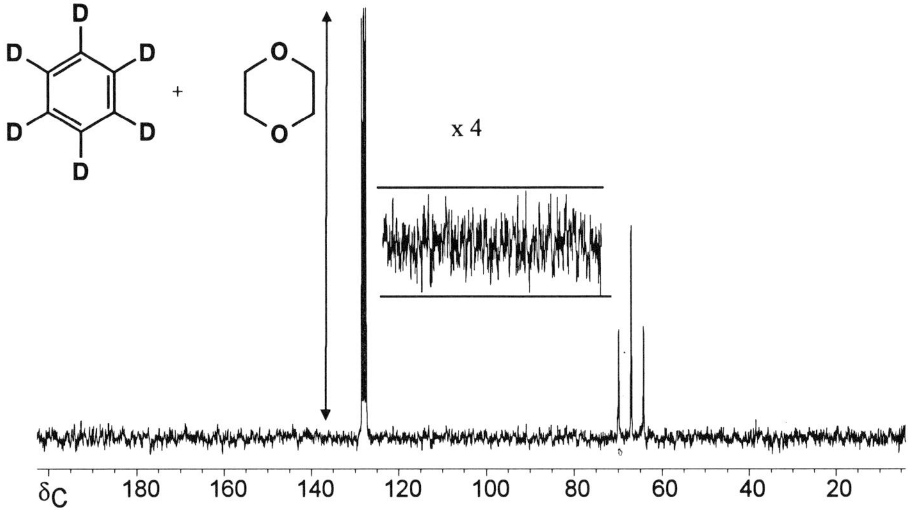

The figure shows the result obtained with a 5 mm probe-head on an ARX-200 spectrometer. The signal height of the benzene triplet was measured to be 70 mm, and the 4 times enlarged peak-to-peak noise was 17.5 mm. From these numbers a signal to rms noise ratio of 40:1 is calculated.

7. Comments

One usually divides the peak-to-peak noise by a factor of 2.5 yielding the rms noise, thus the signal-to-noise ratio S/N is given by Equation (1), where S_H is the signal height and N_{pp} the height of the peak-to-peak noise.

$$S/N = 2.5\ S_H / N_{pp} \tag{1}$$

The resolution can be checked in this test by the splitting of the triplet of benzene which should be visible down to at least 10% of the signal height. If an instrument performs well the ASTM test but badly the ^{13}C sensitivity test with ethyl benzene, check the decoupler settings.

8. Own Observations

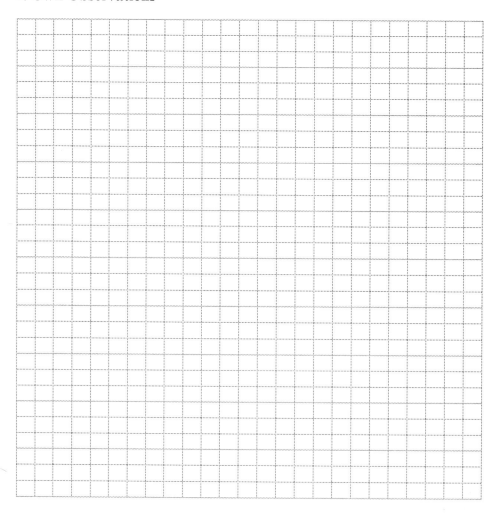

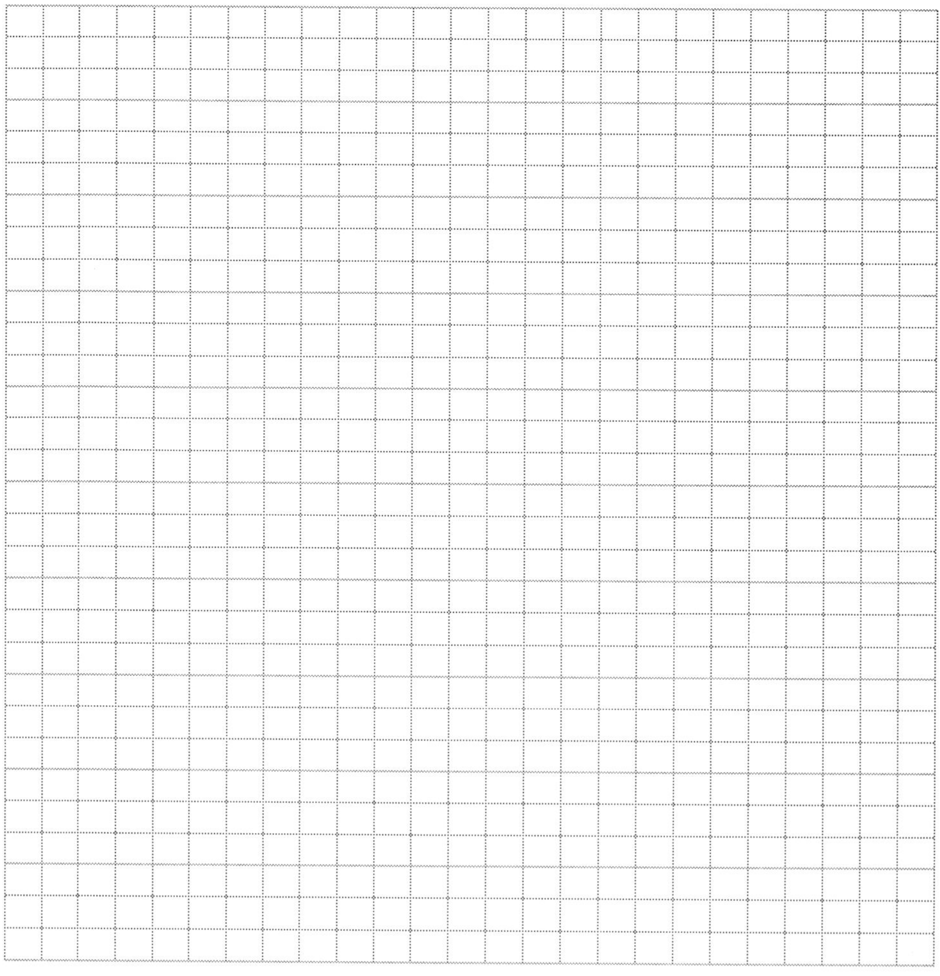

Experiment 3.10

Sensitivity Test for ^{13}C NMR Spectroscopy

1. Purpose

Good ^{13}C sensitivity is one of the most important points concerning the performance of any routine NMR service instrument in its everyday use. Therefore standardized tests have been developed which, to yield meaningful results, must be very critically and honestly performed. For ^{13}C two different tests are in common use. The sensitivity test with ethyl benzene shown here tests the spectrometer performance on both the ^{13}C transmitter and the ^{1}H decoupler channel. In comparison, the ASTM procedure (Exp. 3.9) checks only the ^{13}C performance on the transmitter channel.

2. Literature

[1] A. E. Derome, *Modern NMR Technique for Chemistry Research*, Pergamon, Oxford, **1987**, 50–56.
[2] C. Anklin, D. v. Ow, H. Ruegger, *Standard Test Procedures for Supercon Systems*, Spectrospin AG, Fällanden, **1988**.

3. Pulse Scheme and Phase Cycle

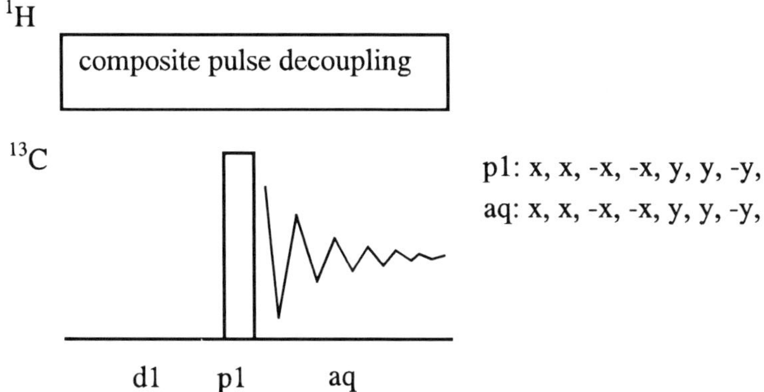

p1: x, x, -x, -x, y, y, -y, -y
aq: x, x, -x, -x, y, y, -y, -y

4. Acquisition

Time requirement: 10 min

Sample: 10% ethyl benzene in $CDCl_3$, *degassed and sealed*

The magnet is shimmed as well as possible (see Ch. 1, Section 1.4). For a meaningful sensitivity test, the ^{13}C line-shape test (Exp. 3.8) should be satisfactory. The CPD pulses for the decoupler should be recalibrated (Exp. 2.3). Load standard ^{13}C NMR parameters. You have to set:

td: 64 k
sw: 200 ppm
o1: middle of ^{13}C NMR spectrum
o2: middle of ^{1}H NMR spectrum
p1: 90° ^{13}C transmitter pulse
d1: 300 s
rg: receiver gain for correct ADC input
decoupler attenuation and pulse width for CPD
ns: 1

5. Processing

Apply standard 1D processing (see exp. 3.2); zero filling to 64 k and an exponential window with lb = 0.3 Hz should be used. The full spectrum should be plotted and the noise between δ_C = 120 and 80 enlarged to allow a correct peak to peak noise measurement.

6. Result

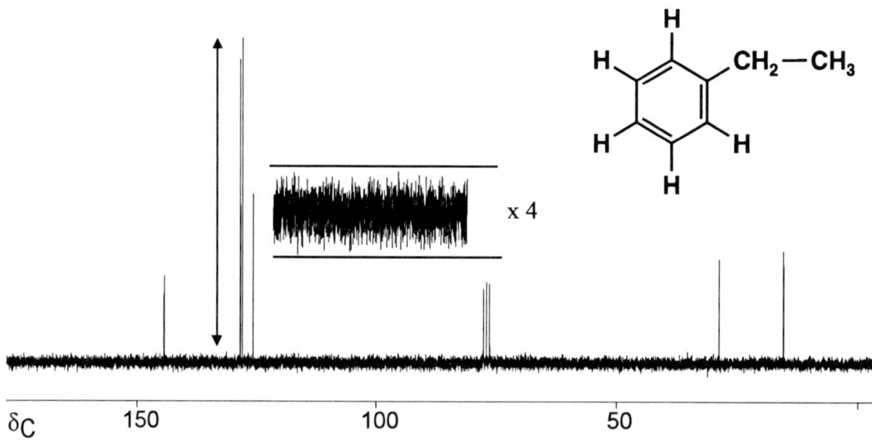

The figure shows the result obtained with a 5 mm dual probehead on an ARX-200 spectrometer. The signal intensity of the *ortho* and *meta* CH groups was measured to be 54 mm, and the 4 times enlarged peak-to-peak noise was 15 mm. From these numbers a signal to rms noise ratio of 36:1 is calculated.

7. Comments

One usually divides the peak-to-peak noise by a factor of 2.5 yielding the rms noise, thus the signal-to-noise ratio S/N is given by Equation (1), where S_H is the signal height and N_{pp} the height of the peak-to-peak noise.

$$S/N = 2.5\, S_H / N_{pp} \qquad (1)$$

There are many factors such as decoupler offset, number of time domain data points, audio filter width, and selection of noise area which influence the result from the ethyl benzene sensitivity test. Very often so-called "optimum conditions" are obtained during an instrument demonstration. For meaningful comparisons keep a test file on the disk of your instrument and perform this test regularly. The results should be recorded in the log-book of the instrument.

8. Own Observations

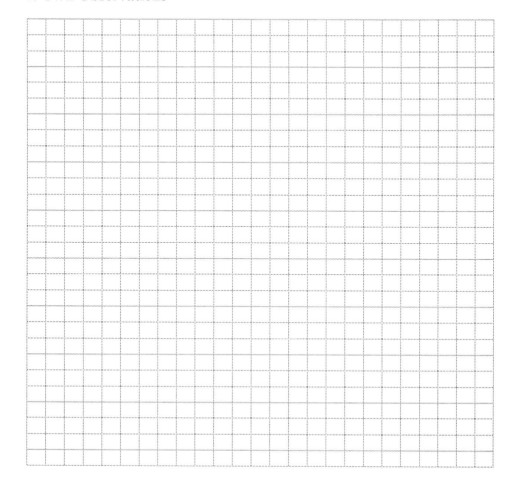

Experiment 3.11

Quadrature Image Test

1. Purpose

NMR signals are usually detected in quadrature mode with two phase detectors which are 90° out of phase (see Ch. 1, Section 1.1.2). These two audio signals are amplified and digitized either sequentially or simultaneously, then stored in different parts of the computer memory. This elegant scheme has many advantages. However, a drawback is that all components of the two channels involved must work identically. Failures which are often encountered include a d.c. offset between the two channels, wrong phase angle difference, and different amplification. Although small deviations can be eliminated by the usual quadrature phase cycle and by applying a baseline correction on the FID, it is important to know how well the two channels are matched to each other. The quadrature image test shown here gives a rapid indication of wrong adjustment.

2. Literature

[1] E. O. Stejskal, J. Schaefer, *J. Magn. Reson.* **1974**, *14*, 160–169.
[2] A. E. Derome, *Modern NMR Techniques for Chemistry Research*, Pergamon, Oxford, **1987**, 77–83.
[3] C. Anklin, D. v. Ow, H. Ruegger, *Standard Test Procedures for Supercon Systems*, Spectrospin AG, Fällanden, **1988**.

3. Pulse Scheme and Phase Cycle

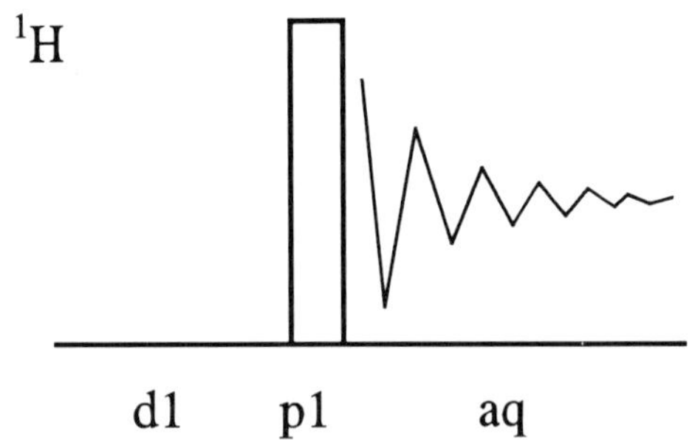

p1: x, x, -x, -x, y, y, -y, -y
aq: x, x, -x, -x, y, y, -y, -y

4. Acquisition

Time requirement: 10 min

Sample: 10% $CHCl_3$ in $[D_6]$acetone

Record a normal ^1H NMR spectrum (see Exp. 3.1) and note the offset of the $CHCl_3$ signal. You have to set:

td: 8 k
sw: 1000 Hz
o1: 250 Hz towards higher frequencies of $CHCl_3$ signal
p1: 90° ^1H transmitter pulse
d1: 1 s
rg: receiver gain for correct ADC input
ns: 1

5. Processing

Use standard 1D processing and exponential window multiplication with lb = 1 Hz. Set the intensity of the $CHCl_3$ signal to 1000 and enlarge the quadrature image signal, which is found 250 Hz towards higher frequencies from the transmitter offset position. This signal should be less than 1% of the main signal after one FID. As an exercise you may perform the experiment with ns = 8. Under these conditions the quadrature image signal should be significantly reduced by the quadrature phase cycle.

6. Result

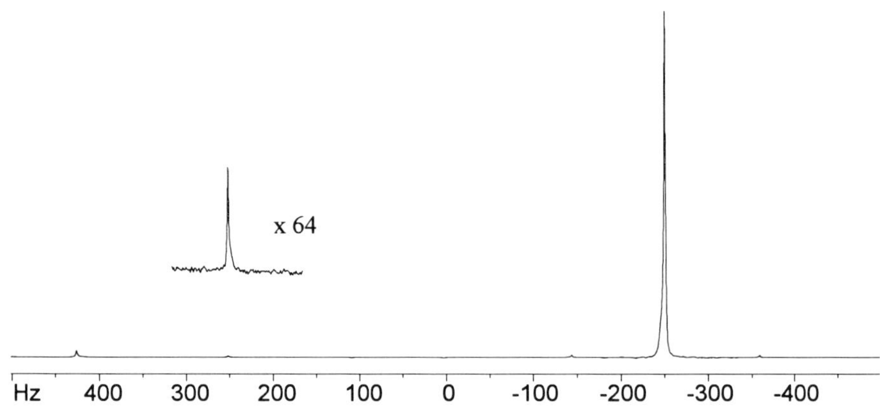

The figure shows the result obtained on an AMX-500 spectrometer. The intensity of the quadrature image peak was 0.4 % of the main signal.

7. Comments

If the quadrature image peak exceeds the 1% limit you can try to correct the problem by adjusting the appropriate potentiometers or capacitors of the phase detection and audio amplification unit. Refer to the schematics provided by the manufacturer. However, these adjustments are a bit tricky and should be performed by experienced personnel only. On very recent instruments a digital quadrature detection has been introduced.

8. Own Observations

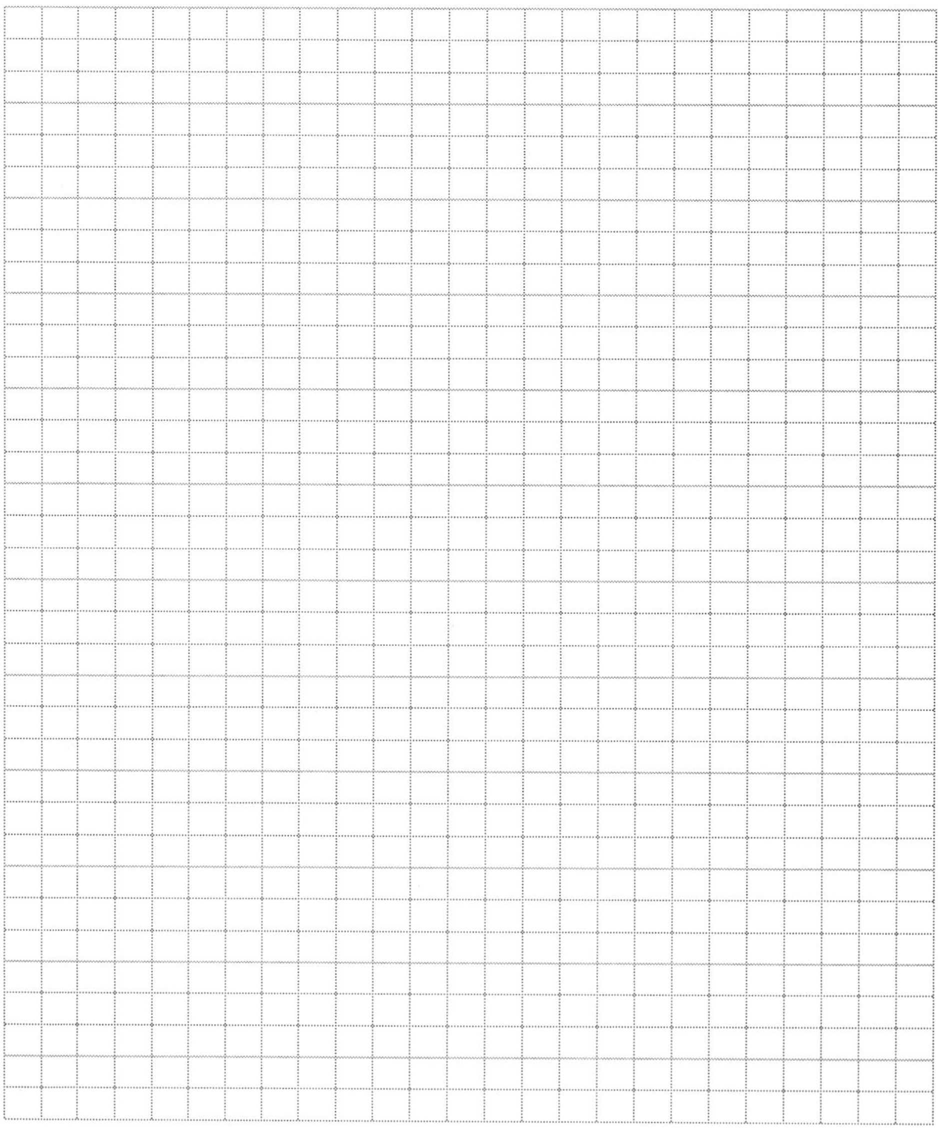

Experiment 3.12

Dynamic Range Test for Signal Amplitudes

1. Purpose

Very often it is necessary to measure rather weak NMR signals in the presence of other strong signals. One example is the detection of the ^1H signals of proteins in normal water. It is therefore useful to check the dynamic range performance of the spectrometer with a standard sample in order to know to what extent small signals can be detected in the presence of a strong signal without distortion.

2. Literature

[1] M. L. Martin, J. J. Delpuech, G. J. Martin, *Practical NMR Spectroscopy* Heyden, London , **1980**, 122–124.

[2] A. E. Derome, *Modern NMR Techniques for Chemistry Research*, Pergamon, Oxford, **1987**, 58–61.

3. Pulse Scheme and Phase Cycle

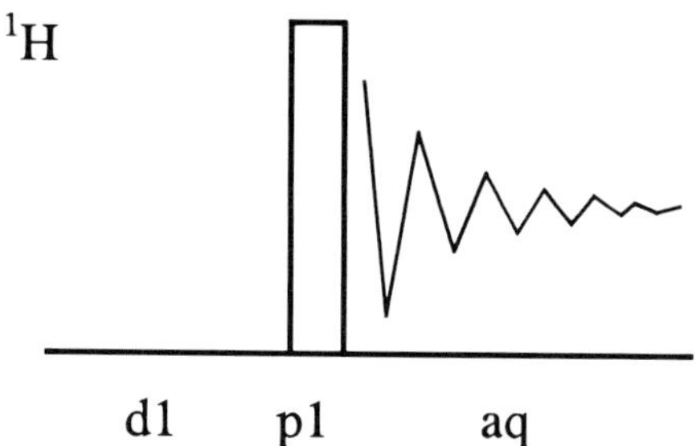

p1: x, x, -x, -x, y, y, -y, -y

aq: x, x, -x, -x, y, y, -y, -y

4. Acquisition

Time requirement: 15 min

Sample: 90% H_2O, 1.056% CH_3OH, 0.136% CH_3CN, 0.008% $(CH_3)_3COH$, 8.8% D_2O (for lock) giving ^1H signal intensity ratios of 10000:100:10:1

Tune the probe-head to the water sample and obtain good shim settings. Load standard ^1H NMR parameters and very carefully adjust the receiver gain to give the optimum input for the analog-to-digital converter. You have to set:

td: 32 k
sw: 10 ppm
o1: middle of the ^{1}H NMR spectrum
p1: 90° ^{1}H transmitter pulse
d1: 5 s
ns: 1

5. Processing

Use standard 1D processing with an exponential weighting (lb = 0.2 Hz), carefully correct the phase of the water signal, try to detect the very small signal of *t*-butanol at δ_H = 1.28, and adjust the phase of this signal as well. Integrate the four relevant signals and check the integrals for consistency with the molar ratios of the four compounds in the sample.

6. Result

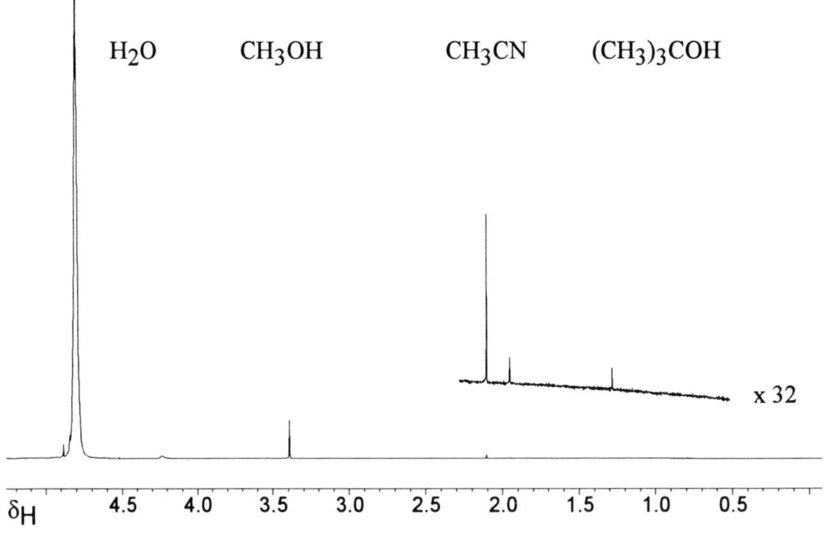

The figure shows the result obtained on an AMX-500 spectrometer equipped with a 16-bit digitizer. Besides some impurities of the sample, the resonances at δ_H = 4.8 (water), δ_H= 3.39 (methanol), and, after enlargement by a factor of 32, at δ_H = 2.1 (acetonitrile) and δ_H = 1.28 (*t*-butanol) can be seen.

7. Comments

The dynamic range is mainly dependent on the digitizer word length and, after accumulation, on the computer word length. The dynamic range behavior of all amplifiers and filter units also comes into account. Contrary to the common belief of many chemists, it does not make sense to accumulate signals endlessly, and newer software even sets a limit which is dependent on the difference between the computer and digitizer word lengths.

8. Own Observations

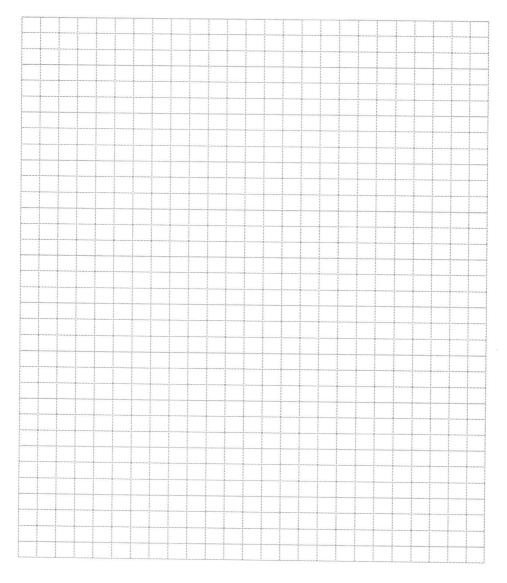

Experiment 3.13

13° Phase Stability Test

1. Purpose

For all multipulse experiments the relative phase between pulses and receiver and the phase between different pulses should be very stable to allow the cancellation of unwanted coherences by phase-cycling procedures. The 13° phase stability test shown here transforms phase stability into signal amplitudes and measures the phase stability between two r.f. pulses. A 1% amplitude variation represents a phase deviation of 0.14°.

2. Literature

[1] *RF Stability of the VXR-500*, Varian Instruments at Work, No. NMR-31, Varian Associates. Palo Alto, Cal. **1987**.
[2] G. A. Morris, *J. Magn. Reson.* **1988**, *78*, 281–291.
[3] G. A. Morris, *J. Magn. Reson.* **1992**, *100*, 316–328.

3. Pulse Scheme and Phase Cycle

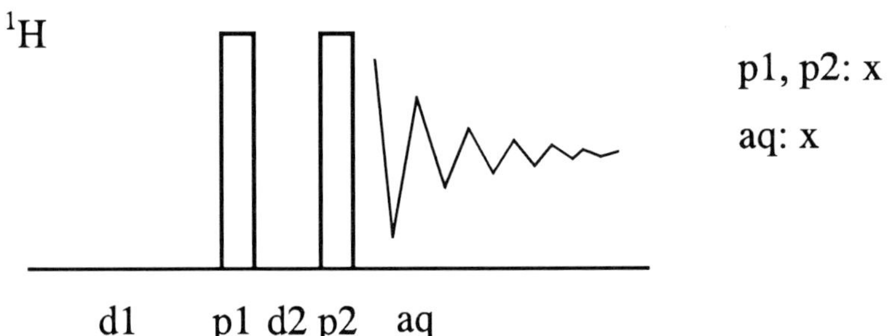

4. Acquisition

Time requirement: 30 min

Sample: 10% $CHCl_3$ in [D_6]acetone with added $Cr(acac)_3$

The probe-head should be tuned to the sample. Load standard ^{1}H parameters, record a normal ^{1}H NMR spectrum and note the offset of $CHCl_3$. *Turn the spinner off to avoid mechanical distortions*. Load the pulse program for the phase stability test. You have to set:

td: 4 k
sw: 500 Hz

o1: 37 Hz to higher frequencies from $CHCl_3$ signal
p1, p2: 90° ^1H transmitter pulse
d1: 20 s
d2: 1 ms
rg: receiver gain for correct ADC input
transmitter attenuation [3 dB]
ns: 1

Record one spectrum and check on all parameters. Use an automation routine which performs this experiment 64 times in sequence in order to have enough data for statistics.

5. Processing

Use standard 1D processing (see Exp. 3.1) applying an exponential window with a line-broadening factor lb = 1 Hz. Adjust the phase of the first spectrum roughly for dispersion and always use the same digital phase correction. The stability can be estimated from the standard deviations of the positive and negative peak heights.

6. Result

The figure on page 88 shows the 64 dispersion signals obtained on an AMX-500 spectrometer with a 5 mm multinuclear probe-head. Note the severe dropout which occurred in spectrum #21, probably due to an external influence, since this was not reproducible. Neglecting this dropout, a statistical evaluation of the positive and negative intensities gave a standard deviation of 0.7% in amplitude, which corresponds to a phase error of 0.1°.

7. Comments

The first 90° pulse p1 aligns the magnetization towards the $-y$ axis. Since the offset of the signal is 37 Hz from the transmitter and d2 was chosen to be 1 ms, the magnetization vector proceeds about 13° from this axis. The second 90° pulse p2, if it comes also exactly from the x-direction, will leave 22.5% (sin 13°) of this signal in the x,y-plane, which is acquired during the acquisition time. In the region of 13° the sine function is rather "linear", and therefore phase instabilities between the two r.f. pulses are well transformed into an amplitude variation of the signal. This is best observed if the signals are displayed in dispersion mode.

Further variations and different processing possibilities of this experiment are given in the literature.

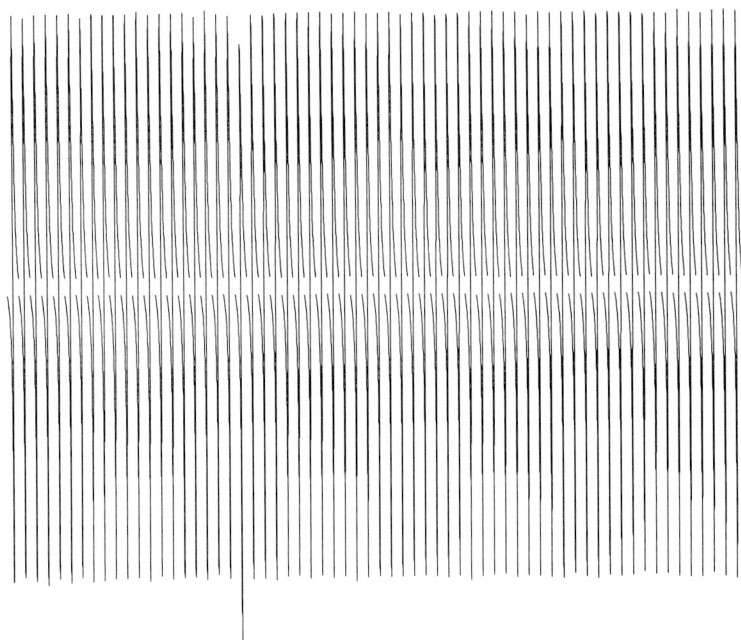

8. Own Observations

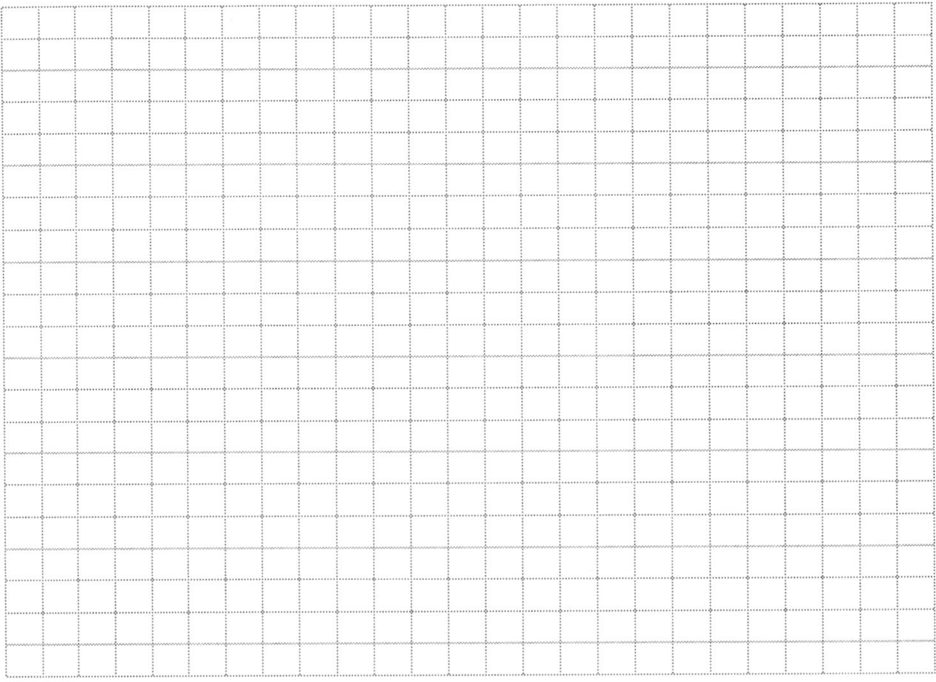

Chapter 4

Decoupling Techniques

In this chapter several basic 1D techniques are described which all use the proton decoupler (^1H broad-band decoupling using CPD is already shown in Exp. 2.3 and Exp. 3.2). Decoupling experiments are among the earliest techniques of NMR spectroscopy, and therefore some of the review literature cited is rather old but is still valid. One can distinguish between *homonuclear* and *heteronuclear* decoupling experiments. In the latter the observing channel is tuned to a heteronuclide X.

For both kinds of experiments it is essential to know the bandwidth of the decoupler for different attenuations. Therefore we provide three introductory experiments to calibrate the decoupler attenuation. It cannot be stressed enough that these experiments should be performed prior to any advanced application. With two exceptions (SPT experiments) the decoupler is used in such a way that no defined short decoupler pulses are required; therefore the experiments described in this chapter can be performed on any older instrument.

The homonuclear decoupling experiments described comprise the basic spin decoupling of protons and an experiment, where decoupling is performed at two frequencies simultaneously. They are followed by the SPT experiment and two variants of NOE difference spectroscopy. These are representative of the experiments which are most often carried out in the routine service applications of any NMR laboratory.

The heteronuclear examples show the different decoupling techniques used in ^{13}C NMR spectroscopy before the advent of multipulse editing sequences and 2D spectroscopy. However, the gated decoupling and inverse gated decoupling experiments are still in routine use, and even the somewhat outdated off-resonance decoupling experiment is of some value in those cases where more modern methods fail.

Many of the 1D experiments described in this chapter have 2D equivalents and these are mentioned in the relevant paragraphs. However, the 1D experiments may provide a quicker answer if only one or two questions are outstanding in the course of a molecular structure determination.

Literature

[1] S. Forsén, R. A. Hoffmann, *Prog. NMR Spectrosc.* **1966**, *1*, 15–204.

[2] J. D. Baldeschwieler, E. W. Randall, *Chem. Rev.* **1963**, *63*, 81–110.

[3] W. McFarlane, *Annu. Rep. NMR Spectrosc.* **1972**, *5A*, 353–393.

[4] W. v. Philipsborn, *Angew. Chem. Int. Ed. Engl.* **1971**, *10*, 472–490.

[5] M. H. Levitt, R. Freeman, T. Frenkiel, *Adv. Magn. Reson.* **1983**, *11*, 47–110.

[6] A. J. Shaka, J. Keeler, *Prog. NMR Spectrosc.* **1987**, *19*, 47–129.

Experiment 4.1

Decoupler Calibration for Homonuclear Decoupling

1. Purpose

In this experiment the decoupler attenuation for homonuclear decoupling is calibrated. In homonuclear decoupling a certain proton signal is irradiated, and the change of the signal patterns of the coupled protons gives structural information immediately. For this experiment the decoupler bandwidth as a function of the decoupler attenuation has to be known. This calibration experiment gives the necessary data to enable the operator to perform a selective decoupling experiment with sufficient decoupler power both for standard homonuclear decoupling and for SPT experiments (see Exp. 4.6). The calibration routine described relies on the Bloch–Siegert shift [2].

2. Literature

[1] M. L. Martin, J.-J. Delpuech, G. J. Martin, *Practical NMR Spectroscopy*, Heyden, London, **1980**, 203–206.

[2] F. Bloch, A. Siegert, *Phys. Rev.* **1940**, *57*, 522–527.

3. Pulse Scheme and Phase Cycle

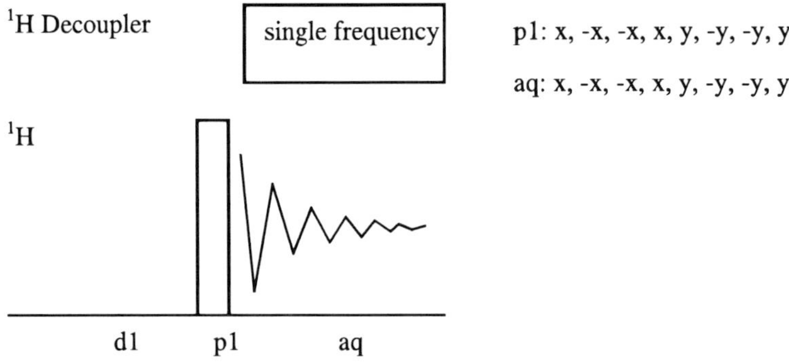

p1: x, -x, -x, x, y, -y, -y, y

aq: x, -x, -x, x, y, -y, -y, y

4. Acquisition

Time requirement: 30 min

Sample: 10% $CHCl_3$ in [D_6]acetone

The probe-head should be tuned to the sample. Load standard ^{1}H parameters, record a normal ^{1}H NMR spectrum and note the offset of $CHCl_3$. Reference the signal to 0 Hz. Switch to the homonuclear decoupling mode of the instrument, adjust the decoupler offset 50 Hz from the $CHCl_3$ signal, and record spectra with different decoupler attenuations. You have to set:

td: 4 k
sw: 500 Hz
o1: on resonance of ^1H signal
o2: 50 Hz towards lower frequency from o1
p1: 45° ^1H transmitter pulse
d1: 2 s
decoupler attenuation, to be varied

Measure the displacement of the signal and calculate the decoupler field strength γB_2 [Hz] from Equation (1)

$$\gamma B_2 = [2(\nu_A - \nu_2)(\nu_{obs} - \nu_A)]^{1/2} \tag{1}$$

where ν_A is the unperturbed resonance frequency, ν_2 the decoupler frequency and $\nu_{obs} - \nu A$ the observed Bloch–Siegert shift. Note that Equation (1) is only valid if Equation (2) holds.

$$(\gamma B_2)^2 \ll (\nu_A - \nu_2)^2. \tag{2}$$

Repeat the experiment for different decoupler offsets.

5. Processing

Use standard 1D processing as described in Experiment 3.1, and use zero-filling to ensure enough data points for the relatively small Bloch–Siegert shifts.

6. Result

A typical set of values for $\nu_A - \nu_2 = 50$ Hz obtained on an ARX-200 spectrometer is given below, from which the graph on page 92 was calculated:

Dec. att. [dB]	70	67	64	61	58	55	52	49	46
$\nu_{obs} - \nu_A$ [Hz]	0.07	0.12	0.24	0.44	0.79	1.5	3.4	8.7	15.7

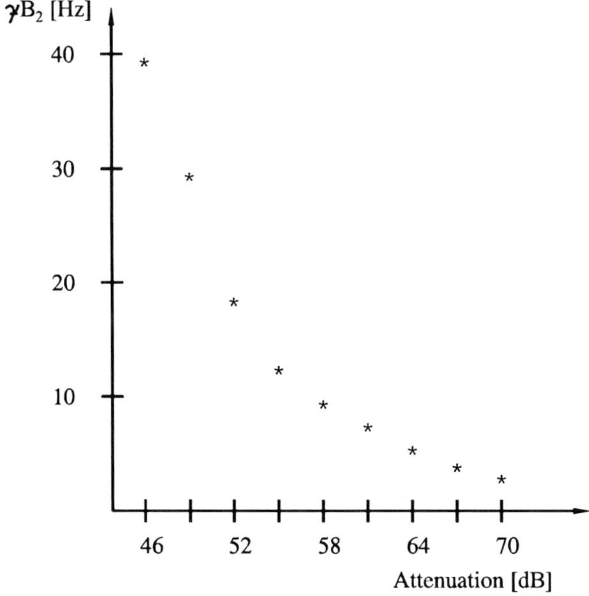

7. Comments

The calibration curve shown is dependent on the probe-head used. It should be determined for all the available probe-heads and documented in the log-book of the instrument. Note that homonuclear decoupling on FT instruments with probe-heads in a single coil arrangement requires a special decoupler mode, since the preamplifier must be protected against the decoupler r.f. power. This is usually done by applying the decoupling power only between the digitization points of the ADC with the preamplifier temporarily blanked.

8. Own Observations

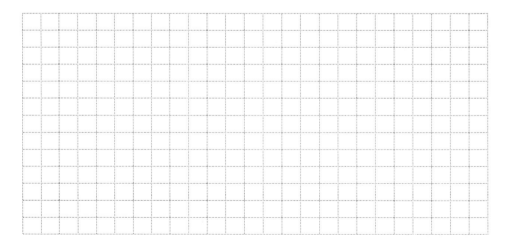

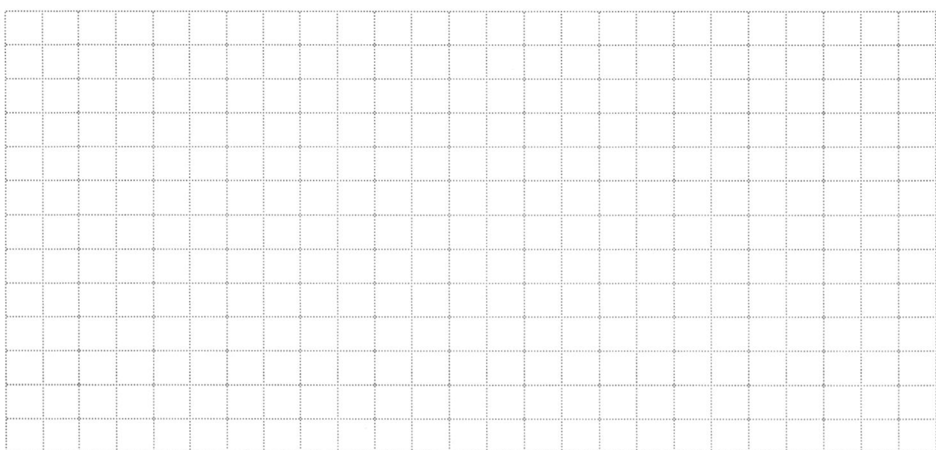

Experiment 4.2

Decoupler Calibration for Heteronuclear Decoupling

1. Purpose

Instead of using 2D NMR techniques such as HETCOR (see Exp. 10.10) it is sometimes more convenient to perform a heteronuclear decoupling experiment. A certain proton signal is irradiated, so that the connected carbon nucleus in the ^{13}C spectrum becomes decoupled. All other carbon signals are in the off-resonance decoupling situation. For this experiment the decoupler bandwidth as a function of the decoupler attenuation has to be known. The calibration experiment gives the necessary data to enable the operator to perform a selective decoupling experiment with the correct power to eliminate ^{1}J(C,H) spin couplings. The experiment also gives information about the decoupler bandwidth under normal broad-band decoupling settings.

2. Literature

[1] S. D. Simova, *J. Magn. Reson.* **1985**, *63*, 583–586.

3. Pulse Scheme and Phase Cycle

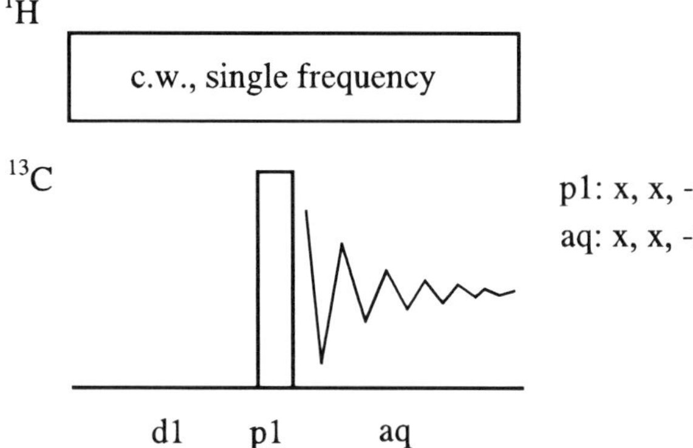

p1: x, x, -x, -x, y, y, -y, -y
aq: x, x, -x, -x, y, y, -y, -y

4. Acquisition

Time requirement: 30 min

Sample: 10% $CHCl_3$ in [D$_6$]acetone

The probe-head should be tuned to the sample. Record a normal ^{1}H NMR spectrum and note the offset of $CHCl_3$. Then load standard ^{13}C parameters with ^{1}H broad-band decoupling, record a normal ^{13}C NMR spectrum and note the offset of $CHCl_3$. Switch the instrument to c.w. decoupling. You have to set:

td: 4 k
sw: 500 Hz
o1: on resonance of ^{13}C signal
o2: 50 Hz offset from the ^{1}H signal
p1: 45° ^{13}C transmitter pulse
d1: 2 s
decoupler attenuation, to be varied
ns: 1

First record the ^{13}C doublet of $CHCl_3$ using 500 Hz spectral width and the transmitter offset adjusted directly on resonance; the decoupler should be switched off for the first experiment. Measure the spin coupling constant from the separation of the two doublet lines. Adjust the decoupler offset 50 Hz from the chloroform ^{1}H resonance and record the ^{13}C spectra with different decoupler attenuations. For certain decoupler offsets and attenuations you will obtain inner and outer doublets; the outer ones have a larger splitting than the normal $CHCl_3$ doublet. Measure the residual splitting within the outer or inner doublet and from this calculate the decoupler field strength γB_2 [Hz] using the Equation (1).

$$\gamma B_2 = [(J\,\Delta/J_R)^2 + 0.25\,(J_R^2 - J^2) - \Delta^2]^{1/2} \tag{1}$$

where J is the unperturbed spin coupling constant (214.8 Hz), J_R the residual splitting of the inner or outer doublett and Δ the offset of the decoupler frequency from the ^{1}H resonance. As an exercise perform the experiment with different decoupler offsets.

5. Processing

Use standard 1D processing as described in Experiment 3.2; use zero-filling to ensure enough data points to obtain accurate values for the residual splittings.

6. Result

A typical set of values for Δ = 50 Hz obtained on an ARX-200 spectrometer is given below, from which the graph on page 95 was calculated:

Dec. att. [dB]	62	59	56	53	50	47	44	41	38
J_R [Hz]	91	90	84	77	61	50	38	29	21

7. Comments

For every probe-head available in the laboratory a corresponding figure should be determined and documented in the log-book of the instrument. Note that for water solutions the tuning of the probe-head might be quite different. Formic acid may be used as a calibration sample for water solutions.

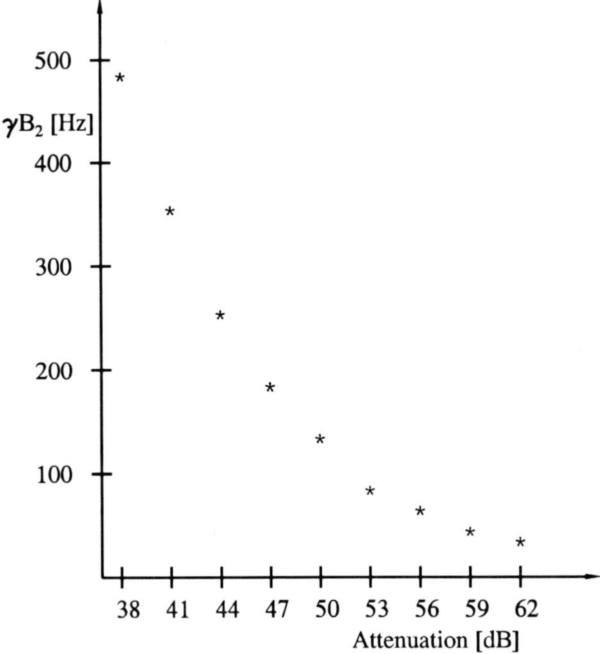

8. Own Observations

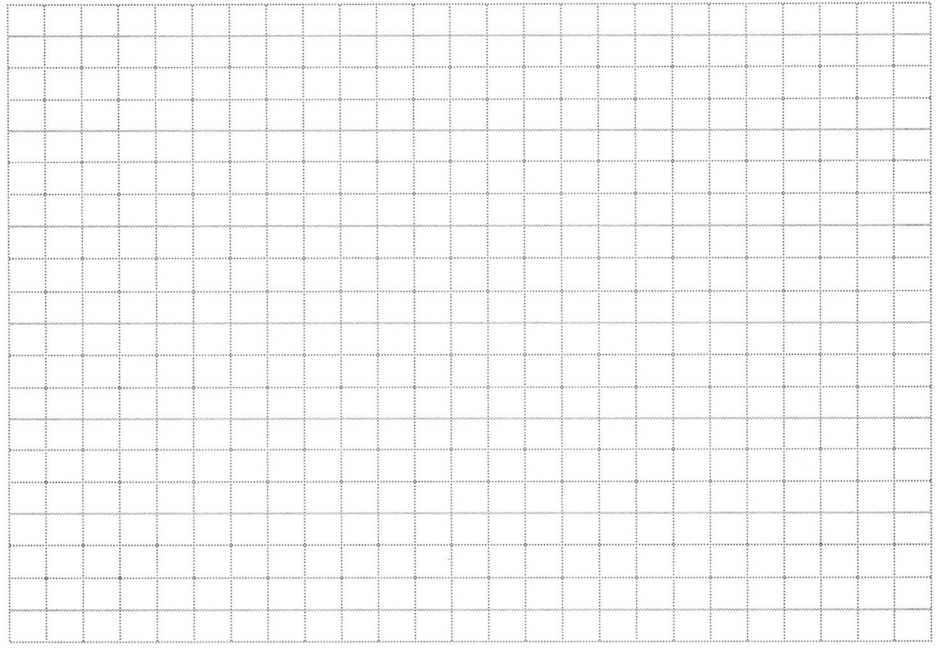

Experiment 4.3

Low Power Calibration for Heteronuclear Decoupling

1. Purpose

For certain applications it is necessary to irradiate a proton resonance with so little power that only the long-range spin couplings to ^{13}C are removed. As in Exp. 4.2 the decoupler bandwidth as a function of the decoupler attenuation has to be known, but in this case the calibration curve should cover γB_2-values from 1 to 40 Hz. Thus, the calibration experiment gives the necessary data to enable the operator to perform such a selective decoupling experiment. Accordingly, instead of $CHCl_3$, acetic acid is used as a calibration sample, where the effect of the decoupler in apparently reducing the ^{2}J(C,H) spin coupling constant is measured. The experiment also gives the necessary information to perform a heteronuclear SPT experiment (see Exp. 4.7).

2. Literature

[1] S. D. Simova, *J. Magn. Reson.* **1985**, *63*, 583–586.

3. Pulse Scheme and Phase Cycle

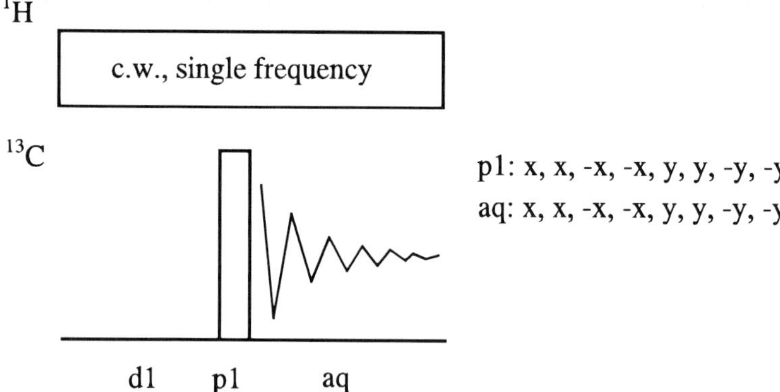

4. Acquisition

Time requirement: 30 min

Sample: 30% acetic acid in D_2O

The probe-head should be tuned to the sample. Record a normal ^{1}H NMR spectrum and note the offset of the methyl group signal of acetic acid. Then load standard ^{13}C parameters with ^{1}H broad-band decoupling, record a normal ^{13}C NMR spectrum, and note the offset of the carboxyl nucleus signal of acetic acid. Switch the instrument to c.w. decoupling. You have to set:

td: 2 k
sw: 100 Hz
o1: on resonance for carboxyl ^{13}C nucleus of acetic acid
o2: 25 Hz offset from the ^1H resonance of the CH_3 group of acetic acid
p1: 45° ^{13}C transmitter pulse
d1: 2 s
decoupler attenuation, to be varied
ns: 1

Record the ^{13}C quartet of the carboxyl signal of acetic acid using 100 Hz spectral width and the transmitter offset directly on resonance; the decoupler should be off for the first experiment. Measure the spin coupling constant from the separation of the two inner lines of the quartet. Adjust the decoupler offset to 25 Hz from the ^1H resonance of the methyl group of acetic acid and record the ^{13}C spectra with different decoupler attenuations. Measure the separation of the two inner lines of the quartet and from this calculate the decoupler field strength γB_2 [Hz] using Equation (1)

$$\gamma B_2 = [(J\,\Delta/J_R)^2 + 0.25\,(J_R^2 - J^2) - \Delta^2]^{1/2} \qquad (1)$$

where J is the unperturbed spin coupling constant (6.63 Hz), J_R the reduced splitting of the inner lines of the quartet and Δ the offset of the decoupler frequency from the ^1H resonance.

5. Processing

Use standard 1D processing as described in Experiment 3.2, with lb = 0.3 Hz, use zero-filling to ensure enough data points for the reduced splittings.

6. Result

A typical set of values for Δ = 25 Hz obtained on an ARX-200 spectrometer is given below, from which the graph on page 98 was calculated:

Dec. att. [dB]	80	77	74	71	68	65	62	59
J_R [Hz]	6.6	6.54	6.4	6.2	5.65	5.05	4.45	3.48

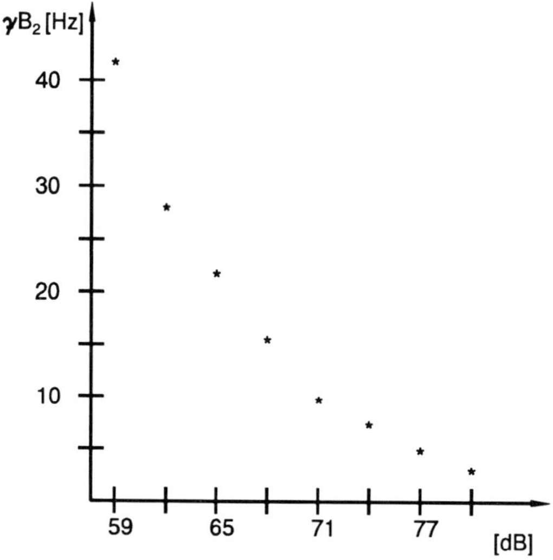

7. Comments

With the sample used here the calibration relates, of course, to a probe-head which was tuned to a sample containing water and thus might be quite different from a typical application with an organic solvent. For these you may use *t*-butyl acetate in $CDCl_3$ or similar compounds. A calibration curve for every probe-head present in the laboratory should be available in the log-book of the instrument.

8. Own Observations

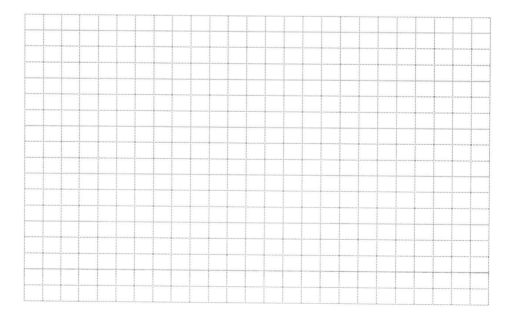

Experiment 4.4

Homonuclear Decoupling

1. Purpose

Complex spin systems can be simplified by homonuclear decoupling. By this technique residual multiplets are obtained in which the spin coupling to the irradiated proton is missing. The signal of the irradiated proton itself cannot be observed during decoupling. From a comparison with the undisturbed multiplet the relevant spin coupling constant can be evaluated. Similar information can also be obtained with the selective COSY technique (Exp. 7.5). Here we show a homonuclear decoupling experiment on ethyl crotonate.

2. Literature

[1] W. A. Anderson, R. Freeman, *J. Chem. Phys.* **1962**, *37*, 85–103.
[2] J. P. Jesson, P. Meakin, G. Kneissel, *J. Am. Chem. Soc.* **1973**, *95*, 618–620.
[3] M. L. Martin, J.-J. Delpuech, G. J. Martin, *Practical NMR Spectroscopy*, Heyden, London, **1980**, 203–206.
[4] J. D. Baldeschwieler, E. W. Randall, *Chem. Rev.* **1963**, *63*, 81–110.
[5] R. W. Dykstra, *J. Magn. Reson. Ser. A* **1993**, *102*, 114–115.

3. Pulse Scheme and Phase Cycle

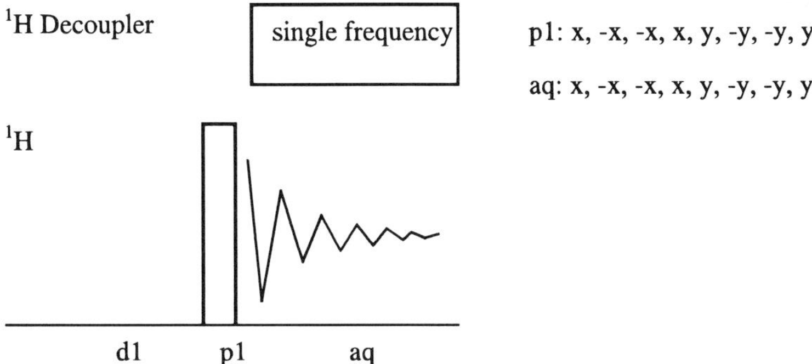

p1: x, -x, -x, x, y, -y, -y, y
aq: x, -x, -x, x, y, -y, -y, y

4. Acquisition

Time requirement: 20 min

Sample: 5% ethyl crotonate in $CDCl_3$

On some instruments you have to connect a directional coupler in order to feed the decoupler frequency into the transmitter line. Load standard ^{1}H parameters, record a normal ^{1}H NMR spectrum, and note the offsets of the signals to be irradiated. Change to the homonuclear decoupling mode of the spectrometer and adjust the decoupling power according to the width of the multiplet (see Exp. 4.1). You have to set:

td : 32 k
sw: 10 ppm
o1: middle of ^{1}H NMR spectrum
o2: on resonance of irradiated proton
p1: 45° ^{1}H transmitter pulse
d1: 1 s
ns: 8
decoupler attenuation for homonuclear decoupling

5. Processing

Use standard 1D processing as described in Exp. 3.1 with exponential multiplication (lb = 0.3 Hz).

6. Result

The figure shows the result for ethyl crotonate obtained on an AM-400 spectrometer. The signal region for the olefinic protons is shown in **a** (not decoupled), while **b** shows the result obtained by irradiation of the methyl group protons H-4 and **c** the result with irradiation of olefinic proton H-2. Note that in **c** some residual splitting is observed.

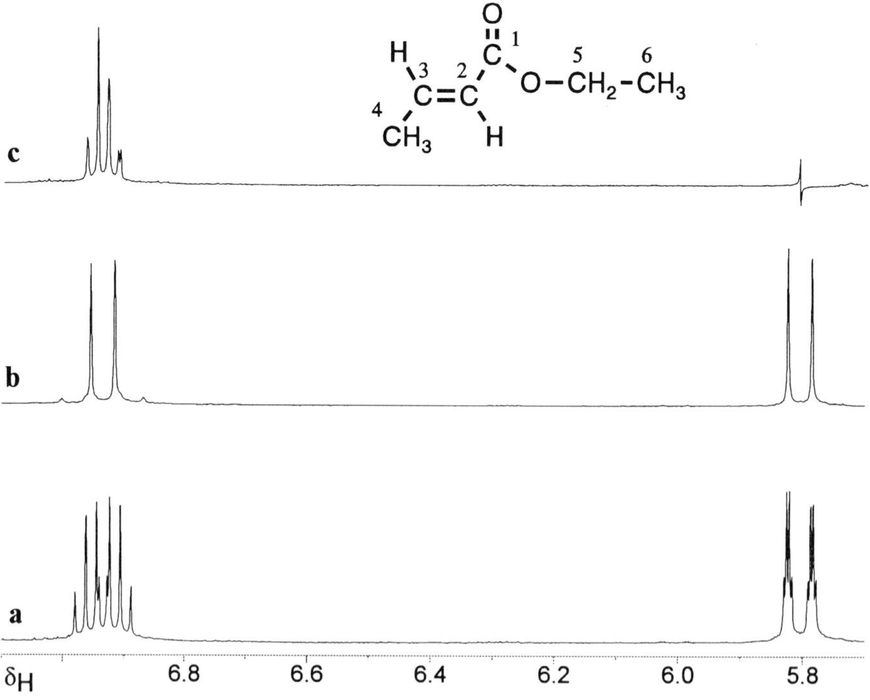

7. Comments

On FT instruments with a single coil probe-head homonuclear decoupling requires a special mode in order to avoid damage to the preamplifier. Thus the decoupling energy is applied in a pulsed mode within the duty cycle of the dwell time and the preamplifier is switched off during the decoupler pulses [2]. Note that homonuclear decoupling causes Bloch–Siegert shifts which can displace the residual multiplets from their original position. Bloch–Siegert shifts also affect the irradiated resonance, so that the best irradiation position is not the center of the unperturbed resonance.

8. Own Observations

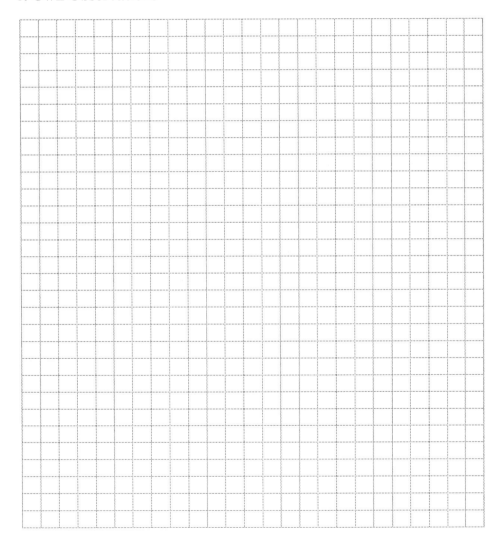

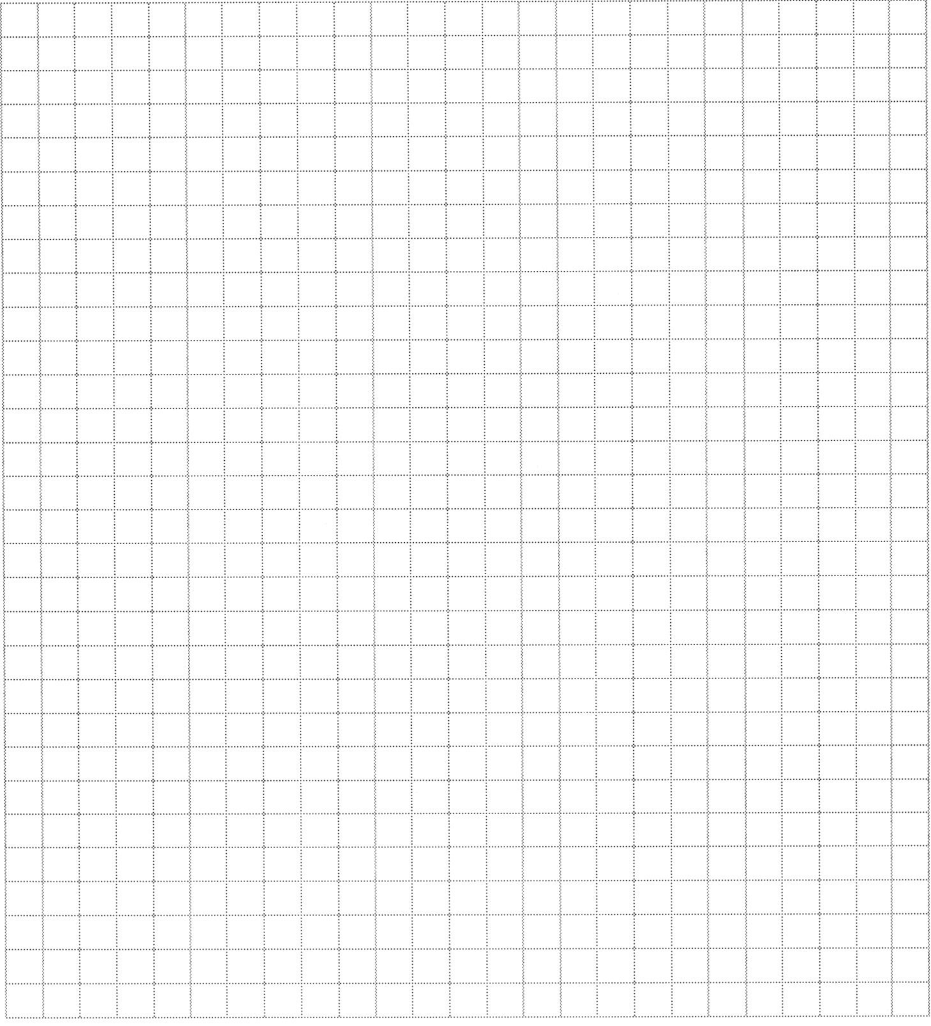

Experiment 4.5

Homonuclear Decoupling at Two Frequencies

1. Purpose

Very often organic molecules have complex spin systems comprised of four or even more protons. Standard homonuclear decoupling as described in Experiment 4.4 simplifies the residual multiplet, but the extraction of the relevant spin-coupling constant may still be difficult. In principle it is possible to decouple more than one proton at the same time and by more recent developments even an area of other protons [3]. In the experiment described here we have chosen a four-spin system AMXY of strychnine demonstrating simultaneous decoupling of protons M and X to observe the spin coupling between A and Y.

2. Literature

[1] J. P. Jesson, P. Meakin, G. Kneissel, *J. Am. Chem. Soc.* **1973**, *95*, 618–620.

[2] M. L. Martin, J.-J. Delpuech, G. J. Martin, *Practical NMR Spectroscopy*, Heyden, London, **1980**, 203–206.

[3] A. Hammarström, G. Otting, *J. Am. Chem. Soc.* **1994**, *116*, 8847–8848.

3. Pulse Scheme and Phase Cycle

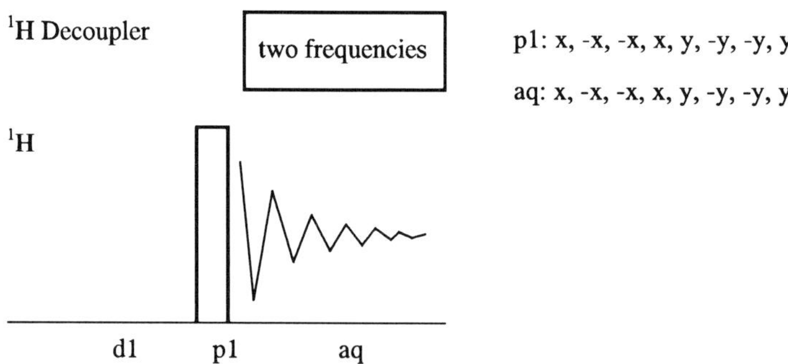

p1: x, -x, -x, x, y, -y, -y, y

aq: x, -x, -x, x, y, -y, -y, y

4. Acquisition

Time requirement: 10 min

Sample: 10% strychnine in $CDCl_3$

On some instruments you have to connect a directional coupler in order to feed the decoupler frequency into the transmitter line. Load standard ^{1}H parameters, record a normal ^{1}H NMR spectrum, and note the offsets of the signals for the protons 11α at $\delta_H = 3.04$ and 11β at $\delta_H = 2.58$. Change to the homonuclear decoupling mode of the spectrometer and obtain first two spectra with single frequency decoupling for protons

11α and 11β. Optimize the decoupling conditions as discussed in Experiment 4.4. Load the pulse program for simultaneous decoupling of two frequencies. You have to set:

td : 32 k
sw: 10 ppm
o1: middle of ^{1}H NMR spectrum
o2: on resonance of irradiated proton 11α
o3: on resonance of irradiated proton 11β
p1: 45° ^{1}H transmitter pulse
d1: 1 s
ns: 1
decoupler attenuation for homonuclear decoupling, may be different for the second and the third channel [25 dB]

5. Processing

Use standard 1D processing as described in Experiment 3.1 with exponential multiplication (lb = 0.1 Hz).

6. Result

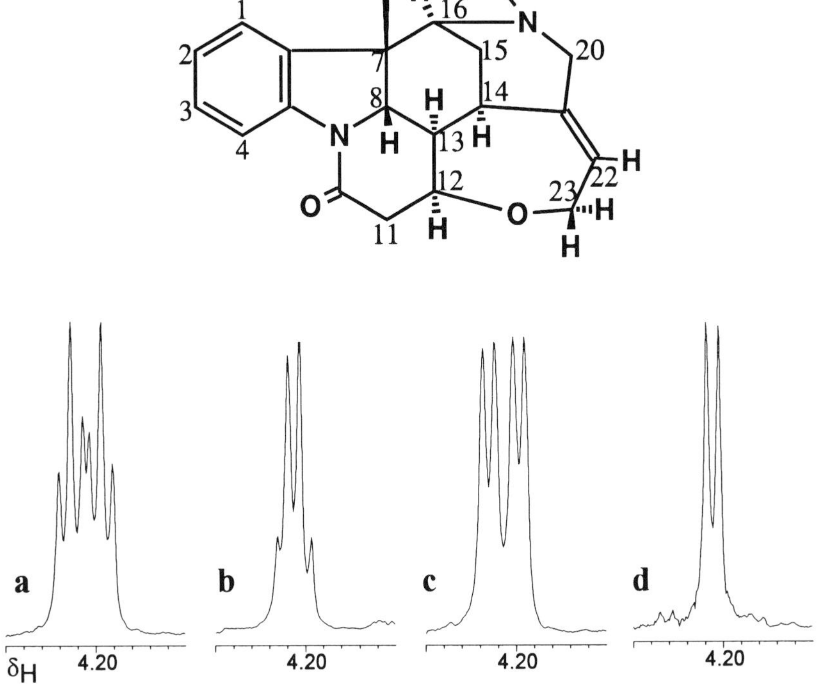

The figure on page 103 shows the result obtained with an AMX-500 spectrometer in an inverse multinuclear probe-head. In **a** the undisturbed multiplet of H-12 due to spin coupling with both H-11α, H-11β and H-13 is shown; in **b** H-11α and in **c** H-11β was decoupled leading to residual multiplets of the AMX type. In **d** the result of the simultaneous decoupling of H-11α and H-11β is shown, where the spin coupling *J*(H-12, 13) = 3.4 Hz can be directly read from the residual doublet.

7. Comments

On FT instruments with a single-coil probe-head, homonuclear decoupling requires a special mode in order to avoid damage to the preamplifier. Thus the decoupling energy is applied in a pulsed mode within the duty cycle of the dwell time, and the preamplifier is switched off during the decoupler pulses [1]. The realization of simultaneous homodecoupling at two frequencies is very instrument dependent and has been performed here by splitting the dwell time to allow decoupling with two different frequency sources (three-channel instrument). In a more recent approach the application of a CPD sequence during the acquisition time was demonstrated, where the CPD sequence consists of a shaped pulse which contains the two irradiation frequencies [3].

8. Own Observations

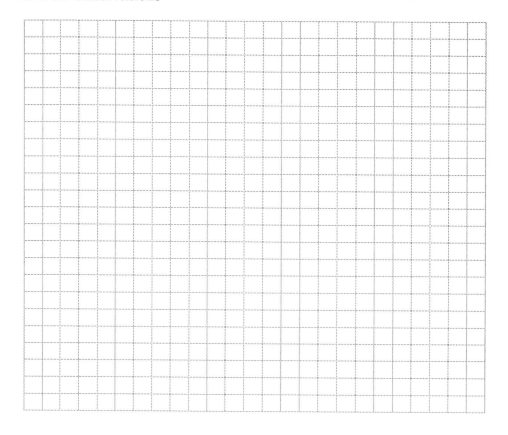

Experiment 4.6

The Homonuclear SPT Experiment

1. Purpose

Spin coupling constants can have either sign, for example, the coupling constant ^2J(H,H) between two diastereotopic methylene hydrogen nuclei is usually negative. A sign determination can be very useful for distinguishing a ^2J from a ^3J coupling constant, as the latter is normally positive. The Selective Population Transfer experiment (SPT) is a simple 1D method which provides this relative sign information [1]. Furthermore, a lot can be learned about the different transitions within a spin system from this experiment. The sign information can also be obtained from a COSY-45 experiment (see Exp. 10.6)

2. Literature

[1] K. G. R. Pachler, P. L. Wessels, *J. Magn. Reson.* **1973**, *12*, 337–339.
[2] M. L. Martin, J. J. Delpuech, G. J. Martin, *Practical NMR Spectroscopy*, Heyden, London, **1980**, 222–226.

3. Pulse Scheme and Phase Cycle

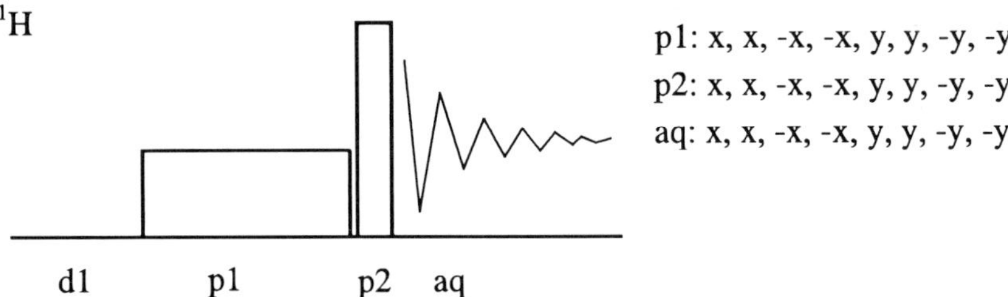

p1: x, x, -x, -x, y, y, -y, -y
p2: x, x, -x, -x, y, y, -y, -y
aq: x, x, -x, -x, y, y, -y, -y

4. Acquisition

Time requirement: 1 h

Sample: 5% 2,3-dibromopropionic acid in [D$_6$]benzene

Depending on the age of the spectrometer this experiment may need to be modified to the available hardware. Older spectrometers have no variable transmitter attenuation; p1 must then be taken from the decoupler and o1 = o2 phase coherence between decoupler and transmitter/receiver should be established. Another possibility is to perform the experiment completely with the decoupler (both p1 and p2) in the inverse mode of the spectrometer. On modern instruments the rectangular pulse p1 can also be replaced by a shaped pulse. In the following description we refer to instruments with transmitter attenuation. You have to set:

td: 8 k
sw: 2.5 ppm
o1: on resonance of a chosen multiplet line of the sample
p1: 180° ^{1}H transmitter pulse at chosen attenuation, here 0.8 s at 90 dB was used, see Exp. 2.9
p2: 30° ^{1}H transmitter pulse with normal attenuation (3 dB), *be sure not to use a 90° pulse*
d1: 5 s
two different transmitter attenuation levels
ns: 1

Record a normal ^{1}H NMR spectrum of the sample. Load a pulse program for the SPT experiment and adjust the power of p1 to $\gamma B_1 \cong 1$Hz, typically in the order of 90 dB (see Exp. 2.9). Adjust o1 to the left-most signal of the sample, record its SPT spectrum and repeat the experiment for all signals of the AMX spin system.

5. Processing

Use standard 1D processing as described in Experiment 3.1

6. Result

In the figure on page 107 the normal ^{1}H NMR spectrum taken on an AMX-500 spectrometer is shown in **a**. In **b** line X_4 was irradiated and in **c** line X_3. Note that in **b** lines A_2 and M_3 are attenuated, whereas A_1 and M_1 are enhanced. In **c** lines M_4 and A_1 are attenuated, M_2 and A_2 are enhanced.

7. Comments

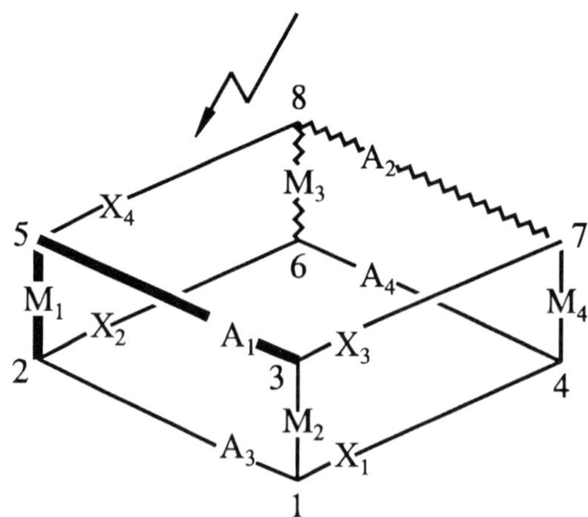

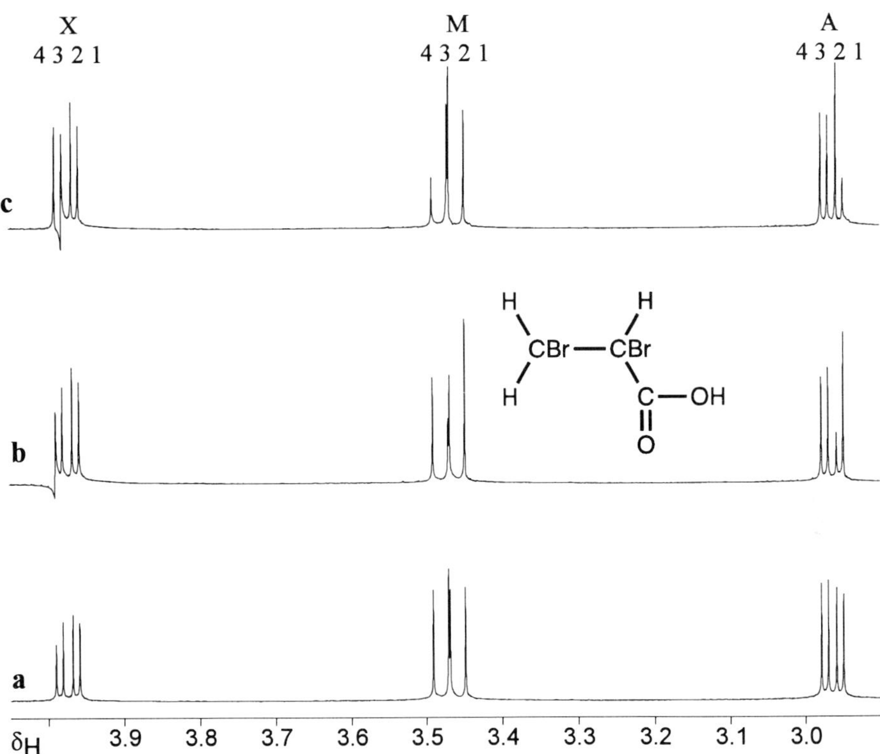

For the AMX spin system of 2,3-dibromopropionic acid the energy levels and transitions can be calculated using a spin simulation program assuming a negative coupling constant between the geminal protons on C-3. From this calculation the level scheme shown on page 106 can be drawn, indicating that the results of the SPT experiment are in agreement with the prediction from the simulation. Thus, irradiation of line X_4 in the spectrum inverts the populations of energy levels 5 and 8. This leads to enhancements of the M_1 and A_1 transitions and to attenuations of the M_3 and A_2 transitions.

8. Own Observations

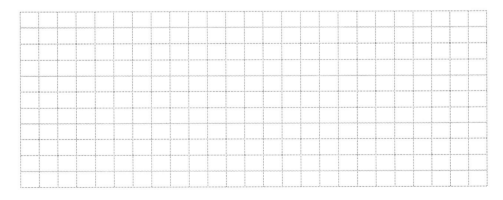

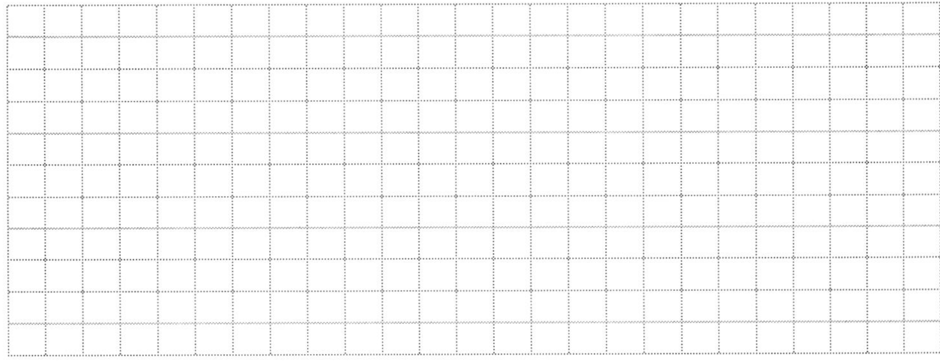

Experiment 4.7

The Heteronuclear SPT Experiment

1. Purpose

The heteronuclear SPT experiment was introduced [1] for the analysis of heteronuclear spin systems H,X with X = ^{13}C or ^{29}Si. As in the homonuclear case (see Exp. 4.6) it is especially able for determining the relative signs of long-range spin coupling constants. In addition to the sign the experiment gives information about whether a particular proton is spin-coupled to the carbon nucleus in question, and thus the experiment is very valuable in the interpretation of long-range C,H multiplets. There are many different versions [2, 3] of the experiment in addition to the basic method described here, such as difference spectroscopy and a version with proton decoupling for assignment purposes only. The theory of the experiment has been described [4, 5].

2. Literature

[1] K. G. R. Pachler, P. L. Wessels, *J. Magn. Reson.* **1973**, *12*, 337–339.
[2] S. K. Sarkar, A. Bax, *J. Magn. Reson.* **1985**, *62*, 109–112.
[3] S. A. Linde, H. J. Jakobsen, *J. Am. Chem. Soc.* **1976**, *98*, 1041–1043.
[4] R. Pachter, P. L. Wessels, *J. Magn. Reson.* **1989**, *81*, 464–473.
[5] M. L. Martin, J. J. Delpuech, G. J. Martin, *Practical NMR Spectroscopy*, Heyden, London, **1980**, 222–226.

3. Pulse Scheme and Phase Cycle

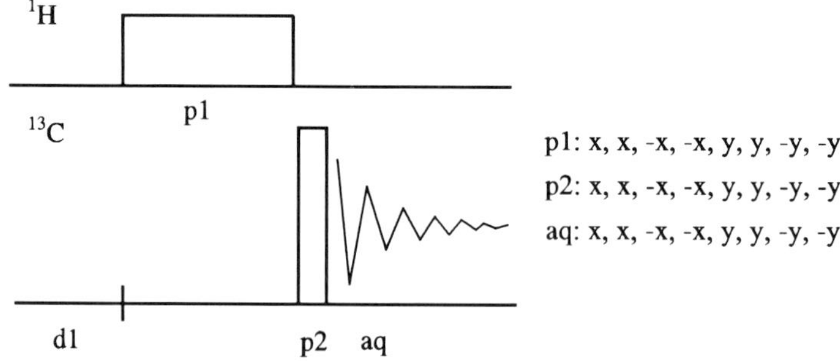

4. Acquisition

Time requirement: 1 h

Sample: 5% 3-chloroacrylic acid in $CDCl_3$

As seen from the pulse sequence above, the experiment consists of a selective 180° proton pulse p1 with a very narrow bandwidth and low power level ($\gamma B_2 \approx 1$Hz) centred on a particular transition of the ^{13}C,H spin system, followed by a hard sampling pulse p2 in the ^{13}C channel. You have to set:

td: 64 k
sw: 200 ppm
o1: middle of the ^{13}C NMR spectrum
o2: exact transition frequency of a ^{13}C satellite in the ^{1}H NMR spectrum
p1: 180° ^{1}H decoupler pulse at chosen attenuation; here 0.4 s at 90 dB was used, see Exp. 2.9
p2: 45° ^{13}C transmitter pulse
d1: 2s
ns =8

Measure a normal ^{1}H NMR spectrum and a proton-coupled ^{13}C spectrum (gated decoupling, see Exp. 4.10) as reference spectra. Load a pulse program for the heteronuclear SPT experiment and adjust the p1 power to $\gamma B_1 \approx 1\text{Hz}$, typically in the order of 90 db (see Exp. 2.9). Adjust o2 to a frequency 2 Hz above that of the left-most line of the proton doublet at δ_H = 7.5, in order to excite the corresponding ^{13}C satellite signal. Measure the SPT spectrum and repeat this for the other satellites of the spin system.

5. Processing

Use standard 1D processing as described in Experiment 3.2.

6. Result

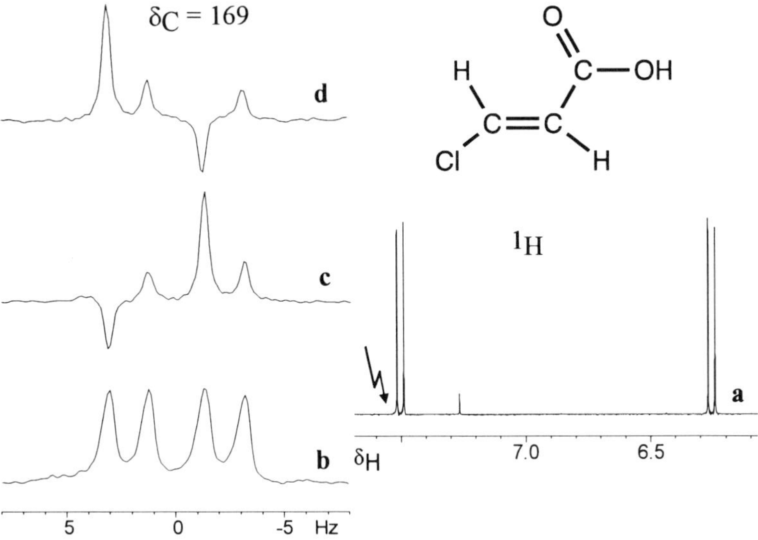

The figure on page 109 shows at **a** the normal ^1H NMR spectrum taken on an AMX-500 spectrometer. In **b** the ^1H-coupled resonance of the carboxyl ^{13}C nucleus at δ_C = 169 is shown. In **c** o2 was set 2 Hz downfield from the left-most proton transition and in **d** 2 Hz upfield from the same transition.

7. Comments

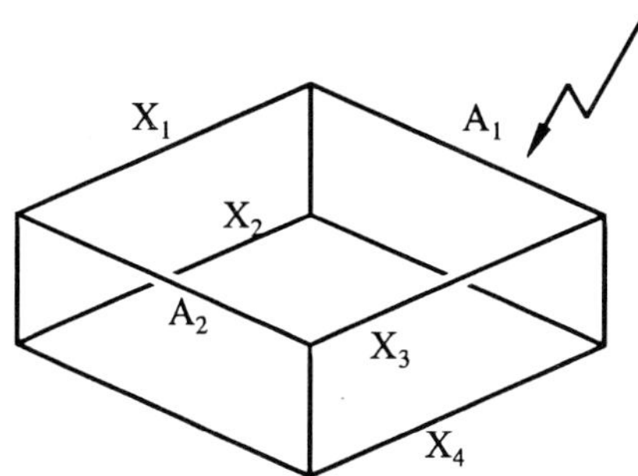

In an AMX spin system, calculated with all coupling constants positive as given in the figure, irradiation of the A_1 transition leads to population changes within the connected energy levels; thus the X_1 transition will be in emission and the X_3 transition in enhanced absorption. The reverse is true, if the A_2 transition is irradiated. Using a spin simulation program the energy levels can be calculated and the transitions are numbered; if the results of the SPT experiment agree with the predictions from the simulation, this confirms the assumed sign of the coupling constants. Note that in contrast with the homonuclear experiment (Exp. 4.6) one obtains true inversions and large enhancements due to the large difference in the γ-values for carbon and hydrogen.

8. Own Observations

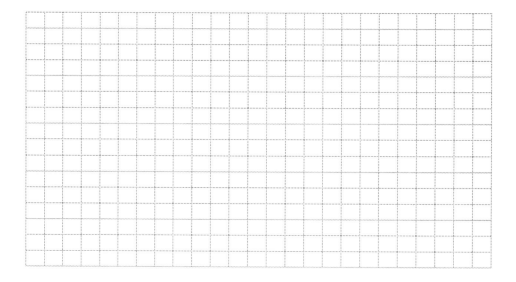

Experiment 4.8

1D Nuclear Overhauser Difference Spectroscopy

1. Purpose

Many problems of stereochemical assignment, for example differentiating between *E* and *Z* double bonds or between *exo* and *endo* groups in bicyclic compounds, cannot be solved using spin coupling constants if no suitable protons are present. In many cases, however, an NOE difference measurement provides an easy and straightforward answer. Irradiation of one group of protons causes a change in the intensities of other signals, which is related to the inverse sixth power of the distance between the spins.

2. Literature

[1] D. Neuhaus, M. Williamson, *The Nuclear Overhauser Effect*, VCH, Weinheim, **1989**.

[2] J. K. M. Sanders, B. K. Hunter, *Modern NMR Spectroscopy*, 2nd Edition, Oxford University Press, Oxford, **1993**, Ch. 6.

3. Pulse Scheme and Phase Cycle

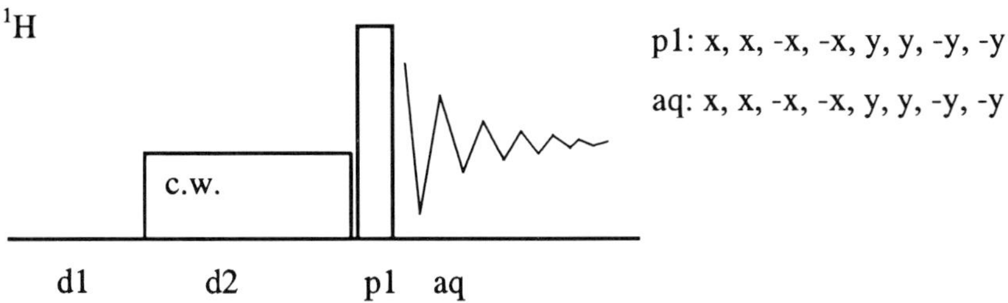

p1: x, x, -x, -x, y, y, -y, -y

aq: x, x, -x, -x, y, y, -y, -y

4. Acquisition

Time requirement: 0.5 h

Sample: 5% ethyl methacrylate in $CDCl_3$, *degassed and sealed sample*

Obtain a normal ^{1}H NMR spectrum and note the frequency of the methyl group signal. Depending on the instrument used, adjust the low power level for pre-irradiation by either decoupler or transmitter to $\gamma B_2 \cong 10$ Hz (see Exp 2.9). Here 70 dB attenuation was used. You should check whether the irradiated signal has disappeared from the ^{1}H NMR spectrum under these conditions. Record two spectra in which first the methyl group resonance and then a reference signal which does not introduce an NOE effect is irradiated. TMS or the residual proton signal of $CDCl_3$ is often used for the latter purpose. For dilute samples these two spectra are accumulated in an interleaved mode. You have to set:

td: 32 k
sw: 10 ppm
o1: middle of ^1H NMR spectrum
o2: on resonance for the methyl protons in the first experiment and on resonance of the residual ^1H signal of $CHCl_3$ in the reference spectrum
p1: 90° ^1H transmitter pulse
d1: 0.1 s
d2: 6 s
decoupler attenuation for selective presaturation
ds: 4
ns:16

5. Processing

NOE difference spectra can be processed in different ways. Since one wants to observe signal intensity changes of 2 to 10%, one should use an exponential window function with lb = 2 Hz to minimize artefacts of subtraction. One can either transform the two spectra separately using a *digitally identical* phase correction and subtract the two spectra, or, more conveniently, subtract the two FIDs directly from each other. In the difference spectrum, adjust the phase of the methyl group signal to be negative and the phase of the reference signal ($CHCl_3$) to be positive. Evaluate only signals which have correct phase and have therefore not been affected by inadequate spectrometer stability.

6. Result

The figure on page 113 shows the results obtained on an ARX-200 spectrometer. In **a** the normal ^1H NMR spectrum is given while in **b** the NOE difference result is shown with an expansion in the olefinic region. Integration of the methyl group signal and comparison with the integral of the *cis* olefinic proton ($\delta_H = 5.54$) gives an NOE effect of 9%, taking into account the threefold number of protons within the methyl group. Note the subtraction artefacts at $\delta_H = 4.2$.

7. Comments

The exact theory of the NOE effect is complicated and is described in detail in the references given. In principle, saturation of one transition W_{1I} in a spin system with dipolar couplings but no indirect (scalar) couplings as given in the figure, first equalizes the populations of the two corresponding energy levels. The system reacts via the relaxation pathways W_2 or W_0, which leads to population changes of those energy levels that are connected by the W_{1S} transition. One should warn against placing too much weight on a quantitative interpretation of the results of these measurements, since the relaxation times of all neighbour protons are important, furthermore three spin effects and other complications are known to play a role. From a qualitative standpoint, however, in most cases, the experiment gives a correct answer to a stereochemical question.

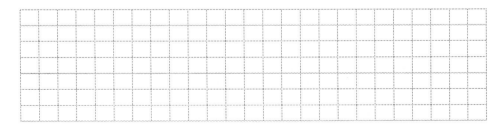

b

H CH3
C=C
H C OCH2CH3
O

a

δH 6.5 6.0 5.5 5.0 4.5 4.0 3.5 3.0 2.5 2.0 1.5

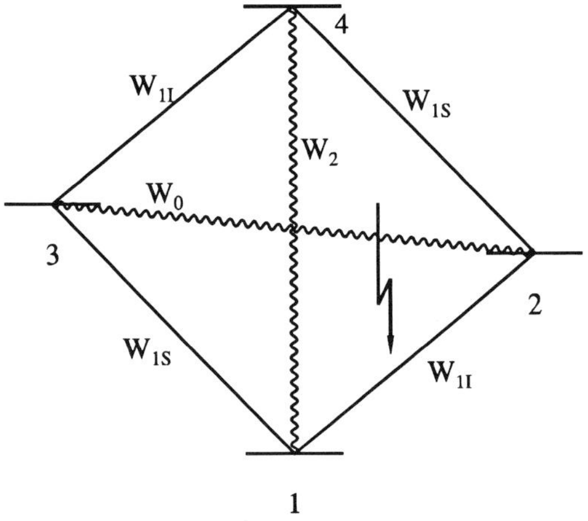

4
W_{1I} W_{1S}
W_2
W_0
3 2
W_{1S} W_{1I}
1

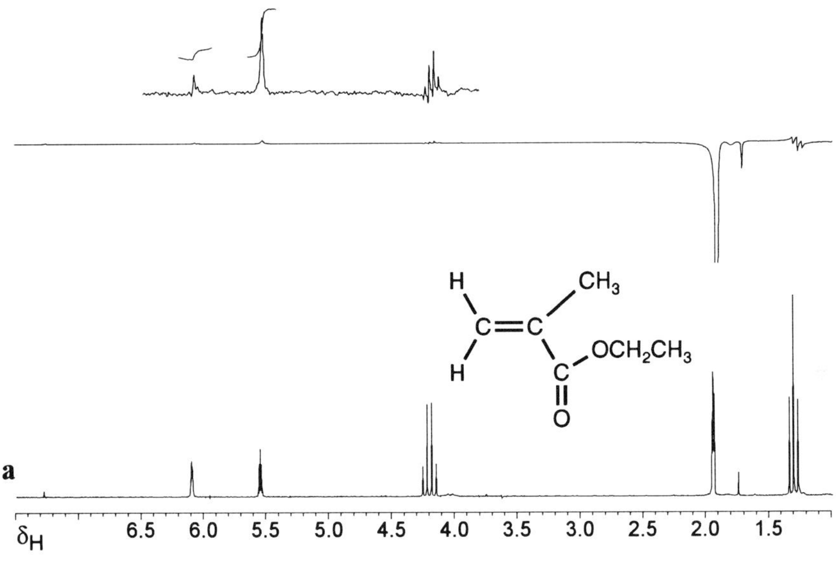

8. Own Observations

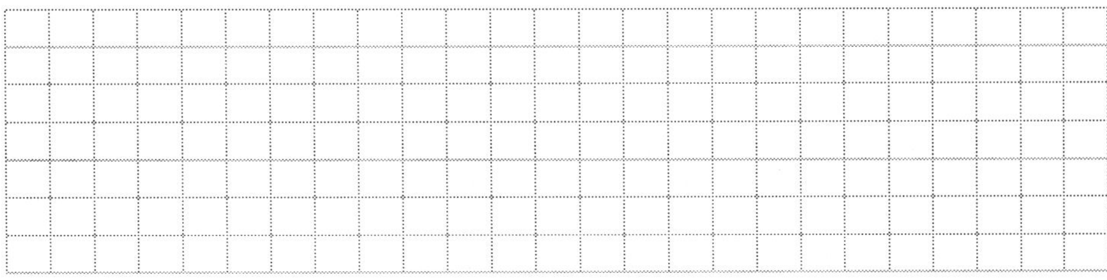

Experiment 4.9

1D NOE Spectroscopy with Multiple Selective Irradiation

1. Purpose

This experiment is a technical variant of Experiment 4.8. In NOE difference spectroscopy one often encounters the problem that the large decoupler bandwidth required to irradiate a broad multiplet spills out to other nearby signals, thus making a correct NOE assignment very difficult. Instead of irradiating the center of a broad multiplet, in the experiment described here each line of the multiplet is irradiated for a short time with a bandwidth of ca. 1–2 Hz and the irradiating frequency is cycled repeatedly in a stepwise manner through the entire multiplet during the pre-irradiation time. In principle, this is a multiple SPT experiment (see Exp 4.6), but here the SPT effects are cancelled and only the NOE effects remain.

2. Literature

[1] M. Kinns, J. K. M. Sanders, *J. Magn. Reson.* **1984**, *56*, 518–520.

[2] D. Neuhaus, M. Williamson, *The Nuclear Overhauser Effect*, VCH, Weinheim, **1989**.

[3] J. K. M. Sanders, B. K. Hunter, *Modern NMR Spectroscopy*, 2nd Edition, Oxford University Press, Oxford, **1993**, Ch. 6.

3. Pulsescheme and Phase Cycle

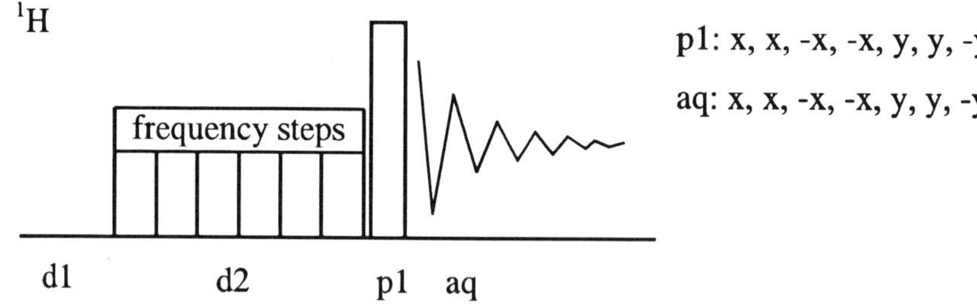

p1: x, x, -x, -x, y, y, -y, -y

aq: x, x, -x, -x, y, y, -y, -y

4. Acquisition

Time requirement: 0.5 h

Sample: 10% strychnine in $CDCl_3$, *degassed and sealed sample*

Obtain a normal ^{1}H NMR spectrum and note the single frequencies to be irradiated within the multiplets. Depending on the instrument used, adjust the low power level for pre-irradiation by either decoupler or transmitter to $\gamma B_2 \cong 1$ to 2 Hz (see Exp 2.9).

Here 85 dB attenuation was used. As in Experiment 4.8 a reference signal which does not introduce an NOE effect is irradiated in a separate measurement. TMS or the residual proton signal of $CDCl_3$ is often used for this purpose. Depending on the spectrometer used, an automatic program may be loaded which organizes the stepping of the irradiation frequency through the multiplets. For dilute samples the spectra are accumulated in an interleaved mode. You have to set:

td: 32 k
sw: 10 ppm
o1: middle of spectrum
o2: lists of frequencies within the multiplets to be irradiated
p1: 90° ^{1}H transmitter pulse
d1: 0.1 s
d2: 7.2 s total pre-irradiation time; in the example multiplets with 6 lines were chosen, each line was irradiated for 400 ms, and the process was performed three times. In addition, the whole sequence for two irradiated multipletts and one reference line was repeated four times and the data averaged.
decoupler attenuation for presaturation (85 dB)
ds: 4
ns: 8

5. Processing

Use processing for NOE difference spectroscopy as described in Experiment 4.8.

6. Result

The figure on page 116 shows the result obtained on an AMX-500 spectrometer. Spectrum **a** shows the normal spectrum in the region between δ_H = 6.0 and 1.0. In spectrum **b** proton 12 at δ_H = 4.2 was irradiated. Three NOE enhancements can be seen, namely for proton 13 at δ_H = 1.18, one of the protons 11 at δ_H = 3.05, and a slight effect at one of the protons 23 at δ_H = 3.98. In spectrum **c** one of the protons 23 at δ_H = 4.05 was irradiated. Note that with this technique the NOE effect of the nearby resonance of the other proton 23 at δ_H = 3.98 can be observed without distortion; the olefinic proton 22 also shows an effect.

7. Comments

See Experiment 4.8 for some short remarks on the mechanism of the NOE effect. Compared with Experiment 4.8 the experiment described here requires more preparations and instrument adjustments, however is less prone to artefacts. An even more advanced version with selective pulses and gradient selection is described in Experiment 11.10.

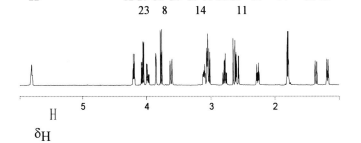

c

b

a

22 12 23 16 20 18 11 18 20 15 17 15 13

23 8 14 11

5 4 3 2

δ_H

8. Own Observations

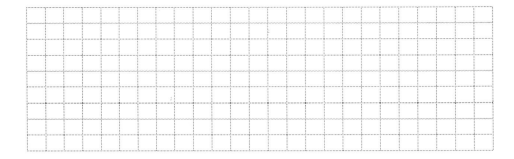

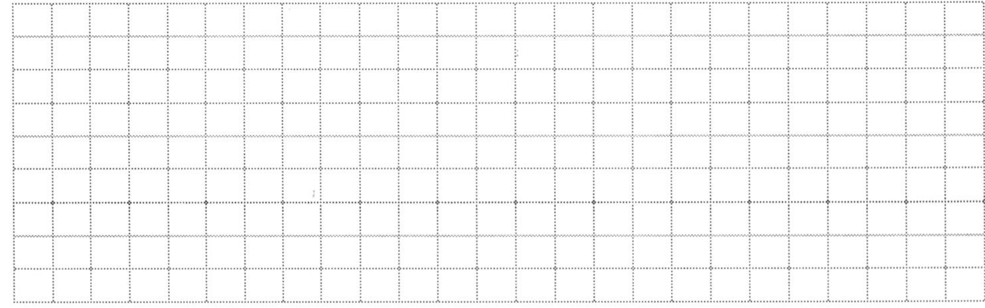

Experiment 4.10

^{1}H Off-Resonance Decoupled ^{13}C NMR Spectra

1. Purpose

This experiment is used to get information about the multiplicity of ^{13}C NMR signals resulting from scalar coupling to the directly bound protons. In certain cases the more modern methods such as APT (see Exp. 6.4) or DEPT (see Exp. 6.9) may fail due to large differences between the C,H spin coupling constants. In these cases the old off-resonance decoupling method yields unambigous results. An additional advantage is the recognition of X-CH_2-CH_2-Y [4] and X-CH=CH-Y [5] groups which can be achieved by this technique. Here we present an ^{1}H off resonance decoupled spectrum of ethyl crotonate.

2. Literature

[1] R. R. Ernst, *J. Chem. Phys.* **1966**, *45*, 3845–3861.
[2] M. L. Martin, J.-J. Delpuech, G. J. Martin, *Practical NMR Spectroscopy*, Heyden, London, **1980,** Ch. 6.2.
[3] H.-O. Kalinowski, S. Berger, S. Braun, *Carbon-13 NMR Spectroscopy*, Wiley, Chichester, **1988**, Ch. 3.3.
[4] R. A. Newmark, J. R. Hill, *J. Am. Chem. Soc.* **1973**, *95*, 4435–4437.
[5] R. Radeglia, H. Poleschner, G. Haufe, *Magn. Reson. Chem.* **1993**, *31*, 639–641.

3. Pulse Scheme and Phase Cycle

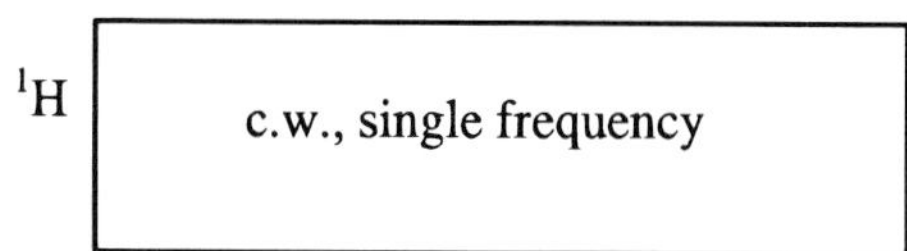

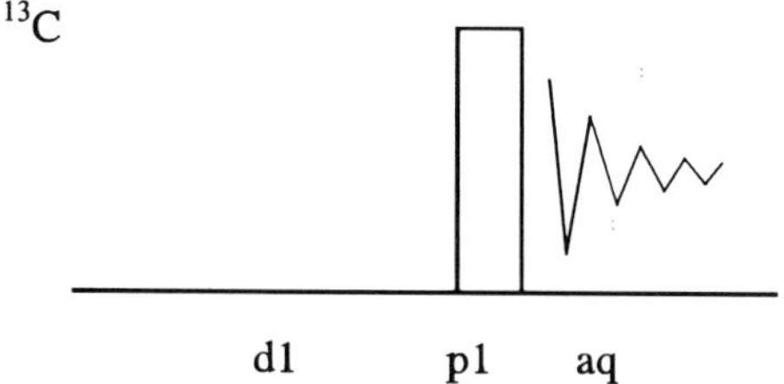

p1: x, x, -x, -x, y, y, -y, -y
aq: x, x, -x, -x, y, y, -y, -y

4. Acquisition

Time requirement: 30 min

Sample: 20% ethyl crotonate in $CDCl_3$

Set up the spectrometer for normal ^{13}C NMR spectroscopy with proton decoupling. Change the decoupler mode to continous wave without broad-band or CPD modulation. You have to set:

td: 64 k
sw: 200 ppm
o1: middle of ^{13}C NMR spectrum
o2: on resonance of ^{1}H TMS signal
p1: 45° ^{13}C transmitter pulse
d1: 0.5 s
decoupler power for c.w. off-resonance irradiation, $\gamma B_2 \cong 3500$ Hz (see Exp. 2.9 and Exp. 4.2)
ns: 512

5. Processing

Use standard 1D processing with lb = 1 Hz as described in Experiment 3.2.

6. Result

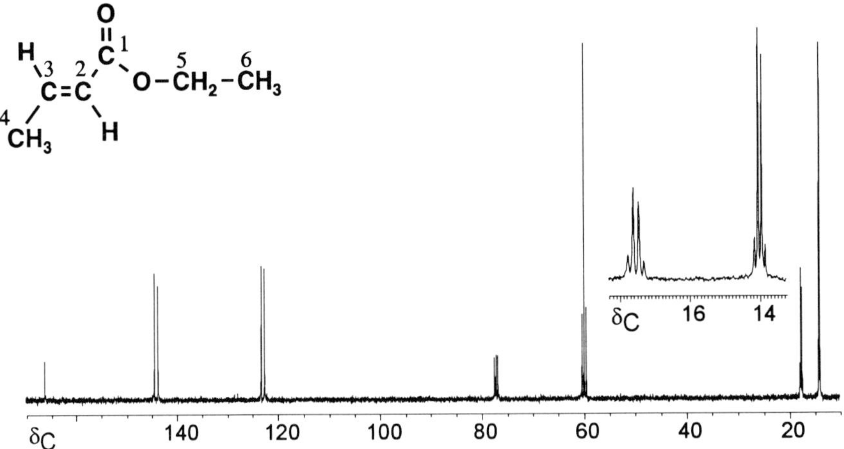

The figure shows the result obtained for ethyl crotonate with an AM-400 spectrometer. Every signal except those of the carboxyl group and the solvent is split into a multiplet according to the number of directly attached protons (CH yields a doublet, CH_2 a triplet and CH_3 a quartet). Note that the residual splitting increases towards higher frequencies, because the decoupler offset was set on the position of the ^{1}H signal of TMS.

7. Comments

Under off-resonance conditions the ^1J(C,H) coupling constants are reduced according to equation (1). J_R is the residual coupling, $\Delta\nu$ is the difference between the proton resonance frequency and the decoupler setting, and γB_2 is the strength of the decoupler

$$J_R = J \cdot \Delta\nu / \gamma B_2 \qquad (1)$$

field. The reduced decoupler bandwidth is usually sufficient to eliminate geminal and vicinal C,H couplings. Therefore only the multiplicity which originates from the directly bonded protons is observed. In some cases additional splittings are observed e.g. in $-CH_2-CH_2-$ or -CH=CH-groups, where the expected triplets and doublets have a fine structure due to higher-order effects.

8. Own Observations

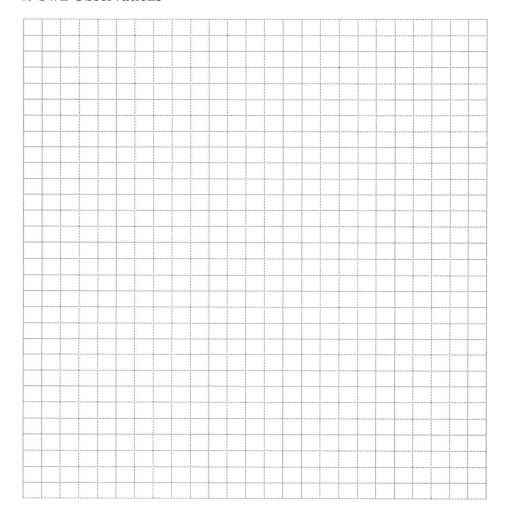

Experiment 4.11

The Gated ^{1}H-Decoupling Technique

1. Purpose

This experiment is used for determining C,H spin–spin coupling constants without losing nuclear Overhauser enhancements. It yields proton-coupled ^{13}C NMR spectra which usually have to be analyzed with the help of spin simulation, since the carbon atoms may often form the X part of relatively complicated A_mB_n...X spin systems. A knowledge of long-range C,H spin coupling constants is very helpful in structural elucidation of organic molecules.

2. Literature

[1] O. A. Gansow, W. Schittenhelm, *J. Am. Chem. Soc.* **1971**, *93*, 4294–4295.

[2] F. W. Wehrli,T. Wirthlin, *Interpretation of Carbon-13 NMR Spectra*, Heyden, London, **1978**, Ch. 3.

[3] M. L. Martin, J.-J. Delpuech, G. J. Martin, *Practical NMR Spectroscopy*, Heyden, London, **1980**, Ch. 6.2.

[4] H.-O. Kalinowski, S. Berger, S. Braun, *Carbon-13 NMR Spectroscopy*, Wiley, Chichester, **1988**, Ch. 2.3.

3. Pulse Scheme and Phase Cycle

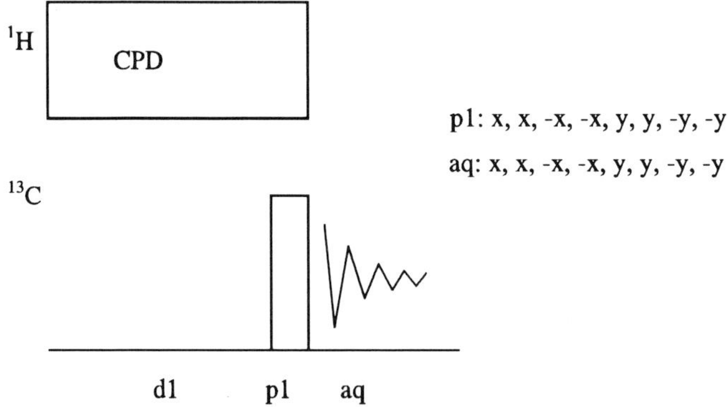

p1: x, x, -x, -x, y, y, -y, -y

aq: x, x, -x, -x, y, y, -y, -y

4. Acquisition

Time requirement: 45 min

Sample: 20% ethyl crotonate in $CDCl_3$

Set up the spectrometer for normal ^{13}C NMR spectroscopy. Load the pulse program for gated decoupling. Use a delay time d1 which is in the same order of magnitude as the acquisition time. You have to set:

td: 64 k
sw: 200 ppm
o1: middle of ^{13}C NMR spectrum
o2: middle of ^{1}H NMR spectrum
p1: 45° ^{13}C transmitter pulse
d1: 2 s
ns: 512
decoupler attenuation and 90° pulse for CPD

5. Processing

Use standard 1D processing with exponential multiplication (lb = 0.3 Hz); you can also use Gaussian multiplication to enhance the resolution.

6. Result

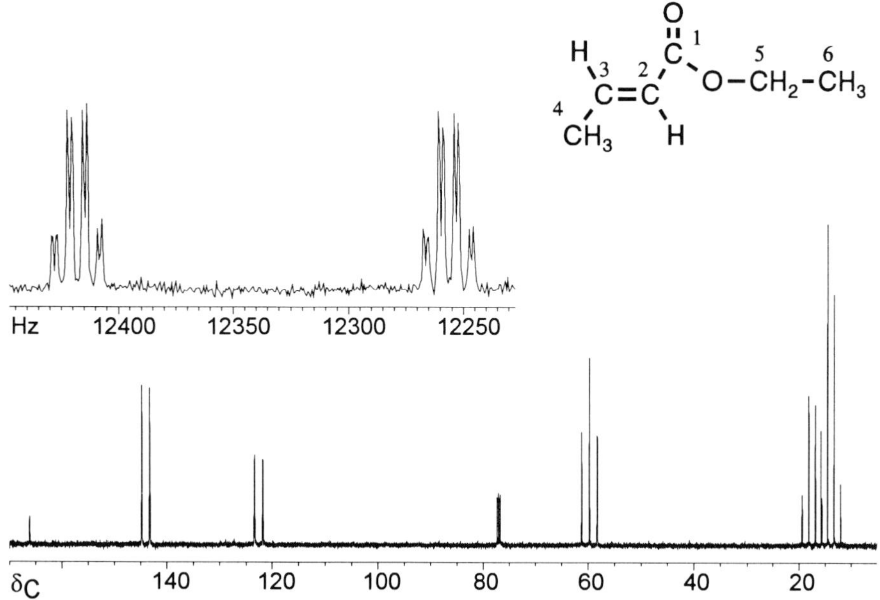

The figure shows the ^{1}H coupled ^{13}C NMR spectrum of ethyl crotonate obtained with an AM-400 spectrometer. Every signal is split into multiplets according to the underlying spin system. The inset is the expanded part of the olefinic region between δ_C = 124 and 121. This multiplet belongs to C-2 of ethyl crotonate showing a doublet of quartets of doublets due to the coupling to H-2, H-4 (CH_3) and H-3, with coupling constants of 161.7, 6.7 and 1.8 Hz. Note that the spectrum is field dependent owing to the fact that both the C-2 and C-3 resonances are the X parts of ABM_3X spin systems.

7. Comments

In this experiment composite pulse decoupling is applied during the delay d1 but not during the acquisition time. Coupling information is present immediately after switching off the decoupling field, whereas the populations of the energy levels decay with the spin–lattice relaxation times. During d1 (same order as the acquisition time) favorable ^{13}C energy level populations become established and coupled spectra with NOE can be obtained. One has to be careful about assuming a first-order interpretation of such spectra since higher order effects can occur. Make sure that the observed splittings are in fact first-order, and use spin simulation programs to analyze the spin systems.

8. Own Observations

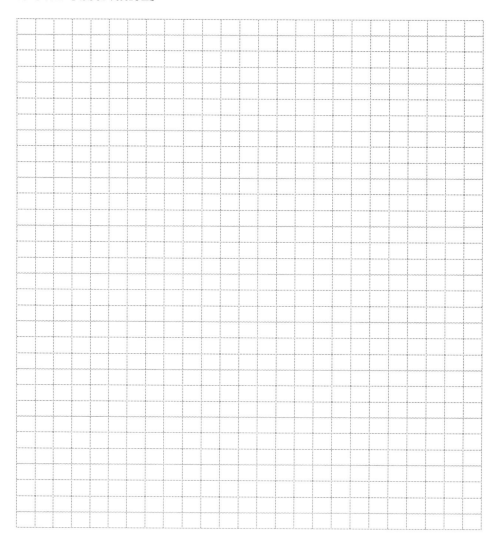

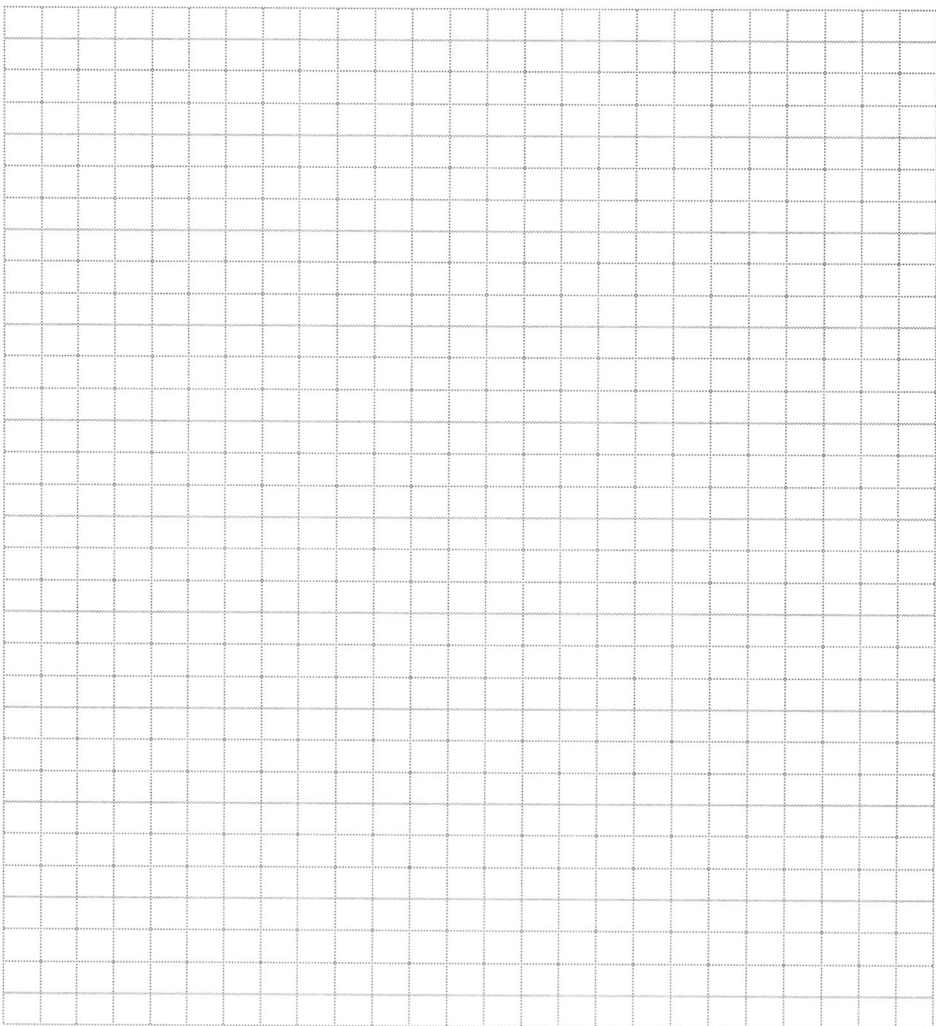

Experiment 4.12

The Inverse Gated ^{1}H-Decoupling Technique

1. Purpose

This experiment yields ^{1}H-decoupled NMR spectra of X nuclei without signal enhancement by the nuclear Overhauser effect. This is important for nuclei with a negative gyromagnetic ratio, where the Overhauser effect can completely suppress some or all signals under certain circumstances. The pulse sequence is also used for quantitative measurements of the Overhauser effect (see Exp. 4.15) and for quantitative ^{13}C NMR spectroscopy (Exp. 8.16) where the Overhauser effect has to be suppressed. Here we show the basic experiment for ethyl crotonate.

2. Literature

[1] R. Freeman, H. D. W. Hill, R. Kaptein, *J. Magn. Reson.* **1972**, *7*, 327–329.

[2] M. L. Martin, J.-J. Delpuech, G. J. Martin, *Practical NMR Spectroscopy*, Heyden, London **1980**, 231–235.

[3] H.-O. Kalinowski, S. Berger, S. Braun, *Carbon-13 NMR Spectroscopy*, Wiley, Chichester, **1988**, Ch. 2.3.

3. Pulse Scheme and Phase Cycle

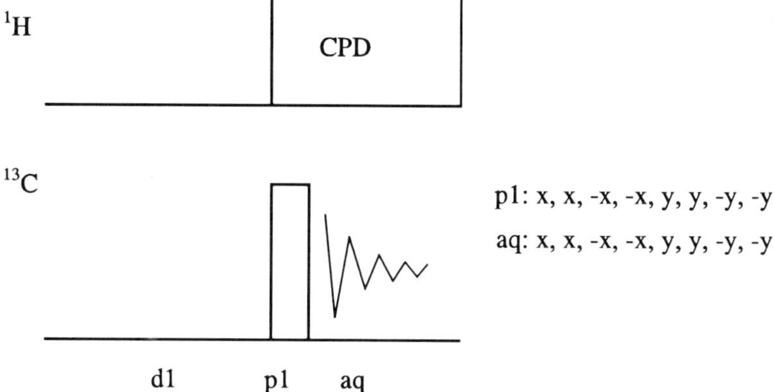

4. Acquisition

Time requirement: 1.5 h

Sample: 20% ethyl crotonate in $CDCl_3$

Set up the spectrometer for normal ^{13}C NMR spectroscopy. Load the pulse program for inverse gated decoupling. Use a delay time d1 which is about ten times the acquisition time. The latter should be rather short to avoid build up of the Overhauser effect during recording of the data. You have to set:

td: 32 k
sw: 200 ppm
o1: middle of ^{13}C NMR spectrum
o2: middle of ^{1}H NMR spectrum
p1: 45° ^{13}C transmitter pulse
d1: 10 s
decoupler attenuation and 90° pulse for CPD
ns: 512

5. Processing

Use standard 1D processing as described in Exp. 3.2. Apply zero filling to 64 k with exponential multiplication (lb = 1 Hz).

6. Result

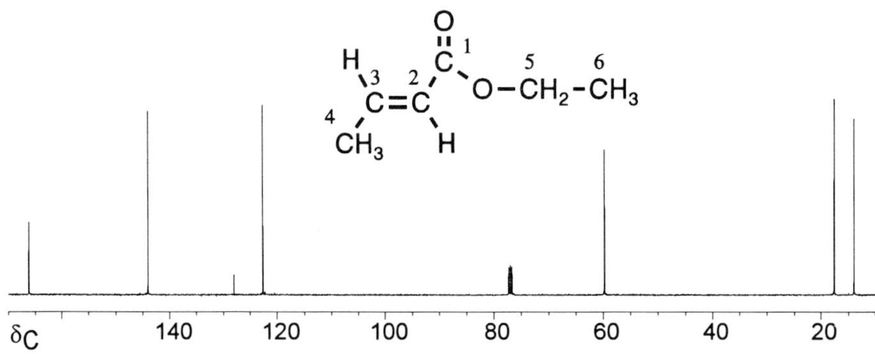

Shown is the ^{1}H decoupled ^{13}C NMR spectrum of ethylcrotonate obtained with an AM-400 spectrometer. The signals of the protonated carbon atoms all have nearly the same height, remaining intensity differences are probably due to different spin–lattice relaxation times which affect most the signal intensity of the carboxyl atom C-1 (see exp. 6.1).

7. Comments

In this experiment composite pulse decoupling is applied only during the short acquisition time and not during the delay d1. Coupling information which is present after the delay is immediatedly eliminated by the decoupling field, whereas the populations of the energy levels and hence NOE enhancements require a build up time in the order of the spin–lattice relaxation times. If the delay d1 is at least 10 times the acquisition time, decoupled spectra without NOE effect can be recorded.

8. Own Observations

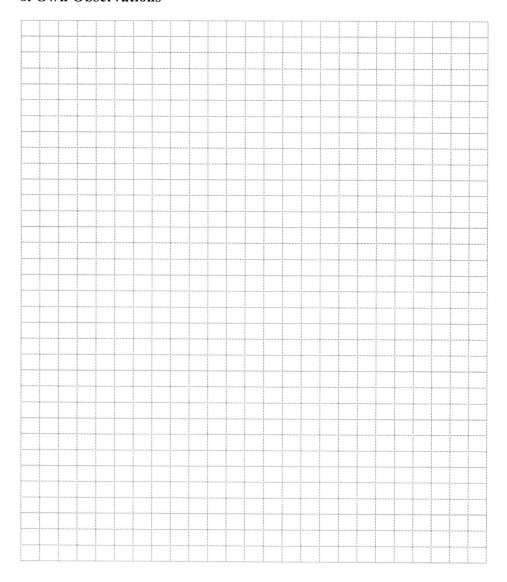

Experiment 4.13

^1H Single Frequency Decoupling of ^{13}C NMR Spectra

1. Purpose

This experiment correlates a chosen ^1H signal with the corresponding carbon signal via $^1J(C,H)$ and is the 1D equivalent of the 2D C,H correlation (Exp 10.10). The experiment runs with ^{13}C as the observed nuclide and can therefore be performed on older instruments without difficulty. The inverse 2D H,C correlation (Exp 10.13) also has a 1D equivalent, called SELINCOR (Exp 7.6). The choice between these four techniques is dictated by the available hardware and the question whether only one specific item of information or the complete C,H correlation is needed. The experiment described here gives in a most straightforward manner the desired connectivity information, provided that the proton signals are sufficiently separated.

2. Literature

[1] M. L. Martin, J.-J. Delpuech, G. J. Martin, *Practical NMR Spectroscopy*, Heyden, London, **1980**, Ch. 6.

3. Pulse Scheme and Phase Cycle

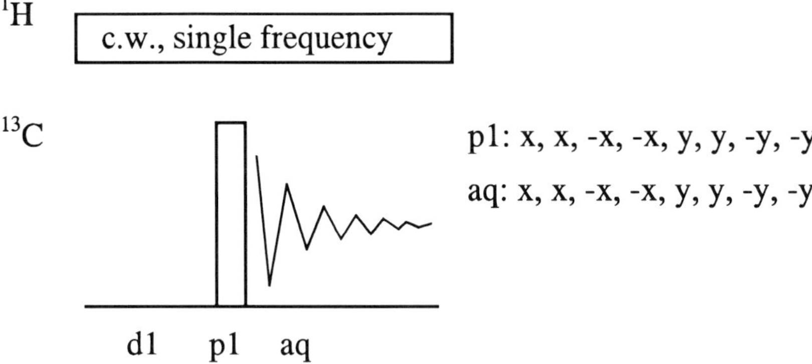

4. Acquisition

Time requirement: 15 min

Sample: 20% ethyl crotonate in $CDCl_3$

Record a normal ^1H and ^{13}C NMR spectrum of ethyl crotonate and note the ^1H frequency of the methyl group attached to the double bond. Load a pulse program for ^{13}C

detection under *continous wave* decoupling. The decoupler power γB_2 should be set to a level appropriate for the ^1J(C,H) coupling constant (Exp.2.9). You have to set:

td: 64 k
sw: 200 ppm
o1: middle of ^{13}C NMR spectrum
o2: center of methyl group ^{1}H resonance at δ_H = 1.6
p1: 45° ^{13}C transmitter pulse
d1: 1 s
decoupler power γB_2 = 150 Hz [45dB was used here]
ns: 8

5. Processing

Use standard ^{13}C NMR processing as described in Exp 3.2.

6. Result

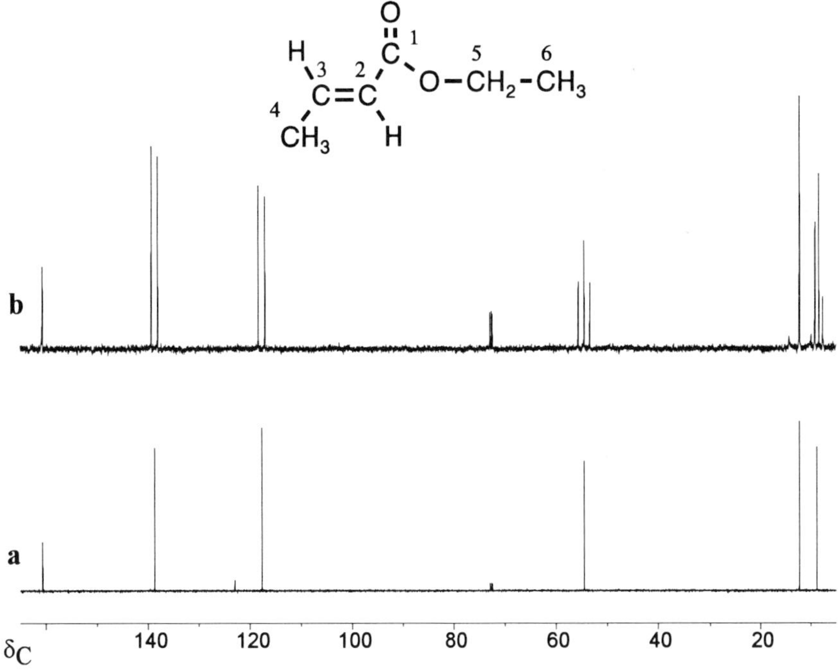

In the figure **a** is the normal ^{1}H-decoupled ^{13}C NMR spectrum of ethyl crotonate and **b** is the result of the single frequency decoupling experiment obtained on an AMX-500 spectrometer. The experiment gives a singlet for C-4, whereas the other signals are multiplets according to the number of attached protons. These multiplets can display an off-resonance pattern (see Exp 4.10).

7. Comments

Single frequency decoupling only works perfectly if all transitions of the spin system are irradiated. A proton-coupled methyl carbon gives a quartet in the ^{13}C NMR spectrum, but the protons show a doublet in the ^{1}H NMR spectrum. The decoupler bandwidth has therefore to match the line separation of the ^{13}C satellite splitting in the ^{1}H NMR spectrum.

8. Own Observations

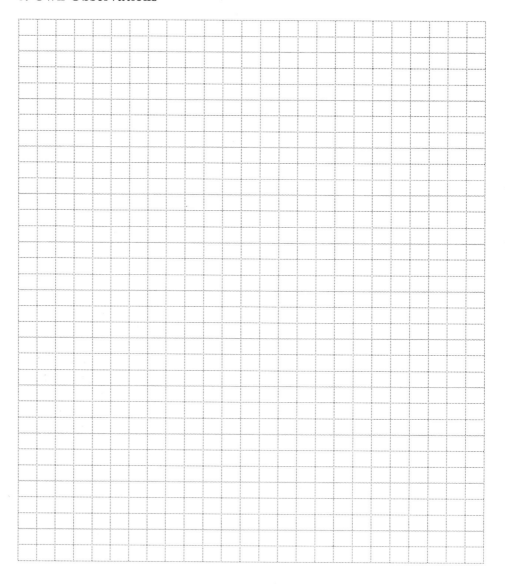

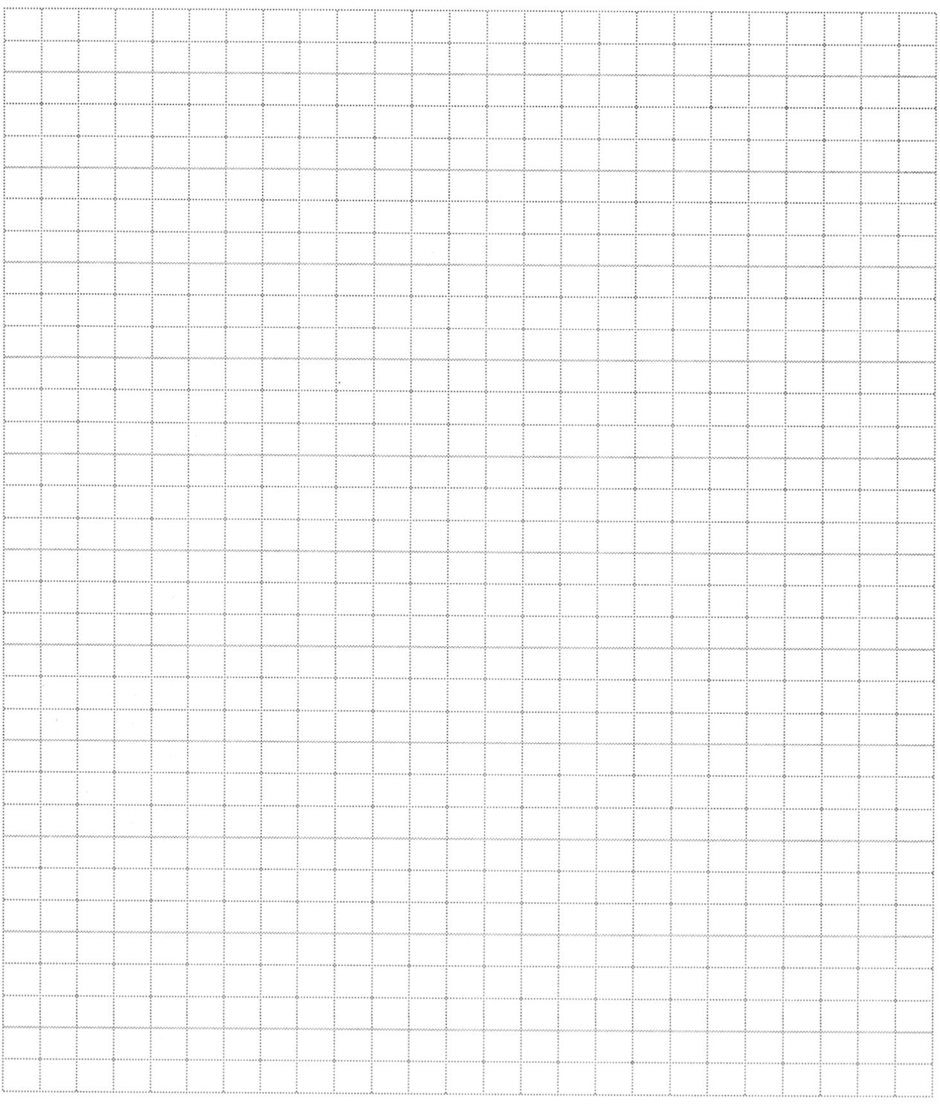

Experiment 4.14

^{1}H Low-Power Decoupling of ^{13}C NMR Spectra

1. Purpose

As in Experiment 4.13, this technique correlates an ^{1}H signal with ^{13}C signals, but in this case the nuclei concerned are separated by two, three or more bonds. As such it is a 1D equivalent of the 2D COLOC technique (Exp. 10.12). Again, this experiment can be performed in the inverse mode either as a 1D or a 2D method (gs-SELINCOR, Exp. 11.11 and HMBC, Exp. 10.15). The main purpose of the experiment described here is to simplify proton-coupled carbon spectra which can not only give assignment information but can make the analysis of the spin system much easier.

2. Literature

[1] M. L. Martin, J.-J. Delpuech, G. J. Martin, *Practical NMR Spectroscopy*, Heyden, London, **1980**, Ch. 6.
[2] K. Bock, C. Pedersen, *J. Magn. Reson.* **1977**, *25*, 227-230.

3. Pulse Scheme and Phase Cycle

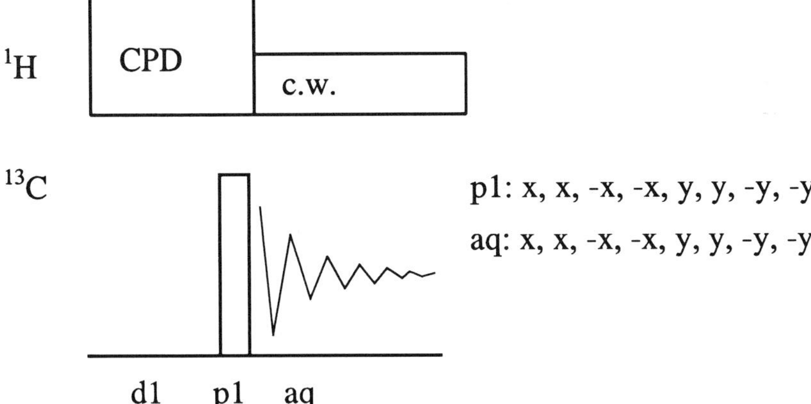

p1: x, x, -x, -x, y, y, -y, -y
aq: x, x, -x, -x, y, y, -y, -y

4. Acquisition

Time requirement: 15 min

Sample: 20% ethyl crotonate in $CDCl_3$

Record a normal ^{1}H NMR spectrum and an ^{1}H-coupled ^{13}C NMR spectrum (gated decoupling, see Exp. 4.11) of ethyl crotonate and note the ^{1}H frequency of the upfield olefinic proton. In order to avoid distortion of intensities due to SPT effects, load a pulse program for ^{13}C detection under *continous wave* decoupling during acquisition and broad-band decoupling during the delay d1. The decoupler power γB_2 during acquisition should be set to a level appropiate for the width of the ^{1}H signal, taking into account the additional splitting by the long-range C,H spin coupling constant of 5–10 Hz. The decoupler power γB_2 during the pulse delay should be set to an appropriate value to maintain the NOE effect. You have to set:

td: 64 k
sw: 200 ppm
o1: middle of ^{13}C NMR spectrum
o2: center of ^{1}H signal of the low frequency olefinic proton
p1: 45° ^{13}C transmitter pulse
d1: 1 s
decoupler attenuation during acquisition γB_2 = 15 Hz [70 dB were used here]
decoupler attenuation and 90° pulse for CPD
ns: 8

5. Processing

Use standard ^{13}C NMR processing as described in Experiment 3.2.

6. Result

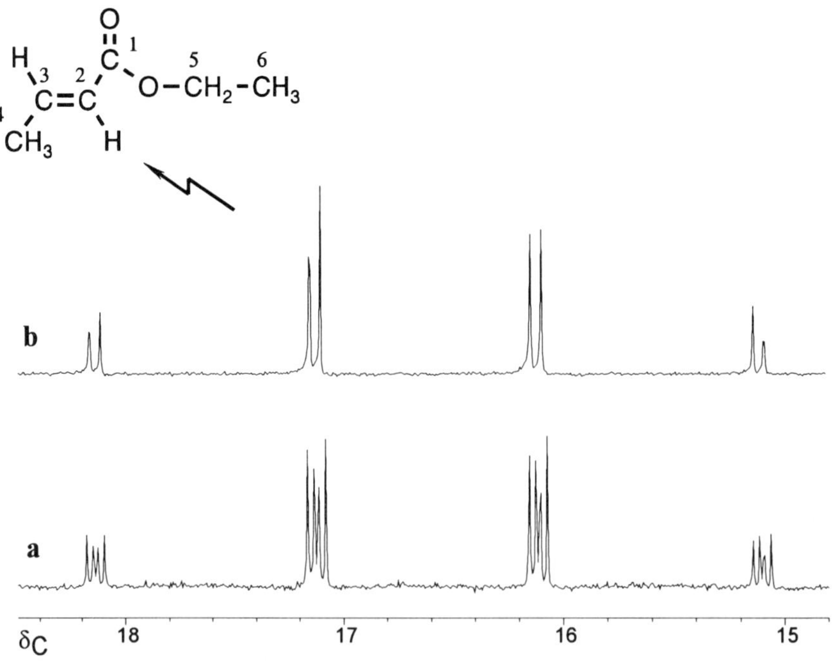

The figure on page 130 shows spectra obtained on an AMX-500 spectrometer. **a** is the expanded methyl carbon region of the ^1H-coupled ^{13}C NMR spectrum; the methyl group C-4 displays two long-range spin coupling constants of 6.5 and 3.6 Hz. In **b** the olefinic proton at C-2 was irradiated, leaving only the coupling to H-3 with $^2J(C,H)$ = 6.5 Hz.

7. Comments

Single frequency decoupling only works perfectly if all transitions of the spin system are irradiated. A proton coupled to ^{13}C forms a doublet in the ^1H NMR spectrum (^{13}C satellites). These satellite lines are separated by the coupling constant $^nJ(C,H)$. The $^1J(C,H)$ couplings and remaining long-range couplings of other protons can be reduced due to off-resonance effects.

8. Own Observations

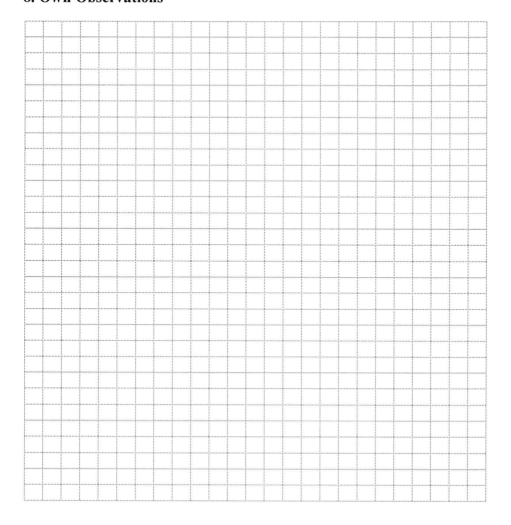

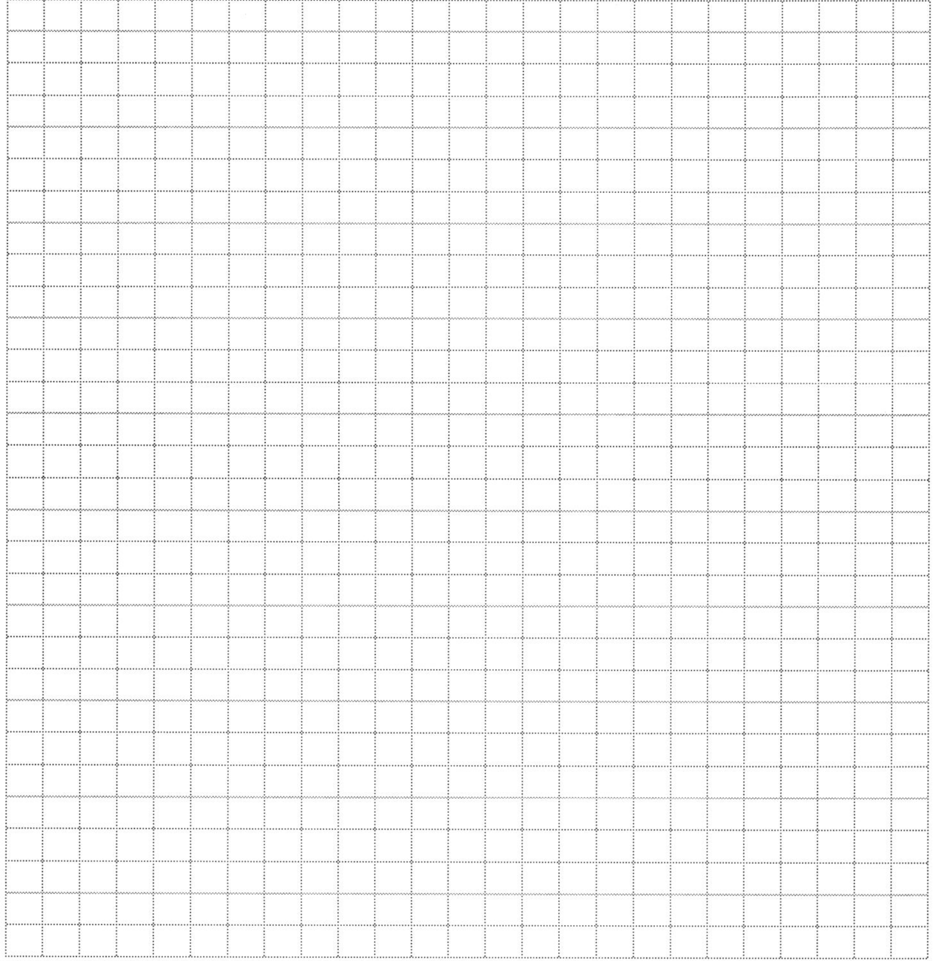

Experiment 4.15

Measurement of the Heteronuclear Overhauser Effect

1. Purpose

To evaluate spin–lattice relaxation data one often needs to know the dipolar contribution $T_{1(DD)}$. This can be obtained by measuring the overall spin–lattice relaxation time $T_{1(exp)}$ (see Exp. 6.1) and the heteronuclear Overhauser effect η as described in this experiment. Selective versions of this experiment can be used for assignment purposes in certain cases [1]. The 2D variant, called HOESY, is described in Experiment 10.21.

2. Literature

[1] K. E. KövER, G. Batta, *Prog. NMR Spectrosc.* **1987**, *19*, 223–266.

[2] D. Neuhaus, M. Williamson, *The Nuclear Overhauser Effect*, VCH, Weinheim, **1989**.

[3] S. Berger, F. R. Kreissl , J. D. Roberts, *J. Am. Chem. Soc.* **1974**, *96,* 4348.

3. Pulse Scheme and Phase Cycle

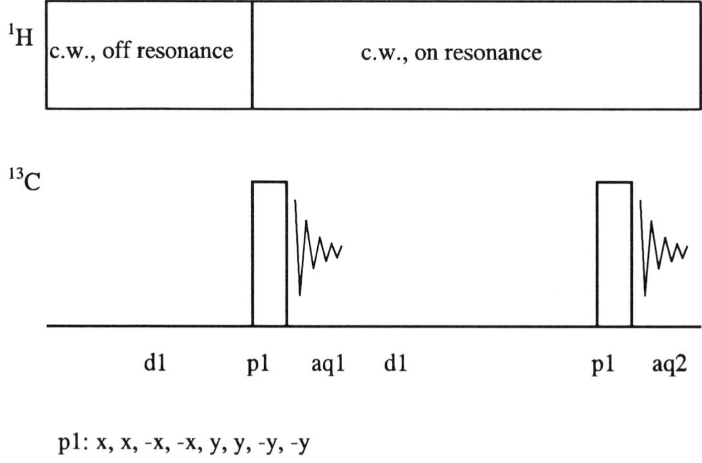

p1: x, x, -x, -x, y, y, -y, -y

aq: x, x, -x, -x, y, y, -y, -y

4. Acquisition

Time requirement: 1 h

Sample: 50% Cyclohexane in $CDCl_3$, *degassed and sealed*

Measure normal ^{1}H and ^{13}C NMR spectra and obtain the offsets. Load a pulse program as shown, which combines measurement with proton irradiation at two different proton offsets and which stores the two FIDs separately. This procedure ensures that the instrument performance is exactly the same in the two experiments. You have to set:

td: 1 k (short aq, to avoid NOE buildup during acquisition)
sw: 500 Hz
o1: on resonance of ^{13}C signal of cyclohexane
o2: provide an ^{1}H frequency list for the pulse program, first value 200 kHz off resonance, second value on resonance of ^{1}H signal of cyclohexane
p1: 90° ^{13}C transmitter pulse
d1: 200 s (10 times the relaxation time of cyclohexane)
decoupler power for c.w. decoupling
ns: 4

5. Processing

The experiment yields two FIDs which must be processed absolute identically. Use a large line-broadening value of lb = 3 Hz to obtain spectra with good signal-to-noise. Measure the two integrals and divide one by the other to obtain $\eta + 1$.

6. Result

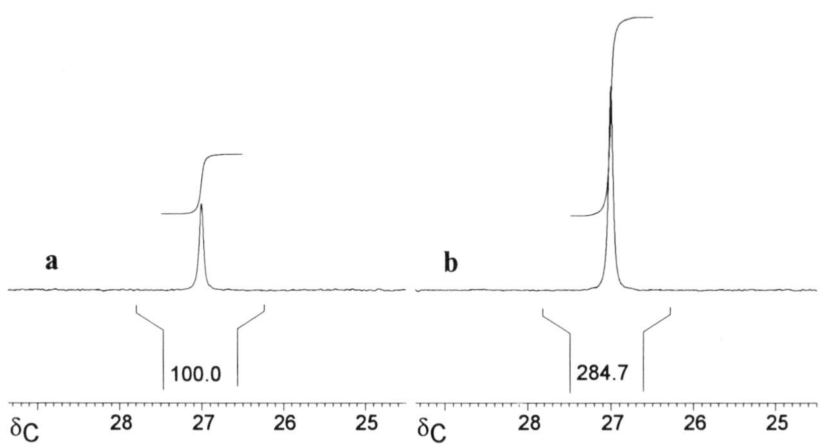

The figure shows the spectra without (**a**) and with NOE (**b**) obtained on an AM-400 spectrometer. From integrals $\eta + 1$ is determined to be 2.84.

7. Comments

Spin–lattice relaxation involves a number of different mechanisms as indicated in Equation (1). The relative importance of these is field dependent.

$$1/T_{1(\mathrm{exp})} = 1/T_{1(\mathrm{DD})} + 1/T_{1(\mathrm{other})} \qquad (1)$$

From the dipolar contribution to the spin–lattice relaxation one can estimate carbon–proton distances. The ratio of the integrals as measured in this experiment yields the NOE effect as defined in Equation (2), where $M_C\{H\}$ is the carbon magnetization with proton decoupling, M_C the carbon magnetization without proton decoupling, σ the cross relaxation rate, and $\rho_C = 1/T_1$.

$$M_C\{H\} / M_C = \sigma\gamma_H/\rho_C\gamma_C + 1 = \eta + 1 \qquad (2)$$

In the extreme narrowing limit it can be shown that with a γ_H/γ_C ratio of approximately 4, $T_{1(\mathrm{DD})}$ can be calculated from Equation (3)

$$T_{1(\mathrm{DD})} = 2\,T_1 / \eta \qquad (3)$$

8. Own Observations

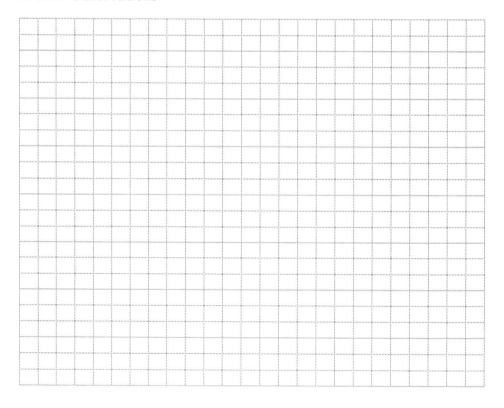

Chapter 5

Dynamic NMR Spectroscopy

In addition to its applications in the determination of static molecular structures, NMR spectroscopy can be used to detect intra- and intermolecular dynamic processes such as hindered rotations about partial double bonds, ring inversions and valence isomerizations. By measuring the temperature dependence of these processes the thermodynamic parameters $\Delta G^{\#}$, $\Delta H^{\#}$ and $\Delta S^{\#}$ can be obtained. The ability of NMR spectroscopy to determine energy barriers in the range from about 20 to 100 kJ mol^{-1} is based on the so-called NMR time-scale. Separate NMR signals are observed for nuclei at two sites A and B only when the site exchange rate constant k is much less than the difference $\Delta\omega$ between the corresponding angular resonance frequencies.

In this short chapter we first provide two basic calibration routines which enable the user to check whether the actual temperature of a sample corresponds to the setting of the temperature unit on the spectrometer. These are very important experiments which have to be performed prior to any dynamic NMR investigation. It cannot be stressed enough that depending on many instrumental factors, such as the position of the thermocouple in the probe-head, there might be quite a difference between actual and indicated temperature.

The chapter also includes a description of a basic dynamic NMR experiment using dimethylformamide as an example, and demonstrates the saturation transfer experiment which can be viewed as the 1D analog of the two-dimensional EXSY technique, given in Experiment 10.23. The chapter concludes with a description of the $T_{1\rho}$ Experiment which extends the range of dynamic NMR measurements into the region of fast exchange.

Literature

[1] G. Binsch, *Top. Stereochem.* **1968**, *3*, 97–192.

[2] G. Binsch, H. Kessler, *Angew. Chem. Int. Ed. Engl.* **1980**, *19*, 411–429.

[3] L. M. Jackman, F. A. Cotton (eds.), *Dynamic NMR Spectroscopy*, Academic Press, New York, **1975**.

[4] J. I. Kaplan, G. Fraenkel, *NMR of Chemically Exchanging Systems*, Academic Press, New York, **1980**.

[5] M. Oki, *Applications of Dynamic NMR Spectroscopy to Organic Chemistry*, VCH, Weinheim, **1985**.

Experiment 5.1

Low Temperature Calibration with Methanol

1. Purpose

There are many NMR experiments which are performed at different temperatures or where the emphasis lies on the measurement of a temperature-dependent effect. Hence, it is most important to know whether the temperature controller of the instrument used gives a correct reading of the actual temperature within the sample. Many different calibration samples have been proposed, working at different temperature ranges and for different nuclides. Here we restrict the description to the most common low temperature standard, methanol, where the chemical shift difference between the OH proton and those of the methyl group is used for the calibration.

2. Literature

[1] A. L. van Geet, *Anal. Chem.* **1970**, *42*, 679–680; ibid. **1968**, *40*, 2227–2229.
[2] H. Friebolin, G. Schilling, L. Pohl, *Org. Magn. Reson.* **1979**, *12*, 569–573.
[3] C. Piccinni-Leopardii, O. Fabre, J. Reisse, *Org. Magn. Reson.* **1976**, *8*, 233–236.
[4] J. Bornais, S. Brownstein, *J. Magn. Reson.* **1978**, *29*, 207–211.
[5] M. L. Martin, J.-J. Delpuech, G. J. Martin, *Practical NMR Spectroscopy*, Heyden, London, **1980**, 330–341.
[6] F. H. Köhler, X. Xie, *Magn. Reson. Chem.* **1997**, *35*, 487–492.

3. Pulse Scheme and Phase Cycle

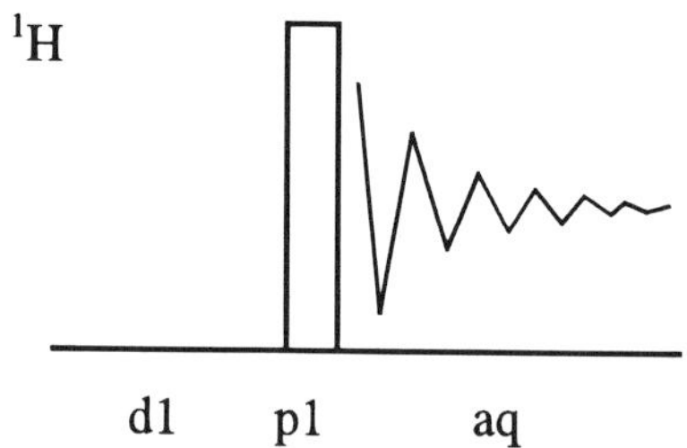

p1: x, x, -x, -x, y, y, -y, -y

aq: x, x, -x, -x, y, y, -y, -y

4. Acquisition

Time requirement: 1 h

Sample: 4% MeOH in [D_4]Methanol containing a trace of HCl

Load standard proton parameters and connect the low temperature equipment for your instrument. Adjust for stable nitrogen flow and set the temperature controller in turn to 193, 223, and 273 K. Let each temperature equilibrate for at least 5 minutes. Measure at each temperature the chemical shift difference $\Delta\delta$ between the two methanol signals. You have to set:

td: 32 k
sw: 8 ppm
o1: middle of ^{1}H NMR spectrum
p1: 45° ^{1}H transmitter pulse
d1: 300 s
ns: 1

5. Processing

Use standard 1D processing as described in Experiment 3.1

6. Result

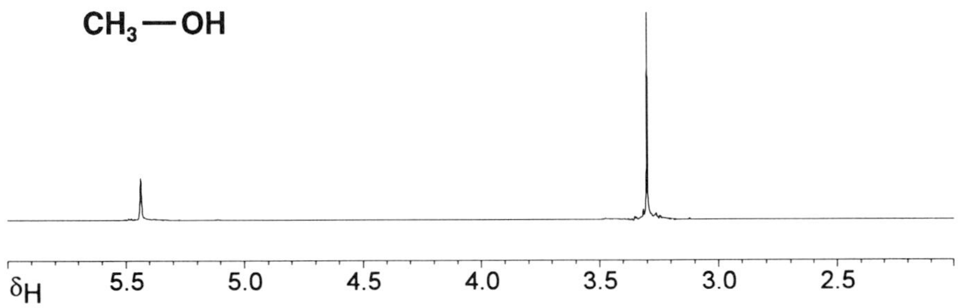

The figure shows the result obtained on an AM-400 spectrometer with the temperature unit set to 223 K. A difference of 2.136 ppm between the two signals was measured. Compare the values with those given in the instrument manufacturer's calibration curve or compute the result with the equations given below. Recent instruments provide temperature calculation programs which detect automatically the chemical shift difference $\Delta\delta$ and compute from this difference the sample temperature. The calibration curve shown on page 138 was drawn with the following equations:

For $\Delta\delta$ 1.4965 to 1.76: $T\,[\text{K}] = -\,114.83\,\Delta\delta + 471.85$

For $\Delta\delta$ 1.76 to 2.08: $T\,[\text{K}] = -\,125\,\Delta\delta + 490$

For $\Delta\delta$ 2.08 to 2.43: $T\,[\text{K}] = -\,140\,\Delta\delta + 521.33$

Outside the temperature range covered by the sample one may use a calibrated thermocouple fixed in an NMR sample.

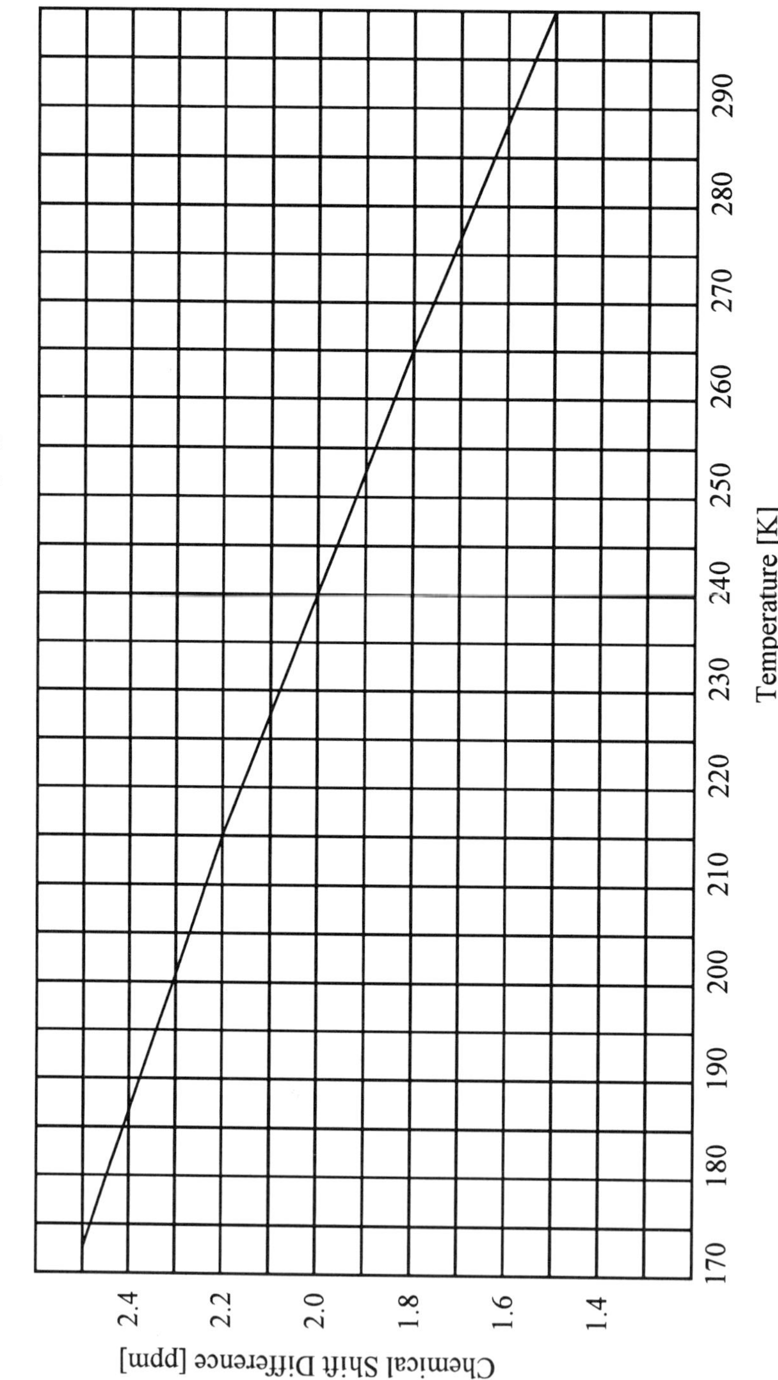
Calibration Curve for 4% Methanol in [D$_4$]Methanol
Temperature [K]
170
180
190
200
210
220
230
240
250
260
270
280
290
Chemical Shift Difference [ppm]
2.4
2.2
2.0
1.8
1.6
1.4

7. Comments

In principle, a long narrow cylinder like an NMR sample surrounded by a gas flow cannot be held at a very accurate and stable temperature compared to the performance of a large temperature bath as used in chemical kinetics. Temperature gradients in the sample are expected. Nevertheless, modern NMR instrumentation gives reasonably reproducible results if enough time is allowed for temperature equilibration. Your temperature readings should not deviate by more than 1–2 K from the calibration curve (page 138) and should be reproducible in repeated measurements.

8. Own Observations

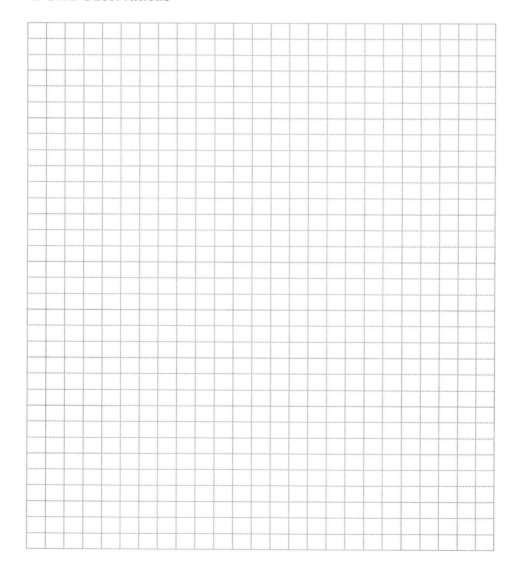

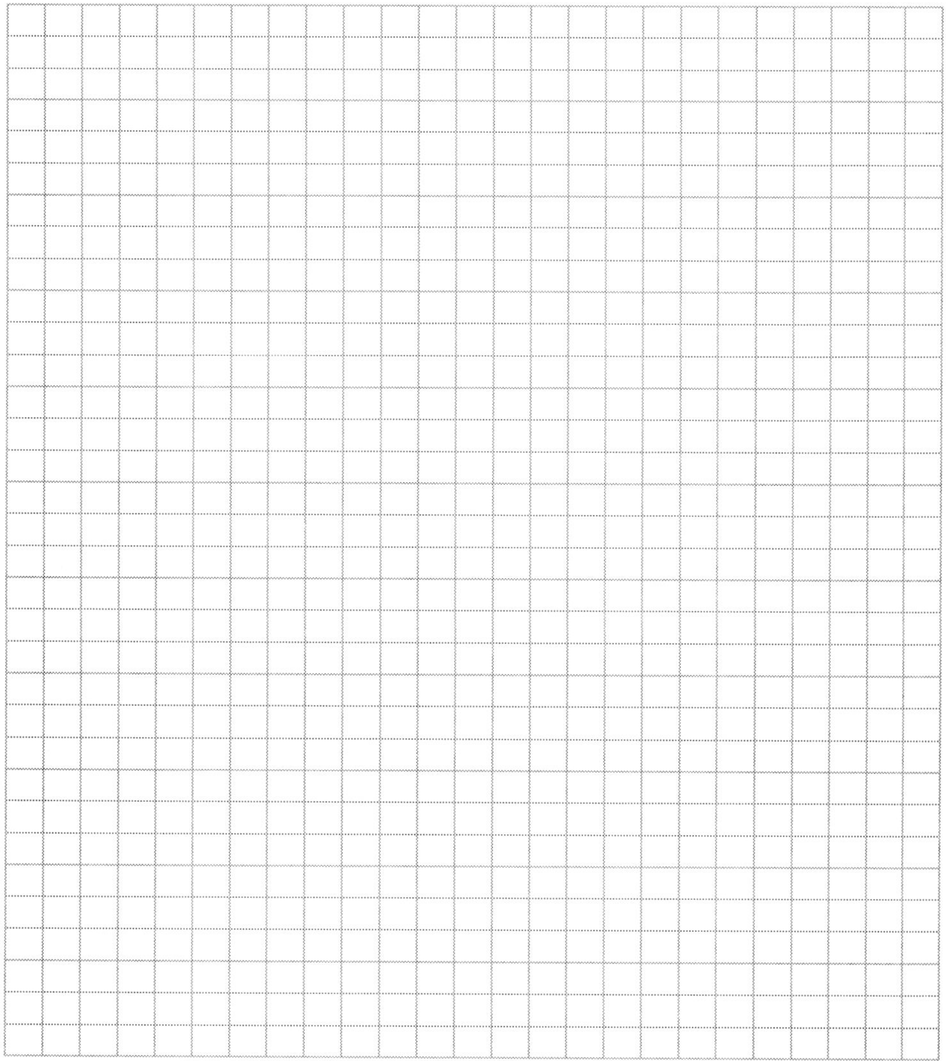

Experiment 5.2

High Temperature Calibration with 1,2-Ethanediol

1. Purpose

There are many NMR experiments which are performed at different temperatures or where the emphasis lies on the measurement of a temperature dependent effect. Hence, it is most important to know whether the temperature controller of the instrument used gives a correct reading of the actual temperature within the sample. Many different calibration samples have been proposed, working at different temperature ranges and for different nuclides. Here we restrict the description to the most common high temperature standard, 1,2-ethanediol, where the temperature-dependent chemical shift difference between the OH protons and those of the methylene groups is used for the calibration.

2. Literature

[1] A. L. van Geet, *Anal. Chem.* **1970**, *42*, 679–680; ibid. **1968**, *40*, 2227–2229.
[2] H. Friebolin, G. Schilling, L. Pohl, *Org. Magn. Reson.* **1979**, *12*, 569–573.
[3] C. Piccinni-Leopardii, O. Fabre, J. Reisse, *Org. Magn. Reson.* **1976**, *8*, 233–236.
[4] M. L. Martin, J.-J. Delpuech, G. J. Martin, *Practical NMR Spectroscopy*, Heyden, London, **1980**, 330–341.
[5] F. H. Köhler, X. Xie, *Magn. Reson. Chem.* **1997**, *35*, 487–492.

3. Pulse Scheme and Phase Cycle

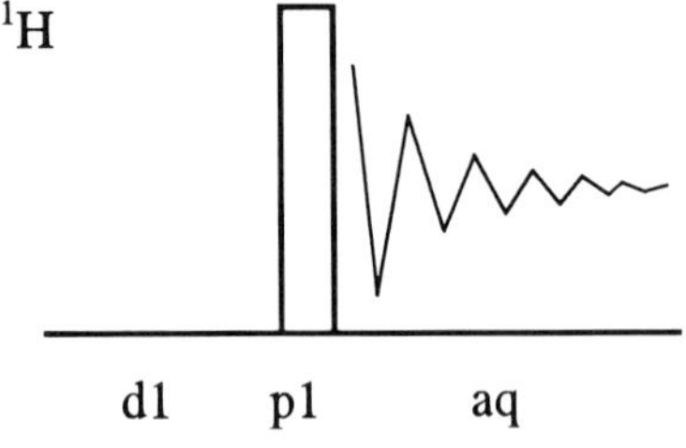

p1: x, x, -x, -x, y, y, -y, -y
aq: x, x, -x, -x, y, y, -y, -y

4. Acquisition

Time requirement: 1 h

Sample: 80% 1,2-ethanediol in [D_6]DMSO

Load standard proton parameters and connect the high temperature equipment for your instrument. Adjust for stable nitrogen flow and set the temperature controller in turn to 300, 330, and 400 K. Let each temperature equilibrate for at least 5 minutes. Measure at each temperature the chemical shift difference $\Delta\delta$ between the two 1,2-ethanediol signals. You have to set:

td: 32 k
sw: 8 ppm
o1: middle of ^{1}H NMR spectrum
p1: 45° ^{1}H transmitter pulse
d1: 300 s
ns: 1

5. Processing

Use standard 1D processing as described in experiment 3.1

6. Result

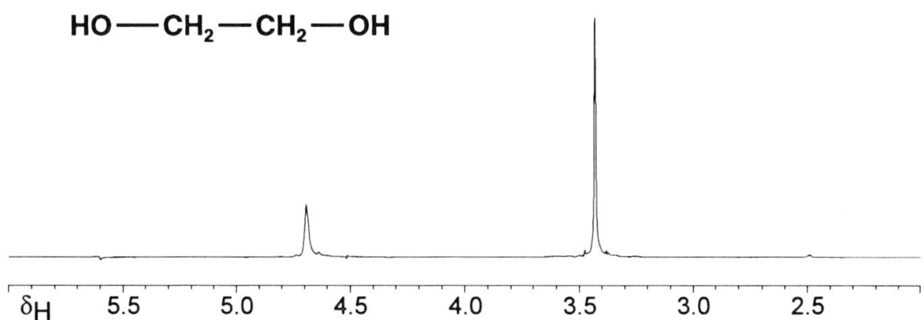

The figure shows the result obtained on an AM-400 spectrometer with the temperature unit set to 330 K. A difference of 1.262 ppm between the two signals was measured.

Compare the values with those given in the instrument manufacturer's calibration curve or compute the result with the equation given below. Recent instruments provide temperature calculation programs which detect automatically the chemical shift difference $\Delta\delta$ and compute from this difference the sample temperature. The calibration curve shown on page 142 was drawn with the following equation:

$$T\,[K] = -\,108.33\,\Delta\delta + 460.41$$

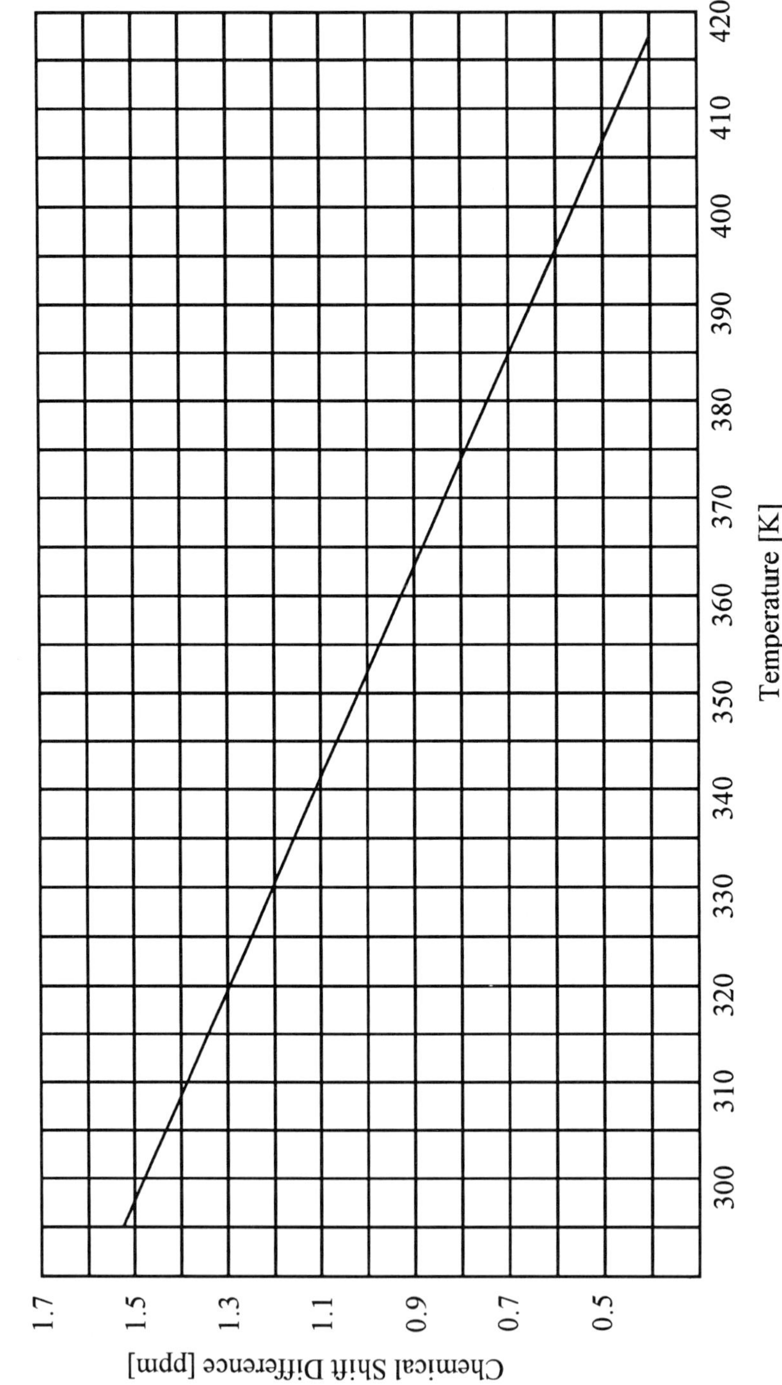
Calibration Curve for 1,2-Ethanediol in [D$_6$]DMSO
Chemical Shift Difference [ppm]
1.7
1.5
1.3
1.1
0.9
0.7
0.5
300
310
320
330
340
350
360
370
380
390
400
410
420
Temperature [K]

7. Comments

In principle, a long narrow cylinder like an NMR sample surrounded by a gas flow cannot be held at a very accurate and stable temperature compared to the performance of a large temperature bath as used in chemical kinetics. Temperature gradients in the sample are expected. Nevertheless, modern NMR instrumentation gives reasonably reproducible results if enough time is allowed for temperature equilibration. Your temperature readings should not deviate by more than 1–2 K from the calibration curve (page 142) and should be reproducible in repeated measurements.

8. Own Observations

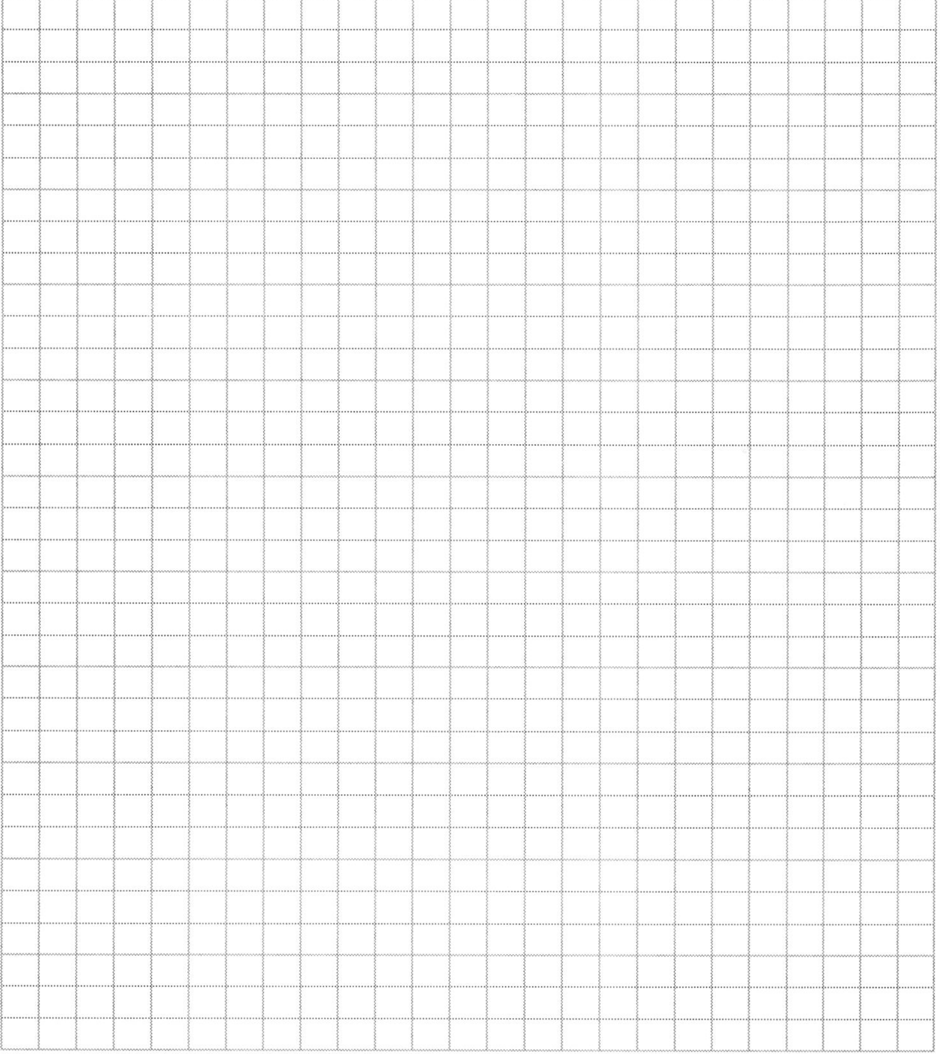

Experiment 5.3

Dynamic ^{1}H NMR Spectroscopy on Dimethylformamide

1. Purpose

The measurement and evaluation of dynamic equilibria and the determination of activation enthalpies and entropies are important tasks which can be performed by high resolution NMR spectroscopy [1-3]. Dimethylformamide (DMF) was one of the earliest molecules investigated when the methodology of dynamic NMR measurements was beeing developed [4-6]. It still provides an easy starting point to learn the procedures involved in performing measurements at different temperatures and evaluating the results for a simple, non-coupled two-site exchange. Many different results were reported in the very early literature, whereas now an agreement seems to have been reached.

2. Literature

[1] G. Binsch, *Top. Stereochemistry*, **1968**, *3*, 97–192.
[2] G. Binsch, H. Kessler, *Angew. Chem. Int. Ed. Engl.* **1980**, *19*, 411–429.
[3] H. Günther, *NMR Spectroscopy*, 2nd Ed., Wiley, Chichester, **1995**.
[4] K. Rabinowitz, A. Pines, *J. Am. Chem. Soc.* **1969**, *91*, 1585–1589.
[5] T. Drakenburg, K. J. Dahlquist, S. Forsén, *J. Phys. Chem.* **1972**, *76*, 2178–2183.
[6] G. J. Martin, M. Berry, D. Le Botlan, B. Mechin, *J. Magn. Reson.* **1976**, *23*, 523–526.

3. Pulsescheme and Phase Cycle

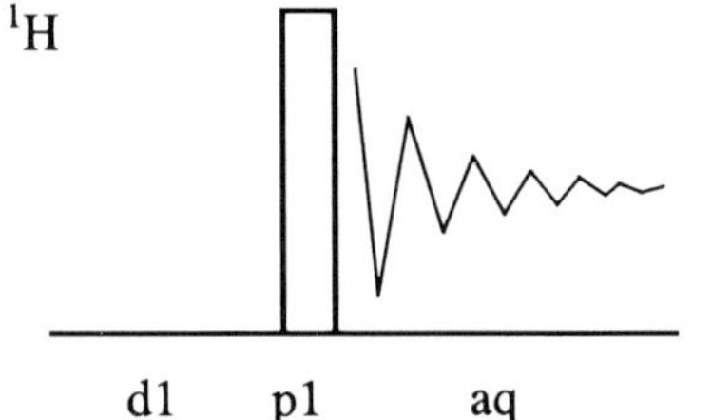

p1: x, x, -x, -x, y, y, -y, -y
aq: x, x, -x, -x, y, y, -y, -y

4. Acquisition

Time requirement: 5 h

Sample: 5% dimethylformamide in $C_2D_2Cl_4$. **Warning: Do not overheat the sample!**

Connect your instrument for high temperature measurements, adjust a reasonable nitrogen flow, set the control to room temperature, and let the sample equilibrate for at least 5 minutes. The spectra should be recorded with homonuclear decoupling of the aldehyde proton in order to simplify the evaluation. Record a normal ^{1}H NMR spectrum, note the position of the aldehyde proton resonance, and set the decoupler offset accordingly. Run the spectrum again under decoupling conditions. The signals of the methyl groups should now have equal heights. Change the temperature in 10 K steps at first and within the actual exchange region in 5 K steps until the signals of the methyl groups coalesce, or with high field instruments to a **maximum** of 430 K. Perform the experiments in reverse order and check for reproducibility. Adjust the decoupler position for the aldehyde proton before every measurement. You have to set:

td: 32 k
sw: 12 ppm
o1: middle of ^{1}H NMR spectrum
o2: on resonance of the aldehyde proton
p1: 45° ^{1}H transmitter pulse
d1: 300 s to equilibrate temperature
decoupler attenuation corresponding to γB_2 = 10 Hz
stable gas flow for temperature regulation
ns: 8

5. Processing

Use standard processing as described in Experiment 3.1, and for each temperature run an expanded plot of the signals of the methyl groups. For comparison with theoretically calculated line-shapes it is helpful to plot the experimental result on a transparent film with an enlargement that corresponds to a filled PC screen. Note for each temperature the line-width of the residual proton signal of the solvent.

6. Result

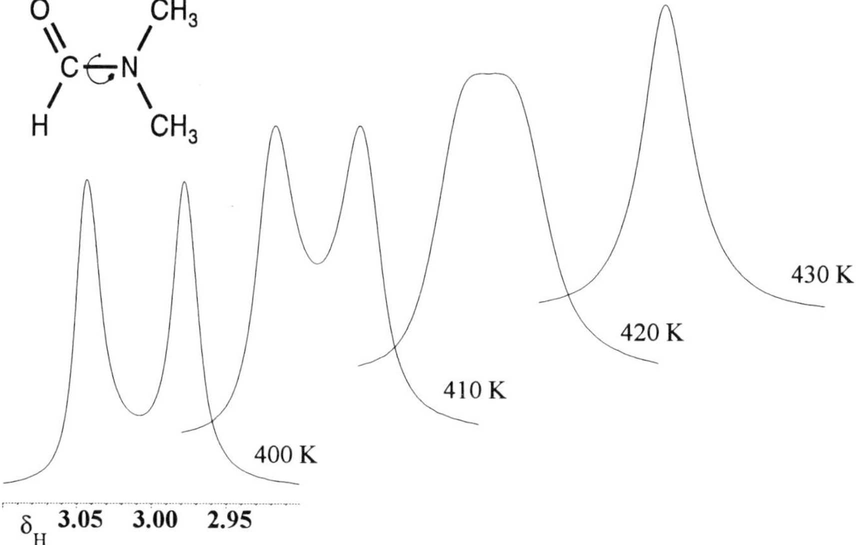

Displayed on page 145 are four typical spectra taken on an AM-400 spectrometer at 400–430 K. From the complete series the result $\Delta G^{\#}(298)$ = 21.4 kcal/mol was obtained; $\Delta S^{\#}$ was essentially zero.

7. Comments

It would be beyond the scope of this book to describe the theory of line-shape investigations in NMR spectroscopy. The interested reader therefore is referred to the fundamental review articles [1–3]. For our purpose here it is sufficient to know that two exchanging sets of nuclei can only be separately observed if the rate constant of the exchange is considerably smaller than their chemical shift difference in Hz (NMR time scale). The goal of the experiment is to derive a table of rate constant vs. temperature. From such a table, using the Eyring equation, you can calculate values of $\Delta G^{\#}$. By plotting $\Delta G^{\#}$ as a function of temperature one can derive $\Delta H^{\#}$ and $\Delta S^{\#}$ for the observed exchange process.

There are many PC-based programs which are able to calculate the theoretical line-shape. These require as input the line separation in the low temperature limit $\Delta\nu$, the ratio of populations of the two sites, the line-width for non-exchanging protons, and the rate constant. From this they calculate a line-shape which must be compared with the experimental result at the corresponding temperature. The rate constant at the temperature of coalescence T_C is given for the simple degenerate two site exchange by Equation (1).

$$k_C = \frac{\pi \Delta \nu}{\sqrt{2}} \qquad (1)$$

Using Equation (1) a formula (2) was derived, by which the $\Delta G^{\#}$ value at the coalescence can be obtained [3].

$$\Delta G^{\#}(T_C) = RT_C[22.96 + \ln(T_C/\Delta \nu)] \qquad (2)$$

8. Own Observations

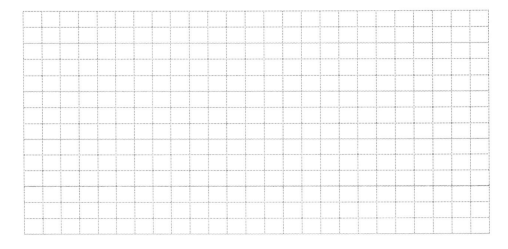

Experiment 5.4

The Saturation Transfer Experiment

1. Purpose

Dynamic NMR experiments such as that described in Experiment 5.3 can detect chemical exchange only if the exchange is fast enough to alter the NMR line-shape. However, slower chemical exchange processes can be detected by the saturation transfer experiment. One signal is irradiated and a change is observed in the intensity of another signal which is connected with the irradiated one by chemical exchange. A similar extension of the NMR time-scale is provided by the 2D EXSY experiment (Exp. 10.23). The FT modification [3] of the original Forsén–Hoffmann method shown here requires modern instruments where the transmitter power can be attenuated; on older instruments a third frequency source is necessary.

2. Literature

[1] S. Forsén, R. A. Hoffmann, *Acta Chem. Scand.* **1963**, *17*, 1787–1788.
[2] S. Forsén, R. A. Hoffmann, *Prog. NMR Spectrosc.* **1966**, *1*, 15–204.
[3] B. E. Mann, *J. Magn. Reson* **1976**, *21*, 17–23.
[4] J. J. Led, H. Gesmar, *J. Magn. Reson* **1982**, *49*, 444–463.
[5] M. L. Martin, J.-J. Delpuech, G. J. Martin, *Practical NMR Spectroscopy*, Heyden, London, **1980**, 315–321.

3. Pulse Scheme and Phase Cycle

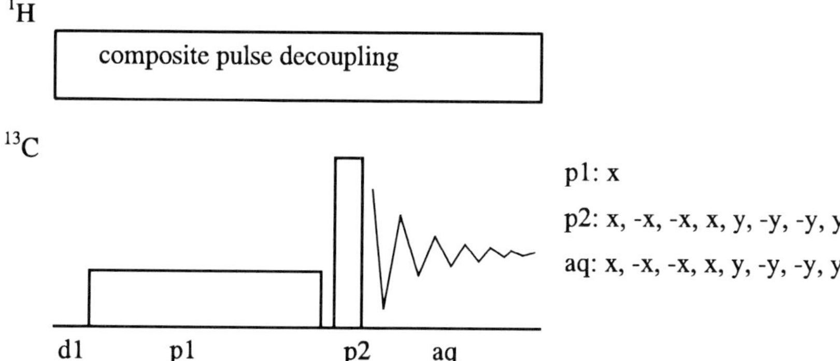

p1: x

p2: x, -x, -x, x, y, -y, -y, y

aq: x, -x, -x, x, y, -y, -y, y

4. Acquisition

Time requirement: 1 h

Sample: 5% dimethylformamide in $C_2D_2Cl_4$

Record a normal ^{13}C NMR spectrum of the sample and note the offsets of the signals of the methyl groups. Connect the high temperature equipment for your instrument, adjust a stable N_2 flow, and set the temperature first to 300 K. Afterwards increase the temperature to 350 K in 10 K steps.You have to set:

td : 8 k
sw: 25 ppm
o1: on resonance of low frequency methyl group signal
o2: middle of ^{1}H NMR spectrum
p1: 25 s pre-irradiation pulse at high transmitter attenuation; the transmitter bandwitdh (see Exp. 2.9) must be small enough in order to saturate only the signal on resonance (70 dB)
p2: 90° ^{13}C transmitter pulse
d1: 0.1 s
decoupler attenuation and 90° pulse for CPD
ns: 8

5. Processing

Use standard 1D processing (see Exp. 3.1) with exponential multiplication (lb = 2 Hz)

6. Result

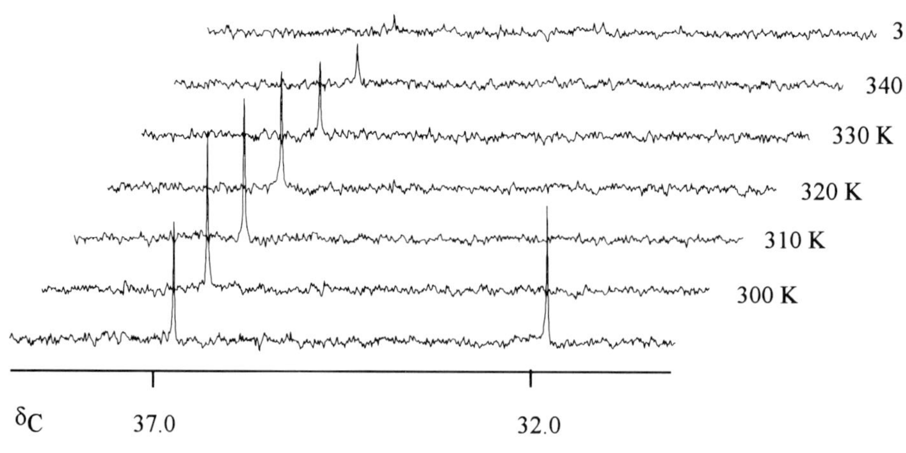

The figure on page 148 shows a series of saturation transfer spectra obtained on an AMX-500 spectrometer. The bottom trace is the normal ^{13}C NMR spectrum of the compound at 300 K. In the other spectra the signal of the low frequency methyl group was pre-irradiated giving nearly complete saturation of the other methyl group signal at 350 K, although no line-broadening can be observed at this temperature and magnetic field strength.

7. Comments

Note that the saturation transfer is field dependent. Performing this sequence with ^{13}C rather than ^{1}H NMR has the distinct advantage that the result is not blurred by Overhauser effects since the exchanging spins are not in the same molecule. The solution consists of a mixture of isotopomers where the ^{13}C atom is located either in the cisoid or transoid methyl group. The experiment gives directly qualitative proof of chemical exchange. However, quantitative treatment is more complicated, since it requires in addition a knowledge of the T_1 relaxation times of the nuclei involved. See the cited literature for the corresponding equations.

8. Own Observations

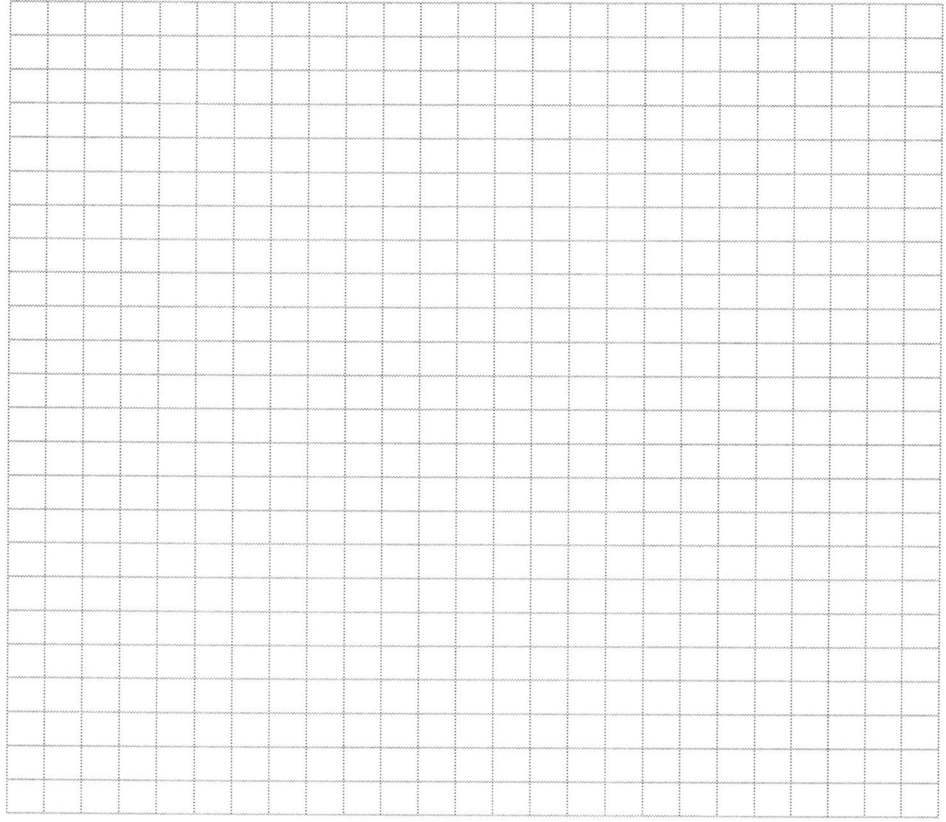

Experiment 5.5

Measurement of the Rotating Frame Relaxation Time $T_{1\rho}$

1. Purpose

Above the coalescence point exchanging AX spin systems form only one line, from which the rate constant k cannot be extracted without additional assumptions. The $T_{1\rho}$ experiment measures the relaxation time in the rotating frame and provides a means to determine the rate constants k and the chemical shift difference $\Delta\nu$ in cases where the low temperature regime cannot be reached. As the saturation transfer experiment (see Exp. 5.4), it extends the range of line shape methods for dynamic NMR, however, into the region of fast exchange. The $T_{1\rho}$ relaxation time becomes further an important parameter in 2D experiments which use a spin-lock, such as TOCSY or ROESY. In the experiment described here we determine the exchange rate for chlorodimethylformamide in the high temperature limit.

2. Literature

[1] I. Solomon, *C. R. Hebd. Séance Acad. Sci. Paris*. **1959**, *249*, 1631–1632.

[2] C. Deverell, R. E. Morgan, J. H. Strange, *Mol. Phys*. **1970**, *18*, 553–559.

[3] T. K. Leipert, J. H. Noggle, W. J. Freeman, D. L. Dalrymple, *J. Magn. Reson.* **1975**, *19*, 208–221.

[4] H. H. Limbach, *NMR-Basic Principles and Progress* **1991**, *23*, 63–164.

3. Pulse Scheme and Phase Cycle

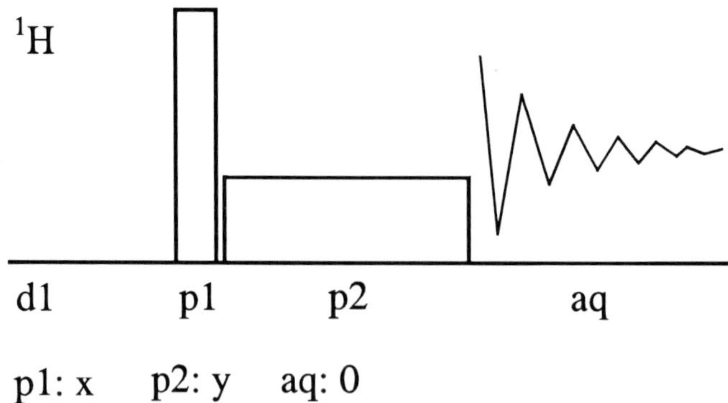

p1: x p2: y aq: 0

4. Acquisition

Time requirement: 2 h

Sample: 5% chlorodimethylformamide in $C_2D_2Cl_4$

This experiment requires an instrument with fast transmitter power switching. Determine the transmitter attenuation corresponding to 90° pulses of 60, 100, 200, 600 and 2000 μs yielding spin-lock fields between 4000 and 100 Hz (see Exp. 2.9). Check whether there is a phase difference between the hard proton transmitter pulse and the attenuated spin-lock pulses and adjust if necessary (see Exp. 7.1). Record a normal ^{1}H NMR spectrum of the sample and note the offset of the signal. Set the temperature to 353 K, which is (for a 500 MHz instrument) just above the coalescence point; let the sample equilibrate and load the $T_{1\rho}$ pulse program. You have to set:

td : 1 k
sw: 1 ppm
o1: on resonance of methyl group signal
p1: 90° ^{1}H transmitter pulse
p2: ^{1}H spin-lock pulse with different spin-lock field [17, 20, 25, 30, 40, 50 dB], create a list for variable spin-lock length; in this experiment 16 p2 values with 0.01, 0.05, 0.1, 0.2, 0.4, 0.8, 1, 1.5, 2, 3, 4, 5, 6, 7, 8 and 10 s have been used.
d1: 15 s
temperature: 353, 363, 373 and 383 K
ns: 1

Determine, at each of the four temperatures and for each of the six spin-lock fields, the rotating frame relaxation time $T_{1\rho}$. In addition determine for each temperature the spin–lattice relaxation time T_1 according to Exp. 6.1.

5. Processing

On recent instruments T_1, T_2 and $T_{1\rho}$ pulse programs usually create 2D NMR files. Apply exponential multiplication in the F_2 dimension and perform the Fourier transformation only in F_2. The series of spectra can then be analyzed by the T_1/T_2 software package of your instrument.

6. Result

The figure on page 152 shows a series of spectra obtained on an AMX-500 spectrometer with an inverse probe-head at 363 K with a spin-lock corresponding to 30 dB transmitter attenuation. A $T_{1\rho}$ value of 1.8 s from this data was calculated; the corresponding T_1 value was 3.1 s.

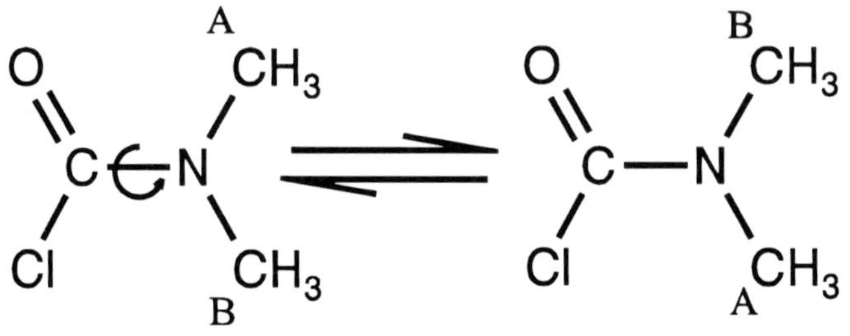

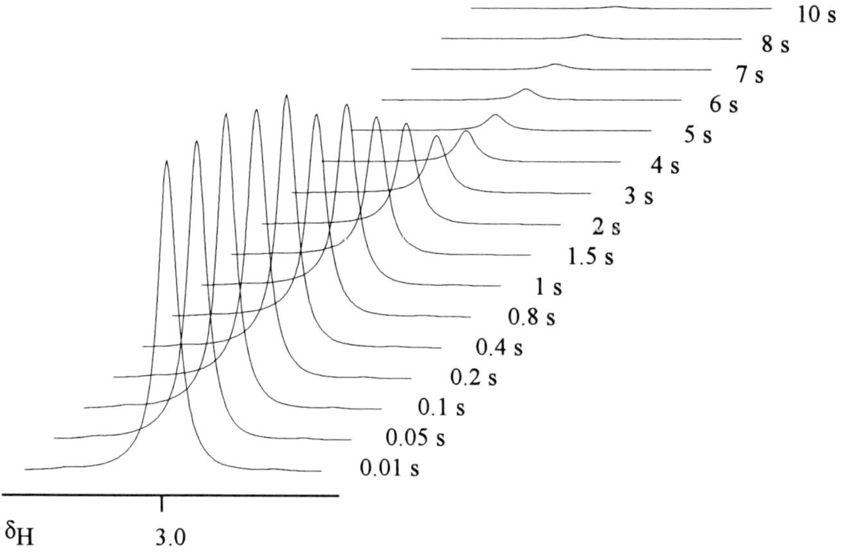

The further evaluation uses the Eq. (1), in which $\Delta\nu$ is the chemical shift difference of the methyl group signals in the slow exchange limit, τ the life time which, for the two site exchange described here, is related to the rate constant by $\tau = 1/2k$;

$$1/T_{1\rho} - 1/T_1 = \pi^2 \Delta\nu^2 \frac{\tau}{1 + \omega_1^2 \tau^2} \qquad (1)$$

$\omega_1 = \gamma B_1$ represents the spin-lock field strength. Thus a plot of $(T_{1\rho} - T_1)$ versus ω_1^2 should give a straight line. From the slope and the y intercept the parameters $\Delta\nu$ and τ can be determined. Indeed, with the data of this experiment a $\Delta\nu$ value close to the experimental value of 49 Hz could be calculated.

7. Comments

The proton transmitter pulse p1 aligns the magnetization along the $-y$ axis. For the spin-lock pulse p2 the phase of the radiofrequency is moved to y and the power is attenuated. Thus, the B_1 field is collinear with M, and this remains locked along the y-axis as long B_1 is applied. The decay of the magnetization during the spin-lock period due to transverse relaxation or, as in this experiment, due to exchange processes is characterized by the relaxation time in the rotating frame, $T_{1\rho}$; it is closely related to the spin–spin relaxation time T_2 (see Exp. 6.2). Since the observed $T_{1\rho}$ time will also have contributions from other mechanisms such as dipolar or spin-rotation contributions, it has to be corrected. This is performed assuming that these contributions are independent of ω_1 and ω_0, thus the difference between $1/T_{1\rho}$ and $1/T_1$ will yield the exchange contribution.

Equation (1) holds only in the absence of spin coupling and in the extreme narrowing limit. For very weak spin-lock fields and small values of τ the equation will become independent of ω_1.

8. Own Observations

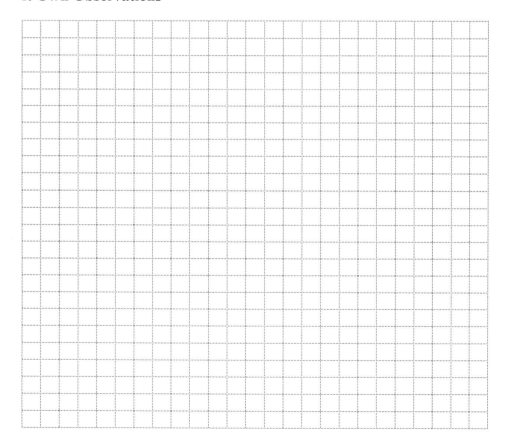

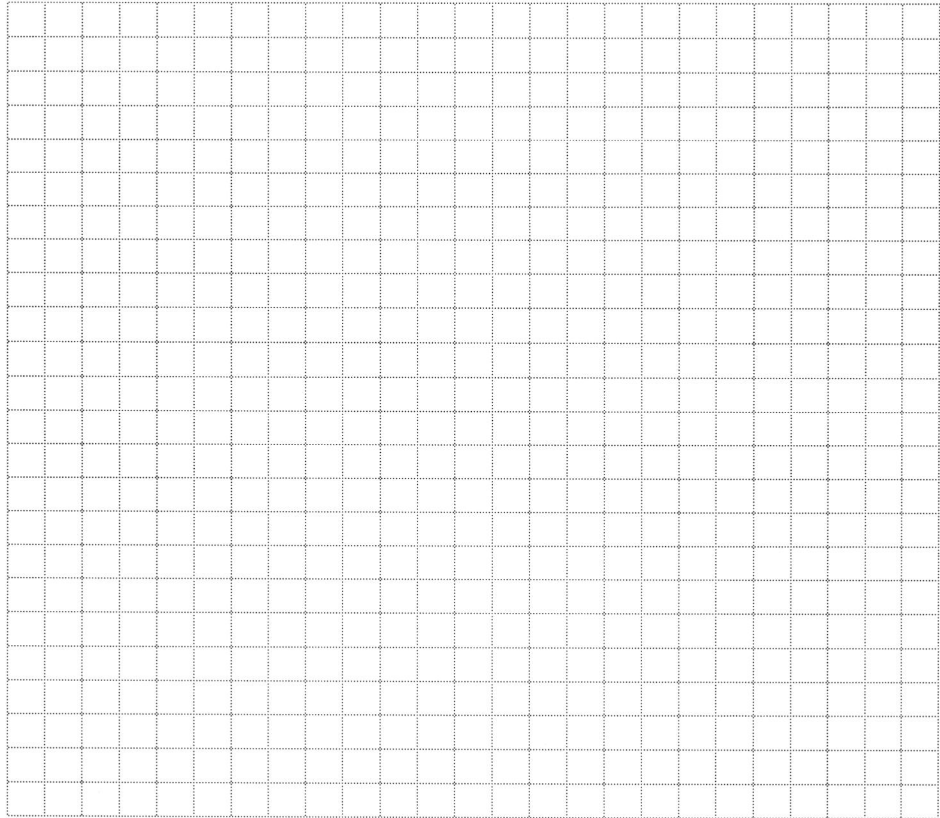

Chapter 6

1D Multipulse Sequences

In this chapter we provide several basic 1D pulse sequences where more than one r.f. pulse is applied. These pulses have to be calibrated for both the transmitter and decoupler channels. Due to error propagation these sequences are sensitive to miscalibration of the r.f. pulses and you will only succeed if these are correct.

Since a knowledge of both spin–lattice and spin–spin relaxation times is very important in multipulse NMR spectroscopy, this chapter begins with the two basic experiments for T_1 and T_2 measurements. A large part (Exp. 6.3–6.10) is then devoted to techniques for multiplicity determination with and without polarization transfer. There have been numerous discussions in the literature on the relative performance of these techniques. We give the basic descriptions of the most often used methods, SEFT, APT, INEPT, DEPT, and the recent PENDANT technique, and leave it to the reader to decide which variety best serves his or her own particular needs.

After an introduction to 1D-INADEQUATE (Exp. 6.11) four purely educational sequences are described: the use of BIRD, TANGO, the double quantum filter, and purging with a spin-lock is introduced. Together with the basic INEPT and reverse INEPT sequences these six experiments are shown for the simple case of the $CHCl_3$ molecule. However, we feel that a lot can be learned about modern multipulse NMR techniques from performing these experiments as shown.

Two methods for suppressing the huge solvent signal of water conclude the chapter. It should also be mentioned that other water suppression techniques using pulsed field gradients are described in Chapter 11.

Literature

[1] C. L. Turner, *Prog. NMR Spectrosc.* **1984**, *16*, 311–370.

Experiment 6.1

Measurement of the Spin–Lattice Relaxation Time T_1

1. Purpose

The longitudinal or spin–lattice relaxation time T_1 is the time constant for re-establishing thermal equilibrium of the *z*-magnetization after an r.f. pulse and must be clearly distinguished from the transversel or spin–spin relaxation time T_2, which describes the decay of the *x,y*-magnetization (see Exp. 6.2). As far as structure determination is concerned, T_1 is not as important a parameter as the chemical shift or the spin–spin coupling. But even for routine work at least a qualitative knowledge of this parameter is essential, e.g. for choosing a reasonable pulse repetition time. Furthermore, T_1-values are important for setting up NOE experiments and for studying molecular motions. Here we describe the inversion recovery experiment as applied to the determination of the ^{13}C NMR T_1-values of ethyl crotonate. Other methods are based on the progressive saturation and saturation recovery experiments.

2. Literature

[1] R. L. Vold, J. S. Waugh, M. P. Klein, D. E. Phelps, *J. Chem. Phys.* **1968**, *48*, 3831–3832.

[2] J. S. Frye, *Concepts Magn. Reson.* **1989**, *1*, 27–33.

[3] D. J. Craik, G. C. Levy, *Top. Carbon-13 NMR Spectrosc.* **1984**, *4*, 239–275.

[4] J. Kowalewski, G. C. Levy, L. F. Johnson, L. Palmer, *J. Magn. Reson.* **1977**, *26*, 533–536.

3. Pulse Scheme and Phase Cycle

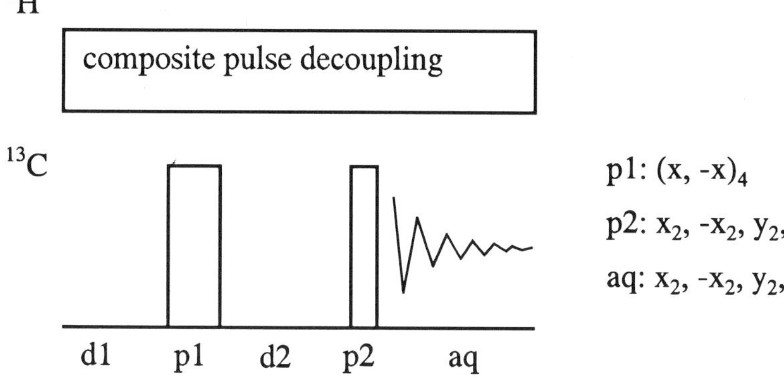

p1: $(x, -x)_4$

p2: $x_2, -x_2, y_2, -y_2$

aq: $x_2, -x_2, y_2, -y_2$

4. Acquisition

Time requirement: 2 h

Sample: 80% ethyl crotonate in $CDCl_3$, *not degassed*

Set up the spectrometer for ^{13}C NMR and load a pulse program for inversion recovery with ^{1}H broad-band decoupling. Newer versions handle the data as 2D data: for example, the results of eight experiments with different delays d2 are stored as rows in a 2D matrix. So you have to create a 2D file and to set:

td1: 8
td2: 32 k
sw: 200 ppm
o1: middle of ^{13}C NMR spectrum
o2: middle of ^{1}H NMR spectrum
p1: 180° ^{13}C transmitter pulse
p2: 90° ^{13}C transmitter pulse
d1: 60 s (> 5 T_1 in order to achieve the equilibrium z-magnetization)
d2: create a list with the follwing values [s]: 0.5, 1, 3, 6, 10, 16, 24, 50
ds: 2
decoupler attenuation and 90° pulse for CPD
ns: 8

5. Processing

If a 2D file has been created, Fourier transformation has to be performed in F_2, using a line broadening factor lb = 2 Hz. In order to adjust the phase, read spectrum number 8 in which all signals have a positive phase, and transfer this phase correction to all other spectra.

6. Result

In the figure on p. 157 the results obtained on an ARX-300 spectrometer (5 mm dual probe-head) are presented as a stacked plot for qualitative inspection. In the first two spectra all signals have a negative intensity, since after the first 180° pulse and the short delays of 0.5 and 1 s all spin vectors are still in the $-z$ direction. After a delay d2 of 3 s spin–lattice relaxation has reduced the intensity of the signal of C-6 nearly to zero, whereas those of all other signals are still negative, showing that C-6 must have the shortest T_1. For a rough estimation of the T_1-values from these spectra you may use Equation (1), where t_{null} is the (interpolated) delay d2 at which the intensity of a signal is zero.

$$T_1 = 1.44\, t_{null} \tag{1}$$

For quantitative analysis apply the T_1/T_2 software which uses either the integrals or the peak heights. The basis for the evaluation is Equation (2) with M_0 = equilibrium *z*-magnetization and M_z = *z*-magnetization after delay τ (d2 above). Replacing *M* by *I* (integral or peak height) yields Equation (3) in which *A* and *B* are constants

$$M_z = M_0\,(1-2e^{-\tau/T_1}) \qquad (2)$$

$$I_t = A + B\,e^{-\tau/T_1} \qquad (3)$$

The recommended procedure is an iterative exponential fitting according to Equation (3). This yields to the following T_1-values [s], based on peak heights and integrals:

	C-1	C-2	C-3	C-4	C-5	C-6
from peak height	(43.1)	8.2	7.5	7.8	7.0	5.3
from integral	(-)	7.6	7.2	8.0	6.7	4.9

It should be noted that the T_1-value for C-1 (C=O) is not reliable since the condition d1 > $5T_1$ is not fulfilled as is the case for the other carbon nuclei. A detailed discussion of the different parameters (length of d1, number and lengths of d2 etc.) is given in Ref. [3, 4]. As an exercise you may design and perform an experiment to determine T_1 for C-1 (about 50 s, not degassed).

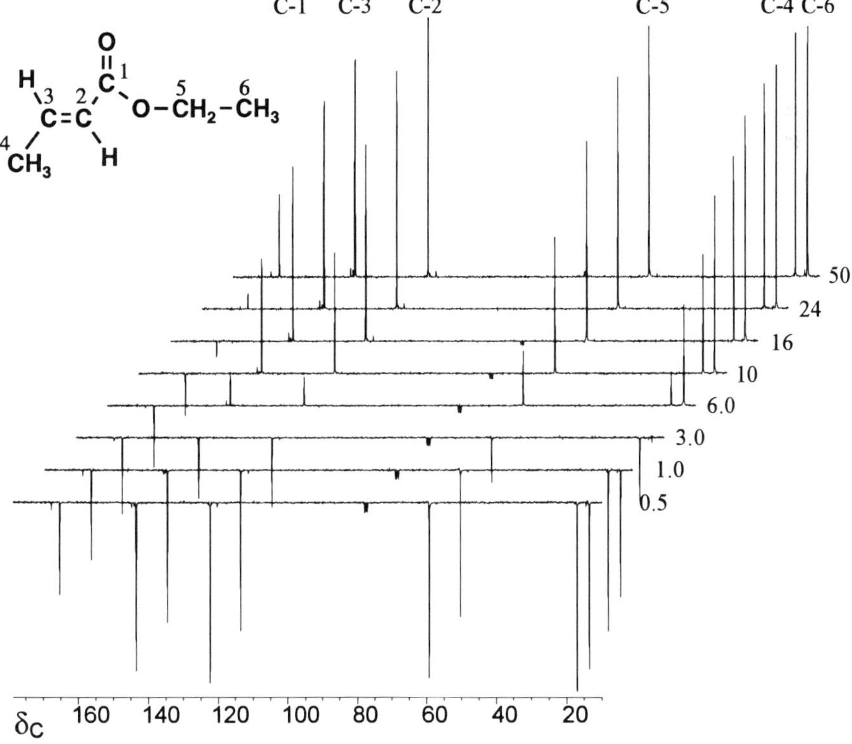

7. Comments

The 180° pulse inverts the magnetization so that it lies along to the $-z$-direction; relaxation then takes place during the delay d2. At the end of d2 the actual magnetization is measured by the 90° read-pulse which transfers z-magnetization into measurable y-magnetization. Note that T_1-values are very dependent on concentration, temperature, oxygen content, and magnetic field strength. In this experiment the sample was not degassed, so as to give relatively short relaxation times which could be more rapidly determined. In scientific applications, however, T_1 measurements should only be performed with carefully degassed samples.

8. Own Observations

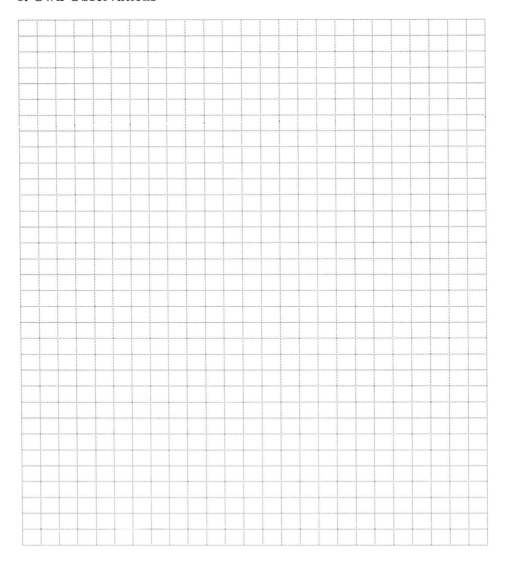

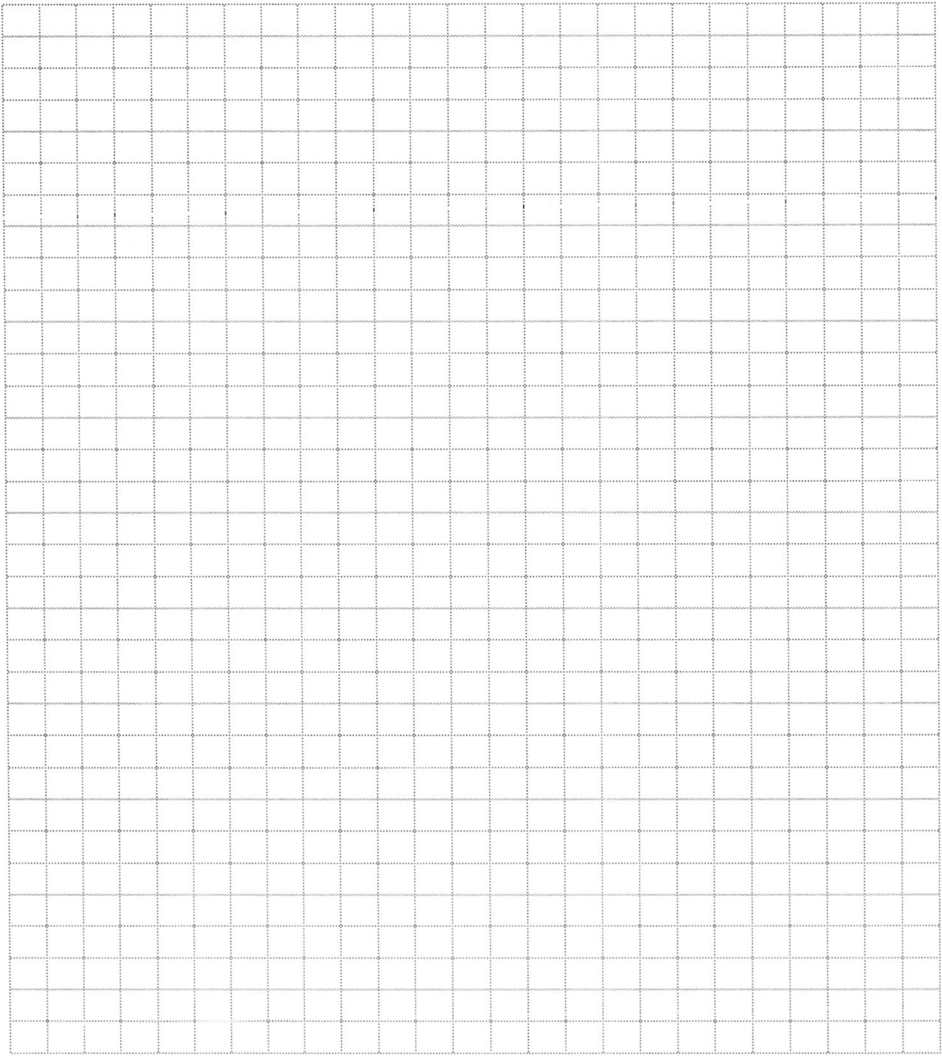

Experiment 6.2

Measurement of the Spin–Spin Relaxation Time T_2

1. Purpose

The transverse or spin–spin relaxation time T_2 determines the decay of the $M_{x,y}$ magnetization and is related to the line-width. It must be clearly distinguished from the longitudinal or spin–lattice relaxation time T_1 (Exp. 6.1) and can be separately measured. Although there is hardly a direct relationship between spin–spin relaxation times and the structure of molecules, a knowledge of its value is important for planning dynamic NMR experiments, investigations on spin diffusion, and generally for devising new pulse sequences, because their evolution periods must not significantly exceed T_2. In the extreme narrowing limit the relationship $T_1 = T_2$ usually holds. The spin-echo method for measuring T_2 is described here.

2. Literature

[1] S. Meiboom, D. Gill, *Rev. Sci. Instrum.* **1958**, *29*, 688–691.

[2] M. L. Martin, J.-J. Delpuech, G. J. Martin, *Practical NMR Spectroscopy*, Heyden, London, **1980**, 280–287.

[3] R. Freeman, *A Handbook of Nuclear Magnetic Resonance*, Longman Scientific Technical, Harlow, **1987**, 262–266.

[4] S. W. Homans, *A Dictionary of Concepts in NMR,* Revised Edition, Clarendon Press, Oxford, **1993**, 310–314.

3. Pulse Scheme and Phase Cycle

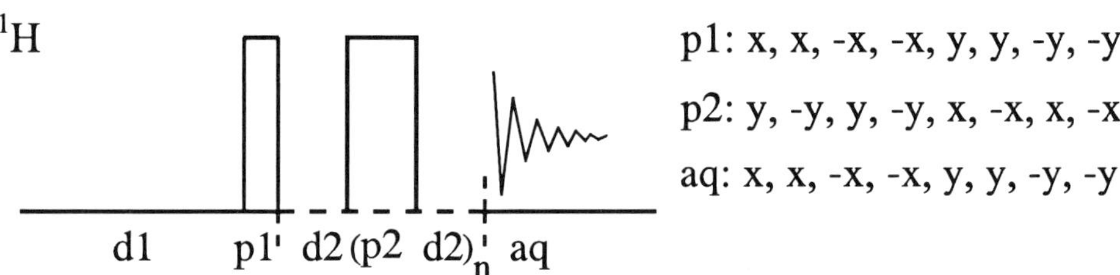

4. Acquisition

Time requirement: 0.5 h

Sample: 3% $CHCl_3$ in [D_6] acetone, *degassed and sealed*

Obtain a normal ^1H NMR spectrum of the sample and adjust the spectral width and the offset. To avoid macroscopic motion, *turn the spinner off*. Load the CPMG

(Carr–Purcell–Meiboom–Gill) pulse sequence and edit a list of ten numbers *n* which define the number of repeated cycles of the p2,d2 period in each of ten different experiments. On modern instruments this sequence is handled in the manner of a 2D experiment, where the ten experiments with the different repetition cycles are stored as rows in the 2D matrix. Therefore you have to create a 2D file prior to the start of the sequence. You have to set:

td: 1 k
sw: 500 Hz
o1: on ^{1}H resonance
p1: 90° ^{1}H transmitter pulse
p2: 180° ^{1}H transmitter pulse
d1: 150 s (5 T_1 of the $CHCl_3$ protons in the sample)
d2: 10 ms
preaquisition delay as short as possible
ns: 1
n-values of 2, 20, 50, 100, 200, 300, 400, 500, 750 and 1000 were used here, leading to delays between the first 90° pulse and start of the acquisition of 0.04, 0.4, 1, 2, 4, 6, 8, 10,15 and 20 s.

5. Processing

Modern software treats this experiment as a 2D file; however, transformation is only performed in the F_2 direction. Use an exponential line broadening of lb = 2 Hz and adjust the phase of the rows. After this, a normal T_1/T_2 software package detects peak integrals or heights from all rows and calculates the T_2 value from the given delays, which the user must provide in a corresponding delay list.

6. Result

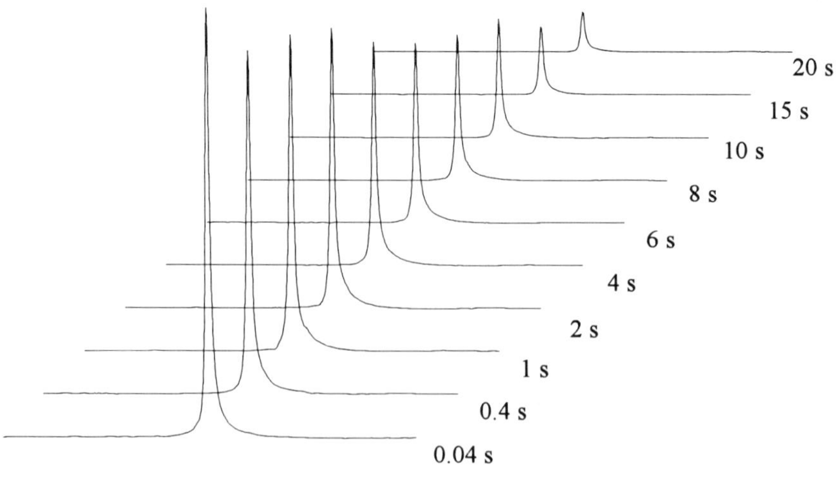

The figure on page 160 shows the ten spectra from this experiment, obtained on an AMX-500 spectrometer. From the intensities a T_2 value of 8.1 s was calculated, which corresponds to a natural linewidth of 0.04 Hz!

7. Comments

In this experiment the intrinsic value of T_2 is measured. This is related to the composite transverse relaxation time T_2^* by equation (1).

$$\frac{1}{T_2^*} = \frac{1}{T_2} + \frac{1}{T_2^{inhom}} \tag{1}$$

The second term on the right hand side describes the effect of the magnetic field inhomogeneity. T_2^* is the decay time constant of the FID and can also be approximated from the line-width using Equation (2) which is based on assuming exponential processes.

$$\Delta \nu_{1/2} = \frac{1}{\pi \cdot T_2^*} \tag{2}$$

For example, if the observed line-width at half height is 0.5 Hz, T_2^* can be calculated as 1.6 s; thus the inhomogeneity of the magnet is predominant for this example. As an exercise you may perform the experiment twice, first with very good resolution and then after poor shimming.

8. Own Observations

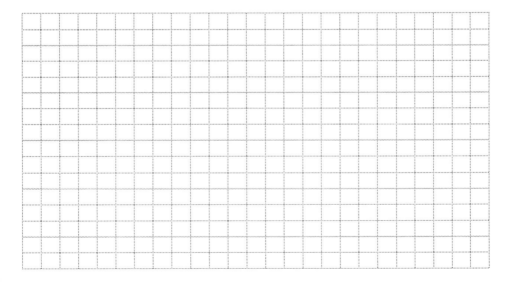

Experiment 6.3

^{13}C NMR Spectra with SEFT

1. Purpose

The SEFT (**S**pin-**E**cho **F**ourier **T**ransform) technique, also known as *J*-modulated spin-echo, is the simplest method of encoding the multiplicity of a ^{13}C signal into the phase of a ^{1}H broad-band decoupled ^{13}C NMR spectrum. From this method APT (Exp. 6.4) was also developed. SEFT can be performed on any instrument, because defined decoupler pulses as for INEPT, DEPT or PENDANT (Exps. 6.5 - 6.11) are not needed. The method does not use polarization transfer as INEPT or DEPT, only the NOE enhancement by broad-band ^{1}H-decoupling is effective.

2. Literature

[1] D. W. Brown, T. T. Nakashima, D. L. Rabenstein, *J. Magn. Reson.* **1981**, *45*, 302–314.

[2] C. Le Cocq, J. Y. Lallemand, *J. Chem. Soc. Chem. Commun.* **1981**, 150–152.

3. Pulse Scheme and Phase Cycle

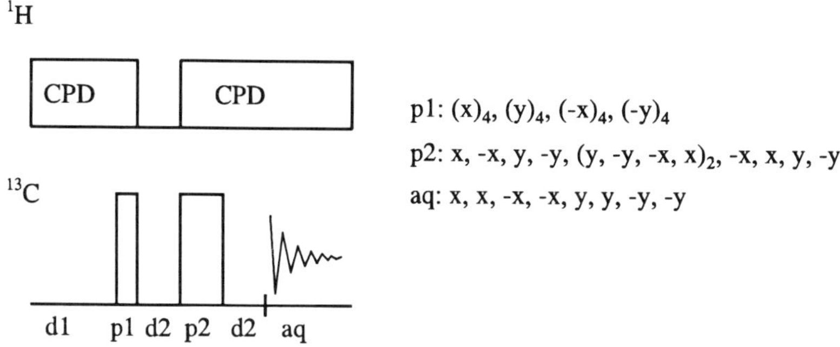

4. Acquisition

Time requirement: 10 min

Sample: 20% ethyl crotonate in $CDCl_3$

Load standard ^{13}C NMR parameters and the pulse program. You have to set:

td: 64 k
sw: 200 ppm
o1: middle of ^{13}C NMR spectrum

o2: middle of ^{1}H NMR spectrum
p1: 90° ^{13}C transmitter pulse
p2: 180° ^{13}C transmitter pulse
d1: 4 s
d2: $1/J$(C,H) = 7 ms, calculated from ^1J(C,H) = 140 Hz
decoupler attenuation and 90° pulse for CPD
preacquisition delay as short as possible
ns: 16

5. Processing

Use standard 1D processing as described in Experiment 3.2. Adjust the phase for the signals of the methyl groups to be positive and of the carboxyl C-atom negative.

6. Result

The figure shows a J modulated spin-echo spectrum obtained on an ARX-200 spectrometer.

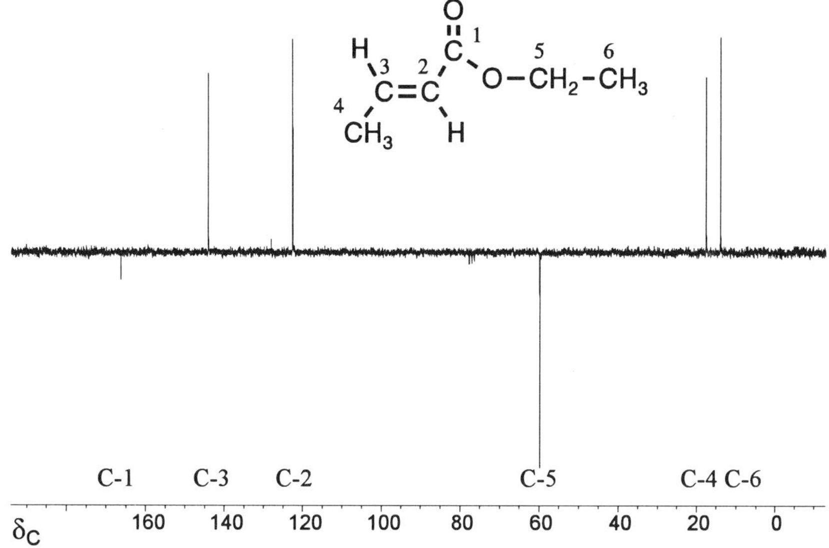

7. Comments

As shown in the figure on page 164, CH_n spin vectors develop differently after a 90° pulse depending on how many hydrogens are bound to the carbon atom. If the delay d2 is set to $1/J$, CH and CH_3 vectors have opposite phase compared with C and CH_2. If the decoupler is switched on at the end of d2 the phases are "frozen" and the corresponding signals have positive or negative sign. The second d2 delay is needed to refocus phase errors caused by the chemical shift evolution. A disadvantage of this sequence is that 90° pulses are used at the start, thus requiring a relatively long relaxation delay. This shortcoming was removed by the development of APT (Exp. 6.4).

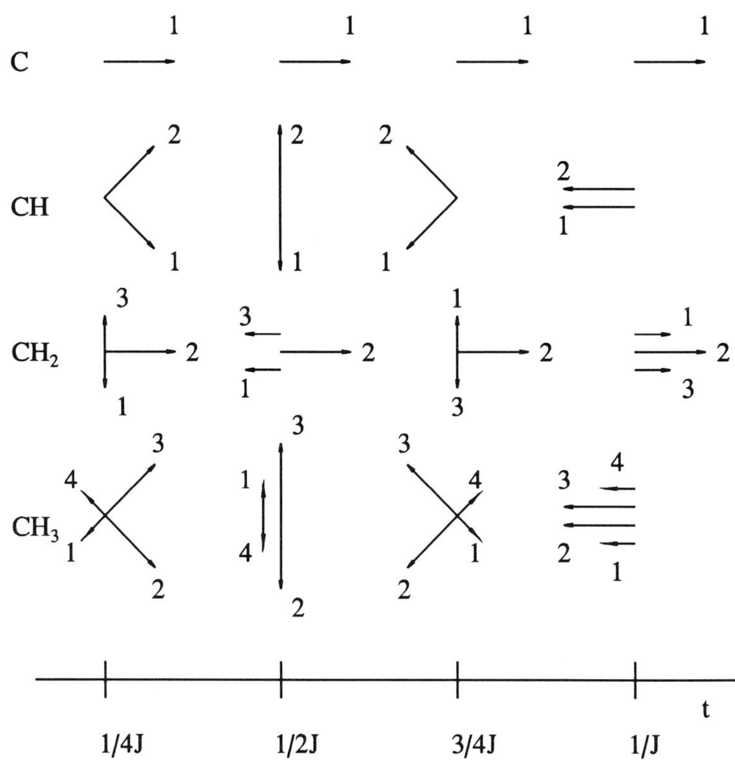

8. Own Observations

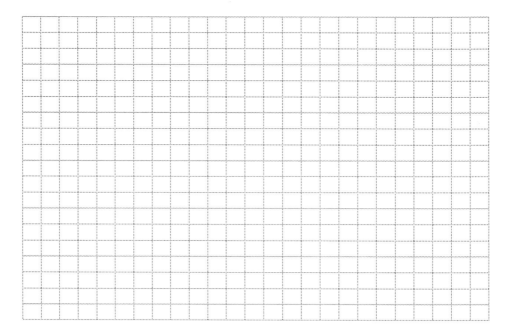

Experiment 6.4

^{13}C NMR spectra with APT

1. Purpose

The APT (**A**ttached **P**roton **T**est) technique is a modification of the SEFT experiment and differentiates between C, CH, CH_2 and CH_3 groups (see Exp. 6.3). The SEFT sequence suffers from the use of a 90° excitation pulse which requires long repetition times. In the APT experiment a shorter excitation pulse is used. Alternative methods which give information about the multiplicities are INEPT, DEPT and PENDANT (see Exps. 6.5 - 6.11), and the old off-resonance ^{1}H-decoupling technique (see Exp. 4.10). Unlike INEPT or DEPT, the APT method yields only NOE enhancements, but also gives information about quaternary carbon atoms. Improved modifications of APT are known [2, 3].

2. Literature

[1] S. Patt, J. N. Shoolery, *J. Magn. Reson.* **1982**, *46*, 535–539.

[2] J. C. Madsen, H. Bildsøe, H. Jakobsen, O. W. Sørensen, *J. Magn. Reson.* **1986**, *67*, 243–257.

[3] A. M. Torres, T. T. Nakashima, R. E. D. McClung, *J. Magn. Reson. Ser. A* **1993**, *101*, 285–294.

3. Pulse Scheme and Phase Cycle

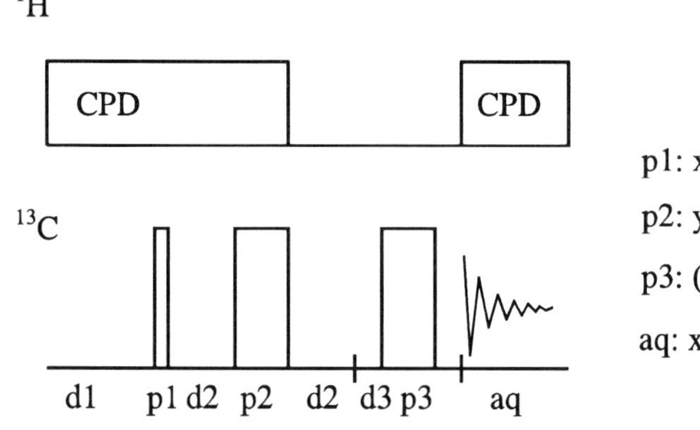

p1: x, x, -x, -x, y, y, -y, -y

p2: y, y, -y, -y

p3: $(y, -y)_2$

aq: x, x, -x, -x, y, y, -y, -y

4. Acquisition

Time requirement: 30 min

Sample: 100 mg cholesteryl acetate in $CDCl_3$

Load standard ^{13}C NMR parameters and the APT pulse program. You have to set:

td: 64 k
sw: 200 ppm
o1: middle of ^{13}C NMR spectrum
o2: middle of ^{1}H NMR spectrum
p1: 45° ^{13}C transmitter pulse
p2, p3: 180° ^{13}C transmitter pulse
d1: 2s
d2: $1/J(C,H)$ = 7 ms, calculated from $^{1}J(C,H)$ = 140 Hz
d3: set d3 equal to preacquisition delay
decoupler attenuation and 90° pulse for CPD
ns = 512

5. Processing

Use standard 1D processing as described in Experiment 3.2. Adjust the phase of the TMS signal positive and that of the carboxyl C-atom negative.

6. Result

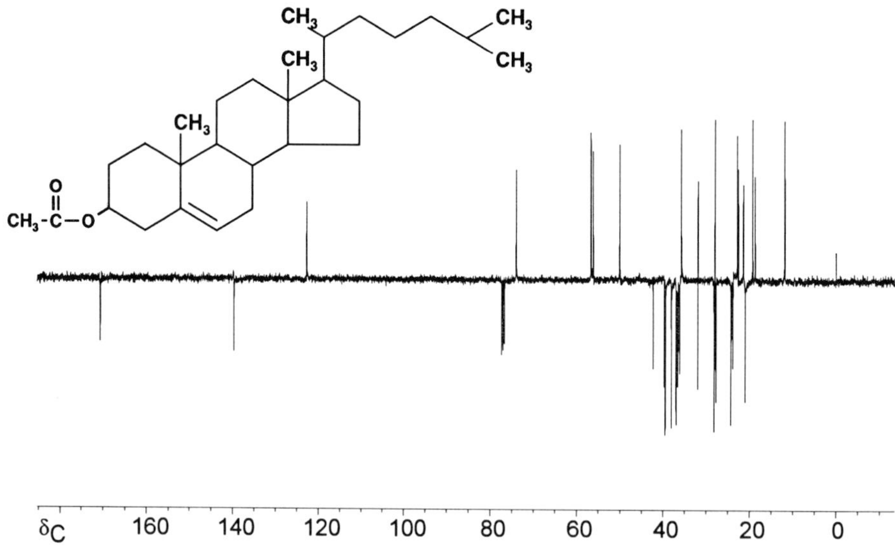

The figure shows the APT spectrum of cholesteryl acetate obtained on an AM-400 spectrometer. Note that the signal of the solvent $CDCl_3$ is negative like the other signals of carbon atoms carrying no protons. Signals of CH and CH_3 groups are positive and signals of CH_2 groups are negative.

7. Comments

The APT sequence is in principle a double spin-echo experiment. By using a 45° or shorter excitation pulse a part of the initial magnetization remains in the *z*-direction and is inverted by the first 180° pulse. This could lead to a cancelling of signals with long spin–lattice relaxation times, but in the second spin-echo period the 180° pulse reinverts the *z*-magnetization, thus eliminating this problem. In comparison with all other editing techniques APT still seems to be the most simple and efficient method, since it gives in one experiment all the necessary information on *all* sorts of carbon atoms. The lower sensitivity compared with polarization transfer methods such as DEPT is in practice not important for the C,H spin pair. APT can be performed on older instruments, since no specific decoupler pulses are required.

8. Own Observations

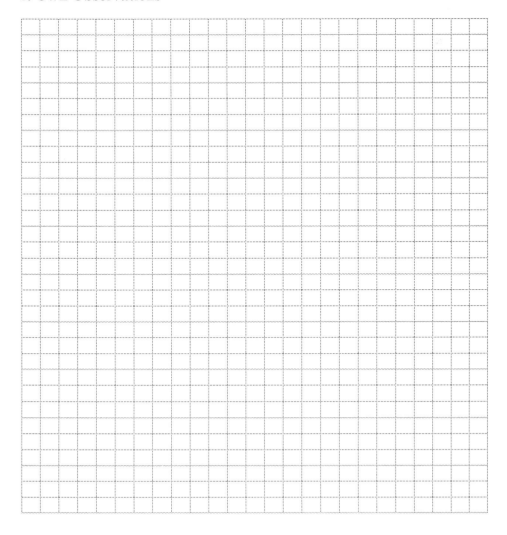

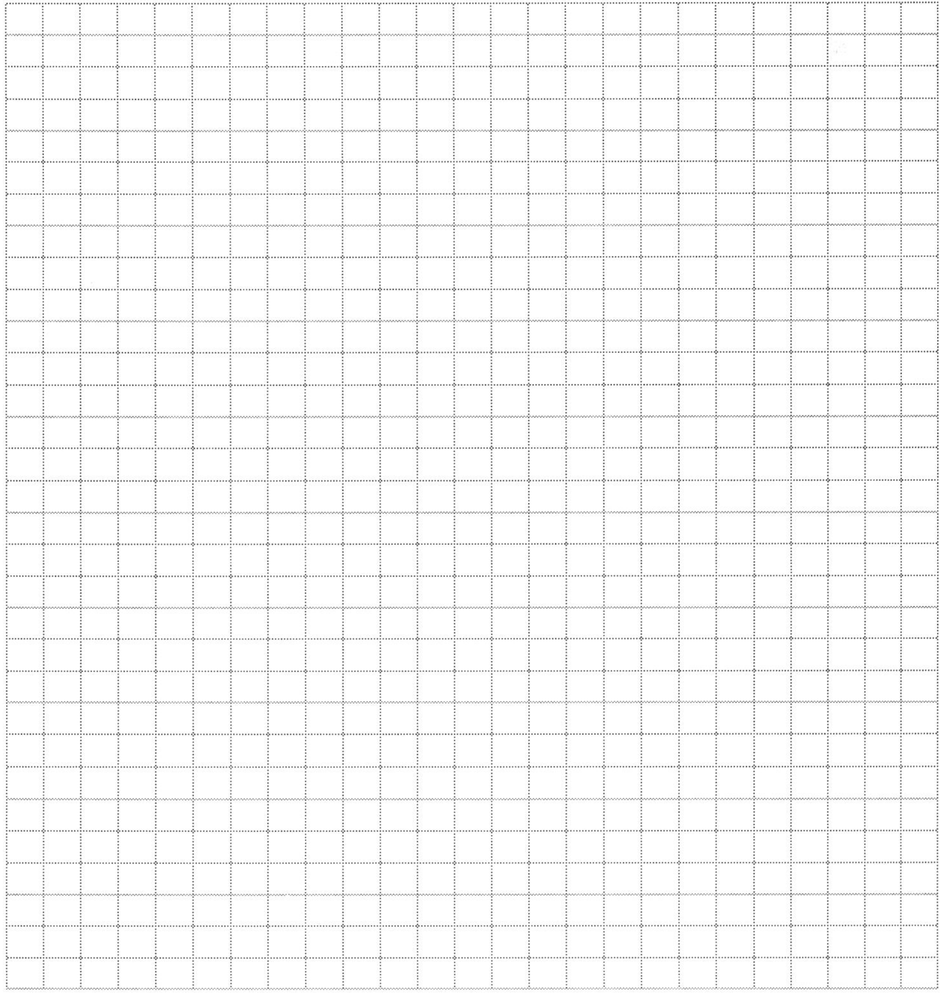

Experiment 6.5

The Basic INEPT Technique

1. Purpose

The INEPT experiment (**I**nsensitive **N**uclei **E**nhanced by **P**olarization **T**ransfer) was developed to increase the signal strength for nuclides with a low gyromagnetic ratio and low natural abundance, such as ^{13}C, ^{29}Si, or ^{15}N. This sensitivity enhancement is usually achieved by polarization transfer from the protons via X,H spin coupling. The increase in sensitivity is γ_A/γ_X , where γ_A represents the gyromagnetic ratio of the nuclide serving as the polarization source, in most cases ^1H, although ^{19}F and ^{31}P can also be used. The polarization transfer experiment delivers larger enhancement factors than the NOE experiment (see Exp. 4.15). The enhancement is independent of the sign of γ. The INEPT sandwich is one of the most frequently used building blocks of modern 2D and 3D sequences. An application to ^{15}N NMR spectroscopy is given in Experiment 9.1. The basic form shown in this educational experiment on $CHCl_3$ is tuned to $^1J(C,H)$ and yields a coupled ^{13}C NMR spectrum.

2. Literature

[1] G. A. Morris, R. Freeman, *J. Am. Chem. Soc.* **1979,** *101*, 760–762.
[2] D. P. Burum, R. R. Ernst, *J. Magn. Reson.* **1980**, *39*, 163–168
[3] O. W. Sørensen, R. R. Ernst, *J. Magn. Reson.* **1983**, *51*, 477–489.

3. Pulse Scheme and Phase Cycle

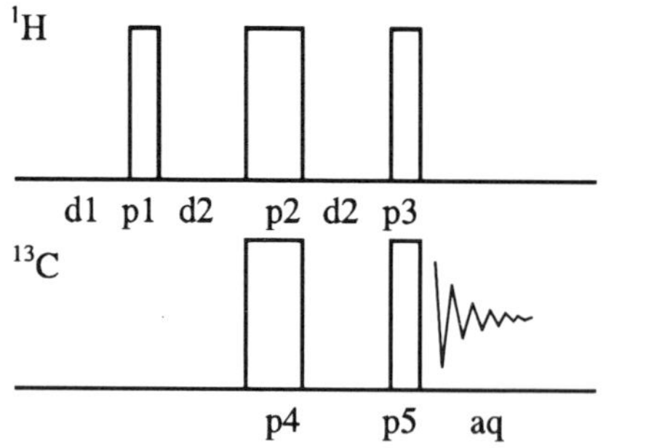

p1: $(x)_8$, $(-x)_8$

p2: x, -x

p3: $(y)_2$, $(-y)_2$

p4: x, -x

p5: $(x)_4$, $(y)_4$, $(-x)_4$, $(-y)_4$

aq: x, x, -x, -x, y, y, -y, -y

4. Acquisition

Time requirement: 10 min

Sample: 80% $CHCl_3$ in $[D_6]$acetone

Record normal ^{13}C and ^{1}H NMR spectra, note the offsets of $CHCl_3$, and load the INEPT pulse program. You have to set:

td: 4 k
sw: 500 Hz
o1: on resonance of ^{13}C NMR signal
o2: on resonance of ^{1}H NMR signal
p1, p3: 90° ^{1}H decoupler pulse
p2: 180° ^{1}H decoupler pulse
p4: 180° ^{13}C transmitter pulse
p5: 90° ^{13}C transmitter pulse
d1: 10 s
d2: $1/[4J(C,H)]$ = 1.18 ms, calculated from $^{1}J(C,H)$ = 212 Hz
decoupler attenuation for hard decoupler pulses
ns: 1 for the first and 4 for the second experiment

5. Processing

Use standard 1D processing as described in Experiment 3.2.

6. Result

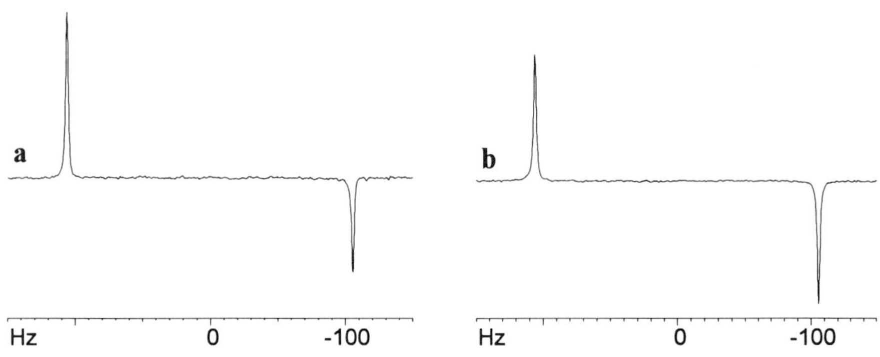

The figure shows two INEPT spectra of $CHCl_3$ obtained on an AM-400 spectrometer. Spectrum **a** was recorded with one scan, spectrum **b** with 4 scans. Due to the phase cycle one obtains in spectrum **b** a 1 : (–1) doublet, whereas in **a** the intensities are in the ratio 5 : (–3). As an additional exercise you may perform the experiment with ethyl crotonate; see Experiment 6.6.

7. Comments

For the product operator formalism we consider a C,H spin pair. The first 90° proton pulse creates transverse magnetization of the protons which develops C,H spin coupling and ^{1}H chemical shift during both delays d2. The chemical shift, however, is refocused by the 180° ^{1}H pulse and therefore for simplicity is not included in the equations. Since a 180° ^{13}C pulse is applied simultaneously with the 180° ^{1}H pulse, the spin-echo after the second d2 delay is modulated by the C,H spin coupling. If the delay τ is set equal to 2·d2 = 1/[2*J*(C,H)] the cosine term becomes zero and the sine term unity, leaving pure antiphase magnetization of the proton with respect to carbon in Equation (1).

$$\gamma_H I_{Hz} \xrightarrow{I_{Hx}} -\gamma_H I_{Hy} \xrightarrow{\pi J \tau 2 I_{Hz} I_{Cz}} -\gamma_H 2 I_{Hx} I_{Cz} \quad (1)$$

This antiphase magnetization is converted into antiphase magnetization of ^{13}C with respect to ^1H by the two simultaneous 90° pulses. During acquisition C,H spin coupling develops again, forming an in-phase ^{13}C magnetization, which however is multiplied by γ_H and thus we obtain a proton-polarized ^{13}C signal. This signal appears in antiphase due to the sine term in Equation (2).

$$-\gamma_H 2 I_{Hx} I_{Cz} \xrightarrow{I_{Hy}, I_{Cx}} -\gamma_H 2 I_{Hz} I_{Cy} \xrightarrow{\pi J \mathrm{aq}\, 2 I_{Hz} I_{Cz}} \gamma_H I_{Cx} \sin\pi J\cdot\mathrm{aq} \quad (2)$$

However, there is another contribution to the signal from the ^{13}C magnetization. I_{Cz} is first inverted by the 180° ^{13}C pulse and then converted into transverse magnetization by the 90° ^{13}C pulse. It develops C,H spin coupling during acquisition, giving an in-phase signal due to the cosine term in Equation (3).

$$\gamma_C I_{Cz} \xrightarrow{180° I_{Cx}} -\gamma_C I_{Cz} \xrightarrow{90° I_{Cx}} \gamma_C I_{Cy} \xrightarrow{\pi J \mathrm{aq}\, 2 I_{Hz} I_{Cz}} \gamma_C I_{Cy} \cos\pi J\cdot\mathrm{aq} \quad (3)$$

This signal has the intensity ratio of 1 : 1, whereas the signal obtained in Equation (2) has the intensity ratio 4: (–4), superposition yields the intensity ratio 5 : (–3) as observed in spectrum **a.** The phase cycle of the INEPT sequence eliminates all signal contributions stemming from initial carbon magnetization; thus in spectrum **b** a 4 : (–4) doublet is seen. Furthermore, all signals from quaternary carbon atoms are suppressed.

In case of the polarization of ^{13}C nuclei by ^1H ($\gamma_H/\gamma_C \approx 4$) the theoretical relative intensities of the multiplets obtained in *one scan* are as follows:

CH	–3 5
CH_2	–7 2 9
CH_3	–11 –9 15 13

This distortion of multiplets is a drawback of the INEPT sequence. Therefore the sequences INEPT+ (see Exp. 6.6) and DEPT (see Exp. 6.9) were developed.

8. Own Observations

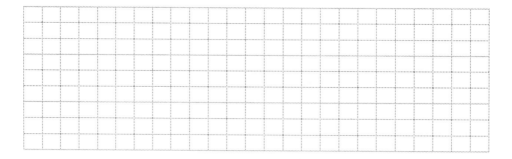

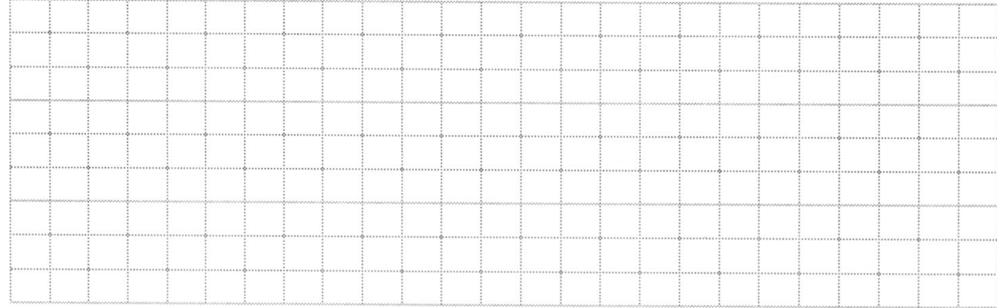

Experiment 6.6

INEPT+

1. Purpose

The disadvantage of the basic INEPT technique is the distortion of the multiplets. By adding a refocusing period with a subsequent additional purging pulse the extended version INEPT+ was developed, which yields coupled polarization-enhanced NMR spectra of X nuclei with correct intensities within the multiplets. Furthermore, the sequence can be tailored to give a different phase for CH_2 groups with respect to the signals of CH and CH_3 groups, providing editing possibilities. Here we describe the ^{13}C INEPT+ technique for ethyl crotonate.

2. Literature

[1] G. A. Morris, R. Freeman, *J. Am. Chem. Soc.* **1979,** *101*, 760–762.
[2] D. P. Burum, R. R. Ernst, *J. Magn. Reson.* **1980**, *39*, 163–168.
[3] O. W. Sørensen, R. R. Ernst, *J. Magn. Reson.* **1983**, *51*, 477–489.

3. Pulse Scheme and Phase Cycle

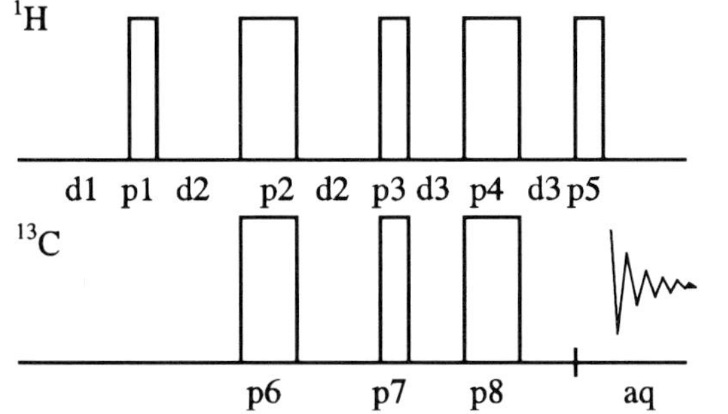

p1, p5: $(x)_8$, $(-x)_8$
p2, p4, p6: x, -x
p3: $(y)_2$, $(-y)_2$
p7: $(x)_4$, $(y)_4$, $(-x)_4$, $(-y)_4$
p8: $(x, -x)_2$, $(y, -y)_2$
aq: x, x, -x, -x, y, y, -y, -y

4. Acquisition

Time requirement: 10 min

Sample: 20% ethyl crotonate in $CDCl_3$

Load standard ^{13}C NMR parameter and the INEPT+ pulse program. You have to set:

td: 64 k
sw: 200 ppm
o1: middle of ^{13}C NMR spectrum
o2: middle of ^1H NMR spectrum
p1, p3, p5: 90° ^1H decoupler pulse

p2, p4: 180° ^{1}H decoupler pulse
p7: 90° ^{13}C transmitter pulse
p6, p8: 180° ^{13}C transmitter pulse
d1: 2 s
d2: $1/[4J(C,H)]$ = 1.78 ms, calculated from $^{1}J(C,H)$ = 140 Hz
d3: $1.5/[4J(C,H)]$ = 2.68 ms, calculated from $^{1}J(C,H)$ = 140 Hz
decoupler attenuation for hard decoupler pulses
ns = 128

5. Processing

Use standard 1D processing as described in Experiment 3.2

6. Result

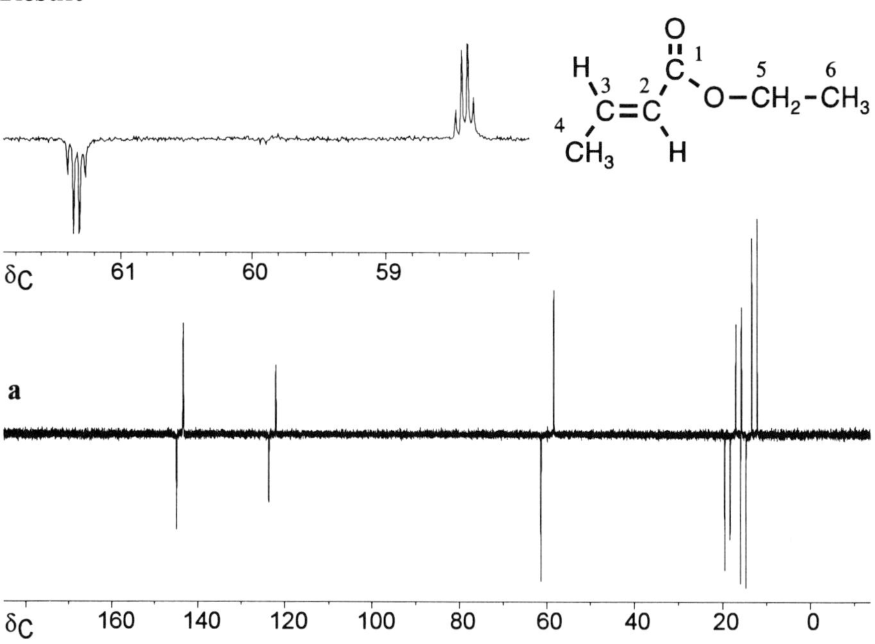

In the figure **a** is the normal INEPT spectrum of ethyl crotonate (sequence of Exp. 6.5) and **b** (page 173) is the INEPT+ modification obtained on an AM-400 spectrometer. Compare the signal patterns of the two spectra. Note that in both spectra the signals of the quaternary carbon nuclei, i.e.those of the C=O group and $CDCl_3$, are missing.

7. Comments

In the INEPT+ sequence an additional refocusing period with two 180° pulses is added to the basic INEPT scheme. It can be shown by the product operator formalism [2] that the polarization factors obtained for CH, CH_2, and CH_3 groups are described by Equations (1).

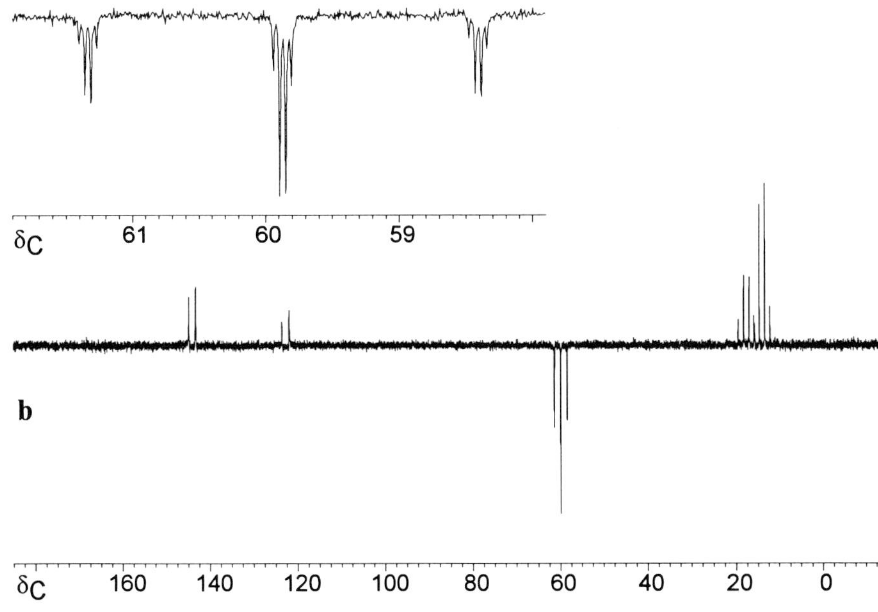

$$
\begin{aligned}
&\text{CH:} \quad \gamma_H/\gamma_C \sin(\pi J\,\text{d2}) \sin(\pi J\,\text{d3}) \\
&\text{CH}_2\text{:} \quad \gamma_H/\gamma_C \sin(\pi J\,\text{d2}) \sin(2\pi J\,\text{d3}) \\
&\text{CH}_3\text{:} \quad 3\gamma_H/4\gamma_C \sin(\pi J\,\text{d2}) [\sin(\pi J\,\text{d3}) + \sin(3\pi J\,\text{d3})]
\end{aligned}
\qquad (1)
$$

Thus choosing delay d3 = 1.5/[4*J*(C,H)], the CH_2 group gives a negative signal. The final purging pulse p5 removes intensity anomalies within the C,H multiplets.

8. Own Observations

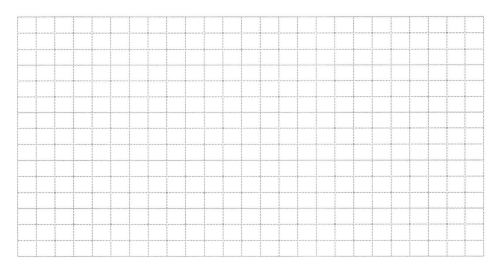

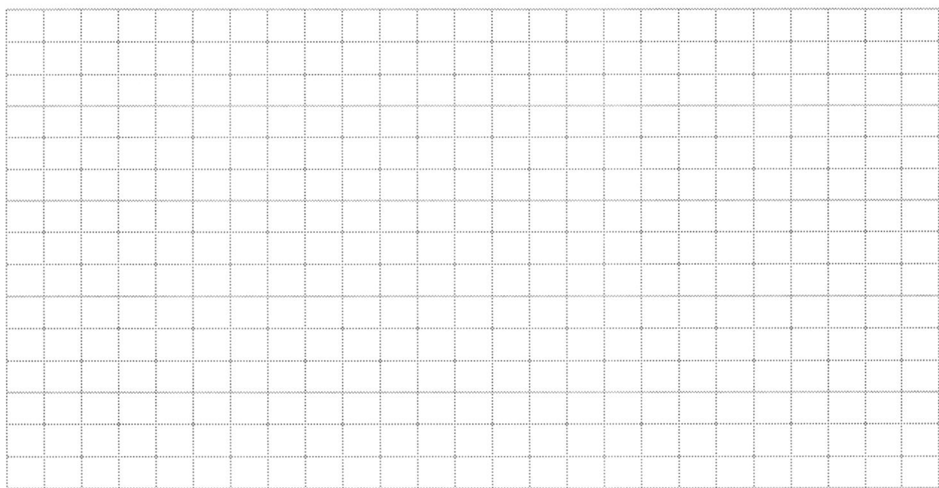

Experiment 6.7

Refocused INEPT

1. Purpose

This variant of INEPT spectroscopy yields proton-decoupled and polarization-enhanced NMR spectra of X nuclei. It is derived from INEPT+ (see Exp. 6.6) by replacing the last proton pulse by broad-band decoupling. The experiment can be tailored to yield different phases of CH_2 groups with respect to CH and CH_3 groups and can therefore be used for editing. Another method yielding the same information is DEPT (see Exps. 6.9 - 6.10). Here we describe the ^{13}C experiment with ethyl crotonate.

2. Literature

[1] G. A. Morris, R. Freeman, *J. Am. Chem. Soc.* **1979,** *101*, 760–762.
[2] D. P. Burum, R. R. Ernst, *J. Magn. Reson.* **1980**, *39*, 163–168
[3] O. W. Sørensen, R. R. Ernst, *J. Magn. Reson.* **1983**, *51*, 477–489.

3. Pulse Scheme and Phase Cycle

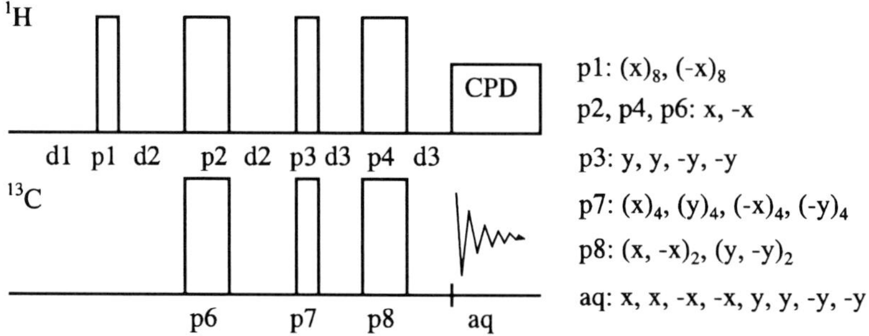

4. Acquisition

Time requirement: 10 min

Sample: 20% ethyl crotonate in $CDCl_3$

Load standard ^{13}C NMR parameter and the pulse program for refocused INEPT. You have to set:

td: 64 k
sw: 200 ppm
o1: middle of ^{13}C NMR spectrum

o2: middle of ^{1}H NMR spectrum
p1, p3: 90° ^{1}H decoupler pulse
p2, p4: 180° ^{1}H decoupler pulse
p6, p8: 180° ^{13}C transmitter pulse
p7: 90° ^{13}C transmitter pulse
d1: 2 s
d2: $1/[4J(C,H)]$ = 1.78 ms, calculated from $^1J(C,H)$ = 140 Hz
d3: $1.5/[4J(C,H)]$ = 2.68 ms, calculated from $^1J(C,H)$ = 140 Hz
decoupler attenuation for hard decoupler pulses
decoupler attenuation and 90° pulse for CPD
ns = 128

5. Processing

Use standard 1D processing as described in Experiment 3.2

6. Result

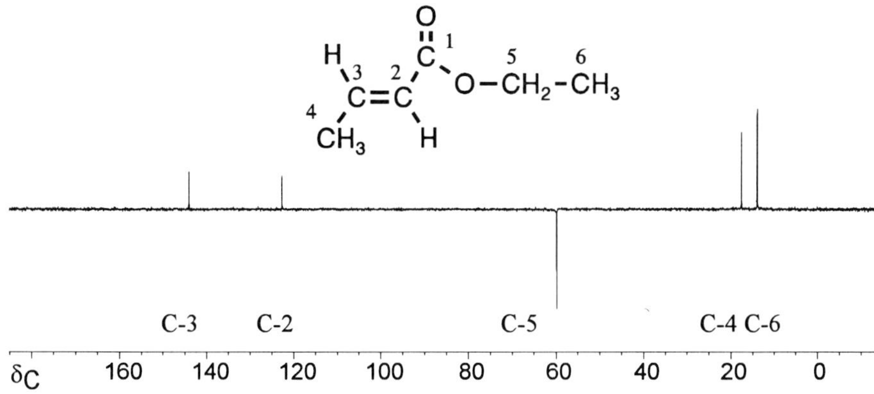

The figure shows the refocused INEPT spectrum of ethyl crotonate obtained on an AM-400 spectrometer. The signals of the quaternary carbon nuclei, i.e. those of the C=O group and $CDCl_3$, are missing.

7. Comments

The polarization and phase factors obtained for the refocused INEPT experiment are the same as in INEPT+ and are given there (see Exp. 6.6). Broad-band ^{1}H decoupling causes collapse of the multiplet lines, which after the refocusing period are all in-phase.

Refocused INEPT is useful for nuclides with low natural abundance and low gyromagnetic ratio. The enhancemant factors η for different nuclides in comparison with the NOE enhancement factors are as follows:

Nuclide	^{13}C	^{15}N	^{29}Si	^{57}Fe	^{103}Rh	^{109}Ag	^{119}Sn	^{183}W
η(NOE)	2.99	−3.94	−1.52	16.48	−16.89	−9.75	−0.41	13.02
η(INEPT)	3.98	9.87	5.03	30.95	31.77	21.50	2.81	24.04

For this reason, INEPT has often been used for the observation of ^{15}N and ^{29}Si. Especially in the case of ^{29}Si special attention has to be payed to the last delay d3, which controls the optimum polarization transfer (see Exp. 9.4).

The advantage of INEPT, like all other polarization transfer methods is that the pulse repetition time of the experiment is dictated by the spin–lattice relaxation time of the protons rather than that of the nuclide under observation, here ^{13}C. The disadvantage of the INEPT sequence is its sensitivity towards both delays d2 and d3. Carbon nuclei with widely different C,H spin coupling constants can give signals with lower intensity or even the wrong sign. Therefore the DEPT sequence is more often used, since the choice of the delays is not as critical (see exp.6.9).

8. Own Observations

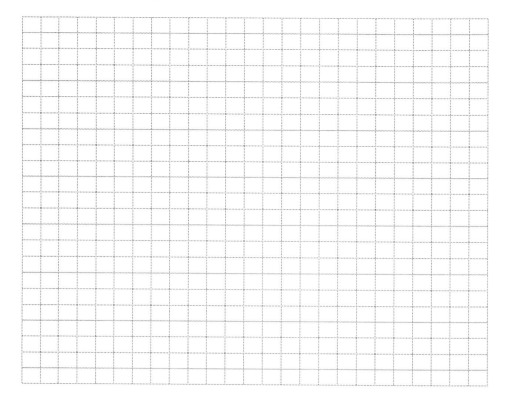

Experiment 6.8

Reverse INEPT

1. Purpose

The INEPT sandwich forms a basic building block in many modern 2D and 3D sequences, such as the HSQC technique (see Exp. 10.17). It transfers proton magnetization to an X nucleus. The reverse transfer is usually also achieved by an INEPT type sandwich, which is shown in the educational experiment given here. Starting from ^{13}C magnetization, the C,H spin coupling is observed by proton detection. Signals of protons bound to ^{12}C are suppressed.

2. Literature

[1] R. Freeman, T. H. Mareci, G. A. Morris, *J. Magn. Reson.* **1981**, *42*, 341–345.

3. Pulse Scheme and Phase Cycle

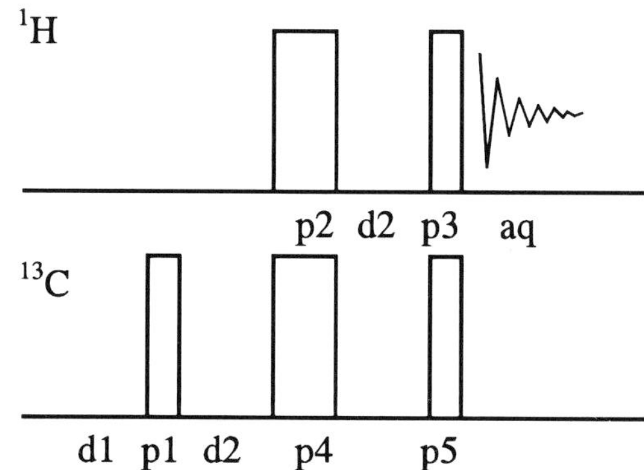

p1: $(x)_8$, $(-x)_8$

p2: x, -x

p3: $(y)_2$, $(-y)_2$

p4: x, -x

p5: $(x)_4$, $(y)_4$, $(-x)_4$, $(-y)_4$

aq: $(x)_2$, $(-x)_2$, $(y)_2$, $(-y)_2$

4. Acquisition

Time requirement: 10 min

Sample: 10% $CHCl_3$ in [D_6]acetone

Record normal ^{13}C and ^{1}H NMR spectra and note the offsets of $CHCl_3$. Set the instrument to ^{1}H observation with ^{13}C decoupling (inverse mode on older instruments) and load the reverse INEPT pulse program. You have to set:

td: 4 k
sw: 500 Hz

o1: on resonance of ^{1}H NMR signal
o2: on resonance of ^{13}C NMR signal
p1, p5: 90° ^{13}C decoupler pulse
p2: 180° ^{1}H transmitter pulse
p3: 90° ^{1}H transmitter pulse
p4: 180° ^{13}C decoupler pulse
d1: 30 s
d2: $1/[4J(C,H)]$ = 1.19 ms, calculated from $^1J(C,H)$ = 214 Hz
decoupler attenuation for hard decoupler pulses
ns: 8

5. Processing

Use standard 1D processing with exponential multiplication (lb = 0.5 Hz) as described in Experiment 3.1

6. Result

The figure shows the result obtained on an AMX-500 spectrometer with an inverse probe-head.

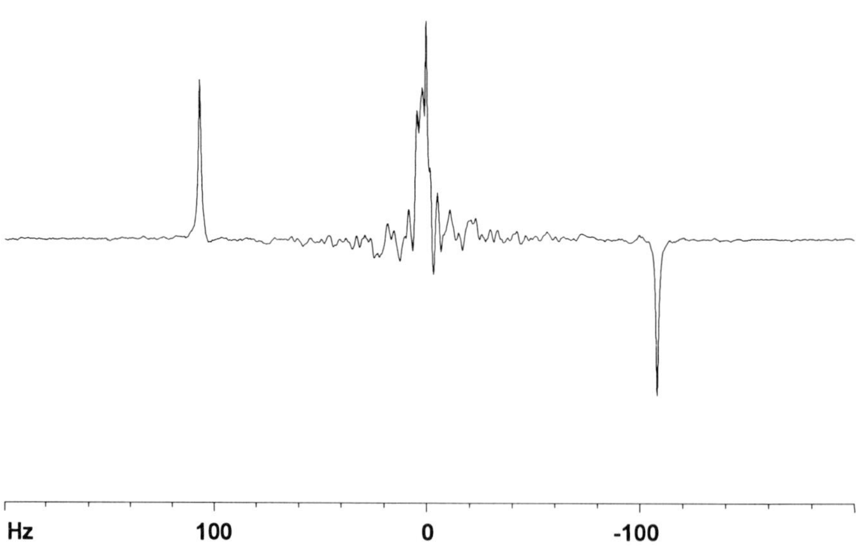

7. Comments

The product operator formalism is exactly the same as given in Experiment 6.5, except that the H and C subscripts of the operator terms have to be interchanged. Note, however, that the reverse INEPT sandwich shown in this experiment differs slightly from the one actually used in 2D experiments; see for example Experiment 10.17. In this educational experiment we have to start with in-phase carbon magnetization, created by the first pulse p1. Antiphase magnetization is developed, chemical shift effects are refocused, and the reverse transfer is achieved by the last two 90° pulses. In an actual 2D or 3D experiment, usually antiphase magnetization is already present after the t_1 evolution. Therefore, first the two 90° reverse transfer pulses are applied and the refocusing period with the two 180° pulses is used after the reverse transfer. Thus the common reverse INEPT building block in 2D or 3D sequences is a pair of 90° pulses followed by a refocusing period with a pair of 180° pulses, as seen in the HSQC sequence of Experiment 10.17. As an exercise you may add to the sequence described here an additional refocusing period with two 180° pulses, which will yield an in-phase signal. The experiment may also be performed with a normal dual probe-head.

8. Own Observations

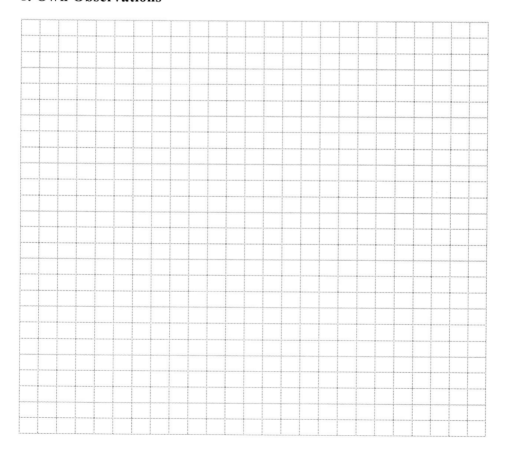

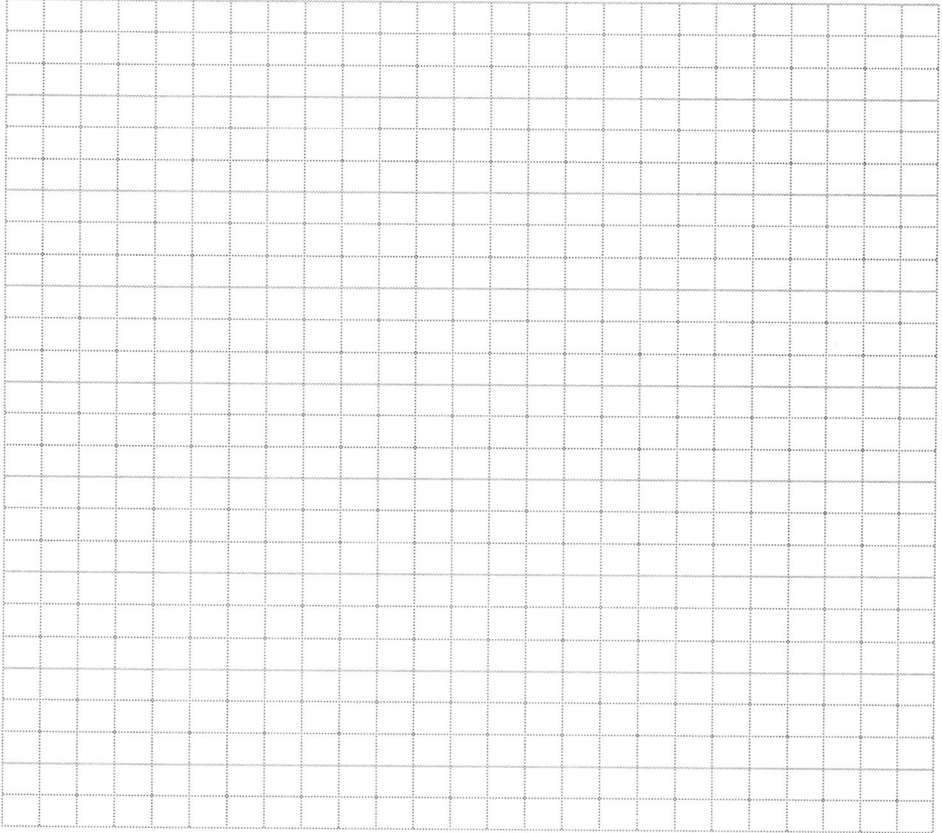

Experiment 6.9

DEPT-135

1. Purpose

The DEPT experiment (**D**istortionless **E**nhancement by **P**olarization **T**ransfer) uses, like the INEPT method (see Exps. 6.5–6.7), a polarization transfer from protons to an X nucleus to increase the signal strength. The experiment may be performed with polarization transfer over one or more bonds, with or without ^{1}H decoupling. It is therefore preferably applied to nuclei with a low γ and a low natural abundance such as ^{15}N or ^{29}Si (see Exps. 9.1–9.2, 9.4). Furthermore, the sequence can also be used for multiplicity determination such as SEFT (Exp. 6.3), APT (Exp. 6.4), refocused INEPT (Exp. 6.7) and the recently developed PENDANT (Exp. 6.11). Described here is the ^{13}C DEPT-135 experiment on cholesteryl acetate.

2. Literature

[1] M. R. Bendall, D. M. Doddrell, D. T. Pegg, *J. Am. Chem. Soc.* **1981**, *103*, 4603–4605.

[2] D. M. Doddrell, D. T. Pegg, M. R. Bendall, *J. Magn. Reson.* **1982**, *48*, 323–327.

[3] K. V. Schenker, W. v. Philipsborn, *J. Magn. Reson.* **1986**, *66*, 219–229.

3. Pulse Scheme and Phase Cycle

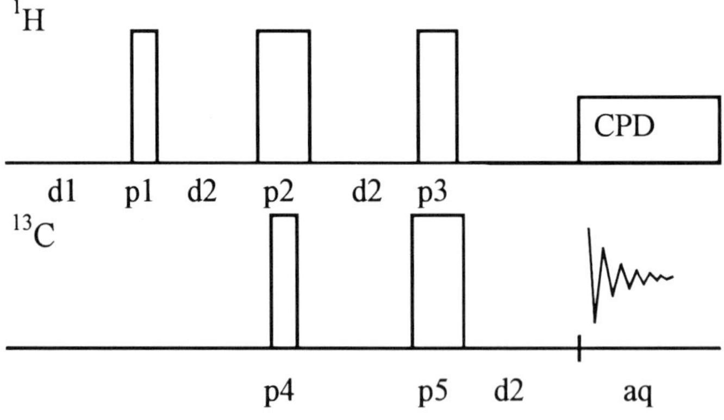

p1: x
p3: $(y)_4$, $(-y)_4$
p5: $(x, -x)_4$, $(y, -y)_4$

p2: x, -x, y, -y
p4: $(x)_8$, $(y)_8$, $(-x)_8$, $(-y)_8$

aq: $(y)_2$, $(-y)_4$, $(y)_2$, $(-x)_2$, $(x)_4$, $(-x)_2$, $(y)_2$, $(y)_4$, $(-y)_2$, $(x)_2$, $(-x)_4$, $(x)_2$

4. Acquisition

Time requirement: 30 min

Sample: 100 mg cholesteryl acetate in $CDCl_3$

Load standard ^{13}C NMR parameters and the DEPT pulse program. You have to set:

td: 64 k
sw: 200 ppm
o1: middle of ^{13}C NMR spectrum
o2: middle of ^{1}H NMR spectrum
p1: 90° ^{1}H decoupler pulse
p2: 180° ^{1}H decoupler pulse
p3: 135° ^{1}H decoupler pulse
p4: 90° ^{13}C transmitter pulse
p5: 180° ^{13}C transmitter pulse
d1: 2 s
d2: $1/[2J(C,H)] = 3.5$ ms, calculated from $^{1}J(C,H) = 140$ Hz
decoupler attenuation for hard decoupler pulses
decoupler attenuation and 90° pulse for CPD
ns = 512

5. Processing

Use standard 1D processing as described in Experiment 3.2. Adjust the phase for the TMS signal positive.

6. Result

The figure on page 182 shows the ^{13}C DEPT-135 spectrum of cholesteryl acetate obtained on an AM-400 spectrometer. Note that there appear no signals of quaternary carbon atoms. This is a disadvantage in comparison with the APT or PENDANT sequences. Compare the signal-to-noise ratio with that of Experiment 6.4, which was obtained under otherwise identical conditions.

7. Comments

For the product operator formalism we consider a C,H spin pair and neglect the effects of chemical shifts, since these are refocused by the 180° pulses. The first 90° proton pulse creates transverse magnetization of the protons which develops C,H spin coupling during the delay d2. If the delay τ is set equal to d2 = $1/[2J(C,H)]$ the cosine term becomes zero and the sine term unity, leaving pure antiphase magnetization of the proton with respect to carbon in the equation. This antiphase magnetization is converted into double quantum magnetization of carbon and proton by the first 90° ^{13}C pulse, see Eq. (1).

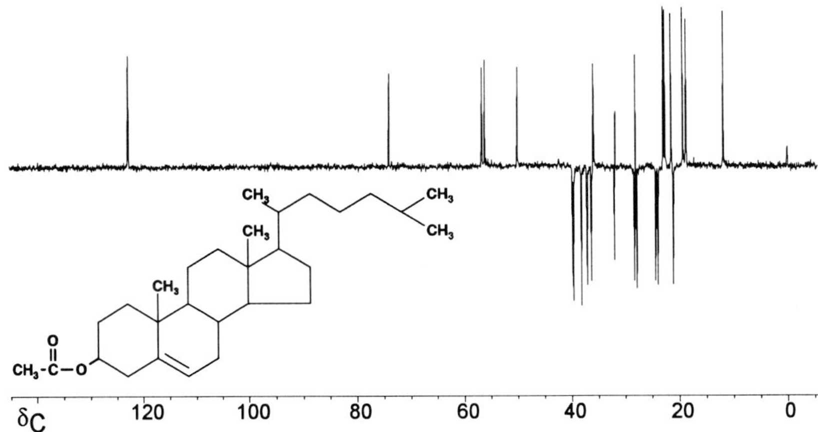

$$\gamma_H I_{Hz} \xrightarrow{I_{Hx}} -\gamma_H I_{Hy} \xrightarrow{\pi J \tau 2 I_{Hz} I_{Cz}} \gamma_H 2 I_{Hx} I_{Cz} \xrightarrow{I_{Cx}} -\gamma_H 2 I_{Hx} I_{Cy} \quad (1)$$

This double quantum term does not further develop spin coupling during the second d2 period. The transfer pulse p3 creates antiphase magnetization of carbon with respect to proton, which during the third d2 delay develops in-phase carbon magnetization I_{Cx}. This is a polarized signal since I_{Cx} is multiplied by γ_H; see Eq. (2). C,H coupling which would develop during acquisition is removed by the decoupling of the protons.

$$-\gamma_H 2 I_{Hx} I_{Cy} \xrightarrow{I_{Hy}} \gamma_H 2 I_{Hz} I_{Cy} \xrightarrow{\pi J \tau 2 I_{Hz} I_{Cz}} \gamma_H I_{Cx} \quad (2)$$

Another signal contribution stemming from carbon magnetization I_{Cz}, which is converted into tranverse carbon magnetization by the first 90° carbon pulse, is removed by the phase cycle.

The adjustment of d2 = $1/2J$ is less crucial compared with INEPT, while the multiplicity selection is performed by the angle of the transfer pulse p3. However, signals of carbon atoms with very widely differing spin coupling constants such as sp-hybridized carbon atoms may display the wrong sign. Often the DEPT sequence is additionally performed with p3 = 90°, yielding only signals for methine carbon atoms and thus distinguishing them from those of CH_3 groups. For complete editing see Exp. 6.10.

8. Own Observations

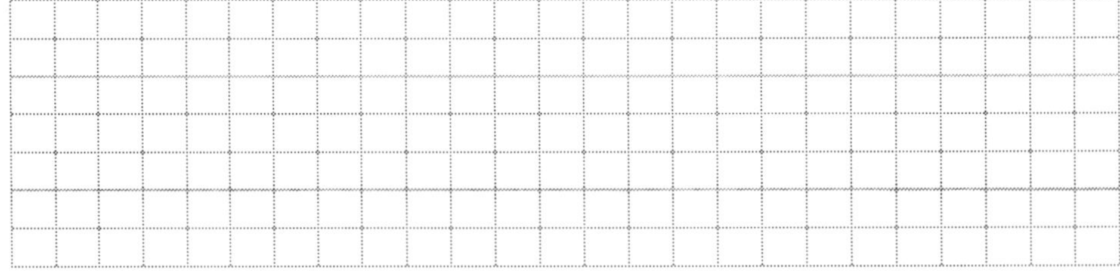

Experiment 6.10

Editing ^{13}C NMR Spectra with DEPT

1. Purpose

The DEPT-135 experiment (**D**istortionless **E**nhancement by **P**olarization **T**ransfer) may be applied as a powerful means for distinguishing CH_3, CH_2, and CH groups, as has been shown in Experiment 6.9. For molecules containing a large number of C-atoms it may be desirable to generate separate subspectra for CH_3, CH_2, and CH groups in order to facilitate the analysis. Described here is the procedure on cholesteryl acetate with complete editing of the three subspectra.

2. Literature

[1] M. R. Bendall, D. M. Doddrell, D. T. Pegg, *J. Am. Chem. Soc.* **1981**, *103*, 4603–4605.

[2] D. M. Doddrell, D. T. Pegg, M. R. Bendall, *J. Magn. Reson.* **1982**, *48*, 323–327.

[1] A. E. Derome, *Modern NMR Techniques for Chemistry Research,* Pergamon, Oxford, **1987**, 145-147.

[4] K. V. Schenker, W. v. Philipsborn, *J. Magn. Reson.* **1986**, *66*, 219–229.

3. Pulse Scheme and Phase Cycle

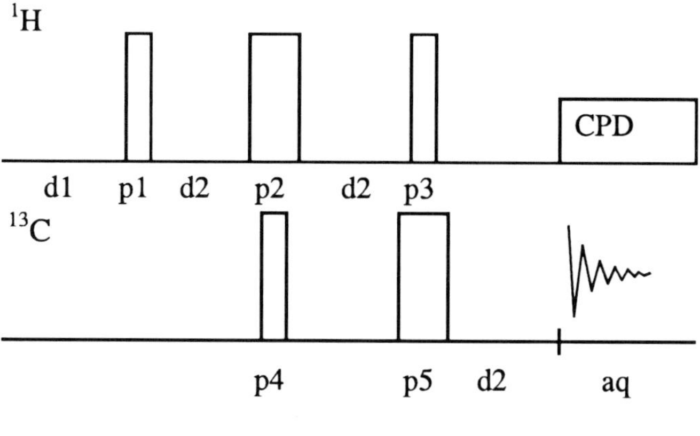

p1: x p3: $(y)_4$, $(-y)_4$ p5: $(x, -x)_4$, $(y, -y)_4$

p2: x, -x, y, -y p4: $(x)_8$, $(y)_8$, $(-x)_8$, $(-y)_8$

aq: $(y)_2$, $(-y)_4$, $(y)_2$, $(-x)_2$, $(x)_4$, $(-x)_2$, $(y)_2$, $(y)_4$, $(-y)_2$, $(x)_2$, $(-x)_4$, $(x)_2$

4. Acquisition

Time requirement: 1.5 h

Sample: 100 mg cholesteryl acetate in $CDCl_3$

Load standard ^{13}C NMR parameters and the DEPT pulse program. You have to set:

td: 64 k
sw: 200 ppm
o1: middle of ^{13}C NMR spectrum
o2: middle of ^{1}H NMR spectrum
p1: 90° ^{1}H decoupler pulse
p2: 180° ^{1}H decoupler pulse
p3: use 45°, 90°, and 135° ^{1}H decoupler pulses for three successive experiments leading to spectra **a**, **b** and **c**. Spectrum **a** will give the signals of CH, CH_2, and CH_3 groups all positive, **b** gives only the signals of CH groups, and the third spectrum **c** gives the signals of CH and CH_3 groups positive and signals of CH_2 groups negative. The second spectrum **b** gives a clear indication of whether the decoupler pulse was determined correctly.
p4: 90° ^{13}C transmitter pulse
p5: 180° ^{13}C transmitter pulse
d1: 2 s
d2: $1/[2J(C,H)] = 3.5$ ms, calculated from $^{1}J(C,H) = 140$ Hz
decoupler attenuation for hard decoupler pulses
decoupler attenuation and 90° pulse for CPD
ns = 512

5. Processing

Use standard 1D processing as described in Experiment 3.2. For editing purposes the three spectra have to be further manipulated. Subtraction of 0.8·**b** from **a** yields spectrum **d**, where the signals of CH_2 and CH_3 groups both remain positive. Subtraction of 0.6·**b** from **c** yields spectrum **e**, where the signals of CH_2 are negative and those of the CH_3 groups remain positive. Subtraction of 1.2·**e** from **d** yields spectrum **f** with only signals of CH_2 groups, whereas addition of 1.2·**e** to **d** yields spectrum **g** with only signals of the CH_3 groups. The factors of 0.8, 0.6, and 1.2 may be finely adjusted according to the exact duration of pulse p3.

6. Result

The figure on page 185 shows the three edited DEPT subspectra **b** (CH groups), **f** (CH_2 groups), and **g** (CH_3 groups), obtained from the three different measurements **a**–**c** on an AM-400 spectrometer and calculated as described.

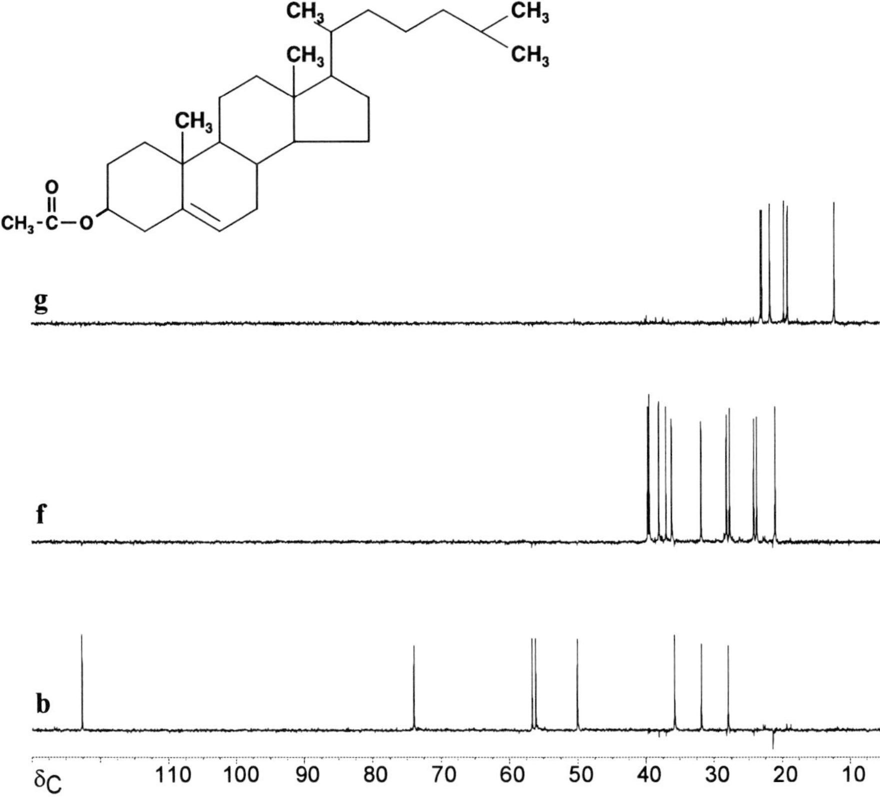

7. Comments

The procedure described in this experiment can also be performed automatically. There are software routines which label after a DEPT-editing experiment all signals of a ^{13}C NMR spectrum with the appropriate characters S, D, T and Q corresponding to the number of protons attached.

8. Own Observations

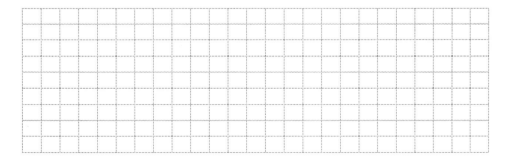

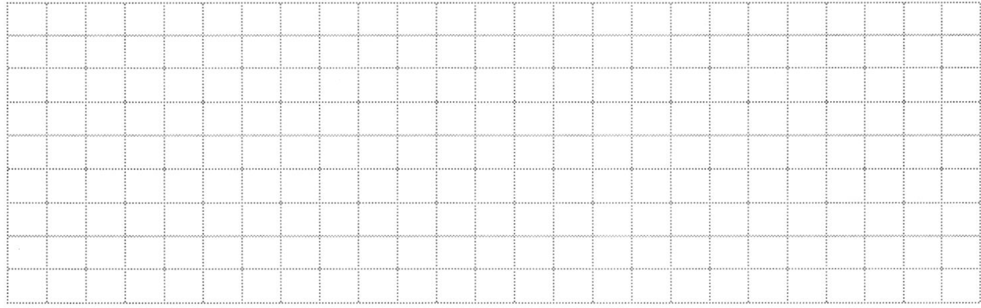

Experiment 6.11

Multiplicity Determination with PENDANT

1. Purpose

There are several methods for distinguishing CH, CH_2, and CH_3 groups in ^1H broadband decoupled ^{13}C NMR spectra. Those most often used are INEPT (Exps. 6.5–6.7), DEPT (Exps. 6.9–6.10) or APT (Exp. 6.4), although all these methods have disadvantages. INEPT and DEPT give no chemical shift information on quaternary carbon atoms, and thus a normal ^{13}C NMR spectrum has to be measured in addition. APT is said to be less sensitive than DEPT, since only the NOE enhancement is operative. The recently introduced PENDANT method (**P**olarization **EN**hancement **D**uring **A**ttached **N**ucleus **T**esting) [1,2] described here is claimed to have the full sensitivity of DEPT and gives signals of quaternary carbon atoms within the same measurement.

2. Literature

[1] J. Homer, M. C. Perry, *J. Chem. Soc. Chem. Commun.* **1994**, 373–374.
[2] J. Homer, M. C. Perry, *J. Chem. Soc. Perkin Trans. 2* **1995**, 533–536.

3. Pulse Scheme with Phasecycle

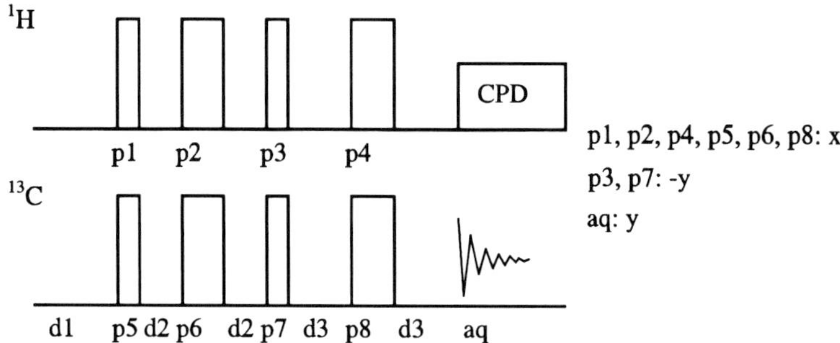

4. Acquisition

Time requirement: 10 min

Sample: 20% ethyl crotonate in $CDCl_3$

Load standard ^{13}C NMR parameters and the pulse program. You have to set:

td: 64 k
sw: 250 ppm

o1: middle of ^{13}C NMR spectrum
o2: middle of ^{1}H NMR spectrum
p1, p3: 90° ^{1}H decoupler pulse
p2, p4: 180° ^{1}H decoupler pulse
p5, p7: 90° ^{13}C transmitter pulse
p6, p8: 180° ^{13}C transmitter pulse
d1: 2 s
d2: $1/[4J(C,H)]$ = 1.72 ms, calculated from $^{1}J(C,H)$ = 145 Hz
d3: $5/[8J(C,H)]$ = 4.31 ms, calculated from $^{1}J(C,H)$ = 145 Hz
decoupler attenuation for hard decoupler pulses
decoupler attenuation and 90° pulse for CPD
de: preaquisition delay as short as possible
ns: 16

5. Processing

Use standard 1D processing as described in Experiment 3.2. Adjust the phase for the signal of the methyl groups to be positive and that of the carboxyl C-atom negative.

6. Result

The figure shows a PENDANT spectrum of ethyl crotonate obtained on an ARX-200 spectrometer.

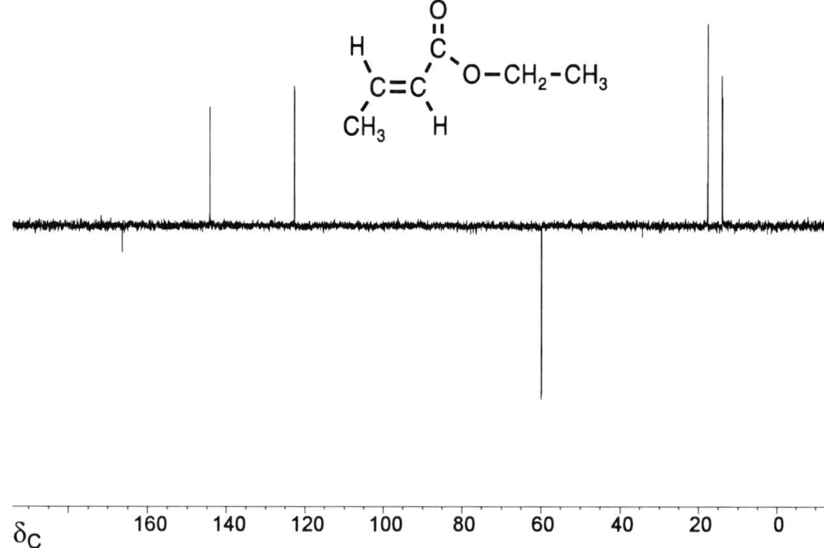

7. Comments

The authors do not provide an explanation in terms of the product operator formalism. One can, however, first consider a quaternary carbon spin I_C, which has no interaction with protons, giving Equation (1).

$$I_{Cz} \xrightarrow{90° I_{Cx}} -I_{Cy} \xrightarrow{180° I_{Cx}} I_{Cy} \xrightarrow{90° I_{C-y}} I_{Cy} \xrightarrow{180° I_{Cx}} -I_{Cy} \quad (1)$$

The first 90° pulse aligns the *z*-magnetization vector along the −y direction. Chemical shift evolution before acquisition can be neglected as the two 180° carbon pulses only change the sign of the operator, and the 90° pulse in the −*y* direction has no effect on the *y*-magnetization, thus the receiver detects a negative signal.

For a CH group we start with $\gamma_H I_{Hz}$, which is transformed into transverse proton magnetization by the first 90° proton pulse, giving Equation (2). During the following two delays d2 C,H spin coupling evolves, which is not refocused due to the simultaneous 180° proton and carbon pulses. Since we set τ equal to d2 = 1/4*J* and use two d2 periods in this sequence, the cosine term becomes zero and the sine term unity. Chemical shift evolution of the protons is refocused by the 180° pulse. Thus at the end of the second delay τ we find antiphase magnetization of the ^{1}H with respect to the ^{13}C.

$$\gamma_H I_{Hz} \xrightarrow{90° I_{Hx}} -\gamma_H I_{Hy} \xrightarrow{\pi J \tau 2 I_{Hz} I_{Cz}} -\gamma_H I_{Hy} \cos(\pi J \tau) + 2\gamma_H I_{Hx} I_{Cz} \sin(\pi J \tau)$$

$$\xrightarrow{180° I_{Hx},\ 180° I_{Cx}} -2\gamma_H I_{Hx} I_{Cz} \xrightarrow{90° I_{H-y},\ 90° I_{Cy}} 2\gamma_H I_{Hz} I_{Cx} \quad (2)$$

The two simultaneous 90° proton and carbon pulses perform the polarization transfer by transforming $-2\gamma_H I_{Hx} I_{Cz}$ into antiphase magnetization of the carbon with respect to the proton. In the final part of the sequence an observable in-phase ^{13}C magnetization is generated. Note that this magnetization bears the factor γ_H, and is thus four times stronger than a normal carbon signal. To choose a suitable d3 value, one has to compromise for CH, CH_2, and CH_3. For a CH group we find that with d3 close to 1/2*J* the sine term disappears, leaving a positive ^{13}C signal at the receiver, giving Equation (3).

$$2\gamma_H I_{Hz} I_{Cx} \xrightarrow{\pi J \mathrm{d3}\ 2 I_{Hz} I_{Cz}} -\gamma_H I_{Cy} \cos(\pi J \mathrm{d3}) + 2\gamma_H I_{Hz} I_{Cx} \sin(\pi J \mathrm{d3}) \quad (3)$$

In summary, the sequence can be viewed as a refocused INEPT with an additional ^{13}C pulse at the beginning.

8. Own Observations

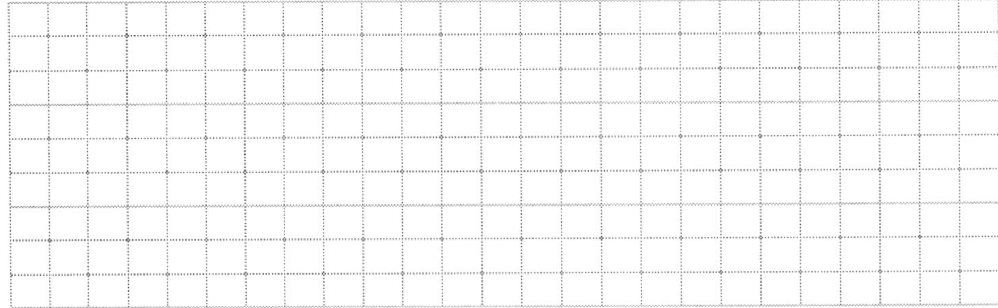

Experiment 6.12

1D-INADEQUATE

1. Purpose

$^{13}C,^{13}C$ spin coupling constants are valuable parameters for the structure elucidation of organic molecules, but are difficult to determine at natural abundance since only every 10 000th molecule contains the necessary two ^{13}C nuclei. Their signals appear as doublets with an intensity 0.5% of that of the singlets from the mono-^{13}C isotopomers, so that only the large ^{1}J(C,C) couplings are accessible under normal conditions. However, the strong singlets can be suppressed by the INADEQUATE (**I**ncredible **N**atural **A**bundance **D**ouble **Q**uantum **T**ransfer **E**xperiment) pulse sequence, so that, at least in principle, even small $^{13}C,^{13}C$ couplings over two or three bonds may be observed. The version given here is the basic experiment tuned to ^{1}J(C,C), leading to antiphase signals [1]. Other variants described elsewhere are SELINQUATE (Exp. 7.7) and 2D-INADEQUATE (Exp. 10.22). 1D variants are refocused INADEQUATE, INEPT-INADEQUATE and DEPT-INADEQUATE [2], as well as a method which uses cross-polarization for signal enhancement [3]. Currently the difficult problem of suppressing the central signal by application of pulsed field gradients is investigated (see Exp. 12.10).

2. Literature

[1] A. Bax, R. Freeman, S. P. Kempsell, *J. Am. Chem. Soc.* **1980**, *102*, 4849–4851.
[2] J. Buddrus, H. Bauer, *Angew. Chem. Int. Ed. Engl.* **1987**, *26*, 625–643.
[3] C. Dalvit, G. Bovermann, *J. Magn. Reson. Ser. A* **1994**, *109*, 113–116.

3. Pulse Scheme and Phase Cycle

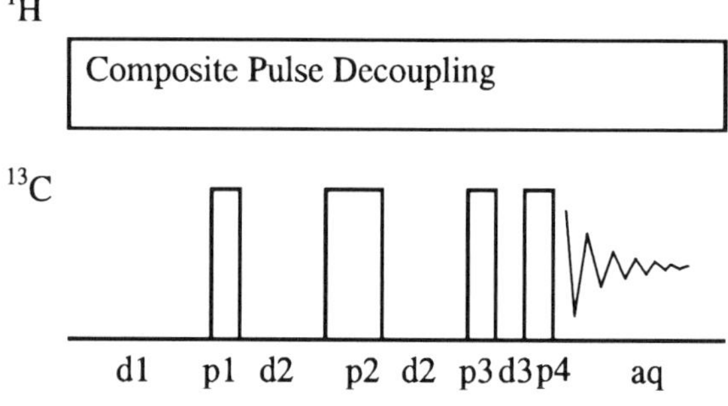

p1: $(x)_4, (y)_4, (-x)_4, (-y)_4, (-x)_4, (-y)_4, (x)_4, (y)_4$

p2: $[(x)_4, (y)_4, (-x)_4, (-y)_4]_2, [(-x)_4, (-y)_4, (x)_4, (y)_4]_2$

p3: $(x)_4, (y)_4, (-x)_4, (-y)_4$

p4: x, -y, -x, y aq: $(x, y, -x, -y, -x, -y, x, y)_2 (-x, -y, x, y, x, y, -x, -y)_2$

4. Acquisition

Time requirement: 1 h

Sample: 90% 1-hexanol in $[D_6]$acetone

Tune the probe-head to the actual sample and record a normal ^{13}C NMR spectrum with ^{1}H broad-band decoupling. Optimize the spectral width and determine the 90° and 180° pulse lengths (see Exp. 2.2) for *this* sample. Load the pulse program and set the following parameters:

td: 32 k
sw: 60 ppm (spectral range for $C_6H_{13}OH$)
o1: 40 ppm downfield from TMS (middle of that range)
o2: middle of ^{1}H NMR spectrum
p1, p3, p4: 90° ^{13}C transmitter pulse
p2: 180° ^{13}C transmitter pulse
d1: 3 s
d2: $1/[4\,J(C,C)]$ = 7.6 ms, calculated from $^{1}J(C,C)$ = 33 Hz
d3: 3 μs
decoupler attenuation and 90° pulse for CPD
ds: 4
ns: 512

5. Processing

Use standard 1D processing as described in Exp. 3.2; apply zero filling to 64 k, use exponential multiplication with lb = 0.5 Hz.

6. Result

The figure on p. 191 shows the ^{13}C-1D-INADEQUATE spectrum of 1-hexanol obtained on an ARX-300 spectrometer using a 5 mm $^{1}H/^{13}C$ dual probe-head. Note the remarkable suppression of the singlets of the mono-^{13}C isotopomers. Closer inspection of the observed splittings (see expansion) shows that some couplings are obviously equal. For instance, the signal at δ_C = 32.4 (C-4) is just one doublet so that an unequivocal assignment based solely on the $^{1}J(C,C)$ values is not possible. Furthermore, doublets may show a"roof effect", characteristic of AB systems (see signal at δ_C = 33.2). The problem of assignment may be overcome by using the 2D version (see Exp. 10.22) in which the various AB spectra are spread out into the second dimension of double quantum frequencies or by using SELINQUATE (Exp. 7.7). As an exercise you may measure the 1D-INADEQUATE spectrum of 2-cyclohexene-1-one, where all C,C couplings are different and clearly resolved. Compare the results with Experiment 7.7.

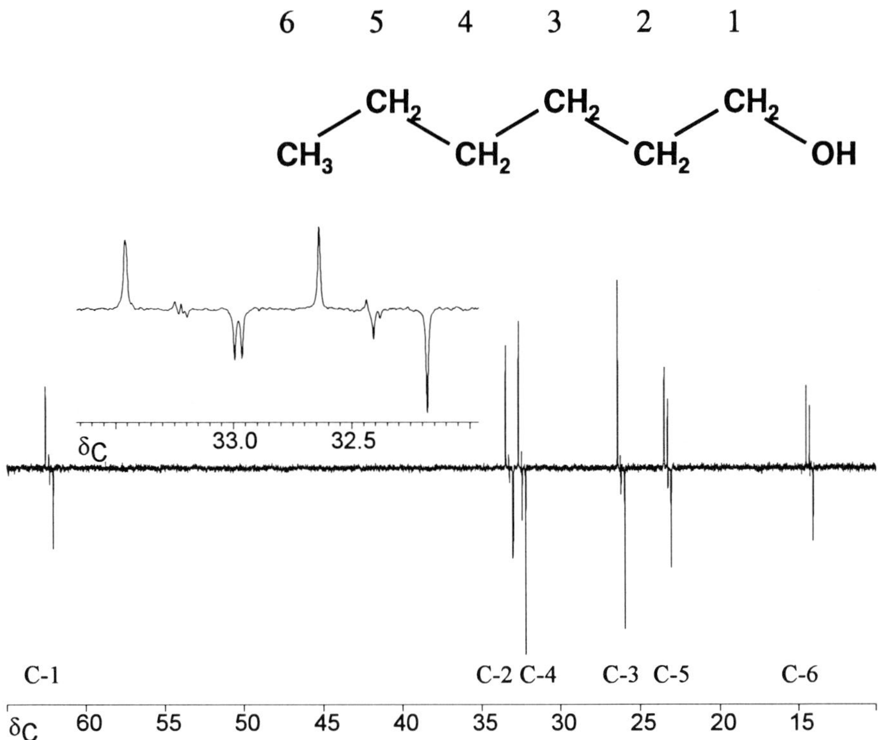

7. Comments

Using the product operator formalism we consider a C,C spin pair. The first pulse p1 creates transverse magnetization, which develops C,C spin coupling during both delays d2. The 180° pulse refocuses the chemical shifts and is for simplicity not shown in the equations. Thus at the end of the spin-echo period we have in-phase and antiphase carbon magnetization multiplied by the respective cosine and sine terms as seen in Equation(1).

$$I_{Cz} \xrightarrow{I_{Cx}} -I_{Cy} \xrightarrow{\pi J \tau 2 I_{Cz} I_{Cz}} -I_{Cy} \cos \pi J \tau \quad +2 I_{Cx} I_{Cz} \sin \pi J \tau \qquad (1)$$

If the delay τ is set equal to 2·d2 = 1/[2*J*(C,C)] the cosine term becomes zero and the sine term unity. The pulse p3 transfers the antiphase magnetization into double quantum magnetization, which is immediatedly transformed back into antiphase magnetization by pulse p4. During acquisition, C,C spin coupling develops again, forming an observable in-phase ^{13}C magnetization which is multiplied by a sine term containing the spin coupling as in Equation (2). Therefore the doublets appear in antiphase.

$$2I_{Cx}I_{Cz} \xrightarrow{I_{Cx}} -2I_{Cx}I_{Cy} \xrightarrow{I_{Cx}} 2I_{Cx}I_{Cz} \xrightarrow{\pi J\,\mathrm{aq}\; 2I_{Cz}I_{Cz}} I_{Cy}\sin\pi J\mathrm{aq} \quad (2)$$

The mechanism for suppressing the central signal is based on the fact that the desired observable coherences must have passed the double quantum filter. This gives them a phase response different from that of ^{13}C signals from molecules which contain only one ^{13}C atom, and thus cannot develop spin–spin coupling. The sequence can therefore be viewed as a spin-echo method with a subsequent homonuclear double quantum filter. For a heteronuclear double quantum filter see Experiment 6.15. It is possible to set the delays d2 according to Equation (3); in cases where the ^{13}C signals have only a small chemical shift difference it is advisable to use n = 1 or 2.

$$d2 = (2n + 1)/[4J(C,C)]\ , n = 0, 1, 2... \quad (3)$$

8. Own Observations

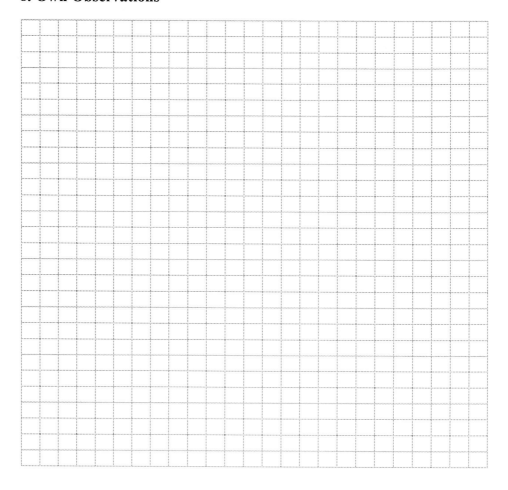

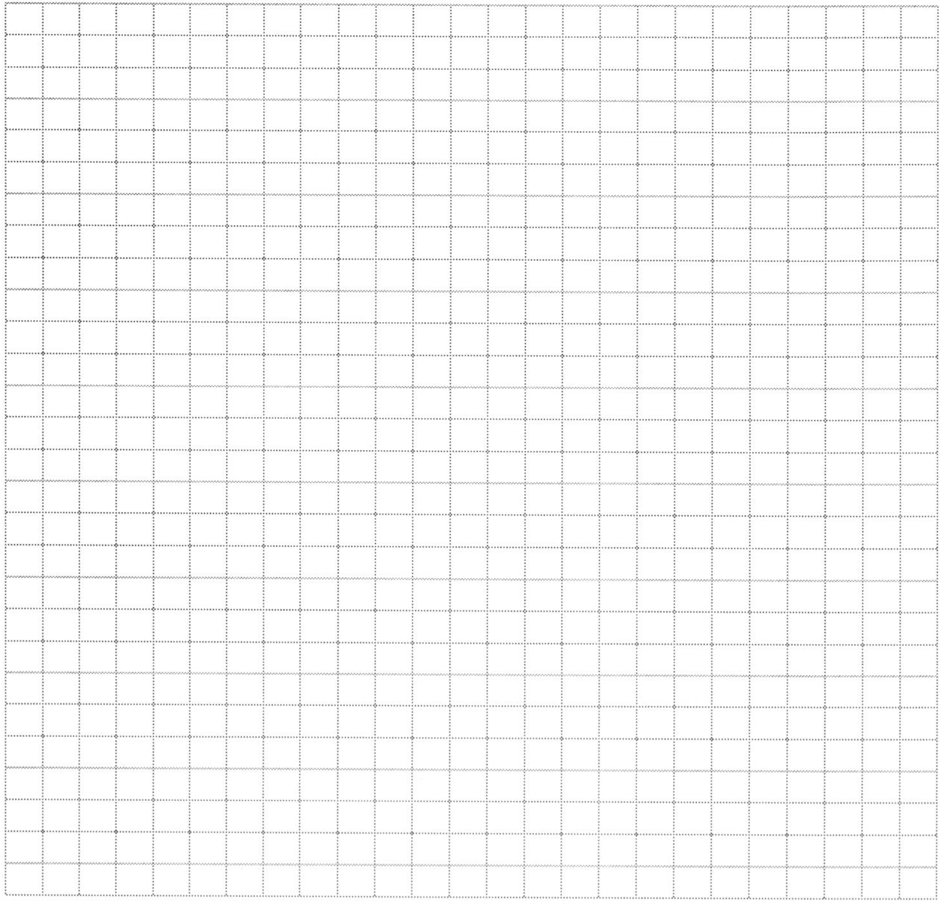

Experiment 6.13

The BIRD Filter

1. Purpose

In many experiments one wants to selectively observe protons that are attached to ^{13}C or ^{15}N. The strong signals of those protons attached to ^{12}C or ^{14}N need to be suppressed prior to the actual pulse sequence in order to be able to adjust the receiver gain for the desired signals only. One method of discriminating between these two kinds of signals is to insert the BIRD (**BI**linear **R**otation **D**ecoupling) sandwich, which rotates the magnetization of the protons attached to ^{12}C into the $-z$-direction of the rotating frame, but leaves the magnetization of the ^{13}C-bound protons unchanged. If one waits a suitable time after the BIRD sandwich, the signals of the former are at the null point and therefore not excited during the following pulse sequence. In this educational experiment the use of the BIRD sandwich is shown for chloroform.

2. Literature

[1] J. R. Garbow, D. P. Weitekamp, A. Pines, *Chem. Phys. Lett.* **1982**, *93*, 504–508.

3. Pulse Scheme and Phase Cycle

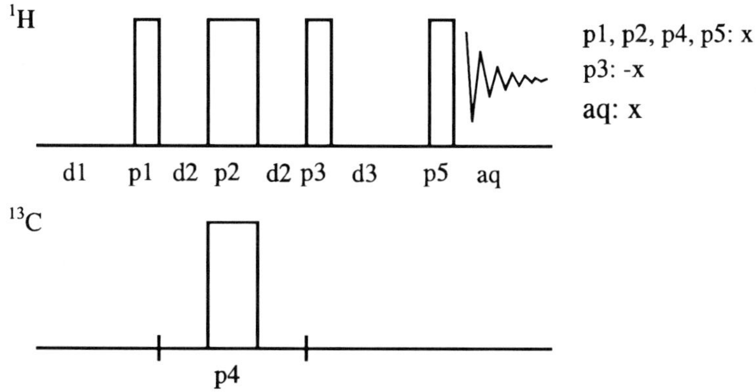

Note that the actual BIRD sandwich consists of the pulses p1 to p4, whereas p5 is only used for detection.

4. Acquisition

Time requirement: 10 min

Sample: 10% $CHCl_3$ in [D_6]acetone

Record normal ^{13}C and ^{1}H NMR spectra and note the offsets of $CHCl_3$. Set the instrument to ^{1}H observation with ^{13}C decoupling (inverse mode on older instruments) and load the BIRD pulse program. You have to set:

td: 4 k
sw: 500 Hz
o1: on ^{1}H resonance
o2: on ^{13}C resonance
p1, p3, p5: 90° ^{1}H transmitter pulse
p2: 180° ^{1}H transmitter pulse
p4: 180° ^{13}C decoupler pulse
d1: 60 s
d2: $1/[2J(C,H)]$ = 2.38 ms, calculated from ^{1}J (C,H) = 214 Hz
d3: 20 s, to be varied
ns: 1

Observe the incoming FID and adjust d3 until you find a minimum intensity, adjust the receiver gain accordingly.

5. Processing

Use standard ^{1}H processing as described in Experiment 3.1 with an exponential multiplication of lb = 1 Hz.

6. Result

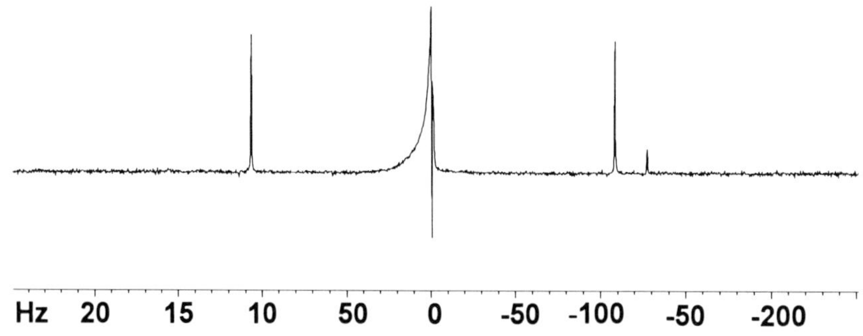

The figure shows the result obtained on an ARX-200 spectrometer with a normal forward dual probe-head. Note that the BIRD filter has suppressed the central line roughly to the height of the ^{13}C satellites. The actual d3 value was 90 s (degassed and sealed sample); it depends greatly on the oxygen content of the sample.

7. Comments

The BIRD sandwich can be understood both either from usual vector diagrams or with the product operator formalism. With the latter we find for a proton bound to ^{12}C:

$$I_{Hz} \xrightarrow{90^\circ I_{Hx}} -I_{Hy} \xrightarrow{180^\circ I_{Hx}} I_{Hy} \xrightarrow{90^\circ I_{H-x}} -I_{Hz} \quad (1)$$

Since this proton develops no spin coupling its magnetization vector reaches the $-z$-direction after the BIRD sandwich as seen in Equation (1). Proton chemical shifts are refocused by the 180° ^{1}H pulse. In contrast, protons bound to ^{13}C develop spin coupling, and due to the two simultaneous ^{1}H and ^{13}C 180° x-pulses this is also refocused. By setting the delay τ equal to d2 = 1/[2J(C,H)] the cosine terms become zero and the sine terms unity:

$$I_{Hz} \xrightarrow{90^\circ I_{Hx}} -I_{Hy} \xrightarrow{\pi J \tau I_{Hz} I_{Cz}} -I_{Hy}\cos(\pi J \tau) + 2I_{Hx}I_{Cz}\sin(\pi J \tau) =$$

$$2I_{Hx}I_{Cz} \xrightarrow{180^\circ I_{Hx}} \xrightarrow{180^\circ I_{Cx}} -2I_{Hx}I_{Cz} \xrightarrow{\pi J \tau I_{Hz} I_{Cz}}$$

$$-2I_{Hx}I_{Cz}\cos(\pi J \tau) - I_{Hy}\sin(\pi J \tau) = -I_{Hy} \xrightarrow{90^\circ I_{H-x}} I_{Hz} \quad (2)$$

As can be seen from Equation (2), the magnetization vector of these protons is returned into the + z-direction.

Note that for typical organic or bioorganic applications the BIRD delay d3 is much shorter than for the degassed example used here; typically one finds d3 values in the order of 0.5 s.

8. Own Observations

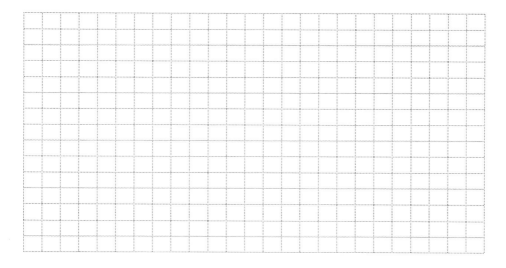

Experiment 6.14

TANGO

1. Purpose

The TANGO sequence (**T**esting for **A**djacent **N**uclei with a **G**yration **O**perator) acts as a 90° pulse for protons bound to ^{13}C, whereas the magnetization vector of protons bound to ^{12}C, which have no or only long-range couplings to ^{13}C remain in the positive *z*-direction. In contrast the BIRD sandwich (see Exp. 6.13) separates protons bound to ^{13}C and protons bound to ^{12}C so that their magnetization vectors are in the positive and negative *z*-direction. Thus, the TANGO sandwich introduces an 90° phase angle between these two sorts of proton spins whereas the BIRD sandwich an angle of 180°. The sequence is used, like BIRD, as a basic building block in many pulse techniques. In this educational experiment we demonstrate the use of the sequence on $CHCl_3$.

2. Literature

[1] S. Wimperis, R. Freeman, *J. Magn. Reson.* **1984**, *58*, 348–353.
[2] T. Parella, F. Sanchez-Ferrando, A. Virgili, *J. Magn. Reson. Ser. A* **1995**, *112*, 241–245.

3. Pulse Scheme with Phasecycle

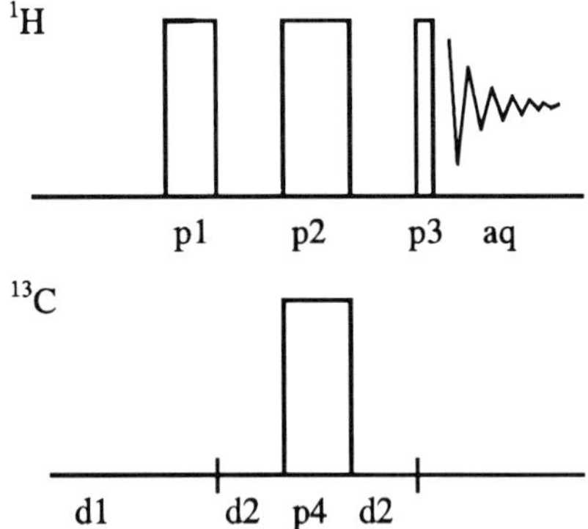

p1, p2, p3: $(x)_2$, $(-x)_2$, $(y)_2$, $(-y)_2$

p4: x, -x, -x, x, y, -y, -y, y

aq: $(x)_2$, $(-x)_2$, $(y)_2$, $(-y)_2$

4. Acquisition

Time requirement: 10 min

Sample: 10% $CHCl_3$ in [D_6]acetone

Obtain first ^{1}H and ^{13}C spectra of the sample and note the offsets of the $CHCl_3$ signals. The instrument must be in the inverse mode, using the proton channel as transmitter and the ^{13}C channel as decoupler. You have to set:

td: 4 k
o1: on ^{1}H resonance
o2: on ^{13}C resonance
sw: 500 Hz
p1: 135° ^{1}H transmitter pulse
p2: 180° ^{1}H transmitter pulse
p3: 45° ^{1}H transmitter pulse
p4: 180° ^{13}C decoupler pulse
d1: 20 s
d2: $1/[2J(C,H)]$ = 2.38 ms, calculated from $^{1}J(C,H)$ = 214 Hz
ns: 8

5. Processing

Use standard ^{1}H processing as described in Experiment 3.1 with exponential multiplication of lb = 1 Hz.

6. Result

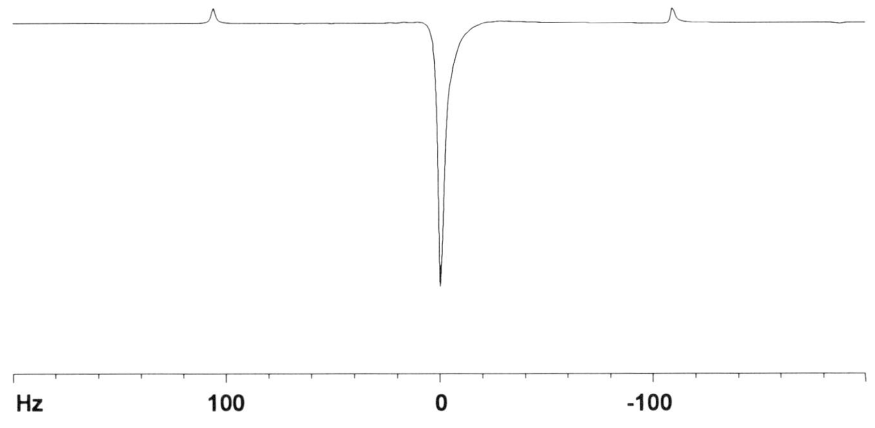

The figure shows the result obtained on an AMX-500 spectrometer with an inverse probe-head. Note that the suppression of the central signal is less efficient compared to the BIRD sequence of Experiment 6.13, since only a slight misadjustment of the pulse-length causes a central signal.

7. Comments

For analysis with the product operator formalism we consider first a proton bound to ^{12}C and thus not being able to develop C,H spin coupling. The effects of chemical shifts are neglected since they are removed by the 180° pulses. The first 135° proton pulse creates transverse magnetization and leaves part of the *z*-magnetization. The 180° proton pulse changes the signs of the terms, then the final 45° proton pulse creates four magnetizations, which simplify due to the different sine and cosine terms to give I_{Hz} unchanged. Thus, for protons without or small C,H spin coupling the TANGO sandwich acts as a 360° pulse, as indicated in Equation (1).

$$I_{Hz} \xrightarrow{135° I_{Hx}} -I_{Hy}\sin(135) + I_{Hz}\cos(135) \xrightarrow{180° I_{Hx}} I_{Hy}\sin(135)$$

$$-I_{Hz}\cos(135) \xrightarrow{45° I_{Hx}} I_{Hy}\sin(135)\cos(45) + I_{Hy}\cos(135)\sin(45)$$

$$-I_{Hz}\cos(135)\cos(45) + I_{Hz}\sin(135)\sin(45) = I_{Hz} \tag{1}$$

In contrast, a proton bound to ^{13}C develops C,H spin coupling after the first 45° pulse. Since both ^{1}H and ^{13}C 180° pulses are applied the spin coupling is not refocused but develops further in the second d2 period. If the delay τ is set equal to $2 \cdot d2 = 1/J$ the corresponding sine terms become zero and the cosine terms −1. Thus, compared with Equation (1), we observe a sign change before the last proton pulse as indicated in Equation (2). This again creates four magnetizations which simplify to $-I_{Hy}$, and therefore the TANGO sandwich acts like a 90° pulse for these protons.

$$I_{Hz} \xrightarrow{135° I_{Hx}} -I_{Hy}\sin(135) + I_{Hz}\cos(135) \xrightarrow{\pi J \tau 2 I_{Hz} I_{Cz}} -I_{Hy}\sin(135)$$

$$+ I_{Hz}\cos(135) \xrightarrow{45° I_{Hx}} -I_{Hy}\sin(135)\cos(45) + I_{Hy}\cos(135)\sin(45)$$

$$-I_{Hz}\cos(135)\cos(45) - I_{Hz}\sin(135)\sin(45) = -I_{Hy} \tag{2}$$

Note that both the BIRD and TANGO sandwiches can be used in a reversed sense by changing the appropriate pulse phases. Then TANGO would act as a 90° pulse for protons bound to ^{12}C and BIRD as a 180° pulse for protons bound to ^{13}C.

8. Own Observations

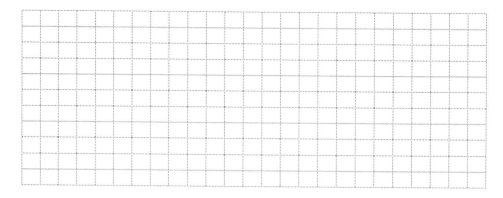

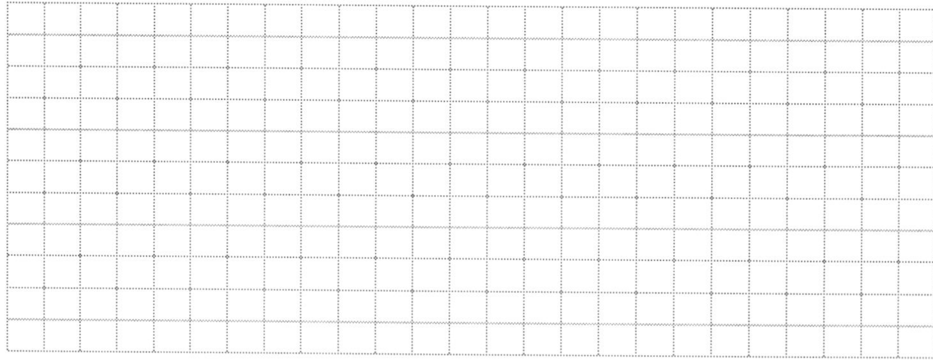

Experiment 6.15

The Heteronuclear Double Quantum Filter

1. Purpose

In many experiments one wants to selectively observe protons that are attached to ^{13}C or ^{15}N. The strong signals of protons attached to ^{12}C or ^{14}N need to be suppressed in order to be able to detect the weak desired signals. One method of discriminating between these two kinds of signals is to use the double quantum filter. It is applied in many pulse sequences and consists essentially of two 90° pulses. Double quantum magnetization passes through this filter, whereas single quantum magnetization is filtered out by the phase cycle. In this educational experiment the use of a heteronuclear double quantum filter is shown for chloroform, an experiment which, in addition, reveals any instability of the spectrometer. In the homonuclear case one type of application is the suppression of a strong solvent signal, e. g. that of water (see Exp 10.7), or of the central signal in 1D-INADEQUATE (see Exp. 6.12).

2. Literature

[1] R. Freeman, *A Handbook of Nuclear Magnetic Resonance*, Longman Scientific Technical, Harlow, **1987**, p. 128–133.
[2] S. W. Homans, *A Dictionary of Concepts in NMR,* Revised Edition, Clarendon Press, Oxford, **1993**, 93–106.

3. Pulse Scheme and Phase Cycle

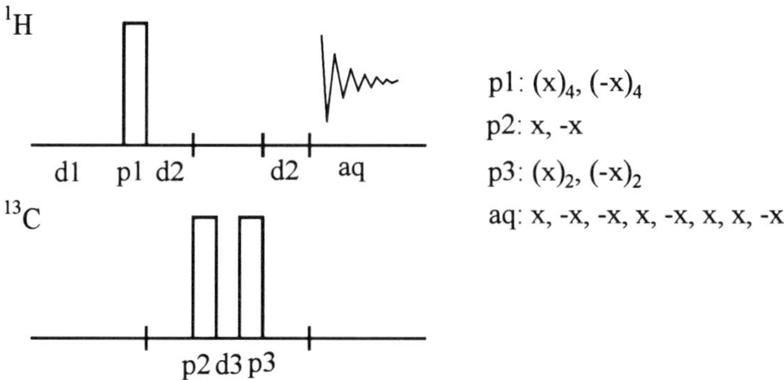

4. Acquisition

Time requirement: 0.5 h

Sample: 10% $CHCl_3$ in $[D_6]$acetone

The instrument must be in the inverse mode, using the proton channel as transmitter and the ^{13}C channel as decoupler. First obtain ^{1}H and ^{13}C spectra of the sample and note the offsets of the $CHCl_3$ signals. You have to set:

td: 4 k
sw: 500 Hz
o1: on ^{1}H resonance
o2: on ^{13}C resonance
p1: 90° ^{1}H transmitter pulse
p2, p3: 90° ^{13}C decoupler pulse
d1: 200 s
d2: $1/[2J(C,H)] = 2.38$ ms, calculated from $^{1}J(C,H) = 214$ Hz
d3: 10 μs
ns: 8

5. Processing

Use standard ^{1}H processing as described in Experiment 3.1.

6. Result

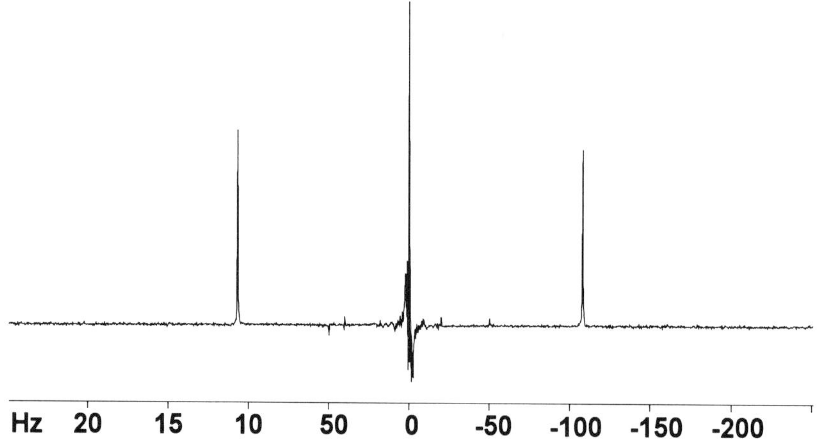

The figure shows the result obtained on an ARX-200 spectrometer with a normal forward dual probe-head. Note that the double quantum filter has suppressed the central line roughly to the height of the ^{13}C satellites.

7. Comments

The double quantum filter can be only understood with the product operator formalism or the density matrix approach. With the former we find for a proton bound to ^{13}C:

$$I_{Hz} \xrightarrow{90° I_{Hx}} -I_{Hy} \xrightarrow{\pi J \tau 2 I_{Hz} I_{Cz}} -I_{Hy}\cos(\pi J \tau) + 2 I_{Hx} I_{Cz} \sin(\pi J \tau) =$$

$$2I_{Hx}I_{Cz} \xrightarrow{90°I_{Cx}} -2I_{Hx}I_{Cy} \xrightarrow{90°I_{Cx}} -2I_{Hx}I_{Cz} \xrightarrow{\pi J \tau 2I_{Hz}I_{Cz}}$$

$$-2I_{Hx}I_{Cz}\cos(\pi J\tau) - I_{Hy}\sin(\pi J\tau) = -I_{Hy} \quad (1)$$

By setting the delay τ equal to d2 = 1/[2*J*(C,H)] the cosine terms become zero and the sine terms unity. As can be seen from Equation (1), the double quantum magnetization $-2I_{Hx}I_{Cy}$ is generated by the first ^{13}C pulse, whereas the second ^{13}C pulse converts this back into antiphase magnetization $-2I_{Hx}I_{Cz}$. Spin coupling evolution within the second d2 period reverts this to an observable single quantum magnetization $-I_{Hy}$. In comparison, a proton bound to ^{12}C does not develop spin coupling, and hence remains after the first proton pulse also as $-I_{Hy}$. However, if in the second scan the phase of the first ^{13}C pulse is changed along with the receiver phase, all signals of protons bound to ^{12}C are cancelled, whereas signals which developed via double quantum magnetization are accumulated.

Note that for typical organic or bioorganic applications the relaxation delay d1 is much shorter than for the sample used here. As an additional exercise you may combine the BIRD filter (Exp 6.13) with the double quantum filter, which should improve the signal suppression further.

8. Own Observations

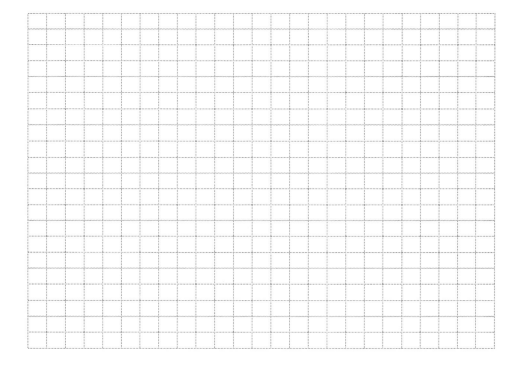

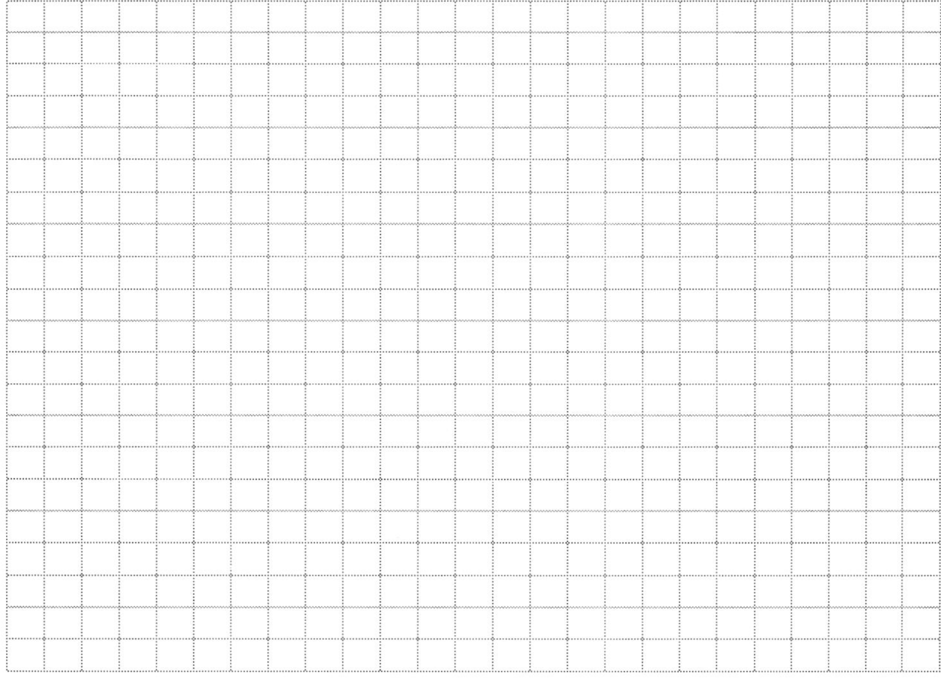

Experiment 6.16

Purging with a Spin-Lock Pulse

1. Purpose

In many experiments one wants to selectively observe protons that are attached to ^{13}C or ^{15}N. The strong signals of protons attached to ^{12}C or ^{14}N need to be suppressed in order to be able to adjust the receiver gain for the desired signals only. It is possible to dephase the undesired signal by the use of a spin-lock purging pulse. This acts like a pulsed field gradient (see Chap. 11 and 12), however not on the main field B_0 but on the r.f. field B_1. Magnetization having the same phase as the spin-lock pulse will be unaffected. The technique is used in many advanced pulse methods, such as in Experiments 12.6 and 12.10, and provides the basis for the PMG method (**P**oor **M**an's **G**radient) as described in Experiment 10.16. In this educational experiment the purging with a spin-lock pulse is shown for chloroform.

2. Literature

[1] G. Otting, K. Wüthrich, *J. Magn. Reson.* **1988**, *76*, 569-574.

[2] J.-M. Nuzillard, G. Gasmi, J.-M. Bernassau, *J. Magn. Reson. Ser. A* **1993**, *104*, 83–87.

[3] P. Mutzenhardt, J. Brondeau, D. Canet, *J. Magn. Reson. Ser. A* **1994**, *108*, 110–115.

3. Pulse Scheme and Phase Cycle

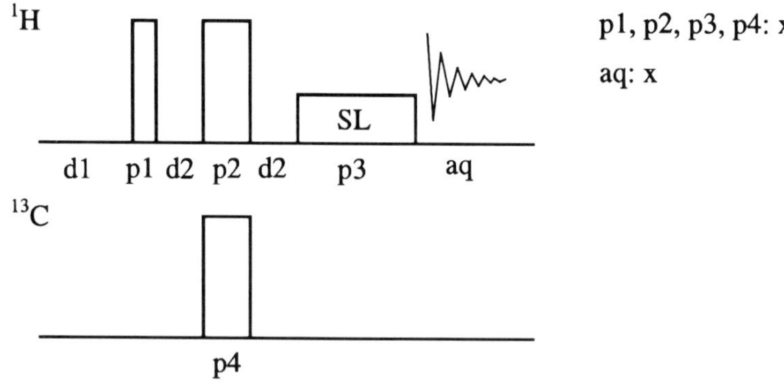

4. Acquisition

Time requirement: 5 min

Sample: 3% $CHCl_3$ in [D_6]acetone

The instrument must be in the inverse mode, using the proton channel as transmitter and the ^{13}C channel as decoupler and must provide fast transmitter power switching. First obtain ^{1}H and ^{13}C spectra of the sample and note the offsets of the $CHCl_3$ signals. For best results you should determine and correct a possible phase difference between the hard transmitter pulses and the spin-lock pulse (see Exp. 7.1). You have to set:

td: 4 k
sw: 500 Hz
o1: on ^{1}H resonance
o2: on ^{13}C resonance
p1: 90° ^{1}H transmitter pulse
p2: 180° ^{1}H transmitter pulse
p3: ^{1}H transmitter spin-lock pulse, 10 ms length at typically 20 dB attenuation
p4: 180° ^{13}C decoupler pulse
d1: 10 s
d2: $1/[4J(C,H)] = 1.16$ ms, calculated from $^{1}J(C,H) = 215$ Hz
transmitter attenuation for hard pulses [3 dB]
ds: 4
ns: 1

5. Processing

Use standard ^{1}H processing as described in Experiment 3.1.

6. Result

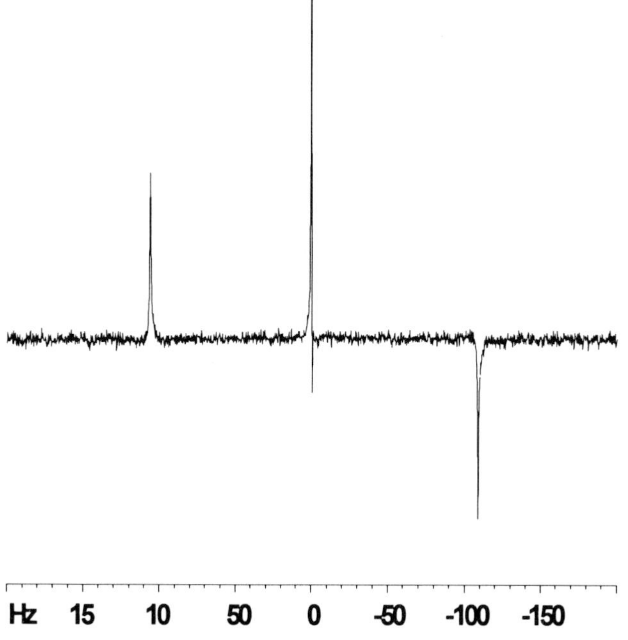

The figure shows the result obtained on an AMX-500 spectrometer with a multinuclear inverse probe-head. Compare the result with the other methods described in this book to achieve a separation between signals of protons bound to ^{13}C and those bound to ^{12}C (see Exp. 6.13–6.15 and 11.7).

7. Comments

The method can best be understood with the product operator formalism. Neglecting the 180° pulses, which refocus the chemical shifts and the heteronuclear spin coupling, we find, for a proton bound to ^{13}C, the result as given in Eq. (1), since by setting the delay 2·d2 = $\tau = 1/[2J(C,H)]$ the cosine term becomes zero and the sine term unity.

$$I_{Hz} \xrightarrow{90° I_{Hx}} -I_{Hy} \xrightarrow{\pi J \tau 2 I_{Hz} I_{Cz}} -I_{Hy}\cos(\pi J\tau) + 2I_{Hx}I_{Cz}\sin(\pi J\tau) = 2I_{Hx}I_{Cz} \quad (1)$$

A proton bound to ^{12}C cannot develop heteronuclear spin coupling and stays as $-I_{Hy}$. A spin-lock pulse with *x*-phase dephases this magnetization depending on the spin-lock strength and length, whereas the wanted magnetization $2I_{Hx}I_{Cz}$ stays spin-locked. During acquisition, in-phase magnetization $I_{Hy}\sin(\pi J aq)$ develops, yielding the anti-phase signals as observed in the figure on page 203.

As an additional exercise you may change the phase of the spin-lock pulse to *y*; the ^{13}C satellites will disappear and only the main signal remains. In recent literature a "hard" spin-lock of 2 ms at 3dB is also often used.

8. Own Observations

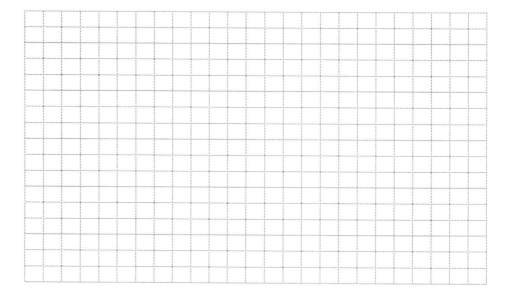

Experiment 6.17

Water Suppression by Presaturation

1. Purpose

For biological and biochemical applications ^{1}H NMR spectra usually have to be recorded in normal water, with the addition of only 10% D_2O to provide the necessary lock signal. Higher D_2O content would cause the signals of the exchangeable NH protons to dissapear. Thus, there is a need to suppress the huge solvent signal. For this purpose a multitude of techniques have been proposed. However, all techniques require that the magnet is well shimmed and a probe-head yielding a reasonable line-shape (see Exp. 3.5) must be available. In this experiment the presaturation method is described. It provides a check on whether the spectrometer set-up is capable of effective water suppression. For other methods see Experiments 6.18, 11.13 and 11.14.

2. Literature

[1] M. Guéron, P. Plateu, M. Decorps, *Prog. NMR Spectrosc.* **1991**, *23*, 135–209.
[2] P. J. Hore, *Methods Enzym.* **1989**, *176*, 64–89.

3. Pulse Scheme and Phase Cycle

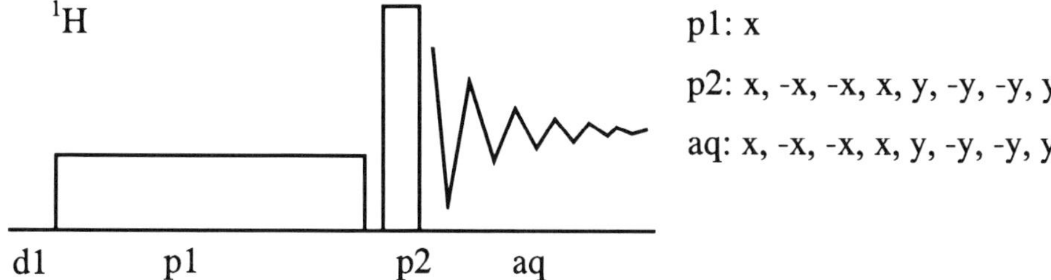

p1: x
p2: x, -x, -x, x, y, -y, -y, y
aq: x, -x, -x, x, y, -y, -y, y

4. Acquisition

Time requirement: 15 min

Sample: 2 mmol sucrose in 90% H_2O / 10% D_2O + 0.5 mmol DSS (2,2-dimethyl-2-silapentane-5-sulfonate, sodium salt) + trace of NaN_3 against bacteria growth

The magnet is well shimmed on the water sample. A normal ^{1}H NMR spectrum is recorded and the offset of the transmitter adjusted on the water resonance. The program for presaturation is loaded, where a weak transmitter pulse p1 of 2 s duration is used for irradiation of the water signal.

On older instruments power switching of the transmitter is often not possible, and therefore the decoupler has to be used as the pre-irradiation source, which may give inferior results. If possible, the decoupler channel should be set to phase coherence with the transmitter. You have to set:

td: 32 k
sw: 10 ppm
o1: on resonance of water signal
p1: 2 s, presaturation pulse at transmitter power level corresponding to $\gamma B_1 \approx$ 25 Hz (90° pulse of ≈10 ms, about 65 dB, see Exp. 2.9)
p2: 90° ^{1}H transmitter pulse
d1: 100 ms
rg: receiver gain for correct ADC input
for inverse probeheads: spinner off
ds: 2
ns: 8

5. Processing

Use standard 1D processing as described in Experiment 3.1, no window multiplication should be used.

6. Result

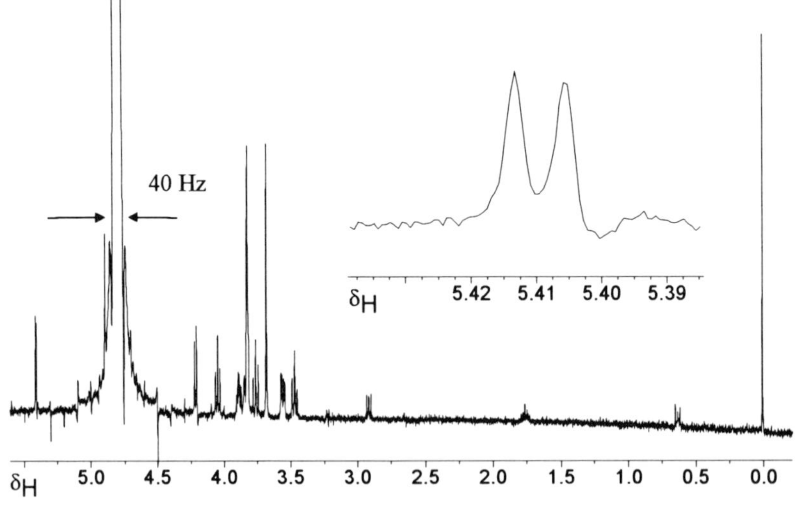

The figure on page 206 shows the result of the above procedure using an inverse probe-head on an AMX-500 spectrometer. The quality of the result is checked by two observations, first the line-width of the residual water signal at half height of the DSS signal and second the signal-to-noise ratio and resolution of the doublet of the anomeric proton under these conditions. The residual water line-width should be below 100 Hz and the splitting of the anomeric signal at $\delta_H = 5.41$ should be visible at least down to 40% of the signal height. Note that an even better supression of the water signal could be obtained, although at the cost of the nearby signal of the anomeric proton.

7. Comments

The water suppression method shown here has the drawback that exchanging NH protons are also saturated and can therefore disappear from the spectrum. There are many other water suppression techniques in the literature; one, the jump and return sequence, is demonstrated in Experiment 6.18. More recent techniques work with coherence selection by pulsed field gradients, see the Experiments 11.13 and 11.14.

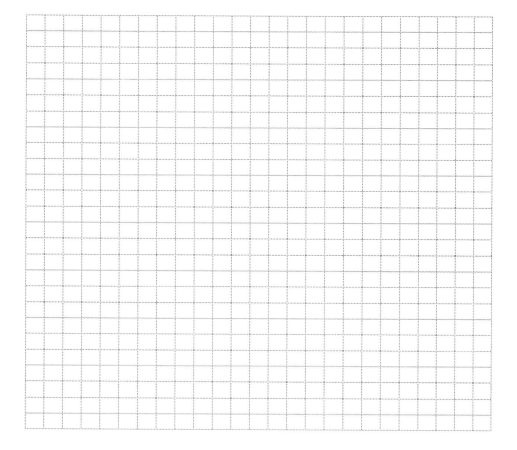

8. Own Observations

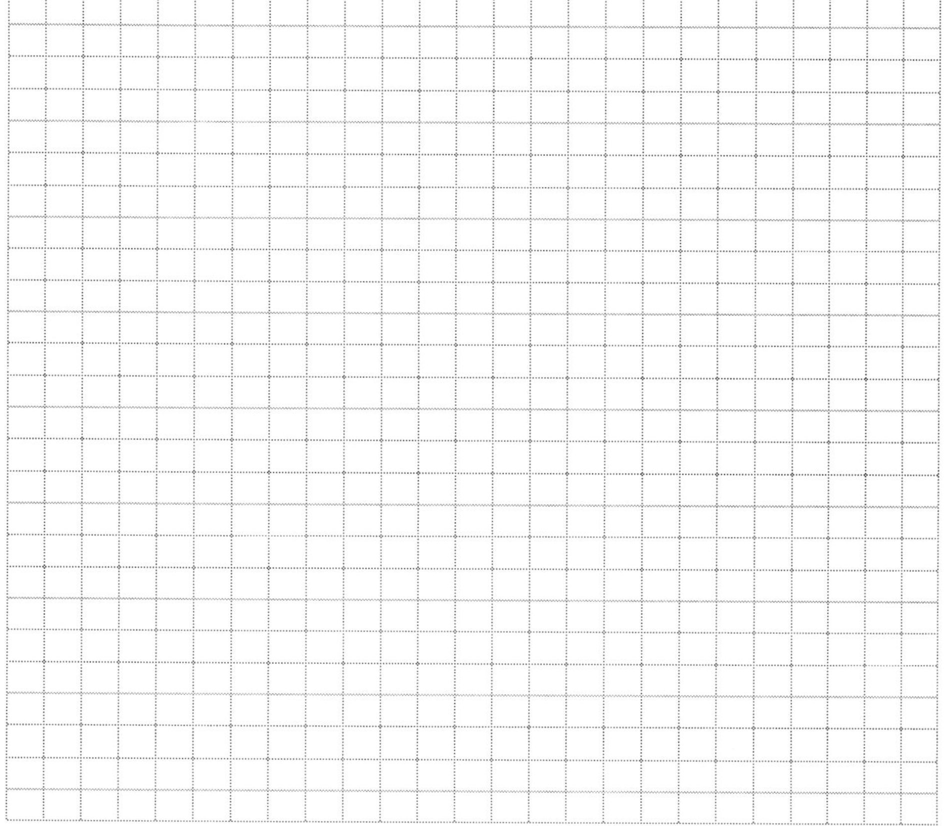

Experiment 6.18

Water Suppression by the Jump and Return Method

1. Purpose

For biological and biochemical applications ^{1}H NMR spectra usually have to be recorded in normal water with the addition of only 10% D_2O to provide the necessary lock signal. Higher D_2O content would cause the exchangeable NH protons to disappear. Thus, there is a need to suppress the huge solvent signal. For this purpose a multitude of techniques have been proposed [1]. The presaturation method (Exp. 6.17) does also affect exchangeable protons, but this is not true for the jump and return technique [2] described here. A method using pulsed field gradients is described in Experiments 11.13 and 11.14.

2. Literature

[1] Guéron, P. Plateu, M. Decorps, *Prog. NMR Spectrosc.* **1991**, *23,* 135–209.
[2] Guéron, P. Plateu, *J. Am. Chem. Soc.* **1982**, *104*, 7310–7311.

3. Pulse Scheme and Phase Cycle

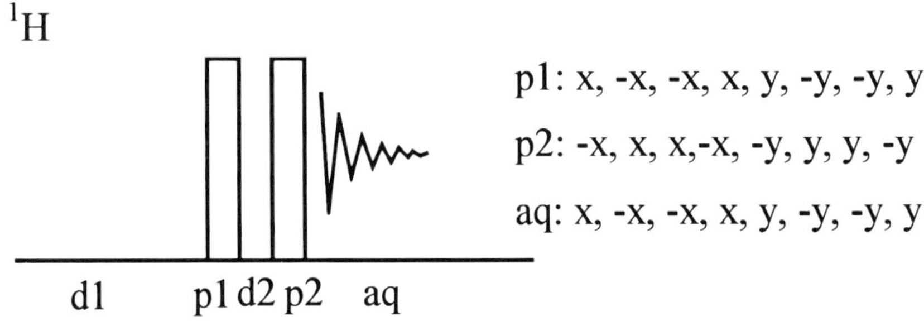

4. Acquisition

Time requirement: 15 min

Sample: 2 mmol sucrose in 90% H_2O / 10% D_2O + 0.5 mmol DSS (2,2-dimethyl-2-silapentane-5-sulfonate, sodium salt) + a trace of NaN_3 (against bacteria growth).

The magnet is well shimmed on the water sample. A normal ^{1}H NMR spectrum is recorded and the offset of the transmitter adjusted on the water resonance. You have to set:

td: 32 k
sw: 10 ppm
o1: on resonance of water signal
p1: 90° ^{1}H transmitter pulse
p2: 90° ^{1}H transmitter pulse
d1: 2 s
d2: 125 μs
rg: receiver gain for correct ADC input
ds: 2
ns: 8

5. Processing

Use standard 1D processing as described in experiment 3.1

6. Result

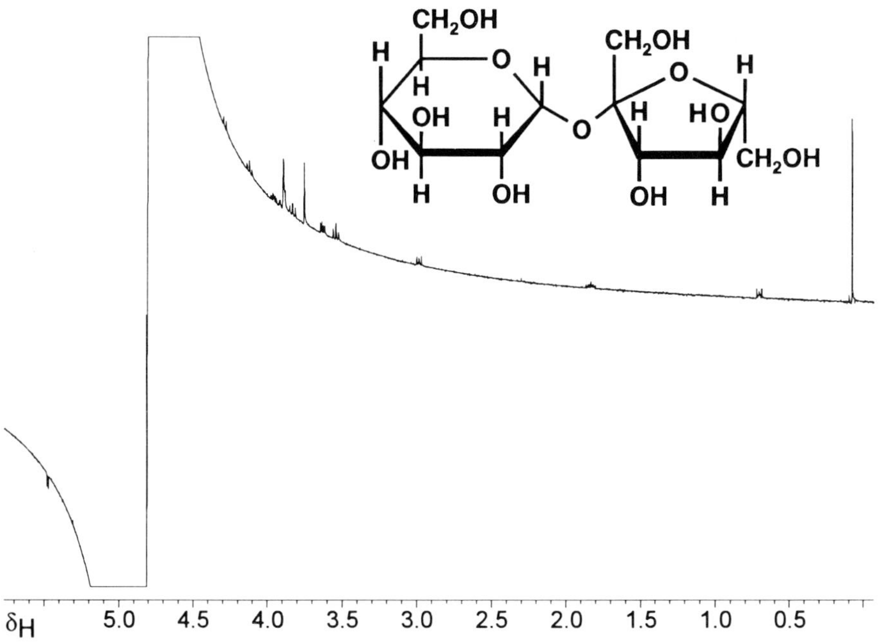

The figure shows the result of the above procedure using an inverse probe-head on an AMX-500 spectrometer. Note that with the jump and return method there is a phase change of 180° at the water resonance position. Compared with the performance of the presaturation method the result is less convincing. For improvement, it has been suggested to increase the pulse length of both pulses to about 20 μs (higher transmitter attenuation) and to decrease the pulse-length of p2 relative to p1 by a small fraction.

7. Comments

The sequence can be easily understood using the classical vector picture. The first pulse aligns all magnetization vectors in the $-y$-direction, where they start to fan out corresponding to their chemical shifts. Only the water signal has no chemical shift with respect to the rotating frame, and thus the second pulse brings it back to the z-direction, yielding in theory no signal during acquisition.

Compare the performance of this sequence with the result of the presaturation method described in Experiment 6.17 and the gradient techniques described in Experiments 11.13 and 11.14.

8. Own Observations

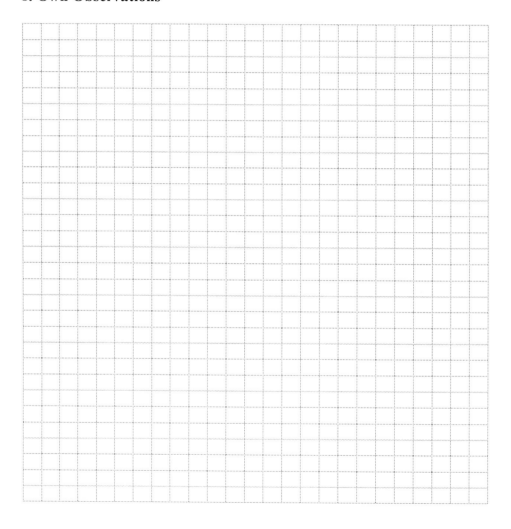

Chapter 7

NMR Spectroscopy with Selective Pulses

The traditional method of continous wave NMR spectroscopy was in principle based on selective excitation. In the field- or frequency-swept instruments of the early days of NMR spectroscopy, each signal of a spectrum was selectively excited in turn when its resonance condition was met. With the advent of pulse Fourier transform spectroscopy these techniques were largely replaced so that all signals of a spectrum are excited non-selectively at the same time by a radiofrequency pulse. According to the Fellgett principle this leads to a much higher sensitivity than could be reached with continous wave-instruments.

However, for some applications it would be extremely useful to be able to excite a particular signal and transfer selected coherences to other spins. Therefore, in the early eighties the use of selective "shaped" or "soft" r.f. pulses of low power and relatively long duration was introduced. Their use in combination with the non-selective "hard" pulses (short rectangular high-power pulses) lead to entirely new possibilities in NMR spectroscopy.

There are some hardware requirements which have to be met before one is able to apply shaped pulses, such as waveform memories, special amplifiers, and the corresponding routers which feed these pulses into the transmitter or decoupler channel. On recent instruments this equipment is already, or will soon become, a standard feature.

Working with selective pulses does, however, require the skills of an experienced spectroscopist. These experiments cannot be performed in the manner of the routine 1D or 2D standard pulse sequences, where meaningful results can be achieved even by robots. The shape of the selective pulses has to be chosen from a large menu of possibilities to obtain the desired action and selectivity. The r.f. power and phase has to be calibrated. Thus, prior to an experiment which uses selective pulses a certain amount of preparation has to be performed. The achievable results, as shown in this chapter, should be worth the greater effort required.

After some calibration methods we show first the DANTE experiment, which can be performed without a pulse shaping unit. Homonuclear experiments such as selective COSY or TOCSY and SELINQUATE follow, and we demonstrate also some heteronuclear applications such as INAPT and SELINCOR. The chapter ends with three procedures in which selective pulses are applied within a 2D sequence, thus reducing a problem which is in principle three-dimensional in a two-dimensional one.

Literature

[1] W. S. Warren, M. S. Silver, *Adv. Magn. Reson.* **1988**, *12*, 247–384.

[2] H. Kessler, S. Mronga, G. Gemmecker, *Magn. Reson. Chem.* **1991**, *29*, 527–557.

[3] L. Emsley, *Meth. Enzym.* **1994**, *239*, 207–256.

[4] T. Parella, *Magn. Reson. Chem.* **1996**, *34*, 329–347.

[5] R. Freeman, *Prog. NMR Spectrosc.* **1998**, *32*, 59–106.

Experiment 7.1

Determination of a Shaped 90° ^{1}H Transmitter Pulse

1. Purpose

Many advanced experiments such as SELCOSY (Exp. 7.5) or selective TOCSY (Exp. 7.8) use "soft" or shaped ^{1}H pulses in the transmitter channel. Prior to these experiments the pulse shapes have to be chosen and their pulse-lengths must be selected in accordance with the desired selectivity. Thus, the 90° shaped pulse must be determined by varying the attenuation of the transmitter and not the pulse-length. After having determined the 90° pulse the relative phase of this pulse with respect to a hard 90° pulse has to be adjusted, since their signal pathways might be quite different. This experiment describes the complete calibration procedure.

2. Literature

[1] H. Kessler, S. Mronga, G. Gemmecker, *Magn. Reson. Chem.* **1991**, *29*, 527–557.

3. Pulse Scheme and Phase Cycle

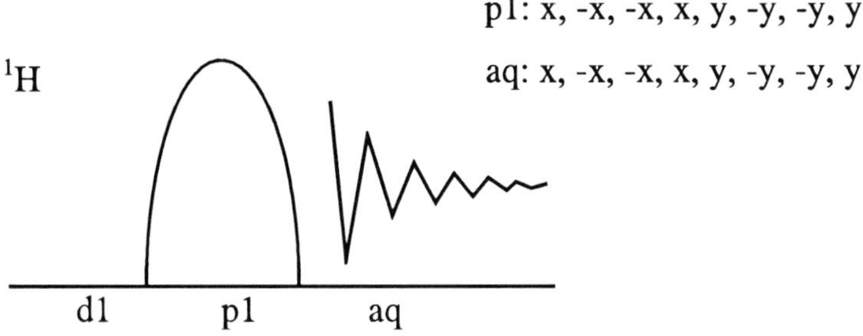

4. Acquisition

Time requirement: 30 min

Sample: 10% $CHCl_3$ in [D_6]acetone; *do not use a degassed and sealed sample*, since that would make the relaxation time of the $CHCl_3$ protons exceedingly long.

The spectrometer should be in normal operation for protons. First the shape and length of the soft pulse has to be selected. For this experiment a Gaussian shape on 1024 data points of 50 ms length was chosen. As a rule of thumb the selectivity in Hz of a shaped pulse corresponds the reciprocal of its length. You have to set:

td: 4 k
sw: 500 Hz

o1: on ^{1}H resonance
p1: shaped ^{1}H transmitter pulse, 50 ms
d1: 20 s
transmitter attenuation, to be varied in steps of 2 dB, initial value 90 dB
rg: receiver gain for correct ADC input
ns: 1

Since it is not allowed to change the pulse length in this experiment you have to determine the signal strength as a function of the transmitter attenuation. This is best done by recording the integrals.

Whenever pulses with different transmitter attenuation are used in a pulse sequence on the same channel one has to make sure that these pulses have the same excitation phase. For very recent instruments with linear amplifiers this check might not be necessary, on older instruments, however, it is mandatory. This applies to all experiments with selective pulses described in this Chapter, but also to spin-lock experiments like TOCSY (Exp. 10.18), ROESY (Exp. 10.20) and others (Exp. 6.16 or 10.16).

Having determined the optimum transmitter attenuation for either the selective or the spin-lock pulses, one has to determine the excitation phase difference between the hard pulse and the attenuated one. For this, a spectrum with a hard 90° pulse is recorded, transformed and phase corrected; the necessary zero order phase correction is noted. Secondly, a spectrum with the attenuated pulse is recorded, transformed and phase corrected. Most likely, the zero order phase constant will differ from that in the first experiment.

The difference between the zero order phase constants obtained in these two experiments is then added to the excitation phase of the attenuated pulse in the corresponding pulse program. Recent software also allows this to be treated as an adjustable parameter of the data set. On repetition, both experiments should then give equally phased spectra using an identical phase correction in the processing routine.

5. Processing

Use standard 1D processing as described in Experiment 3.1

6. Result

The figure on page 214 shows a typical plot of integral values versus transmitter attenuation obtained on a AMX-500 spectrometer, an attenuator setting of ca. 67 dB corresponds to the 90° pulse. Note that the dB scale is logarithmic resulting in a compression of the expected sine curve. In addition, there is often some deviation from the ideal curve.

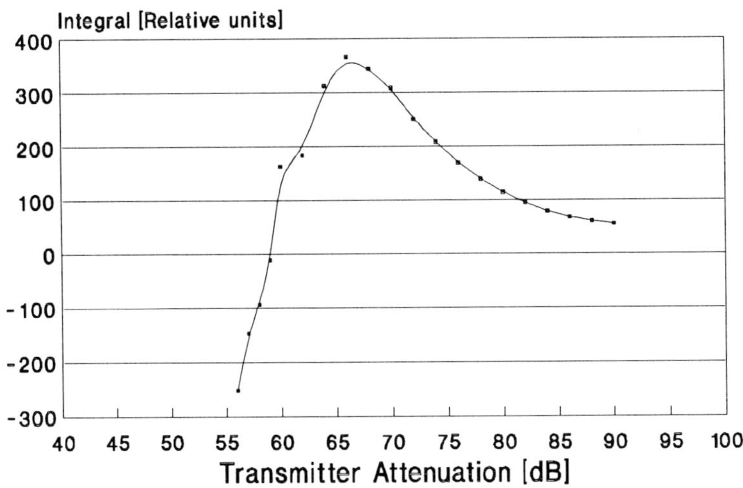

7. Comments

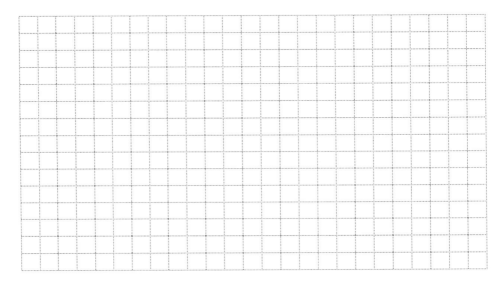

The determination of the relative phase is only necessary if, in the actual pulse sequence used, both hard and soft pulses are applied on the same channel. Recent software allows an offset modulation of the shaped pulse or even multiple excitation. Thus one can excite different signals distant from the offset of the transmitter. Note, however, that the phase of these shaped pulses is dependent on the offset modulation and has to be adjusted for each different offset.

8. Own Observations

Experiment 7.2

Determination of a Shaped 90° ^1H Decoupler Pulse

1. Purpose

Some heteronuclear experiments use shaped pulses in the decoupler channel (see Exp. 7.10). Prior to these experiments the pulse shapes have to be chosen and their pulse-lengths must be selected in accordance with the desired selectivity. Thus, the 90° shaped ^1H decoupler pulse must be determined by varying the attenuation of the decoupler and not the pulse-length. After having determined the 90° shaped decoupler pulse the relative phase of this pulse with respect to a hard 90° decoupler pulse has to be adjusted if in the actual experiment both hard and soft pulses are applied in the decoupler channel, since their signal pathways might be quite different. Note that in Experiment 7.3 the inverse form of this procedure with proton detection and a shaped decoupler pulse on the ^{13}C channel is described.

2. Literature

[1] H. Kessler, S. Mronga, G. Gemmecker, *Magn. Reson. Chem.* **1991**, *29*, 527–557.

[2] J.-M. Bernassau, J.-M. Nuzillard, *J. Magn. Reson. Ser. A* **1993**, *104*, 212–221.

3. Pulse Scheme and Phase Cycle

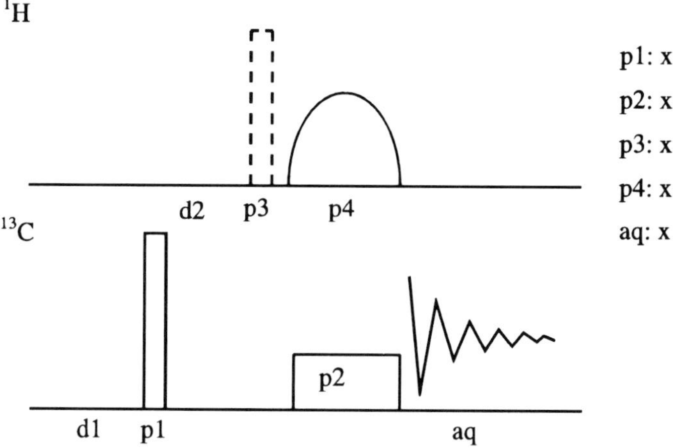

4. Acquisition

Time requirement: 30 min

Sample: 80% $CHCl_3$ in [D_6]acetone; *do not use a degassed and sealed sample*, since that would make the relaxation time of the $CHCl_3$ nuclei exceedingly long.

Obtain normal ^{1}H and ^{13}C spectra of the sample and note the offsets.The spectrometer is set to ^{13}C operation. First the desired pulse shape of the decoupler soft pulse and its length must be selected. For this experiment a Gaussian shape on 1024 data points of 30 ms length was chosen. You have to set:

td: 4 k
sw: 500 Hz
o1: on ^{13}C resonance
o2: on ^{1}H resonance
p1: 90° ^{13}C transmitter pulse
p2: ^{13}C spin-lock pulse for ^{13}C decoupling at 15 dB transmitter attenuation, length = p4, effective phase must be the same as for the hard pulse p1, adjust if necessary.
p3: leave at zero for the determination of the correct attenuation of the shaped pulse. For phase determination set to hard 90° ^{1}H decoupler pulse.
p4: shaped ^{1}H decoupler pulse, 30 ms
d1: 10 s
d2: $1/[2J(C,H)] = 2.36$ ms, calculated from $^1J(C,H) = 212$ Hz
decoupler attenuation for soft pulse, initial value 80 dB, to be varied
rg: receiver gain for correct ADC input
ns: 1

Since it is not allowed to change the pulse length in this experiment you have to determine the signal-to-noise ratio of the signal as a function of the decoupler attenuation. With a very high decoupler attenuation (about 80 dB) adjust the phase of the C,H doublet to an antiphase pattern. Then repeat the experiment with different decoupler attenuations until you get a spectrum with zero intensity of the doublet which corresponds to the soft 90° decoupler pulse. To obtain the correct phase of the soft decoupler pulse one introduces another hard 90° decoupler pulse p3 before the soft pulse in the pulse program. If both have the same phase, they are additive and yield an antiphase pattern with opposite phases to those adjusted before.

5. Processing

Use standard 1D processing as described in Experiment 3.2.

6. Result

The figure on page 217 shows spectra obtained on an AMX-500 spectrometer. **a** is the initial spectrum with high decoupler attenuation; **b** is the spectrum obtained with a shaped 90° decoupler pulse where both signals disappear, and **c** was obtained with an additional hard 90° decoupler pulse p3, where both hard and soft pulses have the same effective phase.

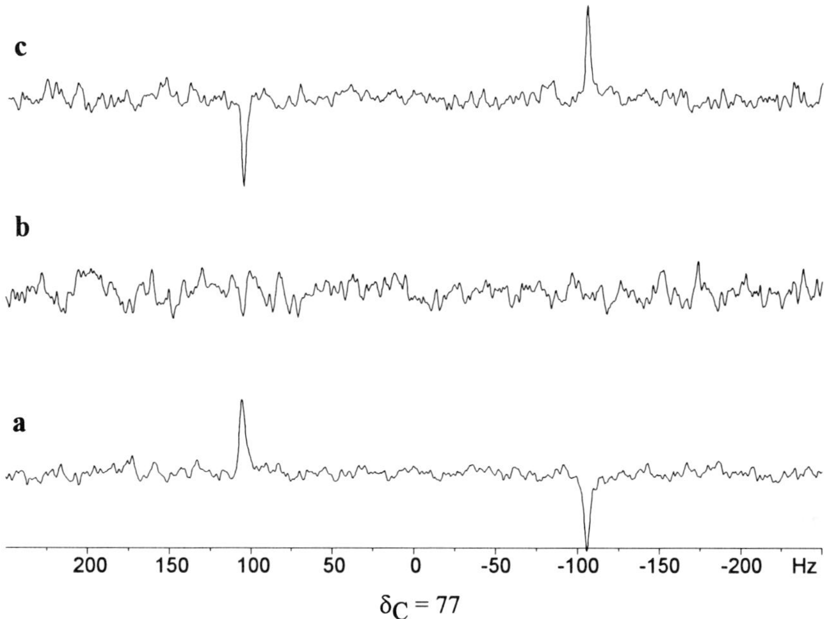

7. Comments

The product operator formalism of the basic mechanism of this experiment has been outlined in Experiment 2.3. This calibration experiment follows the description given in a very recent publication [2]. In earlier work it was found difficult to obtain a correct calibration because of phasing problems. This problem is removed by using the ^{13}C spin-lock pulse p2, which decouples the ^{13}C spins during the application of the shaped ^{1}H decoupler pulse, reducing the ^{13}C satellites of $CHCl_3$ to one singlet. Therefore the shaped pulse can be applied at the center of the proton resonance.

8. Own Observations

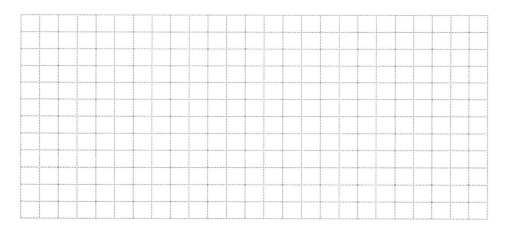

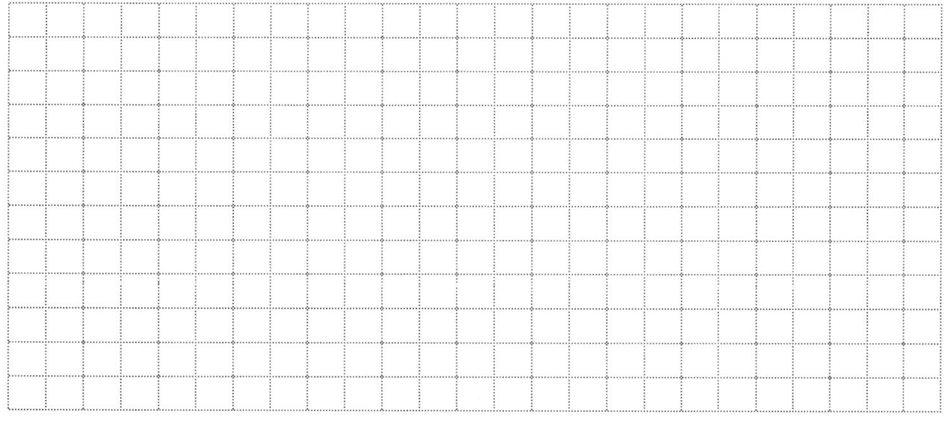

Experiment 7.3

Determination of a Shaped 90° ^{13}C Decoupler Pulse

1. Purpose

Some inverse experiments use shaped pulses in the ^{13}C decoupler channel (Exps. 7.6, 7.11). Prior to these experiments the pulse shapes have to be chosen and their pulse-length selected in accordance with the desired selectivity (reciprocal of pulse-length). Thus, the 90° pulse must be determined by varying the attenuation of the decoupler and not the pulse-length. After having determined the attenuation of the shaped 90° pulse, the relative phase of this pulse with respect to a hard 90° decoupler pulse has to be adjusted. This is necessary if in the actual application both hard and soft pulses are used, since their signal pathways might be different.

2. Literature

[1] J.-M. Bernassau, J.-M. Nuzillard, *J. Magn. Reson. Ser. A* **1993**, *104*, 212–221.

3. Pulse Scheme and Phase Cycle

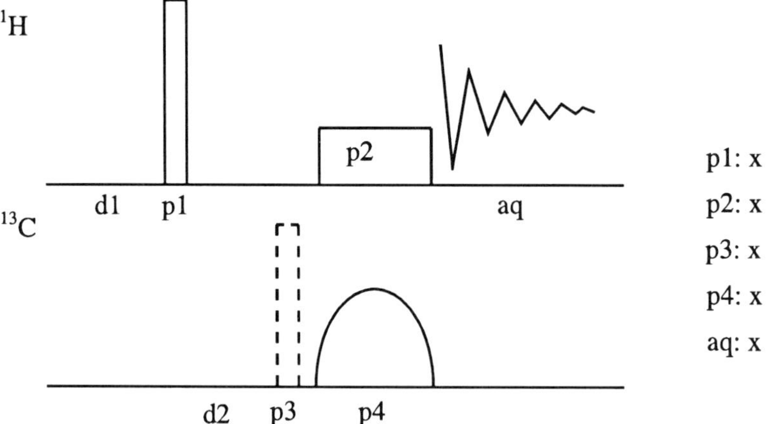

4. Acquisition

Time requirement: 30 min

Sample: 10% $CHCl_3$ in [D_6]acetone; *do not use a degassed and sealed sample*, since that would make the relaxation time of the $CHCl_3$ nuclei exceedingly long.

First the ^1H and ^{13}C offsets of the sample have to be determined.The spectrometer is set up for ^1H observation with ^{13}C decoupling (inverse mode on older instruments).

The desired shape of the decoupler pulse and its length have to be selected. For this experiment a Gaussian shape of 10 ms length was chosen. You have to set:

td: 4 k
sw: 500 Hz
o1: on ^{1}H resonance
o2: on ^{13}C resonance
p1: 90° ^{1}H transmitter pulse
p2: ^{1}H spin-lock pulse for ^{1}H decoupling, same length as p4, typical attenuation 12 dB, effective phase must be the same as for the hard pulse p1, adjust if necessary.
p3: leave at zero for the determination of the correct attenuation of the shaped pulse. For phase determination set to hard 90° ^{13}C decoupler pulse.
p4: shaped ^{13}C decoupler pulse, 10 ms
decoupler attenuation for soft pulse, initial value 80 dB, to be varied
d1: 10 s
d2: $1/[2J(C,H)] = 2.33$ ms, calculated from $^{1}J(C,H) = 215$ Hz
rg = receiver gain for correct ADC input
ns = 1

In order not to change the selectivity you have to determine the effect of the shaped pulse as a function of the decoupler attenuation. With a very high decoupler attenuation (ca. 80 dB) adjust the phase of the C,H doublet to an antiphase pattern. Then repeat the experiment with different decoupler attenuations until the satellites disappear, which corresponds to the soft 90° decoupler pulse. To obtain the correct phase of the soft decoupler pulse one introduces another hard 90° decoupler pulse before the soft pulse in the pulse program. If both have the same phase, they are additive and yield an antiphase pattern with opposite phases as adjusted before.

5. Processing

Use standard 1D processing as described in Experiment 3.1.

6. Result

The figure on page 220 shows spectra obtained on an AMX-500 spectrometer. **a** is the initial spectrum with high decoupler attenuation and in **b** the effect of a shaped 90° decoupler pulse is shown. In **c** p3 was set to 90° and the phase of the shaped pulse was adjusted correctly in order to form, in combination with p3, a 180° pulse.

7. Comments

This calibration experiment follows the description given in a very recent publication [1]. In earlier work, due to phase problems, it was found very difficult to obtain a correct calibration. With the spin lock pulse p2 these problems are removed. It serves two purposes. First, it decouples the protons during the application of the shaped pulse, leaving the ^{13}C resonance as a singlet. Therefore the shaped pulse can be applied at the center of the ^{13}C resonance. Secondly, it purges the signals of all protons bound to ^{12}C, allowing a far easier detection of the ^{13}C satellites. The product operator formalism for this experiment is the same as described in Experiment 2.3 with exchanging the notation for C and H.

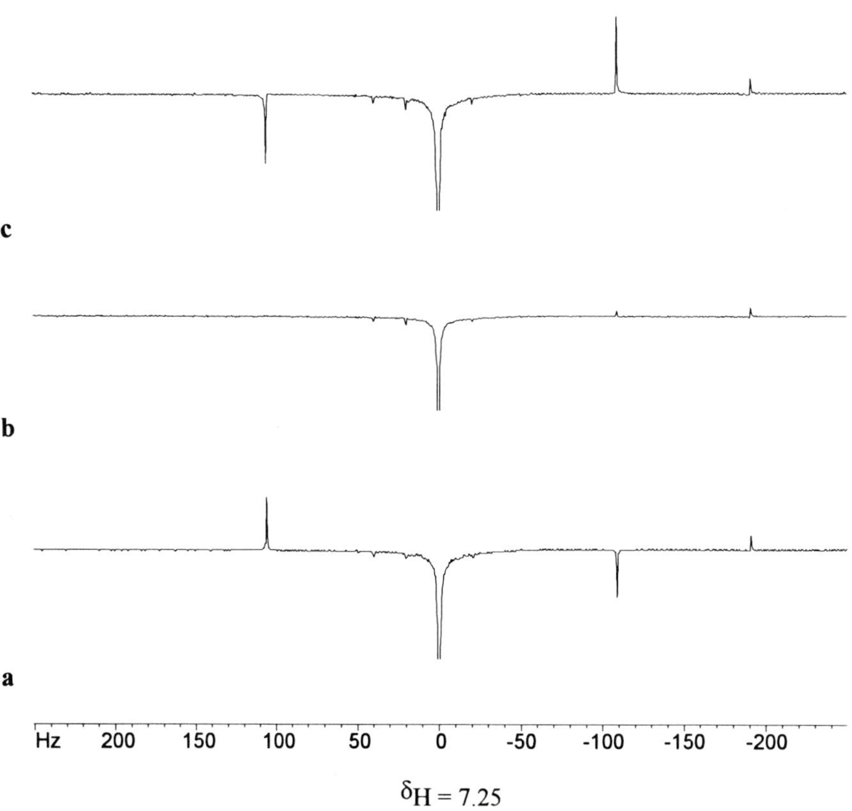

8. Own Observations

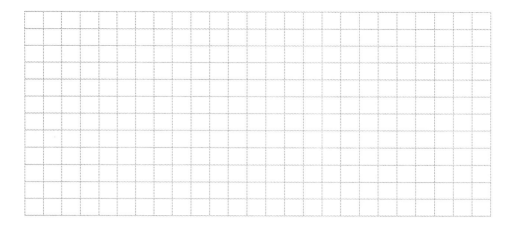

Experiment 7.4

Selective Excitation with DANTE

1. Purpose

One often wants to excite a single resonance selectively. On modern instruments this selective excitation is usually performed with shaped r.f. pulses which require additional hardware. With the DANTE (**D**elays **A**lternating with **N**utation for **T**ailored **E**xcitation) sequence [1] this can be performed on any older instrument. Care must be taken to ensure that the sidebands produced by the DANTE sequence do not excite additional signals. In the experiment presented here selective DANTE excitation is combined with a gated ^{1}H-decoupling ^{13}C experiment (Exp. 4.11), demonstrating that overlapping multiplets can be individually analyzed with this technique [2].

2. Literature

[1] G. A. Morris, R. Freeman, *J. Magn. Reson.* **1978**, *29*, 433–462.
[2] G. Bodenhausen, R. Freeman, G. A. Morris, *J. Magn. Reson.* **1976**, *23*, 171–175.
[3] R. Freeman, *A Handbook of Nuclear Magnetic Resonance*, Longman, Harlow, **1987**, 207–215.

3. Pulse Scheme and Phase Cycle

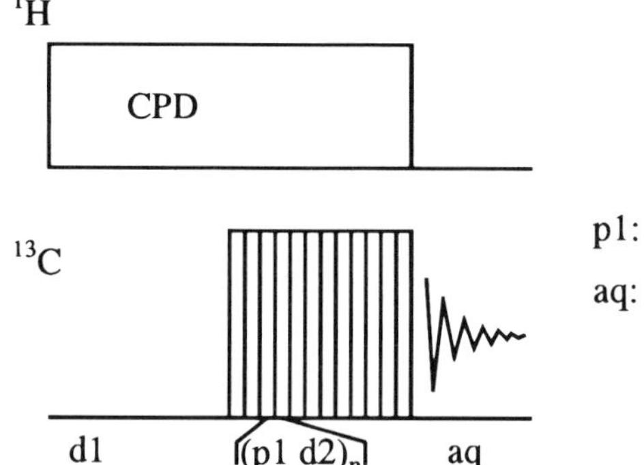

p1: x, x, -x, -x, y, y, -y, -y

aq: x, x, -x, -x, y, y, -y, -y

4. Acquisition

Time requirement: 0.5 h

Sample: 20% ethyl crotonate in $CDCl_3$

First obtain a normal ^{13}C NMR spectrum in order to get the exact resonance positions of the two methyl C-atoms. Then measure a proton coupled ^{13}C NMR spectrum using the gated decoupling method (Exp. 4.11), with the sweep width and offset adjusted to cover the area of the two methyl groups. To record the DANTE spectra you have to attenuate the transmitter power (see Exp. 2.9) until the 90° pulse is about 90 μs, so that a 1° pulse corresponds to about 1 μs. On older instruments, where software control of the transmitter attenuation is not possible, you simply introduce an attenuator box of about 20 dB into the ^{13}C transmitter line. You have to set:

td: 4 k
sw: 10 ppm
o1: on ^{13}C resonance of the selected methyl group
o2: middle of ^{1}H NMR spectrum
p1: 1° ^{13}C transmitter pulse
d1: 2 s
d2: 0.5 ms, yielding a total length of the DANTE excitation of 25.05 ms
n: number of p1 pulses, 50
decoupler attenuation and 90° pulse for CPD
ns: 128

5. Processing

Use standard ^{13}C NMR processing as described in Experiment 3.2 with exponential multiplication (lb = 0.5 Hz).

6. Result

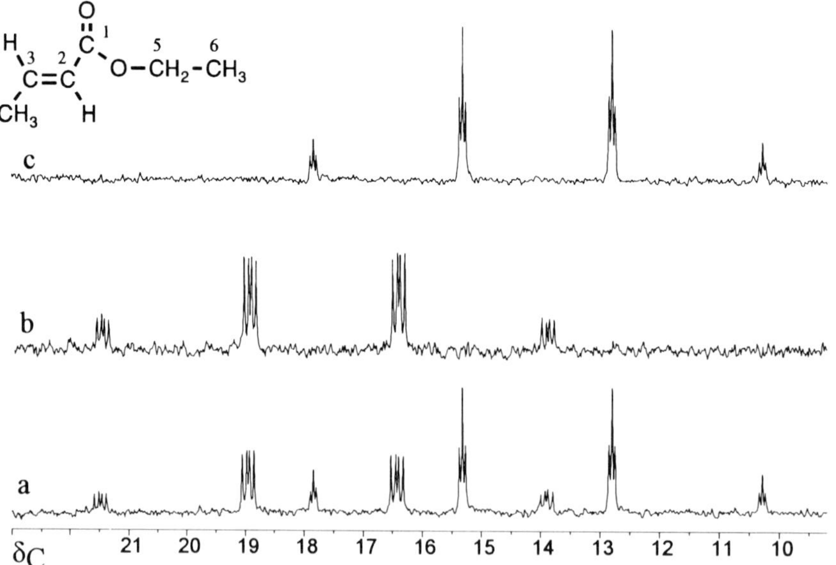

In the figure **a** is the normal ^{1}H coupled ^{13}C NMR spectrum of ethyl crotonate in the area of the two overlapping methyl group quadruplets, obtained on an ARX-200 spectrometer. In **b** and **c** the DANTE spectra of the two selected methyl groups are shown.

7. Comments

The flip angle α caused by an r.f. pulse of duration p and field strength B_1 is given by Equation (1).

$$\alpha = \gamma B_1 p \tag{1}$$

If n very short pulses exactly on resonance are applied in a pulse train, their net effect is given by (2)

$$\alpha = n\gamma B_1 p \tag{2}$$

However, if the frequency is offset from resonance by an amount $\Delta\nu$, then during each pulse cycle the nuclei precess in the rotating frame through an angle $2k\pi + \lambda$ given by Equation (3), where k is an integer, τ is the repetition time of the pulses, and λ is a phase angle less than 2π.

$$2k\pi + \lambda = 2\pi\Delta\nu\,\tau \tag{3}$$

Thus the DANTE sequence produces signal responses at the sidebands k. In quadrature detection it is most convenient to set $k = 0$ with the transmitter directly on the resonance of the desired signal. However, excitation also occurs at $k = 1$ and 2.

Very roughly, one can estimate the selectivity of a selective pulse as the reciprocal of its length; thus the DANTE excitation pulse train of 25 ms length used here corresponds to a selectivity of about 40 Hz. Attenuation of the transmitter is necessary, since the normal pulse programmers are not able to produce very short pulses; therefore the 90° pulse should be in the order of 90 μs to give a 1° pulse angle for a pulse duration of 1 μs.

8. Own Observations

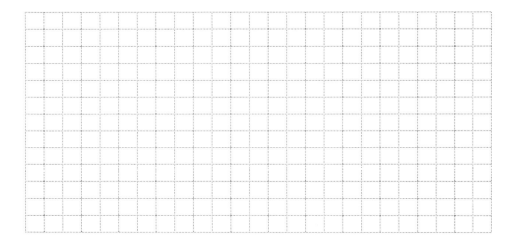

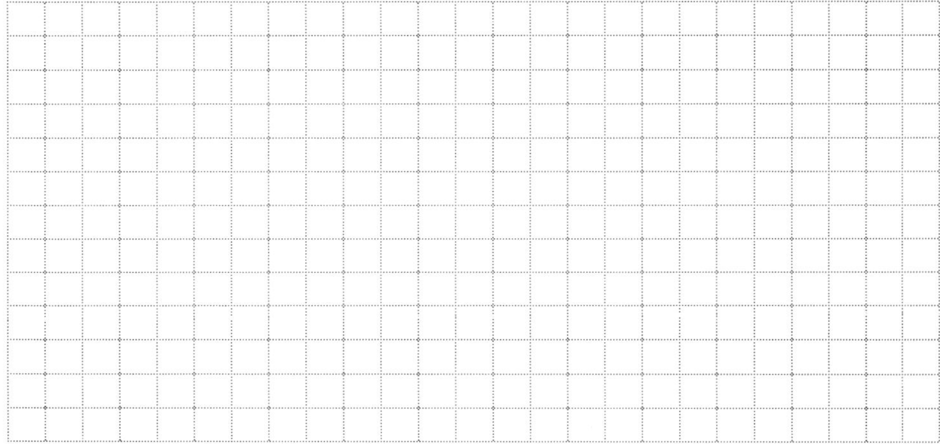

Experiment 7.5

SELCOSY

1. Purpose

This is the 1D variant of the most common 2D experiment. Instead of recording the full 2D matrix, one can simply measure one "row" by replacing the first 90° pulse of the COSY experiment (see Exp. 10.3) with a soft pulse, thus looking only for spin couplings that affect the particular proton excited. The selective COSY method yields the same connectivity information as the homonuclear decoupling technique (Exp. 4.4). In contrast to the latter, however, the multiplets of the coupling partners remain to be seen and can easily be evaluated. Because this is a 1D experiment, it can be performed at high resolution. The most recent extended version of this experiment uses gradient selection [4] and is described in Experiment 11.8.

2. Literature

[1] C. J. Bauer, R. Freeman, T. Frenkiel, J. Keeler, A. J. Shaka, *J. Magn. Reson.* **1984**, *58*, 442–457.

[2] H. Kessler, H. Oschkinat, C. Griesinger, W. Bermel, *J. Magn. Reson.* **1986**, *70*, 106–133.

[3] H. Kessler, S. Mronga, G. Gemmecker, *Magn. Reson. Chem.* **1991**, *29*, 527–557.

[4] M. A. Bernstein, L. A. Trimble, *Magn. Reson. Chem.* **1994**, *32*, 107–110.

3. Pulse Scheme and Phase Cycle

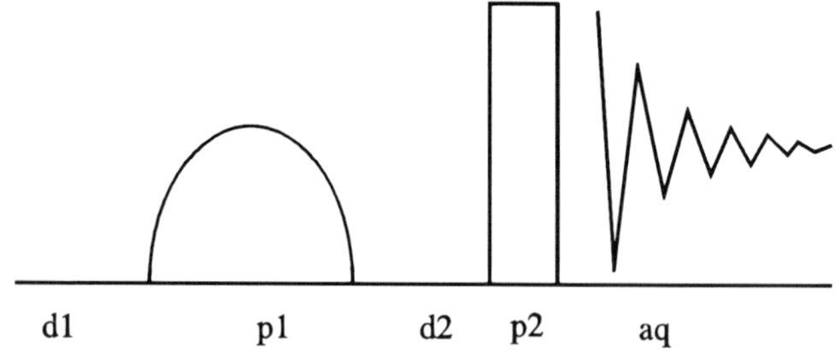

p1: y, -y, -y, y, x, -x, -x, x

p2: x, -x, x, -x, y, -y, y, -y

aq: x, -x, -x, x, y, -y, -y, y

4. Acquisition

Time requirement: 15 min

Sample: 10% strychnine in $CDCl_3$

Record a normal ^1H NMR spectrum of the sample. Determine the 90° pulse width for the hard ^1H transmitter pulse, select a Gaussian pulse shape for the soft pulse and determine the correct attenuation corresponding to a 90° pulse at 50 ms duration (see Exp. 7.1). Determine the phase difference between the hard and the soft pulse and take this into account for the actual phases used in the pulse program. You have to set:

td: 32 k
sw: 10 ppm
o1: on resonance of selected signal. If the software allows offsets for selective pulses, one can also put o1 in the middle of the ^{1}H NMR spectrum. However, the different phases of the selective pulses at different offsets must be determined.
p1: Gaussian shape, 50 ms length, transmitter attenuation corresponding to 90° excitation
p2: 90° ^{1}H transmitter pulse
d1: 2s

$d2 = \frac{1}{2J(\mathrm{H,H})} - \frac{\mathrm{p1}}{2}$, typically 37 ms, calculated from J(H,H) = 8 Hz

ds: 4
ns: 16

5. Processing

Use standard ^{1}H processing as described in Experiment 3.1. Note that the signals of the coupling partners show the active coupling in antiphase.

6. Result

The figure on page 226 shows the result obtained on an AMX-500 spectrometer. In **a** an expanded portion of the normal ^{1}H NMR spectrum is shown, in **b** H-12 was selected, giving the responses of both H-11 and (weakly) of H-13, and in **c** (d2 = 50 ms) H-15β was selected, giving the responses of H-15α, H-14, and H-16. Note that the coupling constant J(H-15β,H-16)) can be measured selectively (4 Hz).

7. Comments

For the COSY part of the sequence exactly the same theory applies as given in Experiment 10.3. Note that the delay d2 determines the intensity of the "cross peak". It may be necessary to perform the experiment twice, for example in order to identify spin coupling partners with both small and large spin–spin coupling constants. In the spectrum **c** d2 was set to 50 ms.

The 90° Gaussian soft pulse used here can be replaced by many other types. You may try a 270° Gaussian, a half Gaussian with or without an additional purge pulse [3], or DANTE excitation (Exp. 7.4).

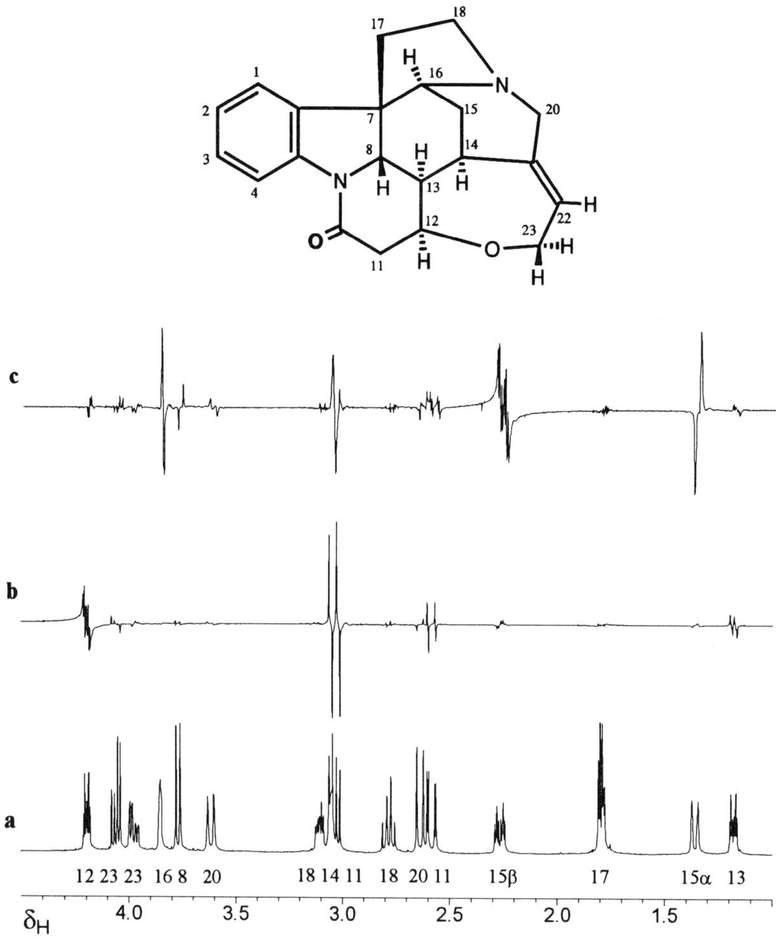

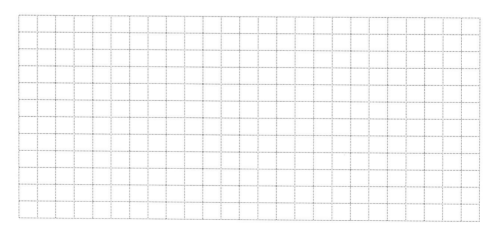

8. Own Observations

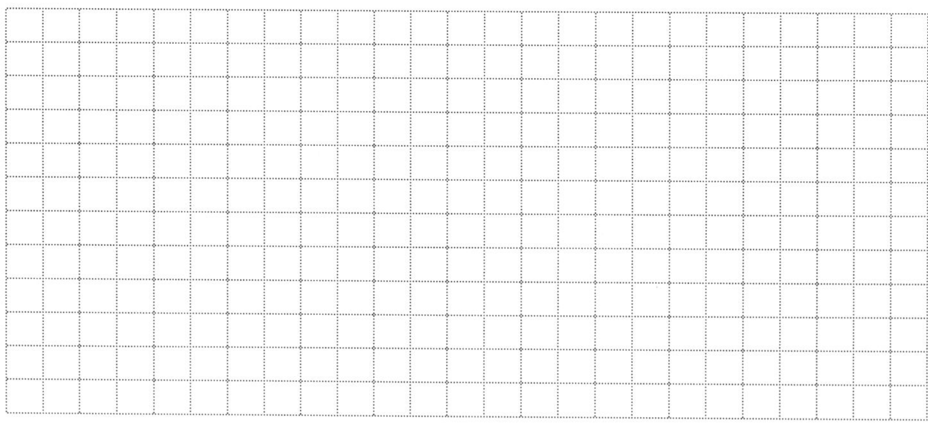

Experiment 7.6

SELINCOR: Selective Inverse H,C Correlation via ^1J(C,H)

1. Purpose

This experiment is the selective 1D version [1] of the two-dimensional inverse H,C correlation (HMQC, see Exp. 10.13). It can also be regarded as the inverse of the single-frequency decoupled ^{13}C NMR spectrum (see Exp. 4.13). The experiment correlates a selected carbon atom with the attached proton via one bond C,H coupling, using proton sensitivity for observation. In the current literature [2–5], there are many modifications and improvements of the basic experiment shown here. With a gradient selected modification (see Exp. 11.11), the sequence can also be used as an initial stage in various advanced ^{13}C-resolved proton experiments [6].

2. Literature

[1] S. Berger, *J. Magn. Reson.* **1989**, *81*, 561–564.
[2] R. C. Crouch, J. P. Shockar, G. E. Martin, *Tetrahedron Letters* **1990**, *31*, 5273–5276.
[3] P. Berthault, B. Perly, M. Petitou, *Magn. Reson. Chem.* **1990**, *28*, 696–670.
[4] J.-M. Bernassau, J.-M. Nuzillard, *J. Magn. Reson. Ser. A* **1993**, *104*, 212–221.
[5] W. Willker, J. Stelten, D. Leibfritz, *J. Magn. Reson. Ser. A* **1994**, *107*, 94–98.
[6] T. Fäcke, S. Berger, *Magn. Reson. Chem.* **1995**, *33*, 144–148.

3. Pulse Scheme and Phase Cycle

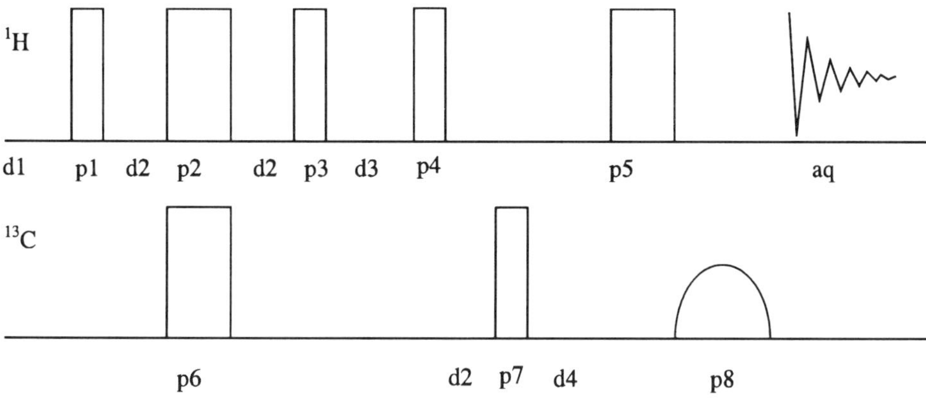

p1, p2, p4, p6: x
p3: -x
p5: $(x)_4$, $(y)_4$, $(-x)_4$, $(-y)_4$

p7: x, -x
p8: x, x, -x, -x
aq: x, -x, -x, x, -x, x, x, -x

4. Acquisition

Time requirement: 0.5 h

Sample: 20% ethyl crotonate in $CDCl_3$

Prior to the experiment you have to choose a Gaussian pulse shape and to determine the attenuation and the relative phase of a selective ^{13}C pulse in the inverse mode of the spectrometer (see Exp. 7.3). Record normal ^1H and ^{13}C NMR spectra and note the offsets of the desired carbon signals. The sequence given above consists of the BIRD filter (see Exp. 6.13) and the actual SELINCOR sequence. You have to set:

td: 32 k
sw: 10 ppm
o1: middle of ^1H NMR spectrum
o2: on resonance of selected ^{13}C nucleus
p1, p3, p4: 90° ^1H transmitter pulse
p2, p5: 180° ^1H transmitter pulse
p6: 180° ^{13}C decoupler pulse
p7: 90° ^{13}C decoupler pulse
p8: Gaussian shaped ^{13}C decoupler pulse, 5 ms length, attenuation corresponding to 90° or 270° excitation.
d1: 1s
d2: $1/[2J(C,H)] = 3.57$ ms, calculated from $^1J(C,H) = 140$ Hz
d3: BIRD relaxation delay, optimized for minimum FID intensity, [2.5 s]
d4: same length as selective pulse p8, 5 ms was used here
ds: 4
ns: 32

5. Processing

Use standard ^1H processing as given in Experiment 3.1. Since the phase of the satellites is not pure process the spectrum in magnitude mode.

6. Result

The figure on page 229 shows spectra obtained on an AMX-500 spectrometer. **a** is the normal ^1H NMR spectrum and **b** to **e** are the SELINCOR spectra, all processed in magnitude mode. In **b** C-2, in **c** C-5, in **d** C-4, and in **e** C-6 was selected. Note that the selective ^{13}C pulse of 5 ms length is able to distinguish between C-4 and C-6, which are separated by 450 Hz at the field strength used, but is broad enough to cover the width of a typical C,H doublet.

O
1
H 3 2 C 5 6
C=C O–CH_2–CH_3
4
CH_3 H

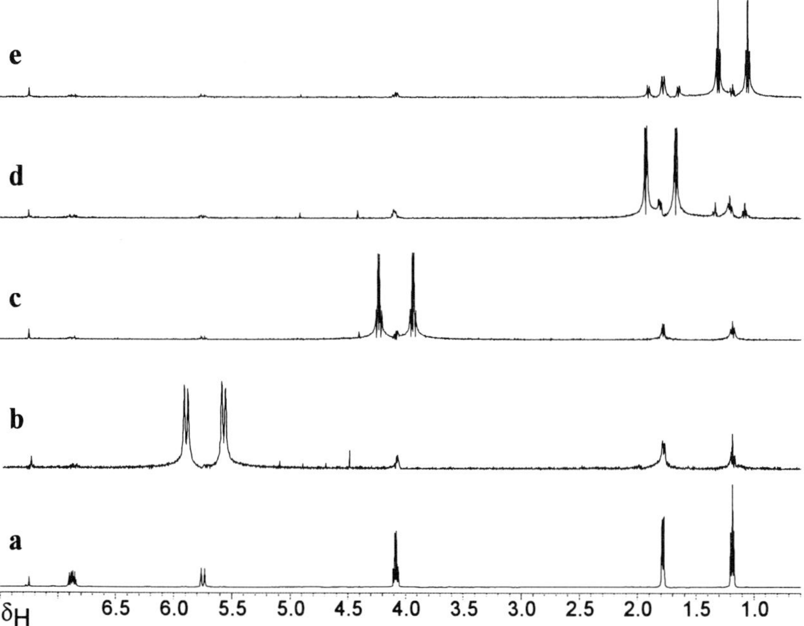

7. Comments

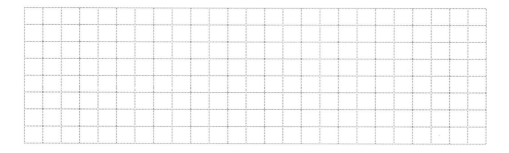

The theory is the same as for the HMQC experiment (Exp.10.13) and is given there. Instead of the evolution of the carbon chemical shifts in t_1 only the double quantum magnetization of the selected carbon nucleus is transformed back into an observable magnetization; all other coherences are suppressed by the phase cycle.

8. Own Observations

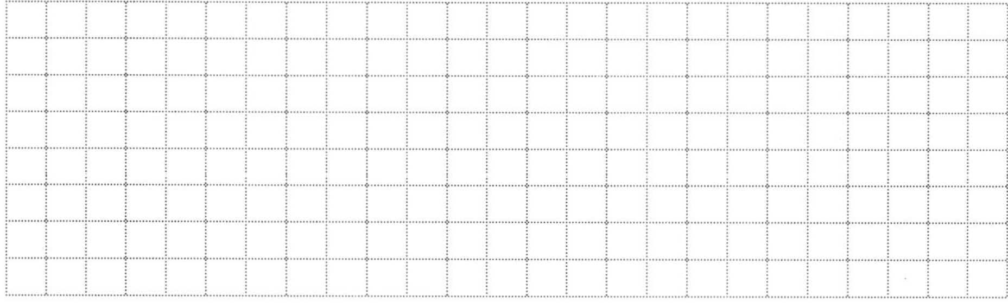

Experiment 7.7

SELINQUATE

1. Purpose

The SELINQUATE method [1] is the selective version of the INADEQUATE [2] sequence. Whereas with 1D-INADEQUATE (see Exp. 6.12) the $^{13}C,^{13}C$ spin coupling constants can yield partially overlapping signals, the 2D version (see Exp 10.22) is very time-consuming and has limited digital resolution [3]. With SELINQUATE it is possible to measure specific $^{13}C,^{13}C$ coupling constants over one or more bonds selectively with the high digital resolution of a 1D method. Thus, the experiment yields connectivity information for the irradiated carbon nucleus and $^{13}C,^{13}C$ spin coupling constants with high accuracy.

2. Literature

[1] S. Berger, *Angew. Chem. Int. Ed. Engl.* **1988,** *27,* 1196–1197.

[2] A. Bax, R. Freeman, S. P. Kempsell, *J. Am. Chem. Soc.* **1980,** *102,* 4849–4851.

[3] A. Bax, R. Freeman, T. A. Frenkiel, M. H. Levitt, *J. Magn. Reson.* **1981,** *43,* 478–483.

[4] J. Buddrus, H. Bauer, *Angew. Chem. Int. Ed. Engl.* **1987,** *26,* 625–643.

3. Pulse Scheme and Phase Cycle

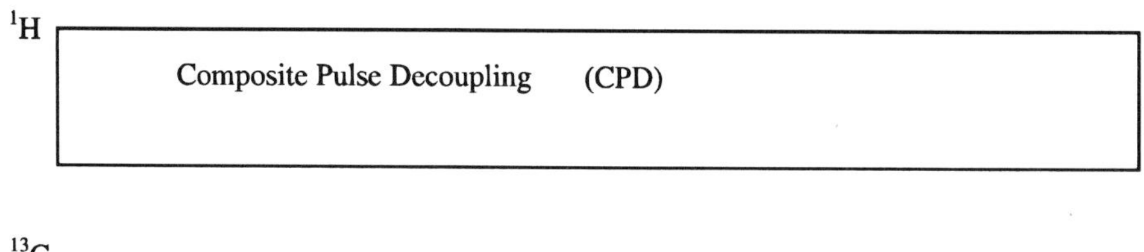

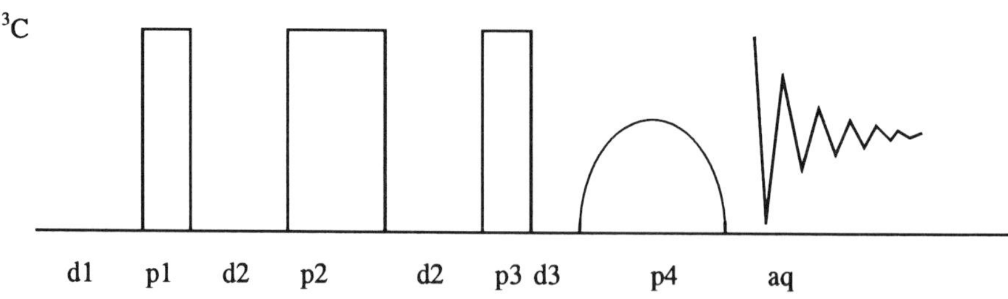

p1: $(x)_4$, $(y)_4$, $[(-x)_4, (-y)_4]_2$, $(x)_4$, $(y)_4$

p2: $[(x)_4, (y)_4, (-x)_4, (-y)_4]_2$, $[(-x)_4, (-y)_4, (x)_4, (y)_4]_2$

p3: $(x)_4$, $(y)_4$, $(-x)_4$, $(-y)_4$

p4: y, x, -y, -x

aq: $(x, y, (-x, -y)_2, x, y)_2$, $(-x, -y, (x, y)_2, -x, -y)_2$

4. Acquisition

Time requirement: 2 h

Sample: 90% 2-cyclohexen-1-one in [D_6]acetone

Prior to the experiment you have to determine the r.f. attenuation and phase for a selective ^{13}C pulse on carbon as described in Experiment 7.1. Record a normal ^{13}C NMR spectrum of the sample and adjust the spectral width to the aliphatic region. Note the offsets for the three signals of the aliphatic carbon nuclei and adjust the frequency of the selective pulse to one of them. You have to set:

td: 16 k
sw: 23 ppm
o1: on resonance of selected signal. If the software allows offsets for selective pulses, one can also set o1 in the middle of the ^{13}C NMR spectrum. However, the different phases of the selective pulses at different offsets must be determined.
o2: middle of ^1H NMR spectrum
p1, p3: 90° ^{13}C transmitter pulse
p2: 180° ^{13}C transmitter pulse
p4: Gaussian shaped ^{13}C transmitter pulse, 10 ms length, attenuation corresponding to 270° excitation, which gives better phase behavior of the satellites than with a 90° pulse.
d1: 4s
d2: $1/[4J(C,C)]$ = 7.6 ms calculated from $^1J(C,C)$ = 33 Hz; for long-range couplings use $J(C,C)$ = 4 Hz, d2 = 62.5 ms
d3: 10 μs
decoupler attenuation and 90° pulse for CPD
ds: 4
ns: 256

5. Processing

Use standard ^{13}C NMR processing as described in Experiment 3.2. Note that the experiment yields ^{13}C satellites in antiphase. The residual signal of the molecules containing only one ^{13}C atom should not be used for phasing.

6. Result

In the figure on page 232 **a** is the normal ^{13}C NMR spectrum of the aliphatic region, obtained on an AMX-500 spectrometer. In **b** C-4 was irradiated, giving a response of C-5 with $J(C,C)$ = 33.4 Hz. In **c** C-5 was irradiated, giving a response of both C-4 and C-6 with $J(C,C)$ values of 33.4 Hz and 31.7 Hz respectively, and in **d** C-6 was irradiated giving a response of C-5 with $J(C,C)$ = 31.7 Hz. In **e** the selective pulse was adjusted to the carbon nucleus of the C=O group, and d2 was adjusted to long-range interaction; the figure shows the signal of C-4 with $^3J(C,C) = 4.8$ Hz. In **f** C-6 was irradiated; the figure shows the signal of C-4 with $^2J(C,C) = 2.5$ Hz.

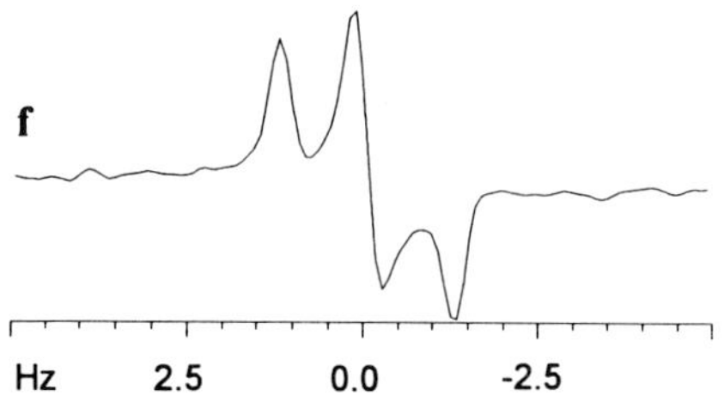

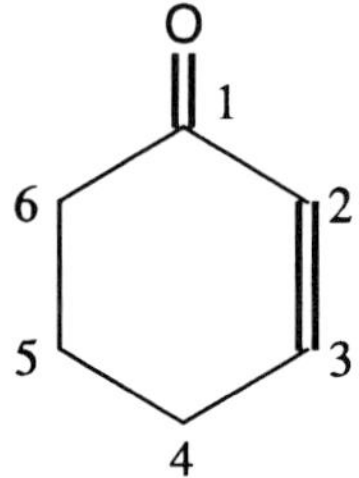

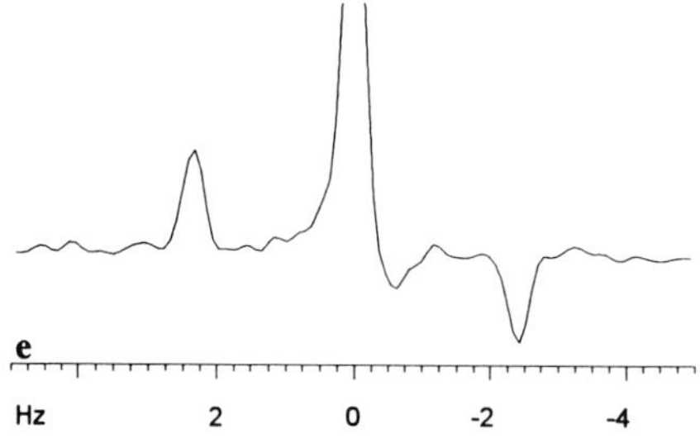

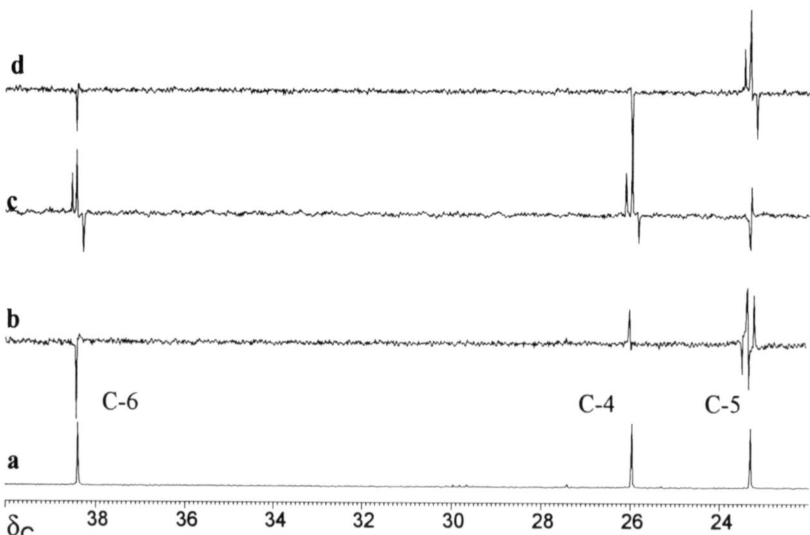

7. Comments

The theory of the experiment is the same as for the 1D-INADEQUATE experiment and is outlined there (see Exp. 6.12). In SELINQUATE only the double quantum coherence of the selectively irradiated carbon atom is transformed back into an observable magnetization. Note that the excitation bandwidth of the selective pulse used must be broad enough to excite both satellites of the carbon signal.

8. Own Observations

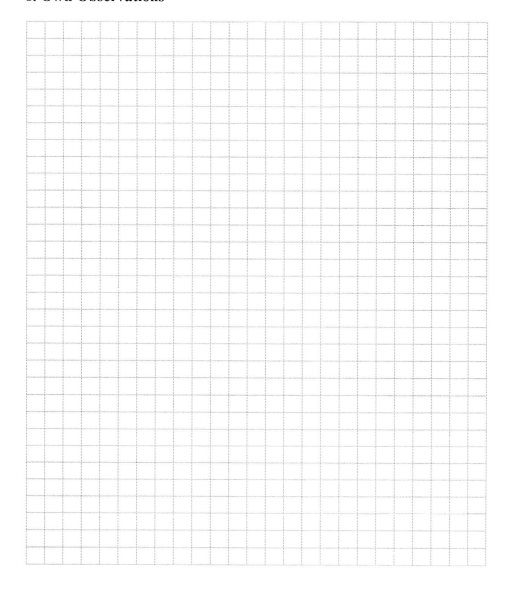

Experiment 7.8

Selective TOCSY

1. Purpose

This experiment is the selective 1D version of the 2D TOCSY (**TO**tal **C**orrelation **S**pectroscop**Y**) method (Exp. 10.18), also known as HOHAHA (**HO**monuclear **HA**rt-mann–**HA**hn). One proton is excited by a shaped pulse and this produces a response from all protons that are connected by spin coupling within a chain. Thus it is possible, for example, to trace all protons in a sugar moiety by starting from the anomeric proton or in one amino acid side-chain by starting from the α proton. Since the publication of the original experiment [1,2] given here, there have been several attempts to improve the performance and to eliminate some shortcomings [3–6], a recent gradient selected version is described in Exp. 11.9.

2. Literature

[1] D. G. Davis, A. Bax, *J. Am. Chem. Soc.* **1985**, *107*, 7197–7198; A. Bax, D. G. Davis, *J. Magn. Reson.* **1985**, *65*, 355–360.

[2] H. Kessler, H. Oschkinat, C. Griesinger, W. Bermel, *J. Magn. Reson.* **1986**, *70*, 106–133.

[3] V. Sklenár, J. Feigon, *J. Am. Chem. Soc.* **1990**, *112*, 5644–5645.

[4] J. P. Shockcor, R. C. Crouch, G. E. Martin, A. Cherif, J.-K. Luo, R. N. Castle, *J. Heterocycl. Chem.* **1990**, *27*, 455–458.

[5] L. Poppe, H. van Halbeek, *J. Magn. Reson.* **1992**, *96*, 185–190.

[6] T. Fäcke, S. Berger, *J. Magn. Reson. Ser. A* **1995**, *113*, 257–259.

3. Pulse Scheme and Phase Cycle

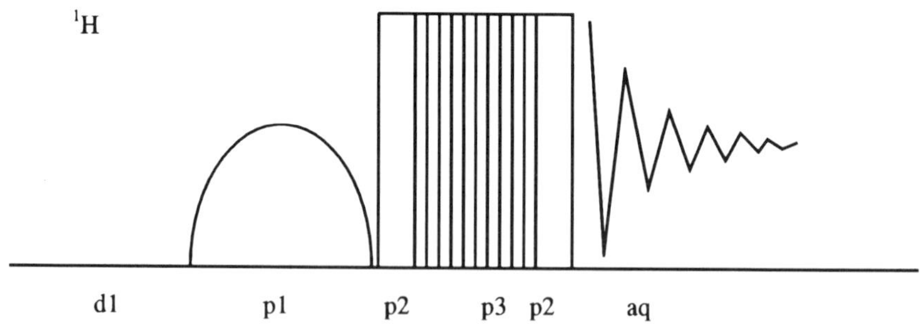

p1: y, $(-y)_2$, y, x, $(-x)_2$, x

p2: $(x, -x)_2$, $(y, -y)_2$ (trim pulses)

aq: y, $(-y)_2$, y, x, $(-x)_2$, x

p3: MLEV-17 spin-lock out of composite 180° pulses (90°, 180°, 90°); sequence:

90 (ph1), 180 (ph2), 90 (ph1)
[90 (ph3), 180 (ph4), 90 (ph3)]$_2$
90 (ph1), 180 (ph2), 90 (ph1)
[90 (ph3), 180 (ph4), 90 (ph3)]$_2$
[90 (ph1), 180 (ph2), 90 (ph1)]$_2$
90 (ph3), 180 (ph4), 90 (ph3)
[90 (ph1), 180 (ph2), 90 (ph1)]$_2$
90 (ph3), 180 (ph4), 90 (ph3)
[90 (ph1), 180 (ph2), 90 (ph1)]$_2$
[90 (ph3), 180 (ph4), 90 (ph3)]$_2$
60 (ph2)

ph1: $(-y, y)_2, (x, -x)_2$
ph2: $(x, -x)_2, (y, -y)_2$
ph3: $(y, -y)_2, (-x, x)_2$
ph4: $(-x, x)_2, (-y, y)_2$

4. Acquisition

Time requirement: 30 min

Sample: 10% strychnine in $CDCl_3$

Prior to the experiment the attenuator setting needed to give a 90° shaped pulse of a chosen duration has to be determined (see Exp 7.1). Its phase relative to the phase of the spin-lock pulses must be known, and the 90° pulse duration and the attenuation of the spin-lock pulses must also be calibrated (see Exp. 2.9). Run a normal ^{1}H NMR spectrum of the sample and note the offsets of the protons to be irradiated. You have to set:

td: 32 k
sw: 10 ppm
o1: on resonance of selected signal. If the software allows offsets for selective pulses, one can also put o1 in the middle of the ^{1}H NMR spectrum. However, the different phases of the selective pulses at different offsets must be determined.
p1: Gaussian shaped ^{1}H transmitter pulse, 50 ms length, transmitter attenuation corresponding to 90° excitation
p2: trim pulse 2.5 ms at power level of spin-lock
p3: series of composite 180° pulses (90°, 180°, 90°) at power level of spin-lock, typically 90° pulse width of 40 μs at 12 dB transmitter attenuation corresponding to an effective spin-lock field of ca. 7000 Hz. Total length of spin-lock set to 200 ms by loop parameter of spin-lock sequence.
d1: 2s
ns: 8
ds: 4

5. Processing

Use standard ^{1}H processing as described in Experiment 3.1

6. Result

In the figure **a** is an expanded region of the normal ^{1}H NMR spectrum of strychnine obtained on an AMX-500 spectrometer. In **b** H-12 was irradiated, giving responses from H-13, both H-11 protons, H-8, and H-14. In **c** H-16 was irradiated, giving responses from both H-15 protons, H-14, H-13, and H-8.

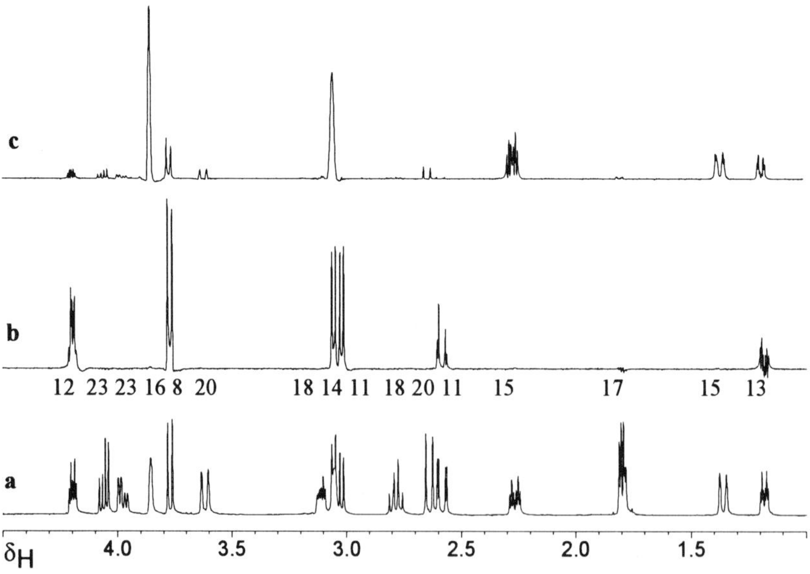

7. Comments

As for the 2D TOCSY experiment there exists a simple picture which explains the result of a spin-lock. The various protons "see" as the effective field only the weak r.f. field of the spin-lock, therefore chemical shift differences vanish and the spin systems are all of higher order leading to a mixing of all spin states. Exciting one proton at the end of a chain connected by spin coupling produces a response from all spins affected by the spin-lock. However, the phase of the response signals is not pure, but a mix of in-phase and antiphase components, thus it is difficult to extract correct spin coupling constants.

8. Own Observations

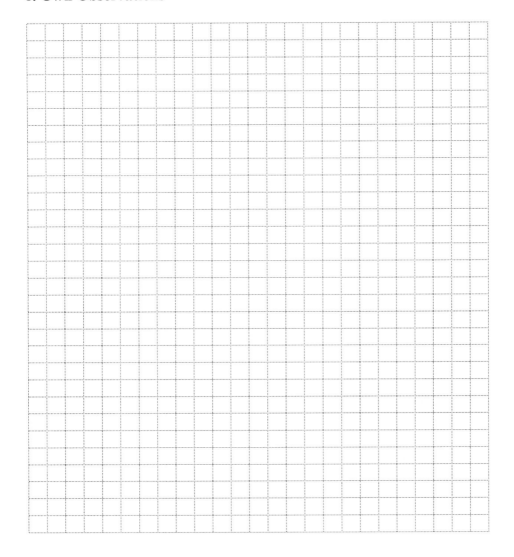

Experiment 7.9

INAPT

1. Purpose

This experiment is the selective version of INEPT (see Exp. 6.7). Here only a particular proton is excited and used for polarization transfer, in order to identify ^{13}C nuclei that are connected to this proton via spin–spin coupling. The experiment is mainly used for detecting long-range interactions and provides a good method for assigning quaternary carbon nuclei. The relative popularity of this experiment in the literature is probably due to the fact that only soft rectangular pulses are used in the proton channel [1,2], and thus it can be implemented on any instrument. Many applications of the INAPT technique have been reported [2–5] and a *J* resolved 2D version is also known [6].

2. Literature

[1] A. Bax, *J. Magn. Reson.* **1984**, *57*, 314–318.

[2] A. Bax, J. A. Ferretti, N. Nahsed, D. M. Jerina, *J. Org. Chem.* **1985**, *50*, 3029–3038.

[3] A. N. Abdel Sayed, L. Bauer, *Tetrahedron* **1988**, *44*, 1883–1892

[4] M. A. Bernstein, *Magn. Reson. Chem.* **1989**, *27*, 659–662.

[5] W. H. Gmeiner, J. W. Lown, *Magn. Reson. Chem.* **1992**, *30*, 101–106.

[6] C. A. Drake, N. Rabjohn, M. S. Tempesta, R. B. Taylor, *J. Org. Chem.* **1988**, *53*, 4555–4562.

3. Pulse Scheme and Phase Cycle

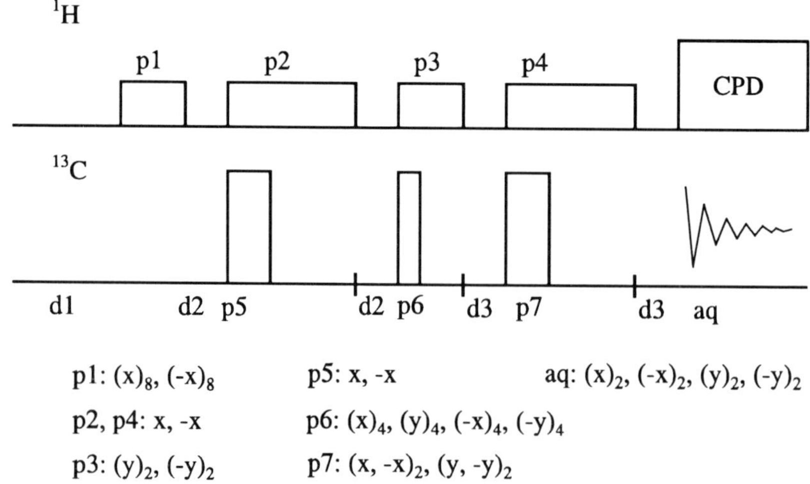

p1: $(x)_8$, $(-x)_8$ p5: x, -x aq: $(x)_2$, $(-x)_2$, $(y)_2$, $(-y)_2$

p2, p4: x, -x p6: $(x)_4$, $(y)_4$, $(-x)_4$, $(-y)_4$

p3: $(y)_2$, $(-y)_2$ p7: $(x, -x)_2$, $(y, -y)_2$

4. Acquisition

Time requirement: 0.5 h

Sample: 2-hydroxynaphthalene, saturated solution in $CDCl_3$

Prior to this experiment the decoupler attenuation for a rectangular soft pulse must be known (see Exp 7.2). Record an ^1H NMR spectrum and note the offsets of the different proton signals. Switch the instrument to ^{13}C operation, record as a reference a normal ^{13}C NMR spectrum, and load the INAPT pulse program. You have to set:

td: 32 k
sw: 55 ppm
o1: center of aromatic region of the ^{13}C NMR spectrum
o2: on resonance of selected ^1H NMR signal
p1, p3: 20 ms rectangular shaped ^1H decoupler pulse at 67 db attenuation
p2, p4: 40 ms rectangular shaped ^1H decoupler pulse at 67 db attenuation
p5, p7: 180° ^{13}C transmitter pulse
p6: 90° ^{13}C transmitter pulse
decoupler attenuation and 90° pulse for CPD
d1: 3 s
d2: 10 ms
d3: 20 ms
ns: 64

5. Processing

Use standard ^{13}C NMR processing as described in Experiment 3.2.

6. Result

In the figure on page 240 **a** is the normal ^1H NMR spectrum recorded on an AMX-500 spectrometer with the assignments obtained by inspection of a NOESY experiment. **b** is the normal ^{13}C NMR spectrum. In **c** the selective proton pulse was adjusted to H-8, giving responses from carbon nuclei C-6, C-10 and C-1, all of which are connected to H-8 via $^3J(C,H)$. Note that the sign of the signals may be positive or negative. In **d** proton H-1 was irradiated, giving responses from C-10, C-8, C-2 and C-3.

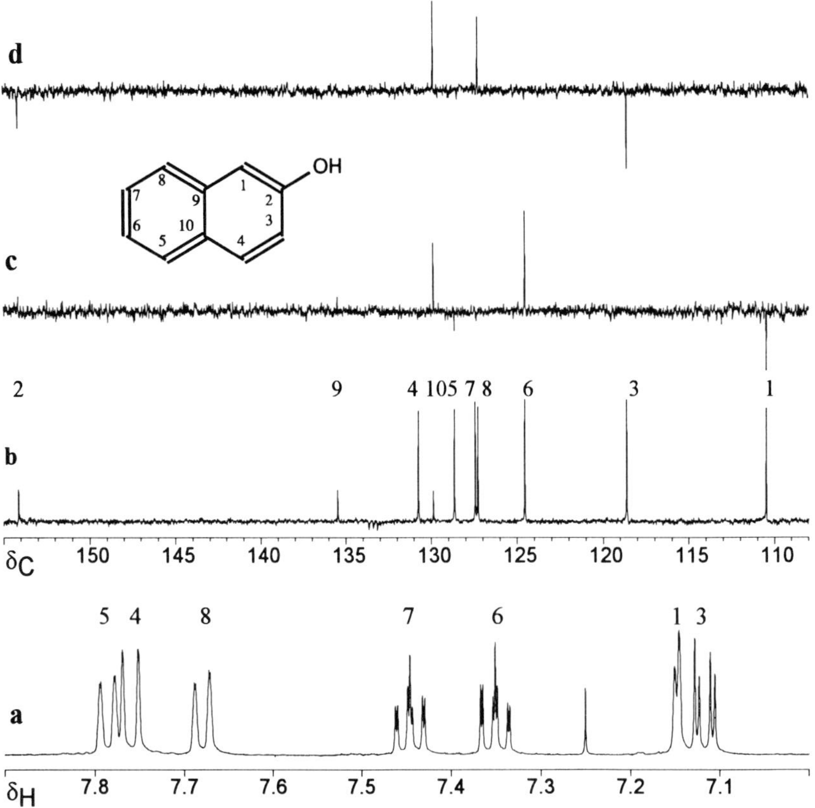

7. Comments

The product operator formalism description is identical with that for the normal refocused INEPT experiment (Exp 6.7) which is given there. Here especially the signals of quaternary carbon nuclei are enhanced by polarization transfer and thus appear with high intensity. Since the experiment can easily be implemented on older instruments it is highly attractive.

8. Own Observations

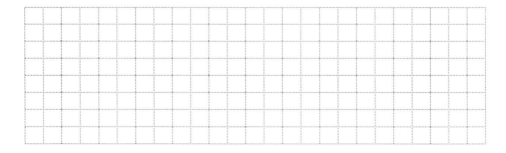

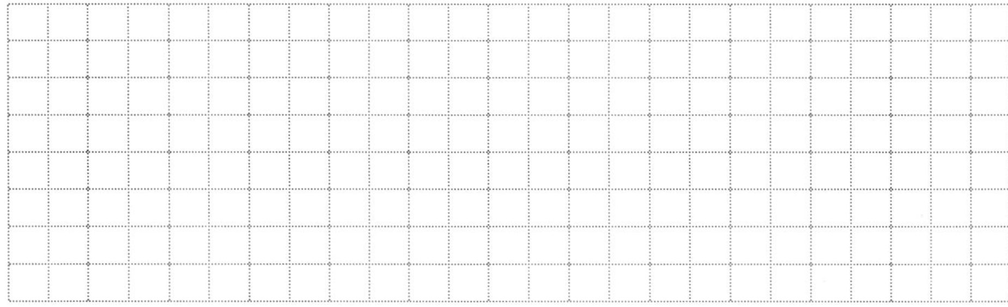

Experiment 7.10

Determination of Long-Range C,H Coupling Constants

1. Purpose

The NMR spectroscopist is often interested in one specific long-range C,H spin coupling constant [1]. There are several methods for simplifying complicated C,H multiplets, such as selective decoupling (see Exp. 4.14), though these methods usually yield residual multiplets which still have to be analyzed by spin simulation. The experiment presented here demonstrates a 2D method related to 2D *J*-resolved spectroscopy and employing a selective pulse. It yields directly the desired spin coupling constant of a chosen C,H pair free of other passive spin couplings. Unlike the original method [2] the pulse sequence given here uses a shaped RE-BURP pulse [3,4].

2. Literature

[1] M. Eberstadt, G. Gemmecker, D. F. Mierke, H. Kessler, *Angew. Chem. Int. Ed. Engl.* **1995**, *34*, 1671–1695.

[2] A. Bax, R. Freeman, *J. Am. Chem. Soc.* **1982**, *104*, 1099–1100.

[3] H. Geen, R. Freeman, *J. Magn. Reson.* **1991**, *93*, 93–141.

[4] T. Fäcke, S. Berger, unpublished results.

3. Pulse Scheme and Phase Cycle

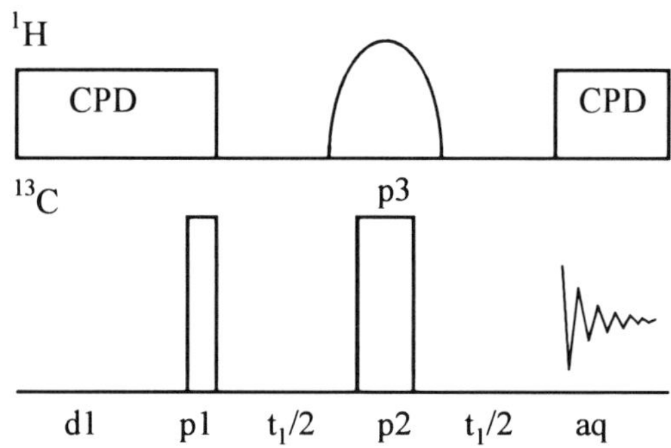

p1: $(x)_4, (y)_4, (-x)_4, (-y)_4$

p2: $x, -x, y, -y, (y, -y, -x, x)_2, -x, x, -y, y$

p3: $x, -x, y, -y, (y, -y, -x, x)_2\ -x, x, -y, y$

aq: $(x)_2, (-x)_2, (y)_2, (-y)_2$

4. Acquisition

Time requirement: 1.5 h

Sample: 20% ethyl crotonate in $CDCl_3$

Record a normal ^{1}H NMR spectrum and note the ^{1}H frequency offset of the methyl group attached to the double bond at $\delta_H = 1.8$. Define and calibrate a RE-BURP pulse for p3 in the ^{1}H channel (see Exp. 7.2). Change to the 2D mode of the spectrometer software. You have to set:

td2: 1 k data points in F_2
td1: 128 data points in F_1
sw2: 200 ppm
sw1: 50 Hz
o1: middle of ^{13}C NMR spectrum
o2: on resonance of the methyl group at $\delta_H = 1.8$
p1: 90° ^{13}C transmitter pulse
p2: 180° ^{13}C transmitter pulse
p3: selective 180° ^{1}H decoupler pulse (40 ms length at 46 dB)
d1: 2 s
initial value for t_1 evolution: 3 μs
increment for t_1 evolution = 1/[2·sw1]
ds: 2
preacquisition delay: as small as possible
ns: 8

5. Processing

Apply zero filling in F_1 to 256 real data points. Use $\pi/2$ shifted sinusoidal windows in both dimensions. Apply complex Fourier transformation corresponding to the N type signal selection using the quadrature off mode in F_1. Phase correction is not necessary, since the data are processed in the magnitude mode.

6. Result

The figure on page 243 shows the result obtained on an AMX-500 spectrometer for the signals of C-2 and C-3. The coupling of these carbon nuclei and the protons of the methyl group C-4 can be seen in F_1 from the splitting of the quartets. Note that in this case the geminal coupling constant ^2J(C-3,H-4) of 7.1 Hz is slightly larger than the vicinal coupling constant ^3J(C-2,H-4) of 6.7 Hz. A reliable assignment of these spin couplings by any other means would be very difficult. Compare the signal patterns with the result obtained in Exp. 10.2.

7. Comments

This method can be thought of as the selective version of the heteronuclear 2D *J*-resolved technique. The selective 180° pulse consisting of p3 and p4 acts only on the chosen protons. At the end of the t_1 period the spin-echo is modulated only by this selected spin coupling. The spin coupling to all other protons is not refocused and is therefore not observable in the final spectrum.

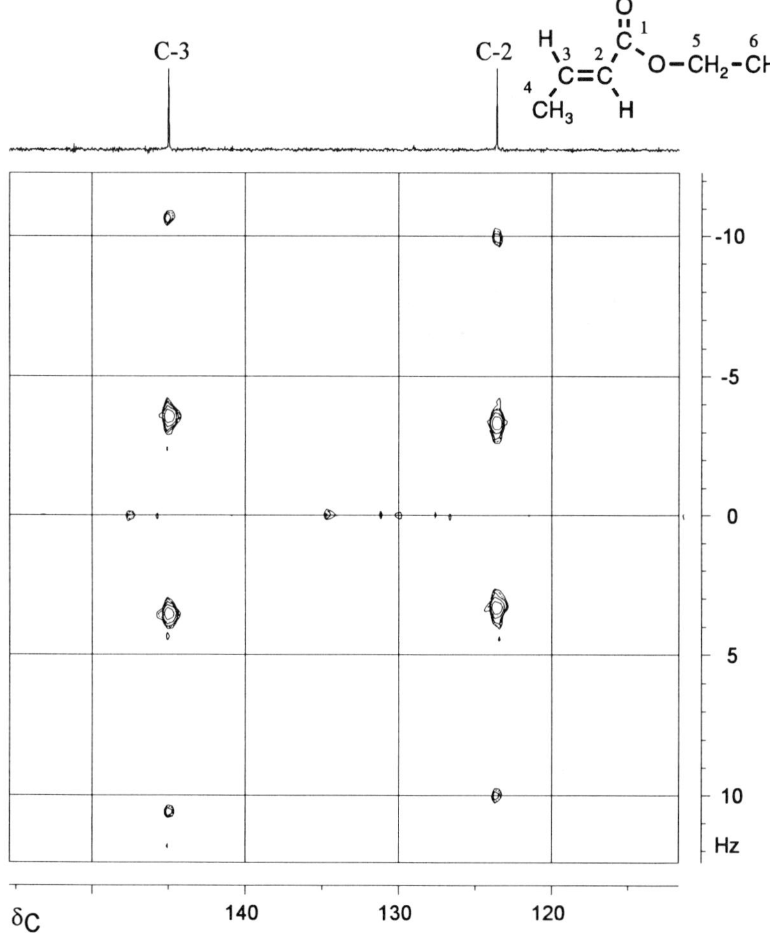

This method provides a unique means of analyzing C,H multiplets without assignment ambiguities, although it is rather insensitive since it is based on a 2D method with ^{13}C detection. The corresponding inverse experiment is described in Experiment 7.11.

8. Own Observations

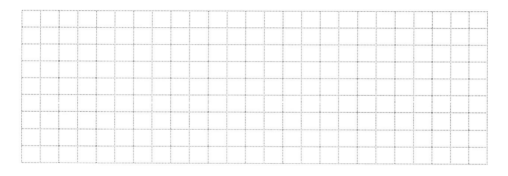

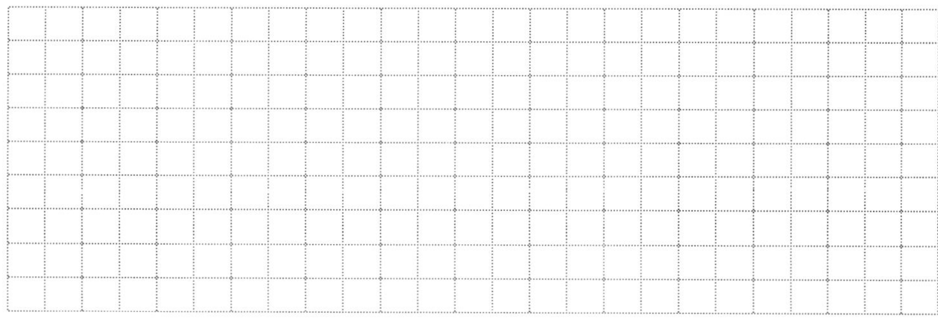

Experiment 7.11

SELRESOLV

1. Purpose

The NMR spectroscopist is often interested in one specific long-range C,H spin coupling constant [1]. There are several methods for simplifying complicated C,H multiplets, such as selective decoupling (see Exp. 4.14), though these methods usually yield residual multiplets which still have to be analyzed by spin simulation. The experiment presented here demonstrates a 2D method related to 2D *J*-resolved spectroscopy and employing a selective pulse [2]. It yields directly the desired spin coupling constant of a chosen C,H pair independent of other passive spin couplings. In contrast to Experiment 7.10, however, the SELRESOLV method is a proton detected experiment and hence more sensitive.

2. Literature

[1] M. Eberstadt, G. Gemmecker, D. F. Mierke, H. Kessler, *Angew. Chem. Int. Ed. Engl.* **1995**, *34*, 1671-1695.

[2] M. Ochs, S. Berger, *Magn. Reson. Chem.* **1990**, *28*, 994–997.

3. Pulse Scheme and Phase Cycle

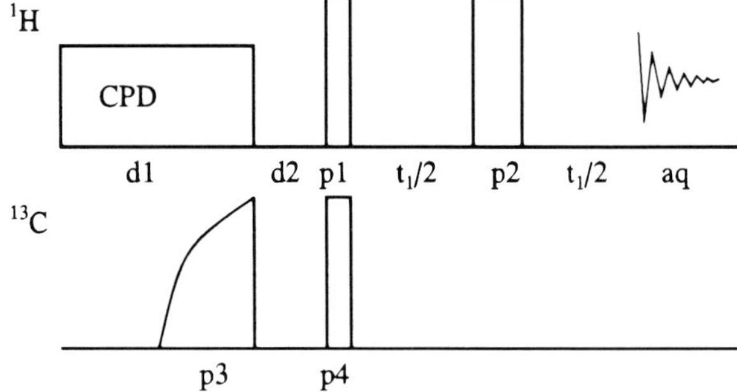

p1: -x, x, x, -x, -y, y, y, -y
p2: $(x)_2$, $(-x)_2$, $(y)_2$, $(-y)_2$
p3: x, -x, -x, x, y, -y, -y, y
p4: $(y)_2$, $(-y)_2$, $(-x)_2$, $(x)_2$
aq: $(x)_2$, $(-x)_2$, $(y)_2$, $(-y)_2$

4. Acquisition

Time requirement: 1.2 h

Sample: 20% ethyl crotonate in $CDCl_3$

Record a normal ^{13}C NMR spectrum and note the offset of the olefinic carbon nucleus C-2 at δ_C = 123.6. Change to ^{1}H observation with ^{13}C decoupling (inverse mode on older instruments) and calibrate a 10 ms soft pulse p3 with half-Gaussian shape as decoupler pulse in the ^{13}C channel. Change to the 2D mode of the spectrometer software. You have to set:

td2: 2 k data points in F_2
td1: 32 data points in F_1
sw2: 1 ppm
sw1: 45 Hz
o1: center of methyl group region in ^1H NMR spectrum
o2: on resonance of the olefinic carbon atom C-2 at $\delta_C = 123.6$
p1: 90° ^1H transmitter pulse
p2: 180° ^1H transmitter pulse
p3: selective 90° ^{13}C decoupler pulse, half-Gaussian shape (10 ms length at 66 dB)
p4: 90° ^{13}C decoupler pulse
d1: 6 s
d2: $1/[2J(C,H)] = 50$ ms, calculated from $^nJ(C,H) = 10$ Hz
^1H transmitter attenuation and 90° pulse width for broad-band presaturation [28 dB, 100 μs]
initial value for t_1 evolution: 3 μs
increment for t_1 evolution = $1/[2 \cdot sw1]$
preacquisition delay: as small as possible
ds: 4
ns: 16

5. Processing

Apply zero filling in F_2 and in F_1 to 2 k and 128 real data points respectively. Use unshifted sinusoidal windows in both dimensions. Apply complex Fourier transformation corresponding to the N type signal selection using the quadrature off mode in F_1. Phase correction is not necessary, since the data are processed in the magnitude mode.

6. Result

The figure on page 246 shows the result obtained on an AMX-500 spectrometer, with the region of the methyl group attached to the olefinic carbon expanded. The ^3J(C-2, H-4) spin coupling of 6.6 Hz can be seen in F_2, whereas the homonuclear couplings to the two olefinic protons are observed in F_1. This long range C,H coupling constant can also be obtained by Experiment 7.10, using ^{13}C rather than proton detection, and is then observed in the F_1 dimension. Note that some axial peak breakthrough at 0 Hz in F_1 is unavoidable; residual signals from protons bound to ^{12}C may also appear, but are outside the region of interest. Along the F_1 axis the normal signal of H-4 is shown.

7. Comments

The method consists of a selective reverse INEPT transfer from ^{13}C to protons followed by a subsequent 2D J-resolved sequence. In this 2D part only the proton selected by the ^{13}C selective pulse is active. The homonuclear spin couplings of this proton are observed in the F_1 dimension, leaving only the heteronuclear spin coupling to the chosen carbon nucleus in F_2. The signals of protons bound to ^{12}C are suppressed by the phase cycle and the presaturation period.

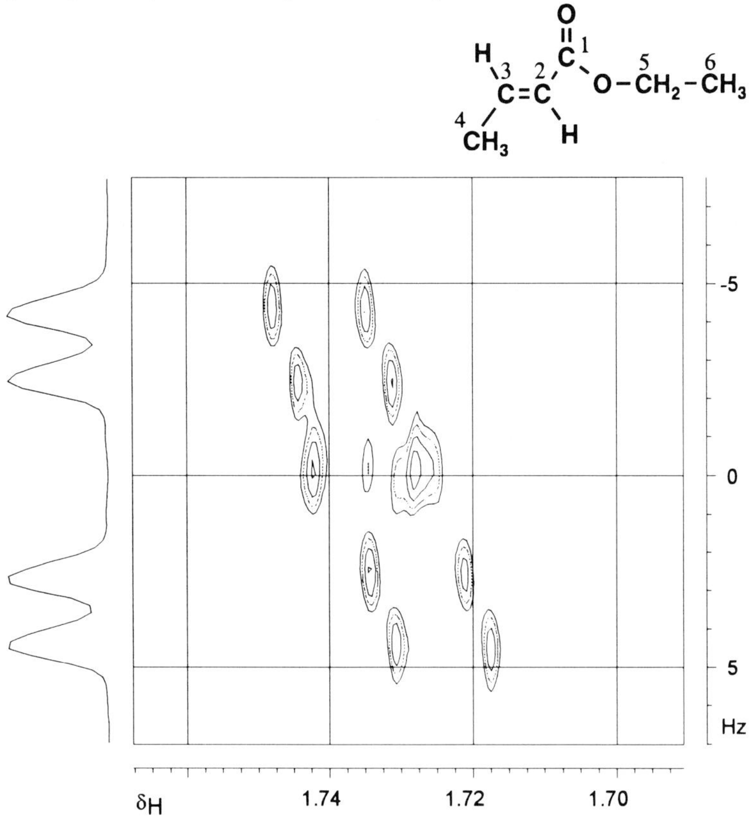

This is a considerable drawback of the method; thus the sensitivity gain obtained through inverse detection is somewhat diminished due to the necessary suppression of unwanted coherences. More effective approaches use pulsed field gradients (Chapters 11 and 12).

8. Own Observations

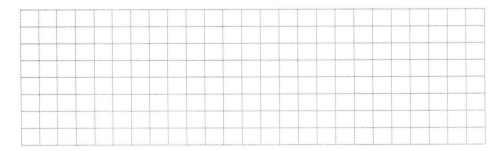

Experiment 7.12

SERF

1. Purpose

The NMR spectroscopist is often interested in determining a specific H,H spin coupling constant [1]. There are several methods for simplifying complicated multiplets, such as homonuclear decoupling (see Exps. 4.4 and 4.5) or selective COSY (see Exps. 7.5 and 11.8); however, these methods usually yield residual multiplets which still have to be analyzed by spin simulation. The SERF (**SE**lective **ReF**ocussing) experiment [2] presented here is a 2D method employing two selective pulses. It directly yields the desired coupling constant of a chosen spin pair without other passive spin couplings.

2. Literature

[1] M. Eberstadt, G. Gemmecker, D. F. Mierke, H. Kessler, *Angew. Chem. Int. Ed. Engl.* **1995**, *34*, 1671-1695.

[2] T. Fäcke, S. Berger, *J. Magn. Reson. Ser. A* **1995**, *113*, 114–116.

3. Pulse Scheme and Phase Cycle

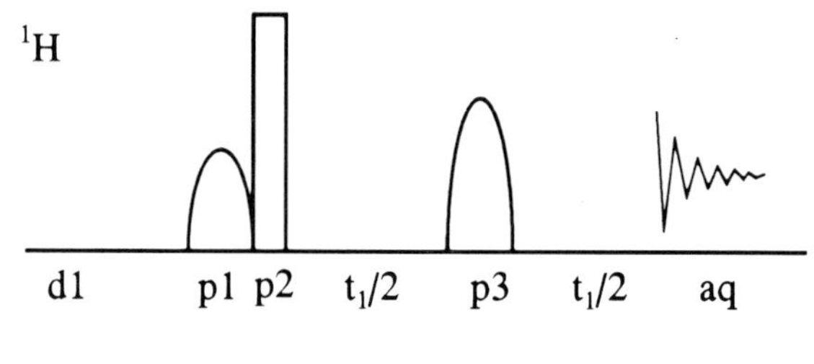

p1: x, -x, -x, x p3: x

p2: y, y, -y, -y aq: x, -x, -x, x

4. Acquisition

Time requirement: 20 min

Sample: 5% ethyl crotonate in $CDCl_3$

Record a normal ^1H NMR spectrum and note the offsets of the olefinic protons and of the methyl group attached to the double bond. Define and calibrate for p1 an E-BURP pulse shape and check its phase relative to the purging pulse p2. Define and calibrate for p3 a double selective RE-BURP pulse shape, so that this pulse acts simultaneously on the olefinic proton at δ_H = 6.9 (H-3) and the methyl group at δ_H = 1.8. For a second spectrum set this pulse to act on both olefinic protons. Change to the 2D mode of the spectrometer software. You have to set:

td2: 1 k data points in F_2
td1: 64 data points in F_1
sw2: 8 ppm
sw1: 50 Hz
o1: middle of ^{1}H NMR spectrum
p1: selective 90° ^{1}H transmitter pulse, E-BURP shape; 50 ms length at 55 dB was used here
p2: 5 ms ^{1}H purging pulse
p3: double selective 180° ^{1}H transmitter pulse, RE-BURP shape; 50 ms length at 45 dB was used here
d1: 2 s
initial value for t_1 evolution: 3 µs
increment for t_1 evolution = 1/[2·sw1]
preacquisition delay: as small as possible
ds: 2
ns: 4

5. Processing

Apply zero filling in F_1 to 128 real data points. Use unshifted sinusoidal windows in both dimensions. Apply complex Fourier transformation corresponding to the N type signal selection using the quadrature off mode in F_1. Phase correction is not necessary, since the data are processed in magnitude mode.

6. Result

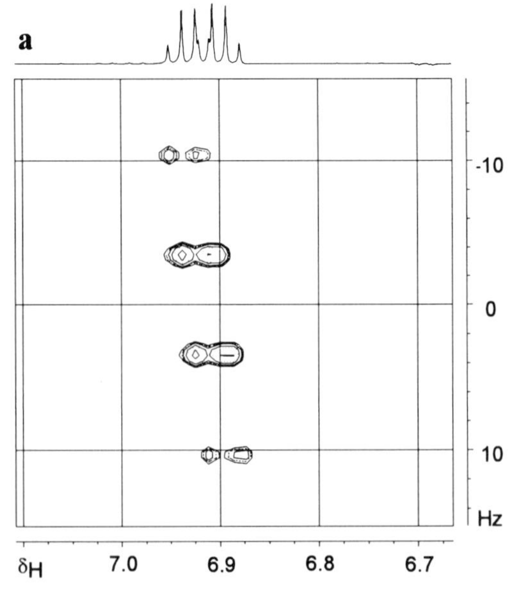

The figures on pages 248 and 249 show the results obtained on an AMX-500 spectrometer. In spectrum **a** the double selective pulse was set to act on the olefinic proton at $\delta_H = 6.9$ (H-3) and on the methyl protons at $\delta_H = 1.8$. Only J(H-3,CH_3) is observed in F_1, whereas J(H-2,CH_3) appears in F_2. In spectrum **b** the double selective pulse was set to act on both olefinic protons, so that only the large olefinic spin coupling constant can be seen in F_1. Compare the signal patterns with the result of Experiment 10.1.

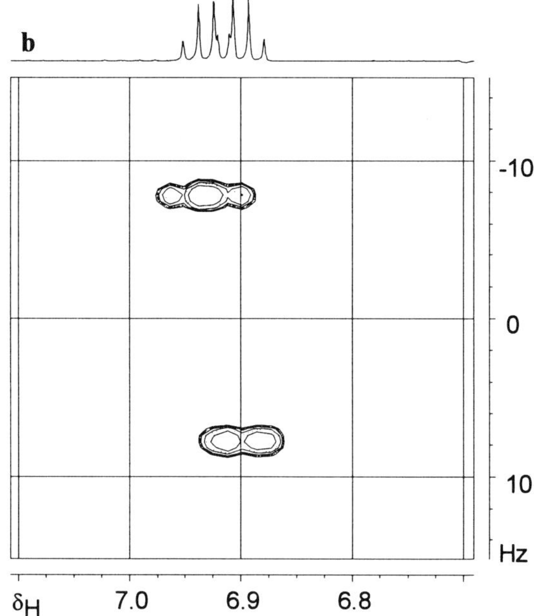

7. Comments

This method can be thought of as the double selective version of the 2D J-resolved technique. After excitation by the selective pulse p1 the selected proton develops spin–spin coupling to all other protons that are coupled to it. The purging pulse p2 suppresses unwanted coherences. The double selective pulse p3 acts only on the chosen spin pair so that at the end of the $t1$ period the spin echo is only modulated by this selected spin coupling.

8. Own Observations

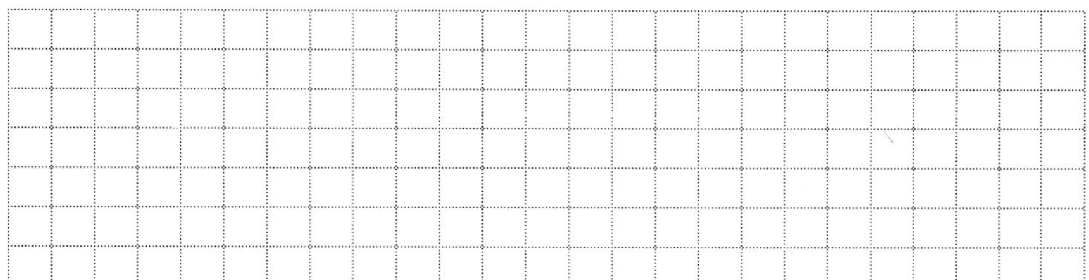

Chapter 8

Auxiliary Reagents, Quantitative Determinations, and Reaction Mechanism

This chapter describes typical applications of routine NMR spectroscopy in organic, inorganic and physical organic chemistry. The emphasis is therefore not on how to perform a special pulse sequence, to set up a certain 2D file, or to tune the spectrometer to a seldom used heteronucleus; instead the NMR experiments shown here mainly use the simple "zero/go" acquisition sequence given in Experiment 3.1, and additionally introduce special reagents such as lanthanide shift reagents or Pirkle's reagent to illustrate how one can obtain meaningful information with these auxiliaries. Various methods how to determine enantiomeric purity and even absolute configuration by NMR are demonstrated. Important effects, like the aromatic solvent induced shifts or the H/D exchange are illustrated. Several experiments involving paramagnetic species are shown, and an important experiment in physical organic chemistry, the CIDNP effect, is described. The analytical application of NMR for various quantitative determinations is also demonstrated.

Literature

[1] M. L. Martin, J. J. Delpuech, G. J. Martin, *Practical NMR Spectroscopy,* Heyden, London, **1980,** Chs. 9 and 10.

[2] L. D. Field, S. Sternhell (eds.) *Analytical NMR*, Wiley, Chichester, **1989**

[3] D. Wendisch, *Appl. Spectrosc. Rev.* **1993**, *28*, 165–229.

[4] H. Duddeck in Houben-Weyl, *Methods in Organic Chemistry, Stereoselective Synthesis*, G. Helmchen, R. W. Hoffmann, J. Mulzer, E. Schaumann (eds.), Thieme, Stuttgart, **1995**, *E21a*, 293–377.

Experiment 8.1

Signal Separation Using a Lanthanide Shift Reagent

1. Purpose

Lanthanide shift reagents are used for simplifying complex NMR spectra. Chiral lanthanide shift reagents can also be used to determine enantiomeric purity (see Exp. 8.2). In the experiment described here tris[1,1,1,2,2,3,3-heptafluoro-7,7-dimethyloctane-4,6-dionato]-europium, Eu(fod)$_3$, is used to separate multiplets of an alkyl chain.

2. Literature

[1] J. Reuben, *Prog. Nucl. Magn. Reson. Spectrosc.* **1975**, *9*, 1–70.
[2] O. Hofer, *Top. Stereochem.* **1976**, *9*, 111–197.
[3] G. R. Sullivan, *Top. Stereochem.* **1978**, *10*, 288–329.
[4] T. C. Morrill (ed.): *Lanthanide Shift Reagents in Stereochemical Analysis*, in *Methods in Stereochemical Analysis*, VCH, Weinheim, **1985**.
[5] J. A. Peters, J. Huskens, D. J. Raber, *Prog. NMR Spectrosc.* **1996**, *28*, 283–350.

3. Pulse Scheme and Phase Cycle

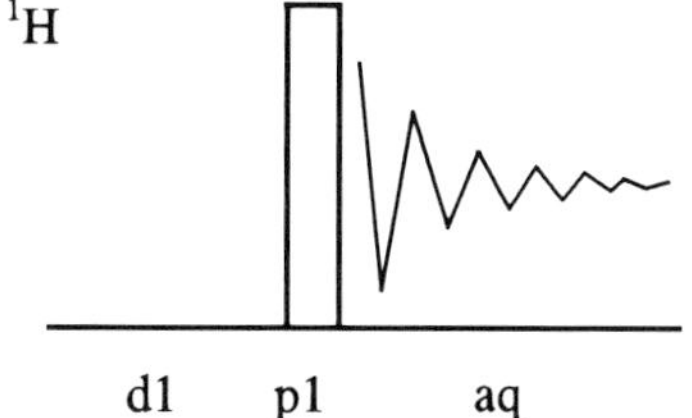

p1: x, x, -x, -x, y, y, -y, -y
aq: x, x, -x, -x, y, y, -y, -y

4. Acquisition

Time requirement: 30 min

Sample: 10 mg 1-octanol in 0.7 ml $CDCl_3$

Load standard ^{1}H NMR acquisition parameters (see Exp. 3.1), set the spectral width to 15 ppm, and record a spectrum of 1-octanol. Add a *very* small portion (about 1 mg) of tris[1,1,1,2,2,3,3-heptafluoro-7,7-dimethyloctane-4,6-dionato]-europium, Eu(fod)$_3$, allow sufficient time to reach equilibrium, and record the spectrum again. Increase the amount of the shift reagent until all the seven CH_2 signals are separated.

For quantitative work it is advisable to prepare substrate and reagent stock solutions. Fill six NMR tubes with 0.1 ml substrate stock solution and varying amounts of reagent solution. Add solvent to each NMR tube to give the same total amount of liquid, so that the substrate concentration remains constant. For experiments with constant susceptibility the reagent concentration is kept constant and the substrate concentration is varied.

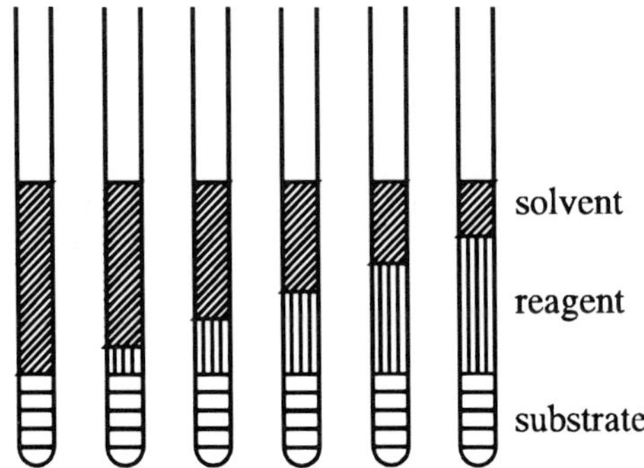

5. Processing

Use standard 1D processing for ^{1}H NMR spectra (see Exp. 3.1).

6. Result

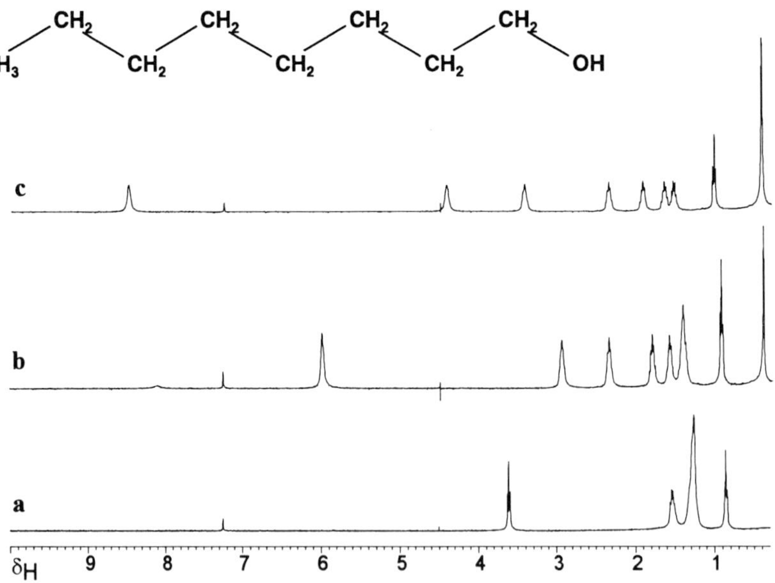

The figure shows in **a** the ^{1}H NMR spectra of 1-octanol without Eu(fod)$_3$ and in **b** and **c** with increasing amounts of Eu(fod)$_3$, obtained on an AM-400 spectrometer. The signals which are shifted most are also severely broadened. The signal of the OH group at $\delta_H = 8.2$ in **b** is no longer visible in **c**. Note that in spectrum **c** the CH_2O group now has a chemical shift of $\delta_H = 8.5$ and all CH_2 groups of the octyl chain are separated. The signal at $\delta_H = 0.5$ stems from the *t*-butyl group of the shift reagent Eu(fod)$_3$.

7. Comments

Lanthanide shift reagents act like an additional magnetic field in the sample and dramatically change the chemical shifts of the signals, especially for nuclei in the vicinity of the complexation site ("the poor man's 1 GHz NMR spectrometer"). Thus, they make it possible to separate signals or to simplify an NMR spin system. The use of lanthanide shift reagents usually gives an unambigious answer to problems such as *cis/trans*, *E/Z*, *endo/exo* or *syn/anti* assignments. The change in chemical shift resulting from the addition of a shift reagent can be expressed as the sum of three components, as given by Equation (1).

$$\Delta = \Delta_{DIA} + \Delta_{CON} + \Delta_{DIP} \qquad (1)$$

The diamagnetic contribution Δ_{DIA} is caused by the complexation of the substrate, whereas the contact contribution Δ_{CON} has its origin in the delocalization of electron spin density from the lanthanide ion to the substrate. Usually both are small and can be neglected. The dipolar contribution Δ_{DIP} depends on the distance r between the lanthanide ion and the nuclear spin being observed, and the angle θ between the principal magnetic dipolar axis of the complex and the distance vector. The complex usually has axial symmetry in solution and the McConnell–Robertson Equation (2) holds:

$$\Delta_{DIP} = K\,(3\cos^2\theta - 1)\, r^{-3} \qquad (2)$$

This equation can be used to give quantitative information about the structure or conformation of a compound, if the bound shifts are known [3]. Lanthanide shift experiments are only successful when the substrate acts as a Lewis base to which complexing can occur. The degree of Lewis basicity decreases in the following order:

$RNH_2 > ROH > RCOR > RCOOR > RCN$

The commercially available lanthanide shift reagents are 1,3 diketone complexes. In contrast to europium, praseodymium complexes shift the signals to lower frequency, while the ytterbium reagents are best used in ^{13}C NMR spectroscopy. Binuclear shift reagents such as Ag(fod)/Eu(fod)$_3$ can be used for unsaturated hydrocarbons.

8. Own Observations

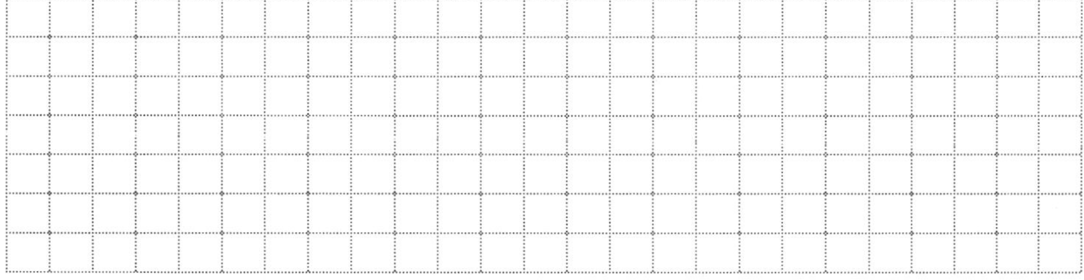

Experiment 8.2

Signal Separation of Enantiomers Using a Chiral Shift Reagent

1. Purpose

Chiral lanthanide shift reagents are used for the determination of enantiomeric purity (for other techniques see Exps. 8.3 to 8.5). In this experiment the chiral tris[3-(heptafluoropropylhydroxymethylene)-*d*-camphorato]-praseodymium complex, $Pr(hfc)_3$, is used, which (unlike europium, Exp. 8.1) results in low frequency rather than high frequency shifts.

2. Literature

[1] J. Reuben, *Prog. Nucl. Magn. Reson. Spectrosc.* **1975**, *9*, 1–70.
[2] O. Hofer, *Top. Stereochem.* **1976**, *9*, 111–197.
[3] G. R. Sullivan, *Top. Stereochem.* **1978**, *10*, 288–329.
[4] W. H. Pirkle, D. J. Hoover, *Top. Stereochem.* **1982**, *13*, 263–331.
[5] T. C. Morrill (ed.): *Lanthanide Shift Reagents in Stereochemical Analysis*, in *Methods in Stereochemical Analysis*, VCH, Weinheim, **1985**.
[6] D. Parker, *Chem. Rev.* **1991**, *91*, 1441–1457.

3. Pulse Scheme and Phase Cycle

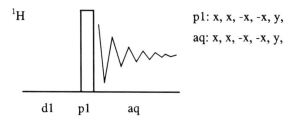

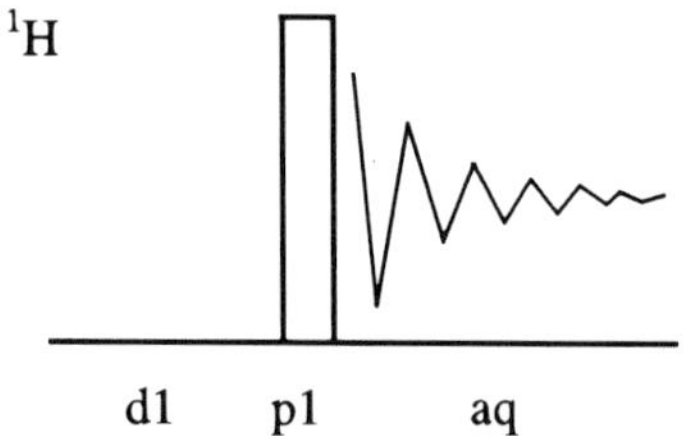

p1: x, x, -x, -x, y, y, -y, -y
aq: x, x, -x, -x, y, y, -y, -y

4. Acquisition

Time requirement: 1 h

Sample: 10 mg rac 1-phenylethanol in 0.7 ml $CDCl_3$

Load standard ^1H NMR acquisition parameters, set the spectral width to 15 ppm and adjust the offset so that signals up to $\delta_H = -3$ are covered. Record a spectrum of 1-phenylethanol, add a very small portion (about 1 mg) of tris[3-(heptafluoropropyl-hydroxy-methylene)-*d*-camphorato]-praseodymium, $Pr(hfc)_3$, allow sufficient time to reach equilibrium, and record the spectrum again. Increase the amount of the chiral shift reagent until the separation of the two quartets of the methine group (initially at $\delta_H = 4.9$) is sufficient for reasonable integration.

5. Processing

Use standard 1D processing for ^{1}H NMR spectra (see Exp. 3.1).

6. Result

The figure shows the ^{1}H NMR spectrum of rac 1-phenylethanol (**a**) in the absence of and (**b**) in the presence of Pr(hfc)$_3$. The signals originally at $\delta_H = 1.5$ (CH_3) and $\delta_H = 4.9$ (CH) are shifted to $\delta_H = -0.3$ and $\delta_H = 1.7$ and 1.9 in spectrum **b**. The methine signals of the two enantiomers are completely separated, but are also significantly broadened. The signals of the methyl group are not yet separated. They begin to separate when the concentration of the added shift reagent is such that the signal is shifted to $\delta_H = -2$ ppm.

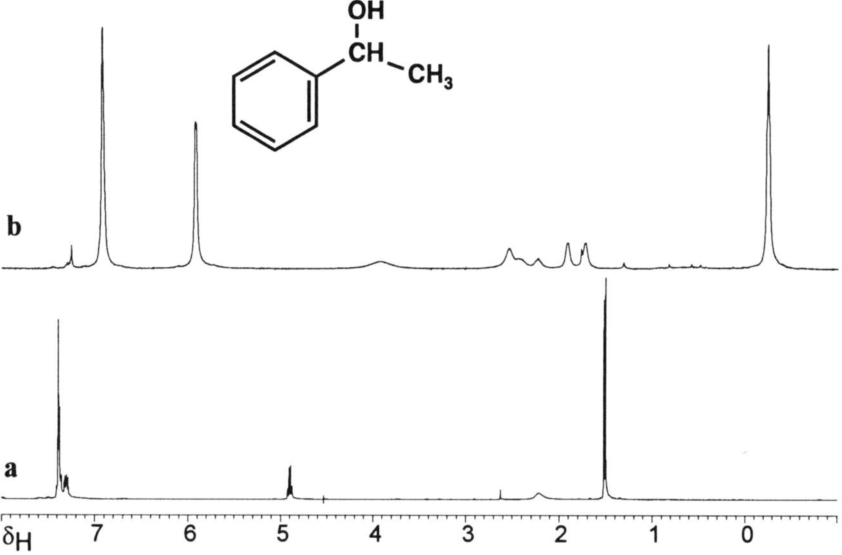

7. Comments

Chiral shift reagents (L_S) form diastereomeric complexes with the substrate molecules S_S and S_R which are in a rapid equilibrium with the uncomplexed species:

$$L_S + S_S \rightleftharpoons [L_S S_S]$$
$$L_S + S_R \rightleftharpoons [L_S S_R]$$

With increasing concentration of the reagent the equilibrium is shifted to the right and the lanthanide-induced shifts increase. Since the shift reagent is paramagnetic one observes significant line broadening at higher concentrations of the shift reagent, especially for nuclei (in this case protons) which are near to the complexation site. Instead of performing several weighing procedures for quantitative work it is better to start with the highest concentration of the lanthanide shift reagent and to dilute by means of a substrate stock solution. Another method is to prepare stock solutions of the reagent and the substrate and to increase the ratio of reagent solution to substrate solution, while always keeping the same total volume of solution by adding solvent (see figure in Exp. 8.1).

Normally the enantiomeric excess *ee* is calculated using to the following equation:

$$ee = (I_1 - I_2) / (I_1 + I_2)$$

where I_1 and I_2 are the integrals of the corresponding signals. The *ee* value is zero for the example shown, because the racemate was investigated. Although it is no problem to separate and integrate the signals of racemates, this method has its limitations if the ratio of the integrals of the enantiomers is about 9 : 1 or greater. The choice of a reagent shifting the signals to lower frequencies, as shown here, compared with one shifting to higher frequencies (see Exp. 8.1) must be carefully considered for each substrate.

8. Own Observations

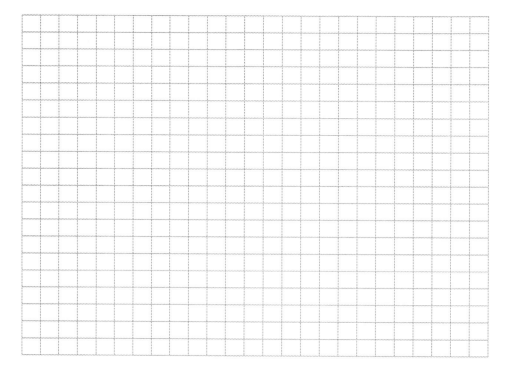

Experiment 8.3

Signal Separation of Enantiomers Using a Chiral Solvating Agent

1. Purpose

This ^{1}H NMR experiment is used to prove the presence of enantiomers or to determine the enantiomeric purity of a compound. A chiral solvating agent, R-(–)-α-acetoxy-phenylacetic acid (R-(–)-O-acetyl-mandelic acid), which is commercially available in pure enantiomeric form is employed [1–3]. A similar application is described in Experiment 8.4.

2. Literature

[1] G. R. Sullivan, *Top. Stereochem.* **1978**, *10*, 288–329.
[2] W. H. Pirkle, D. J. Hoover, *Top. Stereochem.* **1982**, *13*, 263–331.
[3] D. Parker, *Chem. Rev.* **1991**, *91*, 1441–1457.

3. Pulse Scheme and Phase Cycle

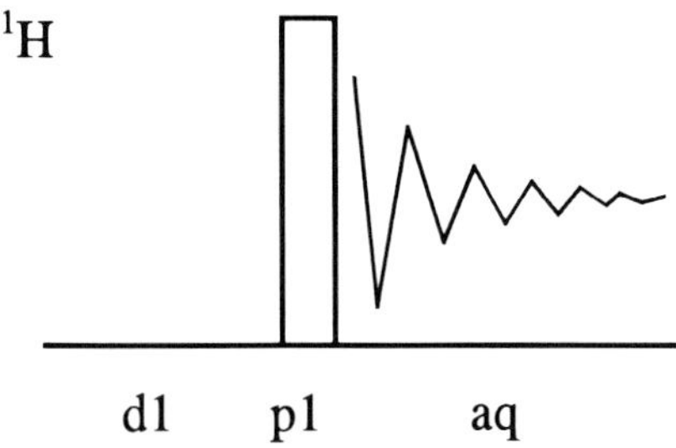

p1: x, x, -x, -x, y, y, -y, -y
aq: x, x, -x, -x, y, y, -y, -y

4. Acquisition

Time requirement: 20 min

Sample: 10 mg rac 1-phenylethylamine in 0.7 ml $CDCl_3$

Load standard ^{1}H NMR acquisition parameters (see Exp. 3.1) and record a spectrum of 1-phenylethylamine. Add about 10 mg of R-(–)-O-acetyl-mandelic acid and record the spectrum again. Increase the amount of the chiral solvating agent until the separation of the two doublets of the methyl group at δ_H = 1.2 is sufficient for integration.

5. Processing

Use standard 1D processing for ^1H NMR spectra (Exp. 3.1).

6. Result

The figure shows the ^1H-NMR spectrum of rac 1-phenylethylamine obtained on an AM-400 spectrometer (**a**) without and (**b**) in the presence of the chiral solvating agent R-(−)-O-acetyl-mandelic acid. The signals at $\delta_H = 1.2$ (CH_3) and at $\delta_H = 4.05$ (CH) are doubled and the CH_3 signals are suitably separated for integration. The phenylethylamine signals in (**b**) are slightly upfield from those in (**a**).

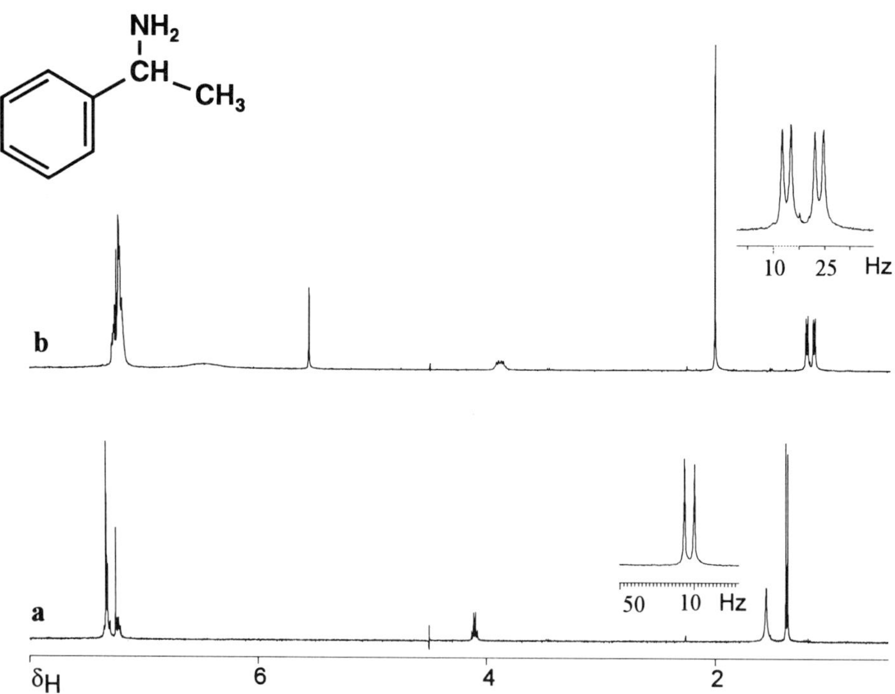

7. Comments

Chiral solvating agents form diastereomeric solvation complexes which are in rapid equilibrium with the uncomplexed species. Solvents with low solvating ability should therefore be used, such as $CDCl_3$, CCl_4 or C_6D_6. Other common commercially available agents are 2,2,2-trifluoro-1-phenylethanol and 1-(9-anthryl)-2,2,2-trifluoroethanol (Pirkle's reagent, see Exp. 8.4). In the example presented here the chiral

auxiliary forms a salt with the basic phenylethylamine. The salt is still in rapid equilibrium with the base and the acid, so this is not a derivatizing agent like, for example, Mosher esters.
For quantitative determinations use stock solutions of the substrate and the chiral solvating agent to make up a series of solutions of constant volume containing a fixed amount of substrate solution and varying amounts of the chiral agent solution (see figure in Exp. 8.1).

The enantiomeric excess *ee* can be calculated from the following equation:

$$ee = (I_1 - I_2) / (I_1 + I_2)$$

where I_1 and I_2 are the integrals of the corresponding signals. The *ee*-value is zero for the example given, as the racemic form was investigated.

8. Own Observations

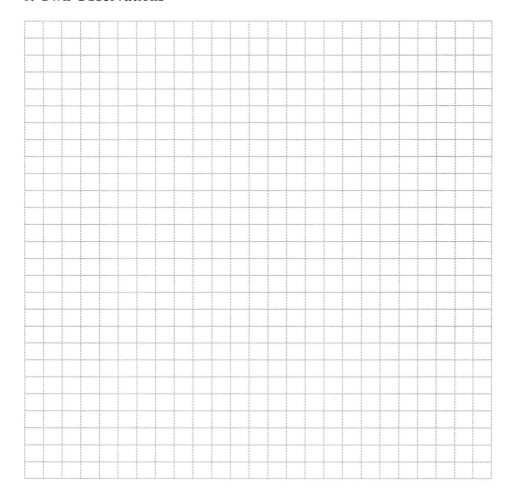

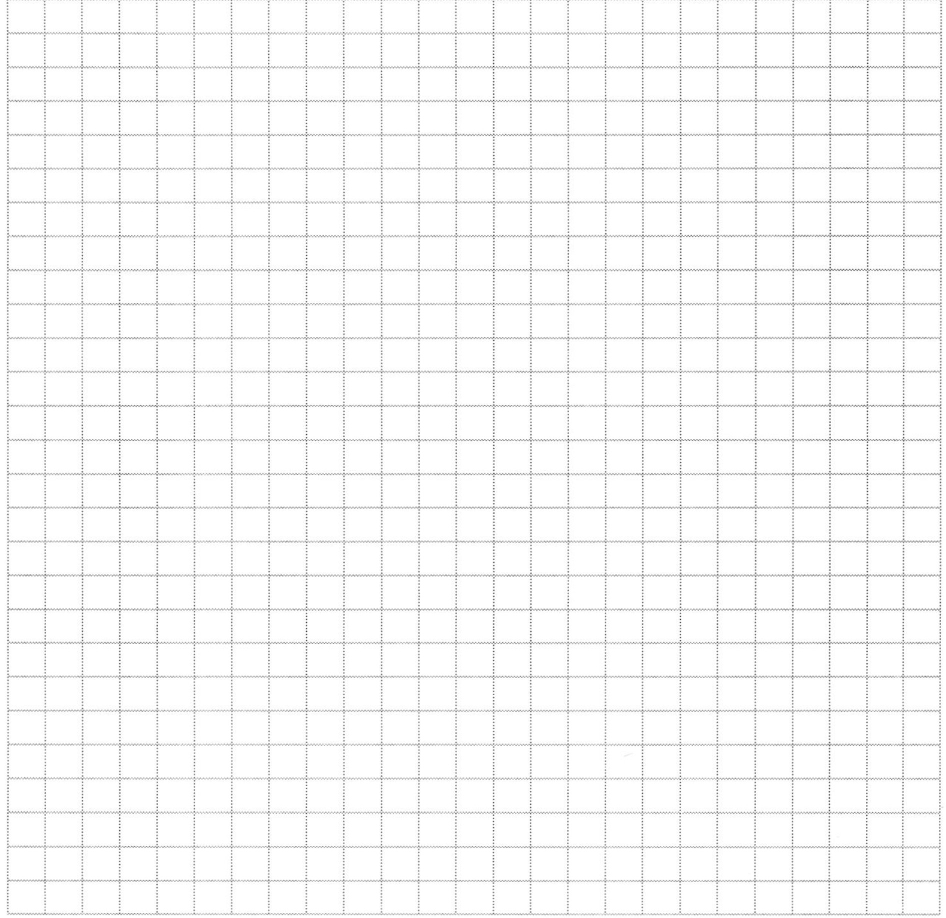

Experiment 8.4

Determination of Enantiomeric Purity with Pirkle's Reagent

1. Purpose

This ^{1}H NMR experiment is used to prove the presence of enantiomers or to determine the enantiomeric purity of a compound, and employs a chiral solvating agent, 1-(9-anthryl)-2,2,2-trifluoroethanol (Pirkle's reagent), which is commercially available in both pure enantiomeric forms [1–3]. A similar application is described in Experiment 8.3.

2. Literature

[1] G. R. Sullivan, *Top. Stereochem.* **1978**, *10*, 288–329.
[2] W. H. Pirkle, D. J. Hoover, *Top. Stereochem.* **1982**, *13*, 263–331.
[3] D. Parker, *Chem. Rev.* **1991**, *91*, 1441–1457.

3. Pulse Scheme and Phase Cycle

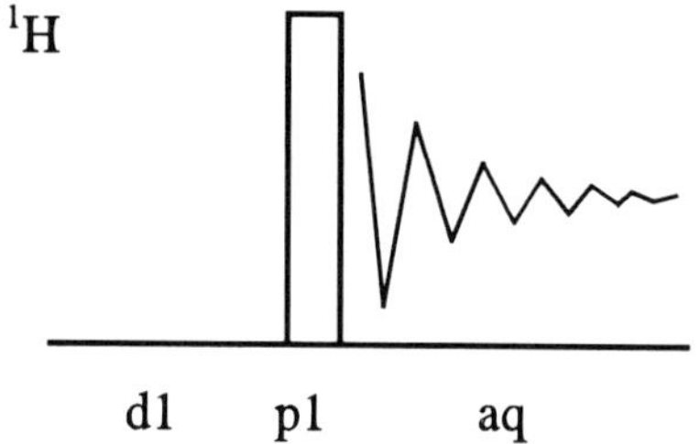

p1: x, x, -x, -x, y, y, -y, -y
aq: x, x, -x, -x, y, y, -y, -y

4. Acquisition

Time requirement: 30 min

Sample: 10 mg rac 1-phenylethanol in 0.7 ml $CDCl_3$

Load standard ^{1}H NMR acquisition parameters (see Exp. 3.1) and record a spectrum of 1-phenylethanol. Add about 40 mg of S-(+)-1-(9-anthryl)-2,2,2-trifluoroethanol and record another ^{1}H NMR spectrum. Increase the amount of the chiral solvating agent until the separation of the two doublets of the methyl group at δ_H = 1.2 is sufficient.

5. Processing

Use standard 1D processing for ^1H NMR spectra (see Experiment 3.1). A Gaussian window (gb = 0.2, lb = –0.5 Hz) was used to process spectrum **b.**

6. Result

The figure shows the ^1H NMR spectrum of rac 1-phenylethanol obtained on an AM-400 spectrometer (**a**) without and (**b**) in the presence of the chiral solvating agent S-(+)-1-(9-anthryl)-2,2,2-trifluoroethanol. The signals at $\delta_H = 1.2$ (CH_3) and at $\delta_H = 4.9$ (CH) are doubled. Note, however, that the separation of the two CH_3 doublets is not sufficient for an integation. This is due to a weaker interaction of the the chiral solvating agent with the 1-phenylethanol, which is less basic than 1-phenylethylamine (see Exp. 8.3). On the other hand the chiral auxiliary used here is not acidic enough to form strong solvation complexes. Better results were achieved for 1-phenylethanol with a chiral shift reagent (see Exp. 8.2).

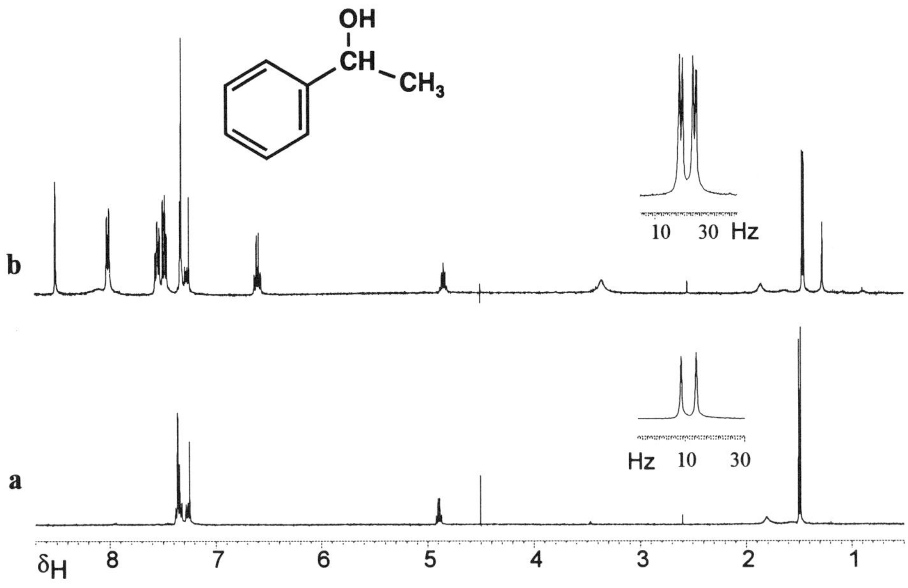

7. Comments

Chiral solvating agents form diastereomeric solvation complexes, which are in rapid equilibrium with the uncomplexed species. Solvents with low solvating ability should therefore be used, such as $CDCl_3$, CCl_4 or C_6D_6. Frequently used commercially available agents are 2,2,2-trifluoro-1-phenylethanol, or R-(−)-O-acetyl-mandelic acid (see Exp. 8.3). It is difficult to predict which chiral solvating agent is the best for a certain compound. The enantiomeric excess *ee* can be calculated from the following relationship:

$$ee = (I_1 - I_2) / (I_1 + I_2)$$

where I_1 and I_2 are the integrals of the corresponding signals. For the example shown here the *ee* value is zero, because the racemate was used.

8. Own Observations

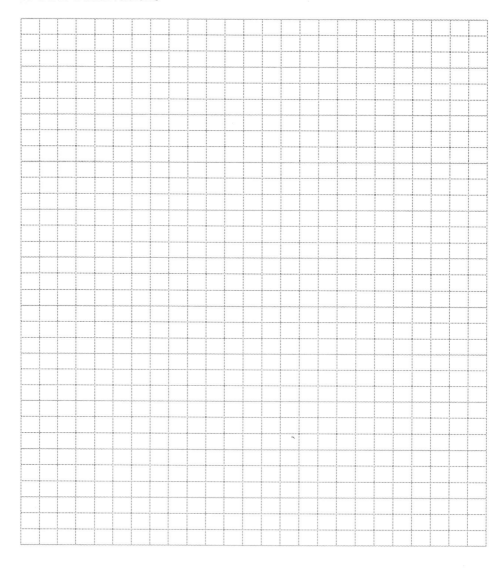

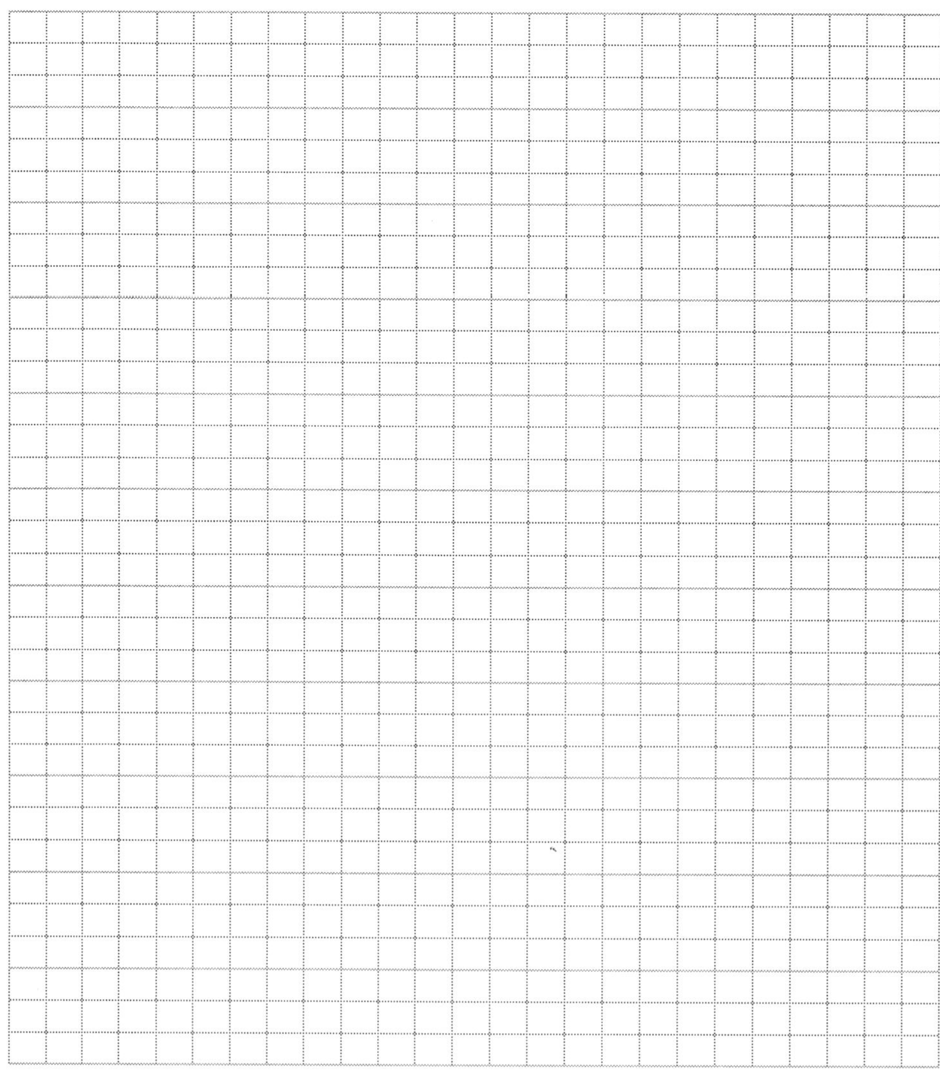

Experiment 8.5

Determination of Enantiomeric Purity by ^{31}P NMR

1. Purpose

Usually the enantiomeric excess (*ee*) is determined by NMR spectroscopy with the aim of *chiral* auxiliary reagents (see Exps. 8.2–8.4). In this experiment the *achiral* auxiliary reagent PCl_3 is used, which forms cleanly and quantitatively dialkylphosphonates on reaction with alcohols, and the ^{31}P NMR spectra are recorded. ^{31}P NMR spectroscopy has the advantage of a large chemical-shift dispersion. The basic idea depends on coupling of enantiomers (R,S) via an achiral reagent A, resulting in diastereomers, a *d,l* pair (RR, SS) and two *meso* compounds according to Eq. (1).

$$\text{R,S} + \text{A} \rightarrow \underbrace{\text{R-A-R} + \text{S-A-S}}_{d,l\ \text{Pair}} + \underbrace{\text{R-A-S} + \text{S-A-R}}_{meso} \qquad (1)$$

Since the *d,l* pair and the *meso* compounds have different ^{31}P chemical shifts, the method can be used to determine the original R/S ratio or the *ee*-value. In the experiment described here we demonstrate the method with *racemic* 2-butanol.

2. Literature

[1] J. P. Vigneron, M. Dhaenens and A. Horeau, *Tetrahedron* **1973**, *29*, 1055–1059.
[2] B. L. Feringa, A. Smardijk and H. Wynberg, *J. Am. Chem. Soc.* **1985**, *107*, 4798–4799.
[3] B. Strijtveen, B. L. Feringa and R. M. Kellogg, *Tetrahedron* **1987**, *43*, 123–130.
[4] D. Parker, *Chem. Rev.* **1991**, *91*,1441–1457.
[5] C. J. Welch, *Tetrahedron Asymmetry* **1991**, *2*, 1127–1132.

3. Pulse Scheme and Phase Cycle

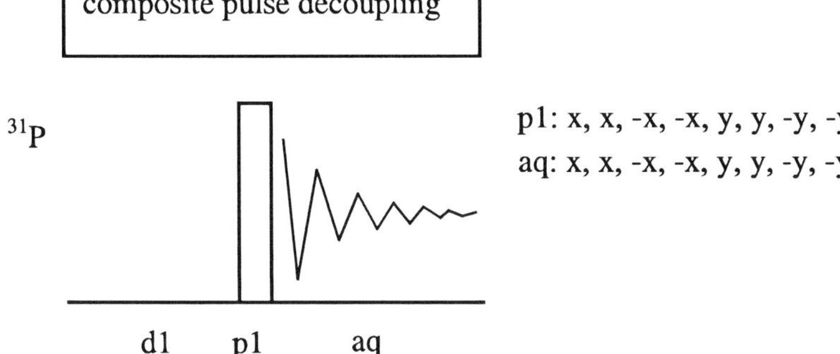

p1: x, x, -x, -x, y, y, -y, -y
aq: x, x, -x, -x, y, y, -y, -y

4. Acquisition

Time requirement: 5 min

Sample: *Racemic* 2-butanol (0.75 mmol) is dissolved in 2 ml of $CDCl_3$, and dry pyridine (0.75 mmol) is added (excess pyridine does not influence the ^{31}P NMR determination). To the stirred solution, 0.25 mmol PCl_3 dissolved in 2 ml $CDCl_3$ is added. The mixture is stirred for 10 min at room temperature and subsequently transferred, without the necessity of any workup or further purification, into an NMR tube.

The instrument is set to ^{31}P detection with composite pulse proton decoupling. You have to set:

td: 16 k
sw: 20 ppm
o1: middle of ^{31}P NMR spectrum
o2: middle of ^{1}H NMR spectrum
p1: 30° ^{31}P transmitter pulse
d1: 2 s
decoupler attenuation and 90° pulse for CPD
ns: 32

5. Processing

Use standard 1D processing as described in Experiment 3.2 using exponential line broadening with lb = 3 Hz. Reference against external 85% H_3PO_4 with $\delta_P = 0$. Apply baseline correction on the spectrum for good integration.

6. Result

The figure on page 265 shows the ^{31}P NMR spectrum obtained on an Avance DRX-400 spectrometer with a multinuclear probe-head. The two smaller signals at $\delta_P = 6.5$ and $\delta_P = 5.75$ stem from the *meso* pair, the components of which are diastereomeric to each other due to the pseudoasymmetric center at phosphorus. The larger signal at $\delta_P = 6.2$ stems from the enantiomeric *d,l* pair. The literature claims that the integrals should give a 1 : 1 ratio of *meso* and *d,l* forms for the racemic 2-butanol; note the deviations given in the figure.

7. Comments

In the reaction, trialkylphosphites are probably first formed, and these are cleaved under the reaction conditions to give phosphonates. An enantiomerically pure alcohol with configuration S will give only SS phosphonate, and the signals of the meso compounds will be missing.
For the *ee*-determination of thiols, methylphosphonic dichloride or in general alkylphosphonic dichlorides are recommended as reagents in the literature. Thiophosphites

$P(SR)_3$ did not give well resolved ^{31}P NMR signals for the diastereo isomers. The advantage of these reagents is that only two equivalents of ROH or RSH are necessary.

```
     Et               Et               Et
     |                |                |
  H—C—Me           H—C—Me           H—C—Me
     |                |                |
     O                O                O
     |                |                |
  H—P=O            H—P=O            O=P—H
     |                |                |
     O                O                O
     |                |                |
 Me—C—H            H—C—Me           H—C—Me
     |                |                |
     Et               Et               Et
```

d,l meso

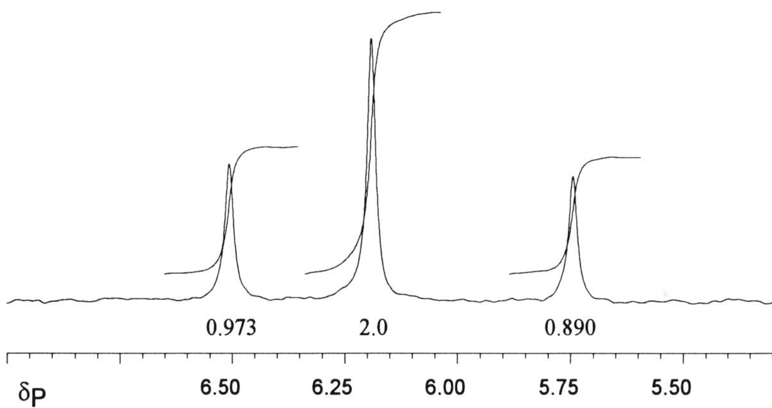

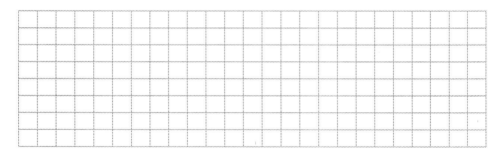

The disadvantage is the longer reaction time of the phosphonate formation. Methylthiophosphonates $CH_3PO(SR)_2$ gave signals at δ_P = 60; the phosphonates $CH_3PO(OR)_2$ show absorptions at δ_P = 30. Note that the method described here only works if the transition states leading to the diastereomeric products are of comparable energy. It is best to test this assumption on the racemate.

8. Own Observations

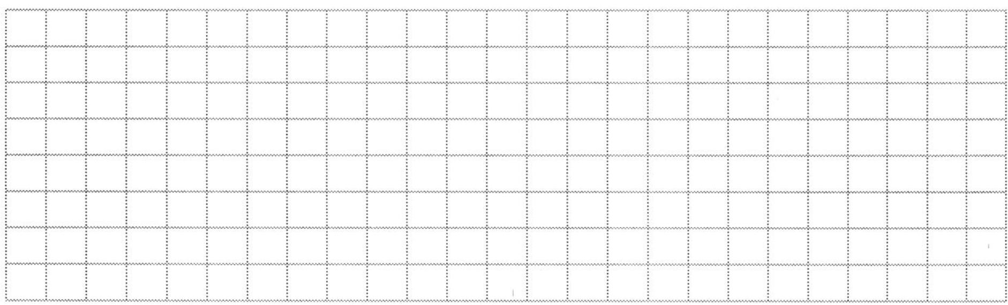

Experiment 8.6

Determination of Absolute Configuration by the Advanced Mosher Method

1. Purpose

The determination of absolute stereochemistry is a most important goal in natural product chemistry. Using high-field instruments, this can be performed by NMR if certain rules are obeyed. An enantiomerically pure alcohol is esterified with both (S)-(+)-2-methoxy-2-(trifluoromethyl)-2-phenylacetic acid chloride (MTPA-Cl) and (R)-(−)-MTPA-Cl. One measures the chemical shift differences $\delta_S - \delta_R$ of *all* protons between the two diastereoisomers obtained. By assuming an idealized conformation, which can be corroborated by molecular mechanics calculation and for some derivatives by NOESY measurements[5], these chemical shift differences can be evaluated to determine the absolute configuration. In this experiment we demonstrate the method using enantiomerically pure menthol of unknown configuration as a substrate.

2. Literature

[1] J. A. Dale, H. S. Mosher, *J. Am. Chem. Soc.* **1973**, *95*, 512–519.

[2] I. Ohtani, T. Kusumi, Y. Kashman, H.Kakisawa, *J. Am. Chem. Soc.* **1991**, *113*, 4092–4096.

[3] G. Uray in Houben-Weyl, *Methods in Organic Chemistry, Stereoselective Synthesis*, G. Helmchen, R. W. Hoffmann, J. Mulzer, E. Schaumann (eds.), Thieme, Stuttgart, **1995**, *E21a*, 253–292.

[4] R. Chinchilla, L. R. Falvello, C. Najera, *J. Org. Chem.* **1996**, *61*, 7285–7290.

[5] A. Heumann, J. M. Brunel, R. Faure, H. Kolshorn, *J.C.S. Chem. Comm.* **1996**, 1159–1160.

[6] M. Kobayashi, *Tetrahedron* **1997**, *53*, 5973–5994.

3. Pulse Scheme and Phase Cycle

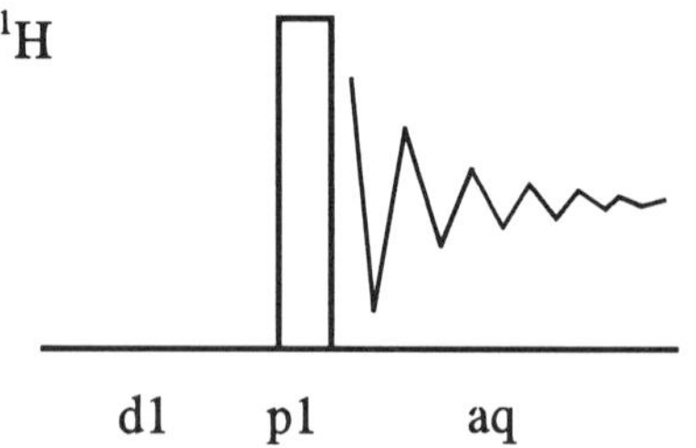

p1: x, x, -x, -x, y, y, -y, -y

aq: x, x, -x, -x, y, y, -y, -y

4. Acquisition

Time requirement: 5 min
Sample: Each 25 mg (S)-MTPA- and (R)-MTPA-ester of one enantiomer of menthol in 0.7 ml $CDCl_3$

Preparation: Dissolve 61.8 mg (0.39 mmol) of one enantiomer of menthol in 0.5 ml dry pyridine, dissolve (**a**) 50 mg (0.2 mmol) (R)-(−)- and (**b**) (S)-(+)-2-methoxy-2-(trifluoromethyl)-2-phenyl acetic acid chloride in each 0.25 ml dry pyridine. Mix each of the acid chloride solutions with 0.25 ml of the menthol solution and let it stand for two days with occasional shaking. Add 20 ml H_2O and a few drops of conc. HCl and extract the solutions with three portions of 20 ml Et_2O. After drying over Mg_2SO_4 the solvent is evaporated and the residue purified by preparative TLC (PE/Et_2O 40:1).

Load standard ^1H NMR acquisition parameters (see Exp. 3.1) and record the spectra of both solutions **a** and **b**.

5. Processing

Use standard 1D processing for ^1H NMR spectra (see Exp. 3.1). The dual display mode is most convenient to extract the chemical shift differences of the two spectra.

6. Result

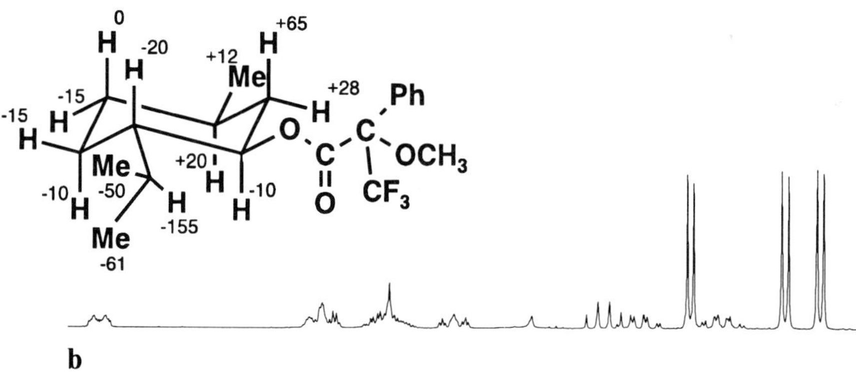

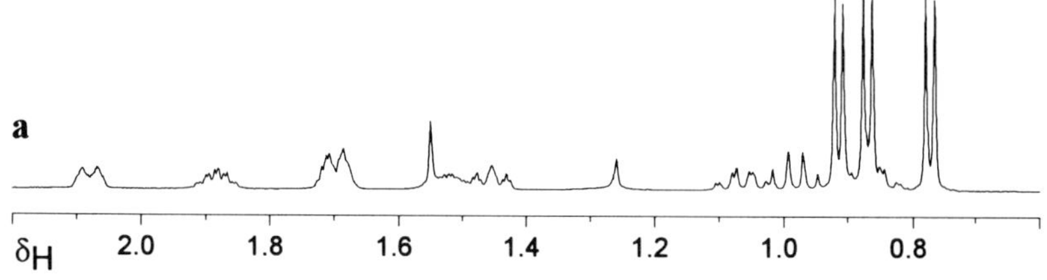

The figure on page 267 shows an expansion of the ^1H NMR spectra obtained on an AMX-500 NMR spectrometer. In **a** the result of the (R)-MTPA ester is given, in **b** the result of the (S)-MTPA ester. The chemical shift differences $\delta_S - \delta_R$ obtained are shown in Hz in the formula. The assignment of the various protons in the menthol moiety must be independently performed using the standard 2D experiments discussed in this book. The further evaluation proceeds as follows: (i) Put the protons with positive $\Delta\delta$ on the right side of the model structure, and those with a negative $\Delta\delta$ on the left side. Construct a molecular model, and confirm that *all* assigned protons with positive and negative $\Delta\delta$ are actually found on the right and left side of the MTPA plane. The absolute values of $\Delta\delta$ must be proportional to the distance from the MTPA moiety. When *all* these conditions are satisfied (do NOT use [D_6]benzene as solvent), the correct configuration can be extracted. For the example shown, the menthol used proved to be (1R,2S,5R)-(−)-menthol.

7. Comments

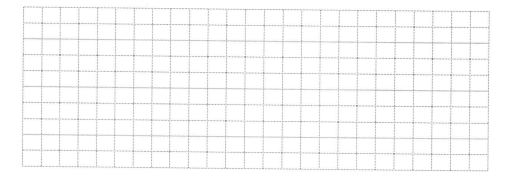

The Mosher method is well known among organic chemists as a method of determining the *relative* ee values in mixtures of enantiomers, mostly using the large chemical shift differences obtained with ^{19}F NMR, by preparing only one MTPA ester. The advantage of the technique shown in this experiment relies mainly on the fact, that $\Delta\delta$ values of *all* protons which show a chemical shift difference are evaluated. Recently, many other reagents have been proposed [3–6], which, in principle, use the same effect, namely the aromatic ring-induced chemical shift differences (see Exp. 8.7).

8. Own Observations

Experiment 8.7

Aromatic Solvent-Induced Shift (ASIS)

1. Purpose

Even in these days of high-field NMR spectroscopy it oftens happens that proton spectra show high-order effects because of small chemical shift differences. In these cases a simple change of solvent, especially from chloroform to aromatic solvents (e.g. benzene or pyridine), can cause a dramatic simplification of the spectrum due to a better separation of the signals. This effect is called aromatic solvent-induced shift (ASIS). The effect is usually strong in ^{1}H NMR spectroscopy but only weak in ^{13}C NMR. In this experiment we demonstrate ASIS with ethyl anthranilate.

2. Literature

[1] P. Lazlo, *Prog. NMR Spectrosc.* **1967**, *3*, 231–402.
[2] J. Ronayne, D. H. Williams, *Annu. Rep. NMR Spectrosc.* **1969,** *2*, 83–124.
[3] F. H. A. Rummens R. H. Krystynak, *J. Am. Chem. Soc.* **1972**, *94*, 6914–6921.
[4] H. Stamm, H. Jäckel, *J. Am. Chem. Soc.* **1989**, *111*, 6544–6550.
[5] K. Nikki, *Magn. Reson. Chem.* **1990**, *28*, 385–388.

3. Pulse Scheme and Phase Cycle

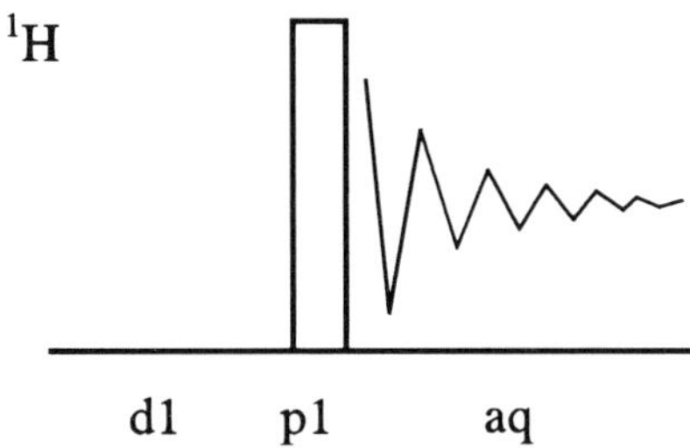

p1: x, x, -x, -x, y, y, -y, -y
aq: x, x, -x, -x, y, y, -y, -y

4. Acquisition

Time requirement: 10 min

Sample: 3 % ethyl anthranilate **a** in $CDCl_3$ and **b** in [D_6]benzene with 0.1% TMS

Load standard ^{1}H NMR acquisition parameters (see Exp. 3.1). Record the spectra of the compound dissolved in both solvents.

5. Processing

Use standard 1D processing for ^{1}H NMR spectra (see Exp. 3.1), reference both spectra to $\delta_H = 0$ and inspect the aromatic region.

6. Result

The figure shows an expansion of the aromatic region obtained on an ARX-200 spectrometer. Spectrum **a** was obtained in $CDCl_3$, whereas spectrum **b** was recorded in [D_6]benzene. In **a** the signals of H-3 and H-5 are overlapping; however, in **b** all signals of the aromatic ABCD pattern can be individually analyzed. The singlet at $\delta_H = 7.16$ stems from the residual protons of [D_6]benzene.

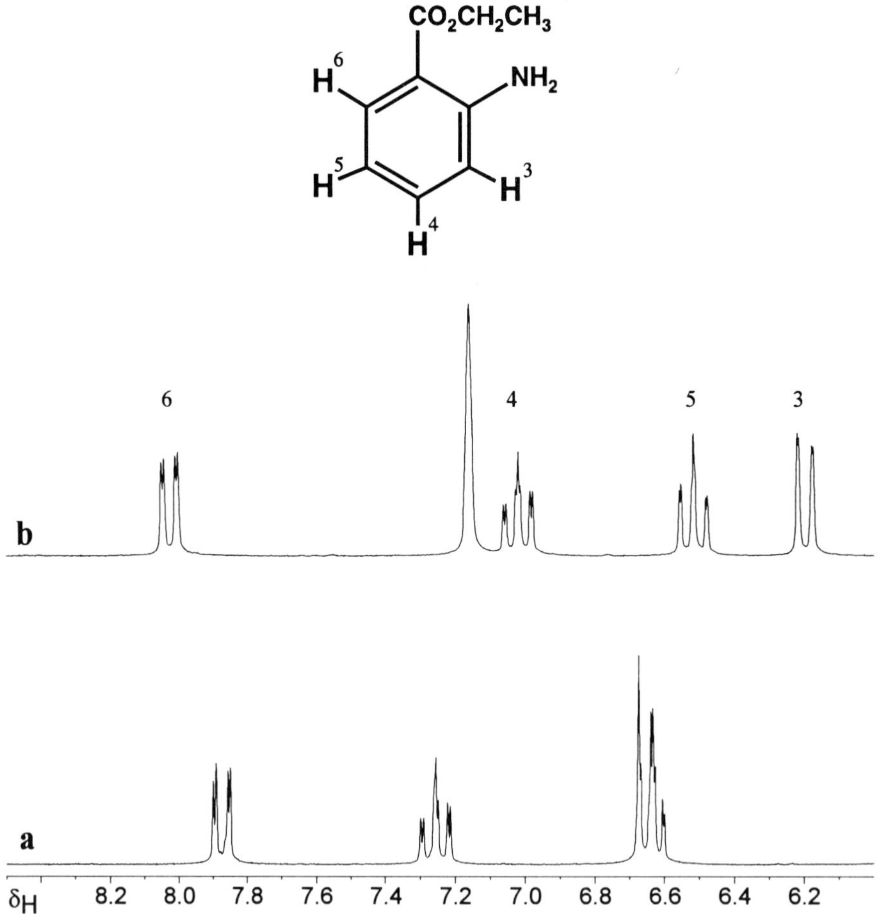

7. Comments

The ASIS technique is the most straightforward approach to simplifying proton NMR spectra and should be tested before other means such as lanthanide shift reagents (see Exp. 8.1) are employed. According to the theory [2], the total shielding σ_t of a proton is composed as described by Eq. (1):

$$\sigma_t = \sigma_g + \sigma_b + \sigma_w + \sigma_a + \sigma_e + \sigma_c \qquad (1)$$

Here, σ_g refers to the chemical shift in the gas phase, and the other contributions stem from the bulk susceptibility of the solvent (σ_b), van der Waals interactions (σ_w), anisotropy effects (σ_a), electric field effects (σ_e), and specific solute–solvent interactions (σ_c). All of these effects together are associated with an interaction energy of about 1 kcal/mole. Quantification of the ASIS, however, seems to be difficult and depends critically on the reference system used [3]. ASIS works best with molecules containing polar groups; sometimes pyridine gives superior results.

8. Own Observations

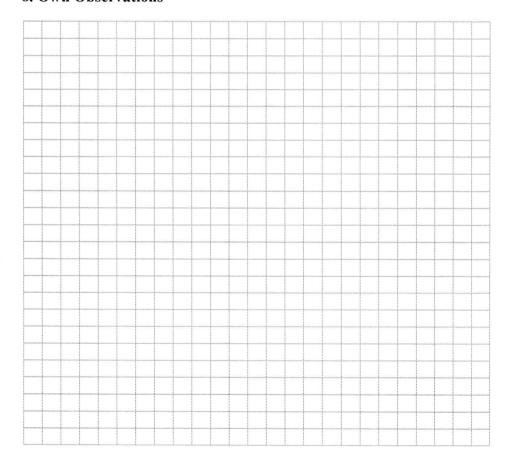

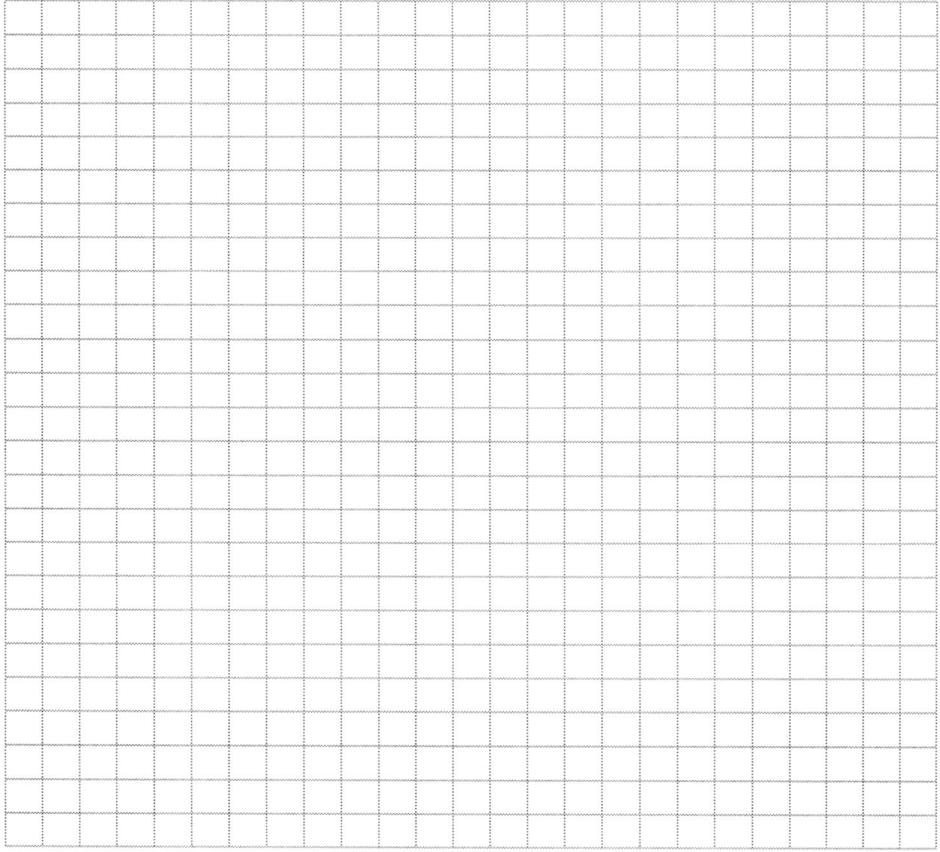

Experiment 8.8

NMR Spectroscopy of OH-Protons and H/D Exchange

1. Purpose

The signals of OH-protons of alcohols as well as of NH-protons of amines are usually assigned by their broadness, exchange with D_2O, or their solvent- and temperature-dependent chemical shifts. This behavior is due to exchange processes. Therefore coupling between OH-protons and adjacent CH-protons are usually not observed. In this experiment we demonstrate the occurrence of OH multiplets when dimethylsulfoxide (DMSO) is used as solvent. In DMSO the exchange processes are so slow that one can observe the OH group of a primary alcohol RCH_2OH as a triplet or the OH group of a secondary alcohol R_2CHOH as a doublet. Thus, these two types of alcohols can be easily distinguished. In addition, one may add a few drops of D_2O or CD_3OD regardless of the solvent used, to exchange the OH-proton by deuterium to confirm the assignment.

2. Literature

[1] K. K. M. Sanders, B. K. Hunter, *Modern NMR Spectroscopy*, 2nd Edition, Oxford University Press, Oxford, **1993**, 220.
[2] D. Martin, A. Weise, H.-J. Niclas, *Angew. Chem. Int. Ed. Engl.* **1967**, *6*, 318–335.

3. Pulse Scheme and Phase Cycle

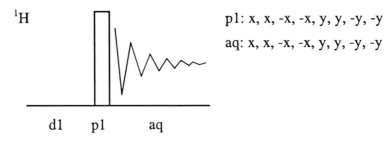

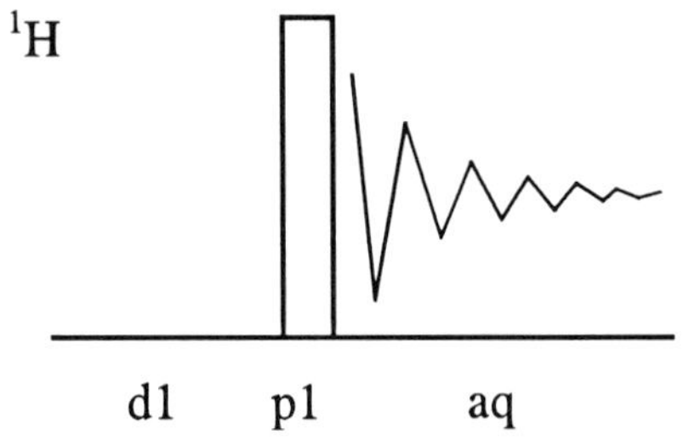

p1: x, x, -x, -x, y, y, -y, -y

aq: x, x, -x, -x, y, y, -y, -y

4. Acquisition

Time requirement: 10 min

Sample: 5% glycerol in [D_6]DMSO

Record a standard ^{1}H NMR spectrum according to Experiment 3.1. Remove the tube from the magnet, add a drop of D_2O, and shake the NMR tube thoroughly. Record again an ^{1}H NMR spectrum.

5. Processing

Use standard 1D processing with exponential line broadening (lb = 0.1) as described in Exp. 3.1.

6. Result

The figure shows the normal ^{1}H NMR spectrum of glycerol (**a**) in DMSO obtained on an AC-300 spectrometer in a dual probe-head. The primary OH group can be easily distinguished from the secondary one, whereas the CH and CH_2 protons exhibit a complex pattern. The reason is that the three types of proton signals, i.e. the methine proton and the diastereotopic methylene protons, have a very small chemical shift difference in the chosen solvent. In **b** the same spectrum is given after addition of D_2O. The signals of the OH groups are strongly reduced and somewhat deshielded because of exchange with the D_2O; the residual HDO signal appears at $\delta_H = 3.42$. Since the coupling with the OH protons is removed, the multiplet between $\delta_H = 3.2$ and 3.45 is somewhat simplified.

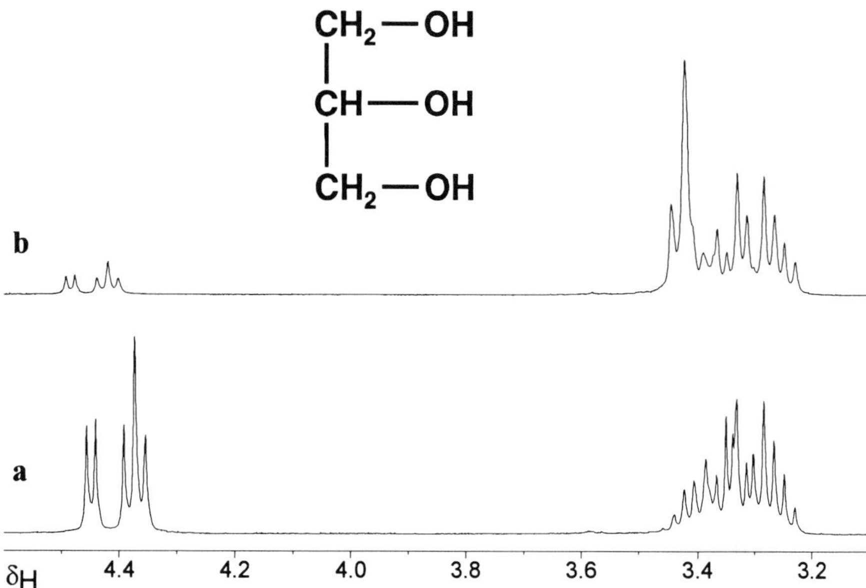

7. Comments

The DMSO used for this experiment was predried over molecular sieves. Otherwise one often obtains a strong water signal in this solvent. The use of DMSO as the solvent has the disadvantage that the sample cannot easily be recovered. Often pyridine can be used as a substitute to show the same effects, and can be removed without difficulty. Although water and $CDCl_3$, the most abundant NMR solvent, are not miscible, the H/D exchange experiment can nevertheless be performed. Thorough shaking of a $CDCl_3$ solution with a drop of D_2O will remove or at least attenuate OH and NH_2 signals. One often wonders about a mysterious singlet in various NMR solvents which usually stems from water. Addition of a tiny drop of normal water will confirm the suspicion, since the signal will increase.

DMSO is a strong acceptor of intermolecular hydrogen bonds [2]. Therefore the exchange processes are slowed down, and the spin couplings of OH-protons can be observed on the NMR time scale.

8. Own Observations

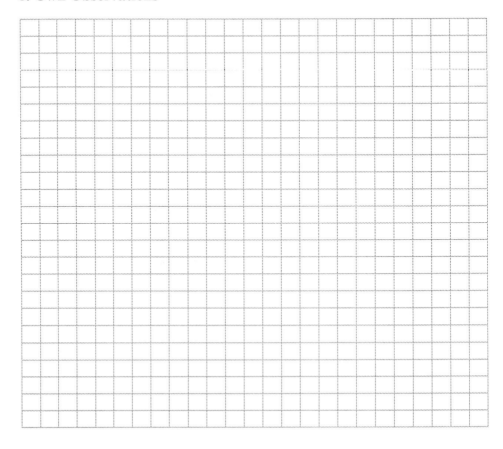

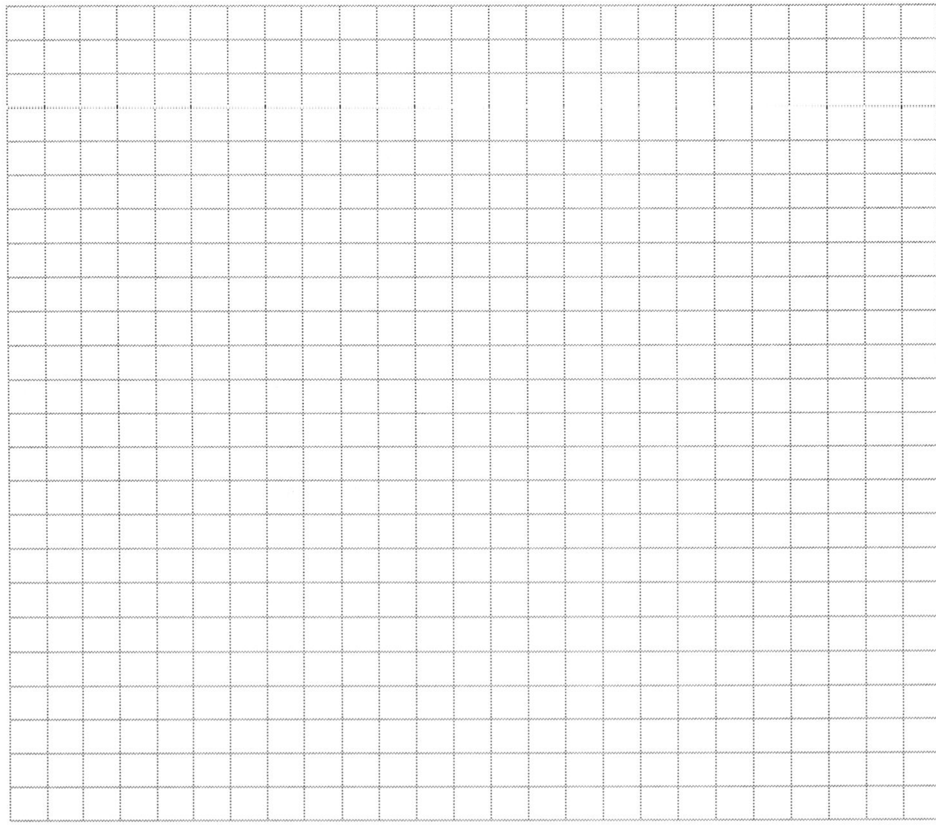

Experiment 8.9

Isotope Effects on Chemical Shielding

1. Purpose

In general, isotopes of an element are considered to have the same electronic environment. This is known as the Born-Oppenheimer approximation. However, because of their different mass, isotopes occupy different vibrational levels within the same electronic potential of a chemical bond, which leads to a somewhat shorter bond length for the heavier isotope. Since NMR spectroscopy averages over the vibrational states, one also finds slightly different chemical shifts for a nucleus X bound to different isotopes of another nucleus Y, such as ^{1}H and ^{2}H. Mostly, but not always, one finds shielding of X if it is bound to the heavier isotope. Isotope effects are expressed by the term $^{n}\Delta X(Y)$, where n depicts the number of chemical bonds between the observed nucleus X and the isotope Y causing the chemical shift. In this experiment, we demonstrate the $^{1}\Delta$ ^{13}C (^{2}H) deuterium isotope effects on the ^{13}C chemical shift.

2. Literature

[1] P. E. Hansen, *Prog. NMR Spectrosc.* **1988**, *20*, 207–255.
[2] C. J. Jameson, *Isotopes in the Physical and Biomedical Sciences, Isotopic Application in NMR Studies*, Elsevier **1991**, *2*, 1–54.
[3] S. Berger, NMR-*Basic Principles and Progress* **1990**, *22*, 1–29.
[4] H. U. Siehl, *Adv. Phys. Org. Chem.* **1987**, *23*, 63–163.

3. Pulse Scheme and Phase Cycle

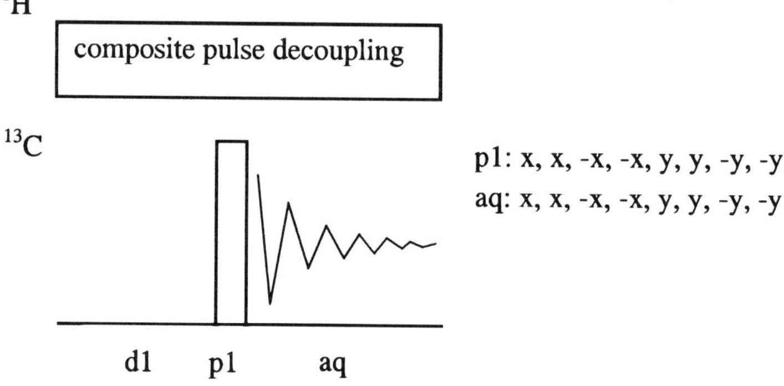

p1: x, x, -x, -x, y, y, -y, -y
aq: x, x, -x, -x, y, y, -y, -y

4. Acquisition

Time requirement: 20 min

Sample: Prepare a mixture of 0.2 ml $CDCl_3$, 0.2 ml [D_2]dichloromethane and 0.2 ml [D_6]acetone. Add for the second measurement 0.2 ml of a 1:1:1 mixture of normal chloroform, dichloromethane and acetone.

Load standard ^{13}C NMR parameters. You have to set:

td: 64 k data points
sw: 200 ppm
o1: middle of ^{13}C NMR spectrum
o2: middle of ^{1}H NMR spectrum
p1: 30° ^{13}C transmitter pulse
d1: 1 s
decoupler attenuation and 90° pulse for CPD
ns: 256

Measure first the mixture of the deuterated solvents alone and inspect the different multiplets caused by the different number of deuterium atoms (I = 1). Then add the mixture of the undeuterated solvents and repeat the measurement.

5. Processing

Use standard 1D processing as described in Experiment 3.2 with exponential multiplication (lb = 0.3 Hz). For better digital resolution apply zero filling to 64 k data points.

6. Result

The figures on pages 277-278 show three expansions of the ^{13}C NMR spectrum obtained from the mixture of deuterated and undeuterated solvents on an AMX-500 spectrometer. For acetone one finds in **a** a septet with $^{1}J(C,D)$ = 19.5 Hz and an $^{1}\Delta$ isotope effect of −758 ppb, for dichloromethane in **b** a quintet with $^{1}J(C,D)$ = 27.5 Hz and an isotope effect $^{1}\Delta$ of −391 ppb, and finally for chloroform in **c** a triplet with $^{1}J(C,D)$ = 32.4 Hz and an isotope effect $^{1}\Delta$ of −192 ppb. For the carbonyl atom of acetone (not shown) a $^{2}\Delta$ of +325 ppb can be detected; note the remarkable sign change. The observed shift measured in Hz for the $^{1}\Delta$ isotope effects is magnetic field dependent; however, the splitting by the coupling to deuterium is not. Therefore the observed overall pattern will change depending on the spectrometer used.

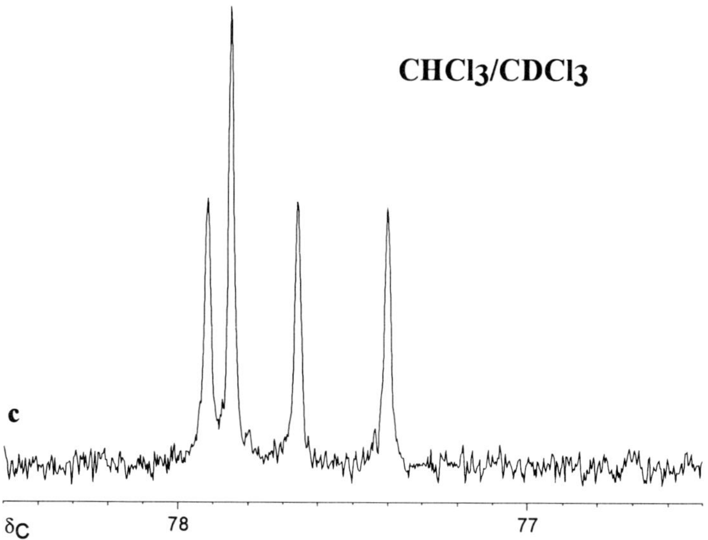
CHCl3/CDCl3
c
δC
78
77

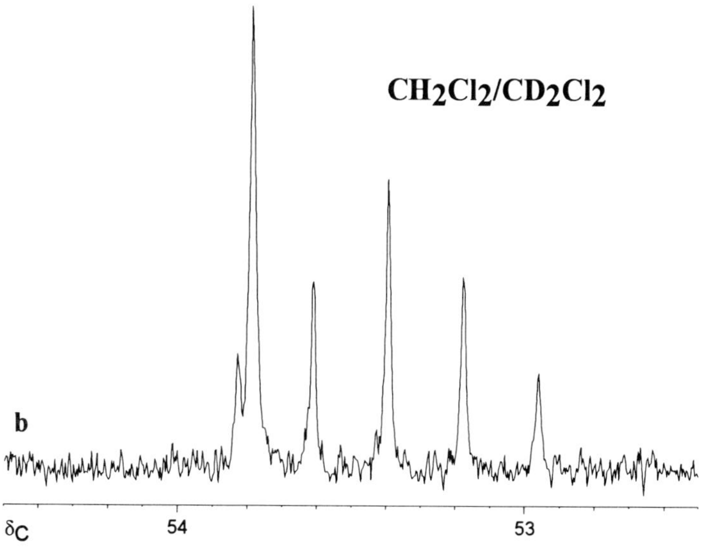
CH2Cl2/CD2Cl2
b
δC
54
53

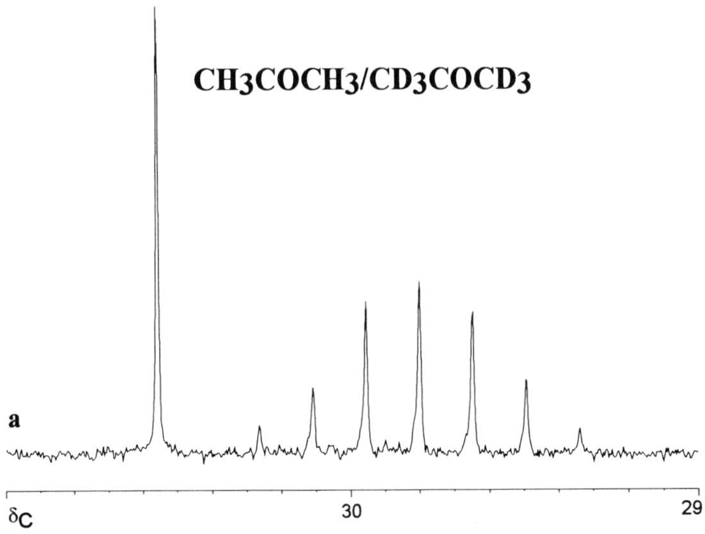

7. Comments

There are numerous applications of the study of deuterium isotope effects in physical organic chemistry [3]. A common feature is the additivity, which can be seen from the result of chloroform and dichloromethane, where the isotope effect in the latter is about twice that in chloroform. The effects shown in this experiment are called *intrinsic*. Other deuterium isotope effects observed by NMR influence a chemical equilibrium; these effects are usually much larger and strongly dependent on the temperature. This field is called "*Isotopic perturbation of equilibrium*" [4].

8. Own Observations

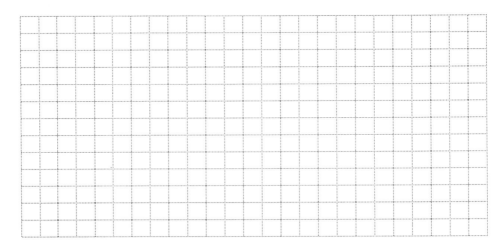

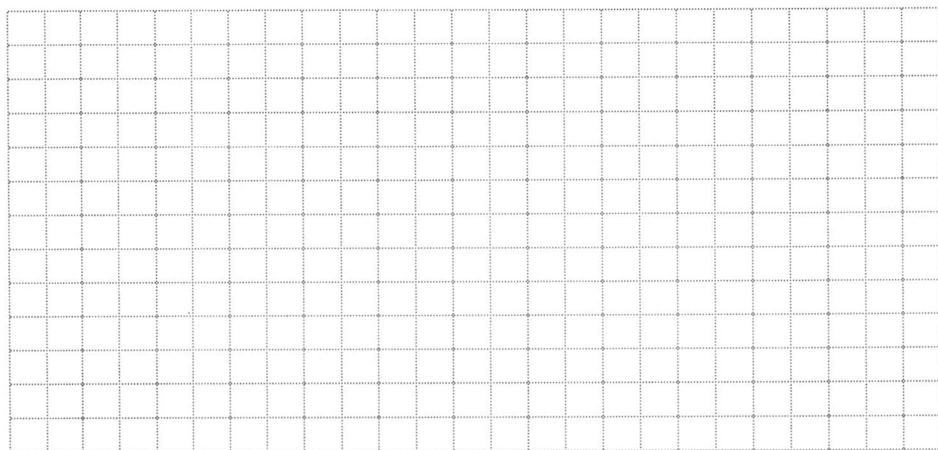

Experiment 8.10

pK_a Determination with ^{13}C NMR

1. Purpose

^{13}C chemical shifts of carboxylic acids are pH dependent as are the ^{15}N and ^{13}C chemical shifts of nitrogen-containing bases. In compounds with more than one acidic group a ^{13}C chemical shift titration immediately reveals which site has the lower pK_a. In this experiment we demonstrate a pKa determination with ^{13}C NMR. Ascorbic acid (Vitamin C) is used as an example since it provides two deprotonation steps. Considerable insight into the chemistry of this important vitamin can be gained from this experiment.

2. Literature

[1] S. Berger, Tetrahedron **1977**, *33*, 1587–1589.
[2] H.-O. Kalinowski, S. Berger, S. Braun, *Carbon-13 NMR Spectroscopy*, Wiley, Chichester, **1988.**
[3] R. E. London, *J. Magn. Reson.* **1980**, *38*, 173–177.
[4] D. H. Holmes, D. A. Lightner, *Tetrahedron* **1996**, *52*, 5319–5338.
[5] D. Farcasiu, A. Ghenciu, *Prog. NMR Spectrosc.* **1997**, *29*, 129–168.

3. Pulse Scheme and Phase Cycle

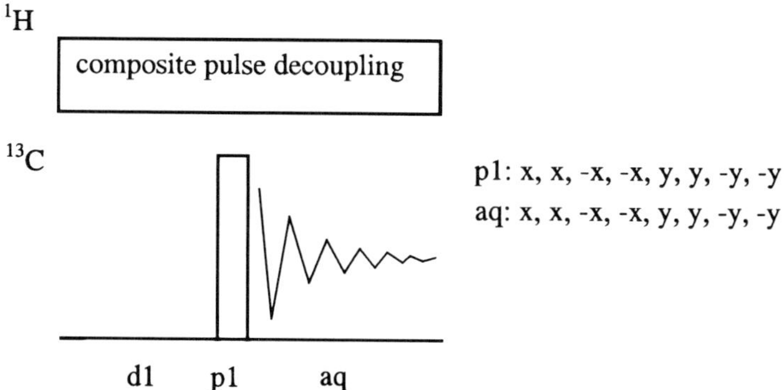

p1: x, x, -x, -x, y, y, -y, -y
aq: x, x, -x, -x, y, y, -y, -y

4. Acquisition

Time requirement: 90 min

Sample: 1 M solution of ascorbic acid in H_2O containing 10% D_2O. Prepare a stock solution and titrate with HCl first to pH = 1.0 using a pH electrode. Remove an aliquot for the first NMR measurement, then adjust the stock solution in steps of 0.5 pH using NaOH, and after each titration remove an aliquot for the measurement. As an internal, p*H*-independant standard 0.1 M 1,4-dioxane (δ_C = 67.6) is used. In alkaline solution vitamin C is very readily oxidized; thus the samples should be measured immediately or kept under an inert atmosphere.

Load standard ^{13}C NMR parameters. Set and control the temperature at 300 K. You have to set:

td: 64 k
sw: 200 ppm
o1: middle of ^{13}C NMR spectrum
o2: middle of ^{1}H NMR spectrum
p1: 30° ^{13}C transmitter pulse
d1: 2 s
decoupler attenuation and 90° pulse for CPD
ns: 32

5. Processing

Use standard 1D processing as described in Experiment 3.2 with exponential multiplication (lb = 2 Hz)

6. Result

The figure shows the ^{13}C NMR spectrum of ascorbic acid at pH 1.2 obtained on an AMX-500 spectrometer with a multinuclear probe-head. The signal labelled D stems from the internal standard dioxane. The complete titration diagram is shown on page 281 as well, indicating the two deprotonation steps, which can best be seen at the signals of C-1, C-2 and C-3. From the inflection points of the curves, the pK_a values were calculated to yield pK_{a1} = 4.05 and pK_{a2} = 11.7. These values compare well with literature results obtained by electrochemical methods considering the relatively high concentration used in this experiment and the influence of temperature.

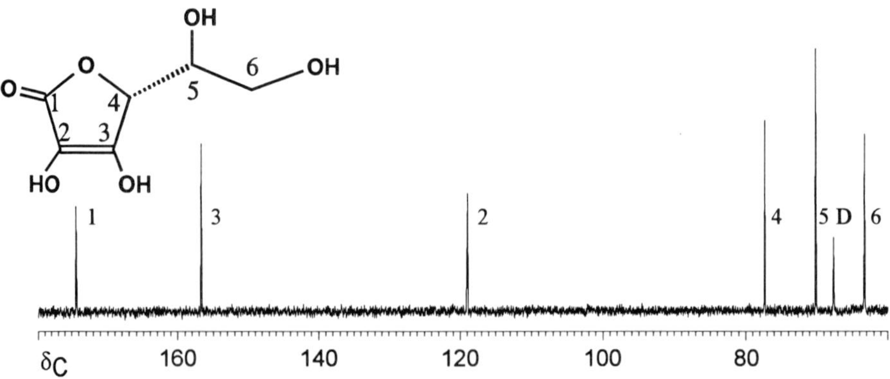

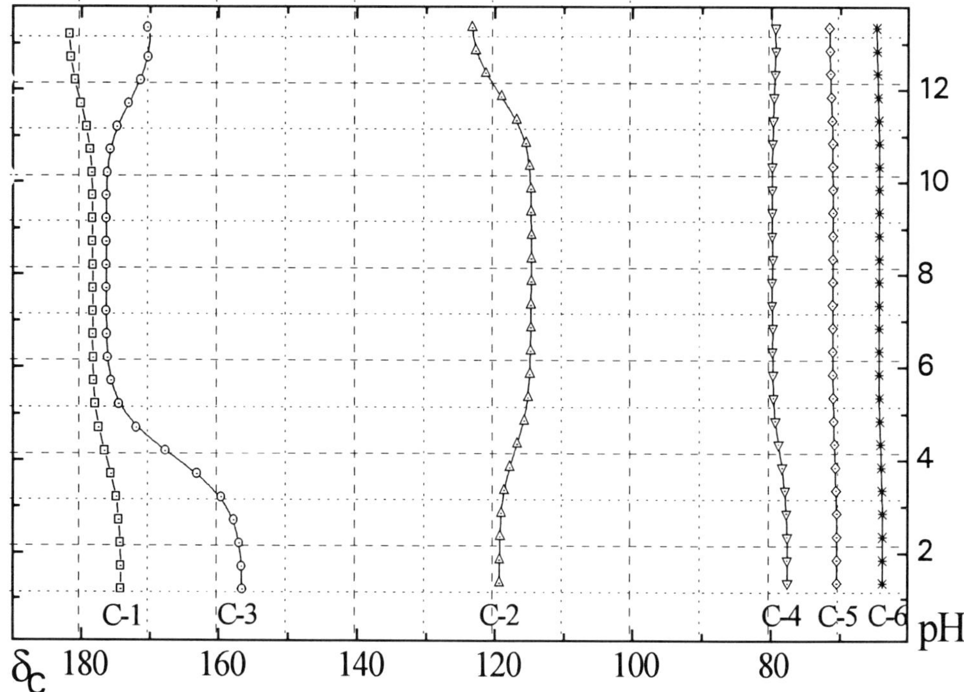

7. Comments

From the titration diagram it can clearly be seen that the first proton is removed from the OH group at C-3. In the second deprotonation step the proton of the OH group at C-2 is removed.

It is interesting to note that, contrary to common "organic feeling", the signal of a carboxylic group is deshielded on deprotonation, although the formal negative charge will be increased. This observation also holds for the endiol moiety of ascorbic acid; the effect, however, is not really understood [4]. Electric field effects, loss of hydrogen bonding, dimer/monomer equilibria and the anisotropy of the C=O double bond have been discussed as possible causes.

8. Own Observations

Experiment 8.11

The Relaxation Reagent Cr(acac)3

1. Purpose

The paramagnetic relaxation reagent $Cr(acac)_3$ is used to increase the intensity of the signals of quaternary ^{13}C nuclei which suffer from long relaxation times T_1. The addition of increasing amounts of the reagent results in a shortening of the relaxation times T_1 and an increase in line-width. The chemical shift is usually unaffected. The relaxation reagent is important for quantitative ^{13}C NMR investigations (see Exp. 8.16). $Cr(dpm)_3$ was reported to be more soluble, however is not commercially available.

2. Literature

[1] M. L. Martin, G. J. Martin, J. J. Delpuech, *Practical NMR Spectroscopy,* Heyden London, **1980**, Ch. 10.3.

[2] G. C. Levy, U. Edlund, *J. Am. Chem. Soc.* **1986**, *97*, 4482–4485.

3. Pulse Scheme and Phase Cycle

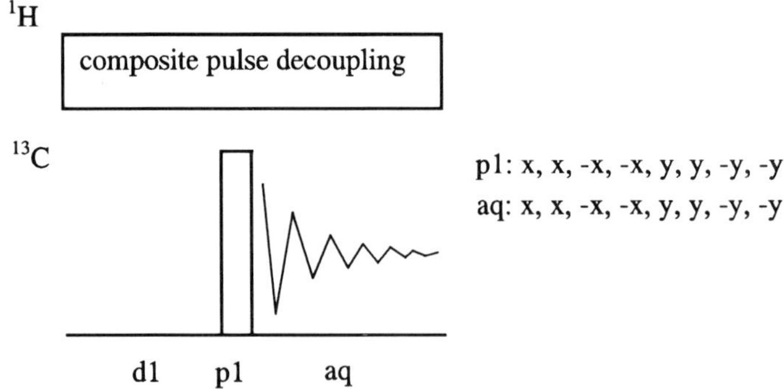

4. Acquisition

Time requirement: 15 min

Sample: 20% ethyl crotonate in $CDCl_3$.

Load standard ^{13}C NMR acquisition parameters (see Exp. 3.2) and record a ^{13}C NMR spectrum. Then add about 20 mg of chromium acetylacetonate, $Cr(acac)_3$. The solution should become slightly purple. A concentration of 0.1 M is ideal for quantitative work. The highest concentration which is reported in the literature to give reasonable results is 0.4 M. At higher concentrations one gets severe line broadening and it is difficult to lock on the solvent signal. You have to set:

td: 64 k
sw: 250 ppm
o1: middle of ^{13}C NMR spectrum
o2: middle of ^{1}H NMR spectrum
d1: 0.5 s
p1: 45° ^{13}C transmitter pulse
decoupler attenuation and 90° pulse for CPD
ns: 64

5. Processing

Use standard 1D processing for ^{13}C NMR (see Exp. 3.2).

6. Result

The figure shows the result obtained on an ARX-200 spectrometer. Note that after the addition of $Cr(acac)_3$ in (**b**) the intensities of all signals of ethyl crotonate are almost the same and the signal of $CDCl_3$ dominates, whereas the spectrum (**a**) without the relaxation reagent shows small intensities for the signal of the carboxyl group of ethyl crotonate and for the solvent signal. However, addition of $Cr(acac)_3$ also causes a reduction in the signal-to-noise ratio due to line broadening.

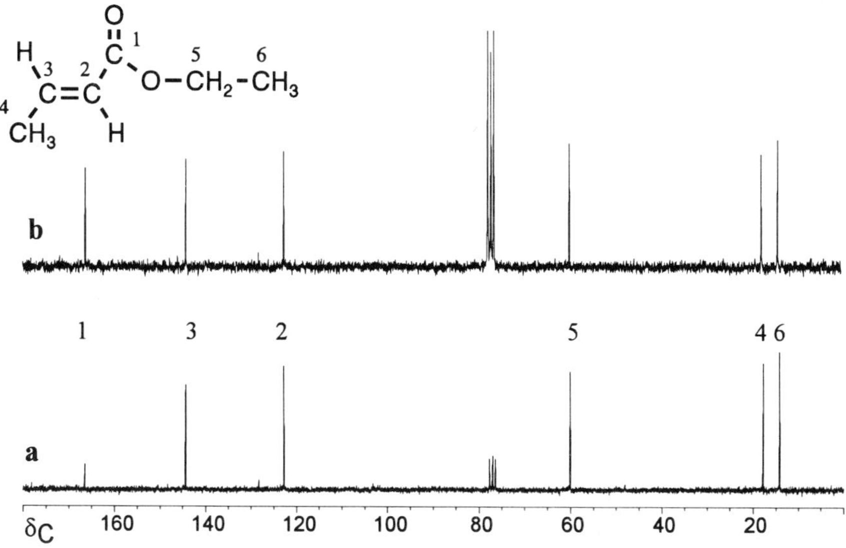

7. Comments

Routine ^{13}C NMR spectra are recorded under conditions which maximize sensitivity, such as using the Ernst angle and ^{1}H broad-band decoupling. This results in reduced intensities for signals of quaternary carbon nuclei, which usually have long relaxation times T_1 and smaller NOE values. The NOEs are dependent on the dipole–dipole interaction between ^{1}H and ^{13}C. The addition of paramagnetic compounds such as $Cr(acac)_3$, $Mn(acac)_2$, $Cu(acac)_2$ or $Gd(acac)_3$ reduces T_1 to less than 1 s for all types of carbon atoms. Because of the large gyromagnetic ratio of the unpaired electrons the mechanism of the relaxation is now an (electron dipole)–(^{13}C dipole) interaction. For excitation one can use 90° pulses and higher pulse repetition rates. Therefore relaxation reagents can also be used for advanced experiments like INADEQUATE (see Exps. 6.12 and 10.22).

8. Own Observations

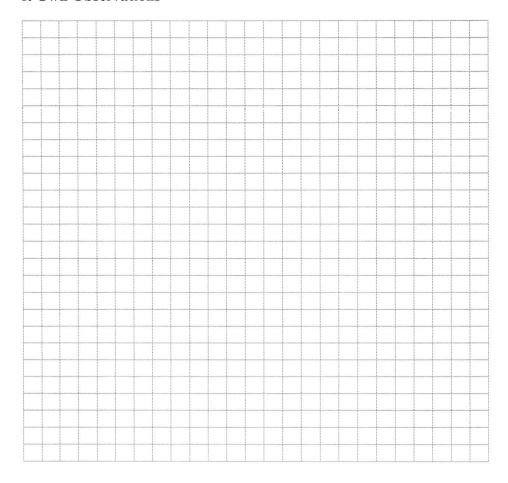

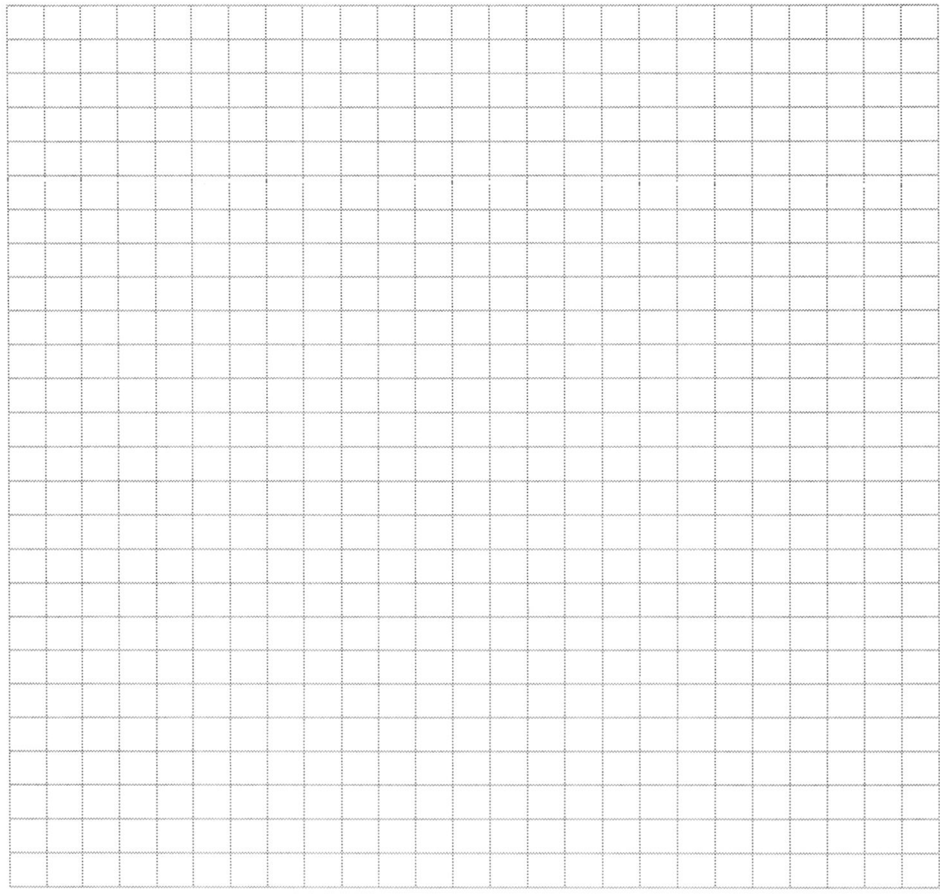

Experiment 8.12

Determination of Paramagnetic Susceptibility by NMR

1. Purpose

Magnetic susceptibilities are traditionally measured using a Gouy balance. This experiment demonstrates how the same information can be gained by a simple NMR measurement. The experiment is based on measuring the shift of the resonance frequency of an indicator compound caused by introducing a known concentration of the paramagnetic compound into the solution. The "doped" solution is prepared in a capillary tube so that the shift can be measured directly.

2. Literature

[1] D. F. Evans, *J. Chem. Soc.* **1959**, 2003–2005.
[2] J. L. Deutsch, S. M. Poling, *J. Chem. Educ.* **1969**, *46*, 167–168.
[3] J. Löliger, R. Scheffold, *J. Chem. Educ.* **1972**, *49*, 646–647.
[4] K. G. Orrel, V. Sik, *Anal. Chem.* **1980**, *52*, 569–572.
[5] A. Furuhashi, I. Ono, A. Yamasaki, *Magn. Reson. Chem.* **1991**, *29*, 1175–1180.
[6] D. H. Grant, *J. Chem. Educ.* **1995**, *72*, 39–40.

3. Pulse Scheme and Phase Cycle

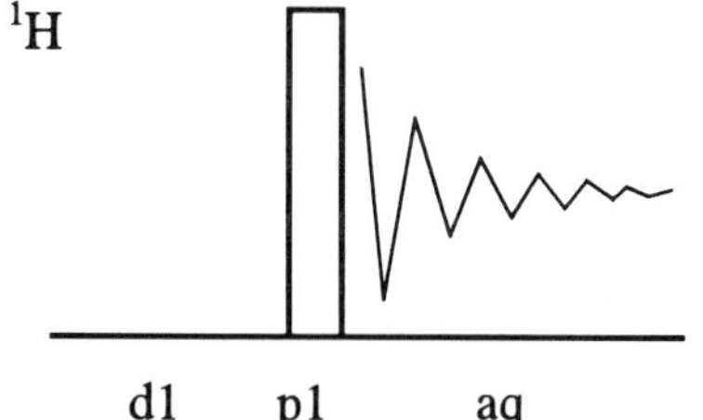

p1: x, x, -x, -x, y, y, -y, -y
aq: x, x, -x, -x, y, y, -y, -y

4. Acquisition

Time requirement: 20 min

Sample: Weigh and dissolve 7 mg $FeSO_4 \cdot 7H_2O$ (p. a.) in 0.5 ml D_2O in a volumetric flask, add 30 µl *t*-butanol and adjust the solution with D_2O to exactly 1 ml. Fill a standard melting-point capillary with this solution and flame-seal the capillary. Prepare an identical solution containing no iron salt and transfer it to a normal 5 mm NMR tube.

Adjust and control the probehead temperature at 298 K. Load standard ^{1}H NMR acquisition parameters, adjust the homogeneity and record a spectrum of the solution in the NMR tube without the capillary. Introduce the capillary and record the spectrum again. The sample should be spun so as to center the capillary in the NMR tube. Measure the frequency difference between the *t*-butanol signals in the two compartments.

5. Processing

Use standard 1D processing as described in Experiment 3.1

6. Result

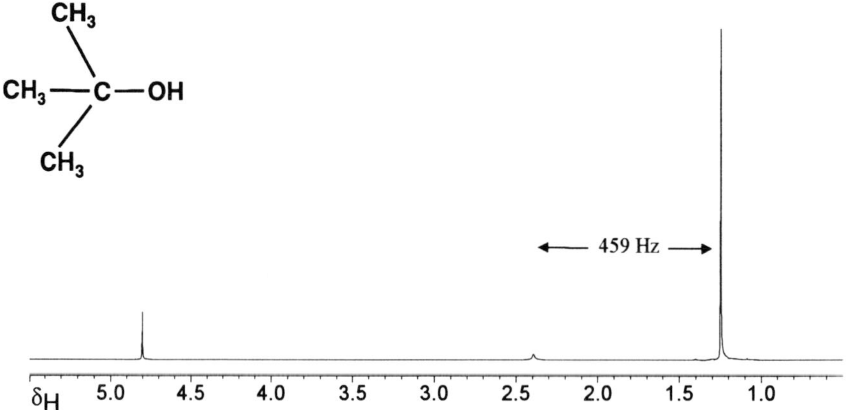

The figure shows the result obtained on an AM-400 spectrometer. The frequency difference between the *t*-butanol signals was 459 Hz using 7.2 mg $FeSO_4{\cdot}7H_2O$ for the solution in the capillary.

Note that the equations and spectra shown in the early literature apply to iron magnets where the magnetic field is perpendicular to the the axis of the NMR tube. For measurements using superconducting magnets the factor $2\pi/3$ must be replaced by $-4\pi/3$; thus the effect is larger and has the opposite sign. The difference between the frequencies of the signals in the outer compartment (ν_o) and in the capillary (ν_i) is related to the volume susceptibilities X by Equation (1).

$$\frac{\nu_o - \nu_i}{\nu_o} = -\frac{4\pi}{3}(X_i - X_o) \tag{1}$$

The volume susceptibilities X are usually replaced by the mass susceptibilities χ, where $\chi = X/\rho$, and the density ρ is equal to the mass m of the paramagnetic compound in 1 ml of solution. Thus (1) rearranges to give Equation (2). Here χ_o is approximately equal to the mass susceptibility of water, which is $-0.72{\cdot}10^{-6}$ cm^3g^{-1}.

$$\chi_i = -\frac{3}{4\pi m}\frac{\nu_o - \nu_i}{\nu_o} + \chi_o \tag{2}$$

The molar susceptibility χ_m is finally obtained by multiplying χ_i by the molar mass, in our case 278. Thus a χ_m-value for $FeSO_4 \cdot 7H_2O$ of $10374 \cdot 10^{-6}$ in c.g.s units was obtained in this experiment from the measured shift difference of 459 Hz, a result which compares reasonably well with the tabulated value of $11200 \cdot 10^{-6}$.

7. Comments

The literature [1–5] gives a variety of methods for measuring paramagnetic susceptibility with NMR. The experiment demonstrated here seems to be the easiest with current NMR instruments. Note that the value obtained is temperature dependent and that the Curie constant can be determined from the temperature dependence.

For the transition metals, where the spin-only approximation for the paramagnetism is valid, the number of unpaired electrons can be calculated from the molar susceptibility. Other indicator compounds may be used, and organometallic paramagnetic compounds can also be investigated.

8. Own Observations

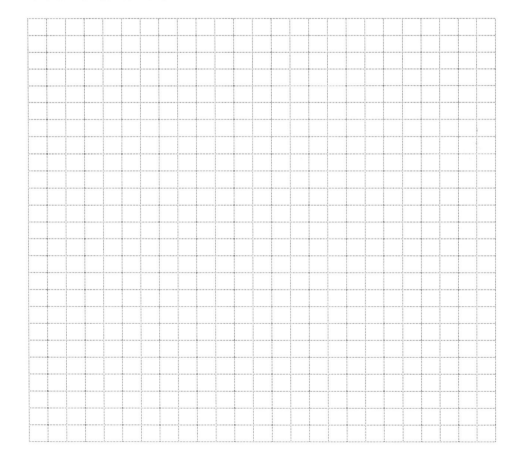

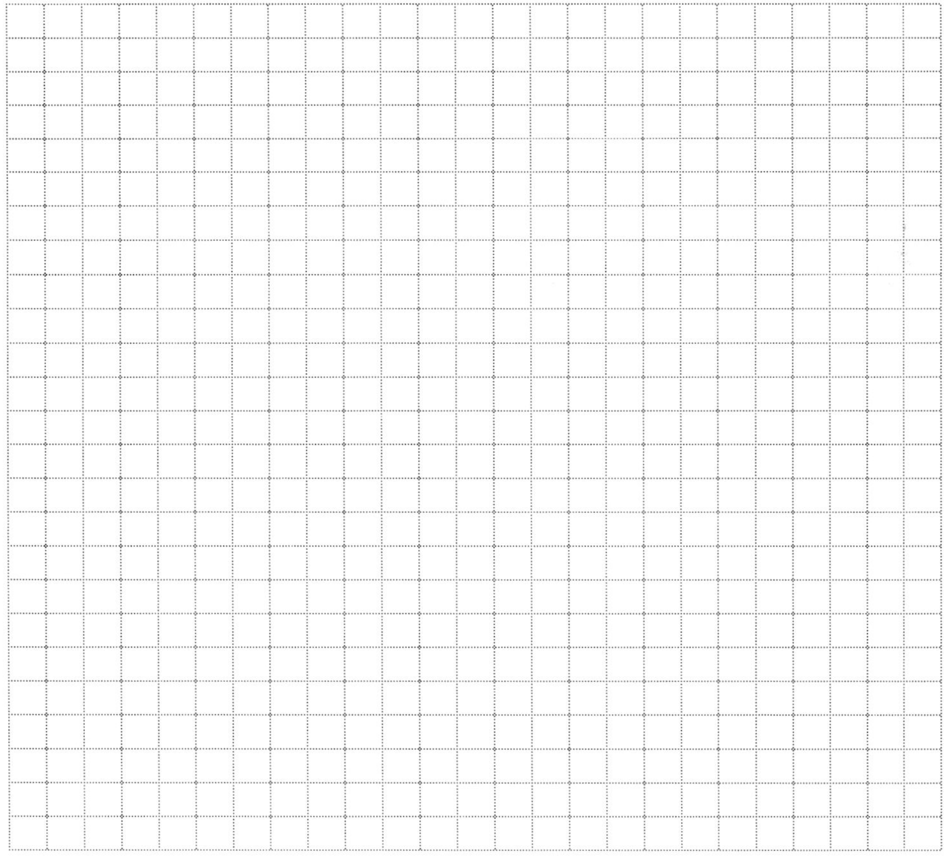

Experiment 8.13

^1H and ^{13}C NMR of Paramagnetic Compounds

1. Purpose

At first glance, NMR and paramagnetism seem to be incompatible. Nevertheless, it is possible to obtain NMR spectra in the presence of unpaired spins, as was demonstrated to be useful in the case of lanthanide shift reagents (see Exps. 8.1–8.2), the relaxation reagent $Cr(acac)_3$ (Exp. 8.11) or in the determination of the magnetic susceptibility (Exp. 8.12). In this experiment, we address the question whether it is possible to obtain NMR spectra from the paramagnetic compounds themselves, for example organometallic complexes, persistent organic radicals or paramagnetic proteins. Indeed, there are many possibilities ranging from severe line broadening, which renders the observation of NMR spectra impossible, to only small effects caused by pseudocontact interaction with the free electron. As an example we have chosen cobaltocene and demonstrate on this compound the large contact shifts observed both in ^1H and in ^{13}C NMR.

2. Literature

[1] H. P. Fritz, H. J. Keller, K. E. Schwarzhans, *Z. Naturforsch.* **1968**, *23b*, 298–302.
[2] G. N. La Mar, W. DeW. Horrocks Jr, R. H. Holm (eds.), *NMR of Paramagnetic Compounds*, Academic Press, New York **1973**.
[3] R. W. Kreilick, *Adv. Magn. Reson.* **1973**, *6*, 141–181.
[4] F. H. Köhler, *J. Organomet. Chem.* **1976**, *110*, 235–246.
[5] N. Hebendanz, F. H. Köhler, F. Scherbaum, B. Schlesinger, *Magn. Reson. Chem.* **1989**, *27*, 798–802.
[6] I. Bertini, C. Luchinat, *Coord. Chem. Rev.* **1996**, *150*, 1–296.

3. Pulse Scheme and Phase Cycle

a)

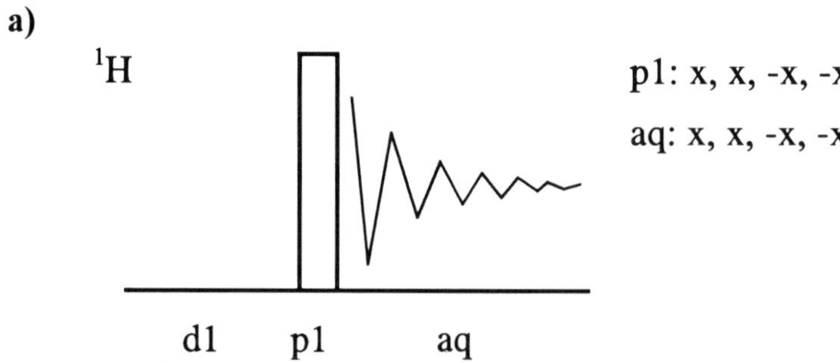

p1: x, x, -x, -x, y, y, -y, -y

aq: x, x, -x, -x, y, y, -y, -y

b)

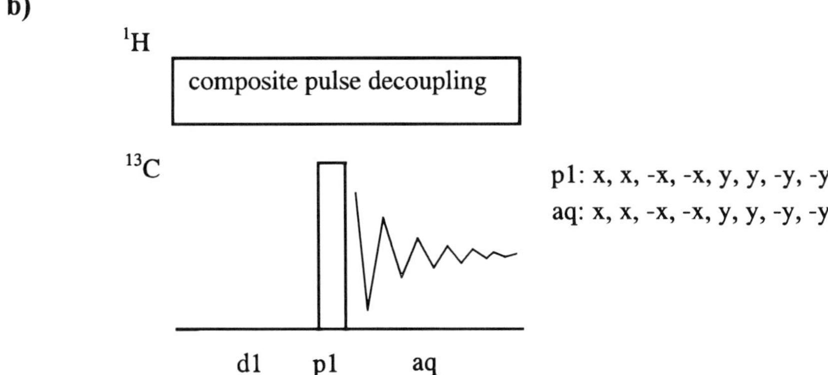

4. Acquisition

Time requirement: 20 min

Sample: ca. 80 mg Cobaltocene in 1 ml [D$_6$]benzene. The preparation of the sample for this experiment is somewhat more elaborate: commercially available cobaltocene is freshly sublimed at 10^{-2} torr / 80°C and afterwards transferred into an NMR tube under strict exclusion of oxygen. This is best performed using an NMR working cross found in many organometallic laboratories or in a glove box. Over sodium-potassium alloy previously dried and O_2 free, [D$_6$]benzene is condensed *in vacuo* into the NMR tube. Finally the NMR tube is sealed *in vacuo*. DO NOT USE ARGON as a protecting gas, since this is easily condensed in the NMR tube which subsequently will explode violently.

For the proton experiment **a** load standard ^{1}H NMR parameters. You have to set:

td: 64 k
sw: 110 ppm
o1: 25 ppm to lower frequencies from TMS signal
p1: 45° ^{1}H transmitter pulse
d1: 100 ms
transmitter attenuation [3 dB]
ns: 8

For the ^{13}C experiment **b** load standard ^{13}C parameters. You have to set:

td: 64 k
sw: 990 ppm
o1: 400 ppm to higher frequencies from TMS signal
o2: on resonance of the previously determined ^{1}H NMR frequency of the cobaltocene signal
p1: 45° ^{13}C transmitter pulse
d1: 100 ms
ns: 2048

5. Processing

Use standard proton and carbon processing as described in Experiments 3.1-3.2. In **a** exponential weighting with lb = 5Hz, in **b** with lb = 100 Hz was applied. Referencing is done relative to an internal solvent peak in order to eliminate bulk susceptibility effects (see Exp. 8.12). Since the TMS scale does not reflect the physically more meaningful paramagnetic signal shifts, these are often calculated using the corresponding signal of an isostructural diamagnetic molecule (in this particular case ferrocene with $\delta_H = 4.1$ and $\delta_C = 69.2$), yielding the paramagnetic signal shifts δ^{para}. These should be reported along with the temperature at which they have been obtained.

6 Result

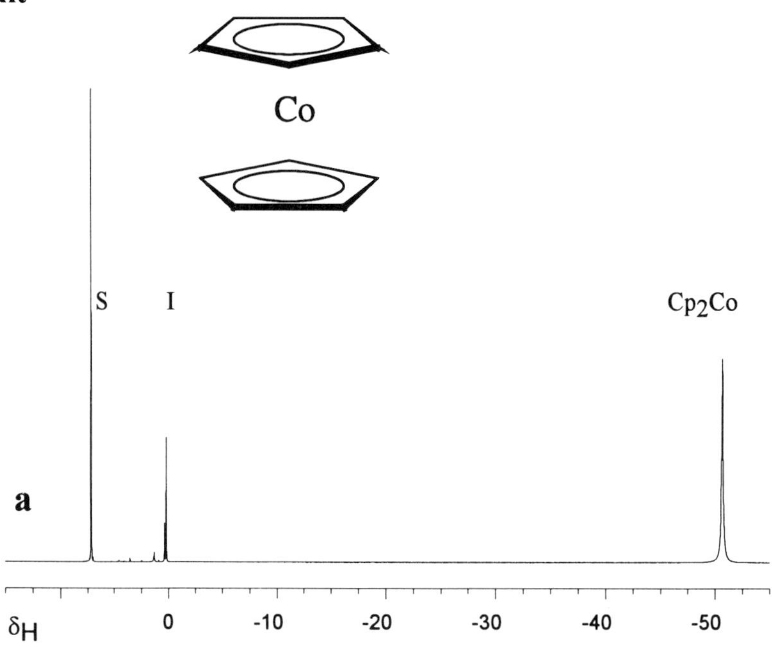

The figures on pages 290 and 291 show expansions of **a** the ^{1}H NMR spectrum (S = solvent, I = impurity) and **b** the ^{13}C NMR spectrum obtained on an AMX-500 spectrometer with a multinuclear probe-head at 300 K. The line-width of the proton signal at $\delta_H = -50.6$ was 80 Hz, and the line width of the ^{13}C signal at $\delta_C = 610$ was 500 Hz.

7. Comments

The paramagnetic shift δ^{para} consist of pseudocontact (see Exp. 8.1) and contact contributions. The contact shift δ^{con} is described by Eq. (1). A is called the contact coupling constant, from which, according to Eq. (2) the spin density ρ of the unpaired spins at the nucleus in question can be calculated; g_e is the electron *g*-factor, and the other constants have the usual meanings.

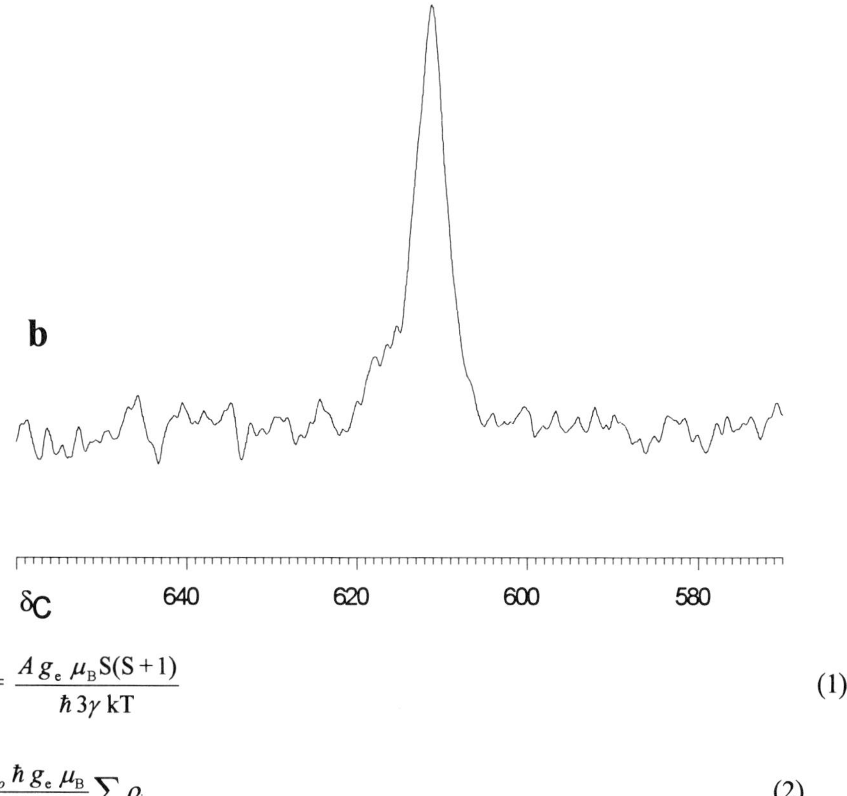

$$\delta con = \frac{A\, g_e\, \mu_B S(S+1)}{\hbar\, 3\gamma\, kT} \tag{1}$$

$$A = \frac{\mu_o\, \hbar\, g_e\, \mu_B}{3S} \sum_i \rho_i \tag{2}$$

MO theoretical calculations of the spin density at the protons and carbons of metallocenes are in reasonable agreement with the NMR results. Note that Eq. (1) predicts a temperature dependence of the contact shift; a concentration dependence has also to be considered.

8. Own Observations

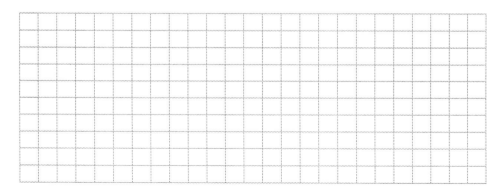

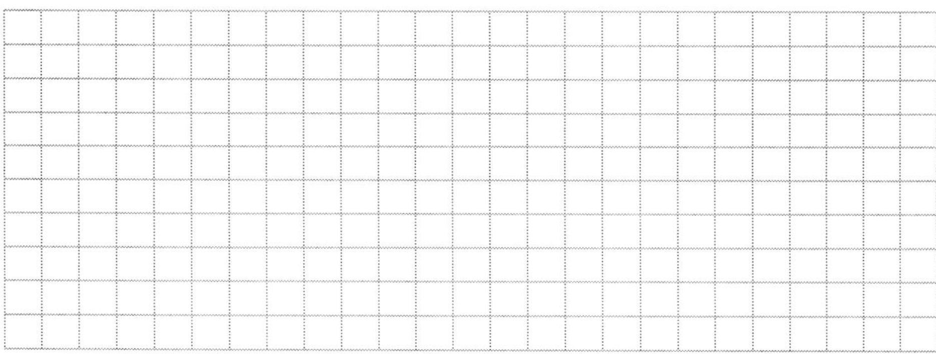

Experiment 8.14

The CIDNP Effect

1. Purpose

This experiment demonstrates the technique used to observe the **C**hemically **I**nduced **D**ynamic **N**uclear **P**olarization. Although the name CIDNP based on an early misinterpretation is somewhat misleading, it has become established in the literature. The effect has been widely used to prove the existence of a radical *pair* intermediate during a chemical reaction. Other applications include the signal assignment of aromatic amino acids in proteins, which uses the photochemical CIDNP technique.

2. Literature

[1] J. Bargon, H. Fischer, U. Johnson, *Z. Naturforsch. A* **1967**, *22*, 1551–1555.

[2] H. R. Ward, R. G. Lawler, *J. Am. Chem. Soc.* **1967**, *89*, 5518–5519.

[3] R. Kaptein, *Adv. Free Radical Chem.* **1975**, *5*, 319–380.

[4] G. L. Closs, R. J. Miller, O. D. Redwine, *Acc. Chem. Res.* **1985**, *18*, 196–202.

[5] P. J. Hore, R. W. Broadhurst, *Prog. NMR Spectrosc.* **1993**, *25*, 345–402.

[6] M. Goez, *Concepts Magn. Reson.* **1995**, *7*, 263–79; *ibid.* 137–52.

3. Pulse Scheme and Phase Cycle

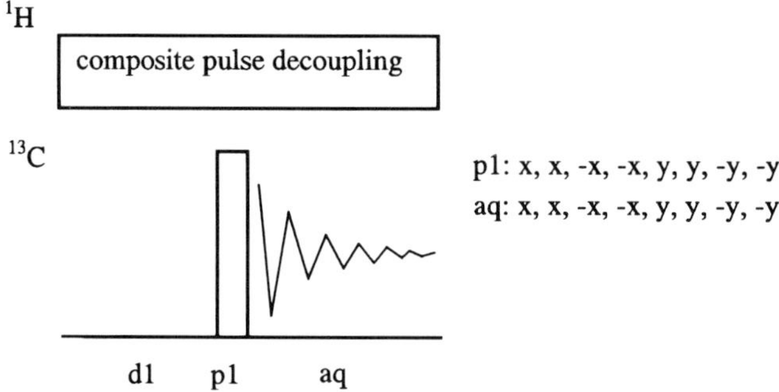

p1: x, x, -x, -x, y, y, -y, -y
aq: x, x, -x, -x, y, y, -y, -y

4. Acquisition

Time requirement: 30 min

Sample: 40 mg benzoyl peroxide in cyclohexanone; *remove the cap from the NMR sample tube*

Use the NMR instrument with the *lowest* available field, as even for protons the *net* CIDNP effect cannot be observed on a 200 MHz instrument. However, the effect can be observed for ^{13}C at this magnetic field strength. Raise the probe-head temperature to 90°C, load standard ^{1}H acquisition parameters (Exp. 3.1), turn off the sawtooth field sweep, and shim the magnet on the incoming FID, since the sample provides no lock signal. Switch to ^{13}C operation and record a standard ^{13}C NMR spectrum as a reference. Load an automatic acquisition routine, which measures and subsequently stores 20 standard ^{13}C spectra with broad-band ^{1}H-decoupling, each with 16 scans. Set the temperature unit to 120°C and immediately start the automatic program. The peroxide decomposes, which results in the CIDNP effect being visible in a few of the recorded spectra. A kinetic diagram can be constructed from the peak heights of the various signals as a function of time.

5. Processing

Use standard 1D processing for ^{13}C NMR as described in Experiment 3.2 with exponential multiplication (lb = 2 Hz).

6. Result

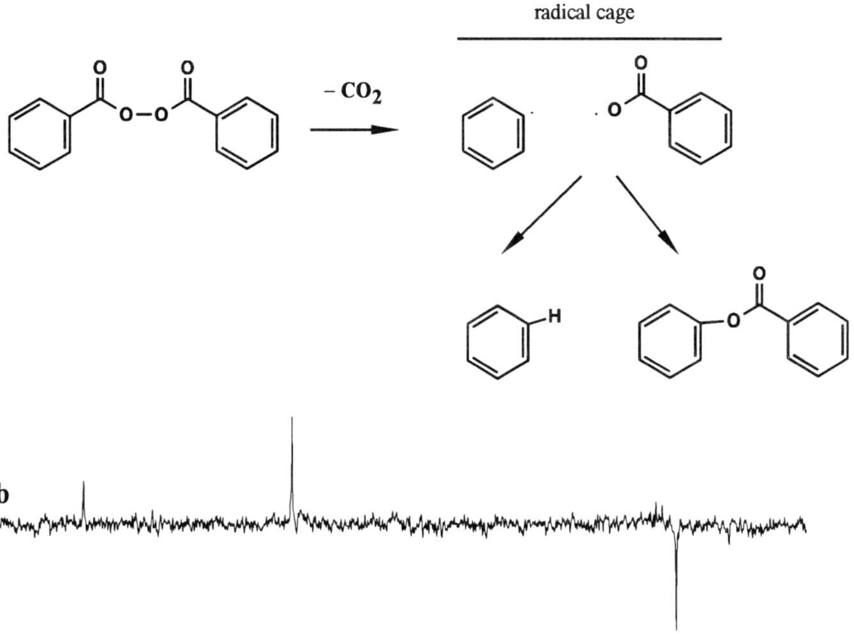

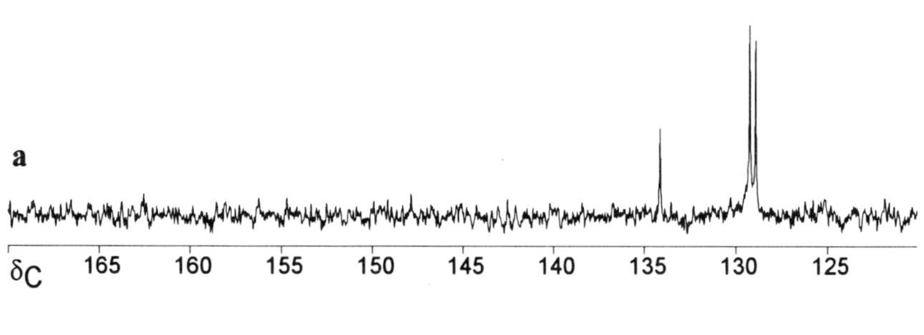

radical cage

$- CO_2$

b

a

δ_C 165 160 155 150 145 140 135 130 125

The figure on page 293 shows the result obtained on an ARX-200 spectrometer. Spectrum **a** is the initial ^{13}C spectrum at 90°C (128 transients). Only the CH signals of the aromatic rings of benzoyl peroxide can be seen under these conditions. Spectrum **b** shows the CIDNP effect at 120°C (16 transients). The emission line of the escape product, benzene, and the two enhanced absorption signals from the carboxyl C-atom and the O-substituted *ipso* carbon atom of the cage recombination product phenyl benzoate are clearly visible. The signals of the initial benzoyl peroxide are hardly visible, which demonstrates the enhancement factor experienced by the other signals.

7. Comments

Other well known examples of the CIDNP effect include organometallic reactions, the decomposition of diazonium salts, 1,2 rearrangements, aromatic nitration, and many photochemically induced reactions. Besides the *net* effect demonstrated in this experiment there exists a *multiplet* effect, where the different lines within one multiplet show enhanced absorption and emission. The complex theory of CIDNP can be studied in the literature cited.

8. Own Observations

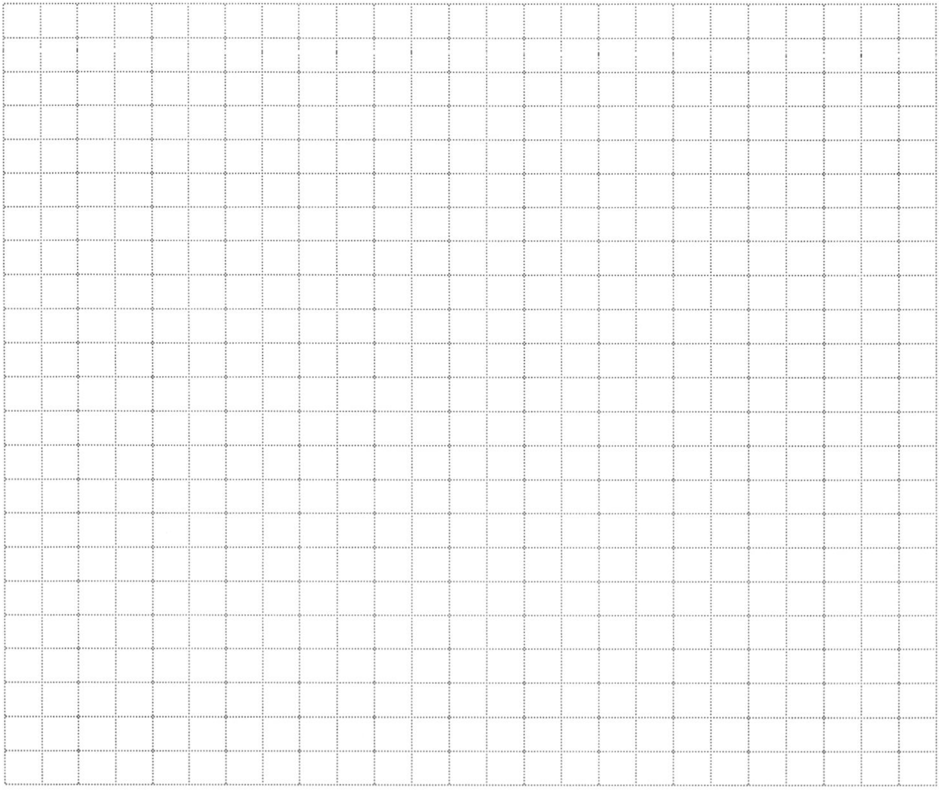

Experiment 8.15

Quantitative ^{1}H NMR Spectroscopy: Determination of the Alcohol Content of Polish Vodka

1. Purpose

In ^{1}H NMR spectroscopy the signal area is normally proportional to the number of nuclei contributing to the signal, provided that saturation is avoided. It is therefore possible to use the integrals of ^{1}H NMR for quantitative determinations in chemistry.

2. Literature

[1] M. L. Martin, G. J. Martin, J. J. Delpuech, *Practical NMR Spectroscopy*, Heyden, London, **1984**, Ch. 9, 350–376.

[2] E. D. Becker, *High Resolution NMR*, Academic Press, New York, **1980**, Ch. 12.

[3] D. D. Traficante, *Concepts Magn. Reson.* **1992**, *4*, 145–160.

[4] D. D. Traficante, L. R. Steward, *Concepts Magn. Reson.* **1994**, *6*, 131–135.

[5] J. Peterson, *J. Chem. Educ.* **1992**, *69*, 843–845.

3. Pulse Scheme and Phase Cycle

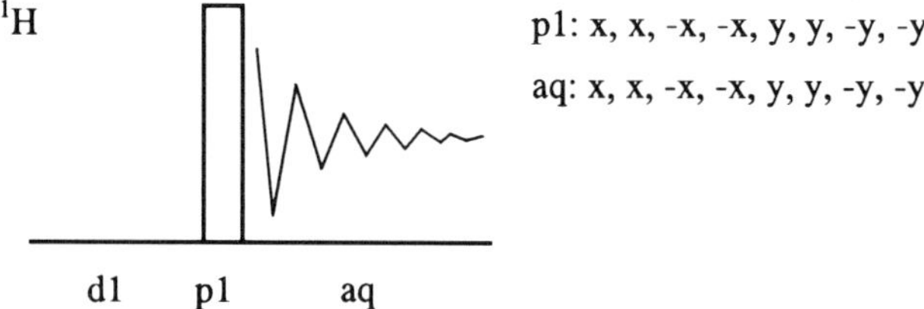

p1: x, x, -x, -x, y, y, -y, -y

aq: x, x, -x, -x, y, y, -y, -y

4. Acquisition

Time requirement: 5 min

Sample: 0.7 ml Polish vodka or any other brand containing a few drops of dry [D$_6$]acetone

Record a standard ^{1}H NMR spectrum with a large data set (10 points/Hz digital resolution). The spectral width should be large enough that the signals at both ends are not affected by the analog audio filter of the spectrometer. Be sure to obtain a good signal-to-noise ratio, at least 35:1. The pulse repetition time must be long enough for complete relaxation ($5T_1$, where T_1 is the longest spin–lattice relaxation time). It is advisable to repeat the experiment with different spectrometer settings and calculate an average of the results. You have to set:

td: 32 k
sw: 10 ppm
o1: middle of ^1H NMR spectrum
p1: 45° ^1H transmitter pulse (for optimization see literature [3])
d1: 5 s
ns: 16

5. Processing

Use standard 1D processing with additional zero-filling to 64 k. Perform a baseline correction on the FID before the Fourier transformation. Phase-adjust the spectrum accurately and perform a baseline correction on the spectrum, then integrate the signals. Ensure the integral limits are far enough apart to give a complete integration. Adjust slope and bias of the individual integrals. This is especially important for the integration of broad signals. In the present case there are three signals (H_2O + OH, CH_2 and CH_3 of ethanol).

6. Result

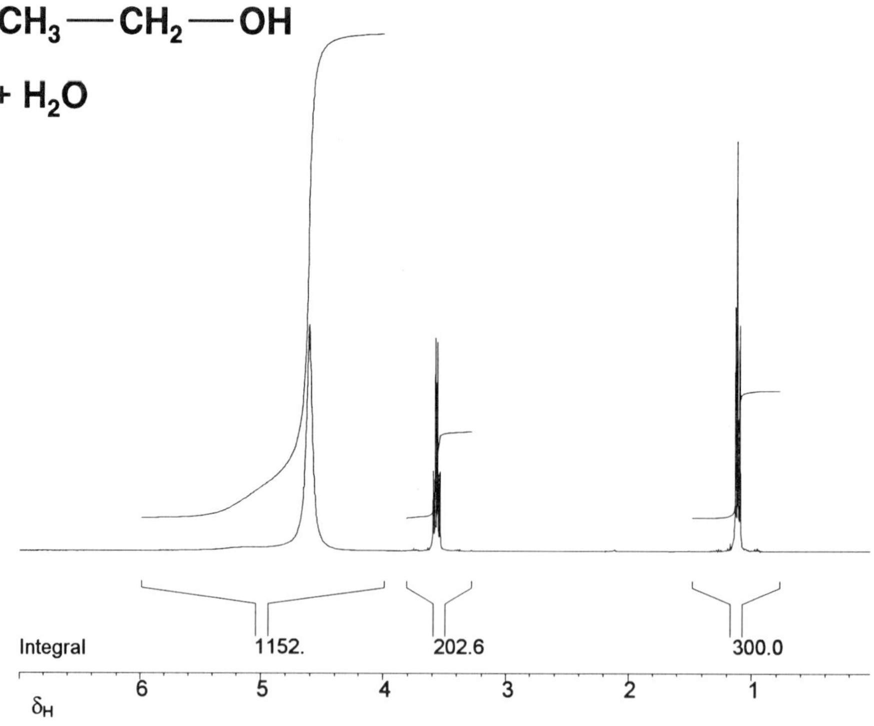

The figure shows the ^1H NMR spectrum of vodka obtained with an AM-400WB spectrometer. The water/alcohol ratio by weight is calculated using the following equation:

$$\frac{Ga}{Gb} = \frac{Fa}{Fb} \cdot \frac{Nb}{Na} \cdot \frac{Ma}{Mb}$$

where G_a and G_b are the parts by weight of the components a and b, in this example water and alcohol respectively and F_a and F_b the areas of the signals of H_2O and CH_2. N_a and N_b are the numbers of nuclei which cause the signals, in this case 2 and 2 respectively, and M_a and M_b are the molecular masses of the two components, 18 and 46 respectively. Using the measured integrals (1152, 202.6) the water/alcohol ratio by weight is calculated to be 2.029/1; in other words the vodka contains 33.0% alcohol by weight or, taking into account the density of the actual ethanol-water mixture, 39.6% by volume. This is in good agreement with the alcohol content given on the label (Bison Brand Vodka 40° by volume). Because the OH signal contains both water and ethanol OH protons, one has to subtract the intensity of one proton from the value of the integral of water before calculating this ratio.

7. Comments

The integrals of NMR signals are relative measures of the numbers of resonating nuclei. If one component is present at much lower concentration than another, the percentage error in measurement of the quantity of this minor component could by quite high (see Exp. 3.12). For quantitative determinations deconvolution and curve-fitting methods have been proposed in the literature, especially for cases where peaks are not fully resolved. Curve-fitting procedures are often included in the commercially available software packages.

8. Own Observations

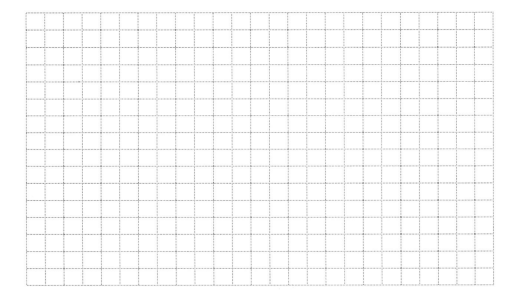

Experiment 8.16

Quantitative ^{13}C NMR Spectroscopy with Inverse Gated ^{1}H-Decoupling

1. Purpose

Quantitative ^{13}C NMR spectroscopy is not as straightforward as quantitative ^{1}H NMR spectroscopy (see Exp. 8.15) as both NOE effects (see Exp. 4.15) and widely varying relaxation times affect the intensity of the signals. For quantitative ^{13}C NMR determinations a relaxation reagent (see Exp. 8.11) should be added and the Overhauser effect suppressed (see Exp. 4.12).

2. Literature

[1] M. L. Martin, G. J. Martin, J. J. Delpuech, *Practical NMR Spectroscopy*, Heyden, London, **1984**, Ch. 9, 350–376.

[2] E. D. Becker, *High Resolution NMR*, Academic Press, New York, **1980**, Ch. 12.

[3] C. H. Sotak, C. L. Dumoulin, G. C. Levy, *Top. Carbon-13 NMR Spectrosc.* **1984**, *4*, 91–121.

[4] L. D. Field, S. Sternhell, *Analytical NMR*, Wiley, Chichester, **1989**, Ch. 3, 41–63.

3. Pulse Scheme and Phase Cycle

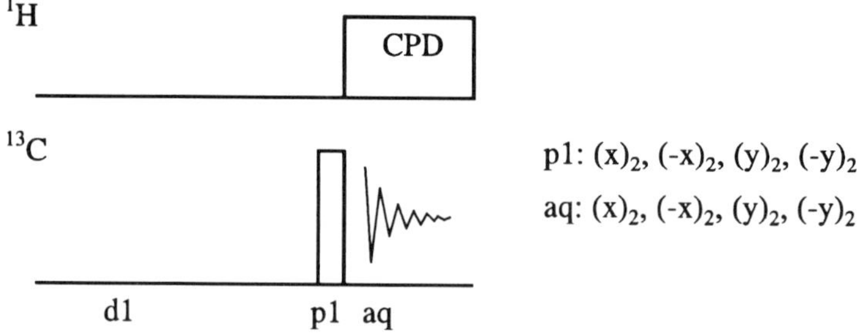

p1: $(x)_2$, $(-x)_2$, $(y)_2$, $(-y)_2$

aq: $(x)_2$, $(-x)_2$, $(y)_2$, $(-y)_2$

4. Acquisition

Time requirement: 30 min

Sample: 156.4 mg naphthalene and 70.0 mg phenanthrene in 1 ml $CDCl_3$. Add 35 mg $Cr(acac)_3$, which corresponds to a 0.1 M solution.

Record a ^{13}C NMR spectrum with the inverse gated decoupling sequence. The spectral width should be large enough that the signals at both ends are not affected by the analog audio filter of the spectrometer. Be sure to obtain a good signal-to-noise ratio, at least 35:1. You have to set:

td: 2 k (short aq to avoid NOE build-up during acquisition)
sw: 20 ppm
o1: middle of aromatic region of the ^{13}C NMR spectrum
o2: middle of aromatic region of the ^{1}H NMR spectrum
p1: 90° ^{13}C transmitter pulse
d1: 10 s
decoupler attenuation and 90° pulse for CPD
ns: 160

5. Processing

Use standard 1D processing as described in Experiment 3.2. Zero-filling to 8 k yields an adequate digital resolution. Perform a baseline correction on the FID before the Fourier transformation. Phase-adjust the spectrum accurately and perform a baseline correction on the spectrum, then integrate the signals. Ensure the integral limits are far enough from both sides of the signal to give a complete integration. In general use as many signal pairs for integration as possible and calculate an average. It is advisable to repeat the experiment with different spectrometer settings and calculate an average of the results.

6. Result

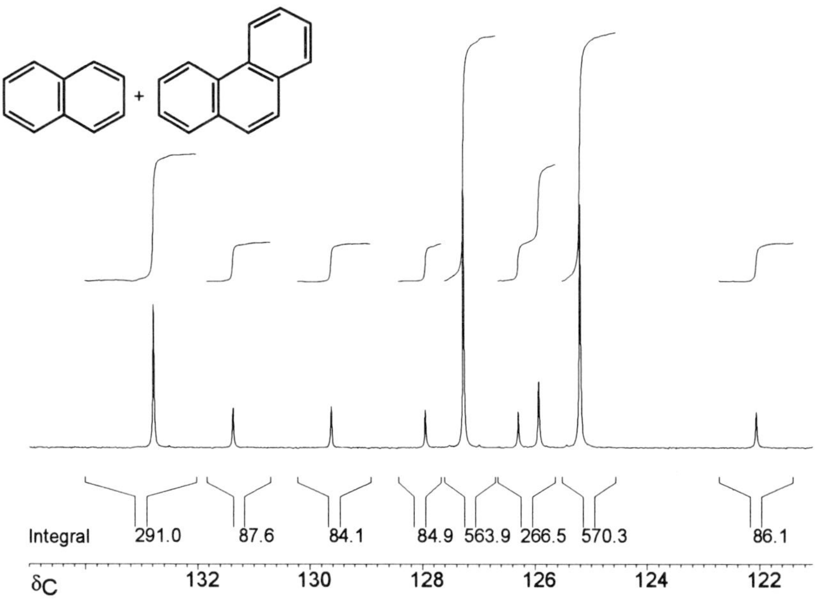

+

Integral 291.0 87.6 84.1 84.9 563.9 266.5 570.3 86.1

δC 132 130 128 126 124 122

The figure on page 299 shows the aromatic region of the ^{13}C NMR spectrum obtained on an AM-400WB spectrometer. The ratio of the components by weight is calculated using the following equation:

$$\frac{Ga}{Gb} = \frac{Fa}{Fb} \cdot \frac{Nb}{Na} \cdot \frac{Ma}{Mb}$$

where G_a and G_b are the parts by weight of the components a (naphthalene) and b (phenanthrene) and N_a and N_b the numbers of nuclei, which cause the signals. M_a and M_b are the molecular masses of the two components, 128.16 and 178.23 respectively. Using the measured averaged integrals F_a for naphthalene (signals at δ_C = 132.8, 127.3, 125.2, relative intensity ratio 2:4:4) and F_b for phenanthrene (signals at δ_C = 131.4, 129.6, 127.9, 126.3. 125.9, 122.1, relative intensity ratio 2:2:2:2:4:2) the naphthalene/phenanthrene ratio by weight is calculated to be 2.27/1, which corresponds to 69.5/30.5 percentage weight. This is in very good agreement with the percentage ratio of 69.1/30.9 corresponding to the weighed amounts used.

7. Comments

The integrals of NMR signals are relative measures of the numbers of nuclei producing that signal. If one component is present at much lower concentration than another, the percentage error in measurement of the quantity of this minor component could by quite high (see Exp. 3.12). It is only when the line shapes and line widths are identical that the relative peak heights are a measure of the relative concentration.

The inverse gated ^{1}H-decoupling experiment is described in Experiment 4.12. Small NOE enhancement build up during the acquisition time, but are dissipated during the relaxation delay which should be at least 10 times longer than the acquisition time. In addition, the paramagnetic relaxation reagent $Cr(acac)_3$ shortens the ^{13}C spin–lattice relaxation times.

8. Own Observations

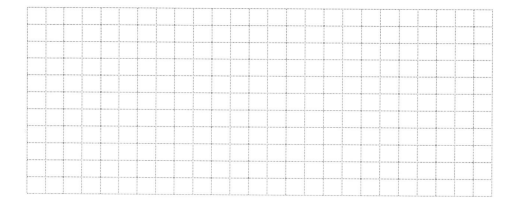

Experiment 8.17

NMR Using Liquid-Crystal Solvents

1. Purpose

If liquid crystals are used as NMR solvents, dissolved molecules no longer tumble isotropically but can be partially oriented along one axis of the liquid-crystal phase. Effects of chemical shift anisotropy, of the dipolar coupling D and of the anisotropy of J become visible; these are normally not present in an isotropic solution. Therefore, from detailed spectral analysis of such spectra, relative internuclear distances, the absolute sign of spin–spin coupling constants and data about molecular reorientation can be obtained. As an example, we show in this experiment the spectrum of benzene in a nematic phase which gives a strikingly complex spectrum for this simple symmetrical molecule.

2. Literature

[1] A. Saupe, G. Englert, *Phys. Rev. Lett.* **1963**, *11*, 462–464.
[2] J. W. Emsley, J. C. Lindon, *NMR Spectroscopy Using Liquid Crystal Solvents*, Pergamon, Oxford, **1975**.
[3] C. L. Khetrapal, A. C. Kunwar, A. S. Tracey, P. Diehl, *NMR Basic Principles and Progress*, **1975**, *9*, 1–85.
[4] A. D. Buckingham, K. A. McLauchlan, *Prog. NMR Spectrosc.* **1967**, *2*, 63–109.
[5] C. L. Khetrapal, A. C. Kunwar, *Adv. NMR Spectrosc.* **1977**, *9*, 301–422.
[6] J. Kaski, J. Vaara, J. Jokisaari, *J. Am. Chem. Soc.* **1996**, *118*, 8879–8886.

3. Pulse Scheme and Phase Cycle

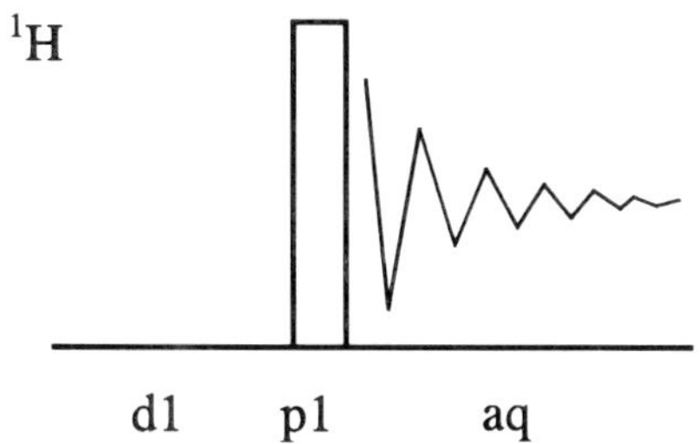

p1: x, x, -x, -x, y, y, -y, -y

aq: x, x, -x, -x, y, y, -y, -y

4. Acquisition

Time requirement: 15 min

Sample: 0.05 ml benzene in 0.5 ml (4-*n*-heptylcyclohexyl)-*p*-cyanobenzene (Merck Licrystal® PCH7)

Load standard ^{1}H NMR parameters. You have to set:

td: 64 k
sw: 22 ppm
o1: on resonance of the ^{1}H benzene signal in isotropic phase
p1: 30° ^{1}H transmitter pulse
d1: 1 s
preacquisition delay: 100 μs to avoid break through of the matrix signal
ns: 8

Set first the probe-head temperature to 330 K. This is above the clearing temperature of the liquid crystal used (T_{cl} = 328 K). Measure at this temperature the spectrum of the liquid crystal alone. Then set the probe-head to 300 K, wait for thermal equilibrium and run the spectrum again. You should not observe any signal. Add the benzene and mix the sample above the clearing temperature. Let the sample equilibrate at 300 K, readjust the receiver gain and measure the signals of the oriented benzene.

5. Processing

Use standard 1D processing as described in Experiment 3.1 with exponential multiplication (lb = 0.3 Hz)

6. Result

The figure shows the ^{1}H NMR spectrum of oriented benzene obtained on an AMX-500 spectrometer. The spectrum is symmetric with respect to the isotropic chemical shift position of benzene and consists of a multitude of lines which become broader at the outer wings. There are no signals from the liquid crystal matrix.

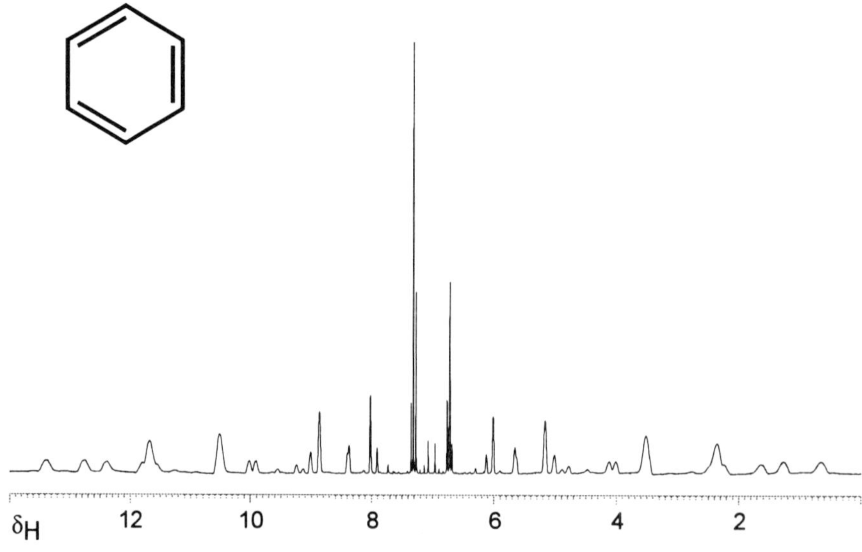

7. Comments

Liquid crystals are usually classified into nematic, cholesteric and smectic phases. Lyotropic phases consist of mixtures of different components. The main application of liquid-crystal NMR is the extraction of relative internuclear distances. For the hexagonal benzene molecule studied in this experiment the distance ratios between the *ortho*, *meta* and *para* protons are expected to be 1: 0.1924 : 0.1250, which can be verified by analysis of the experimental data [2]. Very recently, the anisotropy of the *J* couplings in benzene has been investigated [6]. For high precision work, vibrational averaging has to be considered. In other examples, like *o*-xylene, one can also study the effects of internal motion. For quadrupolar nuclei, liquid-crystal NMR provides a means to determine the quadrupolar coupling constant.

In addition, liquid crystals provide a good means to check the correct temperature setting of the spectrometer (see Ref. [2] in Exp. 5.1). They give a clearly visible indication of whether there is a temperature gradient along the NMR tube.

Often, organic compounds above a certain size tend to arrange in a liquid-crystal like manner which renders the recording of high resolution spectra rather difficult.

8. Own Observations

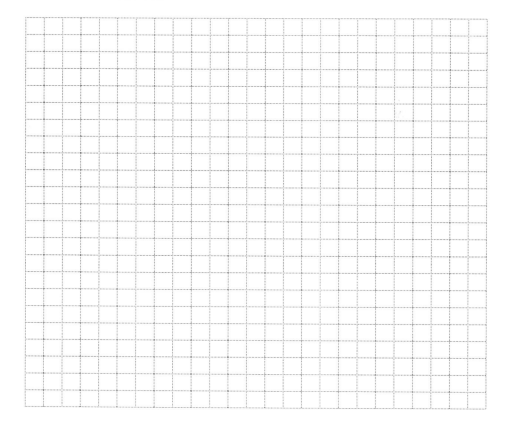

Chapter 9

Heteronuclear NMR Spectroscopy

Nearly all the experiments demonstrated in this book are performed by recording an ^{1}H or a ^{13}C NMR spectrum. However, the world of NMR is much more fascinating if the whole Periodic System is considered, and many of the experiments given can be carried out with other nuclides. Nearly all elements possess at least one isotope with a magnetic moment which is observable, at least in principle, by NMR. However, there are some factors which may prevent routine observations. The main problems arise from low natural abundance and/or a low γ-value giving poor NMR sensitivity. The existence of an electric quadrupolar moment Q for nuclides with $I > 1/2$ may lead to more or less broad lines or even prevents any observations at all, depending on the magnitude of Q. Since the natural abundance and NMR sensitivity of an isotope are constant parameters, they can be combined in a single parameter, the "receptivity", which indicates how difficult it is to obtain a signal in comparison to ^{13}C. In this chapter we provide some basic examples of NMR spectroscopy of the heteroelements to give the beginner an easy start in this field.

For these experiments, the spectrometer must be equipped with a frequency synthesizer to provide the ncccssary frequencies, the appropriate amplifier, and a multinuclear probe-head with a broad-band preamplifier. The following tables give the essential information regarding the resonance frequencies of the NMR-active nuclides, together with the recommended reference compounds.

Referencing, however, is a frequently discussed problem in heteronuclear NMR spectroscopy. Whereas there is an accepted common internal standard for ^{1}H, ^{13}C, and ^{29}Si, namely tetramethylsilane, which is easily removable, comparable universal standards do not exist for the other nuclides. The chemical reactivity often prohibits the use of an internal reference. One way out is to use the reference compound in a capillary inserted into the NMR tube, though it does result in a loss of sensitivity, and a susceptibility correction has to be made.

A common alternative solution is external referencing, where the reference compound is measured before and after the actual experiment and precautions are taken so that the magnetic field is not changed in between. This can work well when using modern instruments, which use digital lock circuits to shift the lock frequency instead of the magnetic field.

A third method uses the Ξ-scale, which does not depend on any nuclide-specific reference compound. Here the chemical shift of TMS is defined to be Ξ_{TMS} = 100.000000 MHz. The Ξ_{X}-value of interest is given by Equation (1).

$$\Xi_{X} = (\Xi_{TMS}/\nu_{TMS})\,\nu_{X} \quad (1)$$

Table of Nuclides with $I = 1/2$

Nuclide	Natural abundance N [%]	Gyromagnetic ratio γ [10^7 rad $T^{-1}s^{-1}$]	NMR frequency[a] Ξ [MHz]	Standard	Receptivity D[b]
^{1}H	99.985	26.7522	100.000 000	Me_4Si	$5.68 \cdot 10^3$
^{3}H	-	28.5335	106.663	Me_4Si	-
^{3}He	$1.3 \cdot 10^{-4}$	−20.378	76.178		$3.26 \cdot 10^{-3}$
^{13}C	1.108	6.7283	25.145 004	Me_4Si	1.00
^{15}N	0.37	−2.7126	10.136 767	$MeNO_2$ (neat)	$2.19 \cdot 10^{-2}$
^{19}F	100.0	25.1815	94.094 003	CCl_3F	4.73
^{29}Si	4.70	−5.319	19.867 184	Me_4Si	2.09
^{31}P	100.0	10.8394	40.480 747	85% H_3PO_4	$3.77 \cdot 10^2$
^{57}Fe	2.19	0.8687	3.237 798	$Fe(CO)_5$	$4.19 \cdot 10^{-3}$
^{77}Se	7.58	5.101	19.091 523	Me_2Se	2.98
^{89}Y	100.0	−1.3163	4.920	$Y(NO_3)_3/H_2O$	0.668
^{103}Rh	100.0	−0.8468	3.17		0.177
^{109}Ag	48.18	−1.2519	4.653 623	Ag^+/H_2O	0.276
^{113}Cd	12.26	−5.99609	22.193 173	$CdMe_2$	7.59
^{119}Sn	8.58	−10.0318	37.290 665	Me_4Sn	25.2
^{125}Te	6.99	−8.453	31.549 802	Me_2Te	12.5
^{129}Xe	26.44	−7.3995	27.661	$XeOF_4$	31.8
^{169}Tm	100.0	−2.21	8.272		3.21
^{171}Yb	14.31	4.7117	17.613		4.44
^{183}W	14.28	1.1283	4.151 888	$W(CO)_6$	$5.89 \cdot 10^{-2}$
^{187}Os	1.64	0.6193	2.282 343	OsO_4	$1.14 \cdot 10^{-3}$
^{195}Pt	33.8	5.8383	21.414 376	$[Pt(CN)_6]^{2-}$	19.1
^{199}Hg	16.84	4.8458	17.910 841	Me_2Hg (neat)	5.42
^{205}Tl	70.50	15.6922	57.633 833	$TlNO_3$/ aq.	$7.69 \cdot 10^2$
^{207}Pb	22.6	5.56264	20.920 597	Me_4Pb	11.8

a) values given in 9 digits as measured for the standard compounds, other data calculated from the γ-values in column 3. b) receptivity relative to ^{13}C

Thus, by measuring the digital frequency of the resonance signal X in question and that of ^{1}H signal of TMS, the corresponding Ξ_X-value can be calculated.

Assume you have detected on a 400 MHz spectrometer the ^{17}O frequency of some compound at $\nu_O = 54.250399$ MHz and you have measured at the same field strength the frequency of the ^{1}H TMS signal at $\nu_{TMS} = 400.130021$ MHz. The calculation according to Eq. (1) then yields a Ξ_O value of 13.558193.

An extension of this procedure is especially common in structural biology, where the ^{15}N chemical shifts of proteins are referenced and calculated using the proton signal of DSS (2,2-dimethyl-2-silapentane-5-sulfonate, sodium salt) [5]. The ratio of the frequency of the common reference compound liquid ammonia (note that in contrast to

Table of Selected Quadrupolar Nuclei with $I > 1/2$

Nuclide	Spin I	Quadrupole moment Q [$10^{-28} m^2$]	Natural abundance N [%]	Gyromagnetic ratio γ [10^7 rad $T^{-1} s^{-1}$]	NMR frequency Ξ [MHz][a)]	Standard	Receptivity D[b)]
^{2}H	1	$2.7 \cdot 10^{-3}$	0.015	4.1064	15.351	$CDCl_3$	$8.2 \cdot 10^{-3}$
^{6}Li	1	$-6.4 \cdot 10^{-4}$	7.42	3.9371	14.717	Li^+ /aq.	3.58
^{7}Li	3/2	$-3.7 \cdot 10^{-2}$	92.58	10.396	38.864	Li^+ /aq.	$1.54 \cdot 10^{3}$
^{9}Be	3/2	$5.2 \cdot 10^{-2}$	100	−3.7594	14.053	Be^{2+} /aq.	78.8
^{11}B	3/2	$3.6 \cdot 10^{-2}$	80.42	8.5827	32.084	$BF_3.OEt_2$	$7.54 \cdot 10^{2}$
^{14}N	1	$1.6 \cdot 10^{-2}$	99.63	1.9324	7.224	CH_3NO_2	5.69
^{17}O	5/2	$2.6 \cdot 10^{-2}$	0.037	−3.6266	13.557	H_2O	$6.1 \cdot 10^{-2}$
^{23}Na	3/2	0.1	100	7.0704	26.429	Na^+ /aq.	$5.25 \cdot 10^{2}$
^{25}Mg	5/2	0.22	10.13	−1.6389	6.126	Mg^{2+} /aq.	1.54
^{27}Al	5/2	0.14	100	6.9762	26.077	$Al(H_2O)_6{}^{3+}$	$1.17 \cdot 10^{3}$
^{33}S	3/2	$-6.4 \cdot 10^{-2}$	0.76	2.0517	7.670	Cs_2SO_4 /aq	$9.7 \cdot 10^{-2}$
^{35}Cl	3/2	$-7.9 \cdot 10^{-2}$	75.53	2.6212	9.798	Cl^- /aq.	20.2
^{37}Cl	3/2	$-6.2 \cdot 10^{-2}$	24.47	2.182	8.156	Cl^- /aq.	3.77
^{39}K	3/2	$-5.5 \cdot 10^{-2}$	93.1	1.2484	4.672	K^+ /aq.	2.69
^{43}Ca	7/2	0.2	0.145	−1.7999	6.728	Ca^{2+}/aq.	$5.3 \cdot 10^{-2}$
^{45}Sc	7/2	−0.22	100	6.4989	24.294	Sc^{3+} /aq.	$1.71 \cdot 10^{3}$
^{47}Ti	5/2	0.29	7.28	−1.5106	5.62476	$TiCl_4$ (neat)	0.864
^{49}Ti	7/2	0.24	5.51	−1.5110	5.638	$TiCl_4$ (neat)	1.18
^{51}V	7/2	$-5.2 \cdot 10^{-2}$	99.76	7.0492	26.350	$VOCl_3$ (neat)	$2.16 \cdot 10^{3}$
^{53}Cr	3/2		9.55	−1.5120	5.652		0.49
^{55}Mn	5/2	0.55	100	6.598	24.67	$KMnO_4$ /aq.	$9.94 \cdot 10^{2}$

^{59}Co	7/2	0.42	100	6.3015	23.727118	$K_3\ Co(CN)_6$	$1.57 \cdot 10^3$
^{63}Cu	3/2	−0.22	69.09	7.1088	26.515473	$[Cu(NCMe)_4]^+$	$3.65 \cdot 10^2$
^{65}Cu	3/2	−0.195	30.91	7.6104	28.403659	$[BF_4]^-$	$2.01 \cdot 10^2$
^{67}Zn	5/2	0.15	4.11	1.6726	6.252		0.665
^{71}Ga	3/2	0.112	39.6	8.1578	30.495	Ga^{3+} /aq.	$3.19 \cdot 10^2$
^{73}Ge	9/2	−0.2	7.76	−0.9332	3.488	$GeMe_4$	0.617
^{75}As	3/2	0.3	100	4.5816	17.127	$KAsF_6$	$1.43 \cdot 10^2$
^{79}Br	3/2	0.3	50.54	6.7021	25.054	Br^- /aq	$2.26 \cdot 10^2$
^{81}Br	3/2	0.28	49.46	7.2245	27.006	Br^- /aq.	$2.77 \cdot 10^2$
^{83}Kr	9/2	0.15	11.55	−1.029	3.848		1.23
^{87}Rb	3/2	0.13	27.85	8.7851	32.839	Rb^+ /aq.	$2.77 \cdot 10^2$
^{87}Sr	9/2	0.36	7.02	−1.1594	4.334	Sr^{2+} /aq.	1.07
^{93}Nb	9/2	−0.32	100	6.5674	24.549	$Nb(O)Cl_3$	$2.74 \cdot 10^3$
^{95}Mo	5/2	$-1.5 \cdot 10^{-2}$	15.72	1.7514	6.547	MoO_4^{2-} /aq.	2.88
^{115}In	9/2	1.16	95.72	5.8622	21.914		$1.89 \cdot 10^3$
^{121}Sb	5/2	−0.5	57.25	6.4016	23.930	$SbCl_6^-$	$5.2 \cdot 10^2$
^{127}I	5/2	−0.7	100	5.3521	20.007	I^- /aq.	$5.3 \cdot 10^2$
^{131}Xe	3/2	−0.12	21.18	2.1935	8.200	$XeOF_4$	3.31
^{133}Cs	7/2	$-3 \cdot 10^{-3}$	100	3.5089	13.117	Cs^+ /aq.	$2.69 \cdot 10^2$
^{137}Ba	3/2	0.28	11.32	2.9729	11.113	Ba^{2+} /aq.	4.41
^{139}La	7/2	0.21	99.911	3.7789	14.126		$3.36 \cdot 10^2$
^{181}Ta	7/2	3	99.988	3.202	11.97	TaF_6^-	$2.04 \cdot 10^2$
^{187}Re	5/2	2.6	62.93	6.0844	22.744	ReO_4^-	$4.9 \cdot 10^2$
^{189}Os	3/2	0.8	16.1	2.0756	7.759	OsO_4	2.13
^{201}Hg	3/2	0.5	13.22	−1.7655	6.600		1.08
^{209}Bi	9/2	−0.4	100	4.2988	16.070		$7.77 \cdot 10^2$

a) values given in 9 digits as measured for the standard compounds, other data calculated from the γ-values in column 3. b) receptivity relative to ^{13}C

the common usage of nitromethane as a reference compound structural biologists prefer NH_3), and the DSS frequency was determined to be 0.101329118. Multiplying this value with the actual DSS frequency of the sample immediately gives the frequency of liquid ammonia which is set to $\delta_N = 0$ by the spectrometer software.

In comparison with ^1H and ^{13}C the chemical shifts of heteronuclides are much more dependent on concentration, temperature, and solvent, which renders this procedure somewhat questionable.

For setting up an NMR experiment with a hetero-nuclide for the first time you have to tune a multinuclear probe-head to the nuclide in question, choosing a frequency according to the tables given here. Since the observation of most hetero-nuclides also requires proton decoupling, the proton channel must also be tuned in order to use ^1H-decoupling or to perform multi-pulse experiments such as DEPT which require ^1H pulses in the decoupler channel. Note that the ^1H pulse-length in the decoupler channel can differ from the setting value used if the observe channel was tuned on ^{13}C.

On modern instruments, these ^1H-decoupler pulses do not differ much from the ^1H pulses in the observe channel, but for older instruments it is best to use special samples which allow the determination of the ^1H-decoupler pulse while the observe channel is tuned to a hetero-nuclide, e.g. formamide for ^{15}N.

The signal of a hetero-nuclide should then be located with a well-known sample yielding a strong signal so as to determine the correct offset and the spectral width for the subsequent measurements on the unknown samples. This procedure might be cumbersome in some cases. To find for the first time the required offset at your instrument it is very helpful to use the above discussed Ξ-values. Using Eq. (1) you calculate ν_X and set the spectrometer offset at this frequency. If no such information is available, it is advisable to record spectra, setting the spectral width to 100 kHz and to shift this spectral window by shifting the offset in 100 kHz steps. Note that the pulse duration used in this process should be $< 90°$ as judged from the situation for other nuclei. Since heteronuclear NMR spectra often consist of only one signal, it is mandatory to check whether the signal is still present without the sample inserted in the probe-head in order to exclude instrumental artefacts.

We have selected for this chapter examples from spin $I = 1/2$ nuclei such as ^{15}N, ^{19}F, ^{29}Si, and ^{119}Sn. The measurement of quadrupolar nuclei is demonstrated with ^2H, ^{11}B, ^{17}O and $^{47/49}Ti$; for a 2D example using ^6Li as the detected nucleus see Experiment 10.21.

Literature

[1] R. K. Harris, B. E. Mann (eds.), *NMR and the Periodic Table*, Academic Press, New York, **1978**.

[2] C. Brevard, P. Granger, *Handbook of High Resolution Multinuclear NMR*, Wiley, Chichester, **1981**.

[3] J. Mason (ed.), *Multinuclear NMR*, 2nd Edition, Plenum Press, London, **1989**.

[4] S. Berger, S. Braun, H.-O. Kalinowski, *NMR Spectroscopy of the Non-Metallic Elements*, Wiley, Chichester, **1997**.

[5] D. S. Wishart, C. G. Bigam, J. Yao, F. Abildgaard, H. J. Dyson, E. Oldfield, J. L. Markley, B.D. Sykes, *J. Biomol. NMR* **1995**, *6*, 135–140.

Experiment 9.1

^1H-Decoupled ^{15}N NMR Spectra with DEPT

1. Purpose

For the observation of nitrogen ^{15}N with I = 1/2 is the isotope of choice, despite its low natural abundance (0.37 %) and low NMR sensitivity; its receptivity as compared to ^{13}C is only 0.022. Additionally one has to take into account that under the conditions of ^1H broad-band decoupling a decrease in intensity may occur due to a negative NOE effect resulting from the negative gyromagnetic ratio of ^{15}N. This problem may be circumvented using the inverse gated ^1H-decoupling technique (see Exp. 4.12). However, the preferred methods are those with polarization transfer, such as the INEPT [1] or DEPT sequences (see Exps. 6.7 and 6.9, respectively), which may be performed with or without ^1H broad-band decoupling and can be tuned either to $^1J(^{15}N,^1H)$ or to $^2J(^{15}N,^1H)$ / $^3J(^{15}N,^1H)$. Here we describe the DEPT experiment on formamide with ^1H broad-band decoupling.

2. Literature

[1] G. A. Morris, *J. Am. Chem. Soc.* **1980**, *102*, 428–429.

[2] W. Witanowski, L. Stefaniak, G. A. Webb, *Annu. Rep. NMR Spectrosc.* **1993**, *25*, 1–480.

[3] S. Berger, S. Braun, H.-O. Kalinowski, *NMR Spectroscopy of the Non-Metallic Elements*, Wiley, Chichester, **1997**.

[4] W. v. Philipsborn, R. Müller, *Angew. Chem. Int. Ed. Engl.* **1986**, *25*, 383–413.

3. Pulse Scheme and Phase Cycle

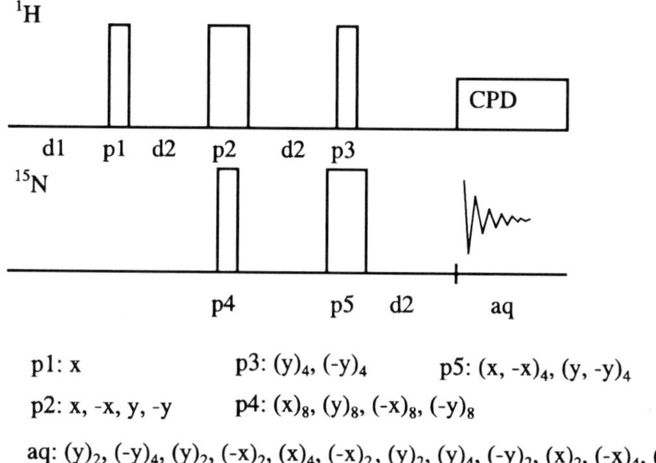

p1: x p3: $(y)_4$, $(-y)_4$ p5: $(x, -x)_4$, $(y, -y)_4$

p2: x, -x, y, -y p4: $(x)_8$, $(y)_8$, $(-x)_8$, $(-y)_8$

aq: $(y)_2$, $(-y)_4$, $(y)_2$, $(-x)_2$, $(x)_4$, $(-x)_2$, $(y)_2$, $(y)_4$, $(-y)_2$, $(x)_2$, $(-x)_4$, $(x)_2$

4. Acquisition

Time requirement: 5 min

Sample: 90% formamide in [D_6]dimethylsulfoxide

Set up your spectrometer to ^{15}N, find the signal, and determine the 90° transmitter pulse-length. The reference compound is CH_3NO_2, and formamide has $\delta_N = -268$. Load the DEPT program with ^1H-decoupling; the settings for a 90° ^1H decoupler pulse must be known. You have to set:

td: 32 k
sw: 350 ppm (chemical shift range of NH-groups)
o1: 220 ppm upfield from CH_3NO_2 (middle of NH region)
o2: middle of ^1H NMR spectrum
p1: 90° ^1H decoupler pulse
p2: 180° ^1H decoupler pulse
p3: 45° ^1H decoupler pulse (optimum for NH_2)
p4: 90° ^{15}N transmitter pulse
p5: 180° ^{15}N transmitter pulse
d1: 2s
d2: $1/[2J(^{15}N,^1H)] = 5.6$ ms, calculated from $^1J(^{15}N,^1H) = 90$ Hz
decoupler attenuation and 90° pulse for CPD
ns: 4

5. Processing

Use standard 1D processing as described in Experiment 3.2

6. Result

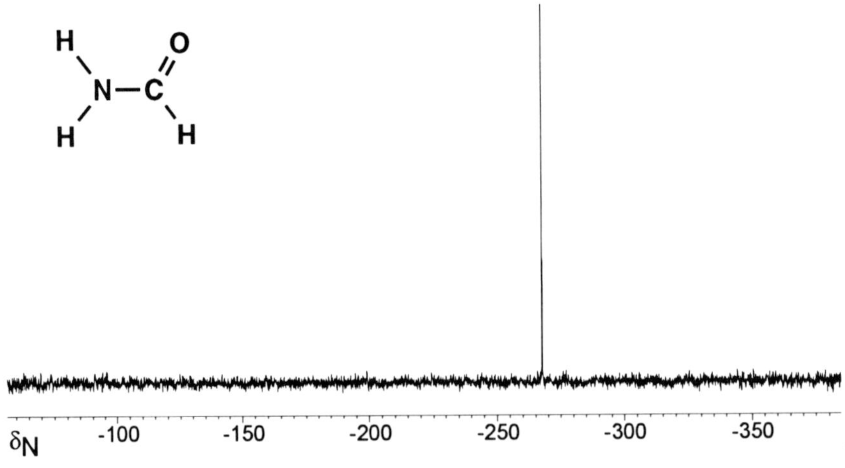

The figure on page 310 shows the 30.4 MHz ^{15}N NMR DEPT spectrum obtained on an ARX-300 spectrometer with a 5 mm multinuclear probe-head. As an exercise you may perform an inverse gated experiment (without NOE, see Exp. 4.12) and one with the normal procedure (Exp. 3.2); theoretically, the gain in intensity using DEPT instead of the inverse gated method is given by $|\gamma(^{1}H)/\gamma(^{15}N)| = 9.87$ (see refocused INEPT, Exp. 6.7).

7. Comments

For a description of the experiment using the product operator formalism see Experiment 6.9; the choice of the pulse duration of p3 (angle α) is discussed in Experiment 9.4. An even greater gain in sensitivity can be achieved by performing a 2D inverse ^{1}H,^{15}N correlation experiment (see Exp. 12.14).

8. Own Observations

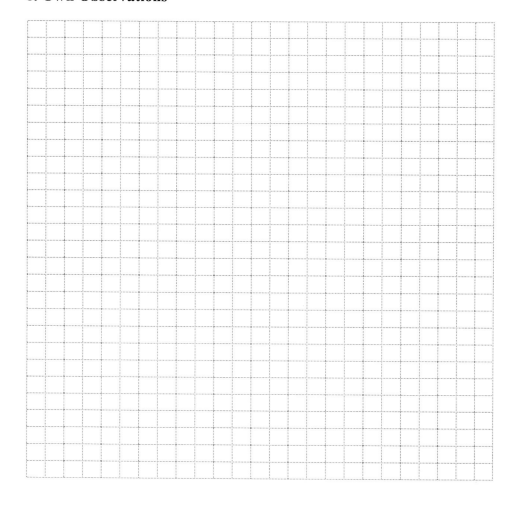

Experiment 9.2

^{1}H-Coupled ^{15}N NMR Spectra with DEPT

1. Purpose

N,H coupling constants are powerful tools in the structure elucidation of nitrogen-containing compounds [1–3] and can be determined by observing ^{15}N at natural abundance using INEPT or DEPT without ^{1}H decoupling. Because of the unusual line intensities associated with the basic INEPT sequence, the use of INEPT$^+$ is recommended. Similarly for DEPT (basic sequence: Exp. 6.9), modifications exist, which are introduced to eliminate spectral distortions [4]. The experiments may be tuned to $^1J(^{15}N,^1H)$, if present, or to $^{15}N,^1H$ couplings over two or three bonds. Here we describe the basic DEPT experiment on formamide without ^{1}H-decoupling.

2. Literature

[1] W. Witanowski, L. Stefaniak, G. A. Webb, *Annu. Rep. NMR Spectrosc.* **1993**, *25*, 1–480.
[2] S. Berger, S. Braun, H.-O. Kalinowski, *NMR Spectroscopy of the Non-Metallic Elements*, Wiley, Chichester, **1997**.
[3] G. C. Levy, R. L. Lichter, *Nitrogen-15 NMR Spectroscopy*, Wiley, New York **1979**.
[4] O. W. Sørensen, R. R. Ernst, *J. Magn. Reson.* **1983**, *51*, 477–489.

3. Pulse Scheme and Phase Cycle

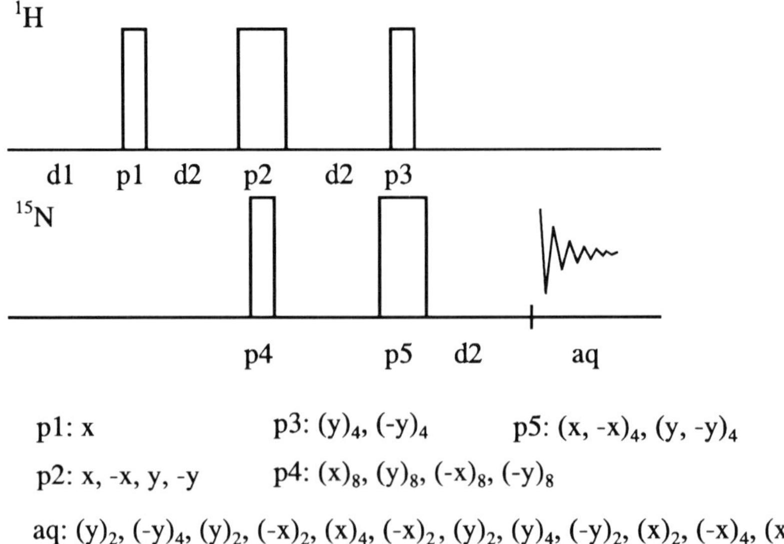

p1: x p3: $(y)_4$, $(-y)_4$ p5: $(x, -x)_4$, $(y, -y)_4$

p2: x, -x, y, -y p4: $(x)_8$, $(y)_8$, $(-x)_8$, $(-y)_8$

aq: $(y)_2$, $(-y)_4$, $(y)_2$, $(-x)_2$, $(x)_4$, $(-x)_2$, $(y)_2$, $(y)_4$, $(-y)_2$, $(x)_2$, $(-x)_4$, $(x)_2$

4. Acquisition

Time requirement: 5 min

Sample: 90% formamide in [D_6]dimethylsulfoxide

Set up your spectrometer to ^{15}N and load the DEPT program without ^{1}H-decoupling. You have to set:

td: 32 k
sw: 350 ppm
o1: 220 ppm upfield from CH_3NO_2
o2: middle of ^{1}H NMR spectrum
p1: 90° ^{1}H decoupler pulse
p2: 180° ^{1}H decoupler pulse
p3: 45° ^{1}H decoupler pulse (optimum for NH_2)
p4: 90° ^{15}N transmitter pulse
p5: 180° ^{15}N transmitter pulse
d1: 2 s
d2: $1/[2J(^{15}N,^{1}H)] = 5.6$ ms, calculated from $^{1}J(^{15}N,^{1}H) = 90$ Hz
ns: 32

5. Processing

Use standard 1D processing as described in Experiment 3.2 with lb = 1 Hz

6. Result

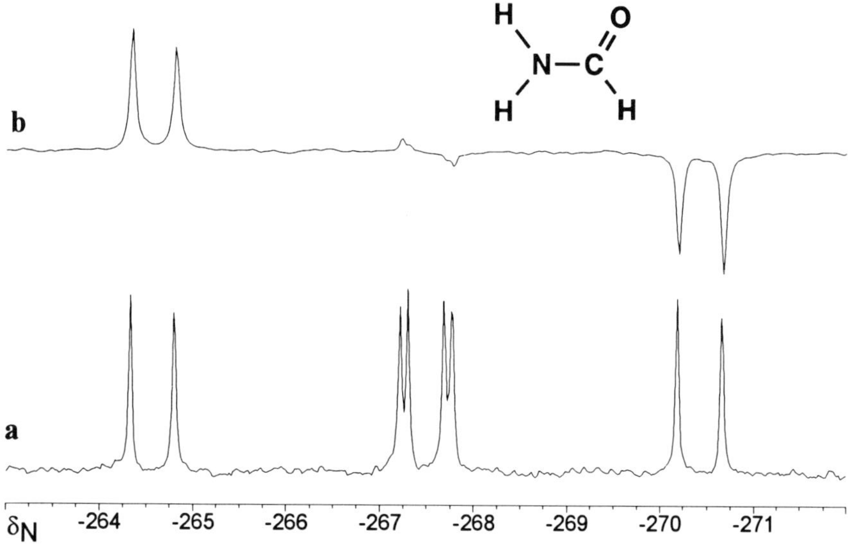

In the figure on page 313 **a** is the 30.4 MHz ^{15}N NMR DEPT spectrum obtained on an ARX-300 spectrometer with a 5 mm multinuclear probe-head. From the pattern, which represents a doublet of doublets of doublets, the following coupling constants can be deduced [2,3]): $^{1}J(^{15}N,H^{syn}) = 87.9$ Hz, $^{1}J(^{15}N,H^{anti}) = 90.3$ Hz, and $^{2}J(^{15}N,H) = 14.0$ Hz. As an exercise you may record the spectrum using INEPT^{+} (Exp. 6.6), leading to the same result. With the basic INEPT (Exp. 6.5), however, you will obtain spectrum **b**.

7. Comments

A description of the DEPT pulse sequence including the product operator formalism is given in Experiment 6.9.

8. Own Observations

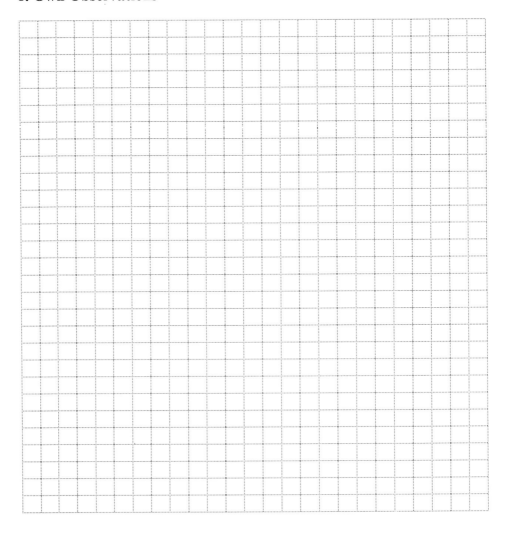

Experiment 9.3

^{19}F NMR Spectroscopy

1. Purpose

The ^{19}F nucleus (I = 1/2, natural abundance 100%) has nearly the same NMR sensitivity as the proton and may occupy the equivalent positions in an organic molecule. However, fluorine is much less widely distributed than hydrogen and hardly occurs at all in natural compounds, and therefore it does not have the same importance for NMR spectroscopy, however. On the other hand, it is exactly for this reason that the nuclide is very well suited for biochemical and medical applications, so that ^{19}F NMR spectroscopy plays an increasing role in these areas [4,5]. Because of the proximity of the resonance frequencies of ^{1}H and ^{19}F it is often possible to tune a proton probe-head to the ^{19}F frequency, so that no special equipment, other than an ^{19}F preamplifier, is needed for the standard experiment described here.

2. Literature

[1] J. W. Emsley, L. Phillipps, *Prog. Nucl. Magn. Reson.* **1971**, *7*, 1–526.
[2] J. W. Emsley, L. Phillipps, V. Wray, *Prog. Nucl. Magn. Reson.* **1976**, *10*, 83–756.
[3] V. Wray, *Annu. Rep. NMR Spectrosc.* **1980**, *10B*, 1–507; ibid. **1983**, *14*, 1–406.
[4] M. J. W. Prior, R. J. Maxwell, J. R. Griffith, *NMR–Basic Principles and Progress* **1992**, *28*, 101–130.
[5] J. T. Gerig, *Prog. Nucl. Magn. Reson.* **1994**, *26*, 293–370.
[6] S. Berger, S. Braun, H.-O. Kalinowski, *NMR Spectroscopy of the Non-Metallic Elements*, Wiley, Chichester, **1997**.

3. Pulse Scheme and Phase Cycle

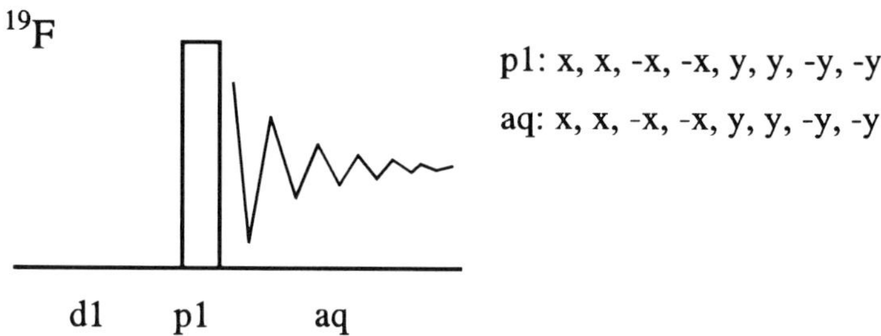

p1: x, x, -x, -x, y, y, -y, -y
aq: x, x, -x, -x, y, y, -y, -y

4. Acquisition

Time requirement: 5 min

Sample: 1% CCl_3F (one drop) in $CDCl_3$

Set up your spectrometer to ^{19}F, find the signal, and determine the 90° transmitter pulse-length. CCl_3F serves as reference compound for ^{19}F NMR. You have to set:

td: 64 k
sw: 300 ppm (typical range for fluorine bound to carbon)
o1: about 100 ppm upfield from CCl_3F (center of that range)
p1: 30° ^{19}F transmitter pulse
d1: 1 s
ns: 1

5. Processing

Use standard 1D processing as described in Experiment 3.1

6. Result

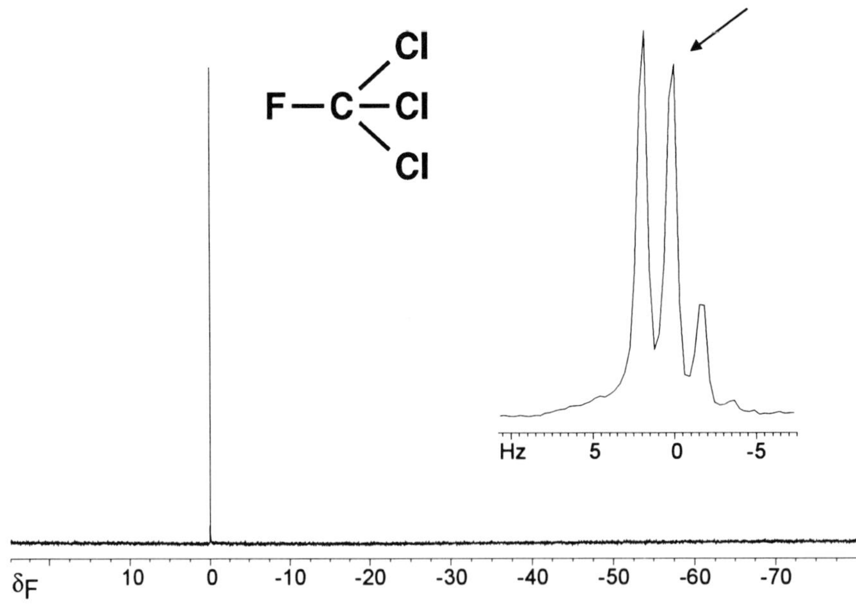

The figure shows the 282.1 MHz ^{19}F NMR spectrum obtained on an ARX-300 spectrometer with a 5 mm $^{1}H/^{13}C$ dual probe-head tuned to the ^{19}F resonance frequency and using a ^{19}F preamplifier for that frequency. In order to achieve better resolution the experiment was repeated with improved digital resolution (see inset). As an exercise you may record an $^{19}F,^{19}F$-COSY spectrum on a mixture of *cis/trans* perfluorodecalin (commercially available) or a 2D *J*-resolved ^{19}F NMR spectrum on 2,4,5-trifluoroaniline. Both experiments may be performed in the above configuration, since they don't need an ^{1}H channel.

7. Comments

The fine structure of the ^{19}F signal of CCl_3F results from the different chlorine isotopomers; $C^{35}Cl_2{}^{37}ClF$ is used for the calibration of high precision ^{19}F NMR spectra. Although isotopes have the same electronic properties within the Born–Oppenheimer approximation, they cause slightly different chemical shifts for an nearby nucleus. This is due to a different ground state vibrational energy which alters the average bond lengths; the heavier isotope usually causes the lower resonance frequency.

8. Own Observations

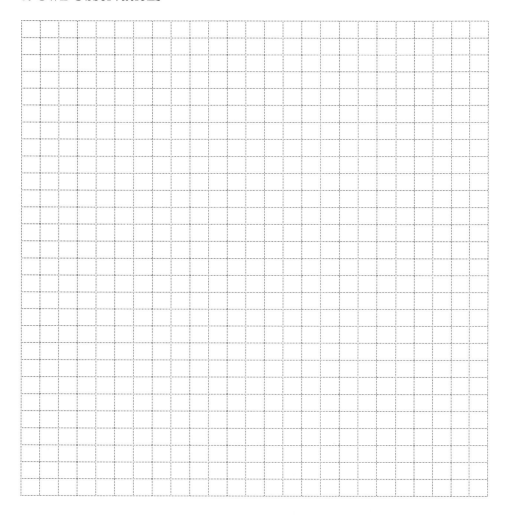

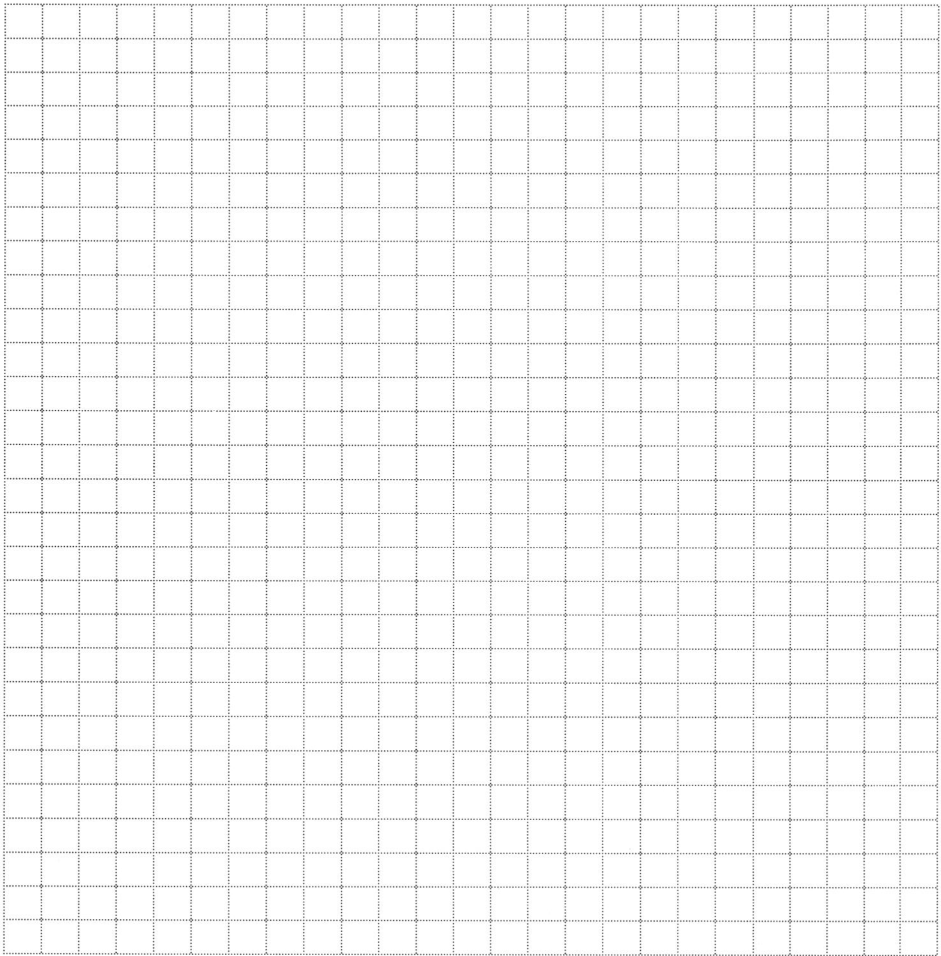

Experiment 9.4

^{29}Si NMR Spectroscopy with DEPT

1. Purpose

^{29}Si (I = 1/2, natural abundance 4.7%) is a nucleus with a small negative gyromagnetic ratio. This means that under normal ^{1}H broad-band decoupling conditions the nuclear Overhauser effect can lead to a reduction in signal intensity or even a cancellation of the signal. It is therefore better to use one of the polarization transfer methods such as INEPT (Exp. 6.5–6.7) or DEPT (Exp. 6.9) which can result in a sensitivity enhancement up to a factor of 5, depending on the number of protons which are responsible for the polarization transfer [1]. In addition the signal of the glassware surrounding the receiver coil is suppressed. Cross-polarization techniques with spin-locking can also be used [5], see Experiment 9.5.

2. Literature

[1] T. A. Blinka, B. J. Helmer, R. West, *Adv. Organomet. Chem.* **1984,** *23,* 193–218.
[2] H. Marsmann, *NMR–Basic Principles and Progress* **1981**, *17*, 65–235.
[3] E. A. Williams, *Annu. Rep. NMR Spectrosc.* **1982**, *15*, 235–289.
[4] J. Schraml, *Prog. NMR Spectrosc.* **1990**, *22*, 289–348.
[5] R. Wagner and S. Berger, *Phosphorus, Sulfur, Silicon, and Rel. Elements* **1994**, *91*, 213–218.

3. Pulse Scheme and Phasecycle

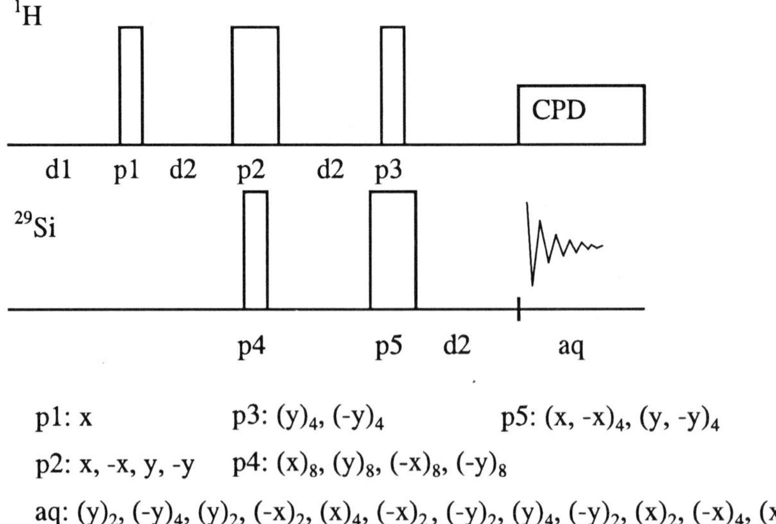

p1: x   p3: $(y)_4$, $(-y)_4$ p5: $(x, -x)_4$, $(y, -y)_4$

p2: x, -x, y, -y   p4: $(x)_8$, $(y)_8$, $(-x)_8$, $(-y)_8$

aq: $(y)_2$, $(-y)_4$, $(y)_2$, $(-x)_2$, $(x)_4$, $(-x)_2$, $(-y)_2$, $(y)_4$, $(-y)_2$, $(x)_2$, $(-x)_4$, $(x)_2$

4. Acquisition

Time requirement: 15 min

Sample: 50% TMS in $CDCl_3$

A multinuclear probe-head is required for the measurement of ^{29}Si spectra. After tuning the probe-head to ^{29}Si on the observe channel and to ^{1}H on the decoupler channel, determine the 90° pulse for ^{29}Si. Load the DEPT pulse sequence (see Exp. 6.9) and record a DEPT spectrum. You have to set:

td: 64 k
sw: 250 ppm
o1: 70 ppm upfield from ^{29}Si signal of TMS
o2: middle of ^{1}H NMR spectrum
p1: 90° ^{1}H decoupler pulse
p2: 180° ^{1}H decoupler pulse
p3: 16.8° ^{1}H decoupler pulse corresponding to the 12 equivalent protons of the sample
p4: 90° ^{29}Si transmitter pulse
p5: 180° ^{29}Si transmitter pulse
d1: 1 s
d2: $1/[2J(\mathrm{Si,H})] = 0.07$ s, calculated from $^{2}J\,(^{29}\mathrm{Si},^{1}\mathrm{H}) = 7$ Hz
decoupler attenuation and 90° pulse for CPD
ds: 2
ns: 32

5. Processing

Use standard 1D processing as described in Experiment 3.2 with exponential multiplication (lb = 3 Hz).

6. Result

The figure on page 320 shows the 79.44 MHz ^{29}Si NMR spectrum obtained for TMS with an AM-400 spectrometer. Spectrum **a** is a normal spectrum, which shows in addition to the TMS signal a broad signal at approximately −110 ppm. This signal is due to the glass NMR tube and the quartz insert surrounding the receiver coil. Spectrum **b** was taken with the DEPT sequence under otherwise identical conditions. Note the improvement in the signal-to-noise ratio. The signal of the glass is suppressed.

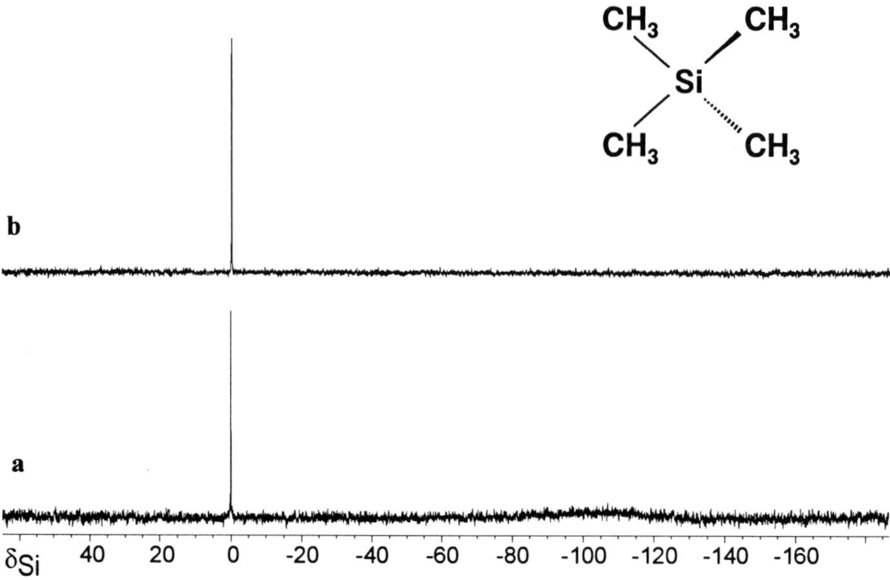

7. Comments

To understand the DEPT sequence you can follow the discussion in Experiment 6.9. In the DEPT experiment the optimal polarization transfer is controlled by the angle α of the last pulse. The optimum pulse angle α_{opt} is independent of J and depends only on the number of coupled nuclei n of the polarization source, usually protons, as given in Equation (2):

$$\alpha_{opt} = \arcsin (n)^{-1/2} \text{ radians} \tag{2}$$

Number of protons n:	1	2	3	6	9	12
Pulse angle α_{opt} (in degrees):	90	45	35	24.1	19.5	16.8

Therefore the DEPT sequence is less sensitive to variations in J. In contrast the NOE enhancement is independent of the number of protons and has a theoretical limit of $\eta = 1 + \gamma_H / 2\gamma_{Si} = -1.5$ for ^{29}Si. Spin-lock experiments as an alternative to INEPT or DEPT are described in Ref. [5].

In the INEPT experiment (see refocused INEPT, Exp. 6.7) the optimum polarization transfer is controlled by the last delay d3. The optimum value of this delay is a function of the scalar coupling constant J and the number of coupled nuclei n which are responsible for the polarization transfer [1] and is given by Equation (1):

$$d3_{opt} = (\pi J)^{-1} \arcsin (n)^{-1/2} \quad (1)$$

As n increases the value of $d3_{opt}$ decreases and the enhancement factor E becomes more sensitive to a variation in d3.

Number of protons n:	1	2	3	6	9	12
Enhancement E:	5.03	5.03	5.82	7.83	9.44	10.82
Delay $d3_{opt}$: (in units of J^{-1})	0.5	0.25	0.196	0.134	0.108	0.093

8. Own Observations

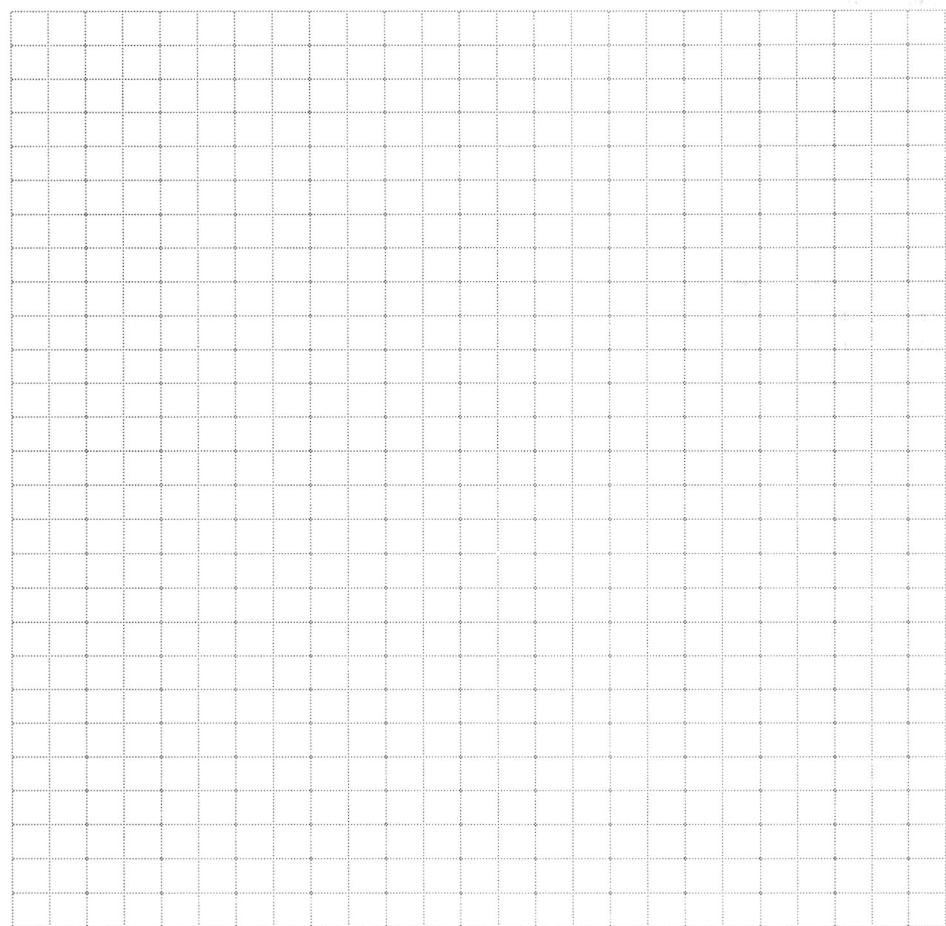

Experiment 9.5

^{29}Si NMR Spectroscopy with Spin-Lock Polarization

1. Purpose

^{29}Si (I = 1/2, natural abundance 4.7%) is a nucleus with a small negative gyromagnetic ratio. Therefore it is traditionally measured with polarization techniques such as INEPT or DEPT (see Exp. 9.4). Because of the various numbers of protons causing the polarization, these techniques are often difficult to optimize. A superior polarization can be achieved with the spin-lock technique, as is commonly used in solid-state NMR spectroscopy by applying the Hartmann Hahn condition (see Exp. 14.3). Thus, in the experiment described here, we demonstrate the application of a heteronuclear spin-lock for the liquid state. This type of polarization transfer in liquids works well for nuclei with no directly attached hydrogens atoms [3], which is most often the case for silicon atoms in organosilicon compounds.

2. Literature

[1] P. D. Murphy, T. Taki, T. Sogabe, R. Metzler, T. G. Squires, B. C. Gerstein, *J. Am. Chem. Soc.* **1979**, *101*, 4055–4058.

[2] G. C. Chingas, R. D. Bertrand, A. N. Garroway, W. B. Moniz, *J. Am. Chem. Soc.* **1979**, *101*, 4058–4059.

[3] M. Ernst, C. Griesinger, R. R. Ernst, W. Bermel, *Mol. Phys.* **1991**, *74*, 219–252.

[4] R. Wagner, S. Berger, *Phosphorus, Sulfur, Silicon, and Rel. Elements* **1994**, *91*, 213–218.

3. Pulse Scheme and Phase Cycle

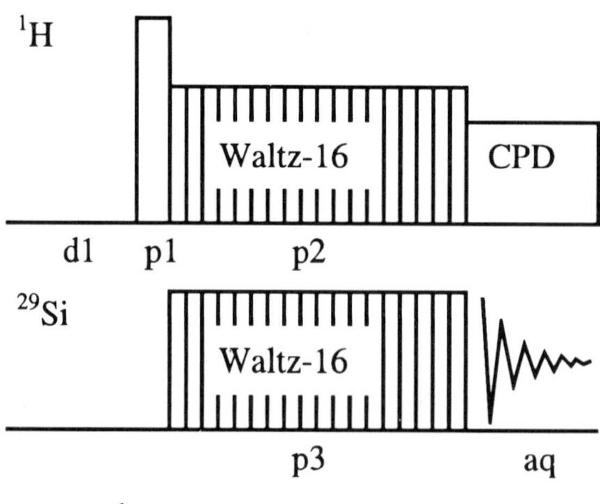

p1: y, -y, -y, y   aq: y, -y, -y, y

waltz-16 spin-lock on both channels consisting of 90°, 180°, 270° and 360° pulses with phases ph1 = −x, x, x, −x and ph2 = x, −x, −x, x:

(270 ph1)	(270 ph2)
(360 ph2)	(360 ph1)
(180 ph1)	(180 ph2)
(270 ph2)	(270 ph1)
(90 ph1)	(90 ph2)
(180 ph2)	(180 ph1)
(360 ph1)	(360 ph2)
(180 ph2)	(180 ph1)
(270 ph1)	(270 ph2)
(270 ph2)	(270 ph1)
(360 ph1)	(360 ph2)
(180 ph2)	(180 ph1)
(270 ph1)	(270 ph2)
(90 ph2)	(90 ph1)
(180 ph1)	(180 ph2)
(360 ph2)	(360 ph1)
(180 ph1)	(180 ph2)
(270 ph2)	(270 ph1)

4. Acquisition

Time requirement: 5 min

Sample: 50% TMS in $CDCl_3$

This experiment requires somewhat advanced preadjustments and can only be performed on instruments providing fast decoupler power switching and variable transmitter attenuation. After tuning the probe-head to ^{29}Si on the observe channel and to ^{1}H on the decoupler channel, determine first the hard 90° ^{1}H decoupler pulse. This can be done as described in Experiment 2.3 but with the TMS sample used here and setting the d2 delay to 76 ms. Repeat the procedure and find a decoupler power level which gives a 90° decoupler pulse of 50 μs for the waltz-16 spin-lock. Check whether the proton channel has a phase difference at these two power levels and adjust if necessary (see Exp.7.1). Finally determine the 90° ^{1}H decoupler pulse and decoupler power level for the usual CPD decoupling. On the ^{29}Si transmitter channel adjust the power level to give a 90° ^{29}Si transmitter pulse of 50 μs. Load the spin-lock polarization pulse sequence. You have to set:

td: 4 k
sw: 500 Hz
o1: on resonance of ^{29}Si signal of TMS
o2: on resonance of ^{1}H NMR signal of TMS

p1: 90° ^{1}H decoupler pulse [3 dB], phase adjustment with respect to spin-lock pulse required
p2: waltz-16 ^{1}H decoupler spin-lock sequence with 50 μs 90° pulse [16 dB]
p3: waltz-16 ^{29}Si transmitter spin-lock sequence with 50 μs 90° pulse [11 dB] length of both spin-lock pulses p2 = p3 = 1/*J*(Si,H) = 152 ms corresponding to ^2J(Si,H) = 7 Hz, determined by loop parameter of waltz-16 spin-lock sequence [l = 32]
d1: 4 s
decoupler attenuation and 90° pulse for CPD [28 dB, 100 μs]
ns: 1

5. Processing

Use standard 1D processing as described in Experiment 3.2 with exponential multiplication (lb = 1 Hz).

6. Result

The figure on page 325 shows the 99.36 MHz ^{29}Si NMR spectrum obtained for TMS with an AMX-500 spectrometer using a multinuclear inverse probe-head. Spectrum **a** is a normal spectrum, obtained with one 90° pulse and CPD decoupling. Spectrum **b** was taken with the spin-lock polarization sequence under otherwise identical conditions. Note the improvement in the signal-to-noise ratio.

7. Comments

For liquids and using pulsed spin-locks it is especially easy to obtain the Hartmann Hahn condition [Eq. (1)], because one simply has to adjust the power levels on both the proton and the X channel so that the 90° pulses have identical lengths. The corresponding equations are outlined in Experiment 2.9.

$$\gamma_H B_1 = \gamma_X B_2 \qquad (1)$$

With the 90° pulse duration of 50 μs used here, a B_1 field of about 5 kHz is covered, which seems to be sufficient in most applications for ^{29}Si. The merits of the technique shown here are its insensitivity to the number of protons causing the polarization transfer, and thus, once adjusted, the method is more robust than INEPT or DEPT and yields better results in routine use. For silanes with directly attached protons the DEPT sequence should be applied.

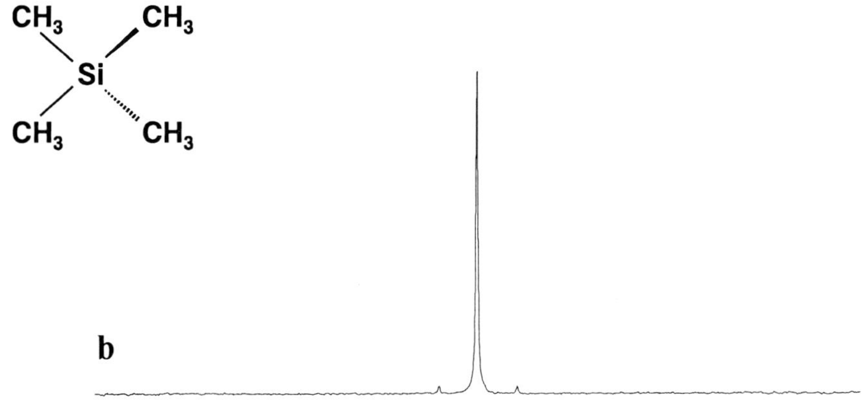

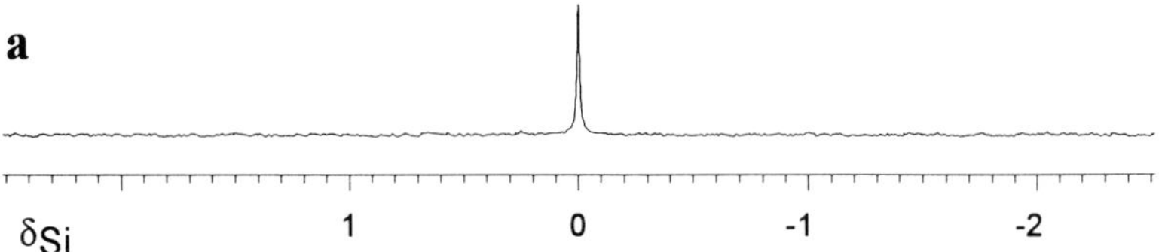

8. Own Observations

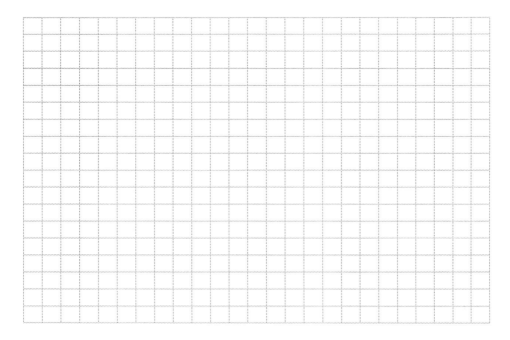

Experiment 9.6

^{119}Sn NMR Spectroscopy

1. Purpose

For the observation of Sn by NMR the isotope ^{119}Sn with $I = 1/2$ and a natural abundance of 8.6% is usually chosen; the alternative is ^{117}Sn with $I = 1/2$, 7.6%. For both isotopes γ has a negative sign, as for ^{15}N and ^{29}Si, so that the comments given in Experiments 9.1 and 9.4 also apply to Sn. Because of the high receptivity of ^{119}Sn (25 relative to ^{13}C) it may be observed by the standard experiment with ^{1}H broad-band decoupling as described here (see Exp. 3.2). For low concentrations the DEPT method is recommended.

2. Literature

[1] B. Wrackmeyer, *Chem. Br.* **1990**, *26*, 48–51.

[2] B. Wrackmeyer, *Annu. Rep. NMR Spectrosc.* **1985**, *16*, 73–186.

[3] V. S. Petrosyan, *Prog. NMR Spectrosc.* **1977**, *11*, 115–148.

3. Pulse Scheme and Phase Cycle

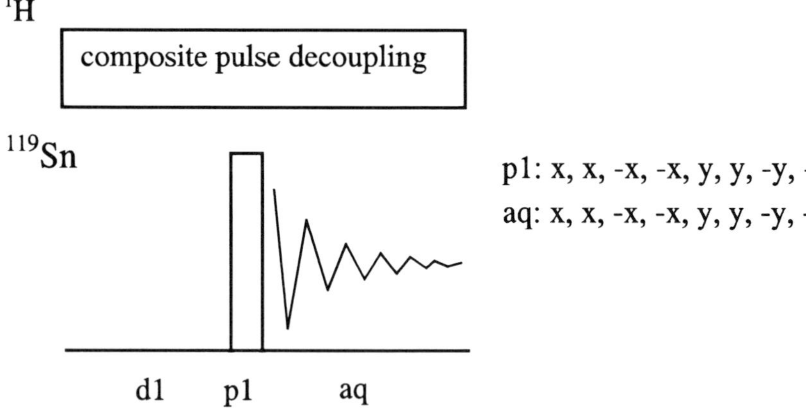

p1: x, x, -x, -x, y, y, -y, -y
aq: x, x, -x, -x, y, y, -y, -y

4. Acquisition

Time requirement: 5 min

Sample: 50% $Sn(CH_3)_4$ in $CDCl_3$

Set up your spectrometer to ^{119}Sn, find the signal, and determine the 90° transmitter pulse. This compound serves as standard, so reference the signal to $\delta_{Sn} = 0$. You have to set:

td: 32 k
sw: 600 ppm (Sn chemical shift range typical for $R_{4-n}SnX_n$)
o1: 100 ppm upfield from $Sn(CH_3)_4$ (center of that chemical shift range)
o2: middle of ^1H NMR spectrum
p1: 30° ^{119}Sn transmitter pulse
d1: 1 s
decoupler attenuation and 90° pulse for CPD
ns: 8

5. Processing

Use standard 1D processing as described in Experiment 3.2 with lb = 3 Hz.

6. Result

The figure shows the 111.9 MHz ^{119}Sn NMR spectrum recorded on an ARX-300 spectrometer; the ^{13}C satellites are clearly visible. For a more precise determination of the coupling constant, a spectrum with a higher digital resolution has been recorded (see inset, $^1J(^{119}Sn,^{13}C) = 336.9$ Hz).

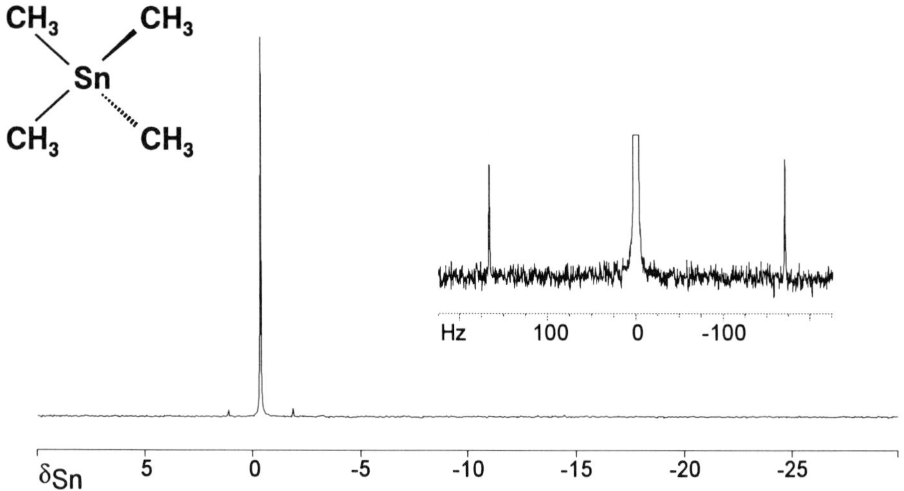

7. Comments

Note the fourfold higher intensity of the ^{13}C satellites (about 2%) because of the four equivalent carbon nuclei.

As an exercise and as an example of what is called multinuclear NMR spectroscopy you may record a standard ^{1}H and a standard ^{13}C NMR spectrum. In these spectra you can determine the ^{1}H and ^{13}C chemical shifts and from the satellites the two-bond couplings of ^{117}Sn and ^{119}Sn with the protons and the one-bond couplings of the two Sn isotopes with the ^{13}C nuclei. It is also interesting to record a ^{1}H-coupled spectrum from the sample to observe the spin coupling pattern caused by the twelve equivalent protons (gated ^{1}H-decoupling, see Exp. 4.11).

8. Own Observations

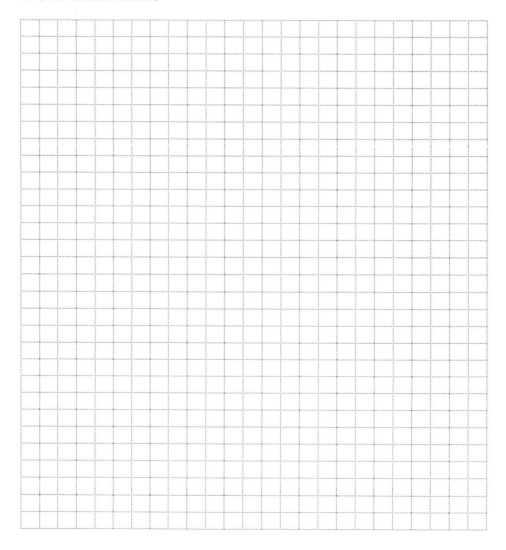

Experiment 9.7

^{2}H NMR Spectroscopy

1. Purpose

This experiment demonstrates the technique used to observe deuterium (^{2}H, $I = 1$) by NMR spectroscopy in natural abundance (0.015%). As an example we have chosen pure ethanol, since this is currently routinely used in food analysis. Since the isotope distribution in ethanol is dependent on the sugar source and its geographic origin, quantitative ^{2}H NMR spectroscopy can be used to detect fraud [3, 4].

2. Literature

[1] C. Brevard, J. P. Kintzinger, in: *NMR and the Periodic Table*, R. K. Harris, B. E. Mann (eds.), Academic Press, London, **1978**, 107–128.

[2] J. W. Akitt, in: *Multinuclear NMR*, J. Mason (ed.), Plenum Press, New York, **1987**, 171–187.

[3] M. L. Martin, G. J. Martin, *NMR–Basic Principles and Progress* **1991**, *23*, 1–62.

[4] G. J. Martin, M. L.Martin, *Annu. Rep. NMR Spectrosc.* **1995**, *31*, 81–104.

3. Pulse Scheme and Phase Cycle

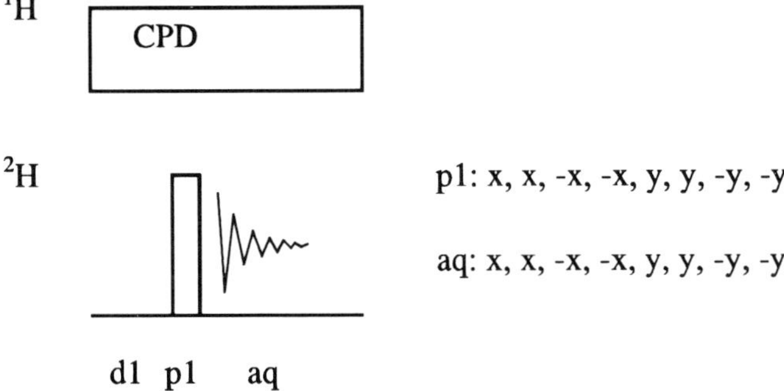

p1: x, x, -x, -x, y, y, -y, -y

aq: x, x, -x, -x, y, y, -y, -y

4. Practical Procedure

Time requirement: 30 min

Sample: pure ethanol

A multinuclear probe-head is required for the measurement of ^{2}H spectra. Disconnect the ^{2}H lock channel and remove any ^{2}H-stop filter from the transmitter line (which is sometimes hidden in the preamplifier). Use an ^{19}F lock, if available; alternatively the magnet must be stable enough to hold its field position for the duration of the measurement. After tuning the multinuclear probe-head to deuterium on the observe channel and to ^{1}H on the decoupler channel, it is best to use a sample of $CDCl_3$ to detect the ^{2}H resonance and to determine the 90° pulse. First record a standard ^{1}H NMR spectrum of the sample, center on the ethanol resonances, and note the offset. This value should be used for the decoupler offset in the following experiment. The chemical shifts of isotopes are essentially identical, and therfore you can reference the ^{2}H resonance of the secondary standard $CDCl_3$ to $\delta_H = 7.25$. You have to set:

td : 8 k
sw: 8 ppm
o1: middle of ^{2}H NMR spectrum
o2: middle of ^{1}H NMR spectrum
p1: 90° ^{2}H transmitter pulse
d1: 100 ms
decoupler attenuation and 90° pulse for CPD
ns: 256

5. Processing

Use standard 1D processing as described in Experiment 3.2 with exponential multiplication (lb = 2 Hz).

6. Result

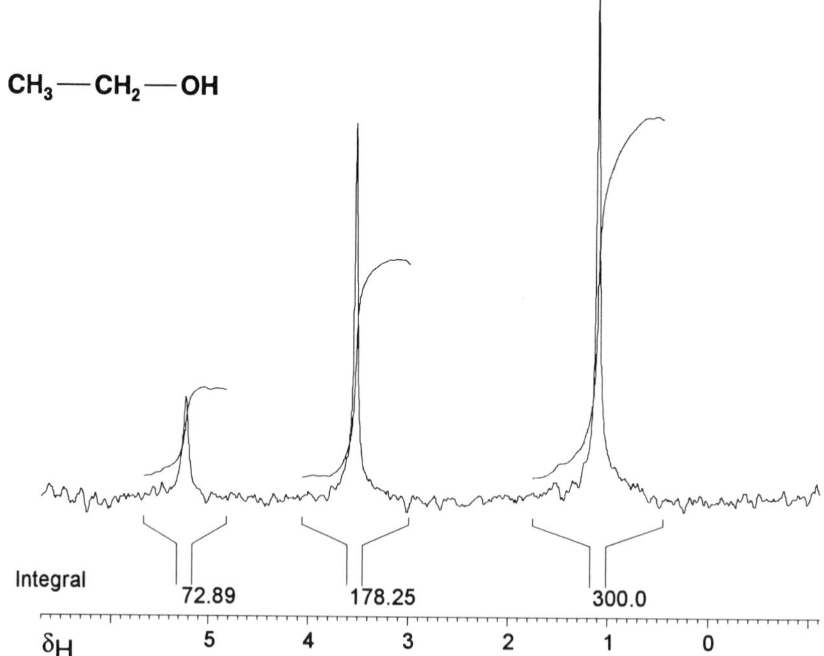

The figure on page 330 shows the result obtained on an AMX-500 spectrometer with a 5 mm multinuclear probe-head. The integrals indicate that the deuterium is not distributed in the expected 3:2:1 ratio. For a reliable quantitative evaluation, however, the signal-to-noise ratio must be much better. Furthermore, a certified standard with a known deuterium content is required. Note that the chemical shift of the OD signal is temperature dependent.

7. Comments

Although deuterium is a quadrupolar nucleus, its relaxation behavior frequently resembles that of spin 1/2 nuclei in small molecules, due to the very small quadrupolar moment. Therefore it is necessary to use a sufficiently long repetition time for quantitative work. As well as the application illustrated here, ^{2}H NMR spectroscopy of labeled compounds is widely used for mechanistic studies in organic chemistry.

8. Own Observations

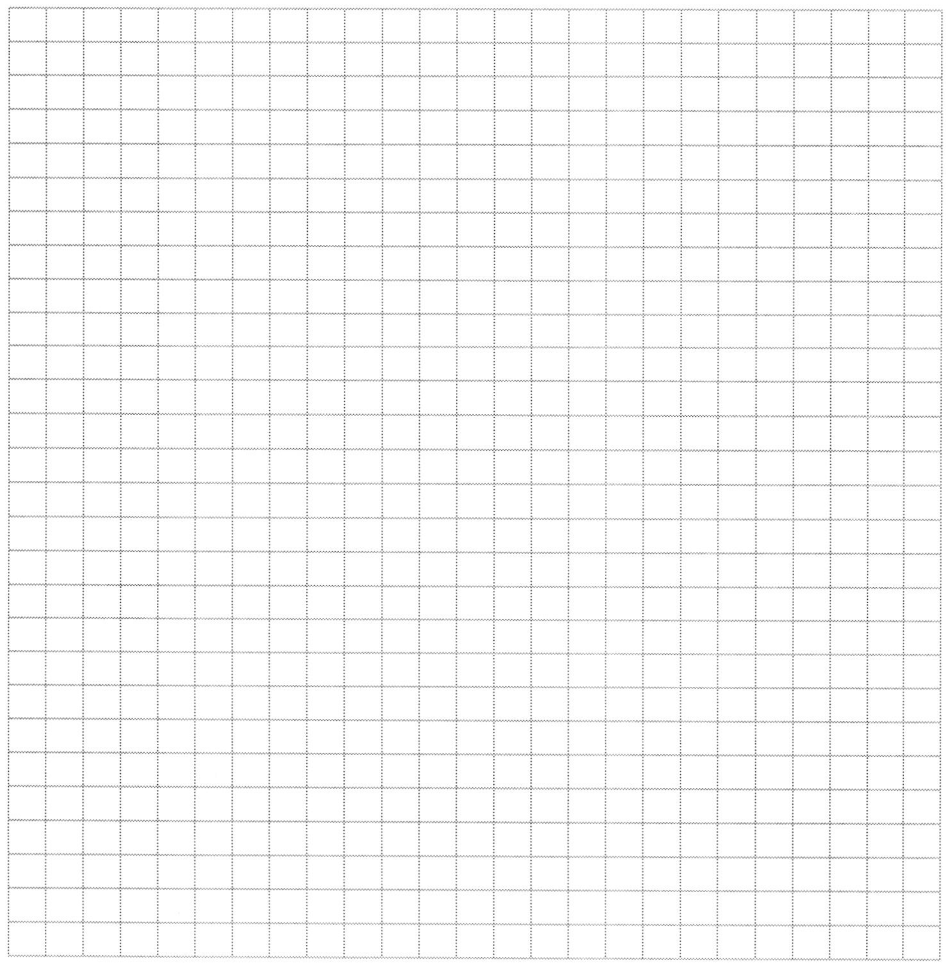

Experiment 9.8

^{11}B NMR Spectroscopy

1. Purpose

This experiment demonstrates the technique used to obtain ^{11}B NMR spectra (I = 3/2, natural abundance 80.42%). As an example we have chosen the commercially available 1,7-dicarba-*closo*-dodecaborane, since this compound gives a spectrum containing four different signals. Furthermore, the spectrum can be recorded with or without proton decoupling; thus a spin–spin coupling ^{1}J(B,H) can be observed.

2. Literature

[1] D. Reed, *Chem. Soc. Rev.* **1993**, 109–116.
[2] B. Wrackmeyer, *Annu. Rep. NMR Spectrosc.* **1988**, *20*, 61–203.
[3] A. R. Siedle, *Annu. Rep. NMR Spectrosc.* **1988**, *20*, 205–314.

3. Pulse Scheme and Phase Cycle

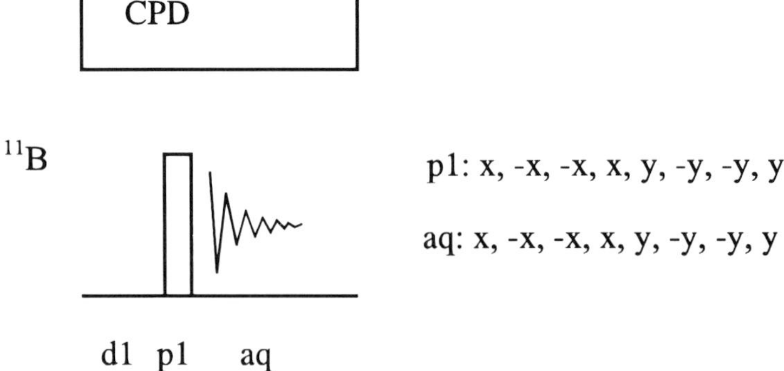

4. Acquisition

Time requirement: 5 min

Sample: 100 mg 1,7-dicarba-*closo*-dodecaborane in 0.7 ml $CDCl_3$

A multinuclear probe-head is required for the measurement of ^{11}B spectra. There are special ^{11}B probe-heads available which don't have a glass insert resulting in a smaller background signal. After tuning the multinuclear probe-head to ^{11}B on the observe channel and to ^{1}H on the decoupler channel, use a sample of $BF_3{\cdot}O(C_2H_5)_2$ to detect the ^{11}B resonance and to determine the 90° pulse. The δ_B value of this standard is referenced to 0. To obtain the spectrum displayed below you have to set:

td : 4 k
sw: 36 ppm
o1: middle of ^{11}B NMR spectrum
o2: middle of ^{1}H NMR spectrum
d1: 100 ms
p1: 90° ^{11}B transmitter pulse
decoupler attenuation and 90° pulse for CPD
ns: 8

5. Processing

Use standard 1D processing with exponential multiplication (lb = 2 Hz) and a baseline correction. For referencing use the external reference of the standard; be sure not to change the magnetic field between the two measurements.

6. Result

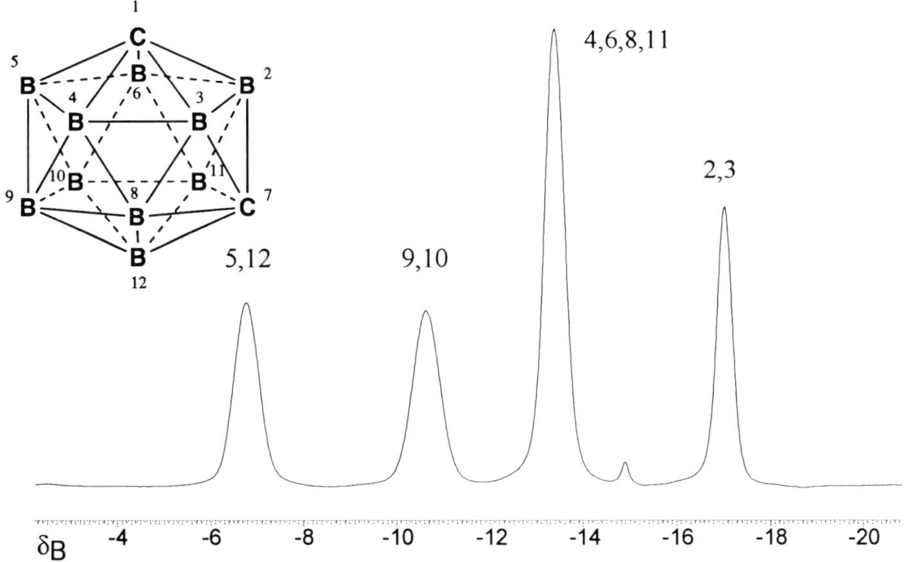

The figure shows the 96.23 MHz ^{11}B NMR spectrum obtained on an AC-300 spectrometer with a 5 mm special boron probehead.

7. Comments

^{11}B is a quadrupolar nucleus and can give very broad signals depending on the molecular environment. The pulse repetition time of the experiment can be selected according to the sample used, so that much faster pulsing than used here is often possible.

To obtain ^{11}B spectra without any background signal both the insert of the probehead and the NMR sample tube must be from boron-free material, such as quartz or teflon.

8. Own Observations

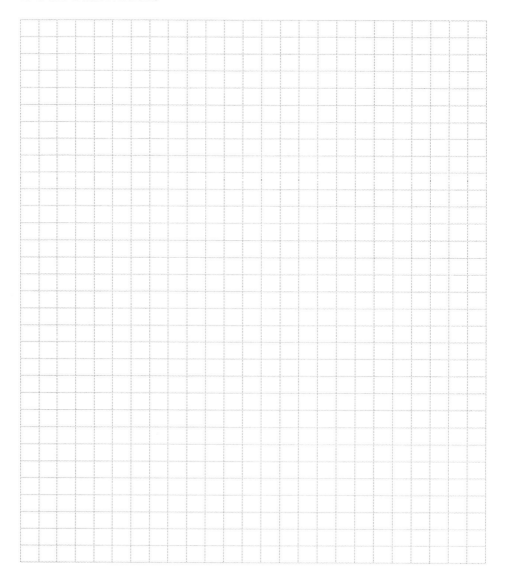

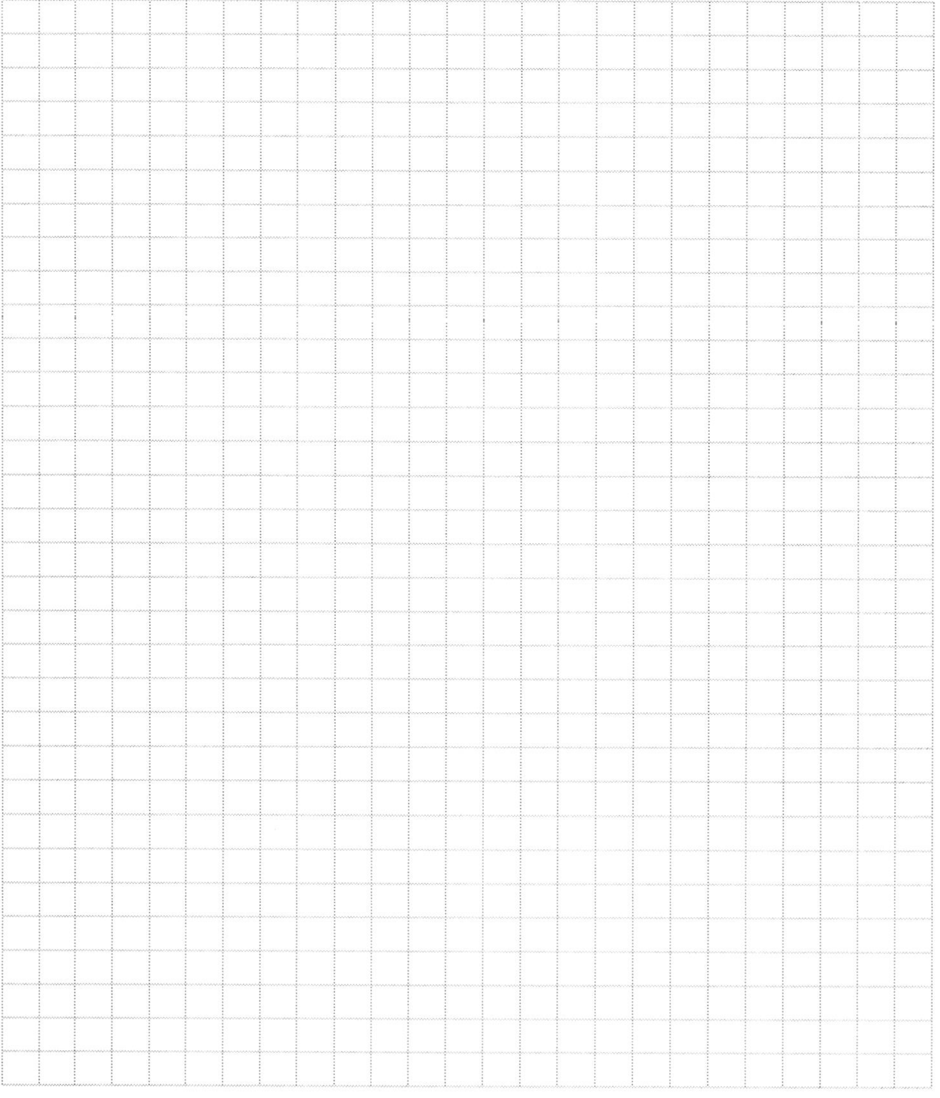

Experiment 9.9

^{17}O NMR Spectroscopy with RIDE

1. Purpose

This experiment demonstrates the technique used to obtain ^{17}O NMR spectra ($I = 5/2$, natural abundance 0.037%). ^{17}O is a quadrupolar nucleus with a relatively low γ-value. Probe-head ringing poses an experimental problem for these types of nuclei, resulting in extensive baseline roll. One possible solution is the RIDE (**RI**ng **D**own **E**limination) pulse sequence which is demonstrated here.

2. Literature

[1] P. S. Belton, I. J. Cox, R. K. Harris, *J. Chem. Soc. Faraday Trans. 2* **1985**, *81*, 63–75.

[2] I. P. Gerothannassis, *Prog. Nucl. Magn. Reson. Spectrosc.* **1987**, *19*, 267–329.

[3] J. P. Kintzinger, *NMR–Basic Principles and Progress*, **1981**, *17*, 1–64.

[4] D. W. Boykin (ed.), *^{17}O NMR in Organic Chemistry*, CRC Press, Boca Raton, Florida, **1991**, 1–325.

[5] S. Berger, S. Braun, H.-O. Kalinowski, *NMR Spectroscopy of the Non-Metallic Elements*, Wiley, Chichester, **1997**, 319–397.

3. Pulse Scheme and Phase Cycle

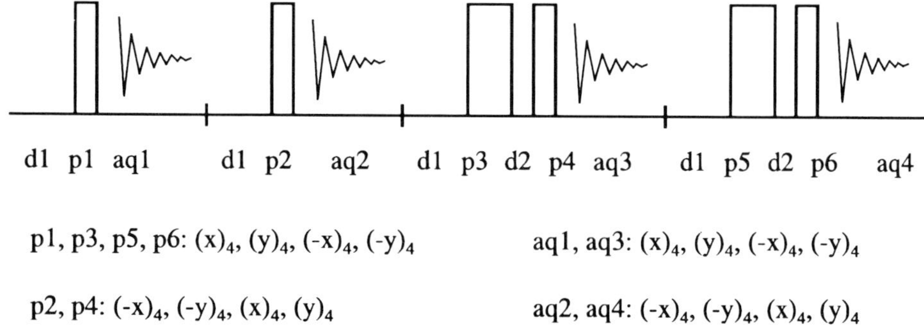

p1, p3, p5, p6: $(x)_4$, $(y)_4$, $(-x)_4$, $(-y)_4$

aq1, aq3: $(x)_4$, $(y)_4$, $(-x)_4$, $(-y)_4$

p2, p4: $(-x)_4$, $(-y)_4$, $(x)_4$, $(y)_4$

aq2, aq4: $(-x)_4$, $(-y)_4$, $(x)_4$, $(y)_4$

4. Acquisition

Time requirement: 5 min

Sample: ethyl crotonate (neat)

A multinuclear probe-head is required for the measurement of ^{17}O spectra. Be sure to remove any ^{2}H stop filter from the transmitter line, since ^{17}O and ^{2}H NMR frequencies are rather similar at lower field strengths. Tune the probe-head to ^{17}O on the observe channel and locate the ^{17}O signal using a sample of D_2O, since the ^{17}O content of D_2O is higher than that of normal water. Determine the 90° pulse for ^{17}O with this sample and use it as reference standard. After loading the RIDE pulse sequence you have to set:

td : 4 k
sw: 500 ppm
o1: 200 ppm downfield from ^{17}O water signal
p1, p2, p4, p6: 90° ^{17}O transmitter pulse
p3, p5: 180° ^{17}O transmitter pulse
d1: 10 ms
d2: 0.5 μs
preacquisition delay: 15 μs
ns: 4·128

5. Processing

Use standard 1D processing as described in Experiment 3.2 with exponential multiplication (lb = 200 Hz)

6. Result

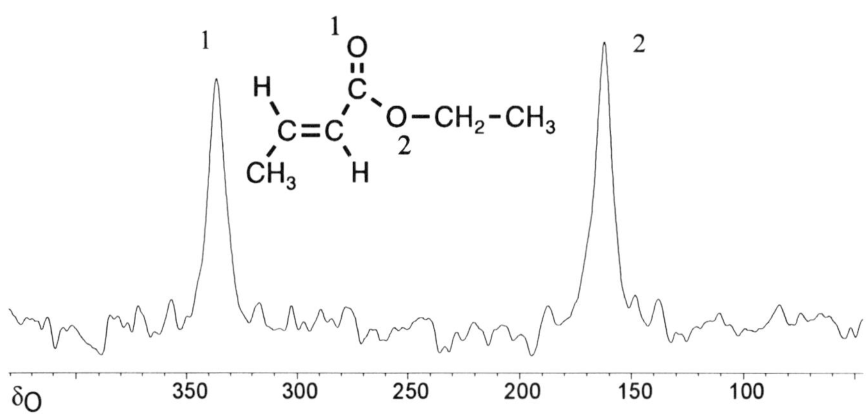

The figure shows the 54.24 MHz ^{17}O NMR spectrum obtained on an AM-400 spectrometer with a 5 mm inverse multinuclear probe-head. A carboxylic ester gives two signals, one for the C=O oxygen (here at δ_O = 336) and one for the C-O oxygen (here at δ_O = 162). As an exercise record a normal ^{17}O NMR spectrum and compare the baseline roll.

7. Comments

One can aim to minimize ring-down effects by using a relatively high preacquisition delay, although this method may also suppress broad signals which decay rapidly. The reasoning behind the RIDE sequence shown here is that probe-head ring-down is dependent of the phase of the r.f. pulses, but independent of any previous pulses. Thus, the ring-down of the first acquisition period is cancelled by the ring-down of the third, which is where the NMR signals first become inverted by an 180° pulse. The ring-down from the second acquisition period is similarily cancelled by the last. Four acquisitions are needed to provide the ringdown elimination both for the 90° and 180° pulses. Note, however, that since the sequence uses 180° pulses it only works well for relatively small spectral widths.

8. Own Observations

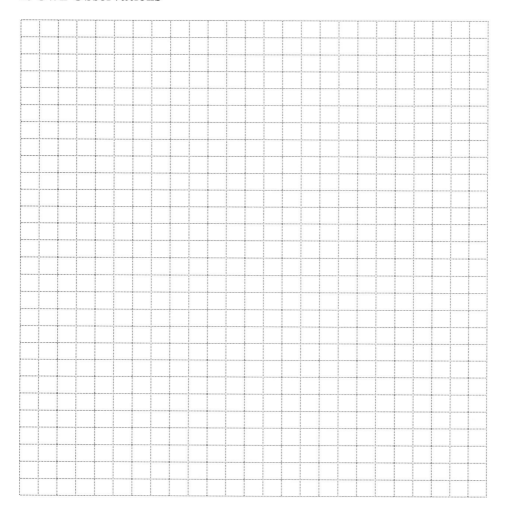

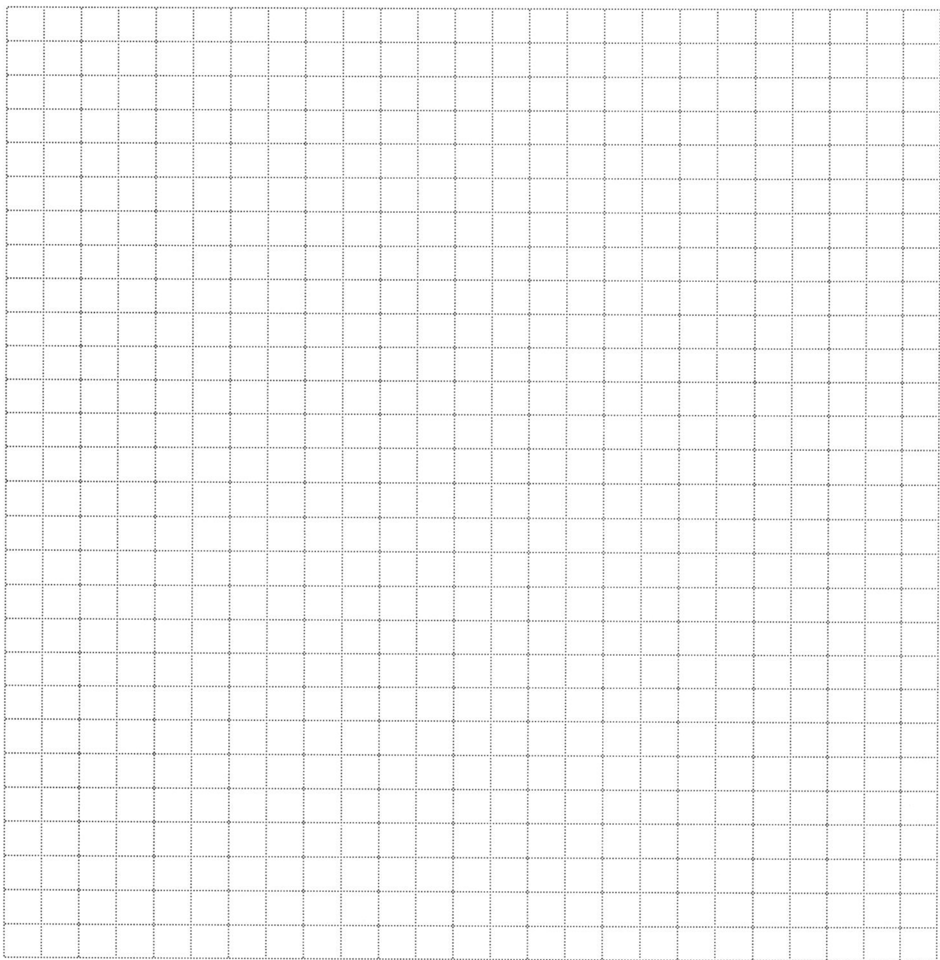

Experiment 9.10

$^{47/49}$Ti NMR Spectroscopy with ARING

1. Purpose

The ^{47}Ti and ^{49}Ti nuclei (I = 5/2 and 7/2, natural abundance 7.28 and 5.51 %) are a curiosity since they are the only isotopes of the periodic table the signals of which appear together in a single NMR spectrum. Therefore one always gets two titanium signals, separated by 266 ppm, even if only one chemical species is present. Furthermore, both nuclides have a quadrupolar moment and a relatively low gyromagnetic ratio, and therefore methods of suppression of probe-head ringing might be tested with these nuclides. Titanium organic catalysts are of prime importance in current synthetic organic chemistry; unfortunately NMR spectroscopy fails in many cases to obtain a signal at all from these nuclei. In this experiment we demonstrate the observation of the titanium signals using $TiCl_4$.

2. Literature

[1] C. D. Jeffries, H. Loeliger, H. H. Staub, *Phys. Rev.* **1952**, *85*, 478–479; C. D. Jeffries, *ibid.* **1953**, *92*, 1262–1263.

[2] N. Hao, B. G. Sayer, G. Denes, D. G. Bickley, C. Detellier, M. J. McGlinchey, *J. Magn. Reson.* **1982**, *50*, 50–63.

[3] D. Rehder in *Multinuclear NMR*, J. Mason (ed.), Plenum Press, New York **1987**, 487–488.

[4] S. Berger, W. Bock, C. F. Marth, B. Raguse, M. T. Reetz, *Magn. Reson. Chem.* **1990**, *28*, 559–560.

[5] S. Berger, W. Bock, G. Frenking, V. Jonas, F. Müller, *J. Am. Chem. Soc.* **1995**, *117*, 3820–3829.

3. Pulse Scheme and Phase Cycle

Experiment **a**

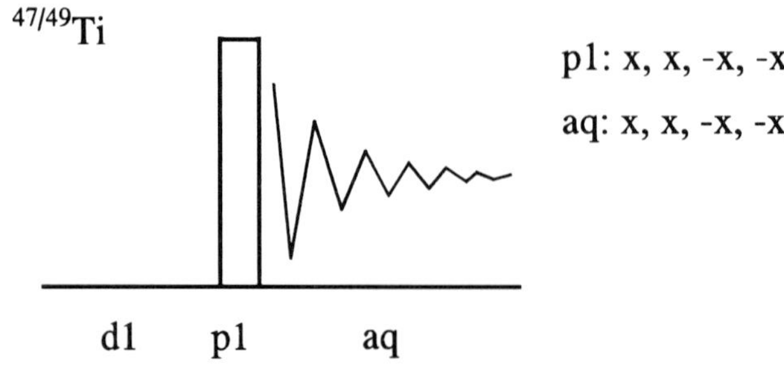

Experiment **b**

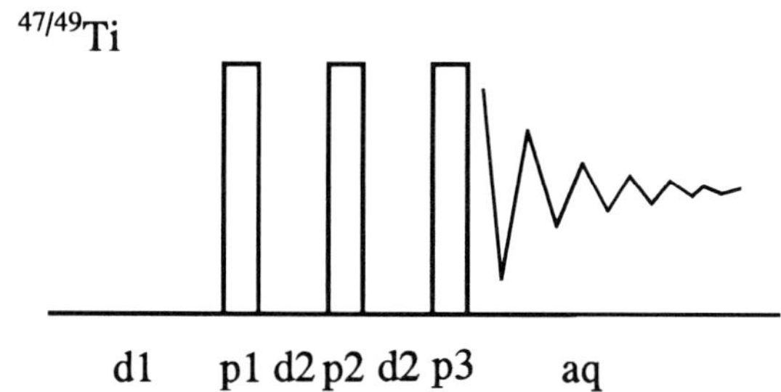

p1: x

p2: -x, x

p3: x, x, -x, -x, y, y, -y, -y

aq: x, -x, -x, x, y, -y, -y, y

4. Acquisition

Time requirement: 10 min

Sample: $TiCl_4$, neat; seal the tube to protect yourself and the spectrometer

Tune the probe head to $^{47/49}$Ti, turn the field sweep unit off since no lock is used; find the signal, and determine the 90° transmitter pulse-length with the pulse sequence shown for experiment **a**. You have to set:

td: 8 k
sw: 600 ppm
o1: middle of the titanium NMR spectrum
p1: 90° $^{47/49}$Ti transmitter pulse
d1: 10 ms
preacquisition delay: 10μs
ns: 8

With the simple sequence of experiment **a** you will likely observe excessive base line roll. Load the pulse sequence shown for experiment **b**. You have to set:

td: 8 k
sw: 600 ppm
o1: middle of the titanium NMR spectrum
p1, p2, p3: 90° $^{47/49}$Ti transmitter pulse
d1: 10 ms
d2: 4μs
preacquisition delay: 10μs
ns: 8

5. Processing

Use standard 1D processing as described in Experiment 3.2, lb = 15 Hz..

6. Result

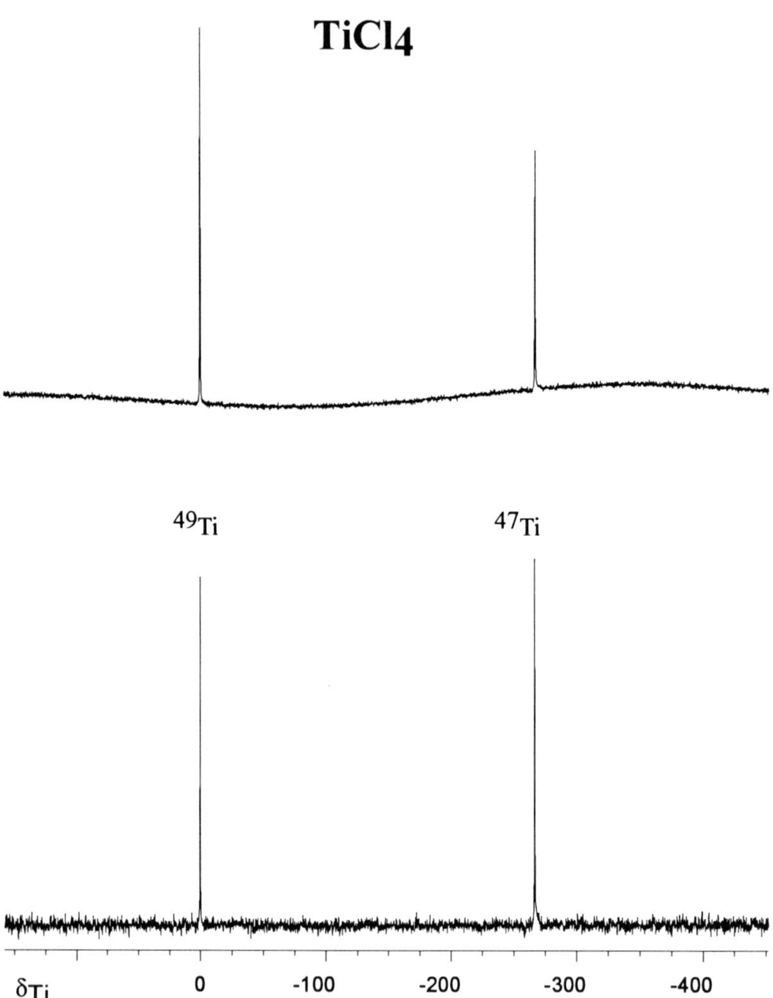

The figure shows the 28.19 MHz $^{47/49}$Ti NMR spectrum obtained on an AMX-500 spectrometer with a 5 mm multinuclear inverse probe-head. In **a** the result with the standard pulse sequence is given, showing considerable base-line roll due to probe-head ringing. In **b** the result of the anti-ring sequence is shown; the base line is completely flat; however, the signal-to-noise ratio is less by a factor of about three.

We use this experiment also to demonstrate the Ξ scale of the chemical shift. The spectral frequency SF for the ^{49}Ti signal was measured to be 28.202631 MHz. For the same digital field position the SF value of TMS in $CDCl_3$ was determined to 500.130204 MHz; thus the Ξ-value for $^{49}TiCl_4$ can be calculated to be 5.639058 MHz. With this value and the SF value of TMS at your instrument, the necessary spectrometer frequency to measure titanium on your instrument can easily be calculated.

7. Comments

As discussed with the RIDE sequence in Experiment 9.9 the acoustic responses of a probe-head are dependent on the phase of the r.f. pulses but, contrary to the magnetization, not on the previous history within one pulse sequence. Looking at the phase cycle given for the ARING sequence above, one can see that, for the first scan, acousting ringing A – A + A = +A is sampled, in the second scan +3A, in the third –A, and in the fourth +A. Since the results of the second and third are subtracted by the receiver phase, this yields a complete cancelling for the acoustic contribution. The magnetization vector moves in the first scan from M_Z to $-M_y$, back to M_Z and again to $-M_y$, in the second scan it ends at + M_y but is subtracted in the receiver, the third and the fourth scan give the same results as the second and first, and thus the final signal will be $-4M_y$ added in the receiver. The sensitivity losses are probably due to pulse imperfections and relaxation losses during the sequence. Since, in comparison to RIDE, no 180° pulses are used, the ARING sequence is less prone to offset effects.

8. Own Observations

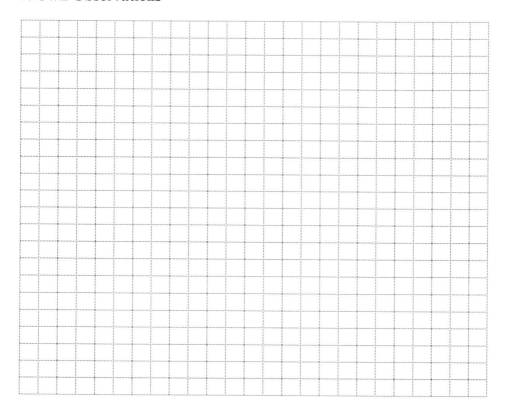

Chapter 10

The Second Dimension

2D NMR spectra are obtained by recording a series of 1D NMR spectra. These individual spectra differ only by a time increment which is introduced within the pulse sequence. It is helpful to distinguish four time periods for these spectra. In the *preparation* period the spin system relaxes and is then excited by at least one r.f. pulse. In the *evolution* period t_1 the chemical shifts and spin–spin couplings evolve; this is the time domain which is incremented during a 2D experiment. In the *mixing* period one or several r.f. pulses are applied and create an observable transverse magnetization. This is recorded in the *detection* period (which in 1D NMR spectroscopy is called the acquisition time); here it is usually labeled with t_2.

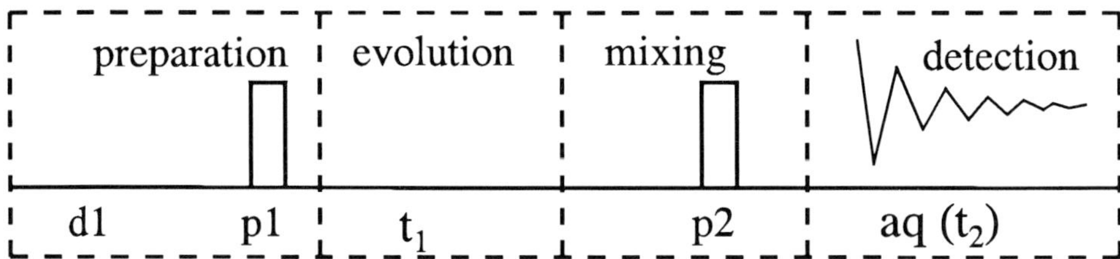

Thus the primary 2D matrix consists of a series of FIDs, from which a set of 1D NMR spectra is obtained by Fourier transformation with respect to t_2. The signals of each transformation may differ in amplitude and/or phase. A second Fourier transformation with respect to t_1 yields the final 2D matrix with frequency axes F_1 and F_2.

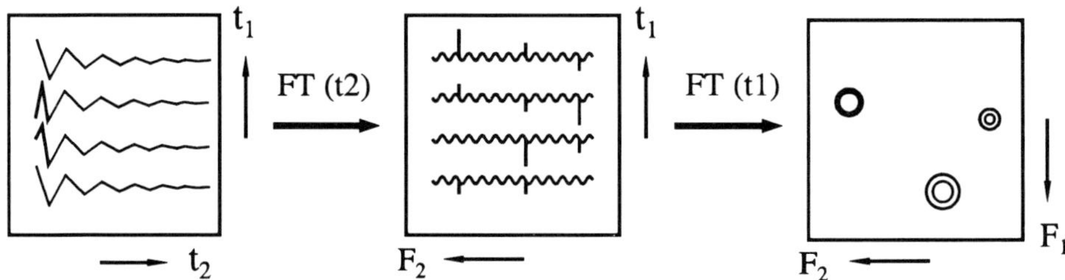

In setting up a 2D experiment one first has to consider the appropriate amount of data that can be acquired and processed. A typical routine set-up for standard 2D spectra in organic chemistry (COSY, C,H correlation) would consist of 128 FIDs, each of 1 k data points, yielding a primary serial file of 128 k data points. The number of FIDs determines the total experiment time and the resolution in F_1, while their data length determines the resolution in F_2.

From both the theoretical and experimental points of view it is important to distinguish whether 2D spectra are recorded and processed in the phase-sensitive mode or not phase-sensitive.

In 1D NMR all spectra are usually taken with quadrature phase detection. The transmitter offset is placed in the middle of the spectrum. The original NMR signal is split and detected by two phase detectors which are 90° out of phase with respect to each other, producing a sine and a cosine component. By either *simultaneous* or *sequential* acquisition of the sine and cosine components of the NMR signal one can then determine the sign of the frequency difference relative to the transmitter offset. This can be visualized by looking at the figure, where the sine and cosine functions, their transforms, and the final result after addition are shown.

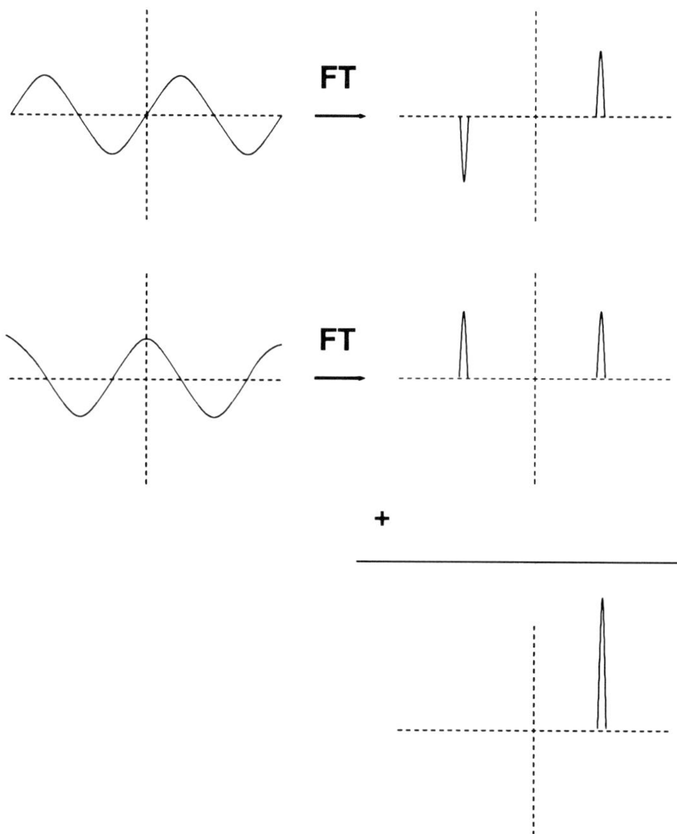

In 2D NMR spectroscopy there is no detector in the F_1 dimension; therefore the signs of the frequencies in F_1 must be determined beforehand during the acquisition in F_2. A specific phase cycling for the individual FIDs obtained in F_2 for each t_1 increment provides the necessary basis for this.

There are two different approaches. In the *phase-sensitive mode* sine and cosine components with respect to the t_1 evolution are created by the phase cycle and stored separately. Subsequent real Fourier transformation detects the signs of the frequencies and the phase of the signals. This is almost equivalent to the procedure used in 1D NMR spectroscopy and can be performed in either the *simultaneous* or the *sequential* mode; the former is called the Ruben–States–Haberkorn procedure and the latter the TPPI (Time **P**roportional **P**hase **I**ncrement) or Redfield method.

The other method is based on subtracting or adding the sine and cosine components (N or P type, or echo/anti-echo selection) created by the phase cycle within the FIDs recorded in t_2; thus they are not stored separately. The subsequent complex Fourier transformation again detects the signs of the frequencies. However, here the signal shapes are skewed, consisting of cosine and sine components in both dimensions. These signals are usually processed in magnitude mode.

There have been numerous debates about the advantages and disadvantages of the two principal approaches. The phase-sensitive approach is clearly the more modern concept and yields Lorentzian lineshapes in both dimensions. However, it requires larger data matrices for processing and, since the full phase information is retained, twice the amount of time. The N or P type mode gives crude information in half the time and requires a smaller data matrix for transformation. However, exact spin coupling constants cannot be extracted from the resulting spectra due to the skewed line shape. Clearly, the choice of method depends upon the information required. In the following examples we use both methods.

Pulsed field gradients can select the coherence pathways directly without the need of phase cycling. If the field gradients are applied during the t_1 period, echo or anti-echo selection results. Various phase sensitive approaches using field gradients also have recently been developed (see Ch. 12).

For a phase-sensitive experiment the NMR spectroscopist has to set the t_1 increment to 1/[2·sw1], where sw1 is the spectral width [Hz] in the F_1 dimension. If the t_1 period is split by a 180° pulse this increment is set to 1/[4·sw1]. For non-phase-sensitive measurements these increments are 1/sw1 and 1/[2·sw1] respectively. In the following experiments (except the *J*-resolved methods) we have given the parameter sw1 in ppm so that it does not depend on the field strength. Of course, the software calculates these increments from sw1 values given in Hz.

Before processing 2D NMR data one first has to consider the data matrix. Usually one performs a zero-filling in F_1 by a factor of at least 2, in order to give nearly symmetrical data matrices. Zero-filling in F_2 is uncommon, it is always better to use more data points in F_2 from the beginning.

The processing of 2D NMR spectra must take into account the phase mode under which the recording was performed and the correct FT procedure has to be chosen. After baseline correction on the FIDs in F_2, the appropriate window functions need to be chosen. For phase-sensitive spectra the usual exponential or gaussian functions should be used; for N or P type spectra sinusoidal windows are popular, as they narrow the skewed line-shape. It is advisable to select the best window function interactively, and this is possible by modern software. After Fourier transformation in F_2, a data column in F_1 (the "FID" in t_1) can be downloaded and inspected as for a 1D spectrum. After a baseline correction the appropriate window function must again be found. Generally the same remarks as for the F_2 dimension hold. N or P type spectra are finally displayed in magnitude mode, so that a phase correction is not necessary.

Phase-sensitive spectra have to be corrected at least in the F_2 dimension, and for some techniques a phase correction in F_1 is also required. In general, the phase in the indirect dimension can also be calculated. This is very helpful, since in crowded 2D spectra the phase correction may become difficult, and, for all 3D methods, calculation of the phases in the indirect dimension is an essential procedure. For 1D spectra the preacquisition delay de is the determining factor for the linear phase shift across the spectrum. For 2D, the equivalent to the preacquisition delay in the indirect dimension is the initial value d0 for t_1 evolution, usually set to 3 μs. If a 180° pulse is placed during the t_1 evolution, the duration p_{180} of this pulse has also to be taken into account, and the initial value for t_1 evolution has to be taken twice for the phase calculation. The chemical shift evolution already starts during the finite length p_{90} of the r.f. pulse which is at the beginning of the t_1 period and spills into the duration of the pulse which is at the end of the t_1 period. This effect is approximated by Eq. (1), which gives the preacquisition delay de1 in the indirect dimension for the phase-sensitive HMQC sequence (see Exp. 10.14) as an example [5].

$$\mathrm{de1} = 4 \cdot p_{90}/2\pi + \mathrm{d0} + p_{180} + \mathrm{d0} + 4 \cdot p_{90}/2\pi \quad (1)$$

With this de1 value, the constants for zero-order and first-order phase corrections in the indirect dimension can be calculated; the details of this computation, however, are dependent on the manufacturers software and whether the simultaneous RSH method or the sequential TPPI method was used to generate the sign discrimination in F_1. On recent instruments software routines are provided, which calculate the phase in the indirect dimension automatically.

The final processing steps of a 2D NMR spectrum include calibration, integration by calculating volume integrals, aligning of the corresponding 1D high-resolution spectra (as shown throughout in this chapter), and choosing correct contour levels for plotting.

This chapter first introduces the basic *J*-resolved methods for the homonuclear and heteronuclear case. A description of the common variants of COSY spectroscopy follows. A large part of the chapter is devoted to the different techniques of C,H correlation, both forward and inverse including correlations via ^1J(C,H) and long-range spin coupling. Pulse sequences using a spin-lock as TOCSY and ROESY are demonstrated and examples for homonuclear and heteronuclear 2D NOE spectroscopy are provided. The 2D-INADEQUATE experiment (Exp. 10.21) is the most time consuming experiment described in this book.

Three 2D experiments using a selective pulse are found in Chapter 7 and those which work with pulsed field gradients are described in Chapter 12.

Literature

[1] R. R. Ernst, G. Bodenhausen, A. Wokaun, *Principles of Nuclear Magnetic Resonance in One and Two Dimensions*, Clarendon Press, Oxford, 1987.

[2] W. E. Hull, in: W. R. Croasmun, R. M. K. Carlson (eds.), *Two Dimensional NMR Spectroscopy*, VCH, Weinheim, **1994**, 67–456.
[3] A. E. Derome, *Modern NMR Techniques for Chemistry Research*, Pergamon, Oxford, **1987**, Chs. 8, 9, 10.
[4] J. K. M. Sanders, B. K. Hunter, *Modern NMR Spectroscopy*, 2nd Edition, Oxford University Press, **1993**, Ch. 4.
[5] A. Bax, M. Ikura, L. E. Kay, G. Zhu, *J. Magn. Reson.* **1991**, *91*, 174–178.

Experiment 10.1

2D *J*-Resolved ^{1}H NMR Spectroscopy

1. Purpose

In a normal 1D ^{1}H NMR spectrum chemical shift and spin–spin coupling information may be obscured by overlapping multiplets. In the 2D *J*-resolved experiment these two parameters are separated and displayed on different axes of the 2D matrix. On the F_2-axis only chemical shift information is present, and on the F_1-axis, only spin–spin coupling information. The projection of the 2D spectrum onto the F_2-axis is effectively a "^{1}H broad-band decoupled" proton spectrum. Another advantage of the experiment is the separation of homonuclear spin couplings from heteronuclear spin couplings (such as couplings to ^{31}P or ^{19}F), since the latter are confined to the F_2-axis.

2. Literature

[1] W. P. Aue, J. Karhan, R. R. Ernst, *J. Chem. Phys.* **1976**, *64*, 4226–4227.
[2] A. E. Derome, *Modern NMR Techniques for Chemistry Research*, Pergamon, Oxford, **1991**, 270–275.
[3] R. Freeman, *A Handbook of NMR*, Longman, Harlow, **1987**, 106–110.
[4] D. D. Traficante, M. D. Meadows, *Concepts Magn. Reson.* **1997**, *9*, 359-384.

3. Pulse Scheme and Phase Cycle

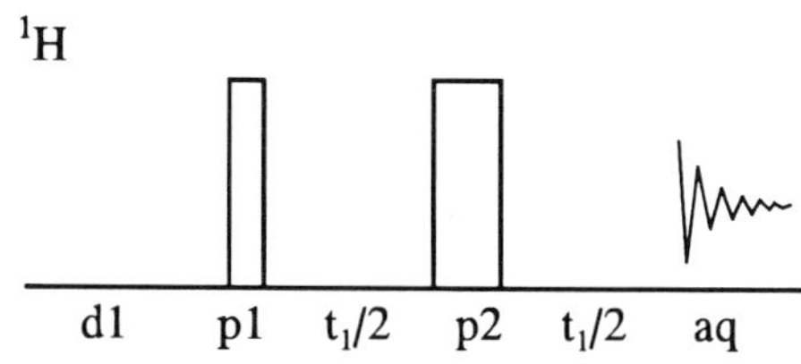

p1: $(x)_4$, $(y)_4$, $(-x)_4$, $(-y)_4$
p2: x, -x, y, -y, $(y, -y, -x, x)_2$, -x, x, -y, y
aq: $(x)_2$, $(-x)_2$, $(y)_2$, $(-y)_2$

4. Acquisition

Time requirement: 20 min.

Sample: 5% ethyl crotonate in $CDCl_3$

Record a normal ^{1}H NMR spectrum and optimize the spectral width. Change to the 2D mode of the spectrometer software and load the pulse program for *J*-resolved spectroscopy. You have to set:

td2: 1 k data points in F_2
td1: 128 data points in F_1

sw2: 8 ppm
sw1: 40 Hz
o1: middle of ^{1}H NMR spectrum
p1: 90° ^{1}H transmitter pulse
p2: 180° ^{1}H transmitter pulse
d1: 2 s
initial value for t_1 evolution: 3 μs
increment for t_1 evolution: 1/[2·sw1]
preacquisition delay: as small as possible
ds: 2
ns: 4

5. Processing

Apply zero filling in F_1 to 256 real data points. Use unshifted sinusoidal windows in both dimensions. Apply complex Fourier transformation corresponding to the N type signal selection using the quadrature off mode in F_1. Phase correction is not necessary, since the data are processed in magnitude mode. After the Fourier transformation the spectrum is tilted, since the signals are also modulated by J in F_2. This tilt can be eliminated by a software command. Finally the data may be symmetrized with respect to the horizontal through the center of F_1.

6. Result

The figure on page 349 shows the 2D spectrum obtained on an ARX-200 spectrometer after the tilt and symmetrization operation. The high resolution spectrum is shown in the F_2 axis, the internal projection (not shown) would give the ^{1}H decoupled proton spectrum. In F_1 the individual proton multiplets are observed.

7. Comments

The sequence is, in principle, identical to that in the spin-echo technique as described in Experiment 6.2, but differs in that now the spin-echo delay 2τ is incremented, thus creating the second dimension. At the end of t_1 the chemical shift information is refocused. However, the echo is modulated by the spin–spin coupling which evolved during t_1, and thus a 2D Fourier transformation will separate these two signal components. Well-resolved multiplet are obtained in F_1, because line broadening due to field inhomogeneity is refocused. On low-field spectrometers artefacts due to higher-order spin systems may occur [4]. There also exist selective and doubly selective variants of this technique, see Experiment 7.12.

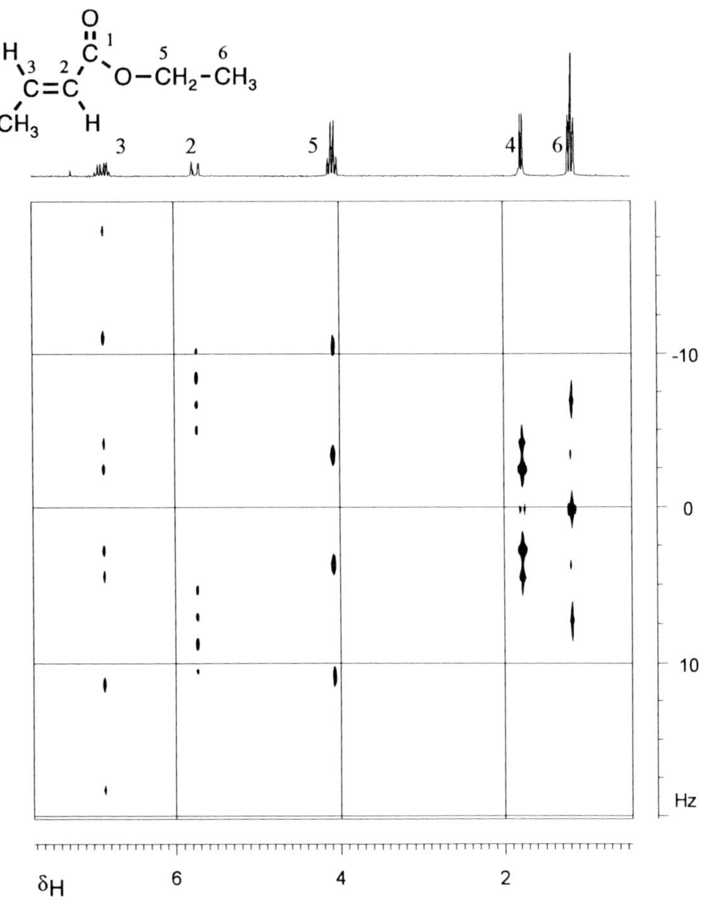

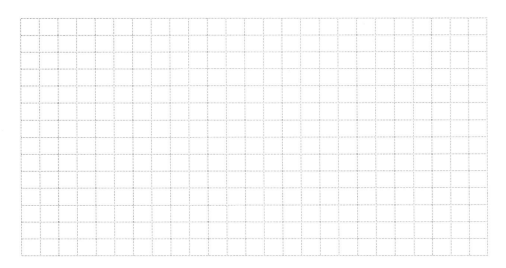

8. Own Observations

Experiment 10.2

2D *J*-Resolved ^{13}C NMR Spectroscopy

1. Purpose

In a ^{1}H coupled ^{13}C NMR spectrum (see Exp. 4.11) chemical shift and spin–spin coupling information may be obscured by overlapping multiplets. In the 2D *J*-resolved experiment these two parameters are separated and displayed on different axes of the 2D matrix. On the F_2-axis only chemical shift information is present, and on the F_1-axis only C,H coupling information. If the experiment is performed with high resolution in F_1, the C,H multiplets can be observed with their natural line-width. There are several variants; here we demonstrate a method in which the decoupler is gated. Note, however, that it results in splittings which are only half the actual spin coupling constants *J*(C,H).

2. Literature

[1] W. P. Aue, E. Bartholdi, R. R. Ernst, *J. Chem. Phys.* **1976**, *64*, 2229–2246.
[2] G. Bodenhausen, R. Freeman, R. Niedermeyer, D. L. Turner, *J. Magn. Reson.* **1977**, *26*, 133–164.
[3] A. E. Derome, *Modern NMR Techniques for Chemistry Research*, Pergamon, Oxford, **1987**, 259–268.
[4] J. K. M. Sanders, B. M. Hunter, *Modern NMR Spectroscopy*, 2nd edition, Oxford University Press, **1993**, 117–120.

3. Pulse Scheme and Phase Cycle

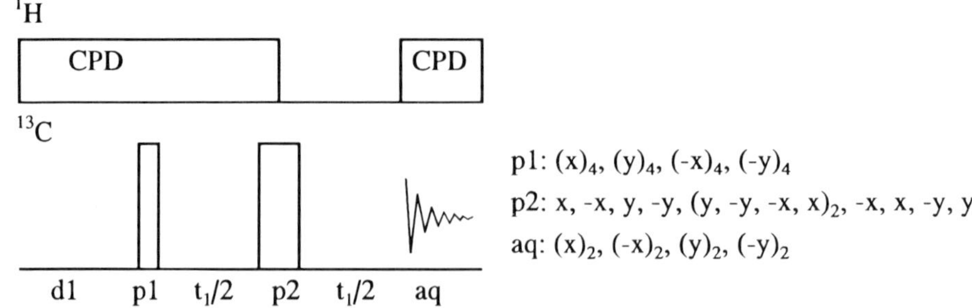

4. Acquisition

Time requirement: 1.5 h

Sample: 20% ethyl crotonate in $CDCl_3$

Record a normal ^{13}C NMR spectrum and optimize the spectral width. Change to the 2D mode of the spectrometer software and load the pulse program for heteronuclear *J*-resolved spectroscopy. You have to set:

td2: 1 k data points in F_2
td1: 64 data points in F_1
sw2: 175 ppm
sw1: 250 Hz
o1: middle of ^{13}C NMR spektrum
o2: middle of ^{1}H NMR spectrum
p1: 90° ^{13}C transmitter pulse
p2: 180° ^{13}C transmitter pulse
d1: 2 s
initial value for $t1$ evolution: 3 μs
increment for $t1$ evolution: 1/[2·sw1]
preacquisition delay: as short as possible
ds: 2
ns: 32

5. Processing

Apply zero filling in F_1 to 256 real data points. Use squared $\pi/2$ shifted sinusoidal windows in both dimensions. Apply complex Fourier transformation corresponding to the N type signal selection using the quadrature-off mode in F_1. Phase correction is not necessary, since the data are processed in magnitude mode. The data can be symmetrized with respect to the horizontal through the middle of the spectrum.

6. Result

The figure on page 352 shows the 2D spectrum obtained on an ARX-200 spectrometer; symmetrization has been performed. Note that the splittings observed in F_1 are only half the actual C,H spin coupling constants.

7. Comments

The sequence is, in principle, identical to the spin-echo technique as described in Experiment 6.3, but differs in that now the spin-echo delay τ is incremented, thus creating the second dimension. At the completion of t_1, the chemical shift information is refocused. The echo, however, contains the spin–spin coupling information, which evolved during the second half of t_1, when the decoupler was switched off. Therefore, these two signal components can be separated by 2D Fourier transformation and the splittings are half of the actual spin coupling constants. There are selective variants of this technique (see Exp. 7.10).

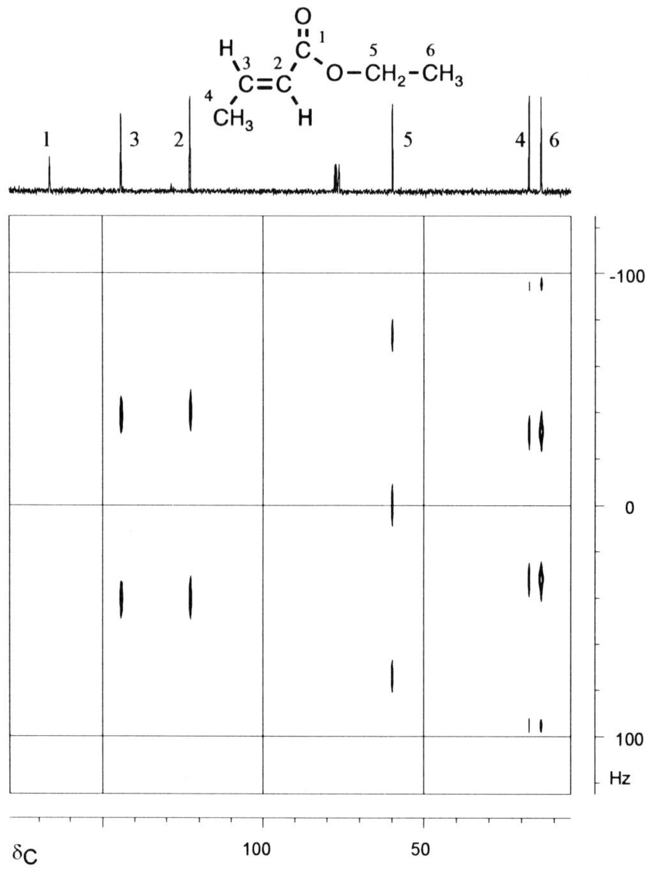

8. Own Observations

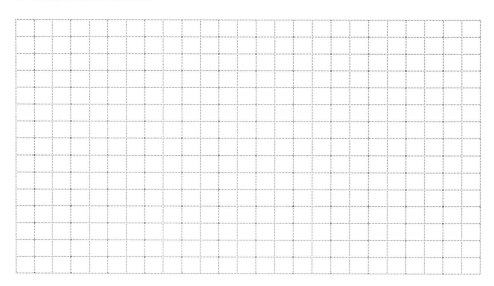

Experiment 10.3

The Basic H,H-COSY Experiment

1. Purpose

The COSY (**CO**rrelation **S**pectroscop**Y**) pulse sequence generates a 2D NMR spectrum in which the signals of a normal ^{1}H NMR spectrum are correlated with each other. Cross-peaks appear if spin coupling is present; thus the COSY sequence detects coupled pairs of protons (or pairs of other nuclei such as ^{19}F or ^{31}P). Since coupled protons are usually separated by two or three bonds, the connectivity and very often a chemical structure can be derived from the COSY spectrum. The COSY sequence is the most important and most frequently used 2D NMR experiment. We describe here the basic COSY technique with two 90° pulses and phase cycling for magnitude processing; other versions are given in Experiments 10.4–10.8 and 12.1.

2. Literature

[1] J. Jeener, *Ampère International Summer School*, Basko Polje, **1971** (proposal).
[2] W. P. Aue, E. Bartholdi, R. R. Ernst, *J. Chem. Phys.* **1975**, *64*, 2229–2246.
[3] A. E. Derome, *Modern NMR Techniques for Chemistry Research,* Pergamon, Oxford, **1987**, 183–234.

3. Pulse Scheme and Phase Cycle

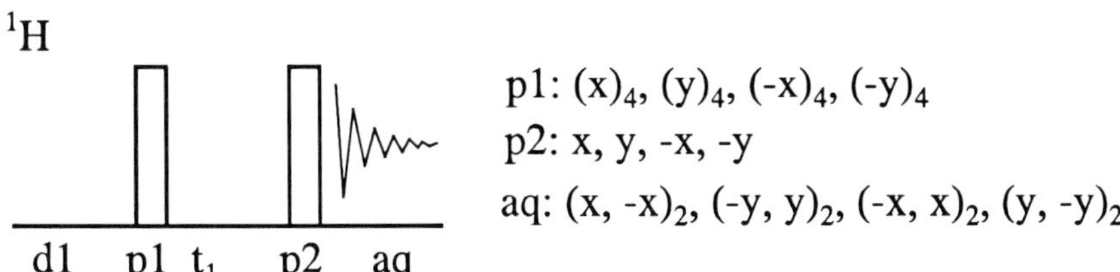

4. Acquisition

Time requirement: 20 min

Sample: 5 % ethyl crotonate in $CDCl_3$

Record a standard ^{1}H NMR spectrum and optimize the spectral width. Change to the 2D mode of the spectrometer software and load the COSY pulse program. You have to set:

td2: 1 k data points in F_2
td1: 128 data points in F_1

sw2: 8 ppm
sw1: 8 ppm
o1: middle of ^{1}H NMR spectrum
p1, p2: 90° ^{1}H transmitter pulse
d1: 2 s
initial value for t_1 evolution: 3 μs
increment for t_1 evolution: 1/sw1
ds: 2
ns: 4

5. Processing

Apply zero filling in F_1 to 512 real data points to obtain a symmetrical matrix of 512·512 real points. Use unshifted sinusoidal windows in both dimensions. Apply complex Fourier transformation corresponding to the N type signal selection using the quadrature-off mode in F_1. Phase correction is not necessary, since the data are processed in magnitude mode. Finally the data may be symmetrized.

6. Result

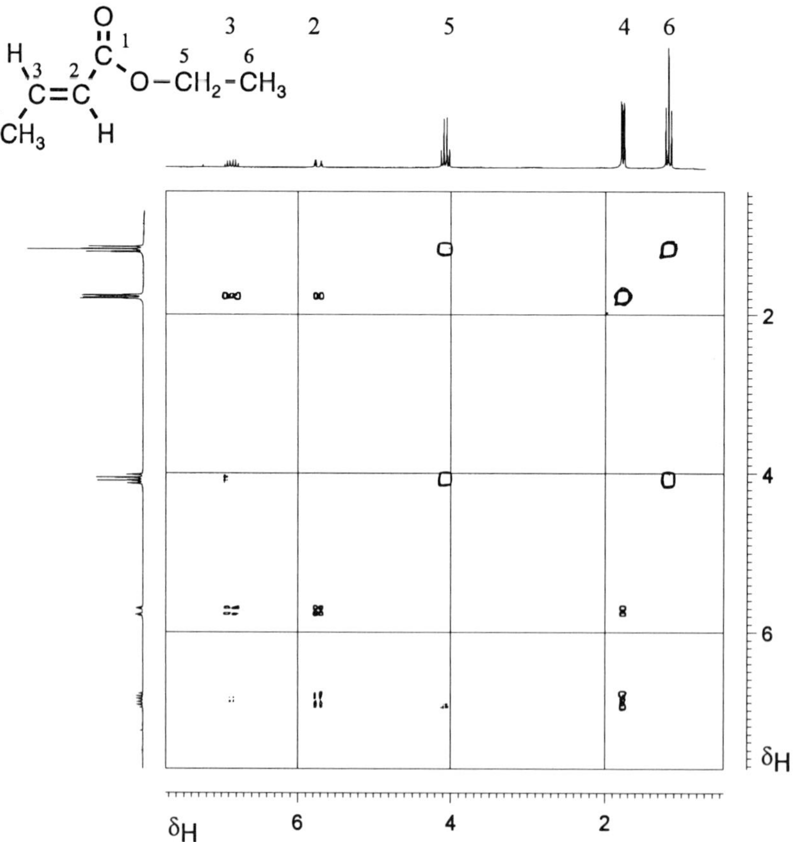

The figure on page 354 shows the H,H-COSY spectrum obtained on an ARX-200 spectrometer with symmetrization. Note that a cross peak connecting the hydrogen nucleus at C-2 with those at C-4 appears, although these hydrogens are not in a vicinal relationship. The cross peak results from an allylic coupling.

7. Comments

For the product operator formalism we consider a spin system of two protons. The key to any COSY protocol is the transformation of antiphase magnetization of proton 1 with respect to proton 2 into antiphase magnetization of proton 2 with respect to proton 1. The first r.f. pulse transforms z magnetization into transversel magnetization. Then the chemical shift develops during t_1, which is written here only for proton 1, giving Equation (1). In addition, spin–spin coupling develops; thus a term $2I_{1y}I_{2z}$ with antiphase magnetization of proton 1 with respect to proton 2 appears, as in Equation (2). Other terms are neglected.

$$I_{1z} + I_{2z} \xrightarrow{I_x} -I_{1y} - I_{2y} \xrightarrow{\Omega_1 t_1 I_{1z}} I_{1x}\sin\Omega_1 t_1 - I_{1y}\cos\Omega_1 t_1 \qquad (1)$$

$$\xrightarrow{\pi J t_1 2I_{1z}I_{2z}} 2I_{1y}I_{2z}\sin\Omega_1 t_1 \sin\pi J t_1 \qquad (2)$$

The second r.f. pulse transforms this into antiphase magnetization $2I_{1z}I_{2y}$, as in Equation (3). During the acquisition time t_2, chemical shift and spin–spin coupling develop once again, giving Equation (4).

$$\xrightarrow{I_x} 2I_{1z}I_{2y}\sin\Omega_1 t_1 \sin\pi J t_1 \qquad (3)$$

$$\xrightarrow{\Omega_2 t_2 I_{1z}} \xrightarrow{\pi J t_2 2I_{1z}I_{2z}} I_{2y}\sin\Omega_1 t_1 \sin\pi J t_1 \sin\Omega_2 t_2 \sin\pi J t_2 \qquad (4)$$

The last expression describes a cross-peak in the COSY matrix.

With the pulse sequence and phase cycling used the sign of the frequencies in F_1 is determined by adding together the sine and cosine terms, which leads to distorted lineshapes. Extraction of spin–spin coupling constants from this type of COSY spectrum is not recommended (see Exp. 10.5).

8. Own Observations

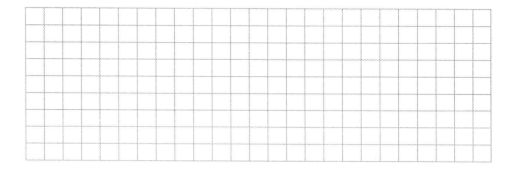

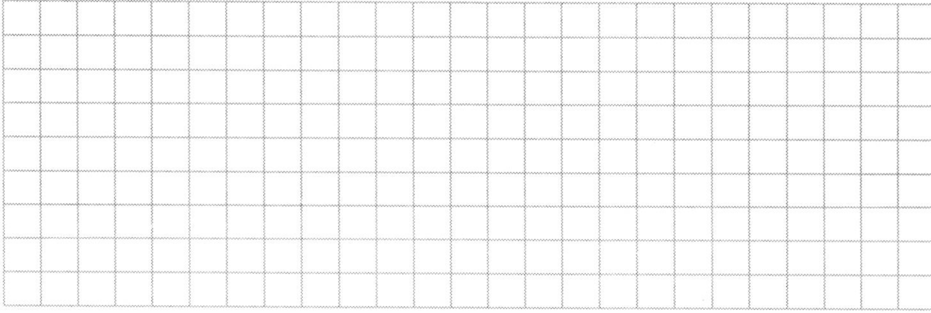

Experiment 10.4

Long-Range COSY

1. Purpose

The standard COSY pulse sequence (see Exp. 10.3) works best for spin–spin coupling constants of 3 to 15 Hz. With the long-range variant it is possible to observe cross-signals between protons which are connected by a very small coupling constant (as in allylic, homoallylic, or W-coupling). The method succeeds even in cases where the spin coupling is not resolved in the normal 1D ^1H NMR spectrum.

2. Literature

[1] A. Bax, R. Freeman, *J. Magn. Reson.* **1981**, *44*, 542–561.
[2] A. E. Derome, *Modern NMR Techniques for Chemistry Research*, Pergamon, Oxford, **1987**, 227–230.

3. Pulse Scheme and Phase Cycle

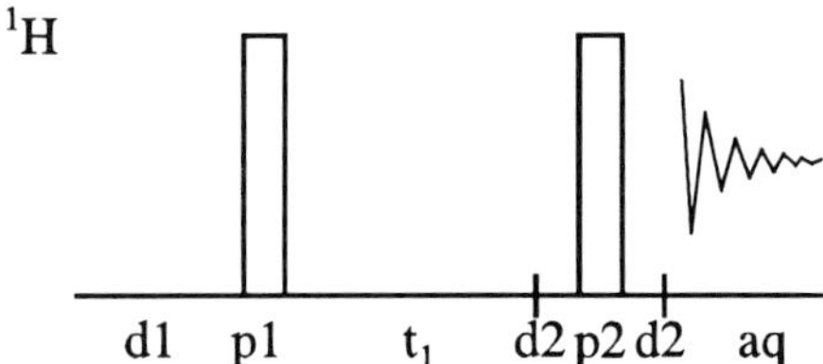

p1: $(x)_4$, $(y)_4$, $(-x)_4$, $(-y)_4$
p2: x, y, -x, -y
aq: $(x, -x)_2$, $(-y, y)_2$, $(-x, x)_2$, $(y, -y)_2$

4. Acquisition

Time requirement: 25 min

Sample: 5% ethyl crotonate in $CDCl_3$

Record a normal ^1H NMR spectrum and optimize the spectral width. Change to the 2D mode of the spectrometer software and load the long-range COSY pulse program. You have to set:

td2: 1 k data points in F_2
td1: 128 data points in F_1
sw2: 8 ppm
sw1: 8 ppm
o1: middle of ^1H NMR spectrum
p1,p2: 90° ^1H transmitter pulse
d1: 2 s

d2: 200 ms
initial value for t_1 evolution: 3 μs
increment for t_1 evolution: 1/sw1
ds: 2
ns: 4

5. Processing

Apply zero filling in F_1 to 512 real data points to obtain a symmetrical matrix of 512·512 real points. Use unshifted sinusoidal windows in both dimensions. Apply complex Fourier transformation corresponding to the N type signal selection using the quadrature-off mode in F_1. Phase correction is not necessary, since the data are processed in magnitude mode. Finally the data may be symmetrized.

6. Result

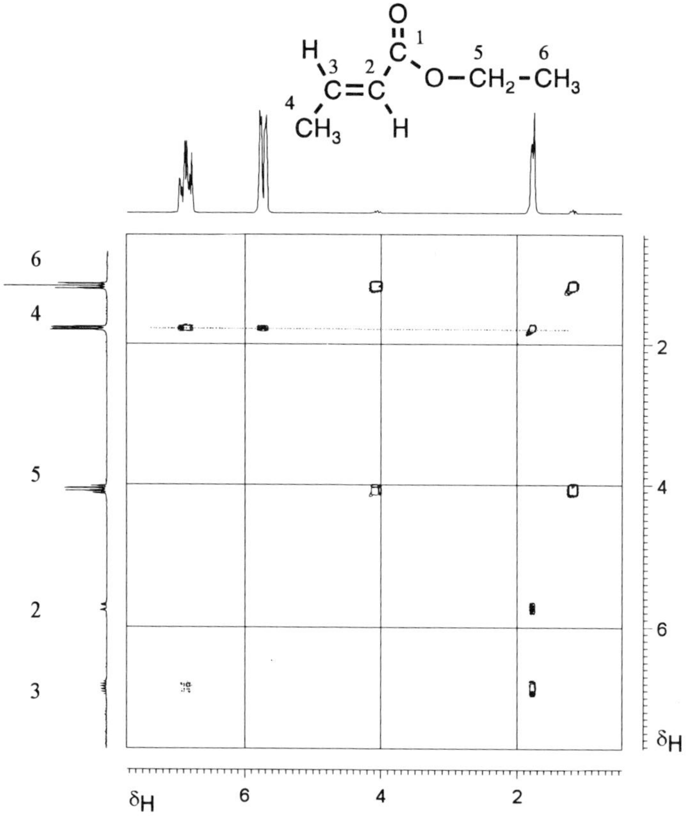

The figure shows a long-range COSY spectrum obtained on a ARX-200 spectrometer with symmetrization of the matrix. Note that the cross peak connecting the hydrogen nucleus at C-2 with those at C-4 is now of similar height to that connecting the hydrogen atoms of C-4 with the proton at C-3. This can best be seen from the corresponding row which was taken from the 2D matrix and displayed on the F_2-axis. Furthermore, the cross signals between the olefinic hydrogen nuclei arising from the large *trans* vicinal coupling virtually disappear at the contour level used.

7. Comments

The sequence differs from the standard version by the insertion of an additional fixed delay before and after the second r.f. pulse. This allows the small spin coupling constants to develop sufficiently to give detectable cross-signals. Values from 0.1 s to 0.4 s may be tried. For sensitivity reasons it is advisable to use two 90° pulses in the long range version of COSY. The first delay d2 can also be inserted directly after the first pulse p1.

8. Own Observations

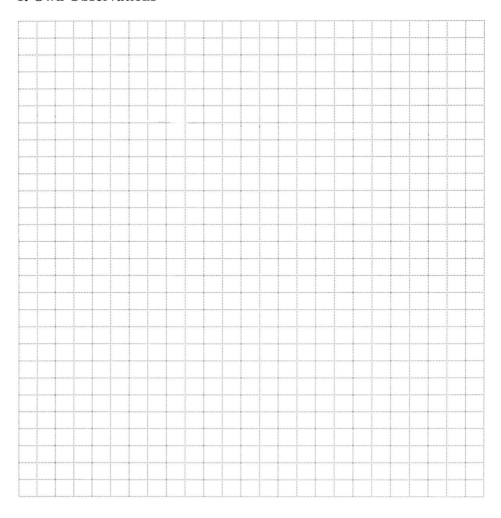

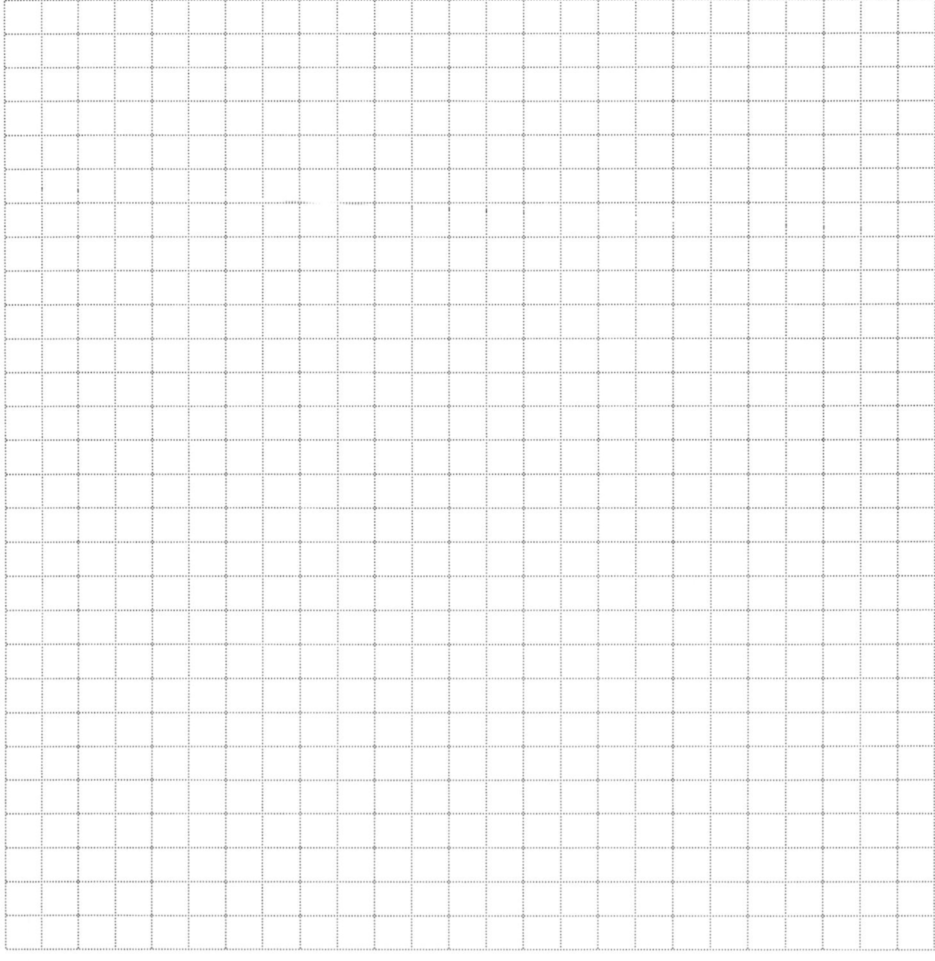

Experiment 10.5

Phase-Sensitive COSY

1. Purpose

The standard COSY experiment (see Exp. 10.3) yields skewed line shapes due to the N type peak selection, arising from the adding together cosine and sine components within the same FID followed by magnitude processing. Therefore spin coupling constants cannot be measured from COSY spectra of this type. To obtain this information it is desirable to have Lorentzian lineshapes in both dimensions; thus the sign of frequencies in F_1 must be determined by separate storing of the sine and cosine components followed by a real Fourier transformation. There are several methods for achieving this goal [1,2]. In the experiment described here the TPPI method of quadrature detection in F_1 is used.

2. Literature

[1] D. Marion, K. Wüthrich, *Biochem. Biophys. Res. Comm.* **1983**, *113*, 967–974.
[2] D. J. States, R. A. Haberkorn, D. J. Ruben, *J. Magn. Reson.* **1982**, *48*, 286–292.
[3] A. E. Derome, *Modern NMR Techniques for Chemistry Research*, Pergamon, Oxford, **1987**, 201–221.

3. Pulse Scheme and Phase Cycle

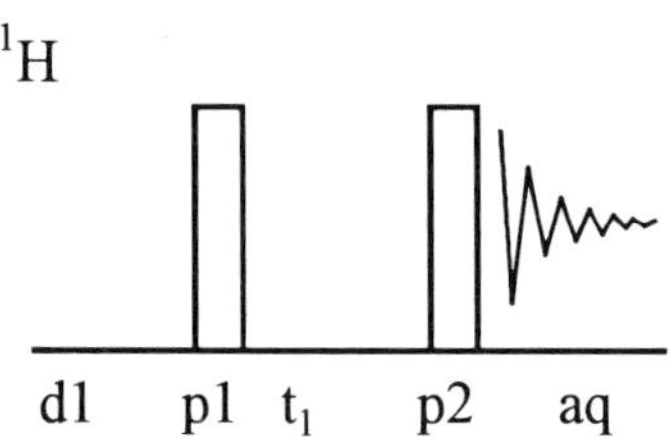

p1: x, -x, -x, x, y, -y, -y, y
p2: x, -x, x, -x, y, -y, y, -y
aq: x, -x, -x, x, y, -y, -y, y
phase cycle for p1 incremented according to TPPI

4. Acquisition

Time requirement: 2 h

Sample: 5% 2,3-dibromopropionic acid in [D$_6$]benzene

Record a standard ^{1}H NMR spectrum and optimize the spectral width. Change to the 2D mode of the spectrometer software and load the pulse program for phase-sensitive COSY with TPPI mode. You have to set:

td2: 2 k data points in F_2
td1: 256 data points in F_1
sw2: 1.5 ppm
sw1: 1.5 ppm
o1: middle of ^{1}H NMR spectrum
p1: 90° ^{1}H transmitter pulse
p2: 90° ^{1}H transmitter pulse
d1: 2 s
initial value for t_1 evolution: 3 μs
increment for t_1 evolution: 1/[2·sw1]
ds: 2
ns: 4

5. Processing

Apply zero filling in F_1 to 1 k real data points to obtain a symmetrical matrix of 1024·1024 real points. Use Gaussian windows in both dimensions. Apply real Fourier transformation corresponding to the TPPI mode of data acquisition in F_1. Phase correction for phase-sensitive COSY spectra can be performed in two ways. If one has measured a 1D ^{1}H NMR spectrum under the same conditions as for the 2D file (same probe-head tuning, same spectral width, time domain and preacquisition delay) one can phase the 1D spectrum and use the phase correction parameters in the F_2 dimension of the 2D file. Otherwise, using the 2D phase correction routines of the NMR software, one adjusts strong diagonal peaks at the left and right of the spectrum in dispersion, which yields the cross signals in pure antiphase.

6. Result

The figure on page 361 shows an expansion of the phase-sensitive COSY spectrum obtained on an AC-300 spectrometer for the two crosspeaks connecting H-3b with H-2 and H-3b with H-3a. The dotted contour lines represent negative signals, the solid contourlines positive signals. In this type of COSY spectrum the active coupling (the one which causes the cross-peak) is in antiphase, whereas the passive one remains in-phase. Thus from the cross peak at δ_H = 3.85 it can be seen that *J*(H-3b,H-2) is small (4.6 Hz), whereas *J*(H-3b,H-3a) is larger (−10.1 Hz). The cross peak at δ_H = 3.35 displays the active coupling *J*(H-3b,H-3a) and the passive coupling *J*(H-3a,H-2) = 11.0 Hz. For the sign determination see Exps. 4.6 and 10.6.

7. Comments

The TPPI (time proportional phase increment) method in 2D resembles somewhat the sequential quadrature detection of 1D spectra. Thus, for each t_1 increment the phase of the first pulse is incremented by 90°, leading to sine, cosine, −sine, −cosine character of the correspinding FIDs. For the same resolution twice the number of FIDs have to be recorded compared with the standard COSY. In the RSH (Ruben–States–Habercorn) method the FID is recorded twice for each t_1 increment with a 90° phase shift of p1. To evaluate spin coupling constants from phase sensitive COSY spectra the digital resolution must be set appropriately high.

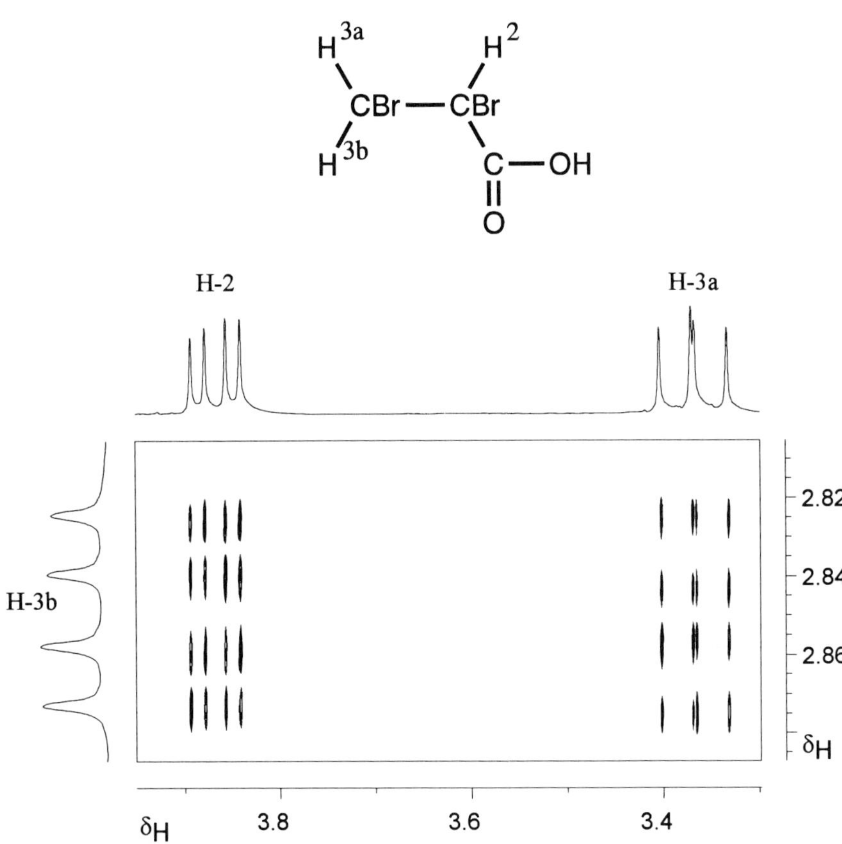

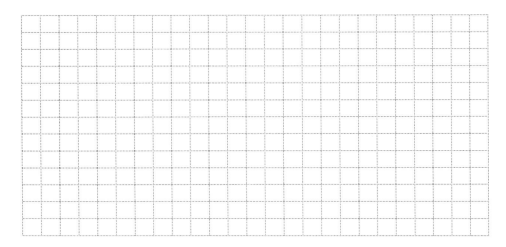

8. Own Observations

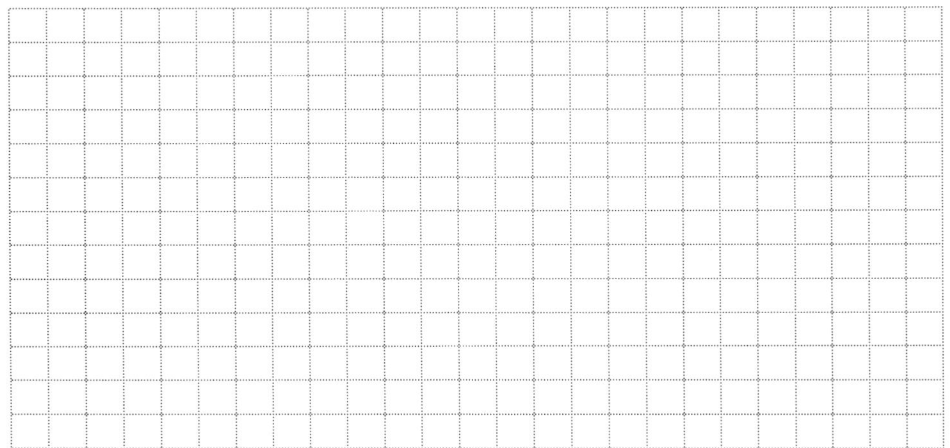

Experiment 10.6

Phase-Sensitive COSY-45

1. Purpose

The second pulse in a COSY experiment can be set to a smaller angle than the usual 90°. Two effects can be achieved by this measure. Firstly, the intensities of the autocorrelation signals, which are the cross-signals *within* a diagonal signal, become smaller; the diagonal will be narrower and cross signals near the diagonal can be observed more easily. Secondly the cross-signals become tilted and from the slope of this tilt the relative signs of spin coupling constants can be derived. Thus the COSY-45 experiment serves to distinguish between a ^2J and a ^3J spin coupling constant. The effect is, of course, best seen if the COSY spectrum is recorded with high digital resolution and in the phase-sensitive mode as shown here.

2. Literature

[1] W. P. Aue, E. Bartholdi, R. R. Ernst, *J. Chem. Phys.* **1975**, *64*, 2229–2246.

[2] A. Bax, R. Freeman, *J. Magn. Reson.* **1981**, *44*, 542–561.

[3] W. E. Hull, in: W. R. Croasmun, R. M. K. Carlson (eds.), *Two Dimensional NMR Spectroscopy*, VCH, Weinheim, **1994**, 301–303.

3. Pulse Scheme and Phase Cycle

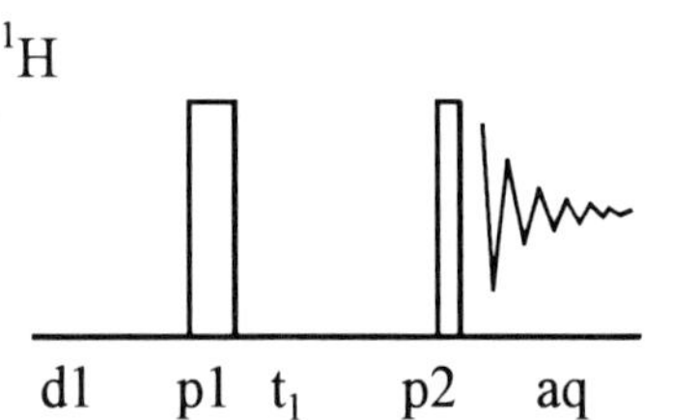

p1: x, -x, -x, x, y, -y, -y, y

p2: x, -x, x, -x, y, -y, y, -y

aq: x, -x, -x, x, y, -y, -y, y

phase cycle for p1 incremented according to TPPI

4. Acquisition

Time requirement: 3 h

Sample: 5% 2,3-dibromopropionic acid in [D_6]benzene

Record a normal ^1H NMR spectrum and optimize the spectral width. Change to the 2D mode of the spectrometer software and load the pulse program for phase-sensitive COSY with TPPI mode. You have to set:

td2: 2 k data points in F_2
td1: 256 data points in F_1
sw2: 1.5 ppm
sw1: 1.5 ppm
o1: middle of ^{1}H NMR spectrum
p1: 90° ^{1}H transmitter pulse
p2: 45° ^{1}H transmitter pulse
d1: 2 s
initial value for t_1 evolution: 3 μs
increment for t_1 evolution: 1/[2·sw1]
ds: 2
ns: 4

5. Processing

Apply zero filling in F_1 to 1 k real data points to obtain a symmetrical matrix of 1024·1024 real points. Use Gaussian windows in both dimensions. Apply real Fourier transformation corresponding to the TPPI mode of data acquisition in F_1. Phase correction as described in Experiment 10.5 for phase-sensitive COSY spectra must be performed.

6. Result

The figure on page 364 shows an expansion of the phase-sensitive COSY-45 spectrum obtained on an AC-300 spectrometer for the same two cross-peaks connecting H-3b with H-2 and H-3b with H-3a as given in Experiment 10.5. Note that specific contours are missing, and the tilt of both cross-peaks has a different slope, since the cross-peak at $\delta_H = 3.85$ is caused by a vicinal spin coupling whereas the cross-peak at $\delta_H = 3.35$ is caused by a geminal spin coupling.

7. Comments

In a two-spin system the intensity of the cross peaks is proportional to $\sin^2\beta$, where β is the pulse angle of the second pulse in the COSY sequence. In multi-spin systems, however, the situation is more complex. In the three-spin system of the example shown the cross-peaks are reduced from a 4·4 matrix to two 2·2 matrices which are offset both in F_1 and F_2 by the passive coupling. Thus, if the sign of active and passive couplings are the same, one obtains a different slope of the cross-peak compared with the situation when the signs differ.

The method shown here is also known as β-COSY, since the pulse angle of the second pulse may be widely varied. Compare the results with the E.COSY technique demonstrated in Experiment 10.7.

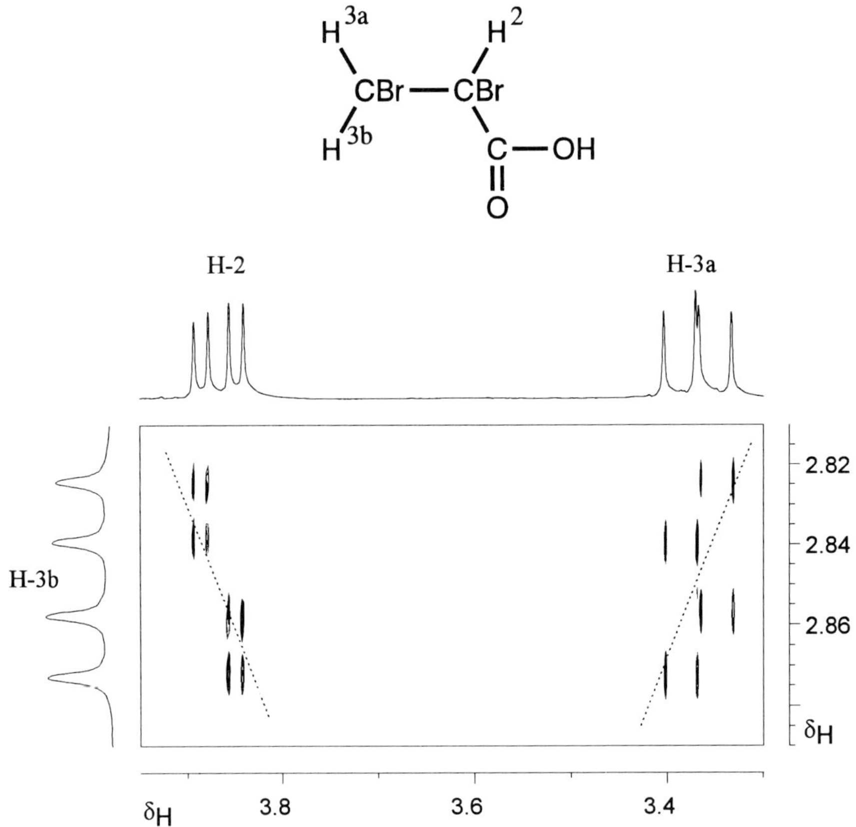

8. Own Observations

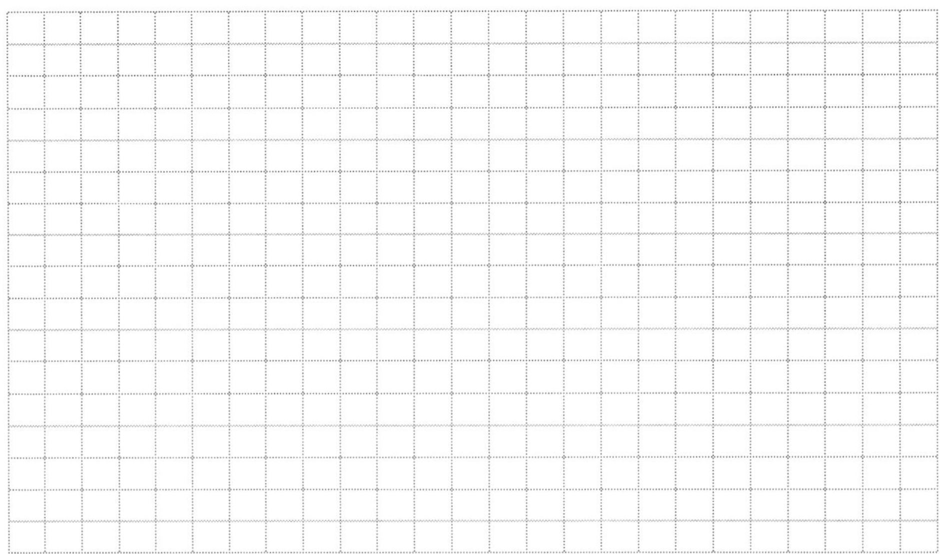

Experiment 10.7

E.COSY

1. Purpose

In the case of more complicated spin systems it is often very difficult to evaluate the cross peak patterns of phase-sensitive or double quantum filtered COSY spectra. The extraction of correct spin coupling constants may be hindered due to mutual cancellation of nearby positive and negative signals. E.COSY (**E**xclusive **CO**rrelation Spectroscop**Y**) provides a solution of this problem, since cross-peak patterns are simplified, displaying only signals of transitions which are directly connected in the energy level diagram, so that signals of the passive spin in a coupling network disappear. The result is very similar to the β-COSY technique (see Exp. 10.6) but more complete, and, furthermore, the diagonal signals are in phase. This facilitates the observation of cross peaks near the diagonal. In principle the E.COSY technique consists of a combination of multiple quantum filtered COSY spectra. Here, we show as an example its application to the three-spin system of 2,3-dibromopropionic acid; the phase cycle for four spin systems is given in ref. [3].

2. Literature

[1] C. Griesinger, O. W. Sørensen, R. R. Ernst, *J. Am. Chem. Soc.* **1985**, *107*, 6394–6396.

[2] C. Griesinger, O. W. Sørensen, R. R. Ernst, *J. Chem. Phys.* **1986**, *85*, 6837–6852.

[3] C. Griesinger, O. W. Sørensen, R. R. Ernst, *J. Magn. Reson.* **1987**, *75*, 474–492.

3. Pulse Scheme and Phase Cycle

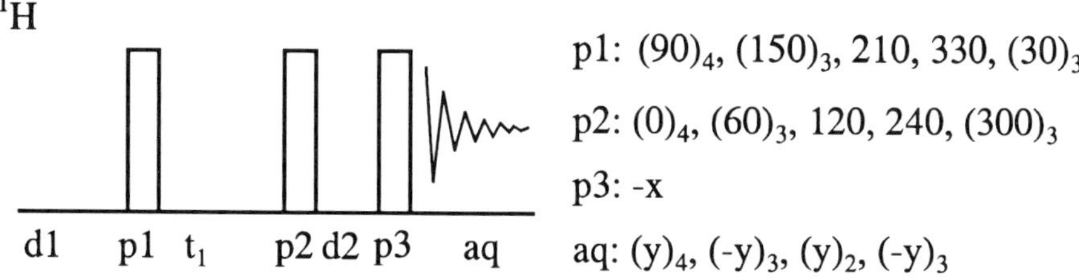

phase cycle for p1 incremented according to TPPI

4. Acquisition

Time requirement: 3 h

Sample: 5% 2,3-dibromopropionic acid in [D_6]benzene

Record a standard ^{1}H NMR spectrum and optimize the spectral width. Change to the 2D mode of the spectrometer software and load the pulse program for phase-sensitive E.COSY with TPPI mode. Note that the pulse sequence shown here can only be performed on spectrometers capable of phase shifts less than 90°. You have to set:

td2: 2 k data points in F_2
td1: 256 data points in F_1
sw2: 1.5 ppm
sw1: 1.5 ppm
o1: middle of ^{1}H NMR spectrum
p1, p2, p3: 90° ^{1}H transmitter pulse
d1: 2 s
d2: 4 μs
initial value for t_1 evolution: 3 μs
increment for t_1 evolution: 1/[2·sw1]
ds: 2
ns: 12

5. Processing

Apply zero filling in F_1 to 1 k real data points to obtain a symmetrical matrix of 1024·1024 real points. Use exponential windows in both dimensions. Apply real Fourier transformation corresponding to the TPPI mode of data acquisition in F_1. Phase correction in F_1 is usually not necessary.

6. Result

The figure on page 367 shows an expansion of the phase-sensitive E.COSY spectrum obtained on an AMX-500 spectrometer displaying the diagonal peak for H-3a (upper left corner) and three cross peaks connecting H-3a with H-2 (lower left corner), H-3b with H-3a (upper right corner) and H-3b with H-2 (lower right corner). The dotted contour lines represent negative signals, the solid contour lines positive signals. In this type of COSY spectrum the active coupling (the one which causes the cross-peak) is in antiphase, whereas the passive one disappears. From the appropriate cross sections the three spin coupling constants *J*(H-3a,H-2) = 11.0 Hz, *J*(H-3b,H-3a) = −10.1 Hz and *J*(H-3b,H-2) = 4.6 Hz can be measured. As in Experiment 10.6, the relative sign of the coupling constants can be taken from the slope of the cross peaks.

7. Comments

A disadvantage of the E.COSY sequence is that it is less sensitive than the COSY-45 procedure. Furthermore, if there are four- and three-spin systems in the same molecule it is better to perform the sequence twice, adapted to the spin system in question. There exists a complementary E.COSY sequence with a different phase cycle yielding in principle the same information, but the cross-peak signals display only the passive coupling. In certain practical cases it may be advantageous to record both varieties. Of rather high current importance is the adaption of the E.COSY scheme to heteronuclear resolved spectra, which allows the extraction of small heteronuclear spin coupling constants.

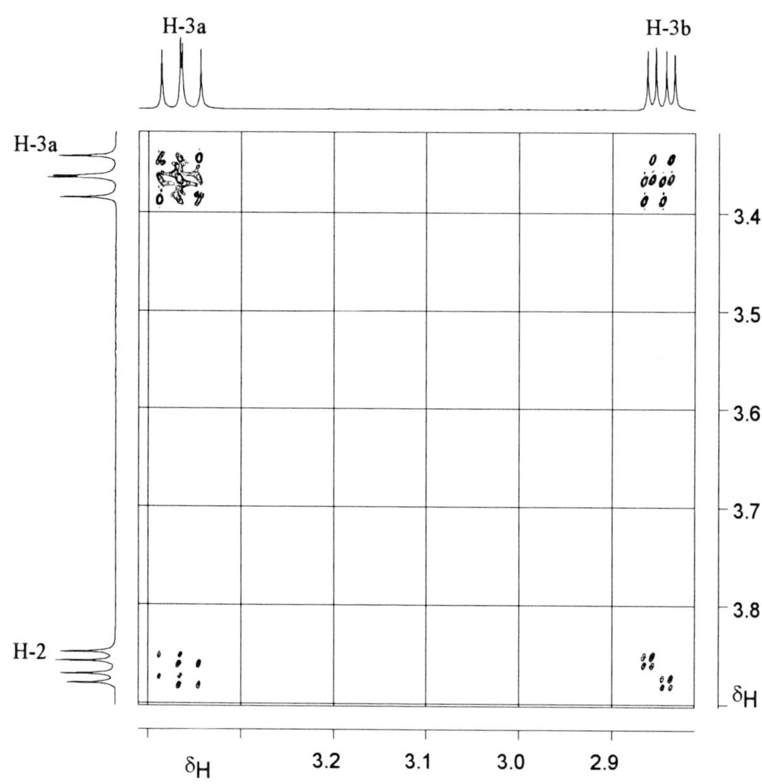

8. Own Observations

Experiment 10.8

Double Quantum Filtered COSY With Presaturation

1. Purpose

The NMR spectra of proteins, peptides, and carbohydrates are usually measured in water solution. Only 10% D_2O is added to provide a lock signal, since otherwise exchangeable NH protons would disappear; thus a huge water signal is present in these samples. In order to get reasonable COSY spectra one first uses a water suppression technique, such as presaturation (see Exp. 6.17) and, in addition, the COSY variant with a double quantum filter. Since water has no double quantum transitions, its signal is further suppressed. Since one wants to analyze the spin systems of the different amino acids in detail, this experiment is usually run in the phase-sensitive mode. An additional asset is the circumstance that the diagonal peaks of the DQF–COSY spectrum can be phased into absorption; thus one avoids having to cope with the tailing of the dispersion diagonal peaks as in the normal phase-sensitive COSY.

2. Literature

[1] U. Piantini, O. W. Sørensen, R. R. Ernst, *J. Am. Chem. Soc.* **1982**, 104, 6800–6801.

[2] M. Rance, O. W. Sørensen, G. Bodenhausen, G. Wagner, R. R. Ernst, K. Wüthrich, *Biochem. Biophys. Res. Commun.* **1983**, *117*, 479–485.

[3] A. E. Derome, M. P. Williamson, *J. Magn. Reson.* **1990**, *88*, 177–185.

3. Pulse Scheme and Phase Cycle

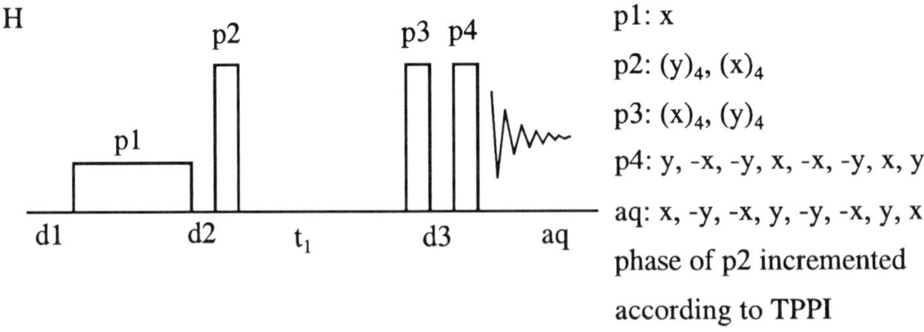

4. Acquisition

Time requirement: 11 h

Sample: 2 mmol sucrose in 90% H_2O/10% D_2O + 0.5 mmol DSS (2,2-dimethyl-2-silapentan-5-sulfonate, sodium salt) including a trace of NaN_3 against bacteria growth.

The probe-head must be tuned to the water sample. Record a normal ^1H NMR spectrum and redetermine the 90° pulse. Record a 1D spectrum with water presaturation and optimize the transmitter power for presaturation (Exp. 6.17). Change to the 2D mode of the spectrometer software and load the pulse program for phase-sensitive double quantum COSY with presaturation. You have to set:

td2: 2 k data points in F_2
td1: 256 data points in F_1
sw2: 10 ppm
sw1: 10 ppm
o1: on resonance of water signal
p1: 2 s at transmitter attenuation corresponding to $\gamma B_1 \approx 25$ Hz, 90° pulse $\approx$ 10 ms, typically 65 dB, see Experiments 2.6 and 6.15
p2, p3, p4: 90° ^1H transmitter pulse
d1: 30 ms
d2: 20 μs
d3: 4 μs
initial value for $t1$ evolution: 3 μs
increment for $t1$ evolution: 1/[2·sw1]
ds: 2
ns: 64

5. Processing

Apply zero-filling in F_1 to 1 k real data points to obtain a symmetrical matrix of 1024·1024 real points. Use Gaussian window in both dimensions. Apply real Fourier transformation corresponding to the TPPI mode of data acquisition in F_1.

6. Result

The figure on page 370 shows an expansion of the DQF-COSY spectrum obtained on an AMX-500 spectrometer. Note that the water signal of the 1D spectrum, which was obtained by presaturation, is much larger than the residual water signal of the 2D file. At the contour level chosen the cross-peak of the anomeric proton at $\delta_H = 5.41$ can be seen only in the upper left part of the matrix.

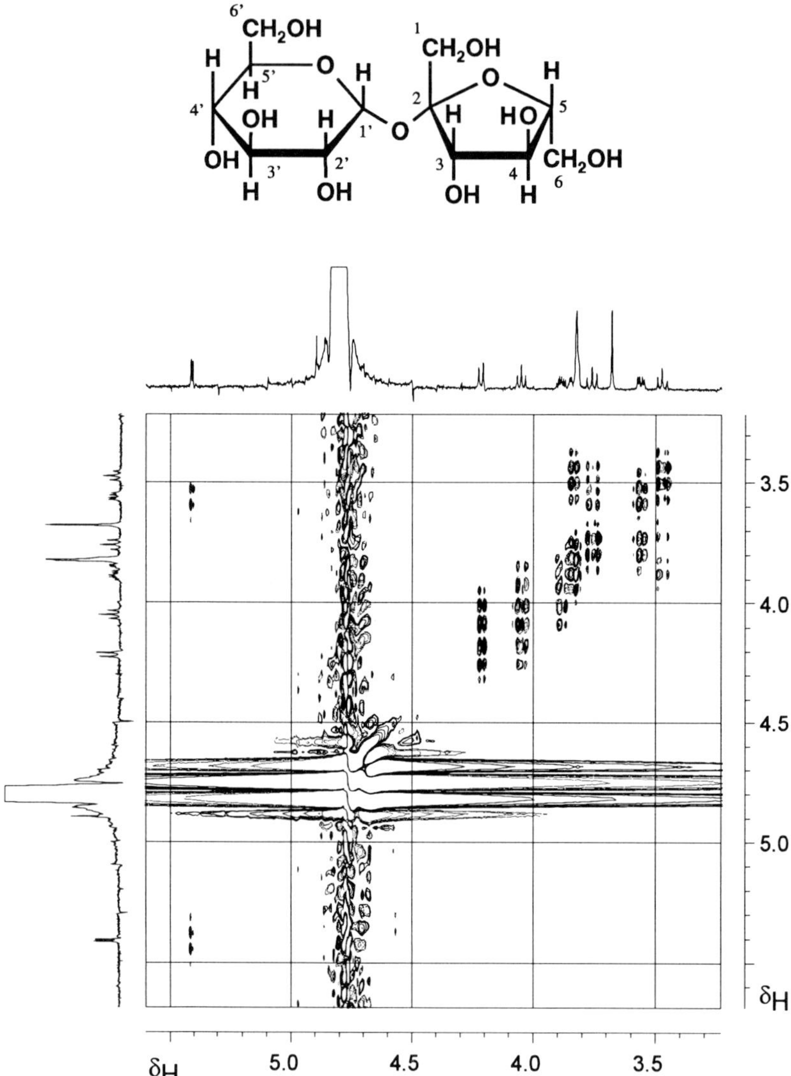

7. Comments

The sequence employs a third 90° pulse acting in combination with p2 as the double quantum filter. The second pulse of a COSY sequence not only generates antiphase magnetization as described in Experiment 10.3, but also creates double quantum terms depending on its phase. In the product operator formalism we can first repeat the findings of Experiment 10.3 with regard to the actual pulse phases used. The first r.f. pulse transforms *z*-magnetization into transverse magnetization. Then the chemical shift develops during t_1, which is written in Equation (1) only for proton 1. In addition, spin–spin coupling develops; thus a term $2I_{1y}I_{2z}$ with antiphase magnetization of proton 1 with respect to proton 2 appears, as indicated in Equation (2). Other terms are neglected.

$$I_{1z}+I_{2z} \xrightarrow{I_y} I_{1x}+I_{2x} \xrightarrow{\Omega_1 t_1 I_{1z}} I_{1x}\cos\Omega_1 t_1 - I_{1y}\sin\Omega_1 t_1 \quad (1)$$

$$\xrightarrow{\pi J t_1 2I_{1z}I_{2z}} -2I_{1x}I_{2z}\sin\Omega_1 t_1 \sin\pi J t_1 \quad (2)$$

The second r.f. pulse now creates double quantum magnetization $2I_{1x}I_{2y}$. This is transformed back into antiphase magnetization of proton 2 with respect to proton 1 by the pulse p4 as indicated in Equation (3). During the acquisition time *t*2 chemical shift and spin–spin coupling develop once again, as indicated in Equation (4). However, due to the full phase cycle used, only those signals are observed which have passed through the double quantum state, whereas all others are suppressed.

$$\xrightarrow{I_x} 2I_{1x}I_{2y}\sin\Omega_1 t_1 \sin\pi J t_1 \xrightarrow{Iy} -2I_{1z}I_{2y}\sin\Omega_1 t_1 \sin\pi J t_1 \quad (3)$$

$$\xrightarrow{\Omega_2 t_2 I_{1z}} \xrightarrow{\pi J t_2 2I_{1z}I_{2z}} I_{2y}\sin\Omega_1 t_1 \sin\pi J t_1 \sin\Omega_2 t_2 \sin\pi J t_2 \quad (4)$$

8. Own Observations

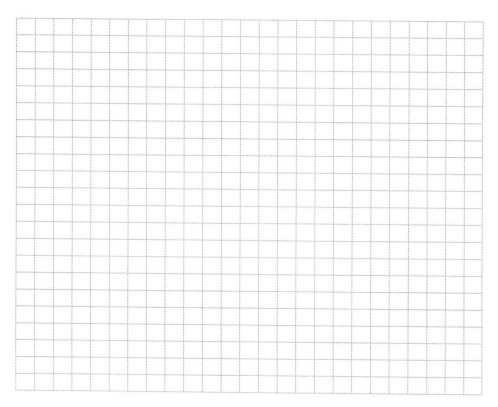

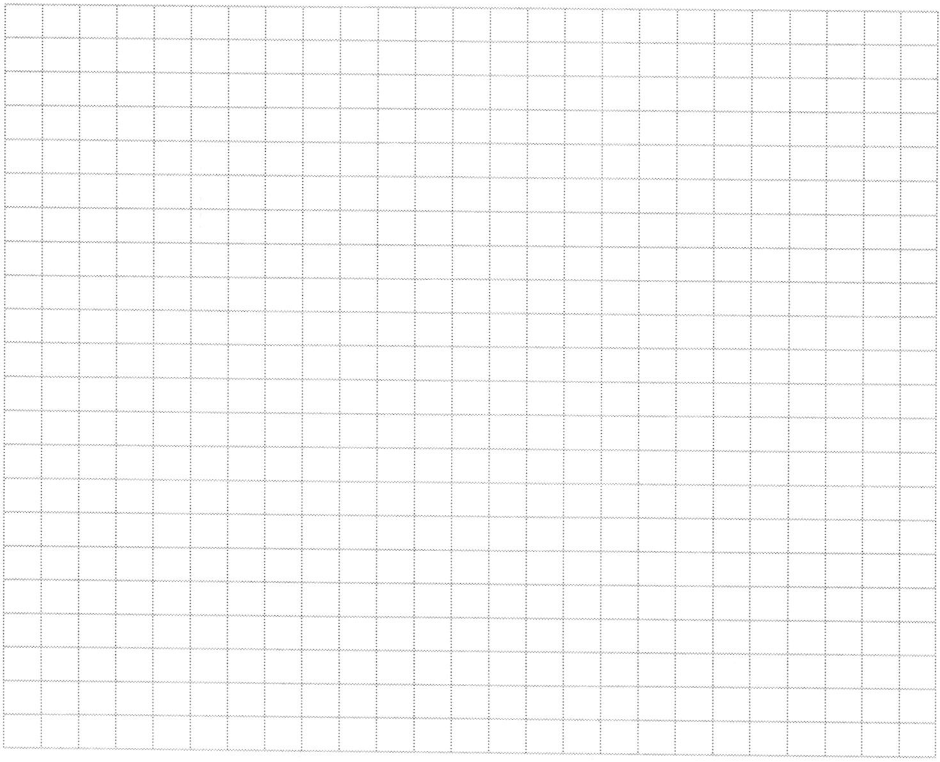

Experiment 10.9

Fully Coupled C,H Correlation (FUCOUP)

1. Purpose

One of the earliest and simplest C,H correlation methods consists only of three r.f. pulses and leads to a 2D spectrum where the C,H spin coupling remains to be seen in both dimensions; therefore it has been called FUCOUP (**FU**lly **COUP**led). The method does not distinguish between ^1J(C,H) and long-range couplings; thus the full information is present in these spectra. Since also the H,H spin coupling is active, the method is very insensitive and gives complex spectra. For practical purposes in structural elucidation it has therefore been replaced by more advanced methods (see Exps. 10.10–10.17). To understand the basics of a C,H correlation, however, this experiment provides an excellent start. In the educational experiment described here we demonstrate the phase-sensitive technique, with chloroform as an example.

2. Literature

[1] G. Bodenhausen, R. Freeman, *J. Magn. Reson.* **1977**, *28*, 471–476.
[2] R. L. Haltermann, N. H. Ngyuyen, K. P. C. Vollhardt, *J. Am. Chem. Soc.* **1985**, *107*, 1379–1387.
[3] G. E. Martin, A. S. Zektzer, *Two-Dimensional NMR Methods for Establishing Molecular Connectivity*, VCH, Weinheim, **1988**, Ch. 3.
[4] W. E. Hull, in: W. R. Croasmun, R. M. K. Carlson (eds.), *Two-Dimensional NMR Spectroscopy*, VCH, Weinheim, **1994**, 382.
[5] W. Bauer, C. Griesinger, *J. Am. Chem. Soc.* **1993**, *115*, 10877–10882.

3. Pulse Scheme and Phase Cycle

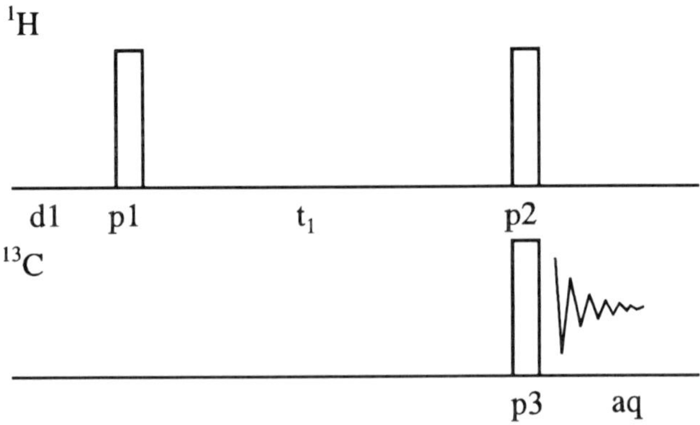

p1, p3: x p2: y, -y aq: x, -x

phase cycle for p1 incremented according to TPPI

4. Acquisition

Time requirement: 25 min

Sample: 50% $CHCl_3$ in $[D_6]$acetone

Record normal ^{13}C and ^{1}H NMR spectra and note the signal positions. Change to the 2D mode of the spectrometer software and load the appropriate pulse program. You have to set:

td2: 512 data points in F_2
td1: 64 data points in F_1
sw2: 500 Hz
sw1: 500 Hz
o1: on resonance of ^{13}C NMR signal of $CHCl_3$
o2: on resonance of ^{1}H NMR signal of $CHCl_3$
p1, p2: 90° ^{1}H decoupler pulse
p3: 90° ^{13}C transmitter pulse
d1: 10 s
initial value for t_1 evolution: 3 μs
increment for t_1 evolution = 1/[2·sw1]
ns: 2

5. Processing

Apply zero filling in F_1 to 128 real data points. Use exponential windows in both dimensions. Apply real Fourier transformation corresponding to the TPPI mode for quadrature detection in F_1. Adjust the phase in F_2 to give antiphase signals; the phase cycle given requires a 90° phase correction in F_1.

6. Result

The figure on page 374 shows an expansion of the 2D spectrum obtained on an ARX-200 spectrometer in a dual probe-head. The dotted contours represent negative signals. Note that the heteronuclear spin coupling is present in antiphase in both dimensions.

7. Comments

Since we are in this experiment on resonance both for ^{1}H and for ^{13}C we do not have to consider chemical shift evolution when using the product operator formalism. The first proton pulse creates $-I_{Hy}$ magnetization, which subsequently develops C,H spin coupling during t_1, as shown in Eq. (1).

$$I_{Hz} \xrightarrow{I_{Hx}} -I_{Hy} \xrightarrow{\pi J t_1\, 2I_{Hz}I_{Cz}} 2I_{Hx}I_{Cz}\sin\pi J t_1 - I_{Hy}\cos\pi J t_1 \qquad (1)$$

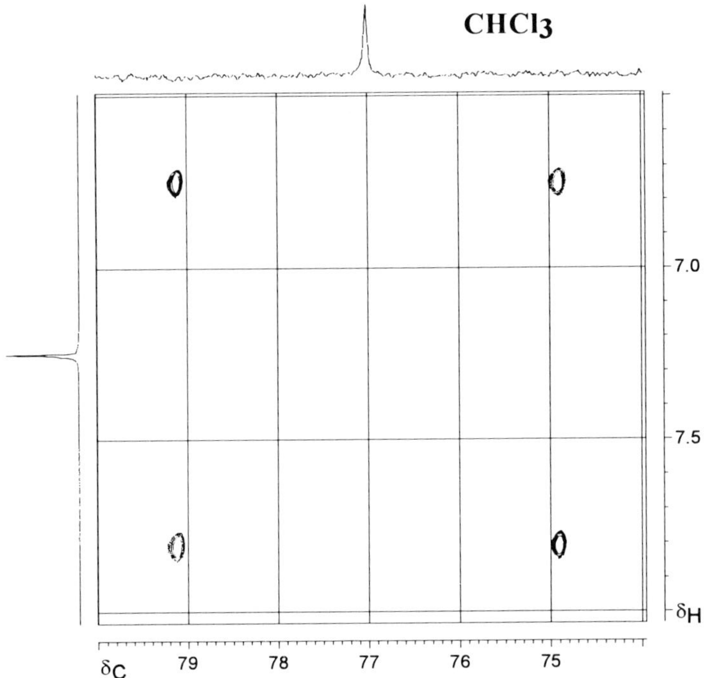

A second proton pulse p2 from the *y*-direction together with the carbon pulse p3 from the *x*-direction will change the antiphase term $2I_{Hx}I_{Cz}$ into the antiphase term $2I_{Hz}I_{Cy}$ as given in Eq. (2).

$$2I_{Hx}I_{Cz}\sin\pi Jt_1 \xrightarrow{I_{Hy},\, I_{Cx}} 2I_{Hz}I_{Cy}\sin\pi Jt_1 \xrightarrow{\pi Jaq\; 2I_{Hz}I_{Cz}}$$

$$-I_{Cx}\sin\pi Jt_1\sin\pi Jaq \tag{2}$$

During acquisition, again C,H spin coupling evolves, giving observable in-phase magnetization $-I_{Cx}$. This signal is both in F_1 and in F_2 modulated with the sine of the spin coupling; thus we observe the antiphase pattern as seen in the figure.

8. Own Observations

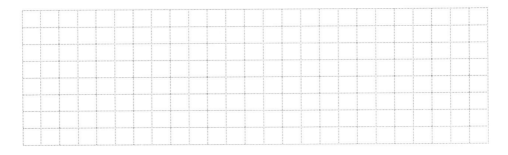

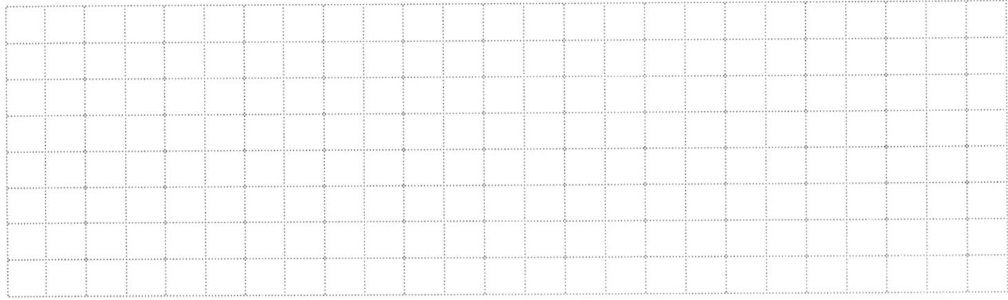

Experiment 10.10

C,H Correlation by Polarization Transfer (HETCOR)

1. Purpose

A two dimensional C,H correlation experiment yields cross signals for all protons and ^{13}C nuclei which are connected by a $^{13}C,^{1}H$ coupling over one bond. The assignment of one member of a spin-coupled pair leads immediately to the assignment of the other. A C,H correlation experiment may be performed in many ways. The experiment described here encodes the proton chemical shift information into the ^{13}C signals and can be performed on most older instruments, since the observed nuclide is ^{13}C. The 1D equivalent of this correlation technique is described in Experiment 4.13.

2. Literature

[1] R. Freeman, G. A. Morris, *J. Chem. Soc. Chem. Commun.* **1978**, 684–686.
[2] A. Bax, G. A. Morris, *J. Magn. Reson.* **1981**, *42*, 501–505.
[3] G. E. Martin, A. S. Zektzer, *Two-Dimensional NMR Methods for Establishing Molecular Connectivity*, VCH, Weinheim, **1988**, Ch. 3.

3. Pulse Scheme and Phase Cycle

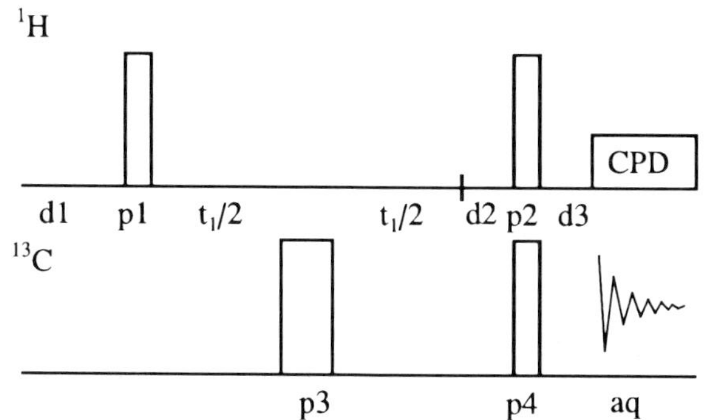

p1: x p3: $(x)_4$, $(-x)_4$

p2: x, -x, y, -y p4: $(x)_8$, $(y)_8$, $(-x)_8$, $(-y)_8$

aq: $(x, -x, y, -y)_2$, $(y, -y, -x, x)_2$ $(-x, x, -y, y)_2$, $(-y, y, x, -x)_2$

4. Acquisition

Time requirement: 2.5 h

Sample: 20% ethyl crotonate in $CDCl_3$

Record normal ^{13}C and ^{1}H NMR spectra and optimize the spectral widths. Change to the 2D mode of the spectrometer software and load the pulse program for X,H correlation. You have to set:

td2: 1 k data points in F_2
td1: 128 data points in F_1
sw2: 175 ppm
sw1: 8 ppm
o1: middle of ^{13}C NMR spectrum
o2: middle of ^{1}H NMR spectrum
p1, p2: 90° ^{1}H decoupler pulse
p3: 180° ^{13}C transmitter pulse
p4: 90° ^{13}C transmitter pulse
d1: 2 s
d2: $1/[2J(C,H)]$ = 3.45 ms, calculated from $^{1}J(C,H)$ = 145 Hz
d3: $1/[3J(C,H)]$ = 2.29 ms, calculated from $^{1}J(C, H)$ = 145 Hz
initial value for t_1 evolution: 3 μs
increment for t_1 evolution = $1/[2 \cdot sw1]$
decoupler attenuation and 90° pulse for CPD
ns: 32

5. Processing

Apply zero filling in F_1 to 256 real data points. Use squared $\pi/2$ shifted sinusoidal windows in both dimensions. Apply complex Fourier transformation corresponding to the N type signal selection using the quadrature-off mode in F_1. Phase correction is not necessary, since the data are processed in magnitude mode.

6. Result

The figure on page 377 shows the 2D spectrum obtained on an ARX-200 spectrometer. Note that the C,H spin coupling is removed in both dimensions. This is achieved by CPD decoupling in F_2 and by the 180° ^{13}C pulse in case of F_1.

7. Comments

The first proton pulse creates $-I_{Hy}$ magnetization, which subsequently develops ^{1}H chemical shift during t_1 as shown in Equation (1). The 180° ^{13}C pulse p3 in the middle of t_1 removes heteronuclear spin coupling during t_1.

$$I_{Hz} \xrightarrow{I_{Hx}} -I_{Hy} \xrightarrow{\Omega_H t_1 I_{Hz}} I_{Hx}\sin\Omega_H t_1 - I_{Hy}\cos\Omega_H t_1 \quad (1)$$

C,H spin coupling evolves during the delay τ, leading to an antiphase magnetization of proton with respect to carbon, as indicated in Equation (2). If the delay τ is set equal to d2 = $1/2J(C,H)$ the corresponding cosine terms become zero.

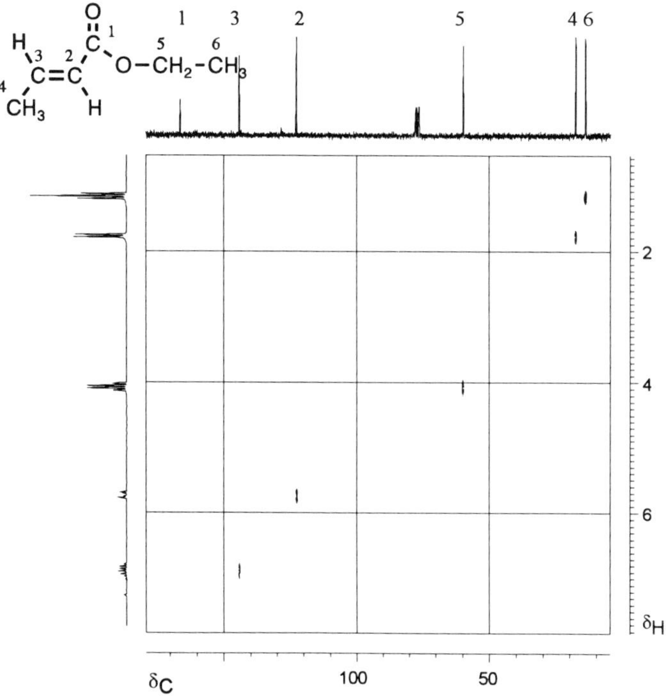

$$\xrightarrow{\pi J \tau 2 I_{Hz} I_{Cz}} 2 I_{Hy} I_{Cz} \sin\Omega_H t_1 + 2 I_{Hx} I_{Cz} \cos\Omega_H t_1 \quad (2)$$

The two simultaneous pulses p2 and p4 transform this into antiphase magnetization of carbon with respect to proton, as shown in Equation (3). Since this magnetization was originally a proton magnetization, bearing γ_H, we call this a polarization transfer.

$$\xrightarrow{I_{Hx}, I_{Cx}} -2 I_{Hz} I_{Cy} \sin\Omega_H t_1 \quad (3)$$

The antiphase magnetization is refocused during the delay d3 to give an observable in-phase magnetization. The delay d3 is chosen so as to obtain the maximum signal for all multiplicities. During acquisition the ^{13}C chemical shift develops, while proton decoupling ensures that no spin coupling appears in F_2.

8. Own Observations

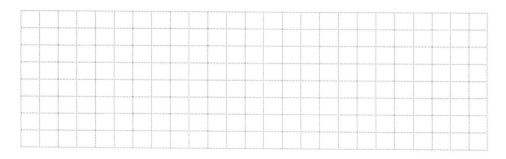

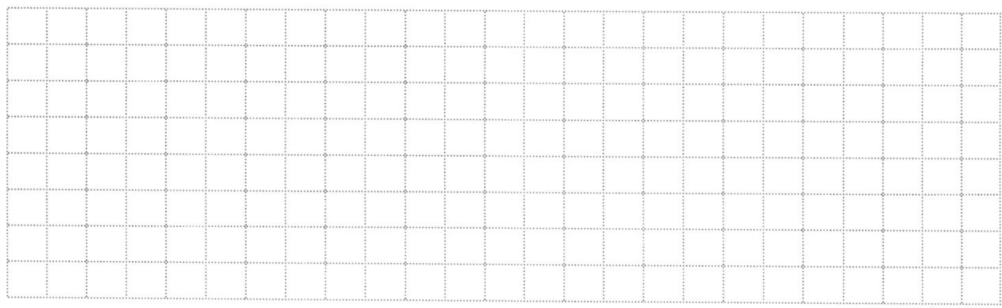

Experiment 10.11

Long Range C,H Correlation by Polarization Transfer

1. Purpose

The normal C,H correlation procedure as described in Experiment 10.10 yields cross-signals for all proton and ^{13}C nuclei that are connected by a one-bond coupling constant $^{1}J(C,H)$. However, it is often desirable to be able to observe cross-signals for C,H spin pairs connected by two- or three-bond couplings $^{2}J(C,H)$ or $^{3}J(C,H)$. This can be achieved with the same pulse sequence by adjusting the appropriate delays. Other alternatives are the COLOC (Exp. 10.12) and HMBC (Exp. 10.16) experiments. The experiment described here incorporates the proton chemical shift information into the carbon signals and can be performed on most older instruments since the observed nuclide is ^{13}C. The 1D equivalent of this correlation technique is described in Experiment 4.14.

2. Literature

[1] C. Bauer, R. Freeman, S. Wimperis, *J. Magn. Reson.* **1984,** *58*, 526–532.
[2] A. S. Zektzer, B. K. John, G. E. Martin , *Magn. Reson. Chem.* **1987**, *25*, 752–756.
[3] G. E. Martin, A. S. Zektzer, *Two-Dimensional NMR Methods for Establishing Molecular Connectivity*, VCH, Weinheim, **1988**, 221–255.

3. Pulse Scheme and Phase Cycle

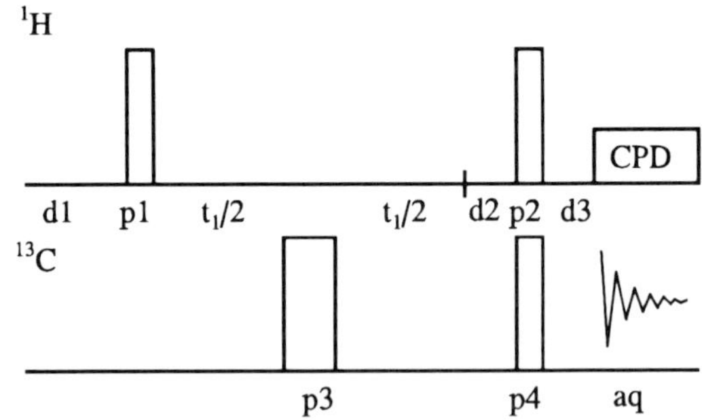

p1: x p3: $(x)_4$, $(-x)_4$

p2: x, -x, y, -y p4: $(x)_8$, $(y)_8$, $(-x)_8$, $(-y)_8$

aq: $(x, -x, y, -y)_2$, $(y, -y, -x, x)_2$ $(-x, x, -y, y)_2$, $(-y, y, x, -x)_2$

4. Acquisition

Time requirement: 5 h

Sample: 20% ethyl crotonate in $CDCl_3$

Record normal ^{13}C and ^{1}H NMR spectra and optimize the spectral widths. Change to the 2D mode of the spectrometer software and load the pulse program for X,H correlation. You have to set:

td2: 1 k data points in F_2
td1: 128 data points in F_1
sw2: 175 ppm
sw1: 8 ppm
o1: middle of ^{13}C NMR spectrum
o2: middle of ^{1}H NMR spectrum
p1, p2: 90° ^{1}H decoupler pulse
p3: 180° ^{13}C transmitter pulse
p4: 90° ^{13}C transmitter pulse
d1: 2 s
d2: $1/[2J(C,H)]$ = 50 ms, calculated from $^{n}J(C,H)$ = 10 Hz
d3: $1/[3J(C,H)]$ = 33 ms, calculated from $^{n}J(C,H)$ = 10 Hz
initial value for t_1 evolution: 3 μs
increment for t_1 evolution = 1/[2·sw1]
decoupler attenuation and 90° pulse for CPD
ds: 2
ns: 64

5. Processing

Apply zero filling in F_1 to 256 real data points. Use squared $\pi/2$ shifted sinusoidal windows in both dimensions. Apply complex Fourier transformation corresponding to the N type signal selection using the quadrature-off mode in F_1. Phase correction is not necessary, since the data are processed in magnitude mode.

6. Result

80 shows the 2D spectrum obtained on an ARX-200 spectrometer. Most of the cross-signals corresponding to $^{1}J(C,H)$ can still be observed. Of interest are the weak cross-signals of H-5, H-3, and H-2 to the carboxyl ^{13}C nucleus C-1, and the large intensity difference between the cross-signals of H-4 to C-2 and C-3. Note, however, that the noise is considerably greater than that obtained for $^{1}J(C,H)$ in Experiment 10.10 and that there is some breakthrough of axial signals.

7. Comments

The product operator description of the experiment is given in Experiment 10.10. The selection of delays d2 and d3 is frequently difficult and is discussed extensively in the literature. A good initial value for d2 is usually 50 ms (corresponding to a coupling constant of 10 Hz), although ^{2}J- and ^{3}J-values are seldom as large as 10 Hz.

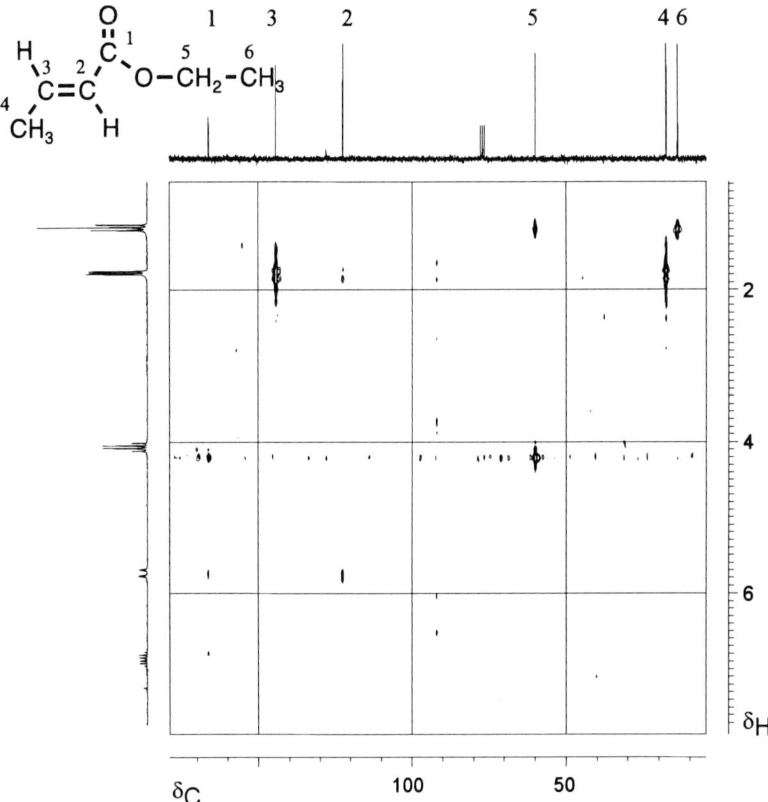

Relaxation times and homonulear spin–spin coupling of protons must also be taken into account.

8. Own Observations

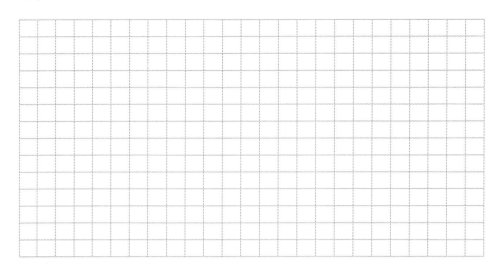

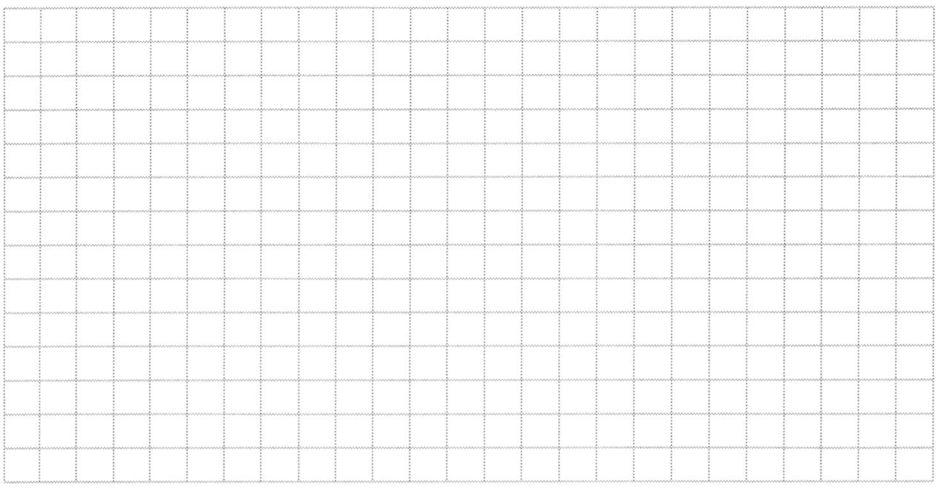

Experiment 10.12

C,H Correlation via Long-Range Couplings (COLOC)

1. Purpose

The long-range C,H correlation procedure as described in Experiment 10.11 is not always satisfactory. With increasing t_1 values homonuclear spin coupling of the protons evolves, and proton relaxation reduces the sensitivity of the experiment. A constant time method called COLOC (**CO**rrelation spectroscopy via **LO**ng range **C**oupling) has therefore been developed. In this experiment the t_1 evolution and the polarization transfer period are combined in one constant time interval and the separation of the two is achieved by incrementally shifted 180° pulses. As in Experiment 10.11, COLOC incorporates the proton chemical shift information into the carbon signals and can be performed on most older instruments since the observed nuclide is ^{13}C. The version shown here is not phase-sensitive.

2. Literature

[1] H. Kessler, C. Griesinger, J. Zarbock, H. R. Loosli, *J. Magn. Reson.* **1984**, *57*, 331–336.

[2] H. Kessler, C. Griesinger, K. Wagner, *J. Am. Chem. Soc.* **1987**, *109*, 6927–6933.

[3] G. E. Martin, A. S. Zektzer, *Two-Dimensional NMR Methods for Establishing Molecular Connectivity*, VCH, Weinheim, **1988**, 255–267.

3. Pulse Scheme and Phase Cycle

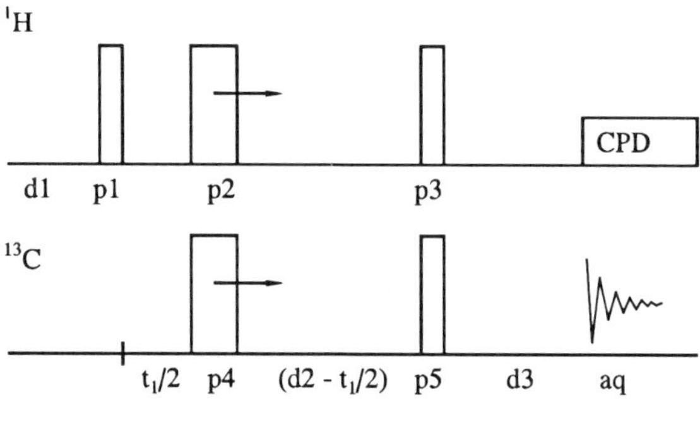

p1: $(x)_{16}$, $(y)_{16}$, $(-x)_{16}$, $(-y)_{16}$ p2, p4: x

p3: $(x, y)_8$, $(-x, -y)_8$ aq: $(x)_2$, $(-x)_2$, $(y)_2$, $(-y)_2$

p5: x, y, -x, -y, y, -x, -y, x, -x, -y, x, y, -y, x, y, -x ,
y, -x, -y, x, -x, -y, x, y, -y, x, y, -x, x, y, -x, -y,
-x, -y, x, y, -y, x, y, -x, x, y, -x, -y, y, -x, -y, x ,
-y, x, y, -x, x, y, -x, -y, y, -x, -y, x, -x, -y, x, y

4. Acquisition

Time requirement: 5 h

Sample: 20% ethyl crotonate in $CDCl_3$

Record normal ^{13}C and ^{1}H NMR spectra and optimize the spectral widths. Change to the 2D mode of the spectrometer software and load the pulse program for X,H correlation. You have to set:

td2: 1 k data points in F_2
td1: 64 data points in F_1
sw2: 175 ppm
sw1: 8 ppm
o1: middle of ^{13}C NMR spectrum
o2: middle of ^{1}H NMR spectrum
p1, p3: 90° ^{1}H decoupler pulse
p2: 180° ^{1}H decoupler pulse
p4: 180° ^{13}C transmitter pulse
p5: 90° ^{13}C transmitter pulse
d1: 2 s
d2: 25 ms
d3: $1/[3J(C,H)]$ = 33 ms, calculated from $^{n}J(C,H)$ = 10 Hz
initial value for t_1 evolution: 3 μs
increment for t_1 evolution: $1/[2 \cdot sw1]$
decrement for d2: $1/[2 \cdot sw1]$; note that d2 must be larger than td1 times $1/[2 \cdot sw1]$
decoupler attenuation and 90° pulse for CPD
ds: 2
ns: 128

5. Processing

Apply zero filling in F_1 to 256 real data points. Use squared $\pi/2$ shifted sinusoidal windows in both dimensions. Apply complex Fourier transformation corresponding to the N type signal selection using the quadrature off mode in F_1. Phase correction is not necessary, since the data are processed in magnitude mode.

6. Result

The figure on page 383 shows the 2D spectrum obtained on an ARX-200 spectrometer. Most of the cross-signals corresponding to $^{1}J(C,H)$ can still be observed. Compared with the result of Experiment 10.11, which was obtained under otherwise identical conditions, there is less noise and the cross-peaks from H-4 to C-2 and C-3, as well the cross peak from H-5 to C-1, are much stronger; however, the correlations of H-3 and

H-2 to C-1 are completely missing. Careful adjustment of d2 and d3 is necessary; this can be optimized by a 1D INEPT experiment.

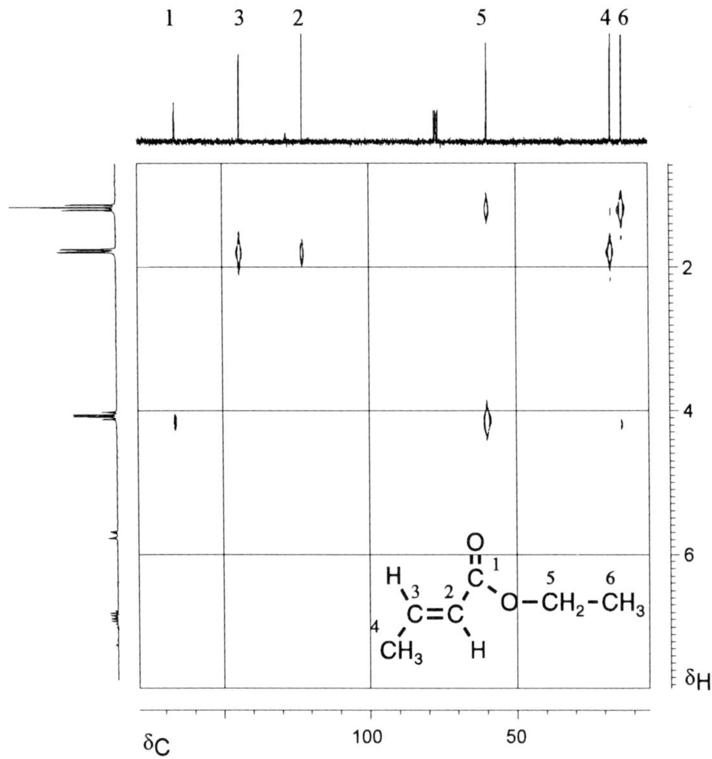

7. Comments

The experiment is of the constant-time type as the time period of the chemical shift evolution is held constant while 180° pulses are moved through this period. The product operator description as given in Experiment 10.10 still holds in principle. The advantages of this approach are that the evolution of homonuclear spin coupling of the protons is held constant during the experiment, and that the protons are given less time to relax, since *t*1 evolution and transfer delay d2 are combined. Several modifications of the original experiment are known [3].

8. Own Observations

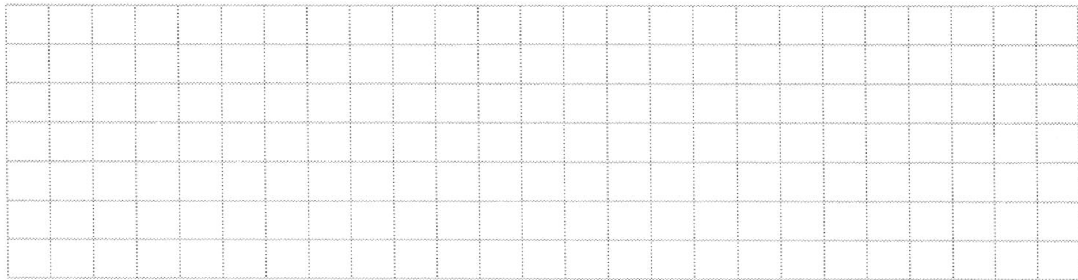

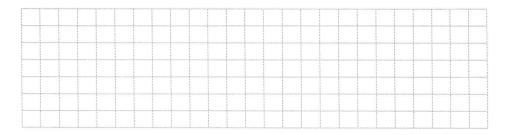

Experiment 10.13

The Basic HMQC Experiment

1. Purpose

The experiment described is the simplest form of an inverse H,X correlation technique. HMQC stands for **H**eteronuclear **M**ultiple **Q**uantum **C**oherence. The suppression of unwanted signals is performed only by the phase cycle; no ^{13}C broad-band decoupling is applied during acquisition and the 2D spectrum is recorded without quadrature detection in F_1, which requires the magnitude mode of data processing. Using only four different r.f. pulses it demonstrates the sensitivity advantage of the inverse detection method.

2. Literature

[1] L. Müller, *J. Am. Chem. Soc.* **1979**, *101*, 4481–4484.
[2] A. Bax, R. H. Griffey, B. L. Hawkins, *J. Magn. Reson.* **1983**, *55*, 301–315.
[3] G. E. Martin, A. S. Zektzer, *Two-Dimensional NMR Methods for Establishing Molecular Connectivity*, VCH, Weinheim, **1988**, 213–221.

3. Pulse Scheme and Phase Cycle

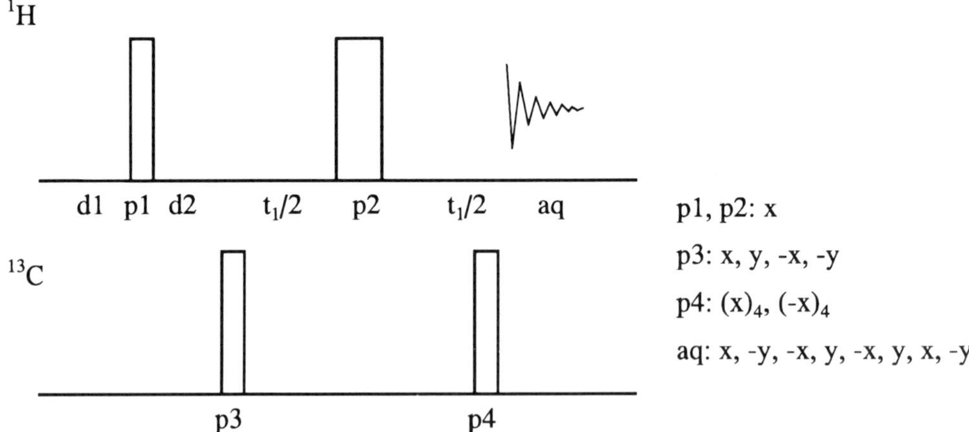

4. Acquisition

Time requirement: 40 min

Sample: 5% ethyl crotonate in $CDCl_3$

Record normal 1D ^{1}H and ^{13}C NMR spectra, optimize the spectral widths, and note the offsets. Switch to the 2D mode of the spectrometer software, load the HMQC pulse program, and, if required, change the spectrometer to the inverse set-up. You have to set:

td2: 1 k data points in F_2
td1: 128 data points in F_1
sw2: 8 ppm
sw1: 175 ppm
offset of ^{1}H frequency: middle of ^{1}H NMR spectrum
offset of ^{13}C frequency: middle of ^{13}C NMR spectrum
p1: 90° ^{1}H transmitter pulse
p2: 180° ^{1}H transmitter pulse
p3, p4: 90° ^{13}C decoupler pulse
d1: 2 s
d2: $1/[2J(C,H)] = 3.5$ ms, calculated from $^1J(C,H) = 145$ Hz
initial value for t_1 evolution: 3 μs
increment for t_1 evolution: $1/[2 \cdot sw1]$
ds: 2
ns: 8

5. Processing

Apply zero filling in F_1 to 256 real data points to obtain a matrix of 512·256 real data points. Use $\pi/2$ shifted squared sinusoidal windows in both dimensions. Apply complex Fourier transformation corresponding to the N type signal selection using the quadrature-off mode in F_1. Phase correction is not necessary, since the data are processed in magnitude mode.

6. Result

The figure on page 386 shows the HMQC spectrum obtained on an ARX-200 spectrometer in a normal forward dual probe-head. This demonstrates that a special inverse probe-head is not an absolute necessity. Note that the spectrum displays in F_2 doublets with the spin coupling constant $^1J(C,H)$, and in addition the homonuclear splittings caused by H,H spin couplings. These also broaden the signals in the F_1 dimension. There is considerable signal breakthrough from the protons bound to ^{12}C, which is seen mostly at the methyl group signals.

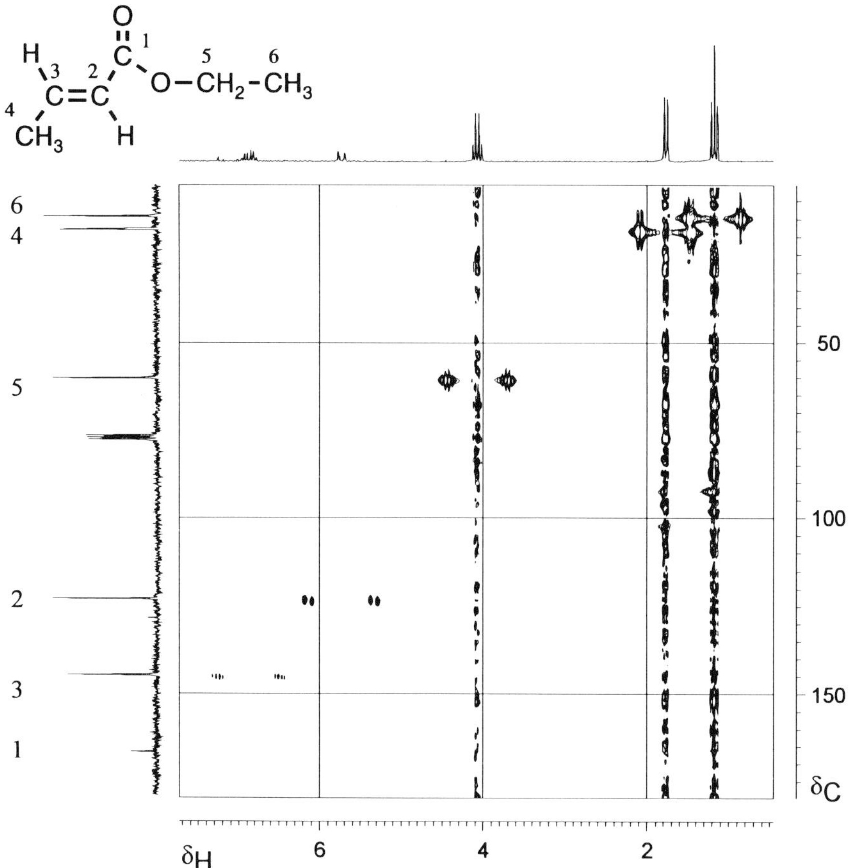

7. Comments

The first 90° proton pulse creates a transverse proton magnetization $-I_{Hy}$ as in Equation (1). During the delay τ = d2 J(C,H) develops and creates antiphase magnetization $2I_{Hx}I_{Cz}$. Since τ was set to $1/[2J(C,H)]$ the cosine term becomes zero and the sine term unity.

$$I_{Hz} + I_{Cz} \xrightarrow{I_{Hx}} -I_{Hy} + I_{Cz} \xrightarrow{\pi J \tau I_{Hz} I_{Cz}} 2I_{Hx}I_{Cz} \tag{1}$$

The first 90° ^{13}C pulse p3 transforms the antiphase magnetization into double quantum magnetization $-2I_{Hx}I_{Cy}$ as in Equation (2). During t_1 this term develops ^{13}C chemical shift as in Equatiuon (3). Of course, ^{1}H chemical shift and H,H spin coupling evolve also during $t1$. The former is removed by the 180° proton pulse p2 which for simplicity is not shown in the equations. Furthermore, this 180° proton pulse interchanges double quantum and zero quantum terms.

$$2I_{Hx}I_{Cz} \xrightarrow{I_{Cx}} -2I_{Hx}I_{Cy} \tag{2}$$

$$\xrightarrow{\Omega_C\, t_1\, I_{Cz}} -2I_{Hx}I_{Cy}\cos\Omega_C t_1 + 2I_{Hx}I_{Cx}\sin\Omega_C t_1 \tag{3}$$

$$\xrightarrow{I_{Cx}} -2I_{Hx}I_{Cz}\cos\Omega_C t_1 + 2I_{Hx}I_{Cx}\sin\Omega_C t_1 \tag{4}$$

The last ^{13}C pulse p4 transforms the double quantum magnetization back into anti-phase terms as shown in Equation (4). During the acquisition time aq C,H spin coupling again evolves, leading to the observable proton signal which is modulated with the ^{13}C chemical shift information during t_1 as in Equation (5).

$$\xrightarrow{\pi J\, \mathrm{aq}\, I_{Hz}I_{Cz}} -I_{Hy}\cos\Omega_C t_1 \sin\pi J\mathrm{aq} \tag{5}$$

8. Own Observations

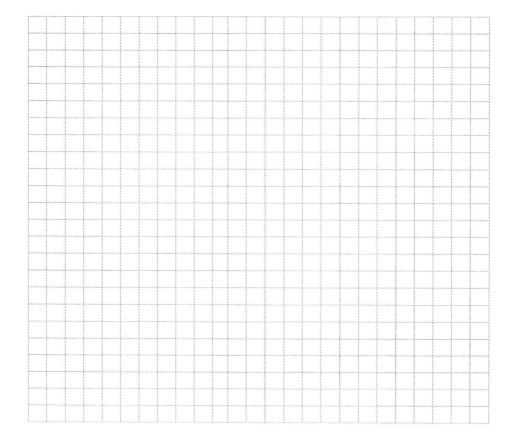

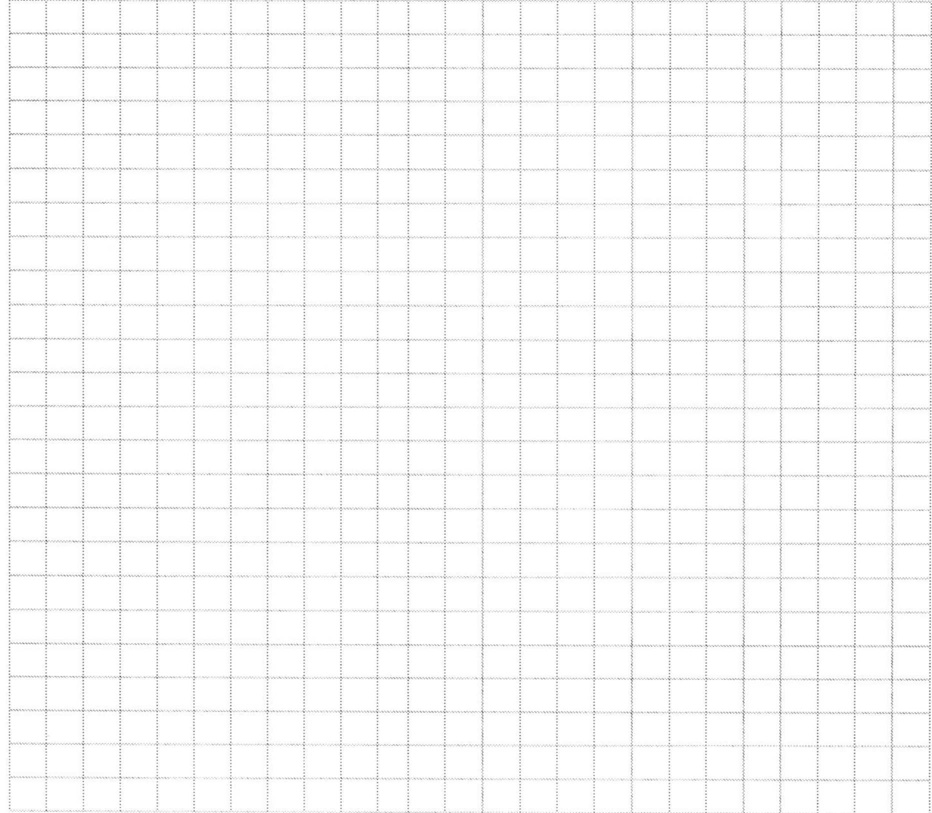

Experiment 10.14

Phase-Sensitive HMQC with BIRD Filter and GARP Decoupling

1. Purpose

The basic HMQC sequence as described in Experiment 10.13 gives rather poor signal suppression for protons bound to ^{12}C or ^{14}N. A considerable improvement [1] can be achieved by using the BIRD sandwich [2] (see Exp. 6.13) prior to the HMQC sequence. Furthermore, decoupling all ^{13}C nuclei with the GARP technique (**G**lobally optimized **A**lternating-phase **R**ectangular **P**ulses) [3] improves the signal-to-noise ratio. The phase-sensitive 2D mode chosen for this example yields Lorentzian line shapes in both dimensions. Prior to the introduction of gradient-selected spectroscopy (see Exp. 12.3), this experiment was the first choice for inverse H,C correlation.

2. Literature

[1] A. Bax, S. Subramanian, *J. Magn. Reson.* **1986**, *67*, 565–569.
[2] J. R. Garbow, D. P. Weitekamp, A. Pines, *Chem. Phys. Lett.* **1982**, *93*, 504–508.
[3] A. J. Shaka, P. B. Barker, R. Freeman, *J. Magn. Reson.* **1985**, *64*, 547–552.

3. Pulse Scheme and Phase Cycle

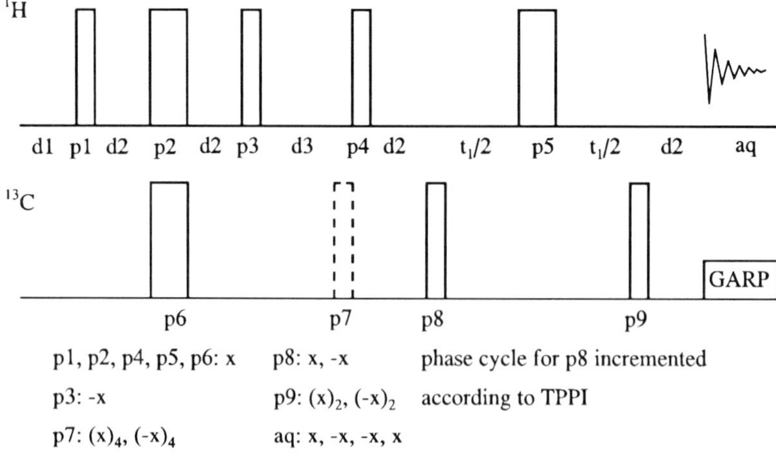

p1, p2, p4, p5, p6: x — p8: x, -x — phase cycle for p8 incremented according to TPPI

p3: -x — p9: $(x)_2$, $(-x)_2$

p7: $(x)_4$, $(-x)_4$ — aq: x, -x, -x, x

4. Acquisition

Time requirement: 40 min

Sample: 5% ethyl crotonate in $CDCl_3$

Record normal 1D ^1H and ^{13}C NMR spectra, optimize the spectral widths and note the offsets. Switch to the 2D mode of the spectrometer software, load the pulse program and, if required, change the spectrometer to the inverse configuration. For the experiment you have to determine the 90° ^{13}C pulse for GARP decoupling (see Exp. 2.5). You have to set:

td2: 1 k data points in F_2
td1: 128 data points in F_1
sw2: 8 ppm
sw1: 175 ppm
offset of ^1H frequency: middle of ^1H NMR spectrum
offset of ^{13}C frequency: middle of ^{13}C NMR spectrum
p1, p3, p4: 90° ^1H transmitter pulse
p2, p5: 180° ^1H transmitter pulse
p7, p8, p9: 90° ^{13}C decoupler pulse
p6: 180° ^{13}C decoupler pulse
d1: 1 s
d2: $1/[2J(C,H)] = 3.5$ ms, calculated from $^1J(C,H) = 145$ Hz
d3: BIRD delay to be optimized for minimum FID, ca 1 s; observe in the set-up mode the incoming FID and adjust d3 for minimum intensity (see Exp. 6.13).
initial value for t_1 evolution: 3 μs
increment for t_1 evolution: 1/[4·sw1]
^{13}C decoupler attenuation and 90° pulse width for GARP
ds: 2
ns: 8

5. Processing

Apply zero filling in F_1 to 256 real data points to obtain a matrix of 512·256 real points. Use exponential or Gaussian windows in both dimensions. Apply real Fourier transformation corresponding to the TPPI type signal selection using the quadrature mode in F_1. Phase correction is required; it is often sufficient to correct the phase only in F_2.

6. Result

The figure on page 390 shows the HMQC spectrum obtained on an ARX-200 spectrometer with a normal dual probe-head. This demonstrates that special inverse probe-heads are not an absolute necessity. Note that the spectrum displays only singlets in F_2 which are further split by the homonuclear spin couplings. These also broaden the signals in the F_1 dimension. The breakthrough of the signals from protons bound to ^{12}C is dramatically reduced in comparison with Experiment 10.13.

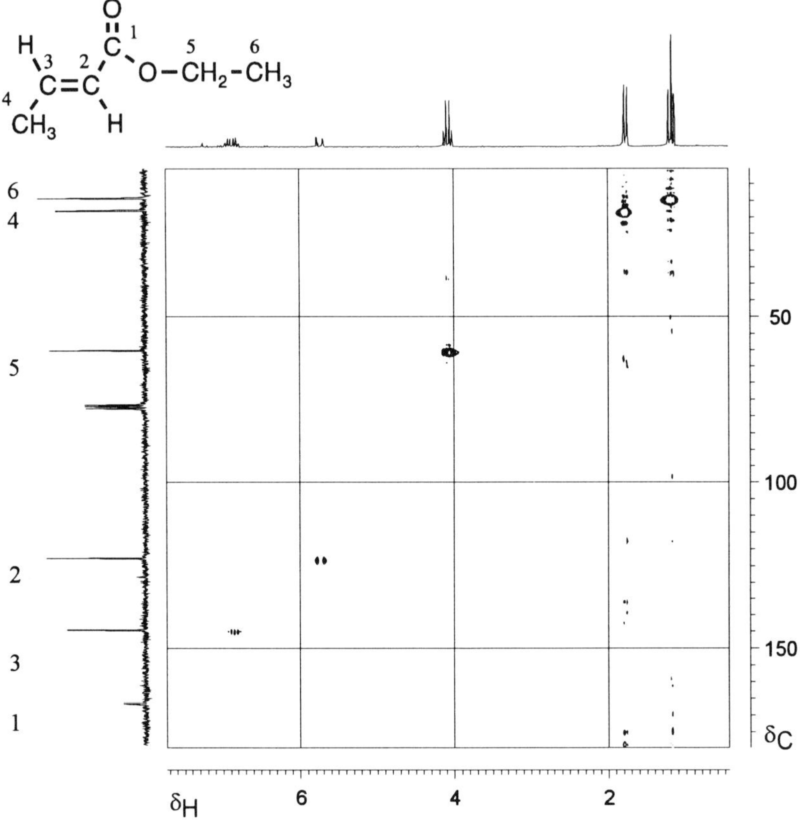

7. Comments

The description of the HMQC and BIRD part of the sequence in the product operator formalism has been given in Experiments 10.13 and 6.13. The dashed pulse p7 in the pulse sequence removes artefacts caused by residual longitudinal ^{13}C magnetization [1]. For the conditions given we have obtained identical results from experiments in which this pulse was used and those in which it was omitted. During the last delay d2 C,H spin coupling is refocused in order to allow GARP decoupling.

8. Own Observations

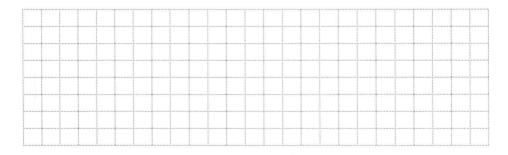

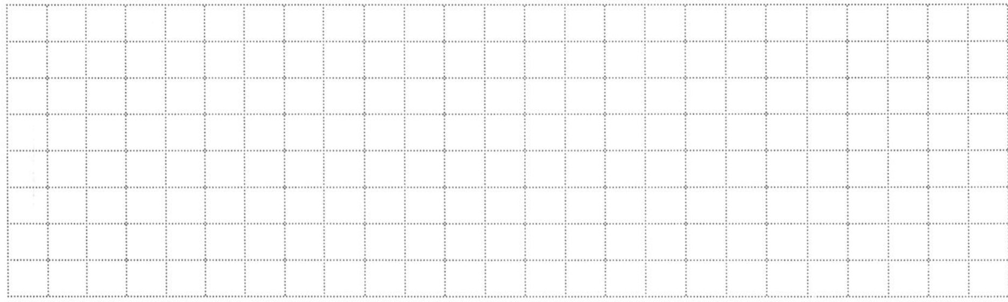

Experiment 10.15

Poor Man's Gradient HMQC

1. Purpose

The basic HMQC sequence as described in Experiment 10.13 gives rather poor signal suppression for protons bound to ^{12}C or ^{14}N. A considerable improvement can be achieved by using the BIRD sandwich (see Exp. 10.14) prior to the HMQC sequence. Furthermore, the use of an additional purging scheme with a spin-lock (see Exp. 6.16) reduces unwanted signals nearly to the level known from pulsed field gradients (see Exp. 12.3) and allows the use of a higher receiver gain. Since this technique can be performed on older instruments not equipped with a field gradient unit it was termed PMG (**P**oor **M**an's **G**radient) [2]. Here we show a phase-sensitive version with ^{13}C-GARP decoupling using ethyl crotonate. The method shown seems currently to be the best for routine H,C correlation under the above-mentioned instrumental restrictions. The basic idea has now also been applied within the HSQC and HMBC pulse techniques [3].

2. Literature

[1] G. Otting, K. Wüthrich, *J. Magn. Reson.* **1988**, *76*, 569–574.
[2] J.-M. Nuzillard, G. Gasmi, J.-M. Bernassau, *J. Magn. Reson. Ser. A* **1993**, *104*, 83–87.
[3] G. Gasmi, G. Massiot, J.-M. Nuzillard, *Magn. Reson. Chem.* **1996**, *34*, 185–190.

3. Pulse Scheme and Phase Cycle

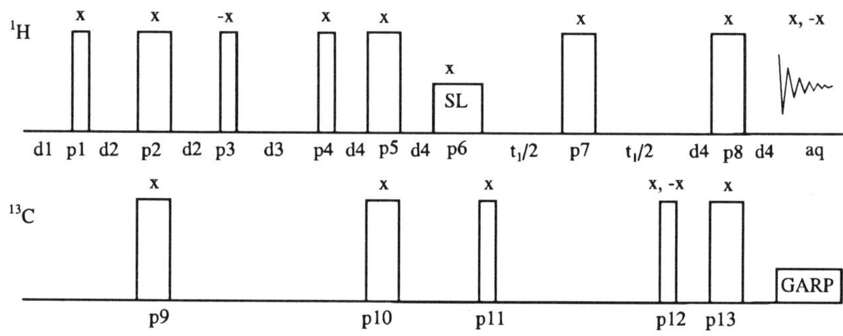

phase cycle for p11 incremented according to TPPI

4. Acquisition

Time requirement: 10 min

Sample: 5% ethyl crotonate in $CDCl_3$

Record normal 1D ^{1}H and ^{13}C NMR spectra, optimize the spectral widths and note the offsets. Switch to the 2D mode of the spectrometer software, load the PMG pulse program and, if required, change the spectrometer to the inverse configuration. For the experiment you have to determine the 90° ^{13}C pulse for GARP decoupling (see Exp. 2.5), and for best results you should determine and correct a possible phase difference between the hard transmitter pulses and the spin-lock pulse (see Exp. 7.1). You have to set:

td2: 1 k data points in F_2
td1: 128 data points in F_1
sw2: 8 ppm
sw1: 175 ppm
offset of ^{1}H frequency: middle of ^{1}H NMR spectrum
offset of ^{13}C frequency: middle of ^{13}C NMR spectrum
p1, p3, p4: 90° ^{1}H transmitter pulse
p2, p5, p7, p8: 180° ^{1}H transmitter pulse
p11, p12: 90° ^{13}C decoupler pulse
p9, p10, p13: 180° ^{13}C decoupler pulse
p6: ^{1}H transmitter spin-lock pulse, 10 ms length at typically 20 dB attenuation
d1: 1 s
d2: $1/[2J(C,H)]$ = 3.5 ms, calculated from $^{1}J(C,H)$ = 145 Hz
d3: BIRD delay to be optimized for minimum FID, ca 1 s; observe in the set-up mode the incoming FID and adjust d3 for minimum intensity (see Exp. 6.13).
d4: $1/[4J(C,H)]$ = 1.75 ms, calculated from $^{1}J(C,H)$ = 145 Hz
initial value for t_1 evolution: 3 μs
increment for t_1 evolution: 1/[4·sw1]
^{1}H transmitter attenuation for hard pulses (3 dB) and for the spin-lock pulse (20 dB)
^{13}C decoupler attenuation and 90° pulse width for GARP
ds: 2
ns: 2

5. Processing

Apply zero filling in F_1 to 256 real data points to obtain a matrix of 512·256 real points. Use exponential or Gaussian windows in both dimensions. Apply real Fourier transformation corresponding to the TPPI-type signal selection using the quadrature mode in F_1. Phase correction is required in both dimensions.

6. Result

The figure on page 393 shows the PMG-HMQC spectrum obtained on an ARX-200 spectrometer with a normal dual probe-head. Due to ^{13}C GARP decoupling, the spectrum displays only singlets in F_2, which are split by the homonuclear spin couplings. The t_1 noise and the signal breakthrough is far less than in Experiments 10.13–10.14, although only two scans per t_1 increment have been used.

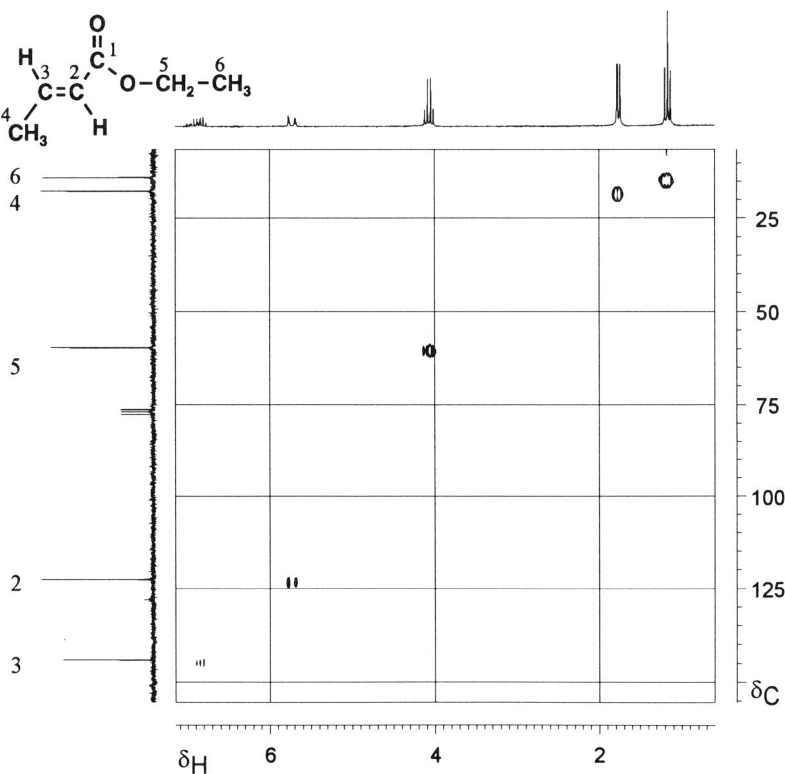

7. Comments

The description of the BIRD part of the sequence in the product operator formalism has been given in Experiment 6.13. The HMQC part of the sequence differs from Experiments 10.13 and 10.14 first by the use of 180° pulses during the development of the CH spin coupling to provide a refocusing of the chemical shifts during both transfer steps. Secondly the spin-lock purging feature is used as described in detail in Experiment 6.16. Note that the receiver gain can be set nearly as high as in experiments with pulsed field gradients and that ns is only 2; in recent literature a 2 ms "hard" spin-lock at 3 dB is also often used.

8. Own Observations

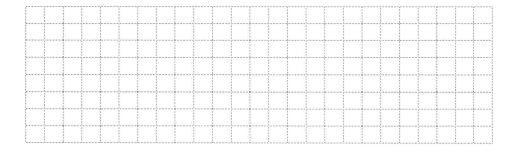

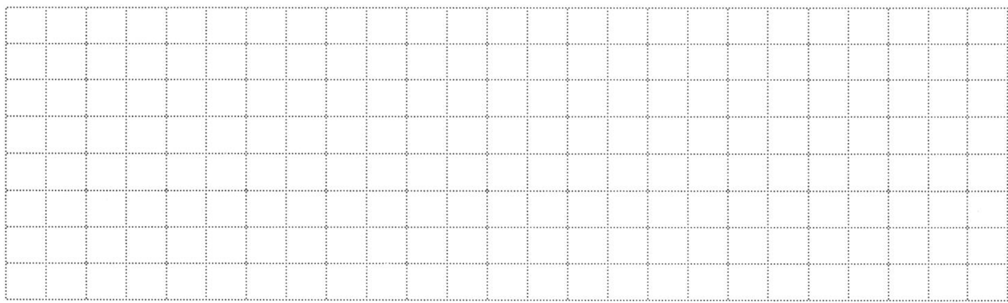

Experiment 10.16

Phase-Sensitive HMBC with BIRD Filter

1. Purpose

The HMQC (**H**eteronuclear **M**ultiple **Q**uantum **C**oherence) sequence as described in Experiments 10.13–10.15 was designed to correlate proton and carbon nuclei via ^{1}J(C,H). To obtain long-range H,C correlations via ^{2}J(C,H) and ^{3}J(C,H) one can simply set the delay d2 to 50 ms, corresponding to a spin coupling constant of 10 Hz. Actually, one uses a special pulse sequence called HMBC (**H**eteronuclear **M**ultiple **B**ond **C**orrelation) [1], the purpose of which is to suppress correlations via ^{1}J(C,H). Since this suppression is not perfect, one usually does not apply ^{13}C-decoupling, so that cross peaks caused by ^{1}J and $^{2/3}J$(C,H) coupling constants can be distinguished. Gradient selected versions are given in Experiments 12.4–12.5 and a 3D version is shown in Experiment 13.4.

2. Literature

[1] A. Bax, M. F. Summers, *J. Am. Chem. Soc.* **1986**, *108*, 2093–2094.
[2] G. E. Martin, A. S. Zektzer, *Two-Dimensional NMR Methods for Establishing Molecular Connectivity*, VCH, Weinheim, **1988**, 267–273.

3. Pulse Scheme and Phase Cycle

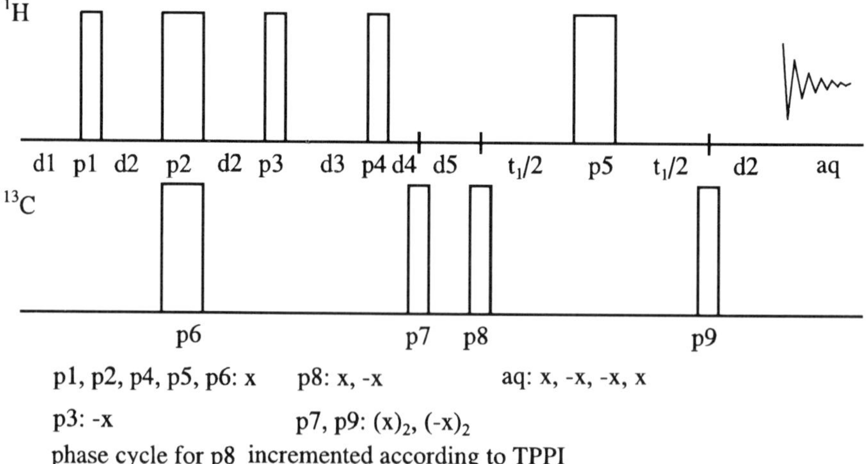

p1, p2, p4, p5, p6: x p8: x, -x aq: x, -x, -x, x

p3: -x p7, p9: $(x)_2$, $(-x)_2$

phase cycle for p8 incremented according to TPPI

4. Acquisition

Time requirement: 9 h

Sample: 5% ethyl crotonate in $CDCl_3$

Record normal 1D ^1H and ^{13}C NMR spectra, optimize the spectral widths and note the offsets. Switch to the 2D mode of the spectrometer software, load the phase-sensitive HMBC pulse program including the BIRD filter, and if required, change the spectrometer to the inverse set-up. You have to set:

td2: 1 k data points in F_2
td1: 128 data points in F_1
sw2: 8 ppm
sw1: 175 ppm
offset of ^1H frequency: middle of ^1H NMR spectrum
offset of ^{13}C frequency: middle of ^{13}C NMR spectrum
p1, p3, p4: 90° ^1H transmitter pulse
p2, p5: 180° ^1H transmitter pulse
p7, p8, p9: 90° ^{13}C decoupler pulse
p6: 180° ^{13}C decoupler pulse
d1: 1 s
d2: $1/[2\,J(C,H)]$ = 50 ms, calculated from $^nJ(C,H)$ = 10 Hz
d3: BIRD delay to be optimized for minimum FID, ca. 1 s; observe in the set-up mode the incoming FID and adjaust d3 for minimum intensity (see Exp. 6.12)
d4: $1/[2J(C,H)]$ = 3.5 ms, calculated from $^1J(C,H)$ = 145 Hz
d5: 46.5 ms (d5 = d2 – d4)
initial value for t_1 evolution: 3 μs
increment for t_1 evolution: 1/[4·sw1]
ds: 2
ns: 128

5. Processing

Apply zero-filling in F_1 to 256 real data points to obtain a matrix of 512·256 real data points. Use exponential or Gaussian windows in both dimensions. Apply real Fourier transformation corresponding to the TPPI-type signal selection using the quadrature mode in F_1. Phase correction in both dimensions is required.

6. Result

The figure on page 396 shows the HMBC spectrum obtained on an ARX-200 spectrometer in a normal forward dual probe-head. This demonstrates that a special inverse probe-head is not an absolute necessity. Note that the 2D spectrum shown is split into two different ranges, since it is often difficult to display cross-peaks of broad multiplets with the same treshold value as used for the sharp singlets. The cross-signals caused by $^1J(C,H)$ of C-6 and C-3 are still observable, but those arising from $^3J(C,H)$ or $^2J(C,H)$ are predominant.

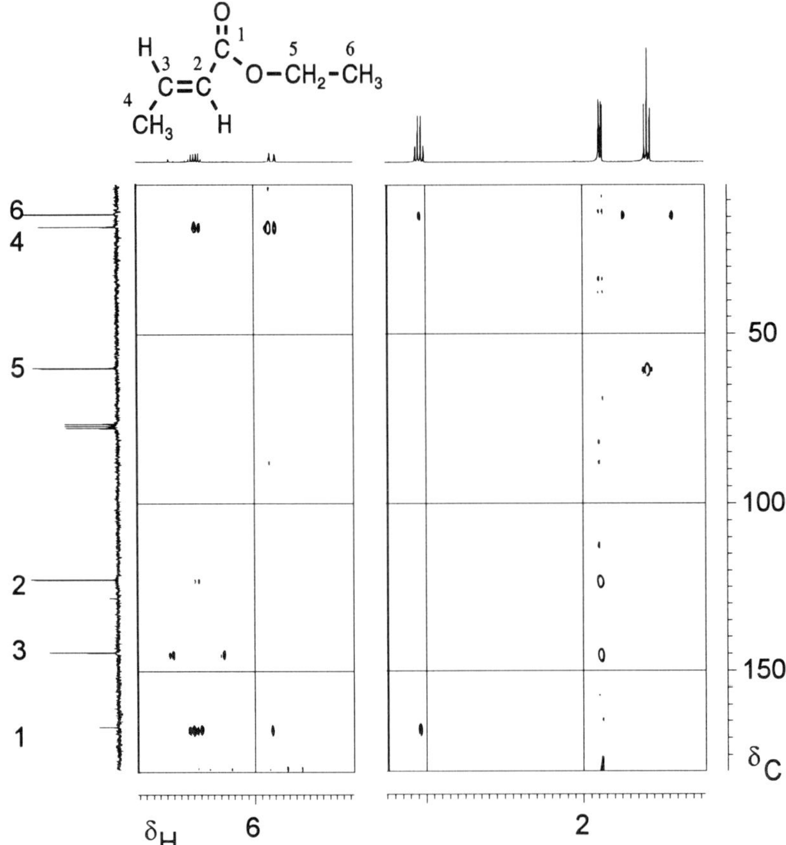

7. Comments

The description of the HMQC and BIRD parts of the sequence in the product operator formalism has been given in Experiments 10.11 and 6.12. The pulse p7 in the pulse sequence acts as a low-pass filter. Coherences stemming from ^{1}J(C,H) are suppressed by the phase cycle of p7. This, however, does not work equally well for all protons; therefore it is advisable not to use GARP decoupling in this sequence, so that one can distinguish between signals arising from ^{1}J and those from ^{2}J or ^{3}J.

8. Own Observations

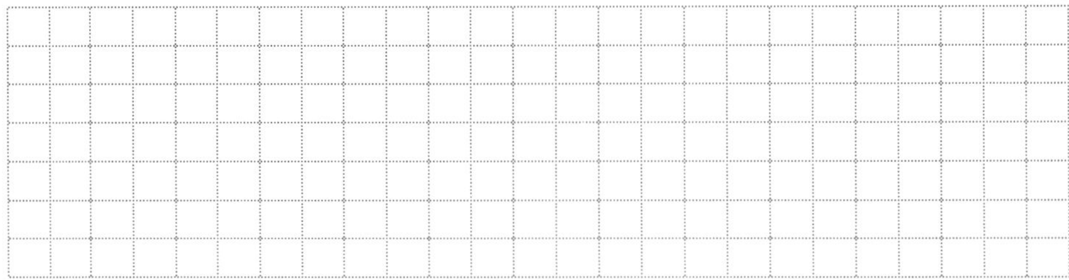

Experiment 10.17

The Basic HSQC Experiment

1. Purpose

Whereas the HMQC experiment 10.13 performs the H,C correlation via the ^{13}C chemical shift evolution of a double quantum coherence, this can also be achieved by the HSQC (**H**eteronuclear **S**ingle **Q**uantum **C**oherence) method. This is sometimes superior in the case of a crowded ^{13}C NMR spectrum, as in this sequence the signals are not broadened by homonuclear H,H coupling in F_1. The HSQC scheme is included as a building block in many 3D sequences. The following example describes this technique, which is also known in NMR jargon as the "Overbodenhausen experiment". A gradient selected phase-sensitive version is given in Experiment 12.6.

2. Literature

[1] G. Bodenhausen, D. J. Ruben, *Chem. Phys. Lett.* **1980**, *69*, 185–188.

3. Pulse Scheme and Phase Cycle

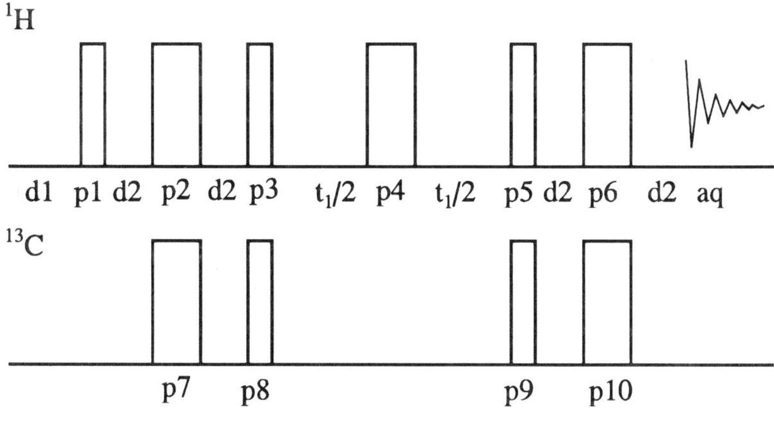

p1, p2, p4, p6: x   p7, p10: $(x)_4$, $(-x)_4$ p9: $(x)_2$, $(-x)_2$

p3, p5: y   p8: x, -x   aq: x, -x, -x, x

phase cycle for p8 incremented according to TPPI

4. Acquisition

Time requirement: 40 min

Sample: 5% ethyl crotonate in $CDCl_3$

Record normal 1D ^{1}H and ^{13}C NMR spectra, optimize the spectral widths and note the offsets. Switch to the 2D mode of the spectrometer software, load the HSQC pulse program, and if required, change the spectrometer to the inverse set-up. You have to set:

td2: 1k data points in F_2
td1: 128 data points in F_1
sw2: 8 ppm
sw1: 175 ppm
offset of ^{1}H frequency: middle of ^{1}H NMR spectrum
offset of ^{13}C frequency: middle of ^{13}C NMR spectrum
p1, p3, p5: 90° ^{1}H transmitter pulse
p2, p4, p6: 180° ^{1}H transmitter pulse
p8, p9: 90° ^{13}C decoupler pulse
p7, p10: 180° ^{13}C decoupler pulse
d1: 2 s
d2: $1/[4J(C,H)] = 1.72$ ms, calculated from $^1J(C,H) = 145$ Hz
initial increment for t_1 evolution: 3 μs
increment for t_1 evolution: $1/[4 \cdot sw1]$
ds: 2
ns: 8

5. Processing

Apply zero filling in F_1 to 256 real data points to obtain a matrix of 512·256 real data points. Use exponential windows in both dimensions. Apply real Fourier transformation corresponding to the TPPI mode signal selection using the quadrature mode in F_1. Phase correction in both dimensions is necessary.

6. Result

The figure on page 399 shows the HSQC spectrum obtained on an ARX-200 spectrometer in a normal forward dual probe-head. This demonstrates that a special inverse probe-head is not an absolute necessity. Note that the spectrum displays in F_2 doublets with the $^1J(C,H)$ spin coupling constant and, in addition, the homonuclear splittings caused by H,H spin couplings. There is considerable breakthrough of signals from the protons bound to ^{12}C since no additional filter was used.

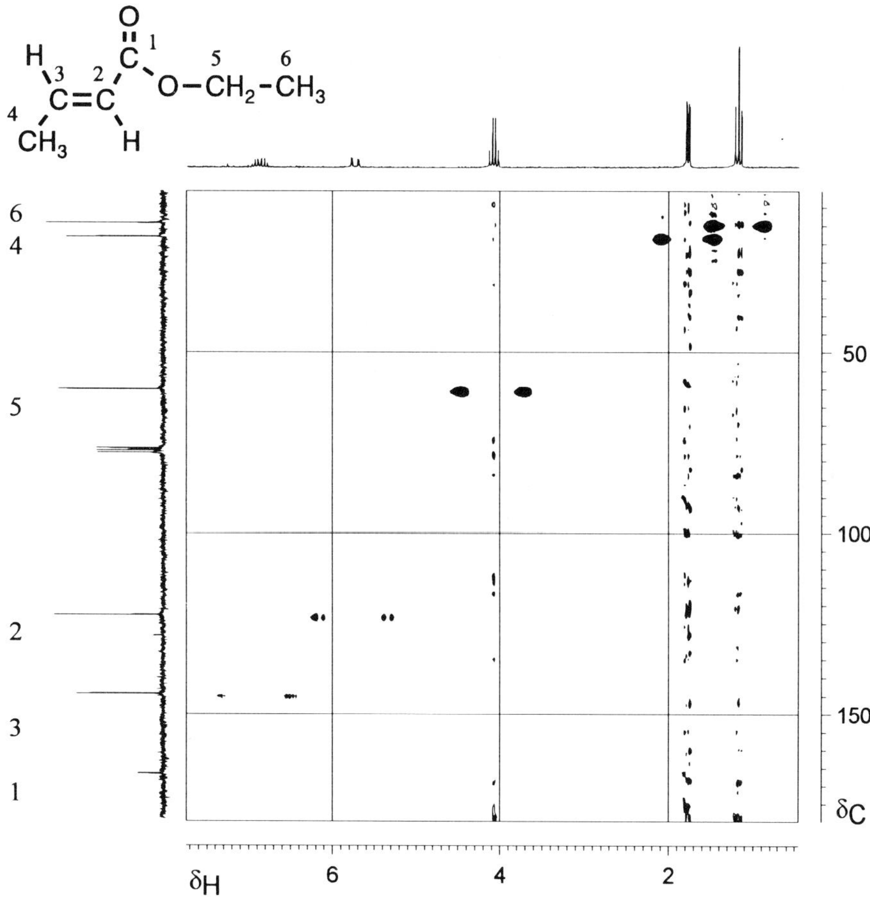

7. Comments

The sequence consists of an INEPT transfer from proton to ^{13}C (see Exp. 6.5), a t_1 period with a 180° pulse on the protons, and a reverse INEPT transfer (see Exp. 6.8) from ^{13}C to proton. As described in Experiment 6.5, the first INEPT transfer results in antiphase magnitization of ^{13}C with respect to proton; thus $-2I_{Hz}I_{Cy}$ single quantum coherence is present. During t_1 this term develops ^{13}C chemical shift as described by Equation (1). The 180° proton pulse p4 (in the middle of t_1) decouples J(C,H) evolution in F_1 and for simplicity is not shown in the equations. The last 90° ^{13}C pulse p9 together with the 90° proton pulse p5 transfers the cosine term back into antiphase of the proton with respect to ^{13}C as in Equation (2). During the subsequent refocussing period, an observable in-phase magnetization develops.

$$-2I_{Hz}I_{Cy} \xrightarrow{\Omega_C t_1 I_{Cz}} -2I_{Hz}I_{Cy}\cos\Omega_C t_1 + 2I_{Hz}I_{Cx}\sin\Omega_C t_1 \quad (1)$$

$$\xrightarrow{I_{Cx}} \xrightarrow{I_{Hy}} -2I_{Hx}I_{Cz}\cos\Omega_C t_1 + 2I_{Hx}I_{Cx}\sin\Omega_C t_1 \quad (2)$$

With regard to sensitivity, the HMQC and HSQC sequences should be identical, since both start with a proton coherence which is transferred to carbon and back to proton. However, HSQC uses 10 r.f. pulses and is therefore more sensitive to experimental error. Nevertheless, in protein research this sequence is of primary importance for N,H correlation, since the correlation signals are not broadened in F_1 by homonuclear coupling.

8. Own Observations

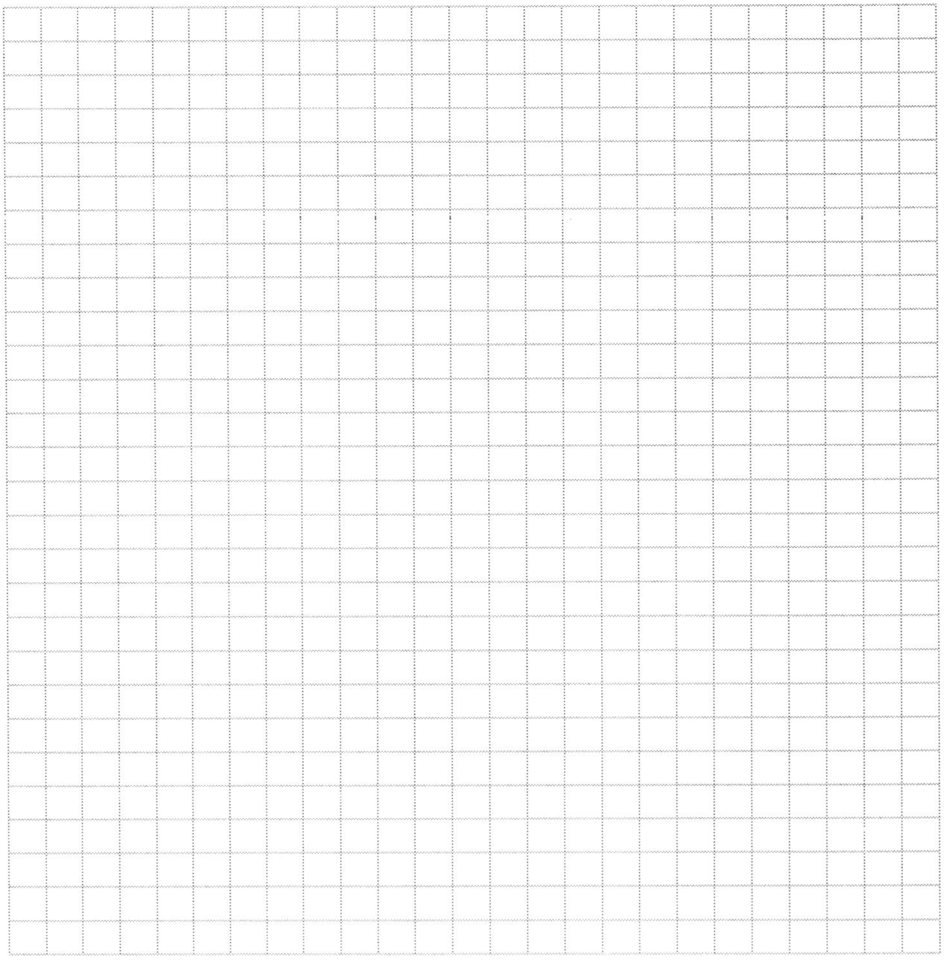

Experiment 10.18

The HOHAHA or TOCSY Experiment

1. Purpose

Homonuclear correlation as described for the COSY technique (see Exps. 10.3–10.8) generally correlates protons via a geminal or vicinal spin coupling. Relayed methods have been proposed to step further along a scalar-coupled spin system. The HOHAHA (**HO**monuclear **HA**rtmann **Ha**hn) or TOCSY (**TO**tal **C**orrelation **S**pectroscop**Y**) method [1,2], can in principle give a total correlation of all protons of a chain with each other. The technique is therefore mostly used for peptides or oligosaccharides, since here it serves for the identification of single residues. In this experiment the basic phase-sensitive method using an MLEV-17 spin-lock is described; many variations, including a gradient-selected method (see Exp. 12.7) and selective methods (see Exps. 7.8 and 11.9), are known.

2. Literature

[1] L. Braunschweiler, R. R. Ernst, *J. Magn. Reson.* **1983**, *53*, 521–528.
[2] A. Bax, D. G. Davis, *J. Magn. Reson.* **1985**, *65*, 355–360.
[3] G. E. Martin, A. S. Zektzer, *Two-Dimensional NMR Methods for Establishing Molecular Connectivity*, VCH, Weinheim, **1988**, 303–316.

3. Pulse Scheme and Phase Cycle

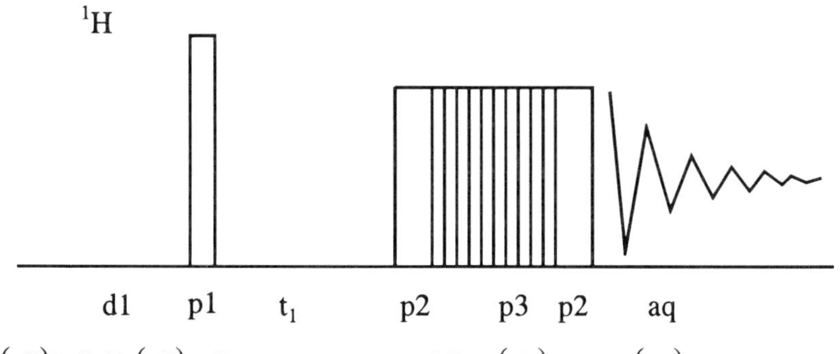

p1: x, $(-x)_2$, x, y, $(-y)_2$, y aq: x, $(-x)_2$, x, y, $(-y)_2$, y
p2: $(x, -x)_2$, $(y, -y)_2$ (trim pulses)
p3: spin-lock out of composite 180° pulses (90°, 180°, 90°); sequence:

90 (ph1), 180 (ph2), 90 (ph1)
$[90\ (ph3), 180\ (ph4), 90\ (ph3)]_2$
90 (ph1), 180 (ph2), 90 (ph1)
$[90\ (ph3), 180\ (ph4), 90\ (ph3)]_2$

ph1: $(-y, y)_2$, $(x, -x)_2$
ph2: $(x, -x)_2$, $(y, -y)_2$
ph3: $(y, -y)_2$, $(-x, x)_2$
ph4: $(-x, x)_2$, $(-y, y)_2$

[90 (ph1), 180 (ph2), 90 (ph1)]$_2$
90 (ph3), 180 (ph4), 90 (ph3)
[90 (ph1), 180 (ph2), 90 (ph1)]$_2$
90 (ph3), 180 (ph4), 90 (ph3)
[90 (ph1), 180 (ph2), 90 (ph1)]$_2$
[90 (ph3), 180 (ph4), 90 (ph3)]$_2$
60 (ph2)

4. Acquisition

Time requirement: 20 min

Sample: 10% strychnine in $CDCl_3$

Prior to the experiment the 90° pulse-width and transmitter attenuation for the spin-lock pulses must be calibrated (see Exp. 2.9). For optimum results one should make allowance for the phase difference between the hard pulse p1 and the spin-lock pulses, either in the pulse program or in the adjustable parameter set if the software allows (see Exp. 7.1). Run a normal ^{1}H NMR spectrum of the sample and optimize the spectral width. The duration of the spin-lock is an adjustable parameter. You have to set:

td2: 1 k data points in F_2
td1: 128 data points in F_1
sw2: 10 ppm
sw1: 10 ppm
o1: middle of ^{1}H NMR spectrum
p1: 90° ^{1}H transmitter pulse
p2: ^{1}H trim pulse, 2.5 ms at transmitter attenuation of spin-lock (12 dB)
p3: series of composite 180° ^{1}H pulses (90°, 180°, 90°) at transmitter attenuation of spin-lock; 90° pulse-width and transmitter attenuation typically in the order of 40 μs and 12 dB, corresponding to an effective spin-lock field of ca. 7000 Hz (magnetic field dependent). Total length of spin-lock set to 200 ms by loop parameter of spin-lock sequence. The loop parameter must be an even number; 76 was used here.
d1: 2 s
initial value for t_1 evolution: 3 μs
increment for t_1 evolution: 1/[2·sw1]
ds: 2
ns: 4

5. Processing

Apply zero filling in F_1 to 512 real data points to obtain a symmetrical matrix of 512·512 real data points. Use exponential windows in both dimensions corresponding to the digital resolution. Apply real Fourier transformation corresponding to the TPPI mode of data acquisition in F_1. Phase correction in both dimensions is necessary.

6. Result

The figure shows the result obtained on an AMX-500 spectrometer. Note that the olefinic proton H-22 displays cross-peaks to many other protons and even to the H-11 pair, although H-12 and H-13 are not reached. The geminal protons H-20 give a weak correlation signal of opposite phase rather than a TOCSY correlation. This is probably a ROESY transfer because of the relatively long spin-lock duration used in this experiment.

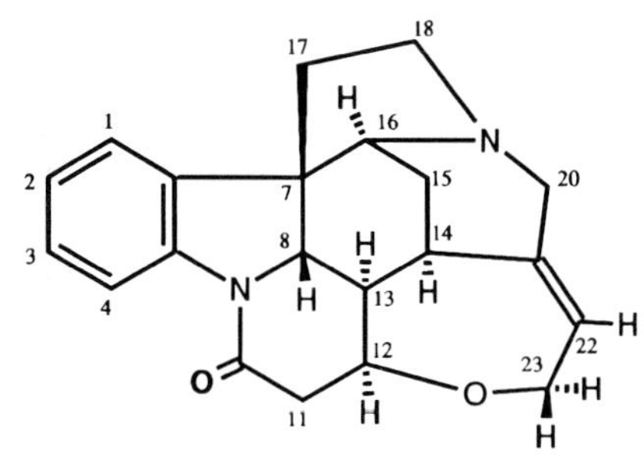

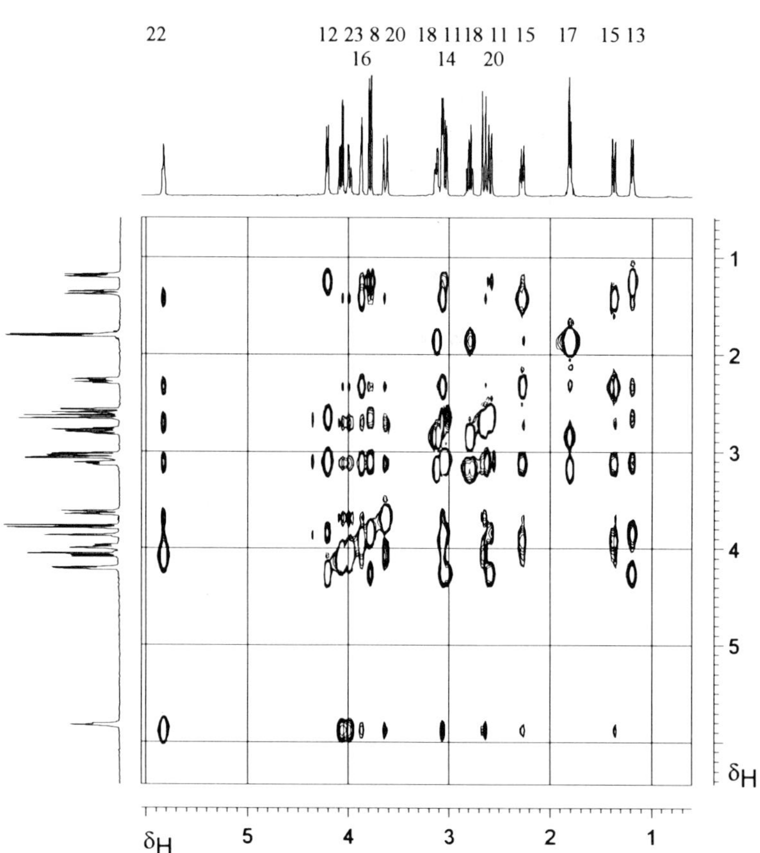

7. Comments

During the spin-lock time the spins "see" only B_1 as the effective field; therefore their chemical shift differences become negligible and the spin systems are all of higher order, leading to cross signals of all protons with each other along a chain of connected XH_n groups. It is possible to adjust the length of the spin-lock for different results. Thus, a rather short spin-lock duration (30 ms) gives roughly the equivalent of a COSY spectrum, intermediate spin-lock times may display results similar to a relayed COSY, and, finally, long spin-lock times result in the desired total correlation. Another adjustable parameter is the individual 90° pulse within the spin-lock defined by the transmitter attenuation which determines the spectral width covered by the spin-lock (see Exp. 2.9).

8. Own Observations

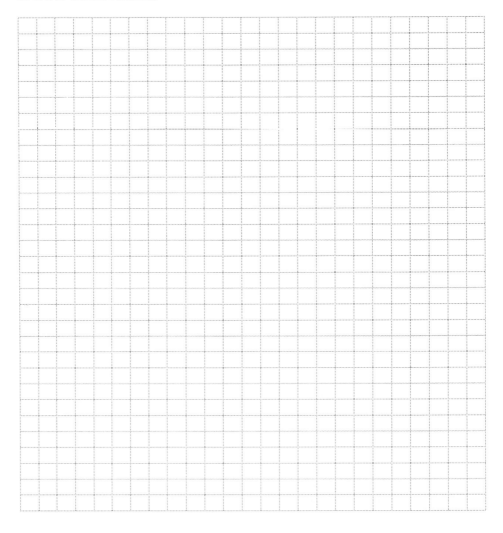

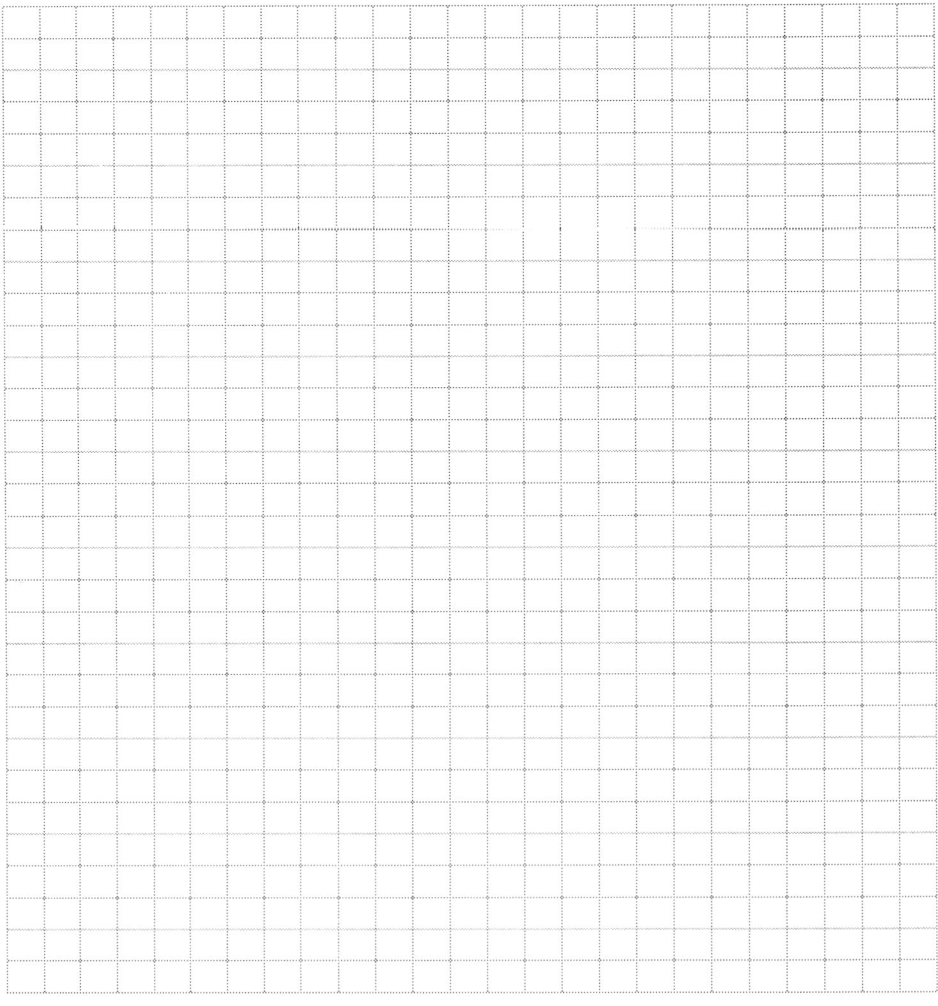

Experiment 10.19

The NOESY Experiment

1. Purpose

The NOESY (**N**uclear **O**verhauser **E**nhancement **S**pectroscop**Y**) experiment is the two-dimensional equivalent of the NOE difference experiment (see Exp. 4.8) and yields correlation signals which are caused by dipolar cross-relaxation between nuclei in a close spatial relationship. The intensities of the cross-peaks are proportional to the sixth power of the proton–proton distances. Quantitatively, however, the results differ from NOE difference spectroscopy, since the latter is a steady-state experiment, whereas NOESY is a transient experiment. In a qualitative way, the NOESY technique gives answers to many stereochemical problems such as *exo*/*endo*, *E*/*Z* and similar assignment questions. In NMR studies of peptides and proteins NOESY is the essential method for determining peptide conformations or tertiary structure of proteins. Here we describe the standard phase-sensitive 2D method, a gradient version is described in Exp. 12.11.

2. Literature

[1] J. Jeener, B. H. Meier, P. Bachmann, R. R. Ernst, *J. Chem. Phys.* **1979**, *71*, 4546–4563.

[2] D. J. States, R. A. Haberkorn, D. J. Ruben, *J. Magn. Reson.* **1982**, *48*, 286–292.

[3] G. Bodenhausen, H. Kogler, R. R. Ernst, *J. Magn. Reson.* **1984**, *58*, 370–388.

[4] D. Neuhaus, M. Williamson, *The Nuclear Overhauser Effect in Structural and Conformational Analysis*, VCH, Weinheim, **1989**, 253–305.

3. Pulse Scheme and Phase Cycle

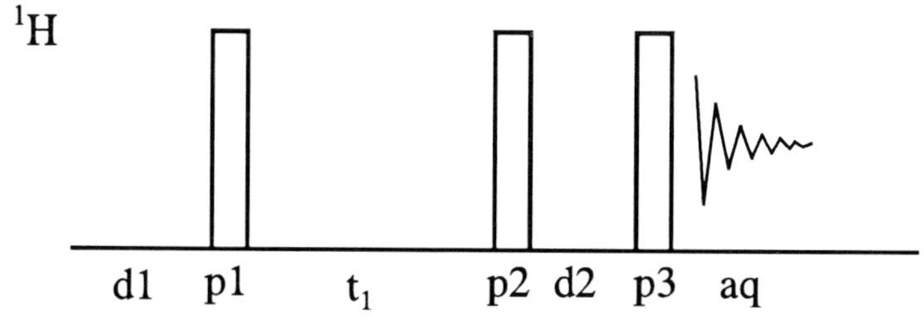

^{1}H

d1 p1 t_1 p2 d2 p3 aq

p1: x, -x p2: $(x)_8$, $(-x)_8$ p3: $(x)_2$, $(-x)_2$, $(y)_2$, $(-y)_2$

aq: x, $(-x)_2$, x, y, $(-y)_2$, y, -x, $(x)_2$, -x, -y, $(y)_2$, -y

phase cycle for p1 incremented according to TPPI

4. Acquisition

Time requirement: 4.5 h

Sample: 10% strychnine in $CDCl_3$; for the best results the sample should be degassed.

Run a normal ^{1}H NMR spectrum of the sample and optimize the spectral width. Change to the 2D mode of the spectrometer software and load the NOESY pulse program. The length of the mixing time d2 is an adjustable parameter. For small organic molecules a trial value of 1–2 s is reasonable. You have to set:

td2: 1 k data points in F_2
td1: 256 data points in F_1
sw2: 10 ppm
sw1: 10 ppm
o1: middle of ^{1}H NMR spectrum
d1: 2 s
d2: 2 s
p1, p2, p3: 90° ^{1}H transmitter pulse
initial value for t_1 evolution: 3 μs
increment for t_1 evolution: 1/[2·sw1]
ds: 2
ns: 16

5. Processing

Apply zero filling in F_1 to 512 real data points to obtain a symmetrical matrix of 512·512 real data points. Use exponential windows in F_2 and F_1. Apply real Fourier transformation corresponding to the TPPI mode of data acquisition in F_1. Phase correction is usually only necessary in F_2. Adjust the phase of the diagonal signals so that they are negative. The NOESY correlation signals will then be positive if the compound has a molar mass below 1000. Correlation signals caused by chemical exchange will have the same phase as the diagonal signals.

6. Result

The figure on page 407 shows the result obtained on an AMX-500 spectrometer. Note that the phase of the diagonal signals is opposite to that of the cross-peaks as can be seen from the dotted contours. There is a wealth of information to be taken from the spectrum which can best be studied using a molecular model. Notice for instance, that only one of the H-20 protons has an NOE contact with one of the H-15 protons, from which a relative assignment of the protons in these methylene groups can be derived.

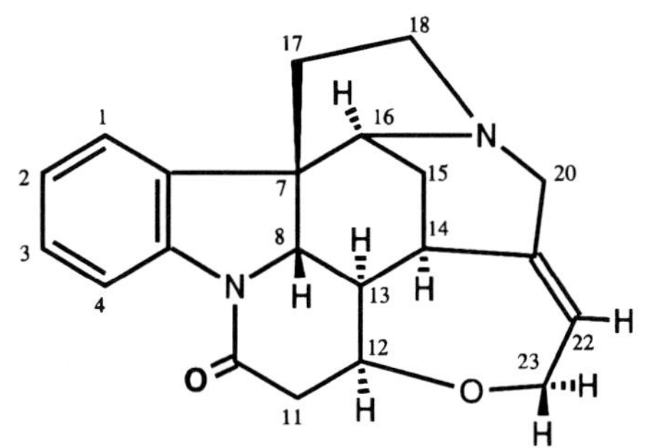
17
18
1
H
16
N
2
7
15
20
8
H
14
3
N
H
H
13
H
4
H
22
12
23
O
O
H
11
H
H

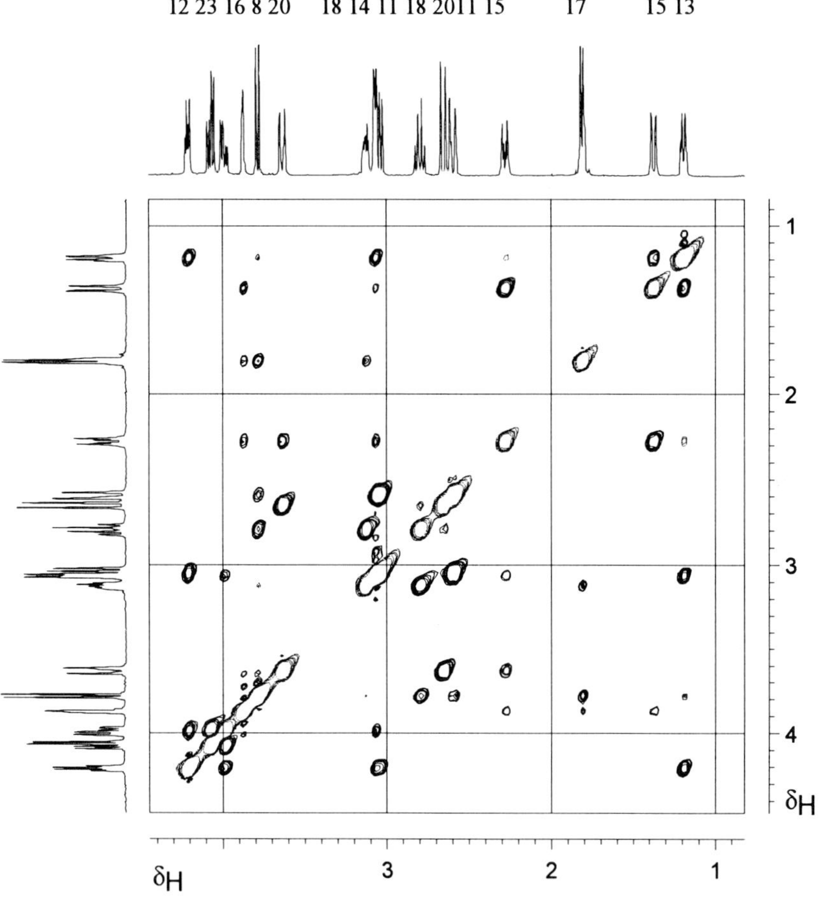
12 23 16 8 20
18 14 11 18 2011 15
17
15 13
δH
3
2
1
δH

7. Comments

The NOESY sequence can be understood from the vector model. We consider two protons with different chemical shifts and no spin–spin coupling. The first pulse of the NOESY sequence aligns all proton magnetization into the *x,y*-plane. After this, chemical shift evolution begins during t_1. The second pulse aligns the two vectors, which are by now labeled with their individual chemical shifts into the negative *z*-direction. During the mixing time d2 both protons are allowed to relax and show cross-relaxation. The final pulse reads the situation at the end of the mixing time and realigns the vectors into the *x,y* plane, where the FID is recorded.

A considerable drawback of the NOESY technique is the dependence of the NOE effect on molar mass and viscosity, which can change its sign and may cause it to disappear for certain conditions. The ROESY technique as described in Experiment 10.20 may be more effective in this case.

8. Own Observations

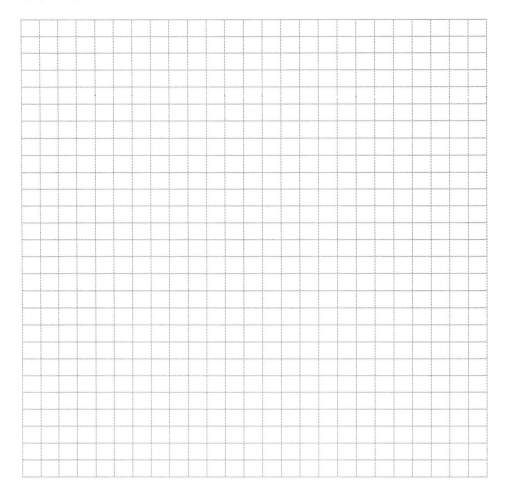

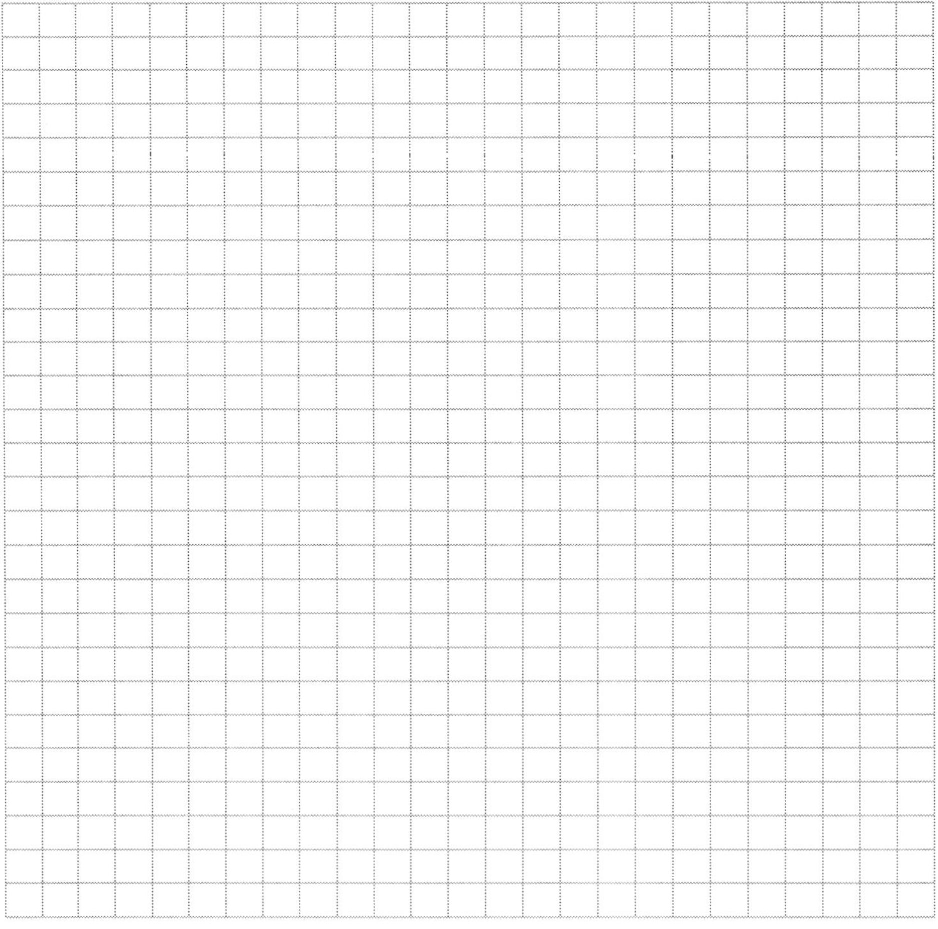

Experiment 10.20

The CAMELSPIN or ROESY Experiment

1. Purpose

The NOESY technique (see Exp. 10.19) has the disadvantage that for molecules with a molar mass in the order of 1000 to 3000 the cross-signals may disappear, since the NOE effect changes its sign depending on the molecular correlation time. However, the nuclear Overhauser effect in the rotating frame under spin-lock conditions is always positive [1,2]. One disadvantage of the ROESY (**R**otating frame **O**verhauser **E**nhancement **S**pectroscop**Y**) experiment is that TOCSY correlations may also break through. This problem has been greatly diminished with a special spin-lock [3,4] which is used here. The experiment described gives identical results to the NOESY technique, but in a shorter time, due to the shorter mixing period during which the spin-lock is used.

2. Literature

[1] A. A. Bothner-By, R. L. Stephens, J.-M. Lee, C. D. Warren, R. W. Jeanloz, *J. Am. Chem. Soc.* **1984**, *106*, 811–813.

[2] A. Bax, D. G. Davis, *J. Magn. Reson.* **1985**, *63*, 207–213.

[3] T.-L. Hwang, A. J. Shaka, *J. Am. Chem. Soc.* **1992**, *114*, 3157–3159.

[4] T.-L. Hwang, M. Kadkhodaei, A. Mohebbi, A. J. Shaka, *Magn. Reson. Chem.* **1992**, *30*, S24–S34.

[5] D. Neuhaus, M. Williamson, *The Nuclear Overhauser Effect in Structural and Conformational Analysis*, VCH, Weinheim, **1989**, 312–327.

3. Pulse Scheme and Phase Cycle

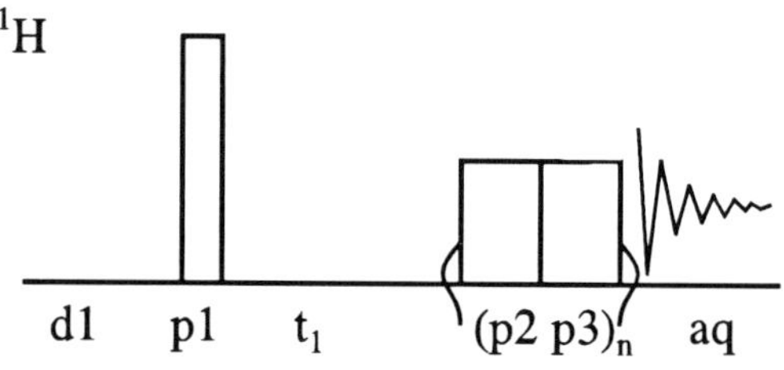

p1: x, $(-x)_2$, x, y, $(-y)_2$, y   p2: $(-y, y)_2$, $(x, -x)_2$

p3: $(y, -y)_2$, $(-x, x)_2$ aq: x, $(-x)_2$, x, y, $(-y)_2$, y

phase cycle for p1 incremented according to TPPI

4. Acquisition

Time requirement: 2.3 h

Sample: 10% strychnine in $CDCl_3$

Prior to the experiment the 90° pulse-width and transmitter attenuation for the spin-lock pulses must be calibrated (see exp. 2.9). For optimum results one should make allowance for the phase difference between the hard pulse p1 and the spin-lock pulses either in the pulse program or in the adjustable parameter set if the software allows (see Exp. 7.1). Run a normal ^1H NMR of the sample and optimize the spectral width. Change to the 2D mode of the spectrometer software and load the ROESY pulse program. The duration of the spin-lock is an adjustable parameter. You have to set:

td2: 1 k data points in F_2
td1: 256 data points in F_1
sw2: 10 ppm
sw1: 10 ppm
o1: middle of ^1H NMR spectrum
p1: 90° ^1H transmitter pulse
p2, p3: series of 180° pulses at transmitter attenuation of spin-lock; 90° pulse width and transmitter attenuation typically in the order of 90 μs and 23 dB. Total duration of spin-lock set to 300 ms by loop parameter *n* of spin lock sequence. The loop parameter must be an even number, 832 was used here.
d1: 2 s
initial value for t_1 evolution: 3 μs
increment for t_1 evolution: 1/[2·sw1]
ds: 2
ns: 16

5. Processing

Apply zero filling in F_1 to 512 real data points to obtain a symmetrical matrix of 512·512 real data points. Use a exponential window in F_2 and a squared $\pi/2$ shifted sinusoidal window in F_1. Apply real Fourier transformation corresponding to the TPPI mode of data acquisition in F_1. Phase correction in both dimensions is necessary. Adjust the phase of the diagonal signals negative so that the ROESY correlation signals are positive. TOCSY breakthrough signals would have the same phase as the diagonal peaks.

6. Result

The figure on page 411 shows the result obtained on an AMX-500 spectrometer. Note that the diagonal signals have a negative phase (dotted contours). No TOCSY breakthrough signals are observed. There is a wealth of information to be taken from the spectrum which can best be studied with a molecular model to hands. Note, for instance, that only one of the H-20 protons is connected with one of the H-15 protons, from which a relative assignment in these methylene groups can be derived.

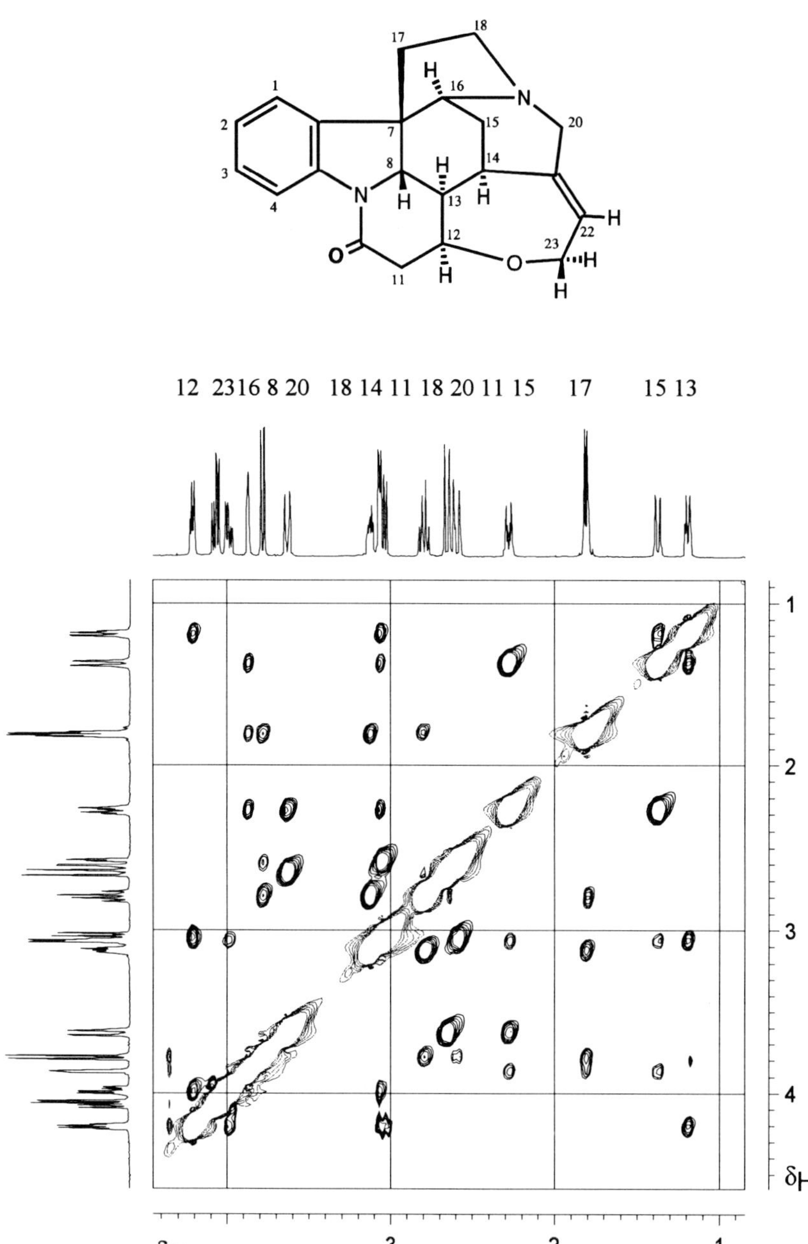

7. Comments

The ROESY sequence is, in principle, identical to the TOCSY sequence as described in Experiment 10.18. After the first pulse and chemical shift evolution during the t_1 period the spins are locked by the spin-lock field B_1, which is considerably weaker in ROESY than in TOCSY. The explanation of the suppression of TOCSY correlations due to the special spin-lock sequence used here is given in the literature [3,4]. In spectra with sharp signals near the offset (e.g. methoxy groups) artefacts have been observed.

8. Own Observations

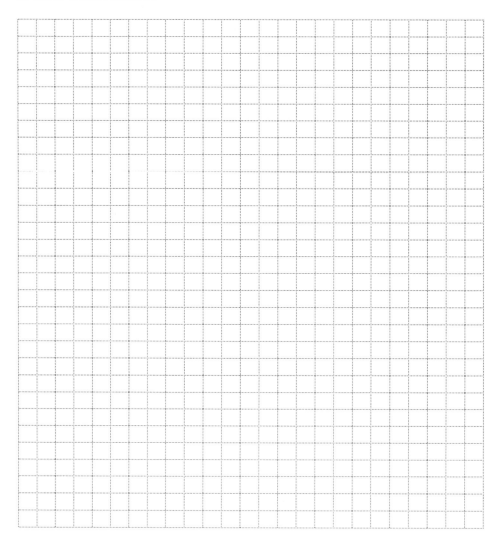

Experiment 10.21

The HOESY Experiment

1. Purpose

The HOESY (**H**eteronuclear **O**verhauser **E**ffect **S**pectroscop**Y**) experiment is the 2D equivalent of Experiment 4.15 and also has many similarities with the NOESY experiment (see Exp. 10.19), yielding information on the spatial relationship between spins in the heteronuclear case. It is therefore used to determine distances between quaternary carbon atoms and protons, especially for cases in which information from spin–spin couplings is unhelpful or unavailable. Although the experiment has been introduced for the C,H and P,H spin pairs, its predominant current application seems to be in the field of organolithium chemistry [5]. The example shown here is therefore taken from this field; the phase-sensitive version is presented. The recent gradient selected version is described in Experiment 12.13.

2. Literature

[1] P. L. Rinaldi, *J. Am. Chem. Soc.* **1983**, *105*, 5167–5168.
[2] C. Yu, G. C. Levy, *J. Am. Chem. Soc.* **1983**, *105*, 6994–6996.
[3] C. Yu, G. C. Levy, *J. Am. Chem. Soc.* **1984**, *106*, 6533–6537.
[4] K. E. Köver, G. Batta, *Prog. NMR Spectrosc.* **1987**, *19*, 223–266.
[5] W. Bauer, P. v. R. Schleyer, *Adv. Carbanion Chem.* **1992**, *1*, 89–175.
[6] W. E. Hull, in: W. R. Croasmun, R. M. K. Carlson (eds.), *Two-Dimensional NMR Spectroscopy*, VCH, Weinheim, **1994**, 406–409.

3. Pulse Scheme and Phase Cycle

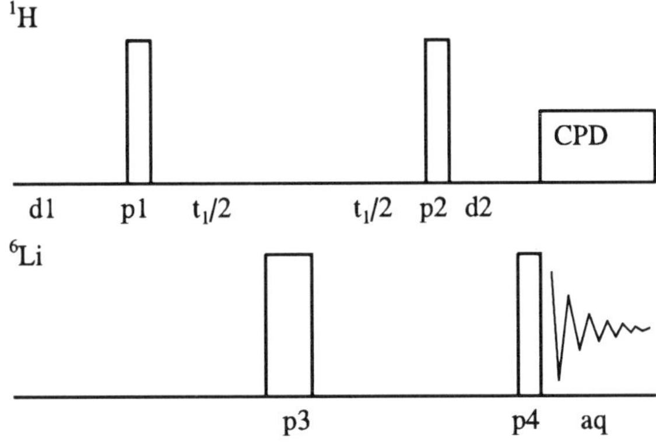

p1: $(x)_2$, $(y)_2$, $(-x)_2$, $(-y)_2$
p2, p4: x, -x, y, -y, -x, x, -y, y
p3: $(x)_8$, $(y)_8$, $(-x)_8$, $(-y)_8$
aq: $(x)_2$, $(y)_2$, $(-x)_2$, $(-y)_2$
phase cycle for p1
incremented according to TPPI

4. Acquisition

Time requirement: 9 h

Sample: commercial 1.4 M n-butyllithium in hexane; add 10% *dry* [D_8]THF for locking purposes. Seal the sample with parafilm. The measurement can be done at room temperature.

Since the ^{2}H and ^{6}Li NMR frequencies are very close, remove any lock stop filter from the transmitter channel at the preamplifier and tune the probe-head for ^{6}Li. Record normal ^{1}H and ^{6}Li NMR spectra, change to the 2D mode of the spectrometer, and load the HOESY pulse program. You have to set:

td2: 512 data points in F_2
td1: 128 data points in F_1
sw2: 4 ppm
sw1: 9 ppm
o1: middle of ^{6}Li NMR spectrum
o2: middle of ^{1}H NMR spectrum
p1, p2: 90° ^{1}H decoupler pulse
p3: 180° ^{6}Li transmitter pulse
p4: 90° ^{6}Li transmitter pulse
d1: 6 s
d2: 1.7 s mixing time
initial value for t_1 evolution: 3 μs
increment for t_1 evolution = 1/[4·sw1]
decoupler attenuation and 90° pulse for CPD
ds: 4
ns: 32

5. Processing

Apply zero filling in F_1 to 256 real data points. Use exponential windows in both dimensions. Apply real Fourier transformation corresponding to the TPPI mode of data acquisition in F_1. Phase correction is usually only necessary in F_2.

6. Result

The figure on page 415 shows the result obtained on an AMX-500 spectrometer with a 5 mm multinuclear probe-head. The cross peaks with the α- and β-protons of butyllithium are clearly visible. The resonance of the β-protons is hidden under one of the resonances of hexane.

7. Comments

The HOESY sequence can be understood from the simple vector analysis. The first pulse creates proton magnetization $-I_{Hy}$. This develops proton chemical shift during

t_1. The 180° Li pulse removes any Li,H spin coupling during t_1 and creates $-I_{Liz}$ magnetization. After t_1 the proton pulse p2 changes the proton transverse magnetization into the negative *z*-direction. Now both spins are in the $-z$-direction, and the proton signal is modulated with its chemical shift information. The spins undergo cross-relaxation during the mixing time d2. The final read pulse p4 creates transverse magnetization of lithium which is detected during t_2. Note that we have chosen a rather long d1 value due to the slow spin–lattice relaxation of ^6Li.

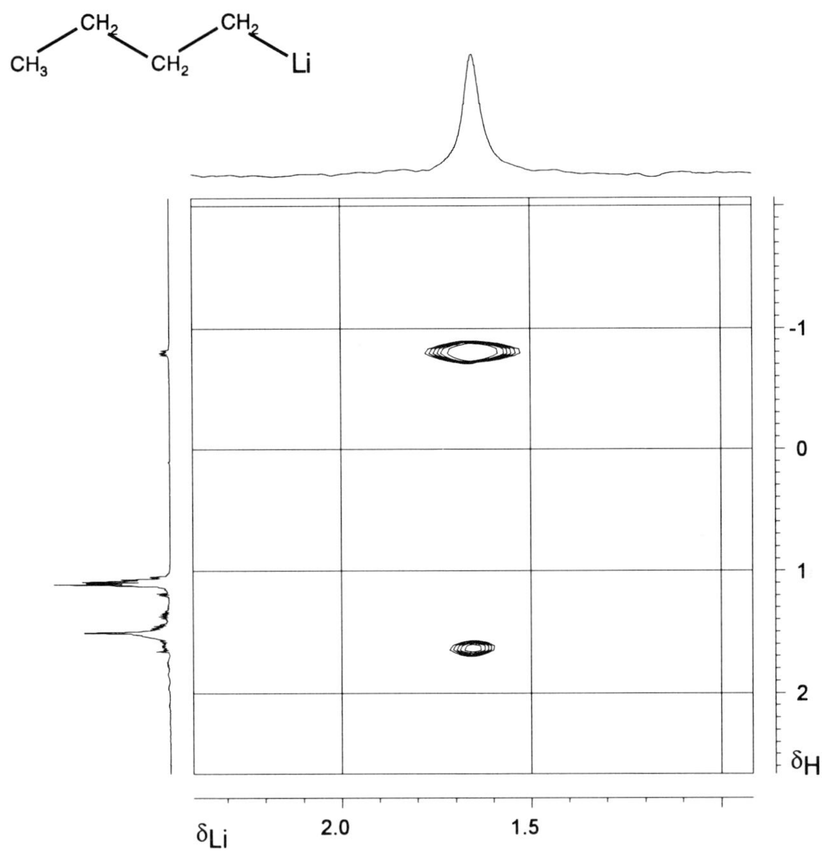

8. Own Observations

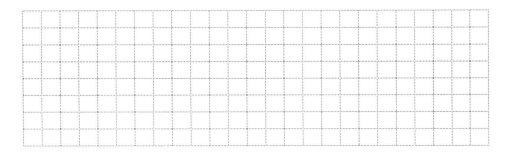

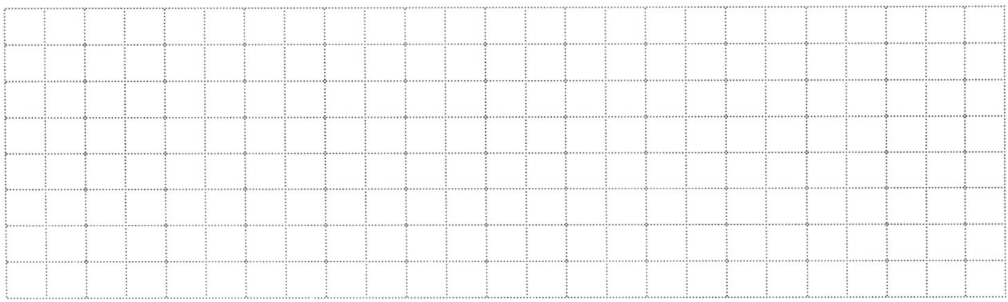

Experiment 10.22

2D-INADEQUATE

1. Purpose

The 2D-INADEQUATE experiment (**I**ncredible **N**atural **A**bundance **D**oubl**E** **QUA**ntum **T**ransfer **E**xperiment) is the two dimensional equivalent of Experiments 6.12 and 7.7. It provides the ultimate form of structure elucidation of organic compounds in solution, since all C,C connectivities can be obtained unequivocally. The main drawback of the method is its very poor sensitivity; many attempts have therefore been made to improve its performance. Nevertheless, there is a rule of thumb by which you can judge whether a 2D-INADEQUATE experiment will be successful. Record a normal ^{13}C NMR spectrum with one transient using a 90° pulse. If the signal-to-noise ratio is better than 30:1 you might invest the time for the experiment. From the many versions known [3] we show here the method with a 90° transfer pulse using 45° steps of phase cycling. Very recently, a gradient selected ^{1}H detected version was introduced, this is shown in Experiment 12.10.

2. Literature

[1] A. Bax, R. Freeman, T. A. Frenkiel, *J. Am. Chem. Soc.* **1981**, *103*, 2102–2104.
[2] J. Buddrus, H. Bauer, *Angew. Chem. Int. Ed. Engl.* **1987**, *26*, 625–643.
[3] W. E. Hull, in: W. R. Croasmun, R. M. K. Carlson (eds.), *Two-Dimensional NMR Spectroscopy*, VCH, Weinheim, **1994**, 353–356.

3. Pulse Scheme and Phase Cycle

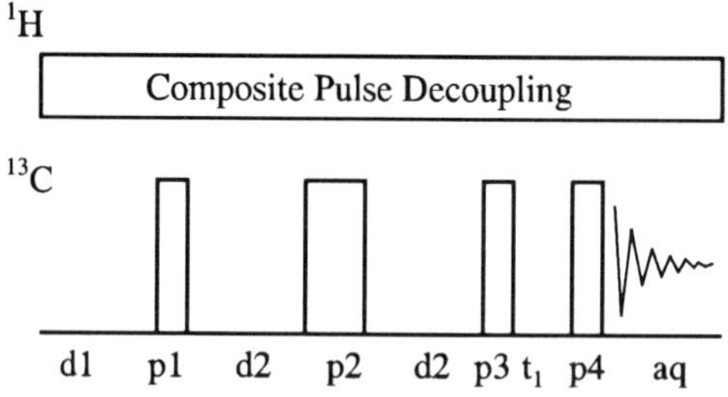

p1, p3: 0, 135, 90, 45, 180, 315, 270, 225, 90, 225, 180, 135, 270, 45, 0, 315, 135, 270, 225, 180, 315, 90, 45, 0, 45, 180, 135, 90, 225, 0, 315, 270

p2: 0, 135, 90, 45, 180, 315, 270, 225, 90, 225, 180, 135, 270, 45, 0, 315, 135, 270, 225, 180, 315, 90, 45, 0, 45, 180, 135, 90, 225, 0, 315, 270 180, 315, 270, 225, 0, 135, 90, 45, 270, 45, 0, 315, 90, 225, 180, 135, 315, 90, 45, 0, 135, 270, 225, 180, 225, 0, 315, 270, 45, 180, 135, 90

p4: x, y, -x, -y aq: $(x)_8$, $(-x)_8$, $(y)_8$, $(-y)_8$

4. Acquisition

Time requirement: 15 h

Sample: 80% 1-hexanol in [D_6]acetone

Tune the probe-head to the actual sample, record a normal ^{13}C NMR spectrum and optimize the spectral width. Determine the ^{13}C observe pulse-width for this sample. For this experiment the instrument must be set so as to obtain optimum performance. Change to the 2D mode of the spectrometer and load the INADEQUATE pulse program. You have to set:

td2: 1 k data points in F_2
td1: 128 data points in F_1
sw2: 60 ppm
sw1: 120 ppm (double quantum frequency)
o1: middle of ^{13}C NMR spectrum
p1, p3, p4: 90° ^{13}C transmitter pulse
p2: 180° ^{13}C transmitter pulse
d1: 3 s
d2: $1/[4J(C,C)] = 7.6$ ms, calculated from ^1J (C,C) = 33 Hz
decoupler attenuation and 90° pulse for CPD
initial value for t_1 evolution: 3 µs
increment for t_1 evolution = 1/sw1
ds: 4
ns: 128

5. Processing

Apply zero filling in F_1 to 256 real data points to obtain a matrix of 512·256 real data points. Use $\pi/2$ shifted sinusoidal windows in both dimensions. Apply complex Fourier transformation corresponding to the N-type signal selection using the quadrature-off mode in F_1. Phase correction is not necessary, since the data are processed in the magnitude mode.

6. Result

The figure on page 418 shows an expansion of the 2D-INADEQUATE spectrum obtained on an ARX-200 spectrometer. Each pair of connected ^{13}C nuclei forms an AX or AB spin system, which is found in the same row of the data matrix; the pairs of doublets are symmetrical with respect to the diagonal (dotted line), and spin coupling constants can be obtained from such a row. If one carbon is connected to more than one other carbon, the corresponding doublets are found at the same chemical shift in F_2, but at another double quantum frequency in F_1. Thus the molecular carbon skeleton can be obtained by a criss-cross progression through the 2D spectrum.

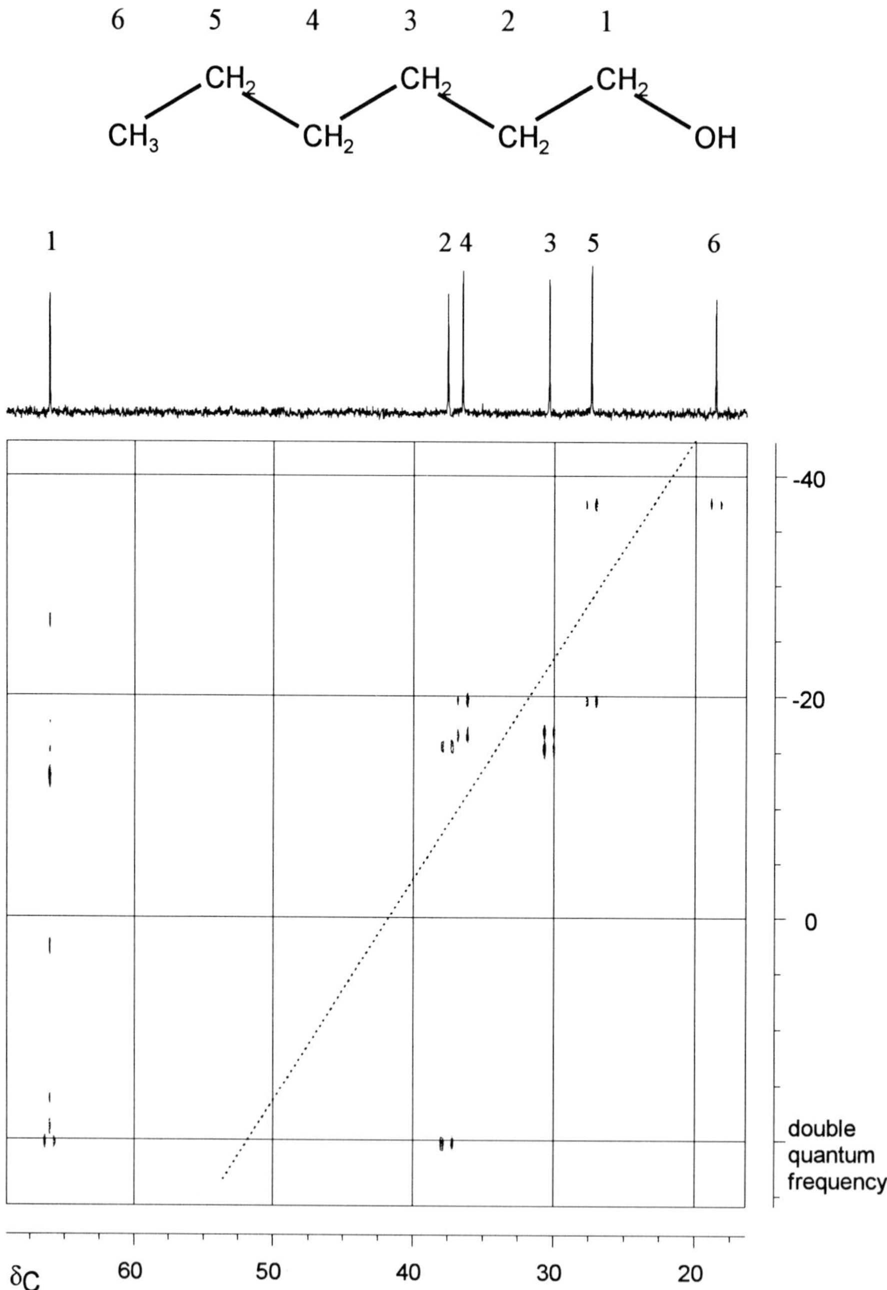
6 5 4 3 2 1
CH2 CH2 CH2
CH3 CH2 CH2 OH
1 2 4 3 5 6
-40
-20
0
double quantum frequency
δC
60
50
40
30
20

7. Comments

The product operator formalism follows the treatment given in Experiment 6.12. After the pulse p3, which creates double quantum magnetization $2I_{1x}I_{2y}$, chemical shift evolution of $2I_{1x}I_{2y}$ during t_1 yields the double quantum frequencies, of which only one typical term is shown in equation (1) for simplicity.

$$2I_{1x}I_{2y} \xrightarrow{\Omega_1 t_1 I_{1z}} \xrightarrow{\Omega_2 t_2 I_{2z}} 2I_{1x}I_{2y}\cos\Omega_1 t_1 \cos\Omega_2 t_1 \qquad (1)$$

The final pulse p4 transforms the double quantum magnetization back into single quantum terms. Evolution of spin coupling between the coupled ^{13}C nuclei creates in-phase magnetization, which is detected during t_2.
On older spectrometers which are not capable of phase cycling in 45° steps, a method equivalent to the experiment shown here uses a 135° pulse for p4 to distinguish between N- and P-type signals.

8. Own Observations

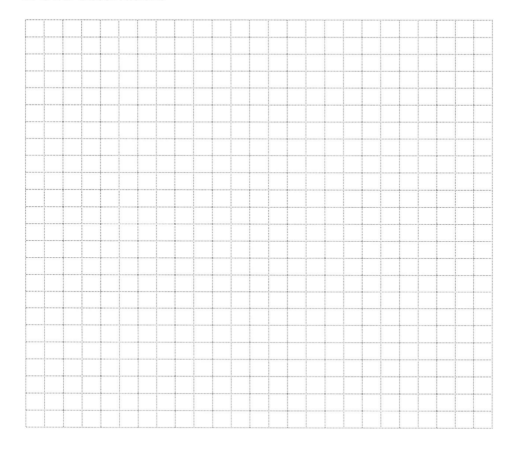

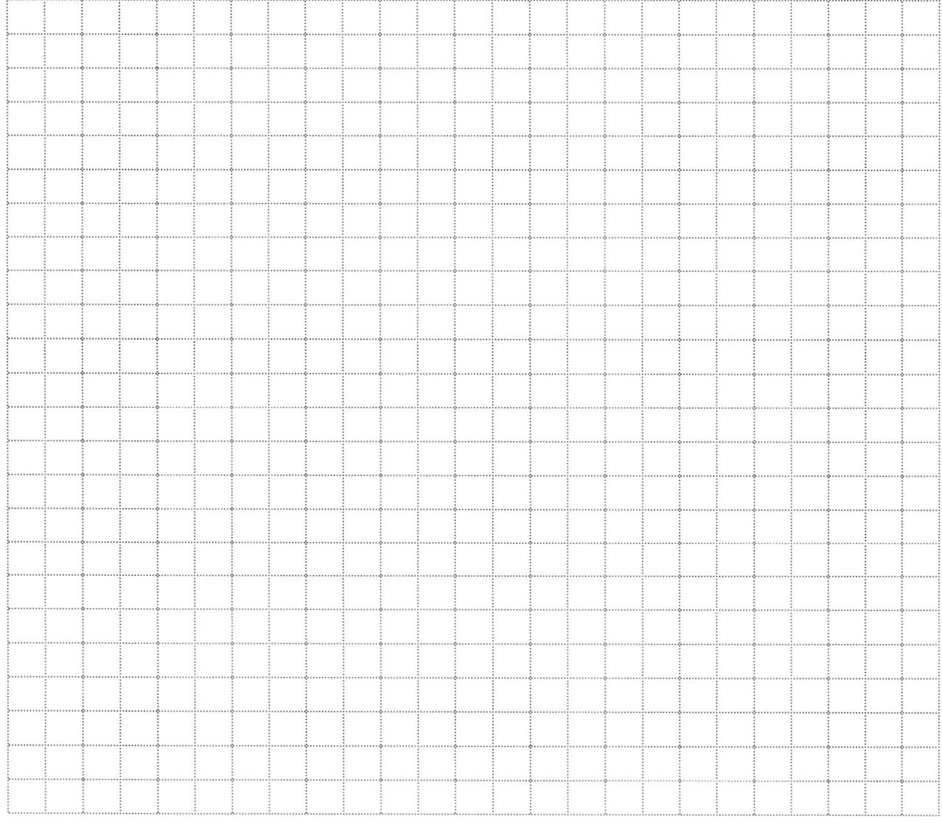

Experiment 10.23

The EXSY Experiment

1. Purpose

For the investigation of dynamic processes 1D spectra are usually recorded at different temperatures (see Exp. 5.3) and the line-broadening and coalescence of signals analyzed. The 2D EXSY (**EX**change **S**pectroscop**Y**) method can indicate chemical exchange before line-broadening occurs. It can therefore be regarded as the 2D equivalent of the saturation transfer experiment (see Exp. 5.4). For multisite exchange it has the important advantage that cross-signals of all exchanging species can be seen and the nature of the exchange process may be clarified by simple inspection. The pulse sequence is exactly the same as that used for phase-sensitive NOESY (Exp. 10.19). Shown here is the EXSY experiment on dimethyl formamide.

2. Literature

[1] J. Jeener, B. H. Meier, P. Bachmann, R. R. Ernst, *J. Chem. Phys.* **1979**, *71*, 4546–4563.
[2] C. L. Perrin, T. J. Dwyer, *Chem. Rev.* **1990**, *90*, 935–967.
[3] S. Macura, W. M. Westler, J. C. Markley, *Methods Enzym.* **1994**, *239*, 106–144.

3. Pulse Scheme and Phase Cycle

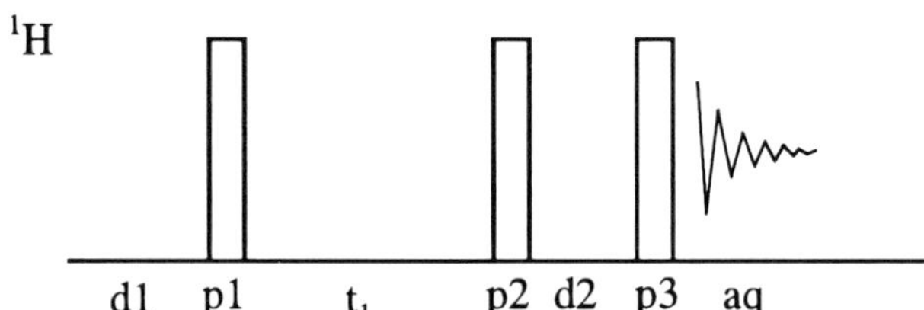

p1: x, -x p2: $(x)_8$, $(-x)_8$ p3: $(x)_2$, $(-x)_2$, $(y)_2$, $(-y)_2$
aq: x, $(-x)_2$, x, y, $(-y)_2$, y, -x, $(x)_2$, -x, -y, $(y)_2$, -y
phase cycle for p1 incremented according to TPPI

4. Acquisition

Time requirement: 0.5 h

Sample: 5% DMF in $C_2D_2Cl_4$

Set up your instrument for high temperature measurement, ensure a reasonable nitrogen flow, set the temperature to 363 K, and let the sample equilibrate for at least 5 minutes. Run a normal ^{1}H NMR spectrum of the sample and optimize the spectral width for the methyl groups only. Change to the 2D mode of the spectrometer and load the NOESY pulse program. The length of the mixing time d2 is an adjustable parameter. You have to set:

td2: 512 data points in F_2
td1: 32 data points in F_1
sw2: 0.7 ppm
sw1: 0.7 ppm
o1: middle of methyl group region
p1, p2, p3: 90° ^{1}H transmitter pulse
d1: 2 s
d2: 1 s
initial increment for t_1 evolution: 3 μs
increment for t_1 evolution: 1/[2·sw1]
ds: 2
ns: 4

5. Processing

Apply zero filling in F_1 to 256 real data points to obtain a symmetrical matrix of 256·256 real data points. Use exponential windows in F_2 and F_1. Apply real Fourier transformation corresponding to the TPPI mode of data acquisition in F_1. Phase correction is usually only necessary in F_2. The cross-signals caused by chemical exchange, unlike the NOESY signals, have the same phase as the diagonal signals.

6. Result

The figure on page 422 shows the result obtained on an AM-400 spectrometer. Note that the cross-signals are very strong, displaying an intensity never reached by NOE signals. In the high-resolution 1D NMR spectrum a dynamic line broadening is not yet present. Note that the coalesence temperature is field-dependent.

7. Comments

Although the occurrence of dynamic processes is easily demonstrated qualitatively by this experiment, the extraction of rate constants is not straightforward. A whole series of EXSY spectra has to be recorded with different mixing times d2 and the volume integrals must be evaluated. Furthermore, the spin–lattice relaxation times of the exchanging spins must be known. From these data a relaxation matrix can be constructed and, using certain assumptions, the rate constants are calculated. For details see Reference [2]. For a qualitative investigation of a two site exchange, the literature gives an optimum mixing time as in Equation (1), where k_{AB} and k_{BA} are the rate constants of the forward and backward reaction.

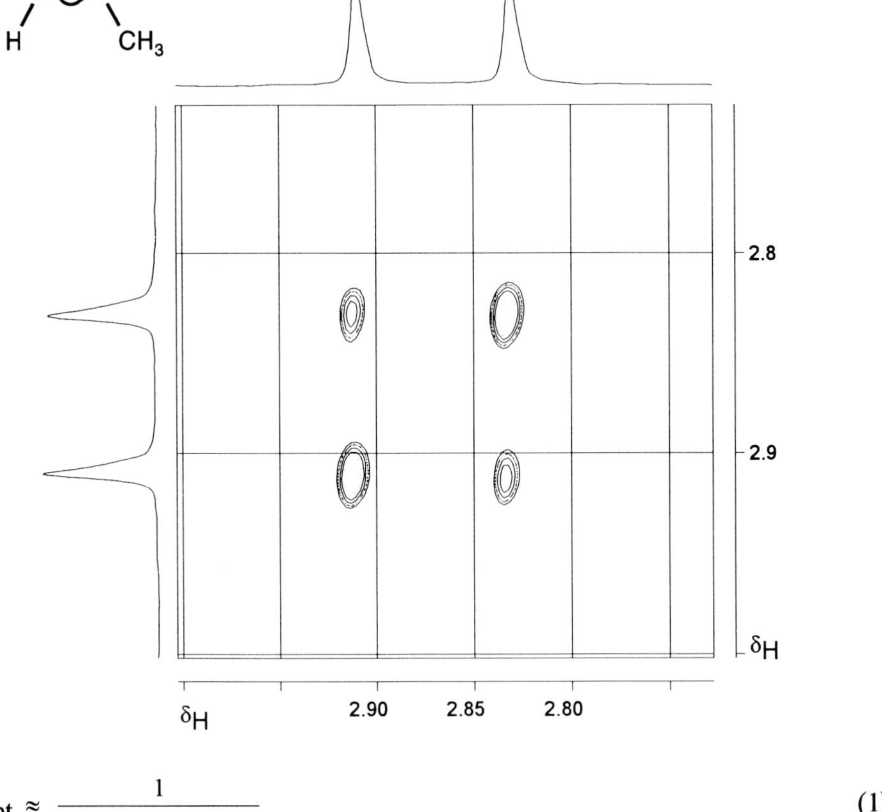

$$\mathrm{d2_{opt}} \approx \frac{1}{T_1^{-1} + k_{AB} + k_{BA}} \quad (1)$$

8. Own Observations

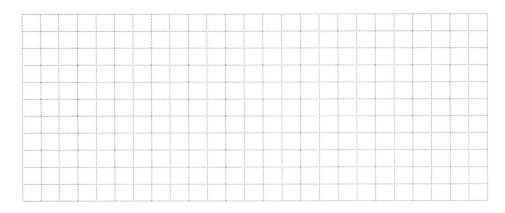

Experiment 10.24

X,Y Correlation

1. Purpose

All 2D correlation experiments described in this book are either H,H or H,X correlations, X mostly being ^{13}C. There is, however, sometimes the need to correlate the signals of hetero-atoms X and Y directly with each other, which is usually performed under complete proton decoupling. For this experiment a triple-resonance probe-head and a three-channel spectrometer are required. Commercially available probe-heads usually have one fixed coil for the nucleus X, e.g. ^{13}C, and one tunable coil for all other frequencies Y. One has to decide which is the detected nucleus and which is the nucleus in the indirect dimension. As an example of the technique, we show in this experiment a ^{13}C,^{31}P correlation on triphenylphosphane as an educational example. Both ^{31}P and ^{13}C detection are described using two different correlation techniques.

2. Literature

[1] L. D. Sims, L. R. Soltero, G. E. Martin, *Magn. Reson. Chem.* **1989**, *27*, 599–602.
[2] P. Bast, S. Berger, H. Günther, *Magn. Reson. Chem.* **1992**, *30*, 587–594.
[3] T. Fäcke, R. Wagner, S. Berger, *Concepts Magn. Reson.* **1994**, *6*, 293–306.
[4] S. Berger, T. Fäcke, R. Wagner, *Magn. Reson. Chem.* **1996**, *34*, 4–13.

3. Pulse Scheme and Phase Cycle

Experiment **a**

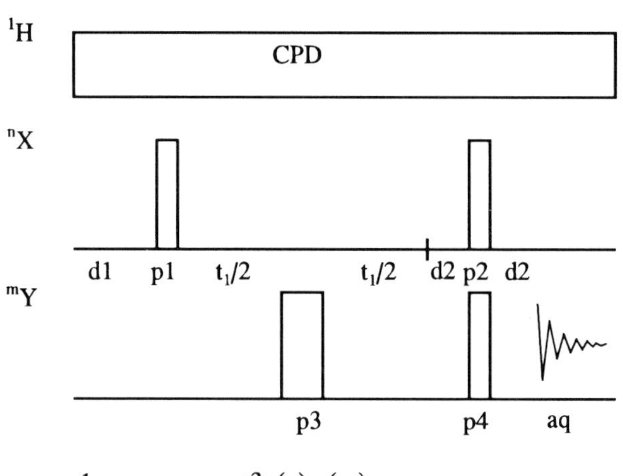

p1: x p3: $(x)_4$, $(-x)_4$

p2: x, -x, y, -y p4: $(x)_8$, $(y)_8$, $(-x)_8$, $(-y)_8$

aq: $(x, -x, y, -y)_2$, $(y, -y, -x, x)_2$, $(-x, x, -y, y)_2$, $(-y, y, x, -x)_2$

Experiment **b**

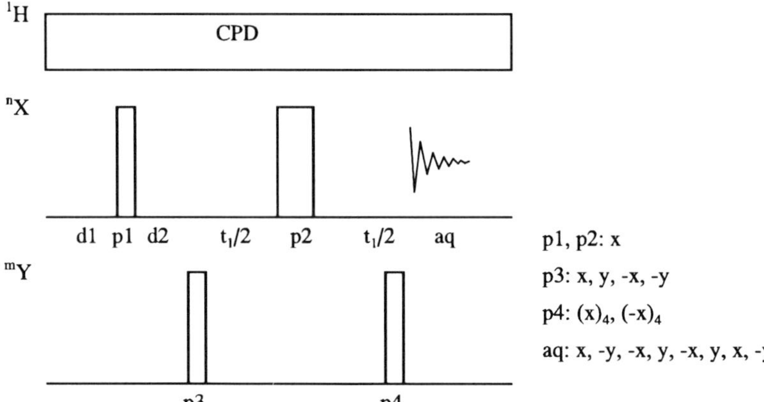

4. Acquisition

Time requirement: 2 x 20 min

Sample: 10% triphenylphosphane in $CDCl_3$

Tune first the fixed ^{13}C coil, then the ^{31}P and finally the ^{1}H coil of the probe-head. Install appropriate pass and stop filters for all three channels. The 90° and 180° pulses in this configuration on all coils must be determined. Record both the ^{13}C and ^{31}P NMR spectra of the sample.

Load the HETCOR-type pulse program according to experiment **a**, where the detected nucleus Y is ^{13}C and the nucleus X in the indirect dimension is ^{31}P. You have to set:

td2: 1 k data points in F_2
td1: 64 data points in F_1
sw2: 12 ppm
sw1: 1 ppm
o1: middle of ^{13}C NMR spectrum
o2: middle of ^{1}H NMR spectrum
o3: middle of ^{31}P NMR spectrum
p1, p2: 90° ^{31}P decoupler pulse
p3: 180° ^{13}C transmitter pulse
p4: 90° ^{13}C transmitter pulse
d1: 2 s
d2: $1/[2J(C,P)]$ = 25 ms, calculated from $^{n}J(C,P)$ = 20 Hz
initial value for t_1 evolution: 3 µs
increment for t_1 evolution = 1/[2·sw1]
proton decoupler attenuation and 90° pulse for CPD
ns: 8

For experiment **b** load the HMQC-type pulse program. You may have to switch r.f. cables and filters since now ^{31}P is the detected nucleus X. You have to set:

td2: 256 data points in F_2
td1: 128 data points in F_1
sw2: 1 ppm
sw1: 12 ppm
o1: middle of ^{31}P NMR spectrum
o2: middle of ^{1}H NMR spectrum
o3: middle of ^{13}C NMR spectrum
p1: 90° ^{31}P transmitter pulse
p2: 180° ^{31}P transmitter pulse
p3, p4: 90° ^{13}C decoupler pulse
d1: 2 s
d2: $1/[2J(C,P)]$ = 25 ms, calculated from $^{n}J(C,P)$ = 20 Hz
initial value for t_1 evolution: 3 μs
increment for t_1 evolution = 1/[2·sw1]
proton decoupler attenuation and 90° pulse for CPD
ns: 4

5. Processing

In both cases apply zero filling in F_1 to 256 real data points to obtain a matrix of 512·256 real data points. Use unshifted sinusoidal windows in both dimensions. Apply complex Fourier transformation corresponding to the N-type signal selection using the quadrature-off mode in F_1. Phase correction is not necessary, since the data are processed in magnitude mode.

6. Result

The figure on page 426 shows the results of both experiments obtained on an AMX-500 spectrometer with an inverse triple resonance probe-head. In both 2D spectra an isotope effect for C-1 of triphenylphosphine can be observed. Compare the signal-to-noise ratio of the two methods by inspecting the rows containing the signals.

3 2
4 1 **P** $(C_6H_5)_2$

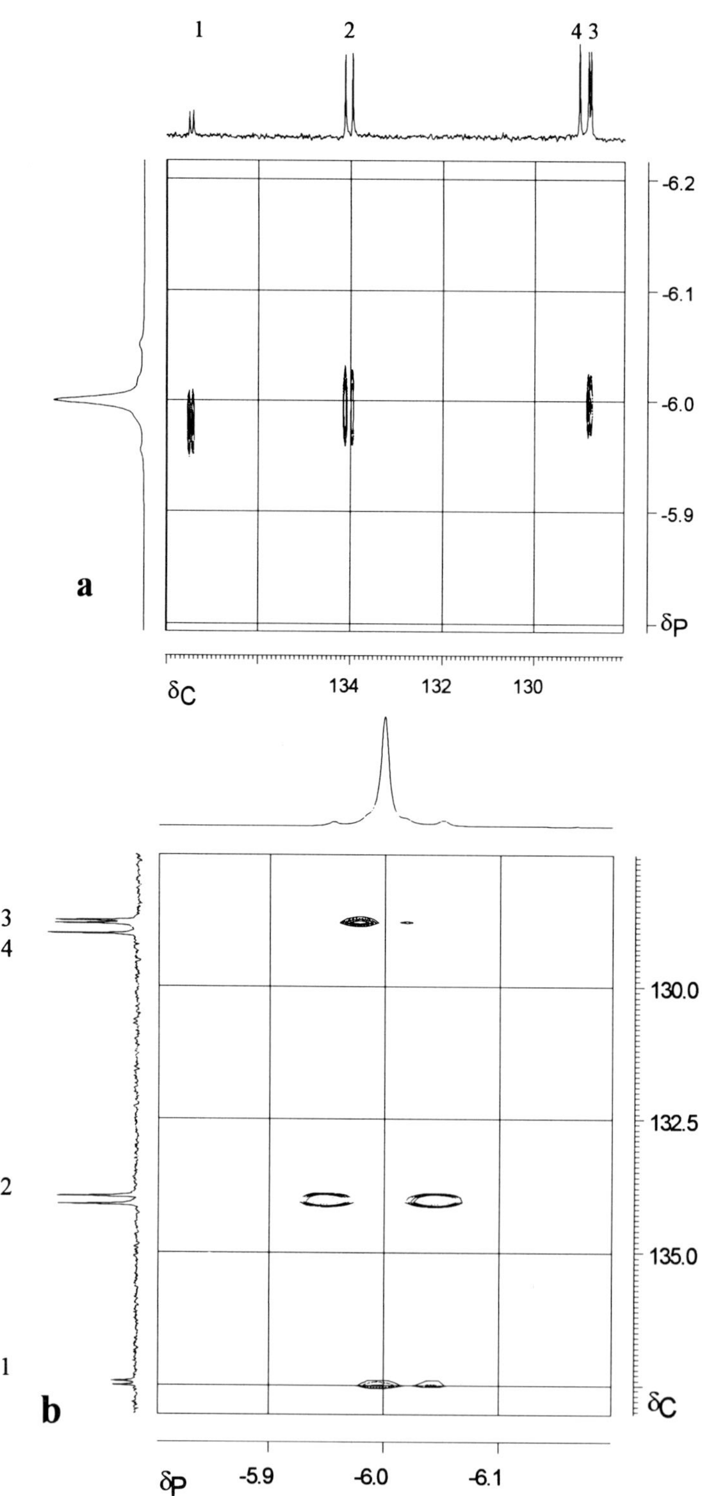

1
2
4 3
-6.2
-6.1
-6.0
-5.9
a
δP
δC
134
132
130
3
4
130.0
132.5
2
135.0
1
b
δC
δP
-5.9
-6.0
-6.1

7. Comments

Note that the delay d2 in these experiments was calculated using a $^{13}C,^{31}P$ spin-coupling constant of 20 Hz, although the actual coupling constants are smaller. The reason is that during the long delay corresponding to small couplings the magnetization is severely diminished because of relaxation.

Contrary to intuitive belief, it is not the natural abundance of a nucleus which decides the choice of the detected nucleus, since in an X,Y correlation experiment the product of both natural abundances is important. In general, one should detect the nucleus with the higher gyromagnetic ratio; however, the situation is rather complex, as discussed at length in the references. Other arguments are the relaxation times of both nuclei, the question of suppression of unwanted signals and the relative spectral width in both dimensions.

For other purposes, such as a $^{15}N,^{31}P$ or a $^{31}P,^{57}Fe$ correlation, a probe-head with a fixed coil tuned to ^{31}P must be available, rendering this approach rather costly if still other pairs of nuclei have to correlated.

8. Own Observations

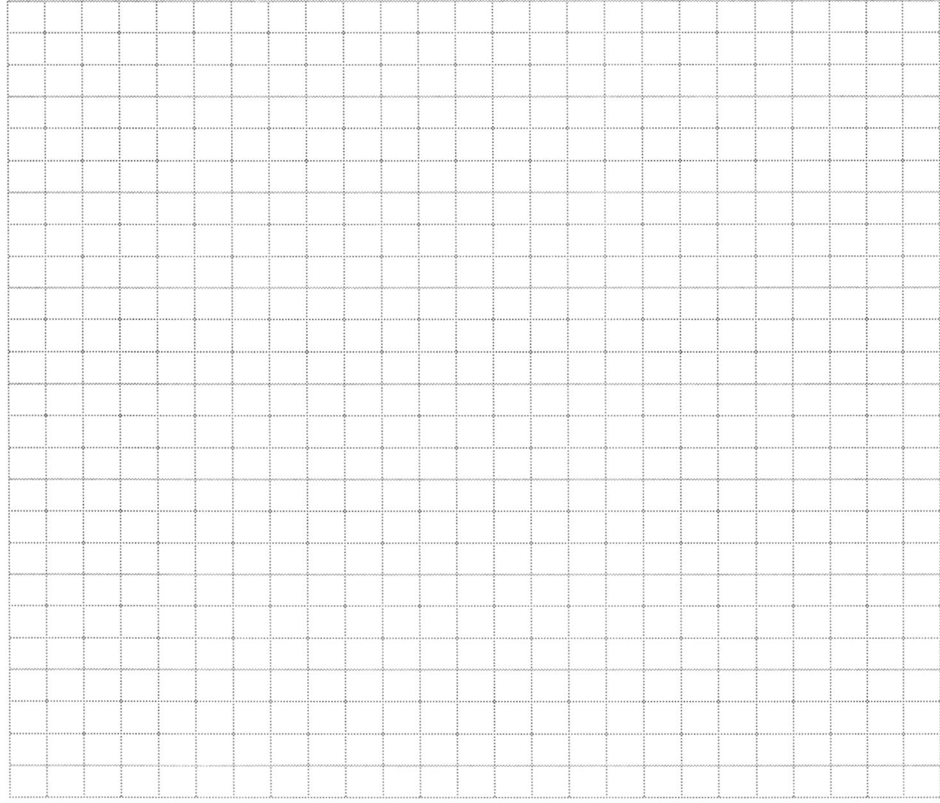

Chapter 11

1D NMR Spectroscopy with Field Gradients

Although pulsed field gradients are used routinely in NMR imaging, in *in-vivo* spectroscopy, and for diffusion measurements, they have only recently been employed extensively in high-resolution NMR spectroscopy. The advantage of experiments with gradient selection is of fundamental importance for homo- and heteronuclear 2D applications. Nearly all 3D experiments currently published use pulsed field gradients. We therefore dedicate two chapters of this book to experiments employing this technology, one for 1D and one for 2D experiments.

Currently these experiments require special hardware for the spectrometer; however, this hardware will soon be or already is regarded as standard equipment. A special probe-head with self-shielded gradient coils and a gradient amplifier are usually required for the experiments described. Since this field is rapidly developing at the time of writing, many experimental details, such as lock and amplifier blanking, gradient ring-down delays and preemphasis, for example, are very dependent on the actual hardware used and must therefore be adapted to the particular instrument of the user. Furthermore, only experiments with *z*-gradients are described, as further developments with three orthogonal gradients are outside the scope of this collection of "basic" experiments.

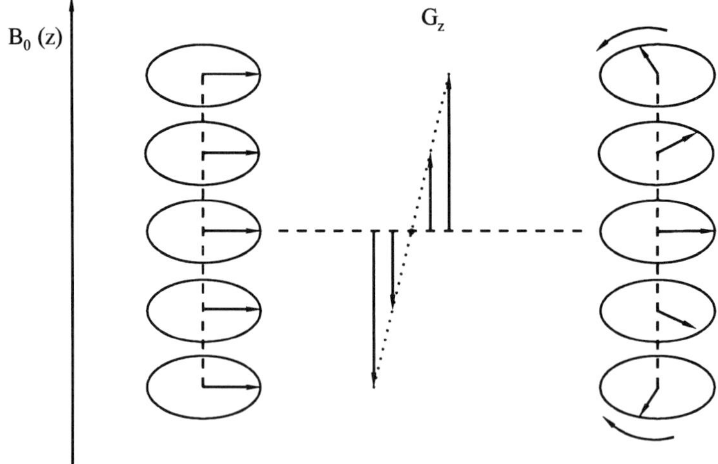

As shown in the figure, a pulsed *z*-field gradient dephases the coherences along the *z*-axis. Gradient-selected experiments rely on the fact that another identical gradient applied at a later stage of the pulse sequence can rephase these coherences if their coherence level was changed, for example by a 180° pulse. Thus one is able to select coherence pathways by combining r.f. pulses and pulsed field gradients in one pulse sequence. To help in understanding the experiments described, the coherence pathway diagram [7] is given below many pulse sequences in this book; the pulsed field gradients are represented with g.

The 1D chapter on pulsed field gradients is meant first to give a working understanding of the equipment in use by providing several calibration experiments (Exps. 11.1–4). Then we show how to measure diffusion constants (Exp. 11.5) and include an educational experiment on a gradient *zz*-filter. Included also are two methods of water suppression, WATERGATE and Excitation Sculpting.

Furthermore, there are several more advanced experiments employing both selective pulses and pulsed field gradients, since this combination seems to be widely used and of practical importance.

Literature

[1] T. J. Norwood, *Chem. Soc. Rev.* **1994**, 59–66.
[2] J. Keeler, R. T. Clowes, A. L. Davis, E. D. Laue, *Methods Enzym.* **1994**, *239*, 145–207.
[3] W. Price, *Annu. Rep. NMR Spectrosc.***1996**, *32*, 55–142.
[4] T. Paralla, *Magn. Reson. Chem.* **1996**, *34*, 329–347.
[5] D. Canet, *Prog. NMR Spectrosc.* **1997**, *30*, 101–135.
[6] S. Berger, *Prog. NMR Spectrosc.* **1997**, *30*, 137–156.
[7] A. D. Bain, *J. Magn. Reson.* **1984**, *56*, 418–427.

Experiment 11.1

Calibration of Pulsed Field Gradients

1. Purpose

For all experiments working with pulsed field gradients the gradient strength has to be known in order to get meaningful results. This experiment describes a calibration routine.

2. Literature

[1] M. Holz, H. Weingärtner, *J. Magn. Reson.* **1991**, *92*, 115–125.
[2] M. I. Hrovat, C. G. Wade, *J. Magn. Reson.* **1981**, *44*, 62–75.
[3] E. Fukushima, S. B. W. Roeder, Experimental Pulse NMR, Addison–Wesley, London, **1981**, 210–215.

3. Pulse Scheme and Phase Cycle

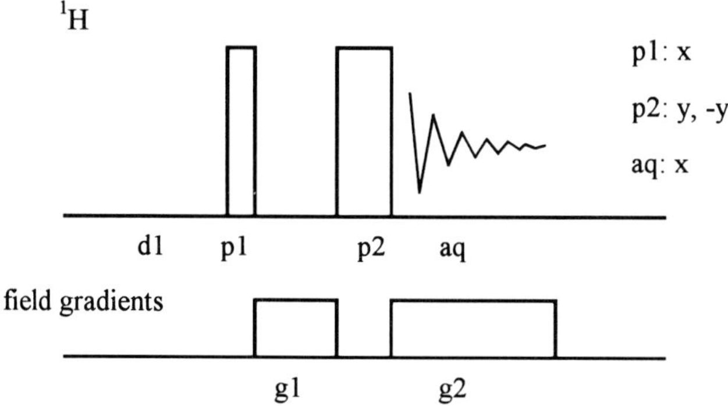

4. Acquisition

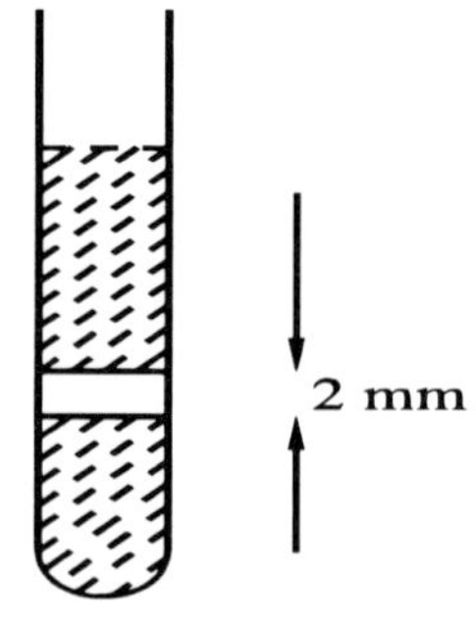

Time requirement: 20 min

Sample: Prepare a special calibration sample as shown. In a 5 mm NMR tube two layers of normal water are separated by a rubber or Teflon disk of 2 mm thickness. The tube should be adjusted in the magnet in such a way that this disk is situated in the center of the r.f. coil. No sample spinning should be applied.

Set the instrument to normal ^{1}H NMR operation and load a pulse program as shown above. You have to set:

td: 2 k
sw: 100 KHz
o1: on resonance of water signal
p1: 90° ^{1}H transmitter pulse
p2: 180° ^{1}H transmitter pulse
d1: 1 s
d2: 10 ms
d3: 8 ms
g1: pulsed field gradient, rectangular shape, duration = 10 ms, strength to be varied
g2: pulsed field gradient, rectangular shape, duration = 20 ms, same strength as g1. Depending on the instrumentation you may in addition have to set a gradient ring-down delay (100 μs), gradient coil blanking switch, and loop counters which define the shaped gradients.
ns: 1

First record a normal ^{1}H spectrum without gradients, then increase the gradient strength in several steps and observe the dip in the water signal.

5. Processing

Use standard ^{1}H processing with an exponential window (lb = 20 Hz), however, apply magnitude calculation. Measure the width of the dip [Hz], and calculate the gradient strength G_Z according to the Equation (1).

$$G_Z = \frac{\Delta\omega}{\gamma \Delta z} = \Delta\nu \cdot 1.17 \cdot 10^{-5} \tag{1}$$

where $\Delta\omega$ is 2π times the width of the dip [Hz], γ is the proton gyromagnetic ratio ($26.751 \cdot 10^7\ T^{-1}s^{-1}$) and Δz is the thickness of the disk (0.002 m). Equation (1) gives the gradient strength in tesla/m, which may be converted to gauss/cm as often used in the literature by multiplying by 100.

6. Result

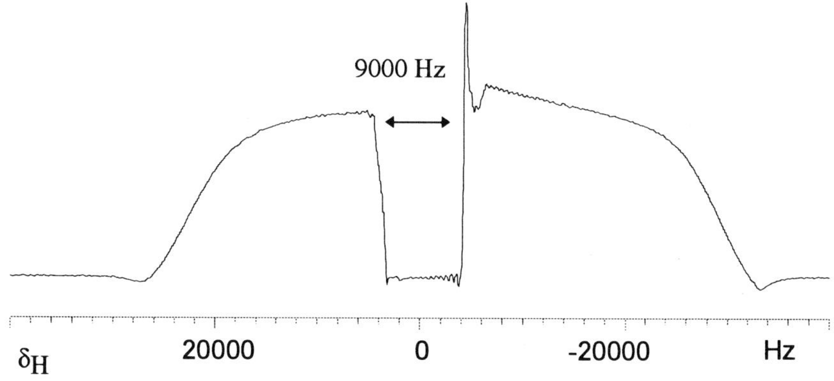

The figure on page 431 shows the result obtained on an AMX-500 spectrometer with a BGU (10 A) gradient unit and an inverse *z*-gradient probe-head. The measurement of the dip is indicated. In this spectrum a gradient of 0.105 T/m was applied.

7. Comments

The experiment as described works only for superconducting magnets in which the direction of the magnetic field B_0 is vertical, i.e. along the axis of the sample tube, and the field gradient is in the same *z*-direction. The resonance frequency of a proton in the presence of a *z*-gradient is given by Equation (2).

$$\omega_Z = \gamma B_{eff} = \gamma [B_0 + B_Z] \qquad (2)$$

Thus the resonance frequency of the proton is dependent on its *z*-position. A field gradient G_Z can be defined as in Equation (3).

$$G_Z = \frac{B_{Z1} - B_{Z2}}{\Delta z} = \frac{\Delta\omega}{\gamma \Delta z} \qquad (3)$$

Using a sample such as the one described above with well-defined points *z*1 and *z*2 determined by the thickness of the Teflon disk, one can measure an image of the water distribution in the tube from a gradient spin-echo experiment and calculate from this image the strength of the gradients.
This experiment can be thought as a normal spin-echo sequence (see Exp. 6.2), where in addition a gradient is switched on after the first 90° pulse. For technical reasons, however, the gradient pulse is divided into two, so that the 180° r.f. pulse can be inserted in between. Each spin is spatially labeled by its resonance frequency, thus giving an image of the water distribution within the sample. The 180° pulse refocuses all magnetizations so that the spin-echo builds up to a maximum at a time d2 after the 180° pulse.

8. Own Observations

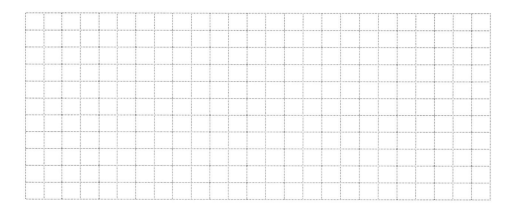

Experiment 11.2

Gradient Preemphasis

1. Purpose

Pulsed field gradients ideally should create magnetic field gradients with a predetermined shape, duration and strength. Because of eddy currents in the surrounding conducting material and because of gradient amplifier imperfections, the actual magnetic field gradient often does not correspond to the programmed shape, leading to undesired long ring-down delays. This is especially true if rectangular gradient pulses are used. One way to compensate for these imperfections is to use gradient preemphasis. With a preemphasis function the shape of the gradient pulse will be changed to anticipate and thus compensate the distortion caused by eddy currents and amplifier rise time. In the experiment described here we demonstrate how to adjust the preemphasis using a sample of chloroform.

2. Literature

[1] J. J. van Vaals, A. H. Bergman, *J. Magn. Reson.* **1990**, *90*, 52–70.
[2] P. Jehenson, M. Westphal, N. Schuff, *J. Magn. Reson.* **1990**, *90*, 264–278.
[3] C. D. Eccles, S. Crozier, M. Westphal, D. M. Doddrell, *J. Magn. Reson. Ser. A* **1993**, *103*, 135–141.
[4] W. E. Hull, in: W. R. Croasmun, R. M. K. Carlson (eds.), *Two-Dimensional NMR Spectroscopy*, VCH, Weinheim, **1994**, 186–193.
[5] M. Czisch, A. Ross, C. Cieslar, T. A. Tolak, *J. Biomol. NMR* **1996**, *7*, 121–130.

3. Pulse Scheme and Phase Cycle

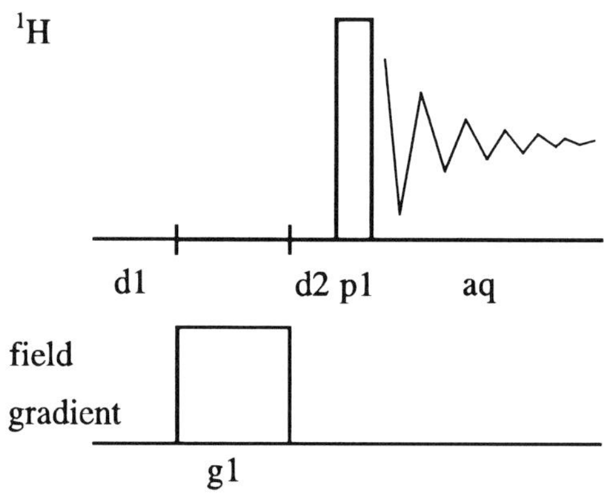

p1: x, x, -x, -x, y, y, -y, -y

aq: x, x, -x, -x, y, y, -y, -y

4. Acquisition

Time requirement: 20 min

Sample: 10% $CHCl_3$ in [D_6]acetone with added Cr(acac)$_3$

Set the instrument to normal ^1H NMR operation and load a pulse program as shown above. You have to set:

td: 4 k
sw: 5000 Hz
o1: 1000 Hz off resonance from $CHCl_3$ signal
p1: 10° ^1H transmitter pulse
d1: 0.1 s
d2: gradient ring-down delay [300 ms - 50 μs]
g1: positive pulsed field gradient, approximately 0.01 T/m, duration 1 ms, rectangular shape
ns: 1

Go into the setup mode of the instrument and display the FID on the screen. Choose a long gradient ring-down delay d2 of 300 ms and note the height and shape of the FID as a reference. According to the description of the manufacturer, shorten the ring-down delay until you observe a significant change in the FID. At this point vary the preemphasis time constants and amplitudes. There are often three sets which work with relatively long, medium and short ring-down delays. The final aim is to observe a maximum FID at the shortest possible d2.

5. Processing

No processing required, since the FID is directly observed

6. Result

The figure on page 435 shows a typical FID obtained on an AMX-500 spectrometer with a BGU (10 A) gradient unit and an inverse gradient probe-head with self-shielded gradients with the parameters described above.

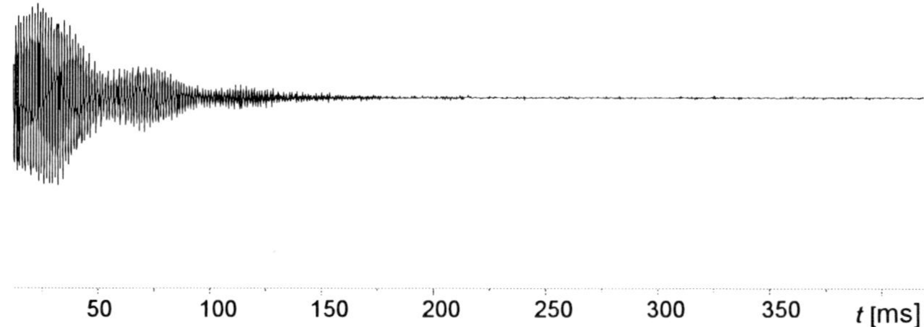

7. Comments

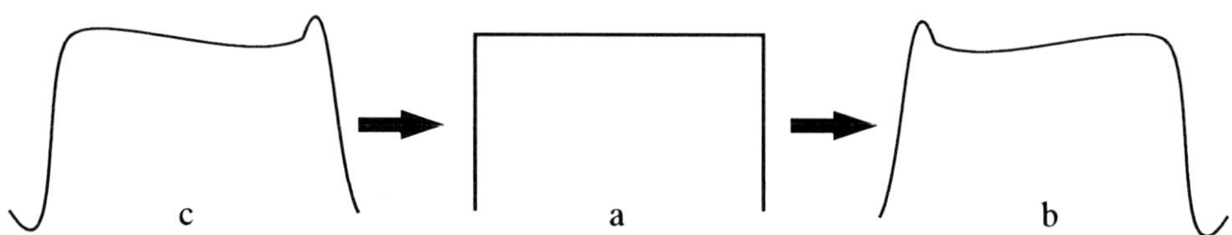

A rectangular waveform of the gradient pulse **a** will typically result in an actual field gradient as given in **b** for a certain configuration of a gradient unit, probe-head and shim coils. The aim of the experiment is to find a preemphasis waveform **c** which gives a rectangular pulsed field in the magnet as a result. Wave form **c** is a complicated function, which in this experiment was approached using three different time constants with different amplitudes to modulate the rectangular shape. For the configuration used (probe-head with *z*-gradient only), the need of preemphasis was not necessary if sinusoidal-shaped gradients were applied. In [4] a special adjustment technique is described in which several FIDs are displayed simultaneously in order to set the different preemphasis constants interactively.

8. Own Observations

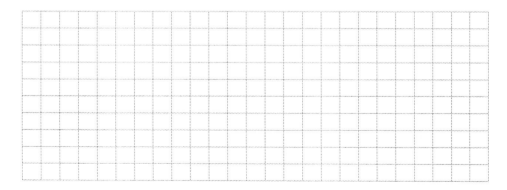

Experiment 11.3

Gradient Amplifier Test

1. Purpose

In all experiments with coherence selection by pulsed field gradients Eq. (1) must be obeyed. In Eq. (1) p_i, the coherence orders present at the instant of the pulsed field gradient, are multiplied by the gyromagnetic ratios γ_i of the corresponding nuclei and the effective field strengths of the gradient pulses G_i.

$$\sum p_i \gamma_i G_i = 0 \tag{1}$$

In order to fulfill this equation, the gradient strengths of either sign must be accurate and reproducible. The simple test provided in the experiment described here checks whether positive and negative gradient pulses have the same effect and thus detects any imbalance of the configuration.

2. Literature

[1] W. E. Hull, in: W. R. Croasmun, R. M. K. Carlson (eds.), *Two-Dimensional NMR Spectroscopy*, VCH, Weinheim, **1994**, 186–193.

[2] J. Keeler, R. T. Clowes, A. L. Davis, E. D. Laue, *Methods Enzym.* **1994**, *239*, 145–207.

[3] M. Czisch, A. Ross, C. Cieslar, T. A. Tolak, *J. Biomol. NMR* **1996**, *7*, 121–130.

3. Pulse Scheme and Phase Cycle

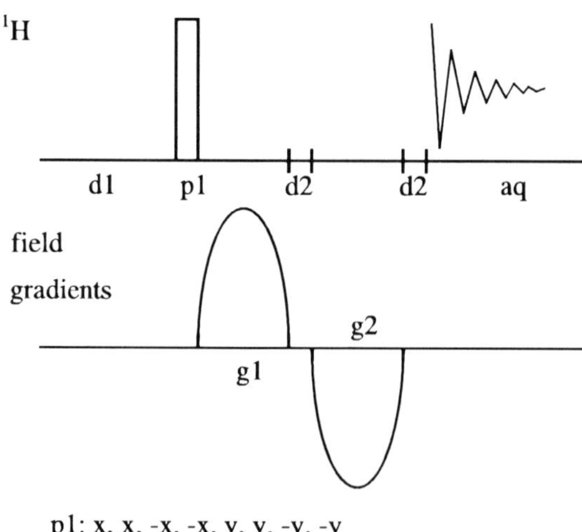

p1: x, x, -x, -x, y, y, -y, -y

aq: x, x, -x, -x, y, y, -y, -y

4. Acquisition

Time requirement: 20 min

Sample: 10% $CHCl_3$ in [D_6]acetone with added $Cr(acac)_3$

Set the instrument to normal ^1H NMR operation, obtain a good homogeneity and load a pulse program as shown above. You have to set:

td: 4 k
sw: 500 Hz
o1: on resonance of $CHCl_3$ signal
p1: 30° ^1H transmitter pulse
d1: 5 s
d2: gradient ring-down delay [100 μs]
g1: positive pulsed field gradient, approximately 0.01 T/m, duration 1 ms, sinusoidal shape
g2: negative pulsed field gradient, approximately 0.01 T/m, duration 1 ms, sinusoidal shape, strength to be varied
ns: 1

First record a ^1H NMR spectrum with identical gradients but of opposite sign. Vary the strength of the second gradient within a ±1% range of the first and note the signal change. Use other strengths and shapes for both gradients to study the influence of these parameters.

5. Processing

Use standard ^1H processing as described in Experiment 3.1 with an exponential window (lb = 2 Hz).

6. Result

The figure on page 438 shows a series of signals obtained on an AMX-500 spectrometer with a BGU (10 A) gradient unit and an inverse gradient probe-head with self-shielded gradients. g1 was set to +50.0 (relative units) and g2 was varied from –49.7 to –50.3. Note that the result is slightly asymmetric and that the zero point of the gradient amplifier has to be more carefully adjusted.

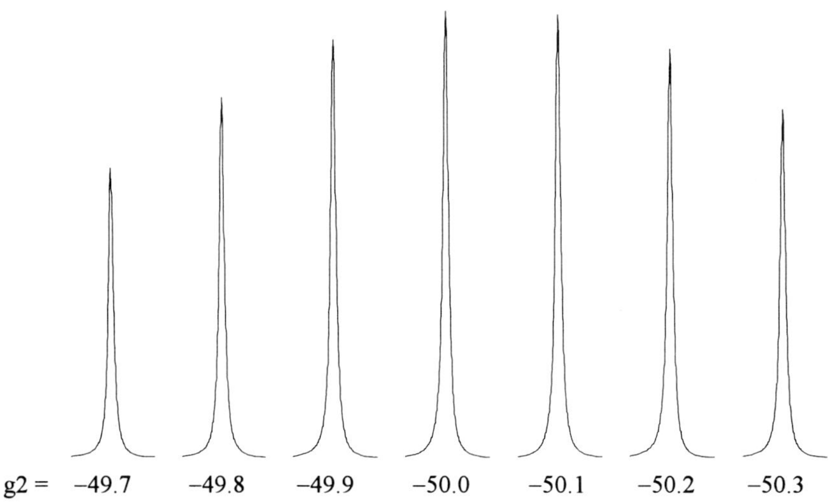

7. Comments

For most coherence selection experiments such as gs-HMQC (see Exp. 12.3) the performance shown is quite sufficient. For an excellent water suppression with gradients or the proton detection of C,C coupling constants (see Exp. 12.10) a rigorous adjustment would seem to be appropriate. Another quite similar test uses both gradients with equal sign, but with an 180° r.f. pulse between the gradients.

8. Own Observations

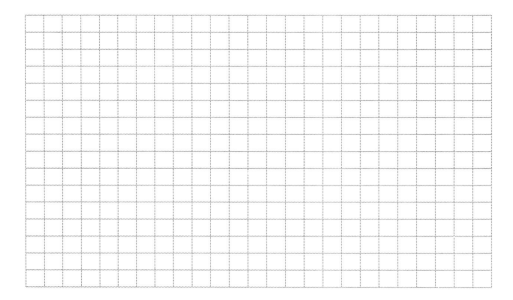

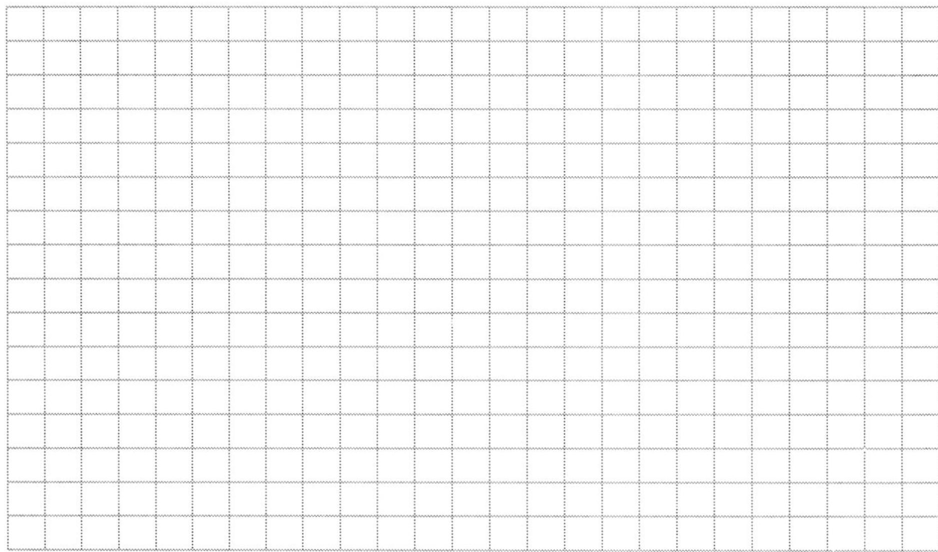

Experiment 11.4

Determination of Pulsed Field Gradient Ring-Down Delays

1. Purpose

Pulsed field gradients cause eddy currents in the surrounding conducting material and thus a certain dead time after the gradient pulse. Within this dead time the signal should not be acquired, nor should other r.f. pulses be applied. The length of the gradient dead time is very much dependent on the design of the gradient coils. The experiment described here demonstrates a calibration routine to define a suitable ringdown delay.

2. Literature

[1] W. E. Hull, in: W. R. Croasmun, R. M. K. Carlson (eds.), *Two-Dimensional NMR Spectroscopy*, VCH, Weinheim, **1994**, 186–193.
[2] J. Keeler, R. T. Clowes, A. L. Davis, E. D. Laue, *Methods Enzym.* **1994**, *239*, 145–207.
[3] P. Mansfield, B. Chapman, *J. Magn. Reson.* **1987**, *72*, 211–223.

3. Pulse Scheme and Phase Cycle

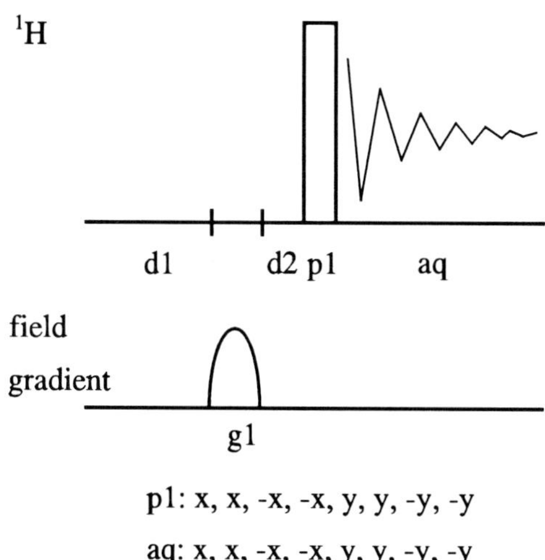

p1: x, x, -x, -x, y, y, -y, -y

aq: x, x, -x, -x, y, y, -y, -y

4. Acquisition

Time requirement: 20 min

Sample: 10% $CHCl_3$ in [D_6]acetone with added $Cr(acac)_3$

Set the instrument to normal ^{1}H NMR operation and load a pulse program as shown above. You have to set:

td: 4 k
sw: 500 Hz
o1: on resonance of $CHCl_3$ signal
p1: 30° ^{1}H transmitter pulse
d1: 5 s
d2: 1 s - 1 µs to be varied
g1: pulsed field gradient, sinusoidal shape on 100 points, duration = 1 ms, strength to be varied
ns: 1

First record a normal ^{1}H NMR spectrum without a gradient, then use a sinusoidal shaped gradient with approximately 0.01 T/m field strength using d2 = 1 s. The signal should have identical intensity to that of the normal ^{1}H NMR spectrum. Decrease d2 until the intensity drops significantly. At this point change the gradient strength and the gradient shape to observe the influence of these parameters. For all further gradient experiments use the shortest possible delay d2 as ring-down delay where the signal is not yet significantly attenuated.

5. Processing

Use standard ^{1}H processing as described in Experiment 3.1 with an exponential window (lb = 2 Hz).

6. Result

The figure on page 441 shows a series of signals obtained on an AMX-500 spectrometer with a BGU (10 A) gradient unit and an inverse gradient probe-head with self-shielded gradients. A ring-down delay of 50 - 100 µs seems to be appropriate for the configuration used. When a rectangular gradient was used, the decrease of the signal was already visible with d2 = 100 µs.

7. Comments

Pulsed field gradients can be generated by gradient coils which are mounted on the shim system or within the probe-head. Recent instruments use self-shielded gradient coils [3], where in the outer part of the assembly a field is generated which opposes the one in the inner part. For this design the ring-down problem is largely attenuated. In general it is better to have the coils mounted within the probe-head; however, this may cause difficulties for very low temperature work.

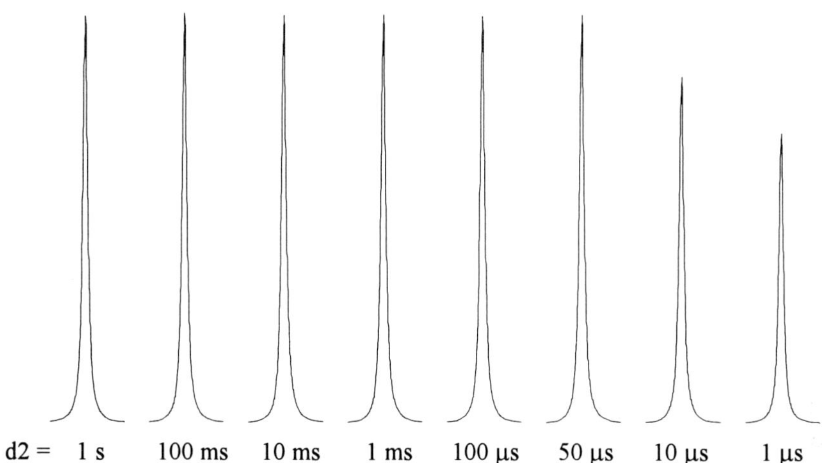

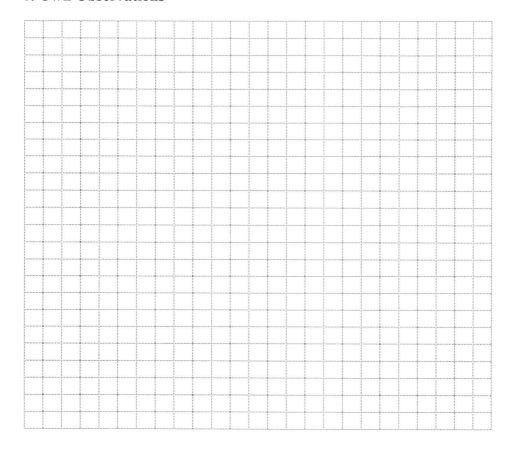

8. Own Observations

Experiment 11.5

The Pulsed Gradient Spin-Echo Experiment

1. Purpose

The PGSE (**P**ulsed **G**radient **S**pin **E**cho) experiment consists of the normal spin-echo experiment (see Exp. 6.2) with additional pulsed field gradients in both half periods before and after the 180° pulse. It was introduced to measure diffusion constants D for restricted and unrestricted diffusion in liquids. It can also be used to determine the strength of field gradients, if the diffusion constant of the sample is accurately known by other means. The experiment provides important insights into the theory and practice of pulsed field gradients and can be viewed as one of the basic physical experiments in NMR. Here the determination of diffusion constants is shown.

2. Literature

[1] E. O. Stejskal, J. E. Tanner, *J. Chem. Phys.* **1965**, *42*, 288–292.
[2] J. R. Singer, *J. Phys. E: Sci. Instrum.* **1978**, *11*, 281–291.
[3] P. Stilbs, *Prog. NMR Spectrosc.* **1987**, *19*, 1–45.
[4] H. Weingärtner, *Z. Phys. Chem. (Neue Folge)* **1982**, *132*, 129–149.
[5] M. Holz, H. Weingärtner, *J. Magn. Reson.* **1991**, *92*, 115–125.

3. Pulse Scheme with Phase Cycle

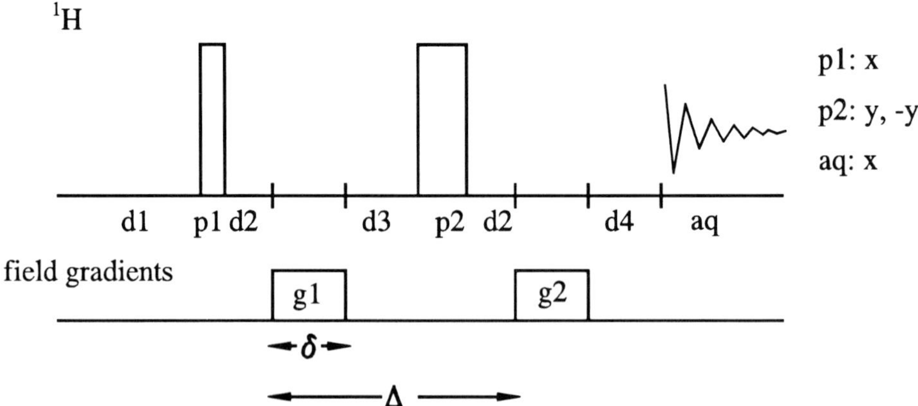

4. Acquisition

Time requirement: 0.5 h

Sample: normal H_2O

Record a normal ^{1}H NMR spectrum of the sample and center the offset at the water resonance; *set and control the temperature at 298 K.* In this experiment one can vary either the gradient strength or the gradient length δ. For observation of restricted diffusion Δ would be varied, since Δ is the time during which the diffusion process occurs. In this example we vary the gradient strength. You have to set:

td: 1 k
sw: 1000 Hz
o1: on resonances of water signal
p1: 90° ^{1}H transmitter pulse
p2: 180° ^{1}H transmitter pulse
d1: 2 s
d2: 1 ms
d3: 10 ms (dependent on gradient ring-down time)
d4: d3 minus preacquisition delay
g1, g2: rectangular shaped field gradients, 4 ms duration and variable strength from 0 to 0.2 T/m in 10 steps, with gradient loop counters, ring-down delays (100 μs), lock blanking and gradient coil blanking switches according to the actual instrumentation used. Gradient strength ratio: 1 : 1.
ns: 1

5. Processing

Process all 10 spectra identically with exponential multiplication using lb = 5 Hz and a baseline correction in order to obtain good integrals. Integrate the water signal in all spectra and refer all integrals to the integral value of the starting spectrum with gradient strength of 0. Compile a table of integral ratios I_g/I_0 vs. gradient strength G used, where the gradient strength is determined as described in Experiment 11.1. Equation (1) relates the integral ratio to the diffusion constant D.

$$\ln (I_g/I_0) = -[\gamma^2 \delta^2 G^2 (\Delta - \delta/3)] D \quad (1)$$

Thus, a plot of $\ln(I_g/I_0)$ versus G^2 yields the diffusion constant from the slope, when the values $\gamma = 2.675 \cdot 10^8$ T^{-1}s^{-1}, $\delta = 0.004$ s, and $\Delta = 0.0151$ s are inserted.

6. Result

The table gives the results obtained on an AMX-500 spectrometer equipped with a BGU(10A) gradient unit and an inverse multinuclear *z*-gradient probe-head, giving a value of $D = 2 \cdot 10^{-9}$ m^2/s, which is reasonably close to the accepted literature value of $2.30 \cdot 10^{-9}$ m^2/s [4, 5].

G [T/m]	0	0.022	0.045	0.065	0.089	0.11	0.13	0.15	0.18	0.19	0.22
I_g/I_0	1	0.96	0.92	0.88	0.82	0.73	0.62	0.52	0.40	0.30	0.22

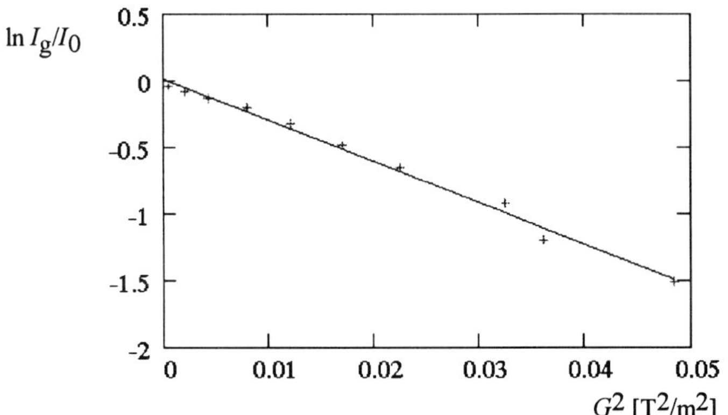

7. Comments

In the normal spin-echo experiment the echo amplitude is dependent on the spin–spin relaxation time and the diffusion constant. If the magnetic field is homogeneous, the latter does not affect the measurement. In the PGSE experiment the pulsed field gradient during the first half period positionally labels the spins with their Larmor frequency. The 180° pulse reverses the coherence order and the second gradient pulse therefore rephases all nuclear spins except those which have diffused during the time period Δ. Thus, the echo amplitude varies strongly with the gradient field strength.

The diffusion constant D is given theoretically by the Stokes–Einstein equation (2), where k_B is the Boltzmann constant, η the viscosity, and r the radius of the molecular sphere.

$$D = \frac{k_B T}{6\pi\eta\, r} \qquad (2)$$

Often a modification of Equation (2) with a factor 4 instead of 6 in the denominator is used when the surrounding particles are of similar size compared to the solute (slip boundary condition).

8. Own Observations

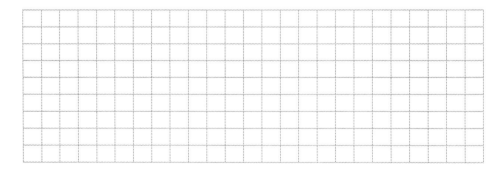

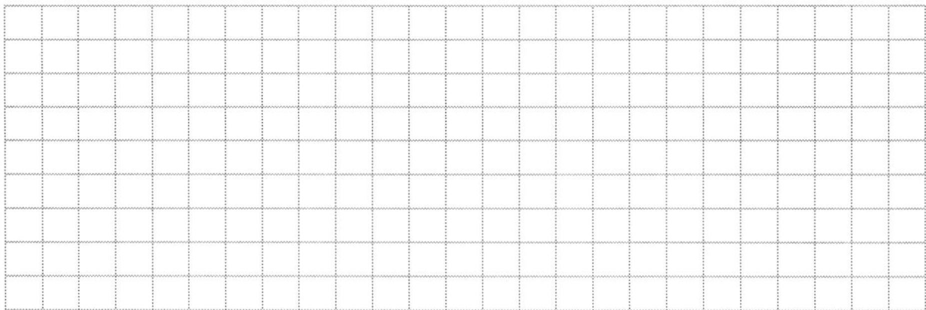

Experiment 11.6

Excitation Pattern of Selective Pulses

1. Purpose

Ideal selective pulses should have a narrow, top-hat-like excitation pattern with equal phase within the excitation regime. In practice, the width of the excitation of a selective pulse corresponds only very roughly to the inverse of its duration. One can determine the excitation profile of a selective pulse by recording many spectra with different offsets; these are moved in small steps through the resonance region of a spectrum consisting of a single line. However, this method is rather time-consuming, and thus we show here two recently developed gradient-selected experiments which produce an image of the excitation pattern in one scan. The experiments are closely related to Experiment 11.1 and provide determinations of the excitation pattern of a 90° and a 180° selective pulse.

2. Literature

[1] V. Belle, G. Cros, H. Lahrech, P. Devoulon, M. Decorps, *J. Magn. Reson. Ser. A* **1995**, *112*, 122–125.

3. Pulse Scheme and Phase Cycle

Experiment **a** (90° selective pulse)

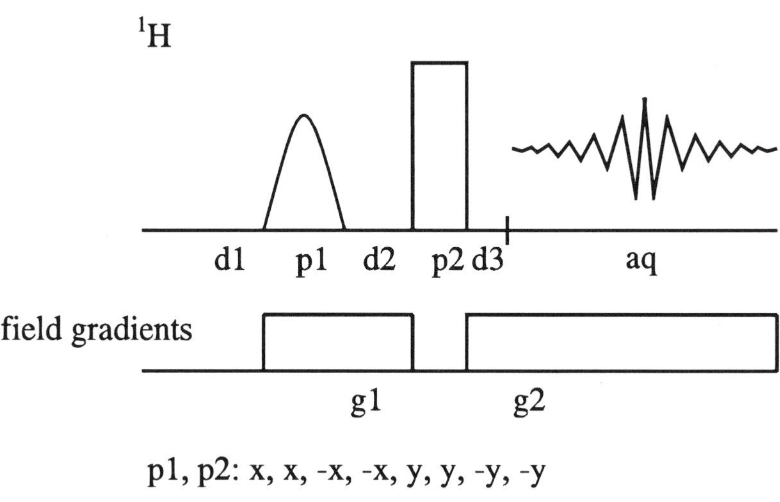

p1, p2: x, x, -x, -x, y, y, -y, -y

aq: x, x, -x, -x, y, y, -y, -y

Experiment **b** (180° selective pulse)

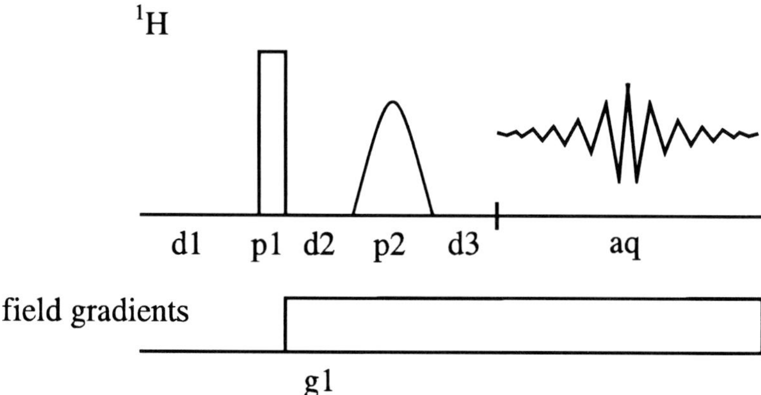

4. Acquisition

Time requirement: 10 min

Sample: 90% H_2O, 10% D_2O

Record a normal ^1H NMR spectrum of the sample. Determine the 90° pulse width for the hard ^1H transmitter pulse, select a Gaussian pulse shape for the soft pulse and determine the correct attenuation corresponding to a 90° and 180° pulse at 50 ms duration (see Exp. 7.1). Load the pulse program according to experiment **a**. You have to set:

td: 1 k
sw: 2500 Hz
o1: on resonance of water signal
p1: 90° Gaussian shape ^1H transmitter pulse, 50 ms length, transmitter attenuation corresponding to 90° excitation [67 dB]
p2: 180° ^1H transmitter pulse
d1: 2s
d2: 4 ms
d3: 900 μs (preacquisition delay), change for fine adjustment of signal oscillation
g1, g2: rectangular field gradients of ca. 0.01 T/m strength, with gradient loop counters, ring-down delays (100 μs), lock blanking and gradient coil blanking switches according to actual instrumentation used. Gradient strength ratio: 1 : 1
ns: 1

For Experiment **b** load the corresponding pulse program and change to:

p1: 90° ^1H transmitter pulse
p2: 180° Gaussian shape ^1H transmitter pulse, 50 ms length, transmitter

attenuation corresponding to 180° excitation [64 dB]

g1: rectangular field gradient of ca. 0.01 T/m strength, with gradient loop counters, ring-down delay (100 μs), lock blanking and gradient coil blanking switches according to actual instrumentation used.

ns: 1

Try other excitation shapes of the selective pulses such as sinc, rectangular or pulses of the BURP family.

5. Processing

Use standard ^{1}H processing as described in Experiment 3.1 using zero filling to 1 k and exponential multiplication with lb = 20 Hz.

6. Result

The figures on pages 447–448 show the result obtained on an AMX-500 spectrometer with a BGU (10 A) gradient unit and an inverse gradient probe-head with self-shielded gradients. In **a** the result of the 90° selective pulse is given. Note the phase change at the center of the pattern. This is a considerable drawback of 90° Gaussian pulses. In contrast to this, the phase of a 180° selective pulse, as shown in **b**, remains constant. Note that the pattern shown in **b** has a width at half height of 75 Hz; thus the 3 dB point for the excitation is ± 37.5 Hz from the center of the resonance.

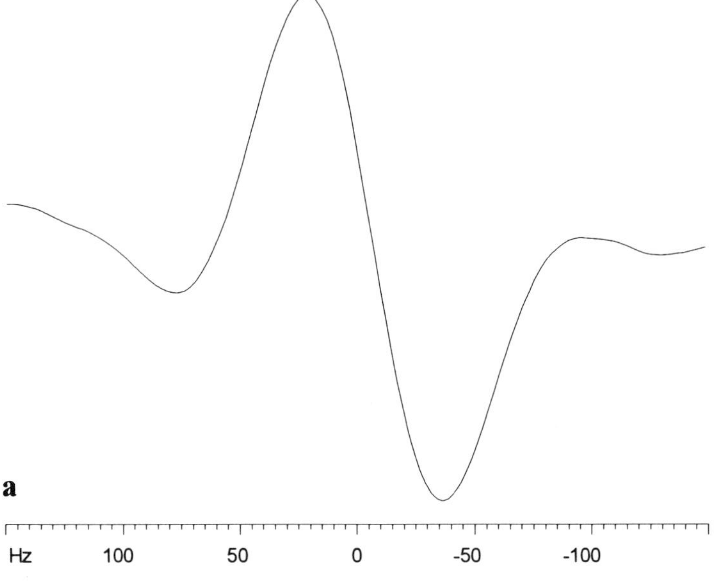

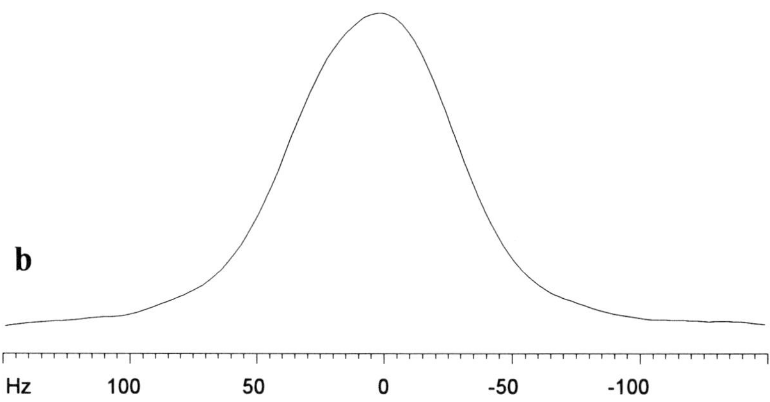

7. Comments

The experiments shown here belong to the class of gradient echo experiments as already discussed in Experiment 11.1. In both experiments **a** and **b**, the field gradient is switched on during the duration of the selective pulse and during the acquisition. The gradient provides a *z*-axis-dependent frequency labelling of the water spins. The frequency-dependent excitation profile of the selective pulses can be imaged, because the 180° pulses in the sequences produce an echo only of those spins which have been affected by the selective pulses.

8. Own Observations

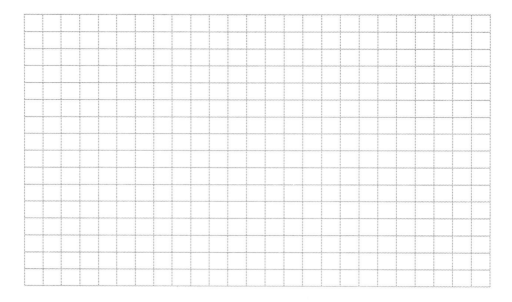

Experiment 11.7

The Gradient *zz*-Filter

1. Purpose

In many experiments one wants to selectively observe protons that are attached to ^{13}C or ^{15}N. The strong signals of protons attached to ^{12}C or ^{14}N need to be suppressed in order to be able to adjust the receiver gain for the desired signals only. One technique to achieve this goal is to dephase unwanted signals with pulsed field gradients [1] after storing the desired magnetization as *z*-magnetization for both proton and carbon atoms [2,3]. The technique is used in several advanced pulse methods. In this educational experiment the application of the gradient *zz*-filter is shown for chloroform.

2. Literature

[1] B. K. John, D. Plant, R. E. Hurd, *J. Magn. Reson. Ser. A* **1992**, *101*, 113–117.
[2] G. Wider, K. Wüthrich, *J. Magn. Reson. Ser. B* **1993**, *102*, 239–241.
[3] G. Otting, K. Wüthrich, *J. Magn. Reson.* **1988**, *76*, 569–574.

3. Pulse Scheme and Phase Cycle

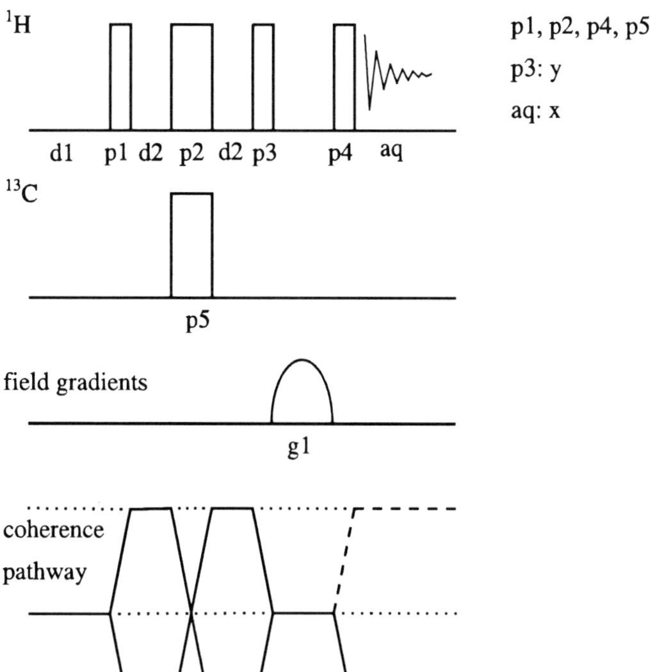

p1, p2, p4, p5: x
p3: y
aq: x

4. Acquisition

Time requirement: 5 min

Sample: 3% $CHCl_3$ in [D_6]acetone

First obtain normal ^{1}H and ^{13}C spectra of the sample and note the offsets of the $CHCl_3$ signals. Change to the inverse mode, using the proton channel as transmitter and the ^{13}C channel as decoupler. Load the pulse program for the gradient *zz*-filter. You have to set:

td: 4 k
sw: 500 Hz
offset of ^{1}H frequency: on ^{1}H resonance
offset of ^{13}C frequency: on ^{13}C resonance
p1, p3, p4: 90° ^{1}H transmitter pulse
p2: 180° ^{1}H transmitter pulse
p5: 180° ^{13}C decoupler pulse
d1: 10 s
d2: $1/[4J(C,H)] = 1.17$ ms, calculated from $^1J(C,H) = 214$ Hz
g1: sinusoidal shaped field gradient with 5% truncation, 1.5 ms duration and ca. 0.01 T/m strength, with gradient loop counter, ring-down delay (100 μs), lock blanking and gradient coil blanking switches according to actual instrumentation used. Gradient strength to be varied for best results.
ds: 4
ns: 1

5. Processing

Use standard ^{1}H processing as described in Experiment 3.1.

6. Result

The figure on page 451 shows the result obtained on an AMX-500 spectrometer with a multinuclear inverse gradient probe-head. Compare the result with the other methods described in this book to achieve a suppression of protons bound to ^{12}C (see Exps. 6.13–6.16).

7. Comments

The method can best be understood with the product operator formalism. Neglecting the 180° pulses, which refocus the chemical shifts, we find for a proton bound to ^{13}C the result as given in Eq. (1), since by setting the delay $2 \cdot d2 = \tau = 1/[2J(C,H)]$ the cosine terms become zero and the sine terms unity.

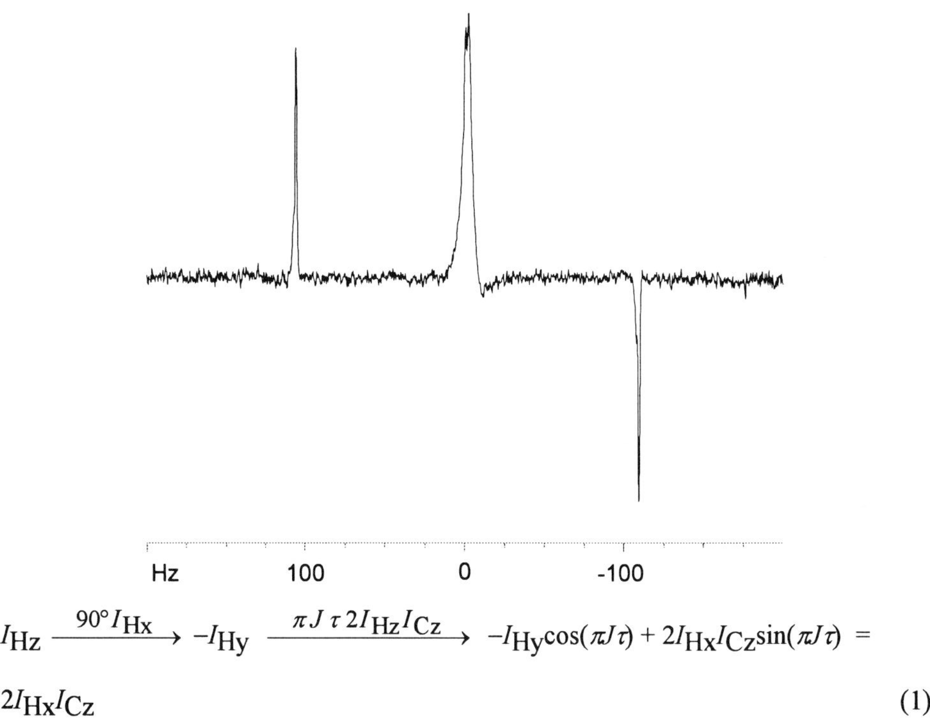

$$I_{Hz} \xrightarrow{90° I_{Hx}} -I_{Hy} \xrightarrow{\pi J \tau 2 I_{Hz} I_{Cz}} -I_{Hy}\cos(\pi J\tau) + 2I_{Hx}I_{Cz}\sin(\pi J\tau) = 2I_{Hx}I_{Cz} \quad (1)$$

A proton bound to ^{12}C cannot develop heteronuclear spin coupling and stays as $-I_{Hy}$. The proton pulse p3 with the phase *y* will not affect this magnetization, but will transform the term $2I_{Hx}I_{Cz}$ into heteronuclear two-spin order $-2I_{Hz}I_{Cz}$. Thus, the desired coherence is stored in the *z*-direction and the gradient pulse g1 only acts upon the magnetization of protons not bound to ^{13}C. The final proton pulse p4 recreates antiphase magnetization $2I_{Hx}I_{Cz}$, which during acquisition develops in-phase magnetization I_{Hy} sin(πJaq) yielding the antiphase signals as observed in the figure.

A very similar sequence can be constructed, which acts as a gradient *z*-filter. Here only in-phase magnetization I_{Hx} is stored as I_{Hz}, an example of this technique is shown for the sensitivity-enhanced HSQC method in Experiment 12.6.

8. Own Observations

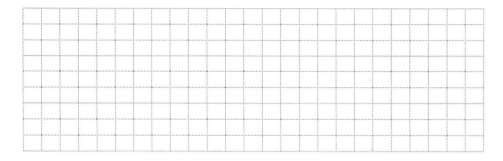

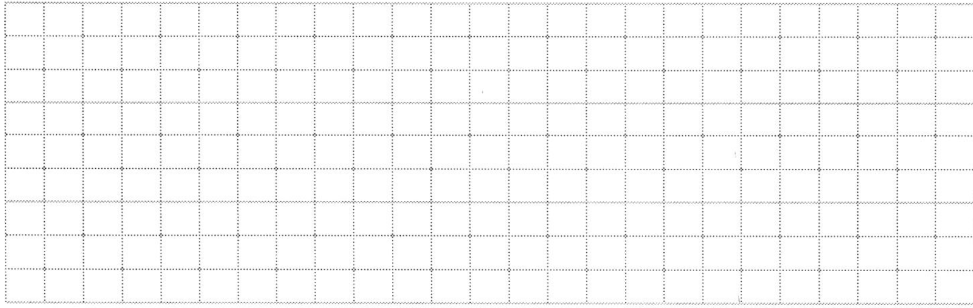

Experiment 11.8

gs-SELCOSY

1. Purpose

This is the advanced 1D variant of the most common 2D experiment. Instead of recording the full 2D matrix, one can simply measure one "row" by replacing the first 90° pulse of the COSY experiment (see Exps. 10.3 and 12.1) with a soft pulse, thus looking only for spin couplings that affect the particular proton excited. Compared with the traditional selective COSY method as described in Exp. 7.5, the gradient-selected method demonstrated here gives excitingly clean results without the need of phase cycling. Shown here is a very recent variant [5] which uses the DPFGSE technique (see Exp. 11.10) to yield a superior frequency selection.

2. Literature

[1] M. A. Bernstein, L. A. Trimble, *Magn. Reson. Chem.* **1994**, *32*, 107–110.
[2] W. Willker, D. Leibfritz, *Magn. Reson. Chem.* **1994**, *32*, 665–669.
[3] C. Dalvit, *J. Magn. Reson. Ser. A* **1995** *113*, 120–123.
[4] C. Dalvit, S. Y. Ko, J. M. Böhlen, *J. Magn. Reson. Ser. B* **1996**, *110*, 124–131.
[5] S. Berger, *Prog. NMR Spectr.* **1997**, *30*, 137–156.

3. Pulse Scheme and Phase Cycle

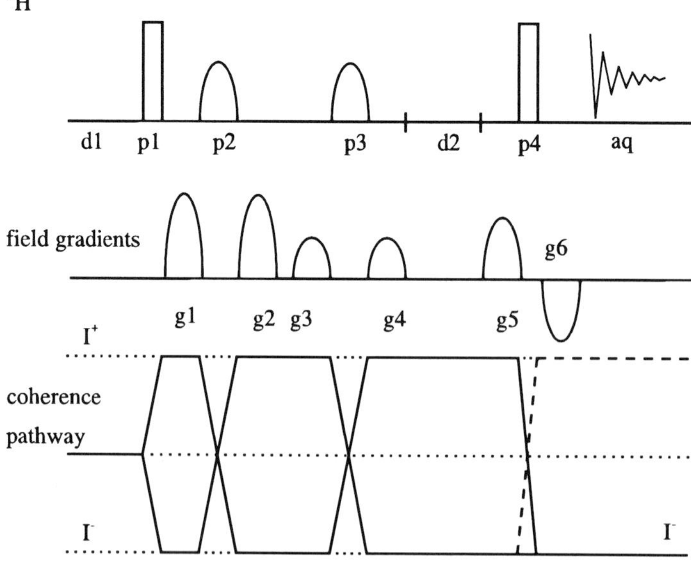

p1, p4: x, -x, -x, x, y, -y, -y, y   p2, p3: x   aq: x, -x, -x, x, y, -y, -y, y

4. Acquisition

Time requirement: 5 min

Sample: 10% strychnine in $CDCl_3$

Record a normal ^1H NMR spectrum of the sample. Determine the 90° pulse width for the hard ^1H transmitter pulse, select a Gaussian pulse shape for the soft pulse and determine the correct attenuation corresponding to a 180° pulse at 50 ms duration (see Exp. 7.1). Determination of the phase difference between the hard and the soft pulse is not necessary. Load the pulse program for gs-SELCOSY. You have to set:

td: 32 k
sw: 10 ppm
o1: on resonance of selected signal. If the software allows offsets for selective pulses, one can also put o1 in the middle of the ^1H NMR spectrum.
p1, p4: 90° ^1H transmitter pulse
p2, p3: 180° Gaussian shape ^1H transmitter pulse, 50 ms length, transmitter attenuation corresponding to 180° excitation [64 dB]
d1: 2s
d2: 30–60 ms, adjusted to ~ $1/[2J(H,H)]$
g1-g6: sinusoidal shaped field gradients with 5% truncation, 2 ms duration and ca. 0.01 T/m strength, with gradient loop counters, ring-down delays (100 μs), lock blanking and gradient coil blanking switches according to actual instrumentation used. Gradient strength ratio: 40 : 40 : 7 : 7 : 20 : –20
ds: 2
ns: 1

5. Processing

Use standard ^1H processing as described in Experiment 3.1. Note that the signals of the irradiated protons are unperturbed and that the signals of the coupling partners show the active coupling in antiphase. For the method shown here you have to adjust the phase for each multiplet individually because of the linear phase shift across the spectrum caused by the finite length of the pulsed field gradients.

6. Result

The figure on page 454 shows the result obtained on an AMX-500 spectrometer with a BGU (10 A) gradient unit and an inverse gradient probe-head with self-shielded gradients. In **a** an expanded portion of the normal ^1H NMR spectrum is shown, in **b** H-12 was selected, giving the responses of both H-11 and (weakly) of H-13, and in **c** (d2 = 50 ms) H-15β was selected, giving the responses of H-15α, H-14, and H-16. Note that the signals of H-14 and H-16 have yet to be correctly phased for analysis. Compare the result with that of Experiment 7.5 and note that all artifacts have disappeared.

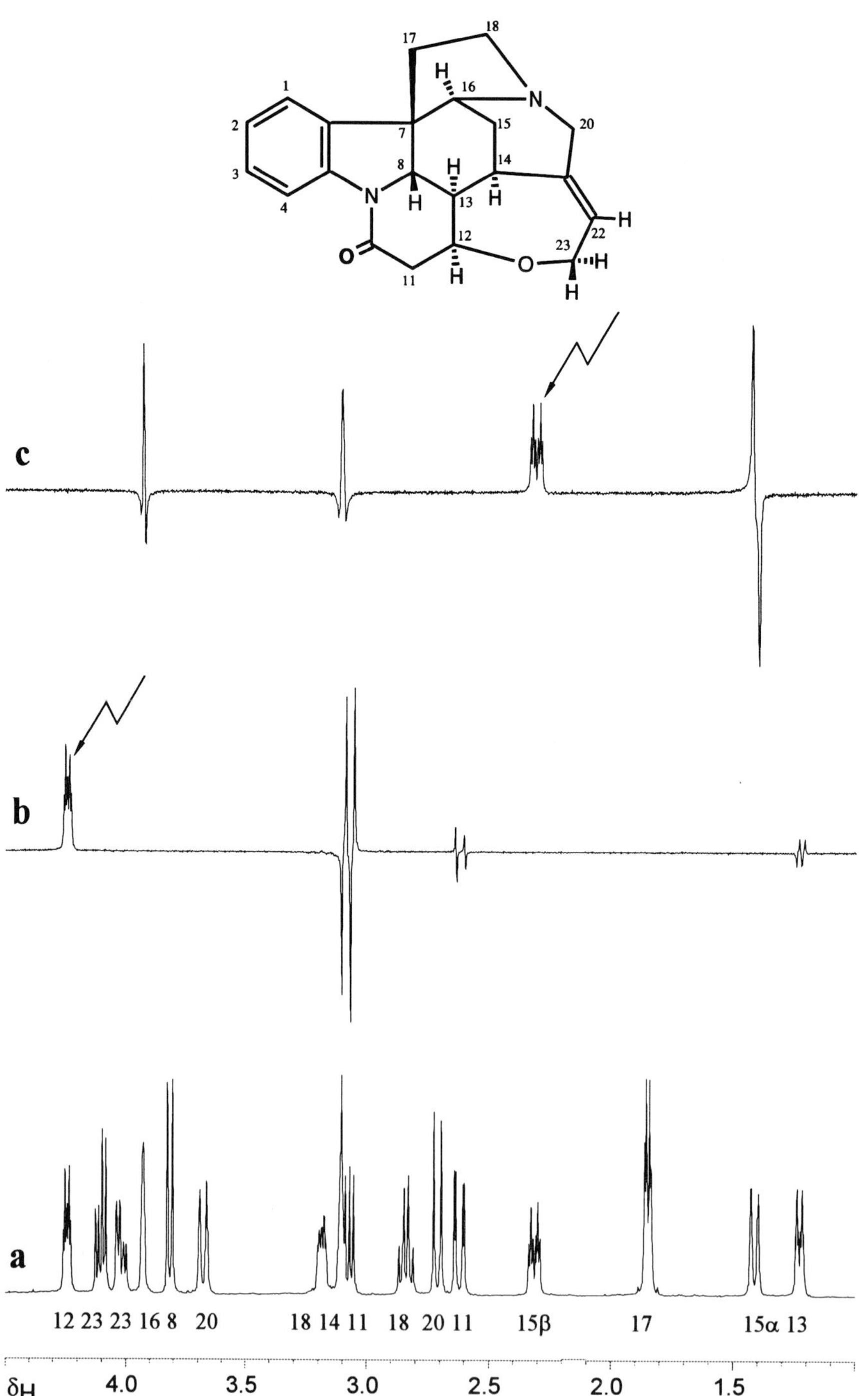

c
b
a
12 23 23 16 8 20
18 14 11 18 20 11
15β
17
15α 13
δH
4.0
3.5
3.0
2.5
2.0
1.5

7. Comments

The first 90° ^{1}H transmitter pulse excites all proton resonances. In the following DPFGSE sandwich (see Exp. 11.10) all these resonances are dephased and only the one chosen by the selective 180° pulse is retained. Since a selective 180° pulse is applied its relative phase with respect to the "hard" pulses does not need to be determined. If one would acquire the signal after the DPFGSE sandwich only the signal excited by the selective pulse could be observed.

For the COSY part of the sequence exactly the same theory applies as given in Experiment 10.3. Note that the delay d2 determines the intensity of the "cross peak". It may be necessary to perform the experiment twice, for example in order to identify spin coupling partners with both small and large spin–spin coupling constants. In the spectrum **c** d2 was set to 50 ms. For multispin systems the delay d2 cannot always be optimized, an alternative is the gs-SELTOCSY method using a short spin-lock (see Experiment 11.9)

8. Own Observations

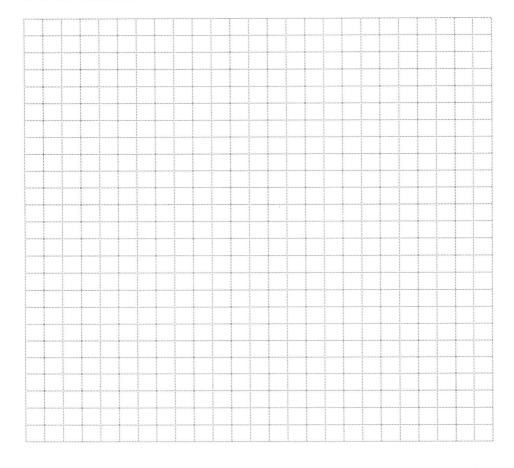

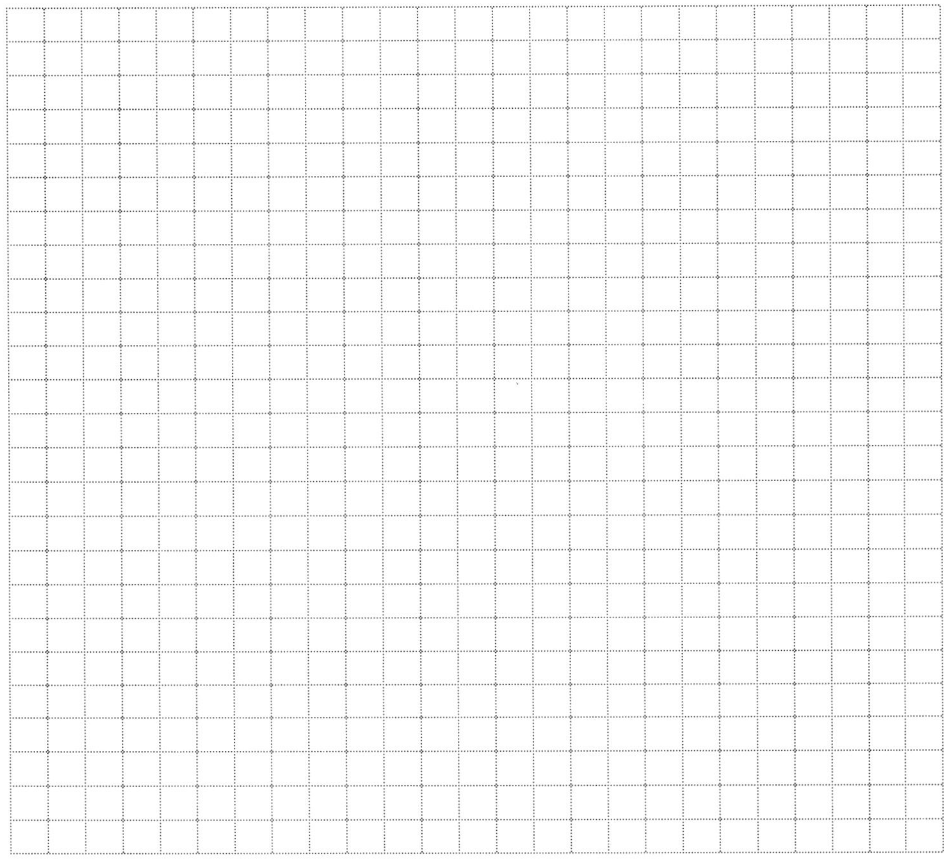

Experiment 11.9

gs-SELTOCSY

1. Purpose

This is the 1D variant of the gs-TOCSY Experiment 12.7. Instead of recording the full 2D matrix, one can simply measure one "row" by selective excitation, thus looking only for spin couplings that affect the particular proton excited. Compared with the traditional selective TOCSY method as described in Experiment 7.8, the gradient-selected method demonstrated here gives clean results without the need of phase cycling, using only one scan. Shown here is a recent variety which uses a 180° selective proton pulse for selective excitation.

2. Literature

[1] T. Fäcke, S. Berger, *J. Magn. Reson. Ser. A* **1995**, *113*, 257–259.
[2] C. Dalvit, S. Y. Ko, J. M. Böhlen, *J. Magn. Reson. Ser. B* **1996**, *110*, 124–131.

3. Pulse Scheme and Phase Cycle

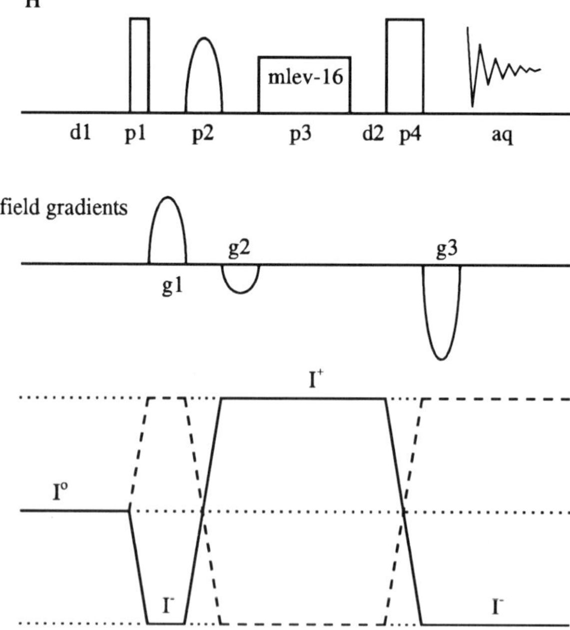

p1, p2, p4: x; aq: x
p3: MLEV-16 spin-lock out of composite 180° pulses (90°, 180°, 90°); sequence:

90 (ph1), 180 (ph2), 90 (ph1)
[90 (ph3), 180 (ph4), 90 (ph3)]2
90 (ph1), 180 (ph2), 90 (ph1)
[90 (ph3), 180 (ph4), 90 (ph3)]2
[90 (ph1), 180 (ph2), 90 (ph1)]2
90 (ph3), 180 (ph4), 90 (ph3)
[90 (ph1), 180 (ph2), 90 (ph1)]2
90 (ph3), 180 (ph4), 90 (ph3)
[90 (ph1), 180 (ph2), 90 (ph1)]2
[90 (ph3), 180 (ph4), 90 (ph3)]2

ph1: $(-y, y)_2, (x, -x)_2$
ph2: $(x, -x)_2, (y, -y)_2$
ph3: $(y, -y)_2, (-x, x)_2$
ph4: $(-x, x)_2, (-y, y)_2$

4. Acquisition

Time requirement: 5 min

Sample: 10% strychnine in $CDCl_3$

Record a normal ^1H NMR spectrum of the sample. Determine the 90° pulse width for the hard ^1H transmitter pulse, select a Gaussian pulse shape for the soft pulse and determine the correct attenuation corresponding to a 180° pulse at 50 ms duration (see Exp. 7.1). Determination of the phase difference between the hard and the soft pulse is not necessary. The 90° pulse duration and the attenuation for the spin-lock pulses must also be known. You have to set:

td: 32 k
sw: 10 ppm
o1: on resonance of selected signal. If the software allows offsets for selective pulses, one can also put o1 in the middle of the ^1H NMR spectrum.
p1: 90° ^1H transmitter pulse
p2: 180° Gaussian shape ^1H transmitter pulse, 50 ms length, transmitter attenuation corresponding to 180° excitation [62 dB]
p3: series of composite 180° pulses (90°, 180°, 90°) at power level of spin-lock, typically 90° pulse width of 40 μs at 12 dB transmitter attenuation corresponding to an effective spin-lock field of ca. 7000 Hz. Total length of spin-lock varied from 250 ms in **b**, 76 ms in **c** to 215 ms in **d** and was adjusted with the loop parameter of the spin-lock sequence.
p4: 180° ^1H transmitter pulse
d1: 2s
d2: equal to the effective length of the pulsed field gradient g3 [1 ms]
g1–g3: sinusoidal shaped field gradients with 5% truncation, 1 ms duration and ca. 0.01 T/m strength, with gradient loop counters, ring-down delays (100 μs), lock blanking and gradient coil blanking switches according to actual instrumentation used. Gradient strength ratio: 7: −3: −10
ns: 1

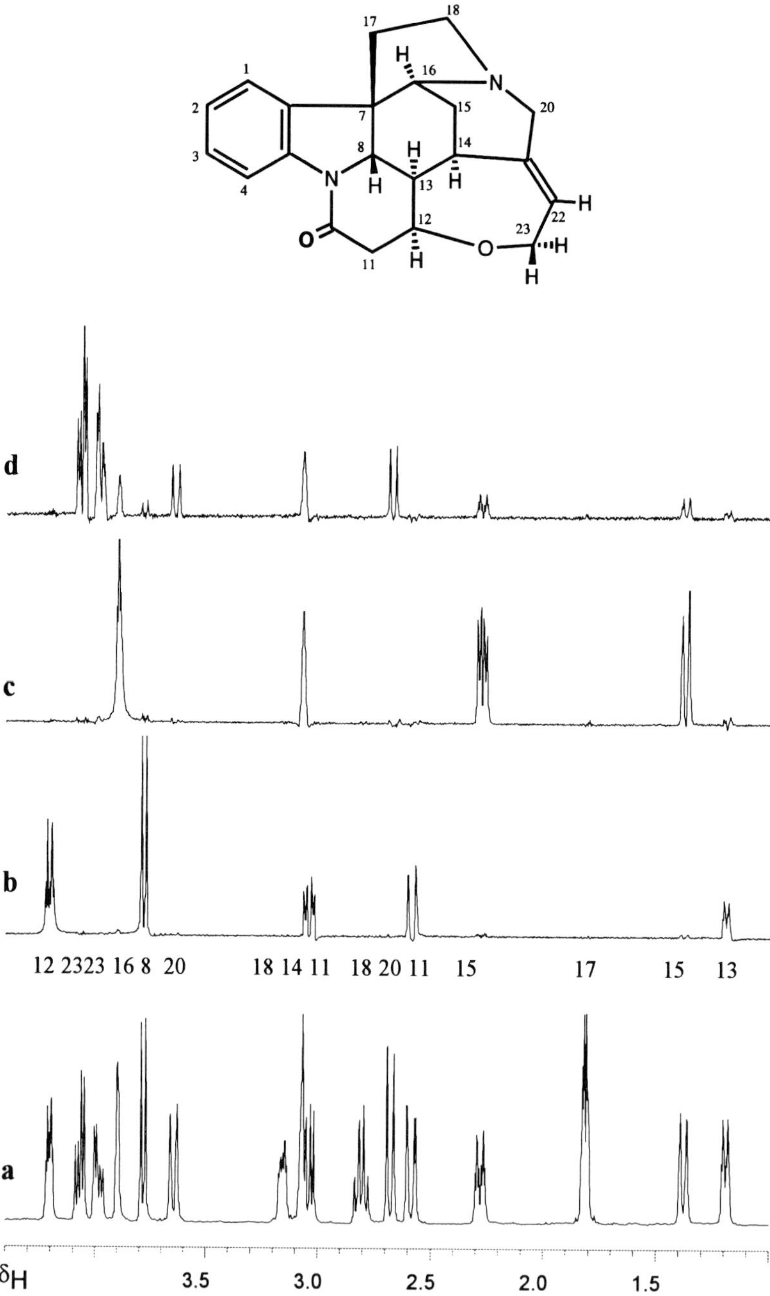
17
18
16
N
1
2
3
4
7
8
15
14
20
13
12
11
22
23
O
H
N
12 2323 16 8 20
18 14 11
18 20 11
15
17
15
13
d
c
b
a
δ_H
3.5
3.0
2.5
2.0
1.5

5. Processing

Use standard ^{1}H processing as described in Experiment 3.1.

6. Result

The figure on page 458 shows the result obtained on an AMX-500 spectrometer with a BGU (10 A) gradient unit and an inverse gradient probe-head with self-shielded gradients. In **a** an expanded portion of the normal ^{1}H NMR spectrum is shown, and in **b** H-12 was selected, giving the responses of both H-11 protons, H-13 and H-8, when a spin-lock of 250 ms was used. In **c** (spin-lock 76 ms), H-16 was selected, giving the responses of both H-15 protons and H-14; in **d** (spin-lock 215 ms), H-22 was selected, giving responses of both H-23 protons, but also of protons 20 and 14 reached via an allylic coupling and further transfers to H-15, H-16 and H-13. Note that no signals have pure phase, but do have components of in- and anti-phase magnetization.

7. Comments

The first 90° ^{1}H transmitter pulse excites all proton resonances. In the following [gradient pulse, selective 180° pulse, gradient pulse] sandwich, all these resonances are dephased and only the one chosen by the selective 180° pulse is retained. Since a selective 180° pulse is applied, its relative phase with respect to the "hard" pulses does not need to be determined. The MLEV-16 spin-lock does not introduce a further change of the coherence level; the solid line in the coherence diagram depicts the coherence chosen by the selective pulse.

Since the experiment requires only one scan, it is ideal to study the influence of the spin-lock length on the number of signal responses as well as their relative intensities and signal shapes.

8. Own Observations

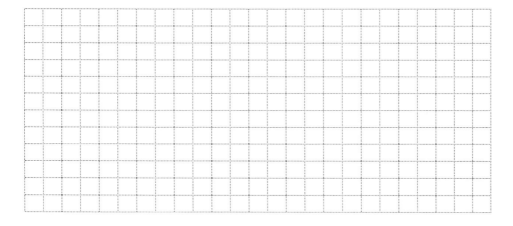

Experiment 11.10

DPFGSE-NOE

1. Purpose

The NOE difference technique (see Exps. 4.8 and 4.9) has to cope with artefacts caused by insufficient spectrometer stability. Very weak NOE effects are often obscured by residual signals. Using pulsed field gradients, unwanted signals can be better suppressed and, with a selective excitation pulse tailored to the multiplet under consideration, the desired NOE effects can be recorded without interference from other signals. This is achieved by the DPFGSE technique, which is a combination of the **D**ouble **P**ulsed **F**ield **G**radient **S**pin **E**cho method [3] and NOE spectroscopy [2,4] and results in spectra without phase distortion. We show here one variant of this method using strychnine as an example. Most likely the technique will replace the traditional nuclear Overhauser difference spectroscopy.

2. Literature

[1] J. Stonehouse, P. Adell, J. Keeler, A. J. Shaka, *J. Am. Chem. Soc.* **1994**, *116*, 6037–6038

[2] K. Stott, J. Stonehouse, J. Keeler, T.-L. Hwang, A. J. Shaka, *J. Am. Chem. Soc.* **1995**, *117*, 4199–4200.

[3] T.-L. Hwang, A. J. Shaka, *J. Magn. Reson. Ser. A*, **1995**, *112*, 275–279.

[4] K. Stott, J. Keeler, Q. N. Van, A. J. Shaka, *J. Magn. Reson.* **1997**, *125*, 302–324.

3. Pulse Scheme and Phase Cycle

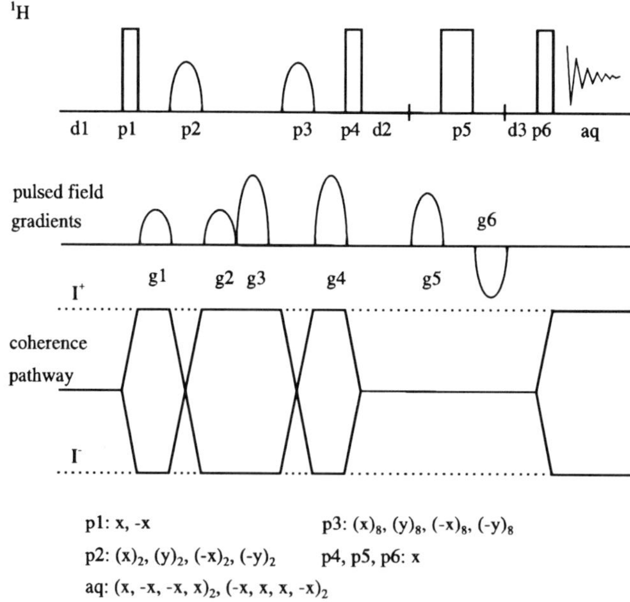

p1: x, -x

p2: $(x)_2$, $(y)_2$, $(-x)_2$, $(-y)_2$

p3: $(x)_8$, $(y)_8$, $(-x)_8$, $(-y)_8$

p4, p5, p6: x

aq: $(x, -x, -x, x)_2$, $(-x, x, x, -x)_2$

4. Acquisition

Time requirement: 15 min

Sample: 10% strychnine in $CDCl_3$

Run a normal ^1H NMR of the sample, optimize the spectral width, and note the offsets of the signals to be irradiated. You have to set:

td: 32 k data points
sw: 10 ppm
o1: middle of ^1H NMR spectrum
p1, p4, p6: 90° ^1H transmitter pulse
p2, p3: selective 180° ^1H transmitter pulse, Gaussian shape, 50 ms length, offset modulated with the difference Δ between o1 and the offset of the signal to be irradiated, transmitter attenuation corresponding to 180° [65 dB]
p5: 180° ^1H transmitter pulse
d1: 2 s
d2: mixing time delay 0.5 s, total mixing time τ_m = d2 + g5 + p5 + g6 + d3
d3: mixing time delay, set to 0.4·d2
g1–g6: sinusoidal shaped field gradients with 1% truncation, 1 ms duration and ca. 0.01 T/m strength, with gradient loop counters, ring-down delays (100 μs), lock blanking and gradient coil blanking switches according to actual instrumentation used. Gradient strength ratio: 70 : 70 : 30 : 30 : 20 : −20.
ds: 4
ns: 32

5. Processing

Use standard 1D processing as described in Experiment 3.1 with exponential multiplication (lb = 0.3 Hz). Adjust a negative phase for the irradiated multiplet.

6. Result

The figure on page 462 shows the result obtained on an AMX-500 spectrometer with a BGU (10 A) gradient amplifier in a 5 mm inverse gradient probe-head. **a** is the normal ^1H NMR spectrum. In **b** the selective pulse was adjusted on the signal of H-16 at δ_H = 3.79; strong NOE effects are observed for one H-18 (δ_H = 2.8), one H-11 (δ_H = 2.58), H-17 (δ_H = 1.8), and H-13 (δ_H = 1.19). Note the very small NOE effect for H-12 (δ_H = 4.2) and the small negative effect for the other H-18 (δ_H = 3.15). In **c** the selective pulse was adjusted to the signal of H-20 at δ_H = 3.65; NOE effects are observed at the other H-20 (δ_H = 2.65) and for one H-15 (δ_H =2.38). A very small negative NOE effect can be observed for the other H-15 (δ_H = 1.36).

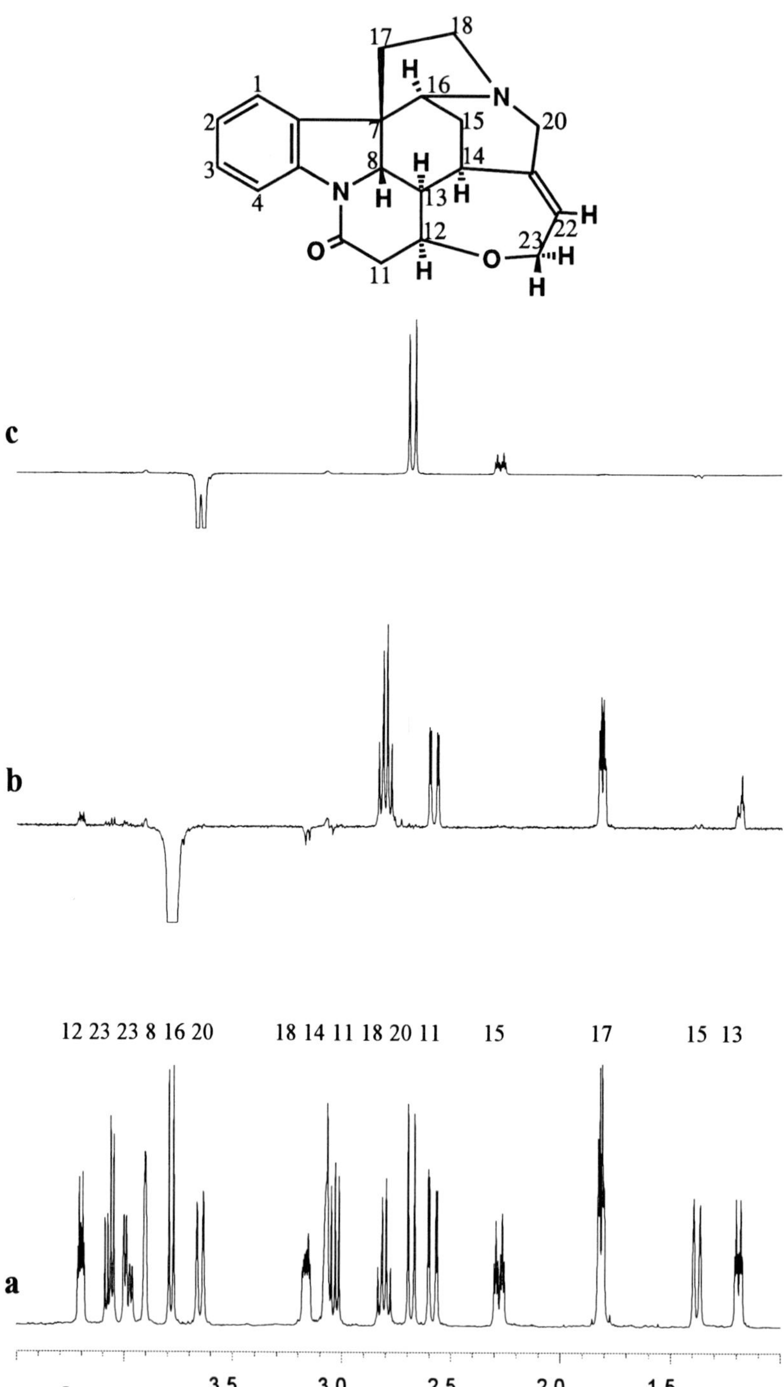
17
18
1
16
N
2
15
20
7
3
8
14
N
4
13
22
12
23
O
O
11
c
b
12 23 23 8 16 20
18 14 11 18 20 11
15
17
15 13
a
3.5
3.0
2.5
2.0
1.5
δ_H

7. Comments

After the first r.f. pulse all spins are dephased by the pulsed field gradient g1. The coherence order is changed only for the signal selected by the shaped r.f. pulse p2, thus only this is rephased by the gradient pulse g2, whereas all other signals are further dephased. This procedure is repeated by the sequence g3, p3, g4. The double gradient spin echo technique provides a distortion-free selective excitation of the desired signal with refocusing of scalar coupling. This method was termed "excitation sculpting". The magnetization is moved into the negative *z*-direction by the r.f. pulse p4. During the mixing time τ_m = d2 + d3 cross relaxation occurs, and the NOE result is transformed in observable magnetization by the read pulse p6. In the mixing time a 180° pulse p5 refocuses the *z*-magnetization caused by relaxation during τ_m; the gradient pulses g5 and g6 remove any *x,y* components caused by an imperfect 180° pulse. The phase cycle provides a difference mode. Note that, in contrast to Experiments 4.8 and 4.9, this experiment is not a steady-state technique but belongs to the transient methods like NOESY, thus quantitatively the results will differ from normal NOE difference spectra. Instead of the Gaussian pulse shape any other pulse shape may be tried; the original literature used hypersecant shapes.

8. Own Observations

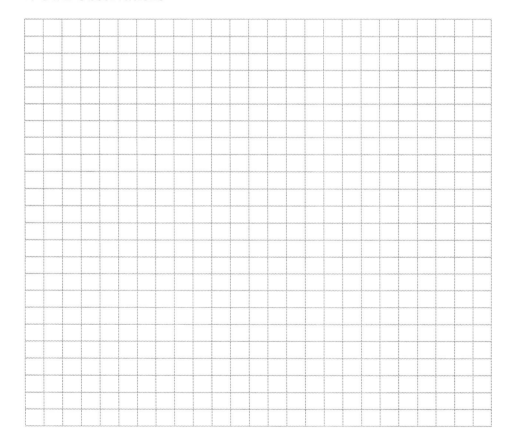

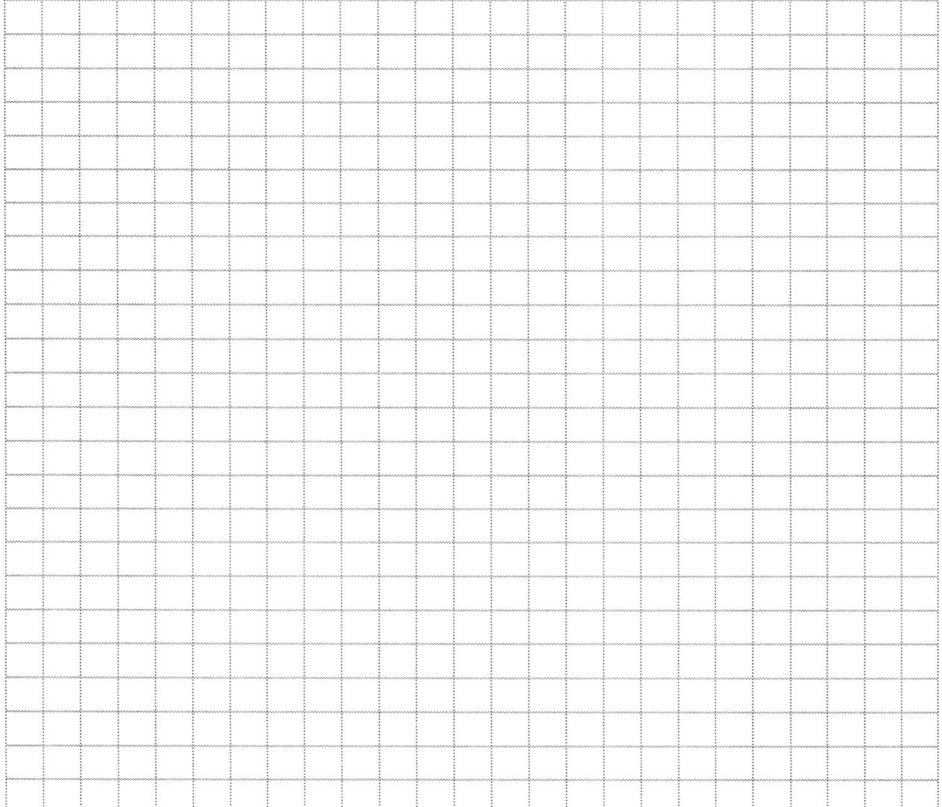

Experiment 11.11

gs-SELINCOR

1. Purpose

This experiment yields 1D proton spectra in which the desired proton signal is selected via a selective pulse on the directly bonded ^{13}C nucleus using the ^{1}J(C,H) spin coupling. In comparison to the normal SELINCOR experiment (Exp. 7.6) the HSQC instead of the HMQC principle is used. The elimination of the signals of protons bound to ^{12}C is achieved by pulsed field gradients and is thus better by an order of magnitude. This pulse scheme can thus be used as an initial building block for a variety of subsequent sequences such as SELINCOR-COSY [1], SELINCOR-TOCSY [2], or 2D *J*-resolved spectroscopy. Here we show a recent gradient-selected SELINCOR version [3] which uses a 180° selective pulse on carbon, and is applied to strychnine as an example.

2. Literature

[1] T. Fäcke, S. Berger, *Magn. Reson. Chem.* **1995**, *33*, 144–148.
[2] T. Fäcke, S. Berger, *Tetrahedron* **1995**, *51*, 3521–3524.
[3] R. Wagncr, S. Berger, *Fresenius Z. Anal. Chem.* **1997**, *357*, 470–472.

3. Pulse Scheme and Phase Cycle

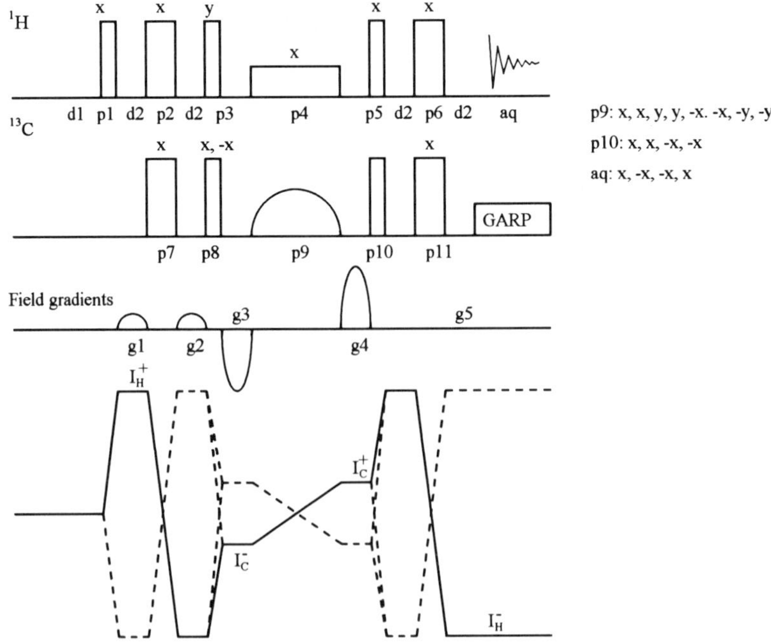

p9: x, x, y, y, -x. -x, -y, -y
p10: x, x, -x, -x
aq: x, -x, -x, x

4. Acquisition

Time requirement: 1 h

Sample: 10% strychnine in $CDCl_3$

Record both normal ^1H and ^{13}C NMR spectra of the sample, and note the offsets of the ^{13}C NMR signals to be irradiated by the selective pulse. You have to set:

td: 32 k
sw: 10 ppm
offset of ^1H frequency: middle of ^1H NMR spectrum
offset of ^{13}C frequency: on resonance of chosen ^{13}C NMR signal
p1, p3, p5,: 90° ^1H transmitter pulse
p2, p6: 180° ^1H transmitter pulse
p8, p10: 90° ^{13}C decoupler pulse
p7, p11: 180° ^{13}C decoupler pulse
p4: ^1H spin-lock pulse, same length as p9 [40 ms , 12 dB]
p9: selective 180° ^{13}C decoupler pulse, Gaussian shape, 40 ms [66 dB]
d1: 2 s
d2: $1/[4J(C,H)] = 1.8$ ms, calculated from $^1J(C,H) \approx 140$ Hz
g1–g5: sinusoidal shaped field gradients, 1 ms duration and ca. 0.02 T/m strength for the largest gradient, with gradient loop counters, ring-down delays (100 μs), lock blanking and gradient coil blanking switches according to actual instrumentation used. Gradient strength ratio: 5 : 5 : –40 : 40 : –20.
^{13}C decoupler attenuation and 90° pulse for GARP (19 dB, 70 μs)
ns: 128

5. Processing

Use standard 1D processing as described in Experiment 3.1, use an exponential window with lb = 0.5 Hz.

6. Result

The figure on page 466 shows the gs-SELINCOR spectra obtained on an AMX-500 spectrometer with a BGU (10 A) gradient unit. **a** is the normal ^1H NMR spectrum. In **b** the selective ^{13}C pulse was adjusted to C-20, in **c** to C-13, and in **d** to C-14. Note that the spectra show correct phases and relative intensities within the multiplets. There are no other signals besides the selected ones.

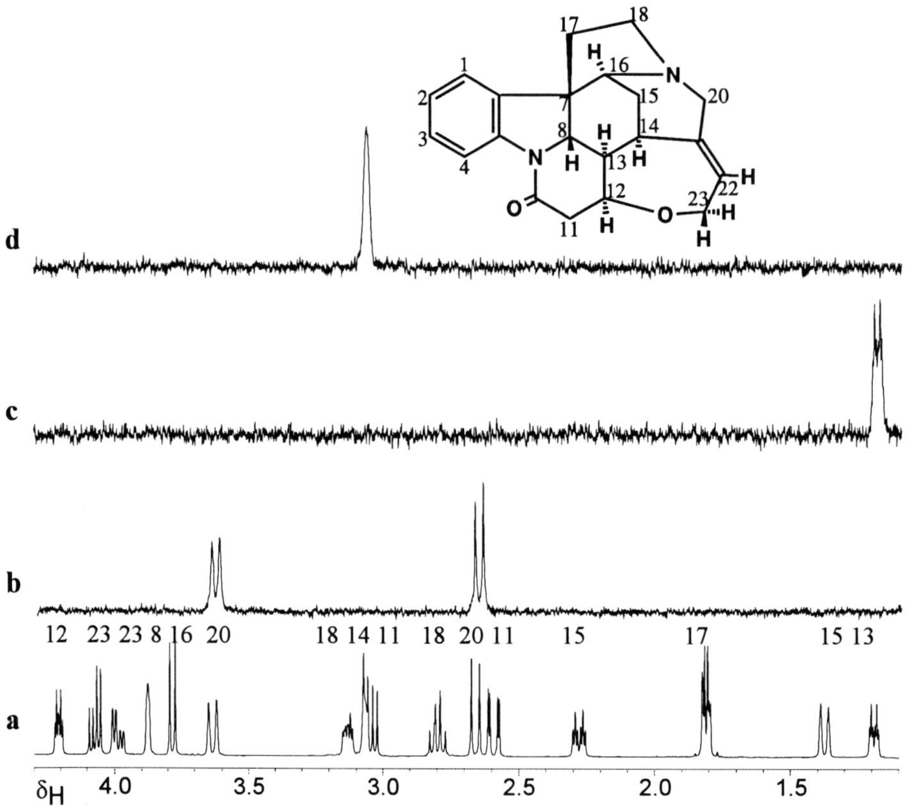

7. Comments

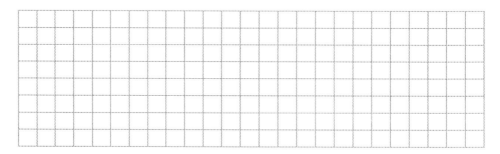

The sequence starts with an INEPT transfer from protons to carbons. A proton spin-lock pulse p4 decouples the carbon nuclei for the duration of the selective pulse. All carbon coherences are dephased by gradients g3 and g4 and only the selected ^{13}C magnetization is retained by the 180° selective pulse. Gradients g1 and g2 only control pulse imperfections. The usual reverse INEPT part of the sequence (see Exp. 6.8) transfers the magnetization back to protons, while the gradient pulse g5 rephases only the desired coherence.

8. Own Observations

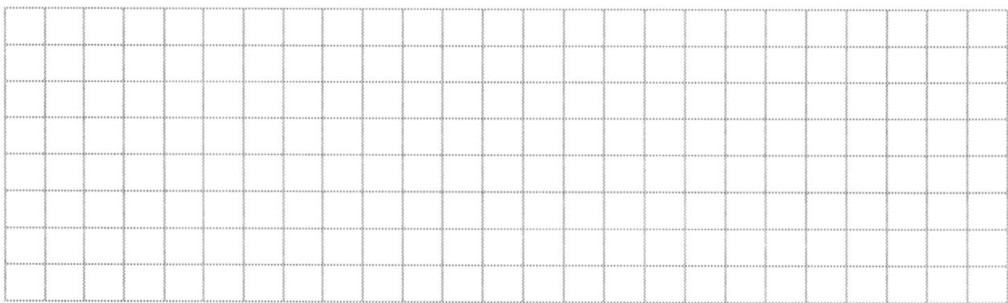

Experiment 11.12

GRECCO

1. Purpose

The GRECCO (**GR**adient **E**nhanced **C**arbon **CO**upling) experiment selectively detects ^{2}J(C,C) and ^{3}J(C,C) carbon carbon couplings, which are useful for a conformational analysis [1]. In principle this task is also performed by SELINQUATE (Exp. 7.7), the selective version of 1D-INADEQUATE (Exp. 6.12). However, whereas the suppression of signals from mono-^{13}C isotopomers is not very important if one is looking for one-bond couplings ^{1}J(C,C), efficient suppression is essential if the small coupling constants ^{2}J(C,C) or ^{3}J(C,C) are to be observed. Otherwise these signals disappear in the foot of the imperfectly suppressed center signal. The method shown here combines three principles of recent NMR developments, namely selective r.f. pulses to choose only the desired carbon signal [1], the use of cross-polarization in liquids for sensitivity enhancement [2], and pulsed field gradients which give efficient suppression of the central signal [3].

2. Literature

[1] T. Fäcke, S. Berger, *J. Am. Chem. Soc.* **1995**, *117*, 9547-9550.
[2] C. Dalvit, G. Bovermann, *J. Magn. Reson. Ser. A* **1994**, *109*, 113–116.
[3] W. Willker, D. Leibfritz, *Magn. Reson. Chem.* **1994**, *32*, 665–669.

3. Pulse Scheme and Phase Cycle

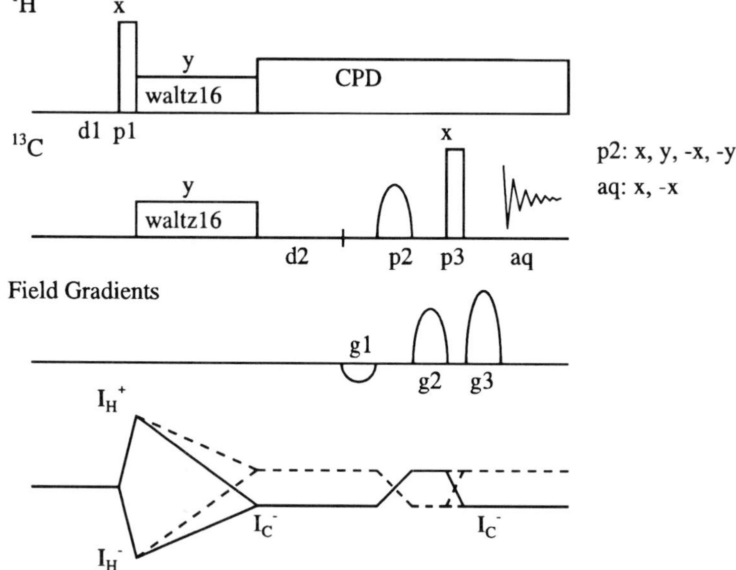

p2: x, y, -x, -y
aq: x, -x

4. Acquisition

Time requirement: 1.5 h

Sample: 90% 2-cyclohexen-1-one in [D_6]acetone

Record normal ^{1}H and ^{13}C NMR spectra of the sample, note the offset of the ^{13}C carbonyl signal and the offset of the β-olefinic hydrogen (H-3) signal. You have to set:

td: 128 k
sw: 450 ppm to avoid folding
o1: on resonance of the carbonyl ^{13}C signal
o2: on resonance of the β-olefinic ^{1}H signal (H-3)
p1: 90° ^{1}H decoupler pulse
p2: selective 180° ^{13}C transmitter pulse [Gaussian shape, 10 ms, 58 dB]
p3: 90° ^{13}C transmitter pulse
^{1}H spin-lock pulse for cross-polarization [1.8 ms, 50 dB]; carefully adjust the phase difference to p1
^{13}C spin-lock pulse for cross-polarization [pulse length must be identical to that of ^{1}H spin-lock pulse]; adjust with ^{13}C transmitter attenuation [1.8 ms, 47 dB]
duration of spin-lock ≈ $1/J$(C,H) = 125 ms, calculated from for ^3J(C,H) ≈ 8 Hz (one cycle of WALTZ-16).
d1: 10 s
d2: $1/[2J$(C,H)] = 100 ms, calculated from ^nJ(C,H) ≈ 6 Hz
g1–g3: sinusoidal shaped field gradients 1.5 ms duration and ca. 0.2 T/m strength for the largest gradient, with gradient loop counters, ring-down delays (100 μs), lock blanking and gradient coil blanking switches according to actual instrumentation used. Gradient strength ratio: −5 : 75 : 80.
decoupler attenuation and 90° pulse for CPD
ns: 512

5. Processing

Use standard 1D processing as described in Experiment 3.1 with zero filling to 128 k and applying an exponential window with lb = 0.5 Hz.

6. Result

The figure on page 469 shows the GRECCO spectra obtained on an AMX-500 spectrometer with a BGU (10 A) gradient unit using an inverse proton-optimized gradient probe-head. Unfortunately this is not the best choice for this experiment, but ^{13}C-optimized gradient probe-heads are not yet available. **a** is the signal of C-5 (J(C1,C5) = 1.6 Hz) and **b** is the signal of C-4 (J(C1,C4) = 4.8 Hz). Note that the strong signal of the carbonyl ^{13}C atom (not shown) is present in these spectra. Compare the excellent suppression of the center line with the result of Experiment 7.7.

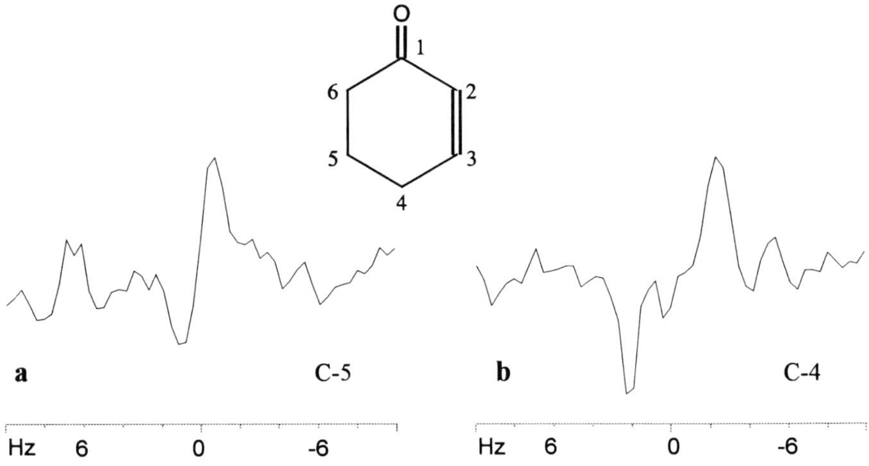

7. Comments

The sequence starts with cross-polarization from protons to ^{13}C nuclei. For this the spin coupling $^3J(C,H)$ of the carbonyl ^{13}C to the β-olefinic hydrogen is used. Thus, at the end of the spin-lock only enhanced magnetization of the carbonyl ^{13}C is present. During the delay d2, antiphase magnetization $2I_{C1x}I_{C2z}$ to the coupled carbon atoms develops and is dephased by the first gradient g1. The selective 180° carbon pulse inverts this coherence, which is then further dephased by gradient g2. The hard 90° carbon pulse p3 converts $2I_{C1x}I_{C2z}$ into $2I_{C1z}I_{C2y}$ which is rephased by gradient g3. During acquisition in-phase magnetization I_{C2x} develops and is detected. All other signals excited by p3 are dephased by gradient g3. Note that the GRECCO technique does not use a double quantum filter, in contrast to all INADEQUATE methods.

8. Own Observations

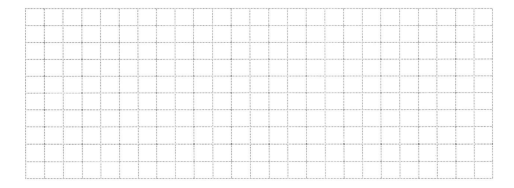

Experiment 11.13

WATERGATE

1. Purpose

The two other water suppression techniques described in this book (see Exps. 6.17 and 6.18) rely on very good shimming. The presaturation technique has the drawback that exchangeable protons may also be saturated. The jump and return method has the disadvantage of the 180° phase shift at the water resonance and the disappearance of signals in the dispersion tail of the residual water peak. The WATERGATE (**WATER** suppression by **GrA**dient **T**ailored **E**xcitation) technique which uses pulsed field gradients is claimed to be independent of line-shape, yielding a superior suppression compared with other methods. Exchangeable protons are not affected and there is no phase jump at the water resonance, although signals very close to the water resonance are also suppressed.

2. Literature

[1] M. Piotto, V. Saudek, V. Sklenar, *J. Biomol. NMR* **1992**, *2*, 661–666.

[2] V. Sklenar, M. Piotto, R. Leppik, V. Saudek, *J. Magn. Reson. Ser. A* **1993**, *102*, 241–245.

[3] L. A. Trimble, M. A. Bernstein, *J. Magn. Reson. Ser. B* **1994**, *105*, 67–72.

[4] T.-L. Hwang, A. J. Shaka, *J. Magn. Reson. Ser.* A **1995**, *112*, 275–279.

3. Pulse Scheme and Phase Cycle

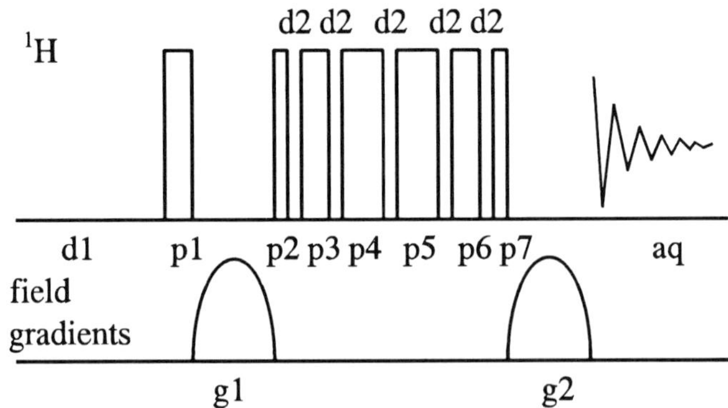

p1, p2, p3, p4: $(x)_2$, $(y)_2$, $(-x)_2$, $(-y)_2$

p5, p6, p7: $(-x)_2$, $(-y)_2$, $(x)_2$, $(y)_2$

aq: $(x)_2$, $(y)_2$, $(-x)_2$, $(-y)_2$

4. Acquisition

Time requirement: 10 min

Sample: 2 mmol sucrose in 90% H_2O/10% D_2O + 0.5 mmol DSS (2,2-dimethyl-2-silapentane-5-sulfonate, sodium salt) + trace of NaN_3 against bacteria growth.

The probe-head must be tuned to the sample used. Record a normal ^1H NMR spectrum and center the offset at the water resonance. Load the WATERGATE pulse program. You have to set:

td : 32 k
sw: 10 ppm
o1: on water resonance
p1: 90° ^1H transmitter pulse
p2, p7: 0.231·p1
p3, p6: 0.692·p1
p4, p5: 1.462·p1
d1: 1 s
d2: 300 μs
g1, g2: sinusoidal shaped field gradients, 2 ms duration and 5% truncation, with gradient loop counters, ring-down delays (100 μs), lock blanking and gradient coil blanking switches according to the actual instrumentation used. Gradient strength must be adjusted experimentally, gradient strength ratio: 1 : 1.
ds: 4
ns: 16

5. Processing

Use standard 1D processing as described in Experiment 3.1 with exponential multiplication (lb = 0.5 Hz) and a baseline correction.

6. Result

The figure on page 472 shows the result obtained on an AMX-500 spectrometer with a BGU (10 A) gradient amplifier in a 5 mm inverse *z*-gradient probe-head. Compared with Experiment 6.15 the advantages claimed for this experiment are less convincing in our hands due to baseline roll and phasing problems which are adressed in Ref. [4].

7. Comments

The sequence is, in principle, a spin-echo experiment in which the 180° pulse is embedded between two pulsed field gradients. After excitation by the first pulse p1 the field gradient g1 dephases all coherences. The pulses p2 through p7 consist of a binomial sequence (3α, 9α, 19α, 19α, 9α, 3α, with 26α = 180°).

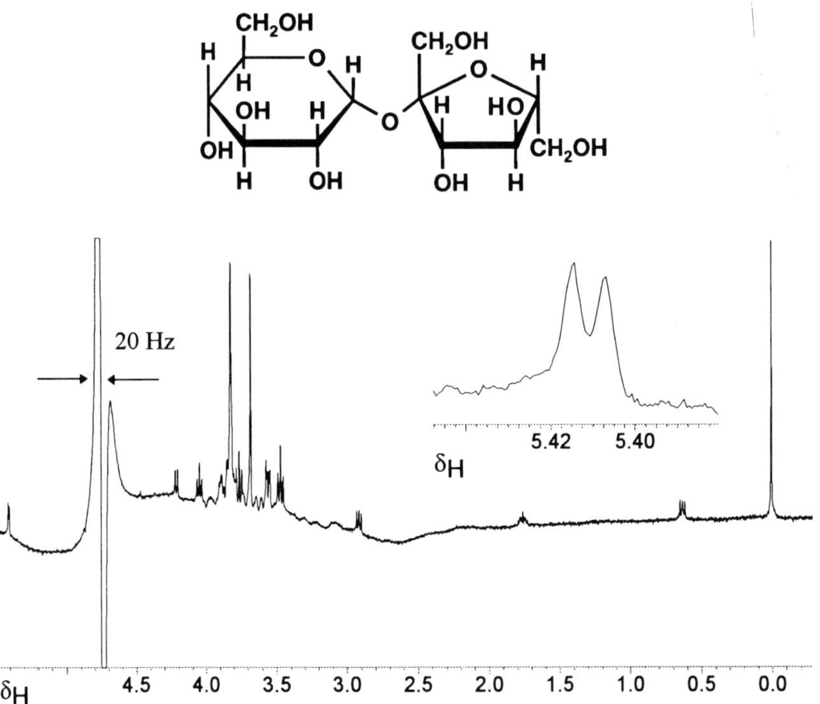

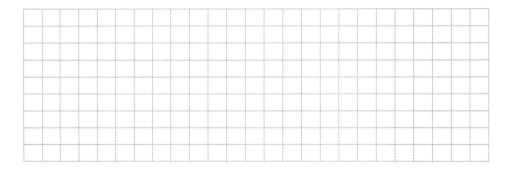

This excites all resonances, excluding those at the offset of the carrier and resonances at positions *k*/d2, where *k* is an integer. Thus the binomial sequence changes the coherence order of all resonances except that of the water. If d2 is set to 300 μs the next zero excitation of the binomial sequence is outside the spectral range. The second gradient dephases the water signal even further, and rephases all other resonances. The binomial sequence can be replaced by a special selective 180° pulse, which leaves the water resonance unchanged [1]. The advantage of the binomial sequence is that it requires less calibration and can be adjusted even if the carrier frequency is not on the water resonance. In Experiment 11.14 the sequence is doubled, leading to a perfect phase behavior.

8. Own Observations

Experiment 11.14

Water Suppression by Excitation Sculpting

1. Purpose

The WATERGATE technique as described in Experiment 11.13 provides a rather good suppression of the water signal, but has problems with baseline roll and signal phasing as can be seen in the figure on p. 472. A new technique, termed DPFGSE (**D**ouble **P**ulsed **F**ield **G**radient **S**pin **E**cho), also being called Excitation Sculpting solves this problem by applying the WATERGATE sequence twice. The DPFGSE technique can even more generally be used with any kind of filter within the two gradient echos; see, for other examples, the gradient NOE difference spectroscopy as described in Experiment 11.10 and gs-SELCOSY in Exp. 11.8. The performance of the method is shown on the 2 mM sucrose sample and is currently the most satisfying water suppression technique available.

2. Literature

[1] T.-L. Hwang, A. J. Shaka, *J. Magn. Reson. Ser.* A **1995**, *112*, 275–279.

3. Pulse Scheme and Phase Cycle

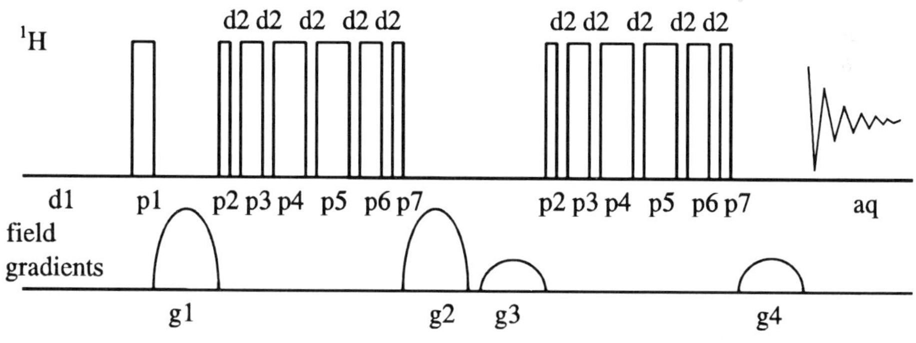

p1: x, -x p2, p3, p4: x p5, p6, p7: -x aq: x, -x

4. Acquisition

Time requirement: 5 min

Sample: 2 mmol sucrose in 90% H_2O/10% D_2O + 0.5 mmol DSS (2,2-dimethyl-2-silapentane-5-sulfonate, sodium salt) + trace of NaN_3 against bacteria growth.

The probe-head must be tuned to the sample used. Record a normal ^1H NMR spectrum and center the offset at the water resonance. Load the DPFGSE pulse program. You have to set:

td : 32 k
sw: 10 ppm
o1: on water resonance
p1: 90° ^1H transmitter pulse
p2, p7: 0.231·p1
p3, p6: 0.692·p1
p4, p5: 1.462·p1
d1: 1 s
d2: 500 μs
g1–g4: sinusoidal shaped field gradients, 1 ms duration and 5% truncation, with gradient loop counters, ring-down delays (100 μs), lock blanking and gradient coil blanking switches according to the actual instrumentation used. Gradient strength ratio: 40 : 40 : 7 : 7.
ds: 4
ns: 16

5. Processing

Use standard 1D processing as described in Experiment 3.1 with exponential multiplication (lb = 0.5 Hz) and a baseline correction.

6. Result

The figure on page 475 shows the result obtained on an AMX-500 spectrometer with a BGU (10 A) gradient amplifier and a 5 mm inverse gradient probe-head. Compared with Experiment 11.13 the method shown here is superior in all respects.

7. Comments

The essence of the water suppression scheme is the same as described in Experiment 11.13. By an elegant matrix treatment given in [1] it can be shown that, by applying the WATERGATE sequence twice, all baseline distortions and phasing problems are eliminated as long as g1 and g2 are not correlated to g3 and g4. As described in Experiment 11.13 the length of the delay d2 decides the frequency positions of zero excitation. As an additional exercise you may try to replace the binomial excitation parts of the sequence by other schemes, such as the Jump and Return technique (Exp. 6.18) or using a selective pulse on the water resonance together with a hard 180° pulse.

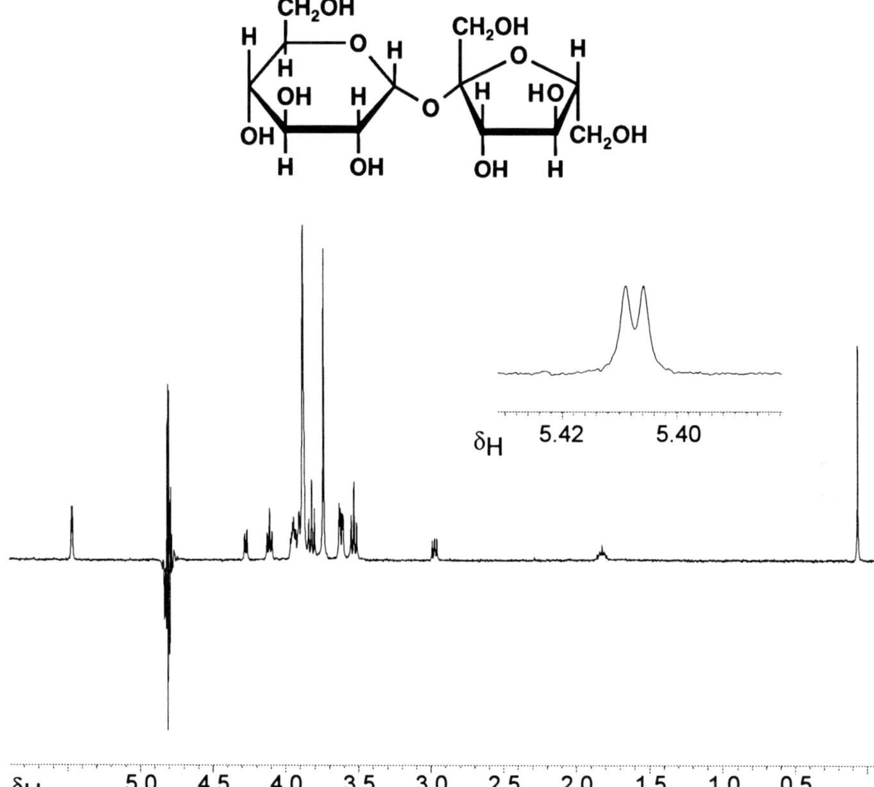

8. Own Observations

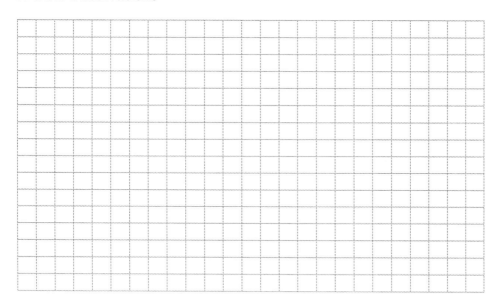

Chapter 12

2D NMR Spectroscopy With Field Gradients

One of the main problems in classical 2D spectroscopy was the differentiation between wanted and unwanted coherences. For example, one had to achieve a frequency discrimination in F_1, or to eliminate axial signals, and to distinguish between protons bound to ^{12}C and ^{13}C. These tasks were previously performed by time-consuming phase cycling. Thus, looking at the experiments in Chap. 10 one finds always the parameter ns ≥ 4. With pulsed field gradients there is a new way to achieve all these tasks. The selection of the desired coherences already happens in the probe-head, and usually, only one single transient with no phase cycling is sufficient, provided that enough substance is available. Since the NMR receiver now detects only the desired signals, its gain can be set much higher. Therefore H,C correlations using pulsed field gradients are performed in a fraction of the time formerly needed.

A considerable drawback of gradient-selected 2D experiments is their N- or P-type signal selection leading to non phase-sensitive 2D spectra when the gradients are applied during the t_1 period. One can circumvent this problem by either not using the gradients for the frequency discrimination in F_1, as shown for the gs-DQF-COSY in Exp. 12.2, or by using the echo/anti-echo technique demonstrated for the sensitivity-enhanced HSQC in Exp. 12.7.

After demonstrating a number of common 2D experiments with gradient-selection, such as gs-COSY, gs-HMQC, gs-HMBC and gs-TOCSY, we also show some advanced experiments including, for example, ^{13}C-edited NOESY, gs-HOESY or the recently published ^1H-detected 2D-INADEQUATE.

In addition, Chap. 13 includes 3D procedures with gradient-selection, one of which uses e.g. the gs-TOCSY (Exp. 12.6) as a building block.

Literature

[1] J. Keeler, R. T. Clowes, A. L. Davis, E. D. Laue, *Methods Enzym.* **1994**, *239*, 145–207.

[2] W. Price, *Annu. Rep. NMR Spectrosc.***1996**, *32*, 55–142.

[3] D. Canet, *Prog. NMR Spectrosc.* **1997**, *30*,101–135.

Experiment 12.1

gs-COSY

1. Purpose

In a 2D experiment it is necessary to distinguish the sign of the frequencies in the F_1 dimension. This is usually achieved by phase cycling, which requires up to four transients per t_1 increment including the suppression of axial peaks. Different phase cycling methods are used to perform the required coherence pathway selection. However, by using pulsed field gradients this coherence pathway selection can be achieved with only one scan per t_1 increment. Thus, if enough substance is available, a typical gs-COSY experiment with 256 time increments can be recorded in 10 minutes.

2. Literature

[1] R. E. Hurd, *J. Magn. Reson.* **1990**, *87*, 422–428.
[2] M. von Kienlin, C. T. W. Moonen, A. van der Toorn, P. C. M. van Zijl, *J. Magn. Reson.* **1991**, *93*, 423–429.

3. Pulse Scheme and Phase Cycle

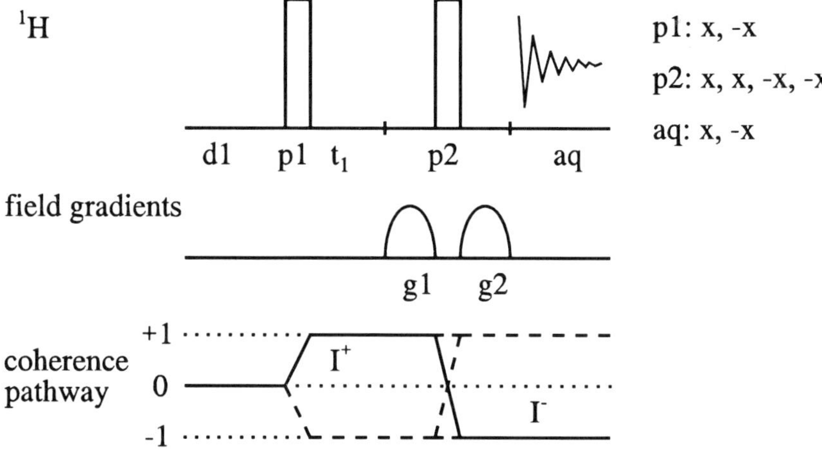

4. Acquisition

Time requirement: 10 min

Sample: 10% strychnine in $CDCl_3$

Record a normal ^1H NMR spectrum of the sample and optimize the spectral width. For the 2D experiment you have to set:

td2: 1 k data points in F_2
td1: 256 data points in F_1
sw2: 10 ppm
sw1: 10 ppm
o1: middle of ^1H NMR spectrum
p1, p2: 90° ^1H transmitter pulse
d1: 2 s
initial value for t_1 evolution: 3 μs
increment for t_1 evolution: 1/sw1
g1, g2: sinusoidal shaped field gradients with 1% truncation, 2 ms duration and 0.01 T/m strength, with gradient loop counters, ringdown delays (100 μs), lock blanking and gradient coil blanking switches according to actual instrumentation used. Gradient strength ratio 1 : 1.
ds: 4
ns: 1

5. Processing

Apply zero filling in F_1 to 512 words in order to have a symmetrical matrix of 512·512 real data points. Use unshifted sinusoidal windows in both dimensions. Apply complex Fourier transformation as for the standard N- or P-type COSY. Instead of phase correction use the absolute value mode; symmetrization of the matrix may be performed.

6. Result

The figure on p. 479 shows the expansion of a 2D spectrum obtained on an AMX-500 spectrometer with an inverse multinuclear z-gradient probe-head and a BGU (10 A) gradient unit. Symmetrization has been used. Note that the intensities of the cross peaks reflect to some extent the magnitude of the spin coupling constants.

7. Comments

Gradient experiments are best understood by using shift operators I^+ and I^- and constructing a coherence pathway diagram like the one shown above. The first pulse of the COSY sequence creates $-I_y$ magnetization, which can be written in shift operator terms as in Equation (1).

$$-I_y = -1/2\, i\, (I^+ - I^-) \tag{1}$$

Both coherence levels, the I^+ and the I^- paths are, after t_1 evolution, dephased by the first gradient g1. The second 90° pulse transforms I^+ and I^- according to Equation (2).

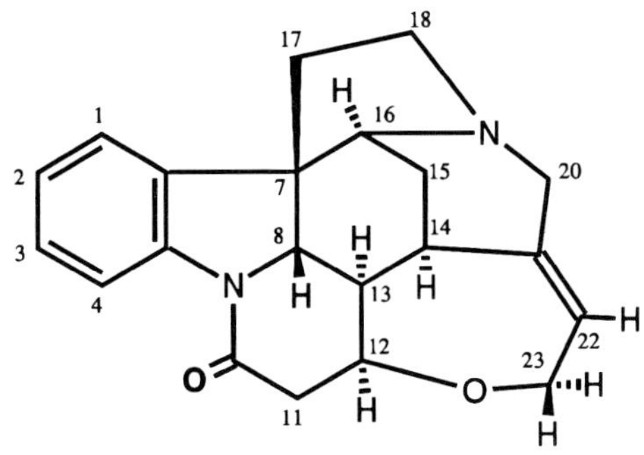

12 23 16 8 20 18 11 18 2011 15 17 15 13
23 14

13
15
17
15
11,20
18
14,11
18
20
8
16
23,23
12

2
3
4
δ_H

δ_H 4 3 2

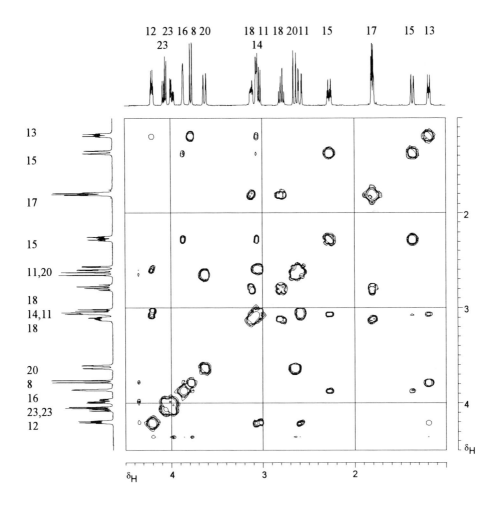

$$I^{+} \xrightarrow{90^\circ I_x} 1/2\ I^{+} + 1/2\, I^{-} + iI_z \ ;\ I^{-} \xrightarrow{90^\circ I_x} 1/2\ I^{-} + 1/2\, I^{+} - iI_z \qquad (2)$$

The second gradient, being identical to the first, further dephases those coherences, which have not changed their coherence order after the second r.f. pulse, but rephases those which changed the sign of the coherence order. Since, by definition, the NMR instruments detect only the I^{-} level, we have with this experiment selected the I^{+} pathway shown in the diagram above.

The chemical shift information developing during t_1 can be written for the I^{+} pathway as in Equation (3).

$$I^{+} \xrightarrow{\Omega t_1 I_z} I^{+}\exp(-i\Omega t_1) = I^{+}(\cos\Omega t_1 + i\sin\Omega t_1) \qquad (3)$$

Therefore, both cosine and sine components are retained and added together as in the standard N-type COSY experiment, leading to phase-skewed lineshapes and requiring complex Fourier transformation in the 2D processing.

8. Own Observations

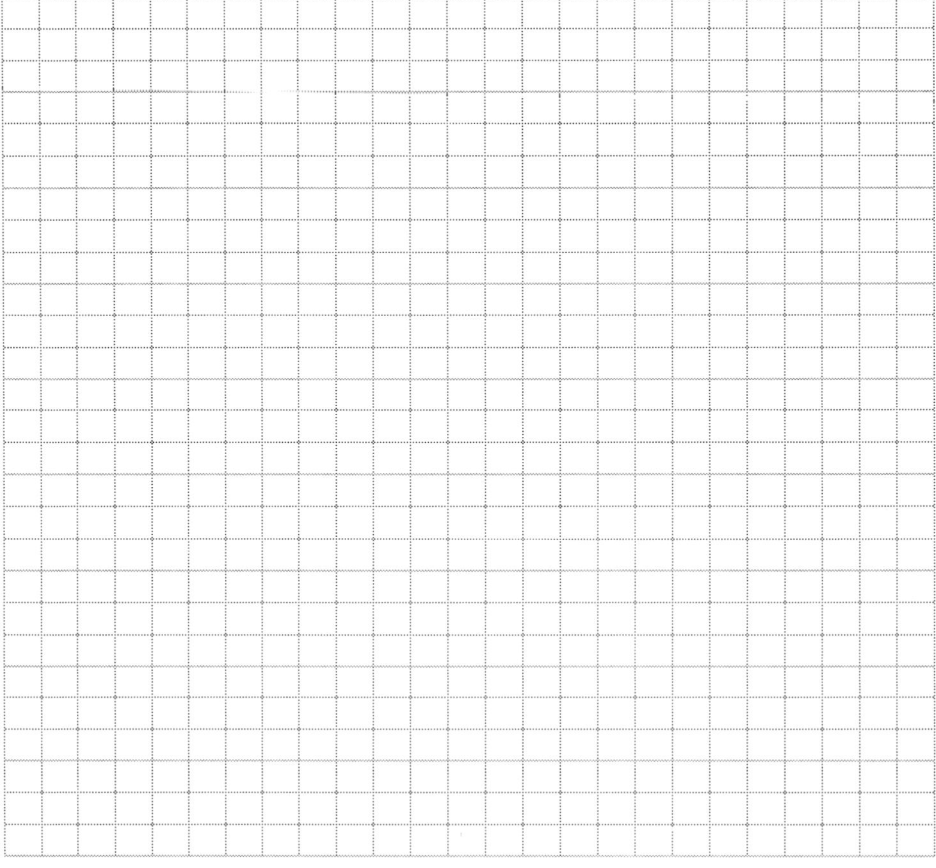

Experiment 12.2

Phase-Sensitive gs-DQF-COSY

1. Purpose

The standard gs-COSY experiment (12.1) solves the problem of distinguishing the sign of the frequencies in F_1 without phase cycling, but gives 2D spectra with phase-skewed line-shapes. The double quantum filtered (DQF) COSY experiment (see Exp. 10.8) can also be performed using gradient pulses, where the gradients only serve as the double quantum filter. In contrast to the gs-COSY, this experiment can be carried out in the phase-sensitive mode, thus giving 2D spectra with correct line-shapes, again with only a single scan per time increment. Since in COSY spectroscopy one is usually interested only in cross-peaks, which have at least double quantum character, this variant of the COSY experiment will probably become the most popular one in future. Furthermore, due to the gradient double quantum filter one can achieve good solvent suppression.

2. Literature

[1] R. E. Hurd, *J. Magn. Reson.* **1990**, *87*, 422–428.

[2] I. M. Brereton, S. Crozier, J. Field, D. M. Doddrell, *J. Magn. Reson.* **1991**, *93*, 54–62.

[3] A. L. Davis, E. D. Laue. J. Keeler, D. Moskau, J. Lohman, *J. Magn. Reson.* **1991**, *94*, 637–644.

[4] A. A. Shaw, C. Salaun, J.-F. Dauphin, B. Ancian, *J. Magn. Reson. Ser. A* **1996**, *120*, 110–115.

3. Pulse Scheme with Phase Cycle

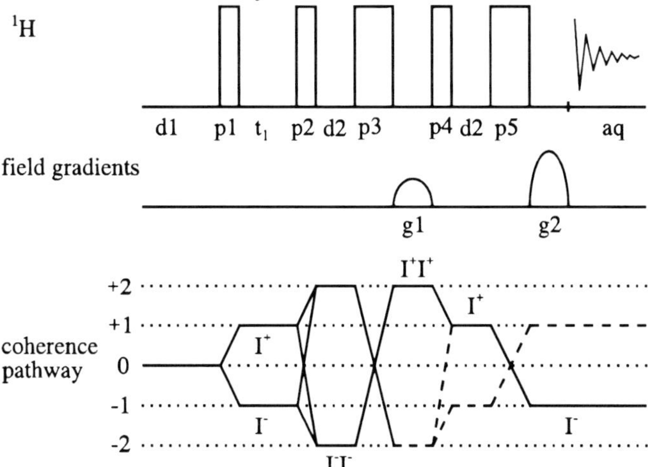

p1: x, -x phase cycle for p1 incremented according to TPPI

p2, p3, p4, p5: x aq: x, -x

4. Acquisition

Time requirement: 1 h

Sample: 10% strychnine in $CDCl_3$

Record a normal ^1H NMR spectrum of the sample and optimize the spectral width. Change to the 2D mode of the spectrometer software and load the pulse program for gs-DQF COSY. You have to set:

td2: 2 k data points in F_2
td1: 512 data points in F_1
sw2: 10 ppm
sw1: 10 ppm
o1: middle of ^1H NMR spectrum
p1, p2, p4: 90° ^1H transmitter pulse
p3, p5: 180° ^1H transmitter pulse
d1: 2 s
d2: equal to effective length of gradient used, here 2 ms
initial value for t_1 evolution: 3 μs
increment for t_1 evolution: 1/[2·sw1]
g1, g2: sinusoidal shaped field gradients with 1% truncation, 2 ms duration and 0.01 T/m strength, with gradient loop counters, ringdown delays (100 μs), lock blanking and gradient coil blanking switches according to actual instrumentation used. Gradient strength ratio: 1 : 2.
rg: One must be very careful in setting the receiver gain for this experiment. The gradient filter allows only the desired coherences to pass into the receiver, however, the double quantum coherences develop only at higher t_1 increments. The receiver gain must therefore be set using a high t_1 increment to avoid overloading.
ds: 4
ns: 4

5. Processing

Apply zero filling in F_1 to 1 k words in order to have a symmetrical matrix of 1024·1024 data points. Use exponential or Gaussian windows in both dimensions corresponding to the Hz/point resolution of your data set. Apply real Fourier transformation in both dimensions. Phase correction in F_2 can be performed after the 2D transformation in order to get clean up/down patterns of the cross peaks. Zero order phase correction of 90° has to be applied in the F_1 dimension.

6. Result

The figure on page 483 shows the expansion of a 2D spectrum obtained on an AMX-500 spectrometer with an inverse multinuclear *z*-gradient probe-head and a BGU (10 A) gradient unit. Note that dotted contours represent negative signals.

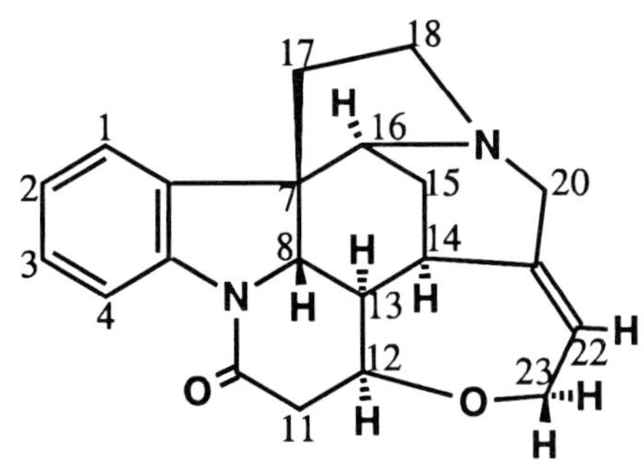

18 11 18 20 11 15β 17 15α
14

15α
17
15β
11,20
18
14,11
18

1.5
2.0
2.5
3.0
δH

δH
2.5
2.0
1.5

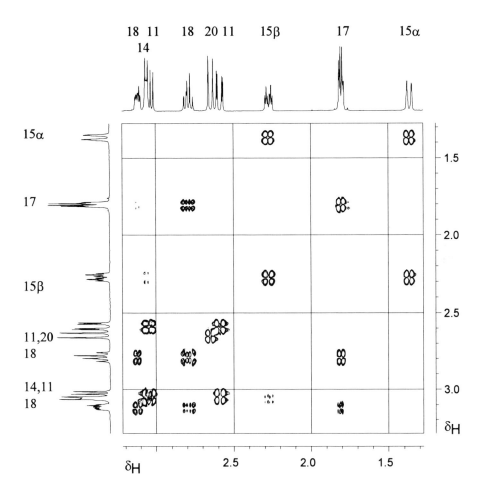

7. Comments

As the normal DQF-COSY procedure (see Exp. 10.8) this experiment uses three 90° pulses where the first two generate double quantum magnetization, whereas the last (reading) pulse transfers it back into observable magnetization. Instead of phase cycling as in Experiment 10.8 the two gradients act as the double quantum filter. The two 180° pulses correct the phase problems introduced by the final length of the gradients; therefore d2 should be set exactly equal to the total gradient length including ring-down time.

As can be seen from the coherence pathway diagram above, the first gradient g1 acts during a period when double quantum magnetization I^+I^+ is present (coherence level +2) , whereas the second acts during a period when single quantum coherence $-I^-$ is present, thus g2 must have twice the gradient strength of g1. All other coherences are further dephased and are not observable. Note that during t_1 both I^+ and I^- are developing chemical shift information; thus the full phase information is retained and can be stored separately for different time increments e.g. using the TPPI mode of phase cycling for the first pulse. Of course, due to the double quantum filter there is a sensitivity loss compared with the gs-COSY procedure described in Experiment 12.1.

8. Own Observations

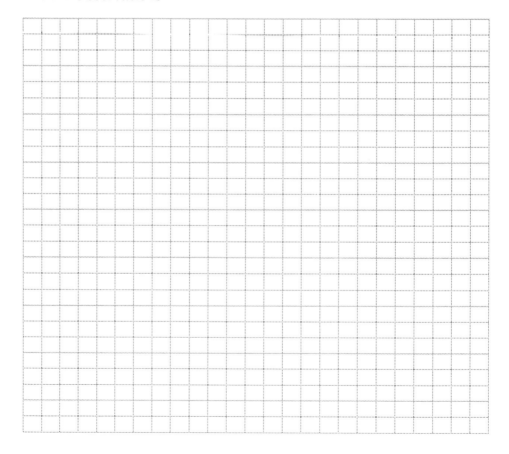

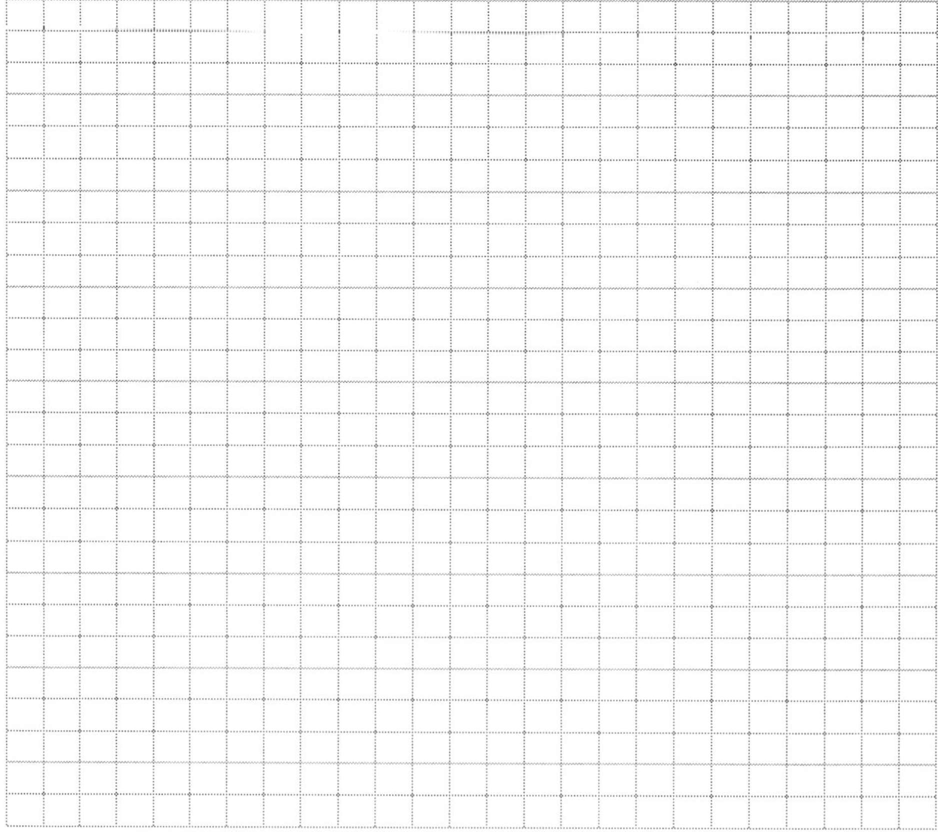

Experiment 12.3

gs- HMQC

1. Purpose

The standard HMQC experiment (see Exp. 10.14) uses the BIRD filter and phase cycling to suppress the undesired signals of protons bound to ^{12}C. With pulsed field gradients the selection of the desired coherences can be drastically improved. This yields artefact free H,C correlation spectra in a fraction of the time needed previously, since the receiver gain of the proton channel can be set to a very high value. The version shown here is not phase-sensitive.

2. Literature

[1] R. E. Hurd, B. K. John, *J. Magn. Reson.* **1991**, *91*, 648–653.

[2] J. Ruiz-Cabello, G. W. Vuister, C. T. W. Moonen, P. van Gelderen, J. S. Cohen, P. C. M. van Zijl, *J. Magn. Reson.* **1992**, *100*, 282–303.

[3] W. Willker, D. Leibfritz, R. Kerssebaum, W. Bermel, *Magn. Reson. Chem.* **1993**, *31*, 287–292.

3. Pulse Scheme and Phase Cycle

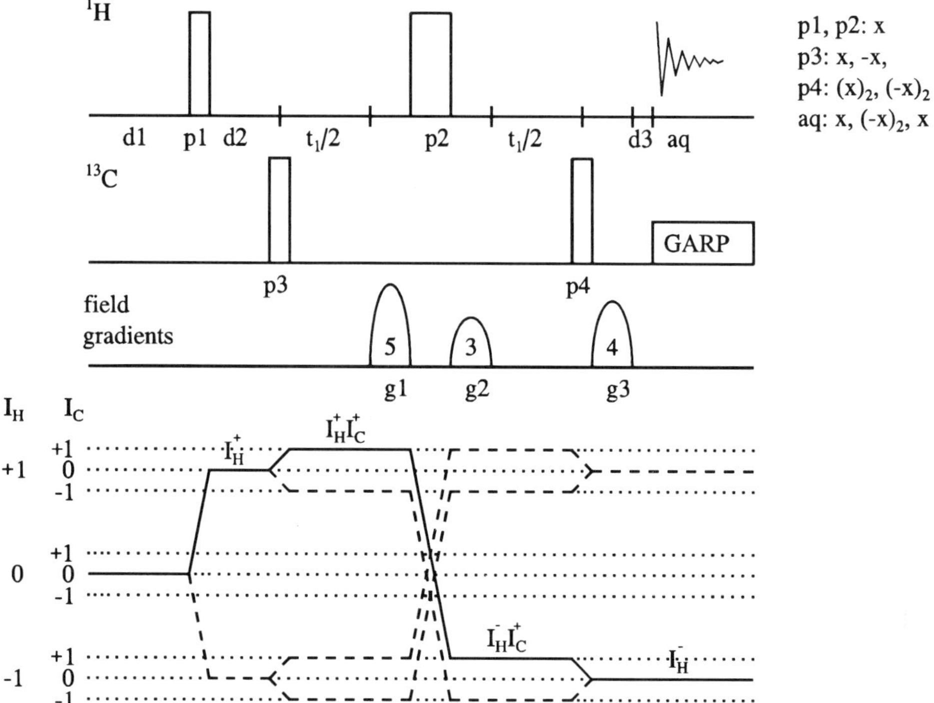

4. Acquisition

Time requirement: 10 min

Sample: 10% strychnine in $CDCl_3$

Record normal ^1H and ^{13}C NMR spectra of the sample and optimize the spectral widths for CH_n signals. Change to the 2D mode of the spectrometer and load the gs-HMQC pulse program. You have to set:

td2: 1 k data points in F_2
td1: 256 data points in F_1
sw2: 10 ppm
sw1: 165 ppm
offset of ^1H frequency: middle of ^1H NMR spectrum
offset of ^{13}C frequency: middle of ^{13}C NMR spectrum
p1: 90° ^1H transmitter pulse
p2: 180° ^1H transmitter pulse
p3, p4: 90° ^{13}C decoupler pulse
d1: 2 s
d2: $1/[2J(C,H)] = 3.57$ ms, calculated from $^1J(C,H) \approx 140$ Hz
d3: set equal to d2 minus gradient length
start increment for t_1 evolution: 3 μs
increment for t_1 evolution: $1/[2 \cdot sw1]$
g1, g2, g3: sinusoidal shaped field gradients with 5% truncation, 2 ms duration and ca. 0.01 T/m strength, with gradient loop counters, ring-down delays (100 μs), lock blanking and gradient coil blanking switches according to actual instrumentation used. Gradient strength ratio: 5 : 3 : 4
^{13}C decoupler attenuation and 90° pulse for GARP (19 dB, 70 μs)
ns: 1

5. Processing

Apply zero filling in F_1 to 512 words in order to have a matrix of 512·512 real data points. Before Fourier transformation use an exponential window in F_2 with lb = 5 Hz and $\pi/3$ shifted squared sine window in F_1. Phase correction is unnecessary, since the spectrum is displayed in the magnitude mode.

6. Result

The figure on page 487 shows the 2D spectrum obtained on an AMX-500 spectrometer with an inverse multinuclear z-gradient probe head and a BGU (10 A) gradient unit. Note that ^{13}C nuclei with attached diastereotopic protons show two different correlation signals.

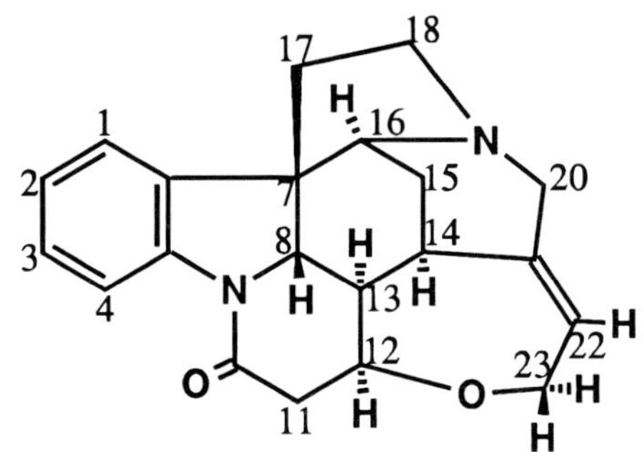

4 3 1 2 22 12 16 18 11 20 15 15 13
23 8 20 14 18 11 17

15
14
11,17
13
18
20
8,16
23
12
4
1
2

40
60
80
100
120
δ_C

δ_H 6 4 2

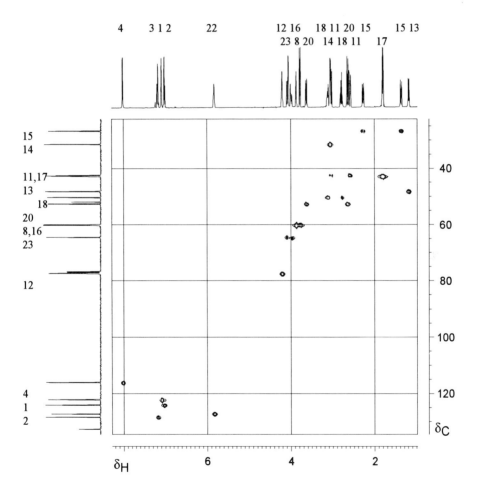

7. Comments

The r.f. pulses are the same as in the basic HMQC sequence, where the first two generate double quantum magnetization, which in the coherence pathway diagram is labeled as $I_H^+ I_C^+$. The relative gradient strength 5 of the dephasing gradient g1 corresponds to this coherence, since $\gamma_H \approx 4\gamma_C$, and the sum of the γ values is the relevant quantity. The 180° pulse in the proton channel transforms the coherence into $I_H^- I_C^+$. At this stage the relative sum of γ values is –3. During acquisition, only I_H^- is present with a relative γ value of –4. Thus, with the gradient strengths used, Equation (1) yields zero only for the selected pathway, whereas all other coherences are effectively dephased. Of course, there are other gradient ratios for which Equation (1) is also fulfilled.

$$g1\,(\gamma_H + \gamma_C) + g2\,(-\gamma_H + \gamma_C) + g3\,(-\gamma_H) = 0 \qquad (1)$$

8. Own Observations

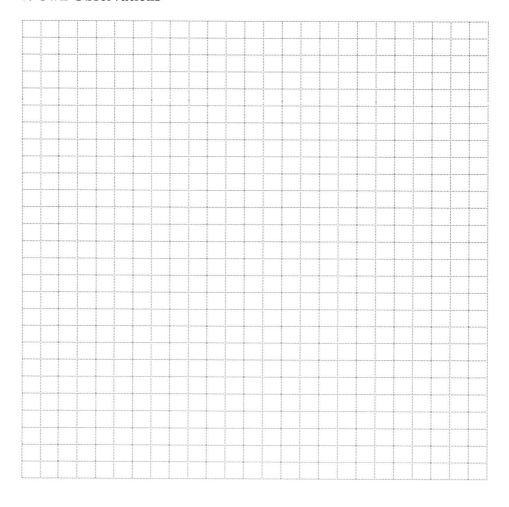

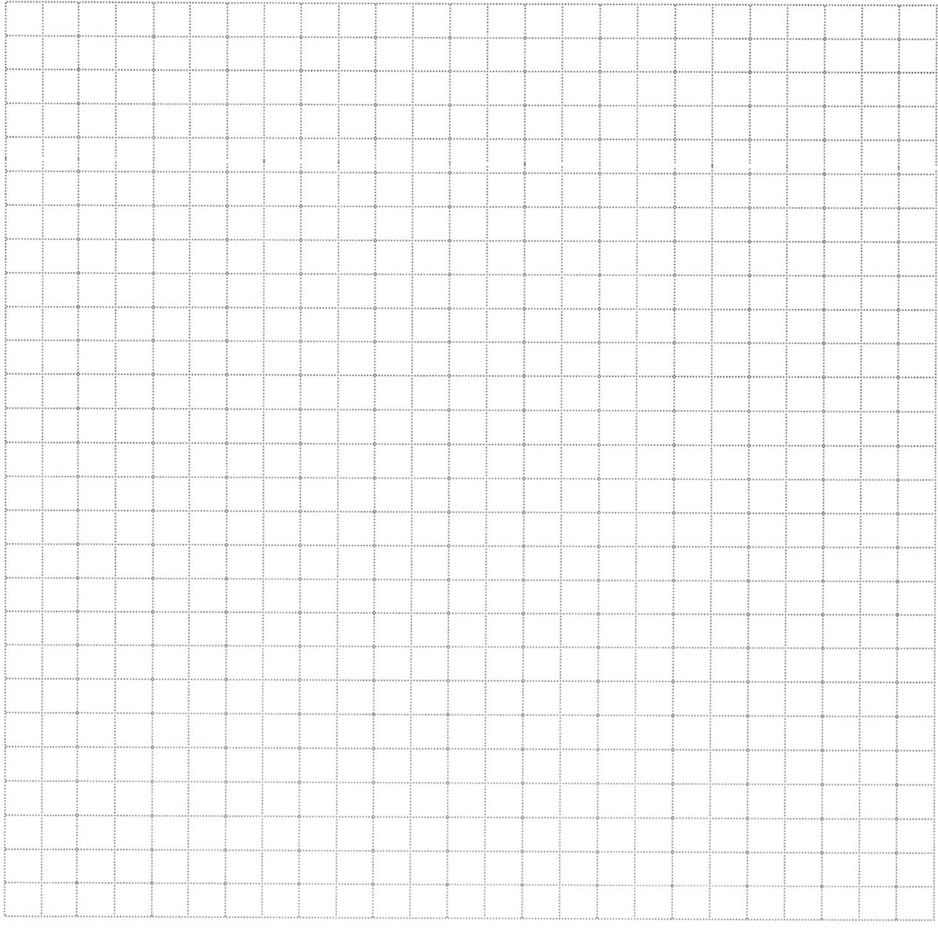

Experiment 12.4

gs- HMBC

1. Purpose

To obtain H,C correlations via ^2J(C,H) and ^3J(C,H) the HMBC pulse sequence was developed which contains a low-pass filter to suppress correlations via ^1J(C,H) (see Exp. 10.16). Here we describe the gradient-selected version [2] which is not phase-sensitive. The experiment is usually performed without GARP ^{13}C decoupling. The sequence allows one to set the receiver gain considerably higher than in the normal HMBC experiment, which leads to far better results in a fraction of the time.

2. Literature

[1] A. Bax, M. F. Summers, *J. Am. Chem. Soc.* **1986**, *108*, 2093–2094.
[2] W. Willker, D. Leibfritz, R. Kerssebaum, W. Bermel, *Magn. Reson. Chem.* **1993**, *31*, 287–292.
[3] J. Ruiz-Cabello, G. W. Vuister, C. T. W. Moonen, P. van Gelderen, J. S. Cohen, P. C. M. van Zijl, *J. Magn. Reson.* **1992**, *100*, 282–303.

3. Pulse Scheme and Phase Cycle

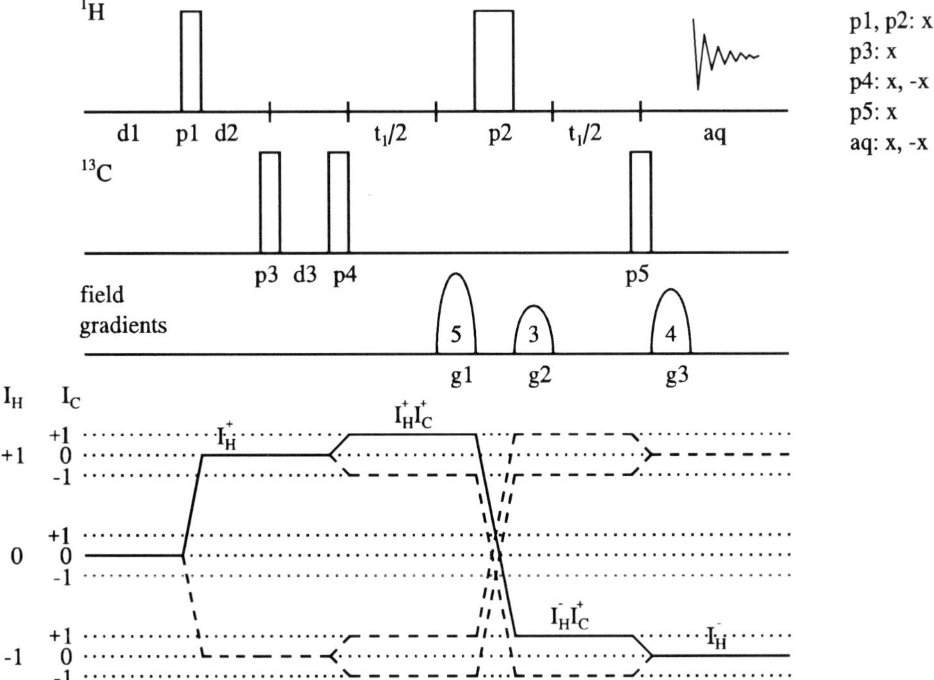

p1, p2: x
p3: x
p4: x, -x
p5: x
aq: x, -x

4. Acquisition

Time requirement: 0.5 h

Sample: 10% strychnine in $CDCl_3$

Record normal ^{1}H and ^{13}C NMR spectra of the sample and optimize the spectral widths. Change to the 2D mode of the spectrometer and load the gs-HMQC pulse program. You have to set:

td2: 1 k data points in F_2
td1: 256 data points in F_1
sw2: 10 ppm
sw1: 165 ppm
offset of ^{1}H frequency: middle of ^{1}H NMR spectrum
offset of ^{13}C frequency: middle of ^{13}C NMR spectrum
p1: 90° ^{1}H transmitter pulse
p2: 180° ^{1}H transmitter pulse
p3, p4, p5: 90° ^{13}C decoupler pulse
d1: 2 s
d2: $1/[2J(C,H)] = 3.57$ ms, calculated from $^{1}J(C,H) \approx 140$ Hz
d3: $1/[2J(C,H)] = 60$ ms, calculated from $^{n}J(C,H) \approx 8$ Hz
start increment for t_1 evolution: 3 μs
increment for t_1 evolution: $1/[2 \cdot sw1]$
g1, g2, g3: sinusoidal shaped field gradients with 5% truncation, 2 ms duration and ca. 0.01 T/m strength, with gradient loop counters, ring-down delays (100 μs), lock blanking and gradient coil blanking switches according to actual instrumentation used. Gradient strength ratio: 5 : 3 : 4
ns: 2

5. Processing

Apply zero filling in F_1 to 512 words in order to have a matrix of 512·512 real data points. Before Fourier transformation use an exponential window in F_2 with lb = 5 Hz and $\pi/3$ shifted squared sine window in F_1. Phase correction is unnecessary, since the spectrum is displayed in the magnitude mode.

6. Result

The figure on page 491 shows the 2D spectrum obtained on an AMX-500 spectrometer with an inverse multinuclear *z*-gradient probe-head and a BGU (10 A) gradient unit. Note the wealth of information obtainable from $^{2}J(C,H)$ and $^{3}J(C,H)$ couplings in this molecule.

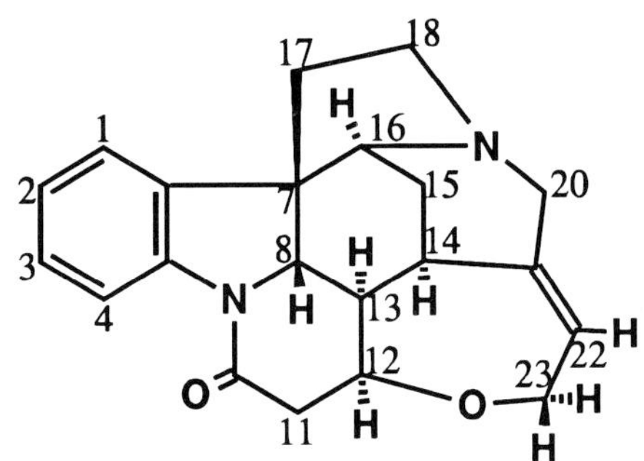

4 3 1 2 22 12 16 18 11 20 15 15 13
23 8 20 14 18 11 17

15
14
11,17
13,18
7,20
8,16
23
12
4
1,2
22,3
6
21
5
10

50
100
150
δ_C

δ_H 6 4 2

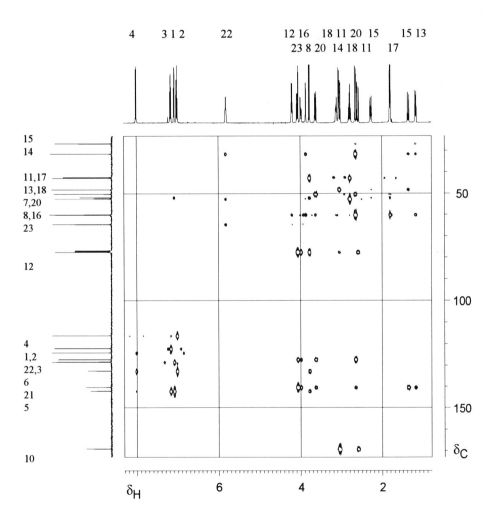

7. Comments

The first ^{13}C pulse serves as a low pass filter, by phase cycling of this pulse the signals of protons experiencing a one-bond coupling ^{1}J(C,H) shall be suppressed. Therefore the action of this pulse is not considered in the coherence pathway diagram. However, this suppression does not work equally well for all proton signals, and despite this low pass filter some correlation signals via ^{1}J(C,H) can be seen in any HMBC spectrum. In order to distinguish these signals from the desired correlations it is advisable not to use GARP decoupling. The second ^{13}C pulse selects proton signals experiencing a long-range C,H coupling. The rest of the sequence is identical to the gs-HMQC sequence as described in Experiment 12.3. Thus the discussion of the coherence pathway diagram is not repeated here. An advanced version which allows GARP decoupling is shown in Experiment 12.5 and a 3D variant is demonstrated in Experiment13.4.

8. Own Observations

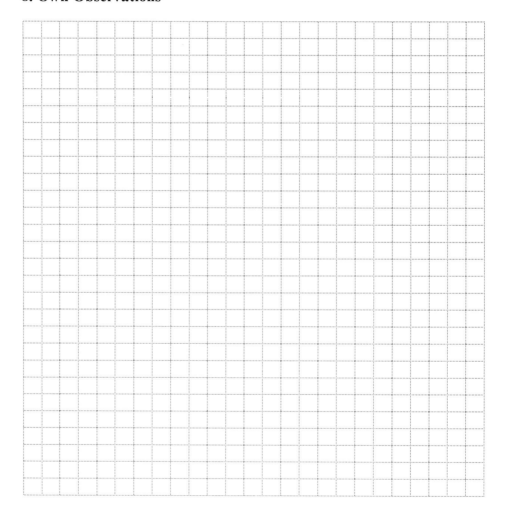

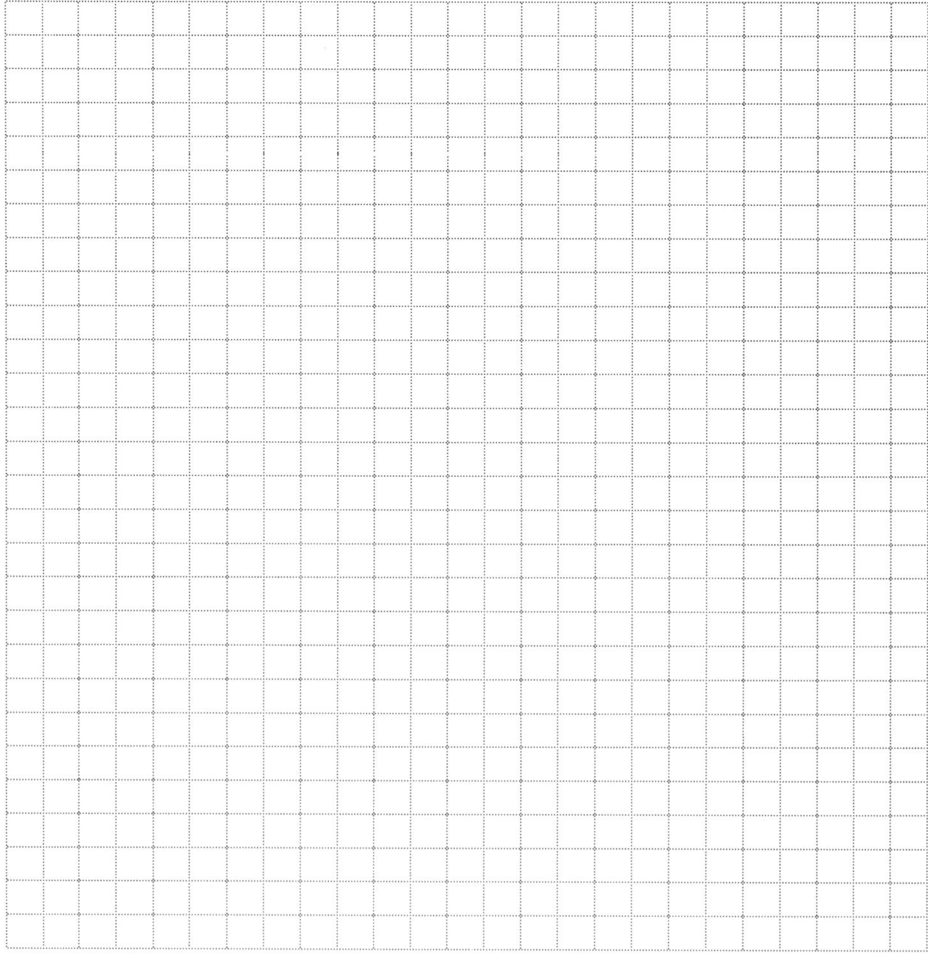

Experiment 12.5

ACCORD-HMBC

1. Purpose

The ACCORD-HMBC method [1] shown in this experiment has two distinct advantages over the standard gradient-selected method outlined in Experiment 12.4. It employs a dual stage low pass filter [2] to suppress effectively all ^{1}J correlation signals. Therefore GARP decoupling can be used without the problem of ambiguity between ^{1}J and $^{2/3}J$ correlations. In addition it uses the ACCORDION principle [3] to sample over a range of $^{2/3}J$ coupling constants, thus more correlation signals will appear compared with the HMBC method with a fixed polarization delay. Here we show the results using strychnine as an example.

2. Literature

[1] R. Wagner, S. Berger, *Magn. Reson. Chem.* **1998**, in press.
[2] G. Bodenhausen, R. R. Ernst, *J. Am. Chem. Soc.* **1982**, *104*, 1304–1309.
[3] H. Kogler, O. W. Sørensen, G. Bodenhausen, R. R. Ernst, *J. Magn. Reson.* **1983**, *55*, 157–163.

3. Pulse Scheme and Phase Cycle

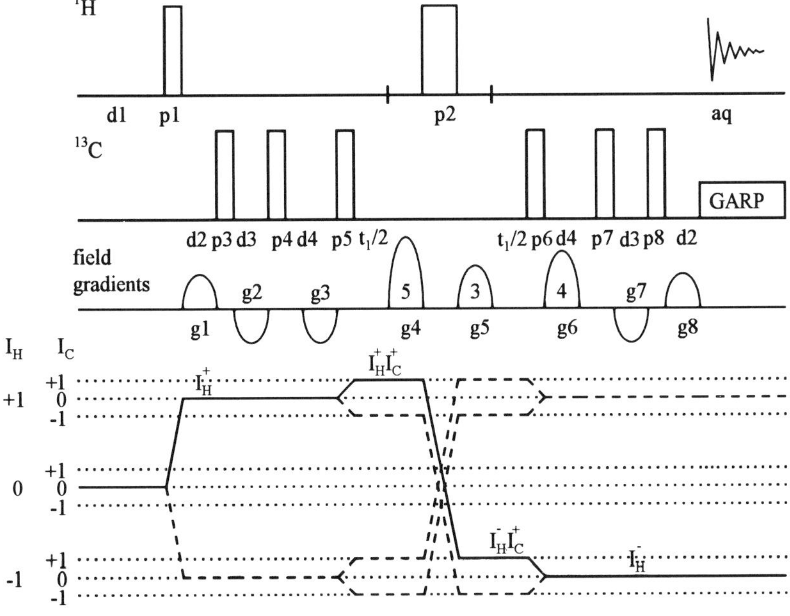

p1: x p2: $(X)_4$, $(-x)_4$ p3: $(x)_2$, $(-x)_2$ p4: x, -x p5: $(x)_8$, $(-x)_8$ aq: $(x, -x)_4$, $(-x, x)_4$

4. Acquisition

Time requirement: 25 min

Sample: 10% strychnine in $CDCl_3$

Record normal ^1H and ^{13}C NMR spectra of the sample and optimize the spectral widths. Change to the 2D mode of the spectrometer and load the ACCORD-HMBC pulse program. You have to set:

td2: 2 k data points in F_2
td1: 256 data points in F_1
sw2: 10 ppm
sw1: 165 ppm
offset of ^1H ferquency: middle of ^1H NMR spectrum
offset of ^{13}C frequency: middle of ^{13}C NMR spectrum
p1: 90° ^1H transmitter pulse
p2: 180° ^1H transmitter pulse
p3–p8: 90° ^{13}C decoupler pulse
d1: 2 s
d2: $1/[2(J(C,H_{min}) + 0.146\cdot(J(C,H_{max}) - J(C,H_{min}))]$ – effective gradient length = 2.7 ms, calculated from $^1J(C,H_{max})$ = 163 Hz and $^1J(C,H_{min})$ = 128 Hz and effective gradient length of 1.05 ms
d3: $1/[2(J(C,H_{max}) - 0.146\cdot(J(C,H_{max}) - J(C,H_{min}))]$ – effective gradient length = 2.1 ms, calculated from $^1J(C,H_{max})$ = 163 Hz and $^1J(C,H_{min})$ = 128 Hz and effective gradient length of 1.05 ms
d4: initial value for long range polarization, 200 ms, calculated from $^{2/3}J(C,H)$ = 2.5 Hz; d4 is decremented during the experiment. The decrement is calculated by the ACCORDION range (200 ms – 20 ms)/ td1 = 0.7 ms corresponding 2.5 to 25 Hz.
start increment for t_1 evolution: 3 μs
increment for t_1 evolution: 1/[2·sw1]
g1–g8: sinusoidal shaped field gradients, 1 ms duration and ca. 0.01 T/m strength, ring-down delays (50 μs), lock blanking and gradient coil blanking switches according to actual instrumentation used. Gradient strength ratio: 15 : –10 : –5 : 50 : 30 : 40 : –5 : 5
^{13}C decoupler attenuation and 90° pulse for GARP [ca. 70 μs at 12 dB]
ns: 2

5. Processing

Apply zero filling in F_1 to 512 words in order to have a matrix of 512·512 real data points. Before Fourier transformation use an exponential window in F_2 with lb = 5 Hz and $\pi/3$ shifted squared sine window in F_1. Phase correction is unnecessary, since the spectrum is displayed in the magnitude mode.

6. Result

The figure shows the 2D spectrum obtained on an Avance DRX-400 spectrometer with an inverse multinuclear *z*-gradient probe-head. Note that the signals contain an additional modulation in F_1 which stems from the varaiable delay d4. The expansion shown demonstrates, in comparison with the result of Experiment 12.4, that additional long range correlations can be seen. For example, the correlations of H-12 with C-10, of both protons H-15 with C-21 and of H-17 and H-16 with C-6 are all observable.

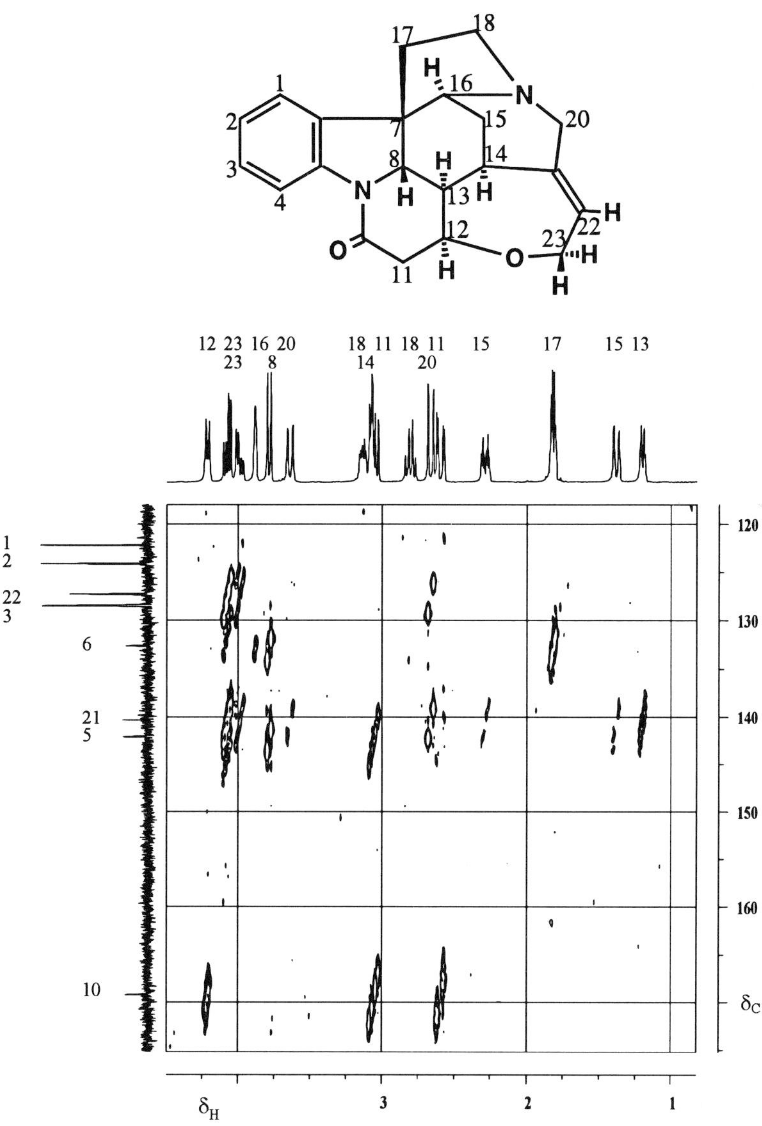

7. Comments

The dual stage low pass filter also employed in Experiment 12.9 is gradient supported and causes a very efficient suppression of correlation signals for one-bond coupling constants. This low pass filter is used twice in the sequence, at the beginning (d2, g1, p3, d3, g2, p4) and at the end (p7, d3, g7, p8, g8, d2). The GARP decoupling introduces a significant gain in sensitivity. To keep the overall length of the sequence to a minimum, the ACCORDION principle is used in a way that during increasing d0, the delay d4 is decreased. Therefore first the correlations due to the small coupling constants and later those due to the larger coupling constants are sampled. A drawback of the sequence is the additional modulation, so that each component of a proton multiplet correlates on a slightly different frequency in F_1 and the corresponding carbon signal bisects this pattern at its center. The idea of sampling different coupling constants during an HMBC experiment can also be performed in a 3D manner, see Experiment 13.4 as an example.

8. Own Observations

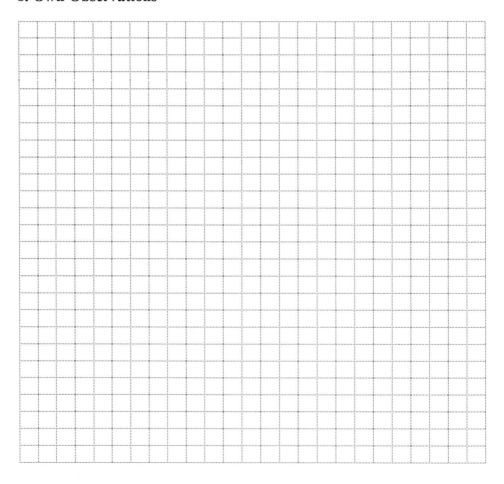

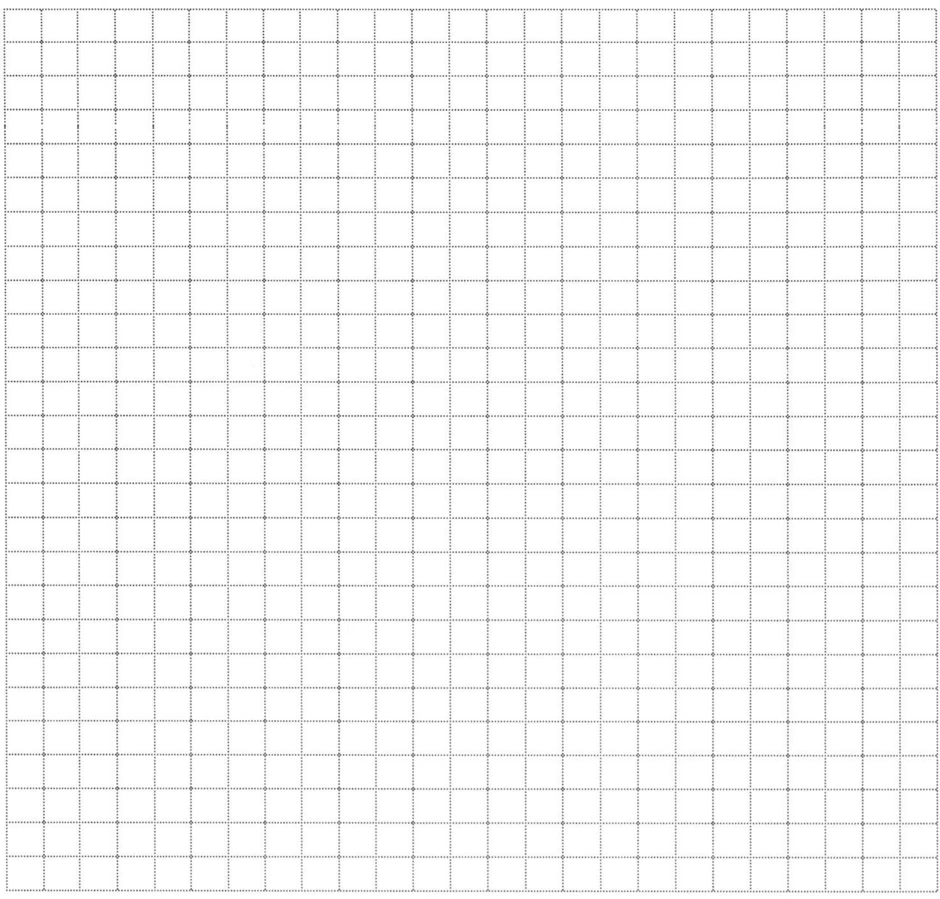

Experiment 12.6

Phase-Sensitive gs-HSQC with Sensitivity Enhancement

1. Purpose

The three gradient-selected heteronuclear correlations given in Experiments 12.3–12.5 are not phase-sensitive. Gradient selected experiments can also be performed in such a way as to yield pure absorption spectra, which give better resolution of signals in crowded regions due to the Lorentzian lineshape. The experiment shown here uses the echo/anti-echo selection method [1]. Another feature of this experiment is a sensitivity enhancement by a factor of $\sqrt{2}$; however, this occurs only for CH groups [2,3].

2. Literature

[1] L. E. Kay, P. Keifer, T. Saarinen, *J. Am. Chem. Soc.* **1992**, *114*, 10663–10665.
[2] A. G. Palmer III, J. Cavanagh, P. E. Wright, M. Rance, *J. Magn. Reson.* **1991**, *93*, 151–170.
[3] G. Kontaxis, J. Stonehouse, E. D. Laue, J. Keeler, *J. Magn. Reons. Ser. A* **1994**, *111*, 70–76.
[4] J. Schleucher, M. Schwendinger, M. Sattler, P. Schmidt, O. Schedletzky, S.J. Glaser, O.W. Sørensen, C. Griesinger, *J. Biomol. NMR* **1994**, *4*, 301-306.

3. Pulse Scheme and Phase Cycle

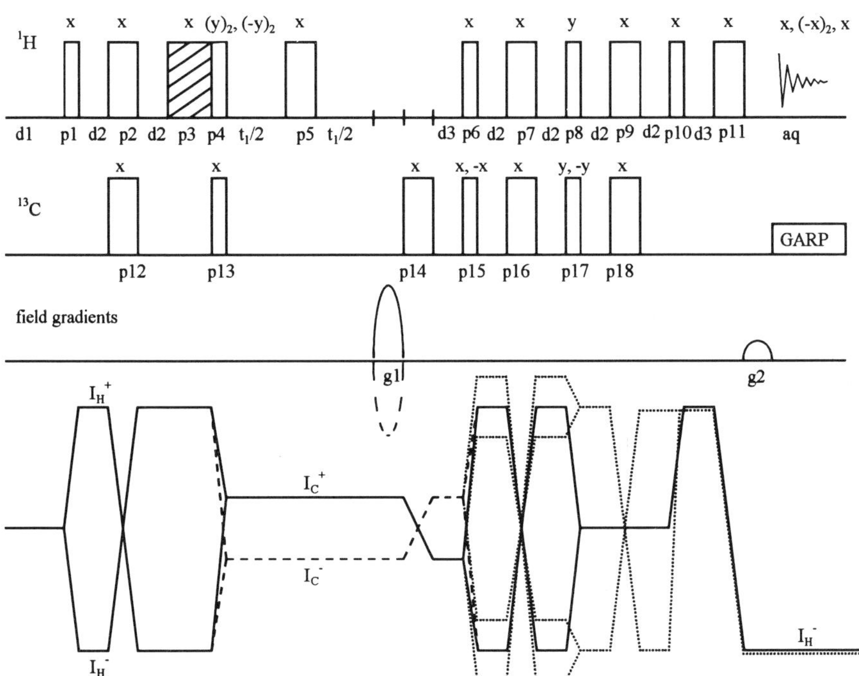

__________ coherence pathway for N type signals
---------- coherence pathway for P type signals
.... coherence pathway for signals leading to sensitivity enhancement

4. Acquisition

Time requirement: 20 min

Sample: 10% strychnine in $CDCl_3$

Record normal ^{1}H and ^{13}C NMR spectra of the sample and optimize the spectral widths for CH_n signals. Change to the 2D mode of the spectrometer and load the phase-sensitive gs-HSQC pulse program with echo/anti-echo acquisition. You have to set:

td2: 1 k data points in F_2
td1: 2 times 64 data points in F_1
sw2: 10 ppm
sw1: 165 ppm
offset of ^{1}H frequency: middle of ^{1}H NMR spectrum
offset of ^{13}C frequency: middle of ^{13}C NMR spectrum
p1, p4, p6, p8, p10: 90° ^{1}H transmitter pulse
p2, p5, p7, p9, p11: 180° ^{1}H transmitter pulse
p13, p15, p17: 90° ^{13}C decoupler pulse
p12, p14, p16, p18: 180° ^{13}C decoupler pulse
p3: 2 ms ^{1}H trim pulse
d1: 2 s
d2: $1/[4J(C,H)] = 1.8$ ms, calculated from $^{1}J(C,H) \approx 140$ Hz
d3: same length as gradient pulse, 1.6 ms used
initial value for t_1 evolution: 3 μs
increment for $t1$ evolution: 1/[2·sw1]
^{13}C decoupler attenuation and 90° pulse for GARP (19 dB, 70 μs)
g1, g2: sinusoidal shaped field gradients with 5% truncation, 1.6 ms duration and ca.0.01T/m strength, with gradient loop counters, ring-down delays (100 μs), lock blanking and gradient coil blanking switches according to actual instrumentation used. Gradient strength ratio: 4 : 1 : –4 : 1
ds: 8
ns: 1

For the phase-sensitive echo/anti-echo scheme in F_1 use simultaneous data mode in F_2. The pulse sequence uses one scan within the go loop. The echoes (gradients: +4, +1) and the anti-echoes (gradients: –4, +1) are stored in different blocks and are sampled with the 180° phase shift of p17. Two further loops cycle the acquisition phase and the phases of p15 and p4. Thus, the experiment shown was performed with 4 scans of 64 echo accumulations and with 4 scans of 64 anti-echo accumulations.

5. Processing

Apply zero filling in F_1 to 512 words in order to have a matrix of 512·512 real data points. Use an exponential window in F_2 with lb = 3 Hz and a Gaussian window in F_1. Choose the echo/anti-echo FT mode of the software corresponding to the acquisition technique. Phase correction is usually only necessary in the F_2 dimension.

6. Result

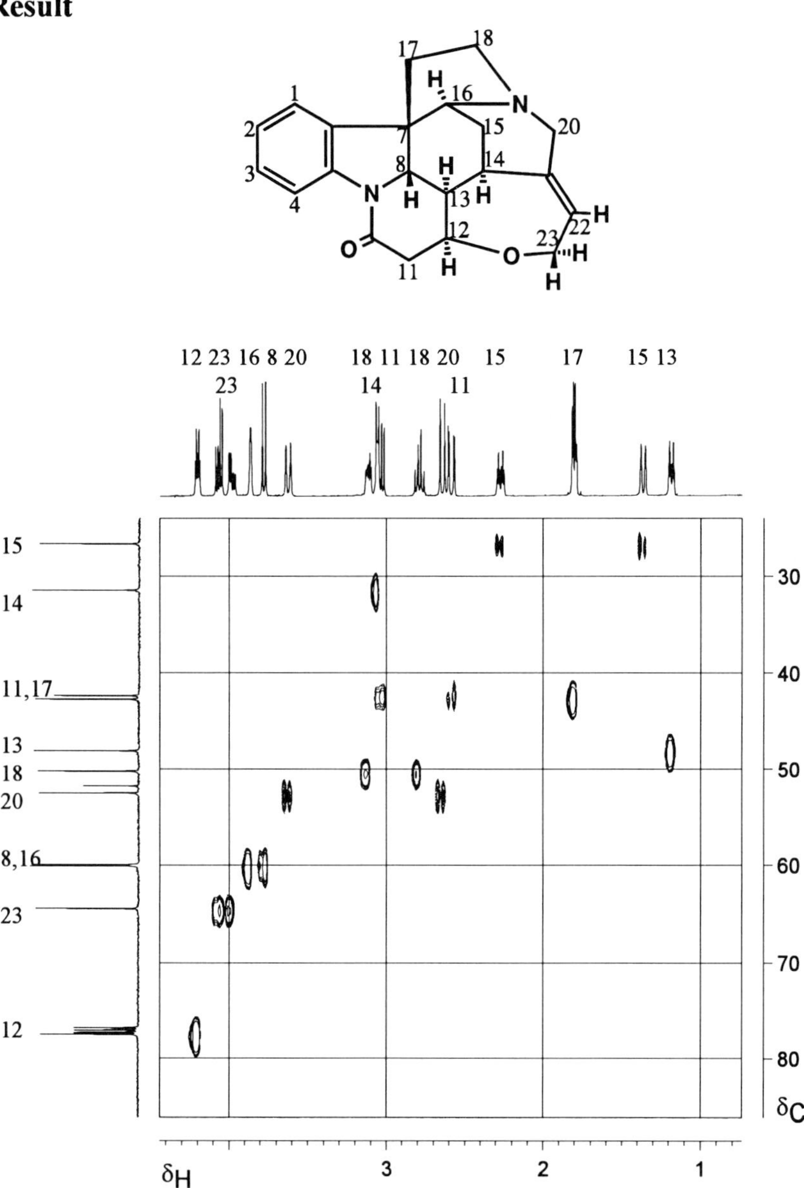

The figure on page 499 shows an expansion of the spectrum obtained on an AMX-500 spectrometer with a BGU (10 A) gradient unit and a multinuclear *z*-gradient probehead.. Note that the correlation signals of CH groups (e.g. H-12, H-13, H-16) have higher intensities than those of the CH_2 groups. This is of significance in protein research, where the CH_α signals are important for the determination of the backbone structure.

7. Comments

There are several modifications to the standard HSQC procedure as described in Experiment 10.14. A proton trim pulse p3 removes unwanted coherences during the first INEPT transfer which arise from imperfect pulses. The two gradients are applied within [delay, 180° pulse, gradient] sandwiches to avoid phase errors due to the finite duration of the gradients. After the first gradient, which acts at a time, when single quantum carbon coherences $2I_{Hz}I_{Cy}$ and $2I_{Hz}I_{Cx}$ are present, a reverse INEPT sandwich transfers the $2I_{Hz}I_{Cy}$ part to in-phase magnetization I_{Hy}. However, $2I_{Hz}I_{Cx}$ is transformed into double quantum magnetization $2I_{Hy}I_{Cx}$ as shown by the dotted line of the coherence pathway diagram. The second reverse INEPT sandwich stores I_{Hy} as *z*-magnetization and transforms $2I_{Hy}I_{Cx}$ to I_{Hx} in-phase magnetization. The proton pulse p10 reconverts the stored *z*-magnetization; thus both components of the proton magnetization which are modulated with carbon chemical shift during t_1 can be observed. The final gradient rephases only the desired coherences. This sequence combines echo/anti-echo selection with the sensitivity enhancement as given by Ref. [2].

8. Own Observations

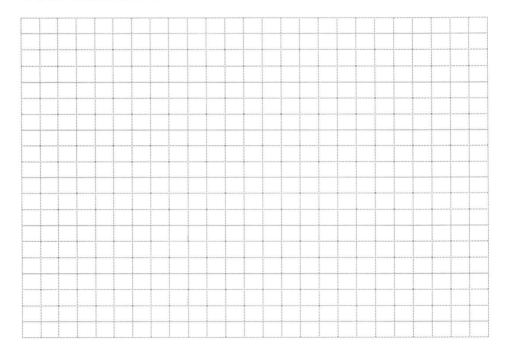

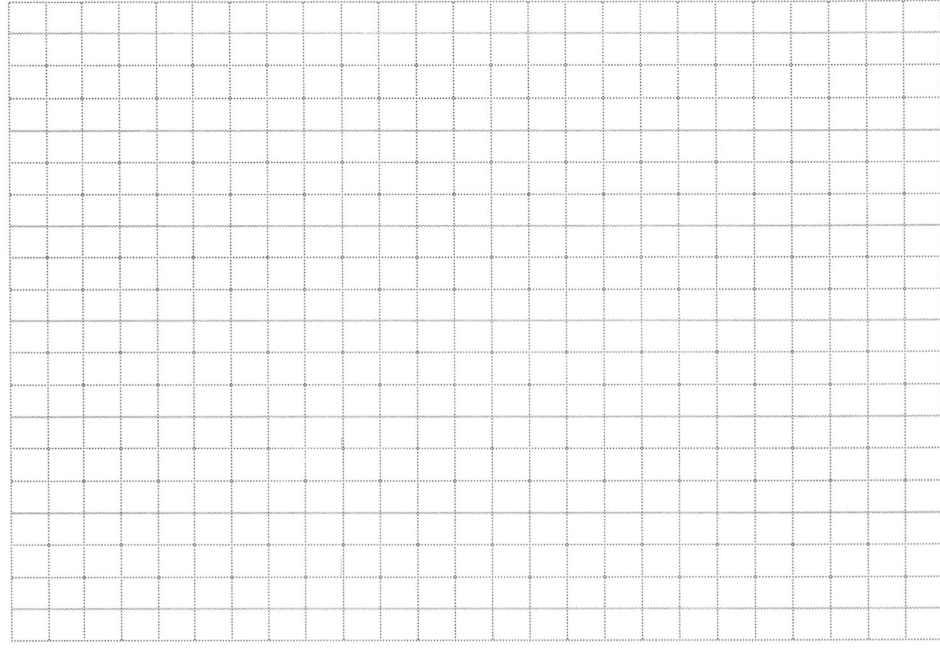

Experiment 12.7

gs-TOCSY

1. Purpose

The TOCSY sequence as described in Experiment 10.18 uses phase cycling to achieve frequency discrimination in F_1. With the necessary suppression of axial peaks it therefore needs a minimum of four transients for each t_1 increment. The gradient-selected method shown here requires only one transient and produces a spectrum which is not phase-sensitive. Since the TOCSY experiment leads to in-phase cross-signals, the magnitude spectrum obtained here is usually sufficient to quickly provide connectivity information. Recently a phase-sensitive version with sensitivity enhancement was communicated [3].

2. Literature

[1] R. E. Hurd, *J. Magn. Reson.* **1990**, *87*, 422–428.
[2] A. Bax, D. G. Davis, *J. Magn. Reson.* **1985**, *65*, 355–360.
[3] K. E. Kövér, D. Uhrín, V. J. Hruby, *J. Magn. Reson.* **1998**, *130*, 162–168.

3. Pulse Scheme and Phase Cycle

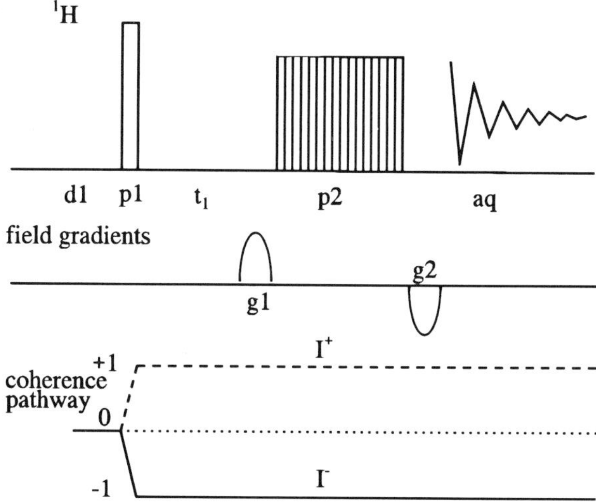

p1: x
p2: spinlock of composite 180° pulses (90°, 180°, 90°) using the MLEV16 sequence:
[90(ph1), 180(ph2), 90(ph1)]$_2$ ph1: x, ph2: y
[90(ph3), 180(ph4), 90(ph3)]$_3$ ph3: -x, ph4: -y
[90(ph1), 180(ph2), 90(ph1)]$_2$
[90(ph3), 180(ph4), 90(ph3)]$_3$ aq: x
[90(ph1), 180(ph2), 90(ph1)]$_3$

$[90(ph3), 180(ph4), 90(ph3)]_2$
$[90(ph1), 180(ph2), 90(ph1)]_1$

4. Acquisition

Time requirement: 10 min

Sample: 10% strychnine in $CDCl_3$

Run a normal ^1H NMR of the sample and optimize the spectral width. Change to the 2D mode of the spectrometer and load the gs-TOCSY pulse program. The 90° pulse-width and attenuation of the spin-lock pulses must be calibrated prior to the experiment (see Exp. 2.9). For optimum results one should take into account the phase difference between the hard pulse p1 and the spin-lock pulses either in the pulse program or in the adjustable parameter set if the software allows (see Exp. 7.1). The duration of the spin-lock is an adjustable parameter. You have to set:

td2: 1 k data points in F_2
td1: 256 data points in F_1
sw2: 9 ppm
sw1: 9 ppm
o1: middle of ^1H NMR spectrum
p1: 90° ^1H transmitter pulse
p2: series of composite 180° pulses (90°, 180°, 90°) at transmitter attenuation of spin-lock; 90° pulse-width and transmitter attenuation typically in the order of 40 μs and 16 dB, corresponding to an effective spin-lock field of ca. 7000 Hz (magnetic field dependent). Duration of spin-lock set to 100 ms by loop parameter of spin-lock sequence. The loop parameter must be an even number (38 was used here).
d1: 2 s
initial value for t_1 evolution: 3 μs
increment for t_1 evolution: 1/sw1
g1, g2: sinusoidal shaped field gradients with 5% truncation, 2 ms duration and ca. 0.01 T/m strength, with gradient loop counters, ring-down delays (100 μs), lock blanking and gradient coil blanking switches according to actual instrumentation used. Gradient strength ratio: 1 : −1
ds: 4
ns: 1

5. Processing

Apply zero filling in F_1 to 512 real data points to obtain a symmetrical matrix of 512·512 real data points. Use unshifted sinusoidal windows in both dimensions. Apply complex Fourier transformation corresponding to the quadrature-off mode of data acquisition in F_1. Since magnitude data are calculated, no phase correction is necessary. Since the P-type coherence pathway is selected, one needs frequency reversal in the F_1 dimension.

6. Result

The figure shows the result obtained on an AMX-500 spectrometer using a BGU (10 A) gradient unit and a multinuclear *z*-gradient probe-head. A short spin-lock (100 ms) was used, in contrast to Experiment 10.18, and since only one transient per t_1 increment is required, twice the number of t_1 increments were recorded in half the time.

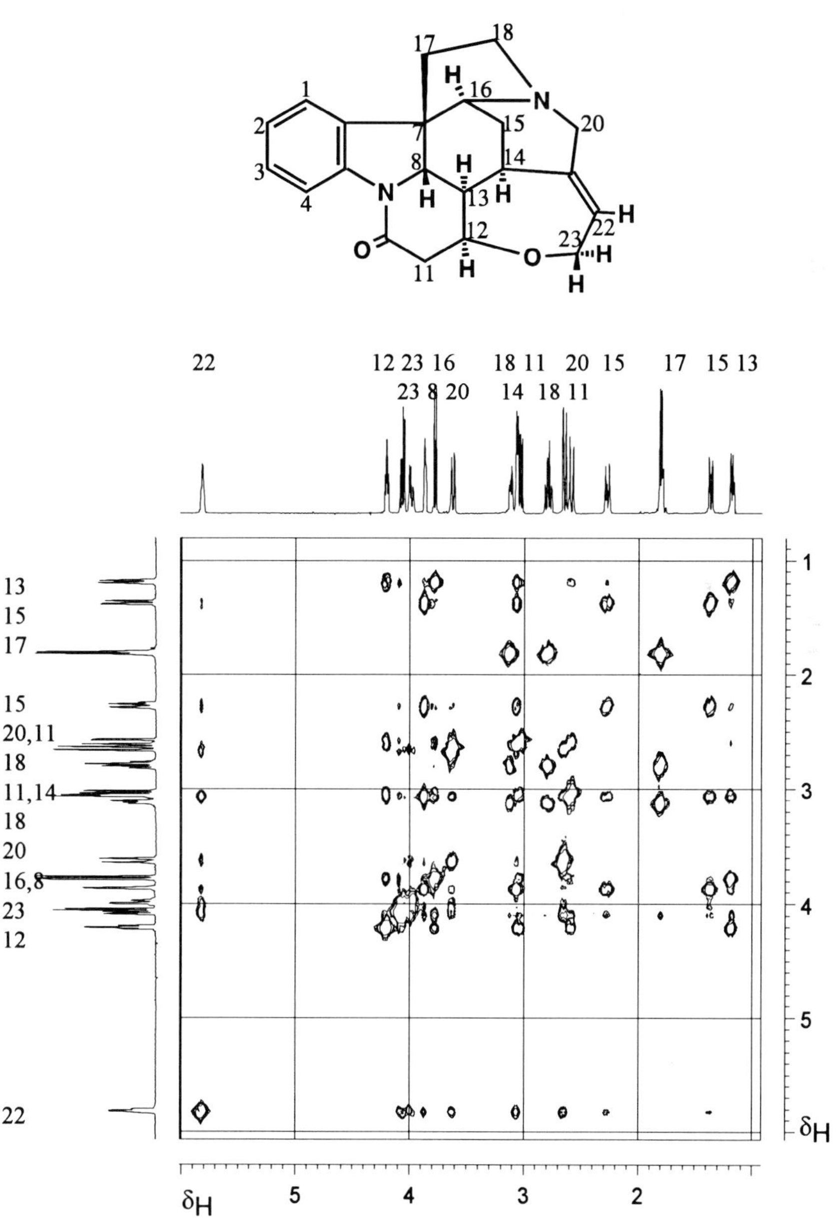

7. Comments

The MLEV-16 spin-lock consists of an even number of composite 180° pulses, so that the coherence level is not changed during its action. Since, by convention, I_{H}^{-} is detected, the pair of oppositely signed gradients selects P-type signals during t_1. The signal distortions mentioned in Ref. [1] were not observed. If a MLEV-17 spin-lock sequence is used, the gradient ratio should be 1 : 1. Phase-sensitive versions using the echo/anti-echo procedure are also known.

8. Own Observations

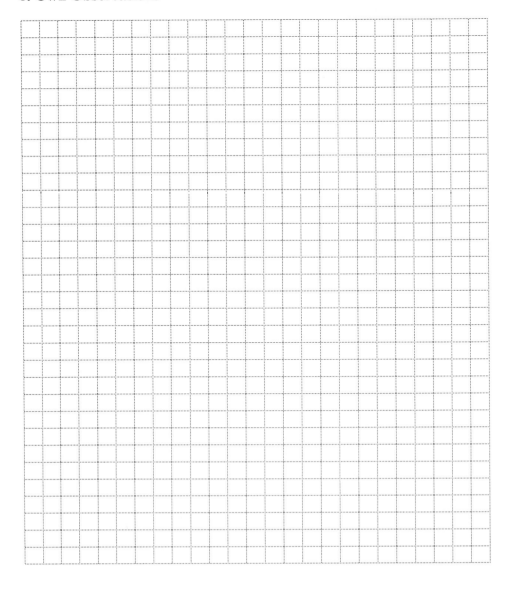

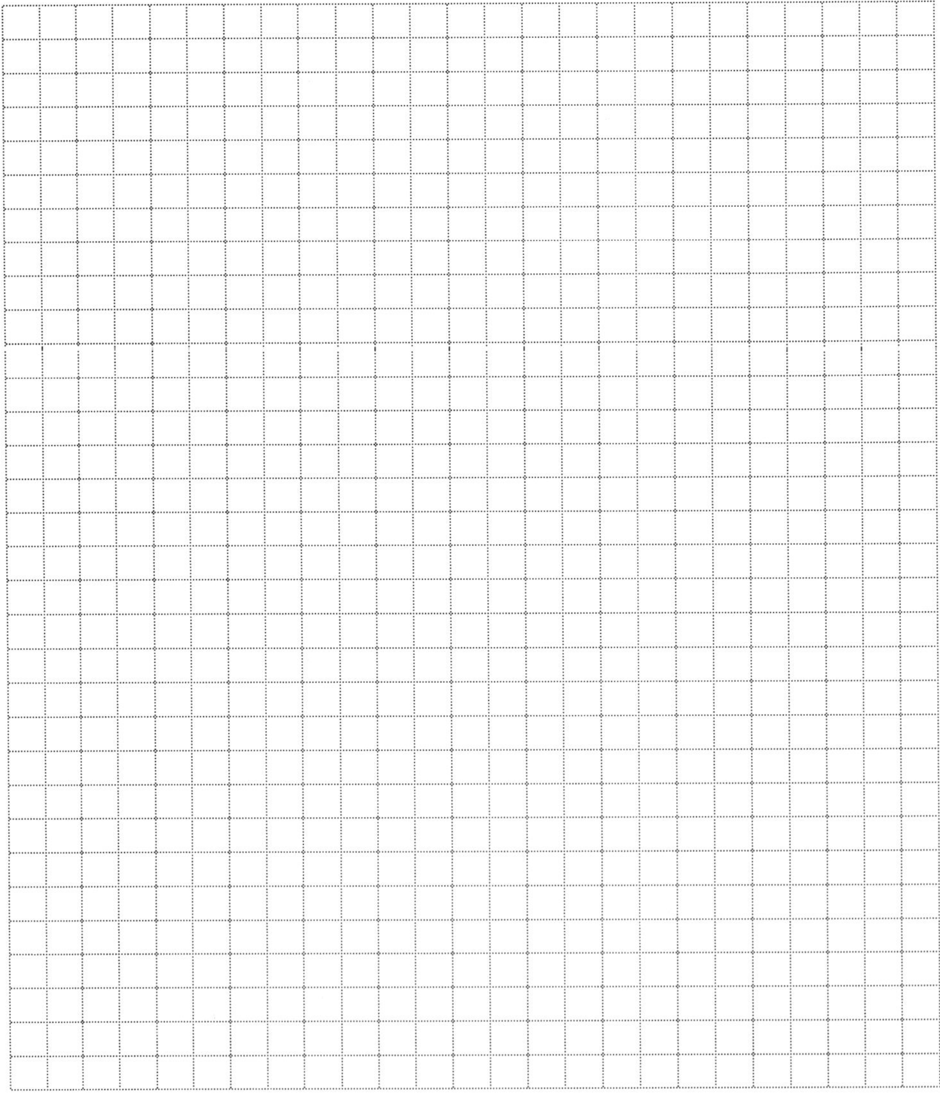

Experiment 12.8

gs-HMQC-TOCSY

1. Purpose

The combination of the HMQC method with the TOCSY sequence leads, in principle, to a 3D technique (see Chapter 13). However, if the evolution period of the TOCSY part is omitted, one obtains a 2D sequence which provides a ^{13}C edited TOCSY spectrum. Starting from each HMQC cross-signal one finds in the same row in F_1 additional signals which are caused by a TOCSY transfer. This is very helpful for structural elucidation since normal TOCSY spectra may often be rather crowded. Compared with true 3D sequences the digital resolution is far better using significantly less measuring time. There are many variants known, here we show a non phase-sensitive gradient-selected method, which does not need a BIRD filter.

2. Literature

[1] L. Lerner, A. Bax, *J. Magn. Reson.* **1986**, *69*, 375–380.
[2] T. Domke, *J. Magn. Reson.* **1991**, *95*, 174–177.
[3] G. E. Martin, T. D. Spitzer, R. C. Crouch, J.-K. Luo, R. N. Castle, *J. Heterocyclic Chem.* **1992**, *29*, 577–582.
[4] B. K. John, D. Plant, S. L. Heald, R. E. Hurd, *J. Magn. Reson.* **1991**, *94*, 664–669
[5] W. Willker, D. Leibfritz, R. Kerssebaum, W. Bermel, *Magn. Reson. Chem.* **1993**, *31*, 287–292.

3. Pulse Scheme and Phase Cycle

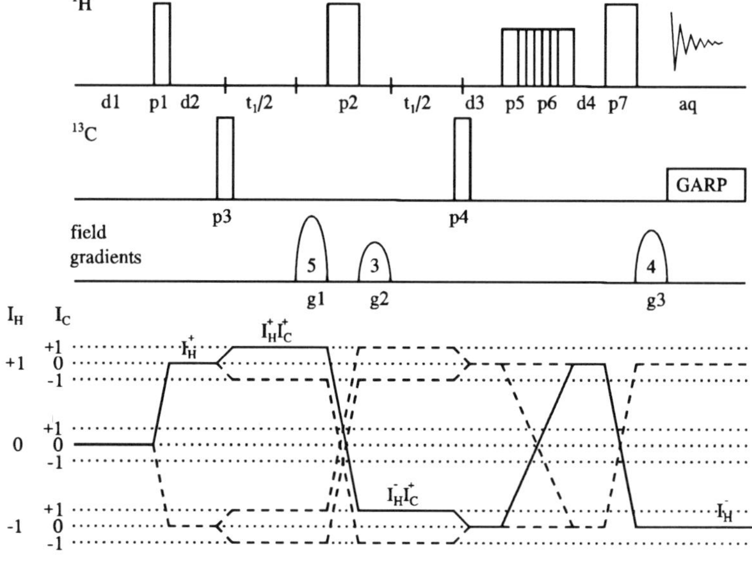

p1: x p2: $(x)_2$, $(-x)_2$ p3: x, -x p4: $(x)_4$, $(-x)_4$
p5: y p7: x aq: $(x, -x)_2$, $(-x, x)_2$
p6: spinlock of composite 180° pulses (90°, 180°, 90°)
using the MLEV-17 sequence:
$[90(ph1), 180(ph2), 90(ph1)]_2$ ph1: x, ph2: y
$[90(ph3), 180(ph4), 90(ph3)]_3$ ph3: -x, ph4: -y
$[90(ph1), 180(ph2), 90(ph1)]_2$
$[90(ph3), 180(ph4), 90(ph3)]_3$
$[90(ph1), 180(ph2), 90(ph1)]_3$
$[90(ph3), 180(ph4), 90(ph3)]_2$
$[90(ph1), 180(ph2), 90(ph1)]_1$
[60(ph2)]

4. Acquisition

Time requirement: 1.2 h

Sample: 10% strychnine in $CDCl_3$

Record normal ^1H and ^{13}C NMR spectra of the sample and optimize the spectral widths for CH_n signals. Change to the 2D mode of the spectrometer and load the gs-HMQC-TOCSY pulse program. You have to set:

td2: 1 k data points in F_2
td1: 256 data points in F_1
sw2: 10 ppm
sw1: 165 ppm
offset of ^1H frequency: middle of ^1H NMR spectrum
offset of ^{13}C frequency: middle of ^{13}C NMR spectrum
p1: 90° ^1H transmitter pulse
p2, p7: 180° ^1H transmitter pulse
p3, p4: 90° ^{13}C decoupler pulse
p5: 2.5 ms ^1H trim pulse
p6: series of composite 180° pulses (90°, 180°, 90°) at transmitter attenuation of spin-lock; 90° pulse-width and transmitter attenuation typically in the order of 40 μs and 16 dB, corresponding to an effective spin-lock field of ca. 7000 Hz (magnetic field dependent). Last pulse 60° according to MLEV-17 scheme. Duration of spin-lock set to 81.8 ms by loop parameter of spin-lock sequence. The loop parameter must be an even number (30 was used here).
d1: 2 s
d2: $1/[2J(C,H)] = 3.57$ ms, calculated from $^1J(C,H) \approx 140$ Hz
d3: d2 minus length of gradient pulse
d4: equal to length of gradient pulse
start increment for t_1 evolution: 3 μs
increment for t_1 evolution: $1/[2 \cdot sw1]$

g1, g2, g3: sinusoidal shaped field gradients with 5% truncation, 2 ms duration and ca. 0.01 T/m strength, with gradient loop counters, ring-down delays (100 μs), lock blanking and gradient coil blanking switches according to actual instrumentation used. Gradient strength ratio: 5:3:4

^{13}C decoupler attenuation and 90° pulse for GARP (19 dB, 70 μs)

ds: 8

ns: 4

5. Processing

Apply zero filling in F_1 to 512 words in order to have a matrix of 512·512 real data points. Before Fourier transformation use sinusoidal windows both in F_2 and in F_1. Phase correction is unnecessary, since the spectrum is displayed in the magnitude mode.

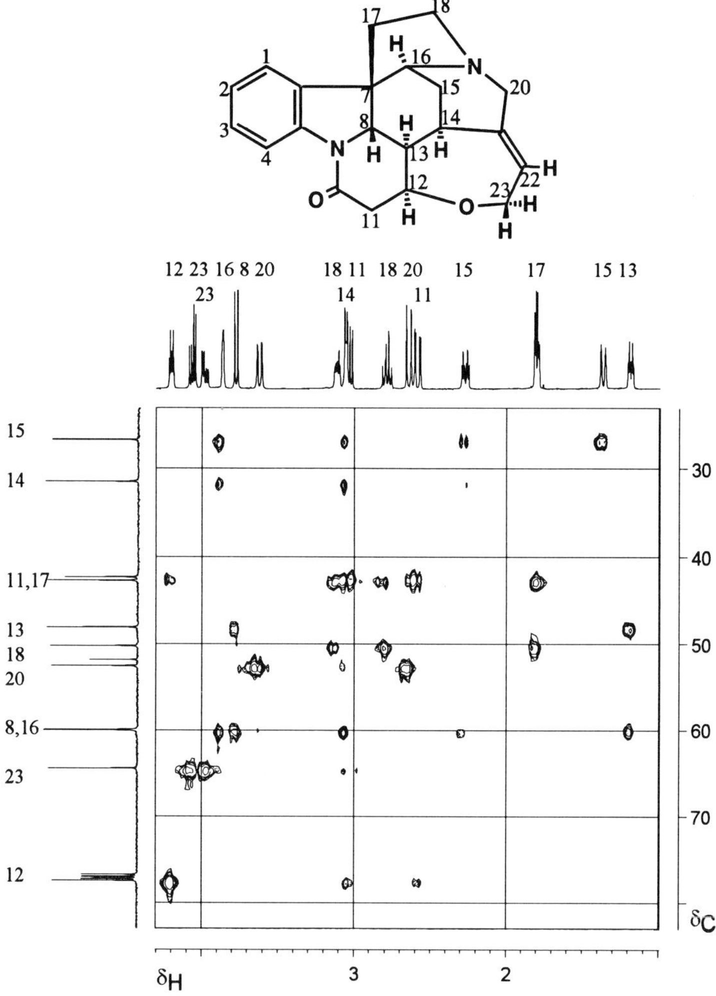

6. Result

The figure on page 507 shows an expansion of the HMQC-TOCSY spectrum obtained on an AMX-500 spectrometer using a BGU (10 A) gradient unit and a multinuclear *z*-gradient probe-head. Since a relative short spin-lock of 82 ms was used, mainly correlations over two bonds, starting from a ^{13}C nucleus, are present. Note e.g. the connectivities C-12 – H-12 – H-11α and H-11β ($\delta_C = 78$) or C-15 – H-15α and H-15β to H-14 and H-16 ($\delta_C = 25$).

7. Comments

The HMQC and TOCSY parts of the sequence have been described in Experiments 10.13 and 10.18, their gradient-selected variants in Experiments 12.3 and 12.7. Note that the TOCSY part contains no evolution time as in 3D sequences. The gradients are set similar to Experiment 12.3 for the gradient-selected HMQC, only the final refocusing gradient is set after the spin-lock. As a spin-lock a MLEV-17 sequence is used; the last 180° proton pulse inverts the coherence level once more in order to finally detect I^-. Similar information has been provided earlier by the heteronuclear relayed methods.

8. Own Observations

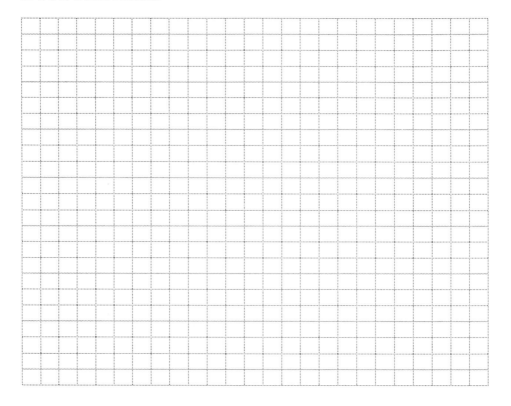

Experiment 12.9

2Q-HMBC

1. Purpose

This experiment detects long-range carbon–carbon connectivities. Like Experiment 12.10, it starts from ^1H magnetization and detects ^1H magnetization. It differs from Experiment 12.10 by the circumstance that the transfer from protons to ^{13}C uses a $^3J(C,H)$ or a $^2J(C,H)$ instead of a $^1J(C,H)$ coupling and that it is not dependent on the C,C coupling constants [1]. Therefore carbon–carbon relationships can also be detected, where the C,C spin coupling constant is close to zero. Thus, the experiment detects long-range interactions both between proton and carbon and between carbon atoms; together with (n,n)ADEQUATE [2], it is therefore a supplement for Experiment 12.10 in structural elucidation of organic compounds, especially when proton signals overlap. In the experiment described here we demonstrate the phase-sensitive version using the echo/anti-echo approach with salicylaldehyde as an example.

2. Literature

[1] A. Meissner, D. Moskau, N. C. Nielsen, O. W. Sørensen, *J. Magn. Reson.* **1997**, *124*, 245–249.

[2] B. Reif, M. Köck, R. Kerssebaum, H. Kang, W. Fenical, C. Griesinger, *J. Magn. Reson. Ser. A* **1996**, *118*, 282–285.

3. Pulse Scheme and Phase Cycle

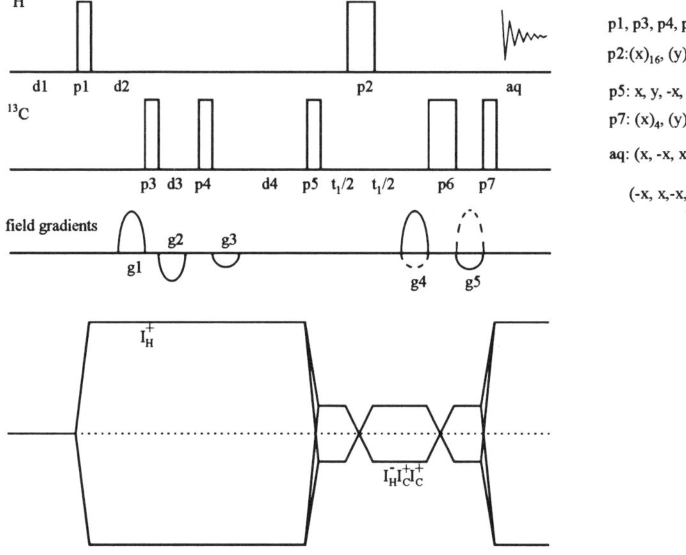

p1, p3, p4, p6: x

p2: $(x)_{16}$, $(y)_{16}$, $(-x)_{16}$, $(-y)_{16}$

p5: x, y, -x, -y

p7: $(x)_4$, $(y)_4$, $(-x)_4$, $(-y)_4$

aq: $(x, -x, x, -x, -x, x, -x, x)_2$,

$(-x, x, -x, x, x, -x, x. -x)_2$

4. Acquisition

Time requirement: 5 h

Sample: 0.5 ml salicylaldehyde with 0.2 ml [D_6]DMSO

Record normal ^{1}H and ^{13}C NMR spectra of the sample, note the required spectral widths and note the offset of the middle of each spectrum. Change to the 2D mode of the spectrometer and load the 2Q-HMBC pulse program. You have to set:

td2: 1 k data points in F_2
td1: 128 data points in F_1
sw2: 5.2 ppm
sw1: 206 ppm (C,C double quantum frequency)
offset of ^{1}H frequency: center of ^{1}H NMR spectrum
offset of ^{13}C frequency: center of ^{13}C NMR spectrum
p1: 90° ^{1}H transmitter pulse
p2: 180° ^{1}H transmitter pulse
p3, p4, p5, p7: 90° ^{13}C decoupler pulse
p6: 180° ^{13}C decoupler pulse
d1: 4 s
d2: $1/[2(J(C,H_{min}) + 0.146\cdot(J(C,H_{max}) - J(C,H_{min}))]$ – effective gradient length = 1.936 ms, calculated from $^1J(C,H_{max})$ = 179 Hz and $^1J(C,H_{min})$ = 159 Hz
d3: $1/[2(J(C,H_{max}) - 0.146\cdot(J(C,H_{max}) - J(C,H_{min}))]$ – effective gradient length = 1.689 ms, calculated from $^1J(C,H_{max})$ = 179 Hz and $^1J(C,H_{min})$ = 159 Hz
d4: $1/[J(C,H)$ – effective gradient length = 57.6 ms, calculated from $^3J(C,H) \approx$ 8.5 Hz
initial value for t_1 evolution: 3 μs
increment for t_1 evolution: $1/[2\cdot sw1]$
g1–g5: sinusoidal shaped field gradients with 5% truncation, 1 ms duration and ca. 0.02 T/m strength, with gradient loop counters, ring-down delays (100 μs), lock blanking and gradient coil blanking switches according to actual instrumentation used. Gradient strength ratio: 3 : −2 : −1 : 3 : −1 (echo) and 3 : −2 : −1 : −3 : 1 (anti-echo)
ns: 32

For the phase-sensitive echo/anti-echo scheme in F_1 use simultaneous data mode in F_2.

5. Processing

Use zero filling in F_2 to 1 k real data points and to 256 real data points in F_1. Use a π/4 shifted sinusoidal window in F_2 and a π/2 shifted squared sinusoidal window in F_1

Choose the echo/anti-echo FT mode of the software corresponding to the acquisition technique. The spectrum shown is displayed in magnitude mode.

6. Result

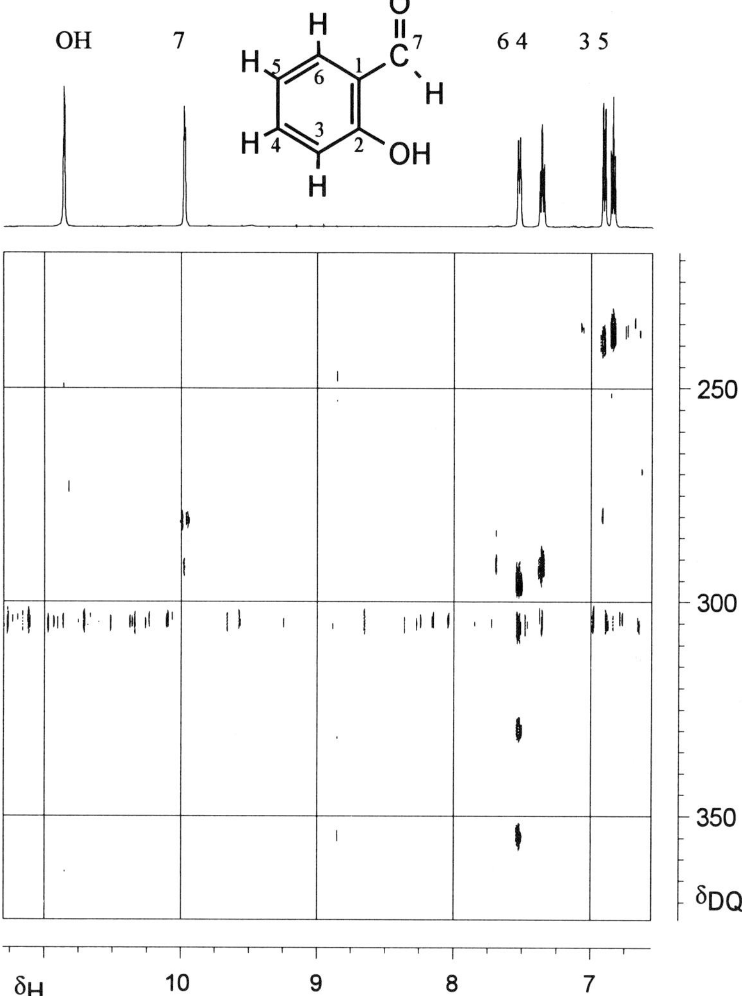

The figure shows the ^{1}H-detected 2Q-HMBC spectrum obtained on an AMX-500 spectrometer using a BGU (10 A) gradient unit and a multinuclear *z*-gradient probe-head. H-5 displays a double quantum signal at δ_{DQ} = 237.1, which corresponds to $\delta_{C\text{-}3}$ = 116.4 + $\delta_{C\text{-}1}$ = 120.7. H-3 yields a DQ signal at δ_{DQ} = 239.5 [$\delta_{C\text{-}5}$ (118.8) + $\delta_{C\text{-}1}$]. H-4 shows the DQ signal at δ_{DQ} = 291.4 [$\delta_{C\text{-}2}$ (160.2) + $\delta_{C\text{-}6}$ (131.2)], whereas H-6 gives three DQ signals. The first at δ_{DQ} = 295.8 connects C-2 with C-4 (135.6), the second at δ_{DQ} = 329.6 connects C-4 with C-7 (194.0), and the third at δ_{DQ} = 354.2 connects C-2 with C-7. Finally for H-7 a DQ signal for the connection of

C-1 with C-2 can be seen (280.9). Note that there are considerable axial peaks at δ_{DQ} = 305.

7. Comments

The very simple looking but rather elegant pulse sequence consists in principle only of the standard HMBC sequence using a gradient double quantum filter detecting protons which "see" two ^{13}C atoms in the same molecule. Another feature already described in Experiment 12.5 is a two-step low-pass filter, consisting of the first two 90° carbon pulses, which is also gradient-supported. The 3 : −2 : −1 ratio of the first three gradient pulses dephases all but the long-range $2I_{Hx}I_{Cz}$ coherences. Pulse p5 directly creates 2Q HMBC relations of the type $4I_{Hx}I_{Cy}I_{Cy}$ which develop double quantum chemical shift information during t_1. After the 180° pulse on protons we therefore have a coherence level of −4 + 1 + 1 = −2 which is dephased by gradient g4 of relative strength 3. The 180° carbon pulse p7 changes the coherence level to −6; thus the last gradient pulse of relative strength −1 rephases just these coherences which are transformed back to proton magnetization by the last 90° carbon pulse. Note that the action of the low-pass filter pulses is not considered in the coherence pathway diagram.

8. Own Observations

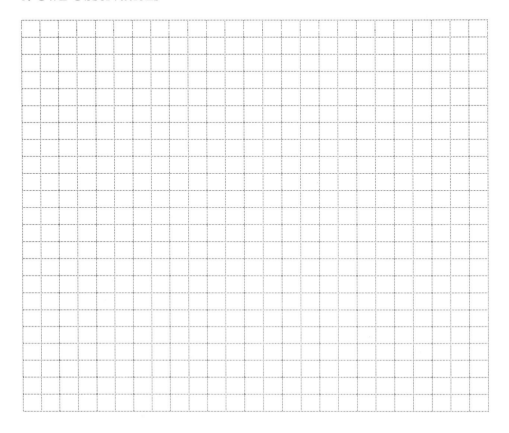

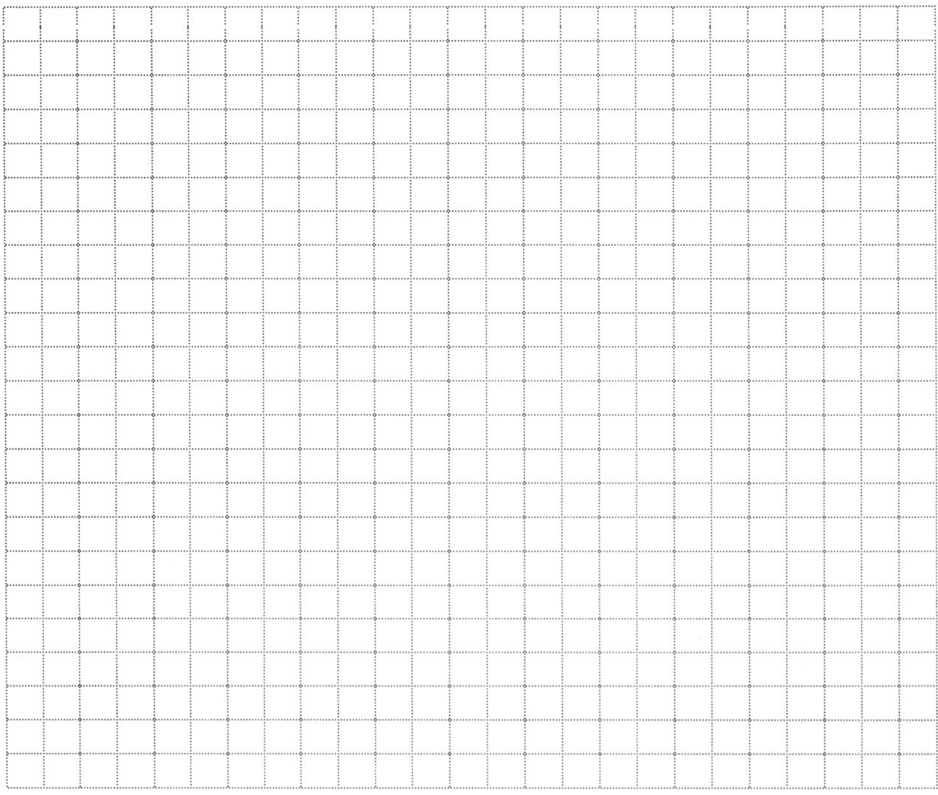

Experiment 12.10

^{1}H-Detected 2D INEPT-INADEQUATE

1. Purpose

This experiment detects carbon–carbon connectivities. In contrast to the standard 2D INADEQUATE experiment 10.22, which starts from ^{13}C magnetization and detects ^{13}C magnetization, the experiment described here starts from ^{1}H magnetization and detects ^{1}H magnetization. It is estimated to be about a factor of 13 times more sensitive than Experiment 10.22. The formidable task of suppressing protons bound to ^{12}C (1: 10 000) and protons in molecules containing only one ^{13}C nucleus (1:100) is achieved by the use of pulsed field gradients with additional phase cycling. The method lacks the generality of the normal 2D INADEQUATE since connectivities between two quaternary carbon atoms C_q–C_q cannot be detected, however it is possible to see a C_q–CH_n moiety.

2. Literature

[1] J. Weigelt, G. Otting, *J. Magn. Reson. Ser. A*. **1995**, *113*, 128–130.
[2] B. Reif, M. Köck, R. Kerssebaum, H. Kang, W. Fenical, C. Griesinger, *J. Magn. Reson. Ser. A* **1996**, *118*, 282–285.

3. Pulse Scheme and Phase Cycle

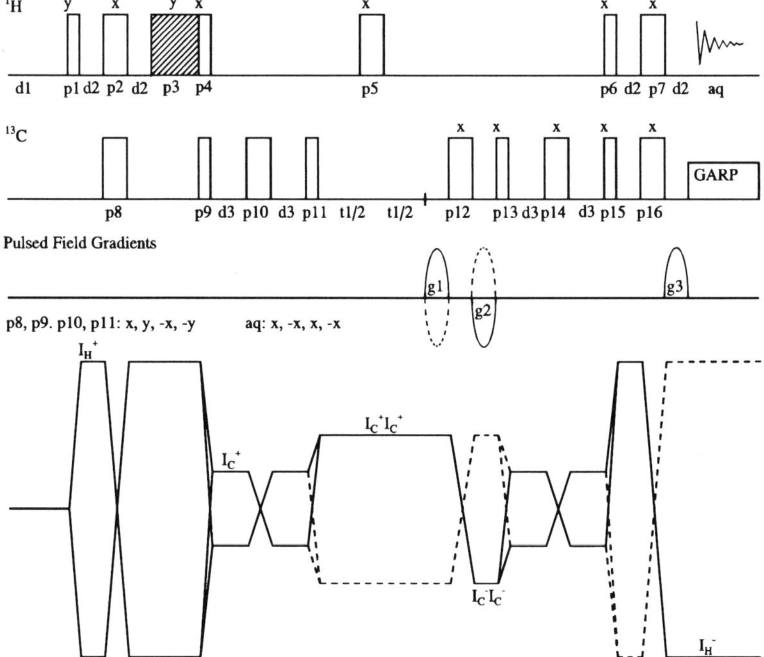

4. Acquisition

Time requirement: 6.5 h

Sample: 2 M sucrose in D_2O

Record normal ^1H and ^{13}C NMR spectra of the sample, note the required spectral widths and note the offset of the middle of each spectrum. Change to the 2D mode of the spectrometer and load the required pulse program. You have to set:

td2: 1 k data points in F_2
td1: 2 times 512 data points in F_1
sw2: 3.7 ppm
sw1: 80 ppm (C,C double quantum frequency)
o1: center of ^1H NMR spectrum
o2: center of ^{13}C NMR spectrum
p1, p4, p6: 90° ^1H transmitter pulse
p2, p5, p7: 180° ^1H transmitter pulse
p9, p11, p13, p15: 90° ^{13}C decoupler pulse
p8, p10, p12, p14, p16: 180° ^{13}C decoupler pulse
p3: 2 ms ^1H spin-lock purging pulse
d1: 1.5 s
d2: $1/[4J(C,H)] = 1.8$ ms, calculated from $^1J(C,H) \approx 140$ Hz
d3: $1/[4J(C,C)] = 5$ ms, calculated from $^1J(C,C) \approx 50$ Hz
initial value for t_1 evolution: 3 μs
increment for t_1 evolution: $1/[2 \cdot sw1]$
g1, g2, g3: shaped field gradients with a sinusoidal start reaching a plateau after 32 data points and falling off sinusoidally with 5% truncation, 1 ms duration and ca. 0.02 T/m strength, with gradient loop counters, ring-down delays (200 μs), lock blanking and gradient coil blanking switches according to actual instrumentation used. Gradient strength ratio: 3.97 : −3.97 : 4.0 (exact γ-ratios)
Decoupler attenuation and 90° pulse for GARP (19 dB, 70 μs)
ns: 12 (see below)

For the phase-sensitive echo/anti-echo scheme in F_1 use simultaneous data mode in F_2. The pulse sequence used here for the AMX spectrometer uses one scan within the go loop. The echoes (gradients: 3.97 : −3.97 : 4) and the anti-echoes (gradients: −3.97 : +3.97 : 4) are stored in different blocks. Two further loops cycle the acquisition phase and the phases of p8–p11. Thus, the result shown was obtained with 12 scans of 512 echo and 12 scans of 512 anti-echo accumulations.

5. Processing

Use an exponential window in F_2 with lb = 6.5 Hz and a $\pi/2$ shifted squared sinusoidal window in F_1. Choose the echo/anti-echo FT mode of the software corresponding to the acquisition technique.

6. Result

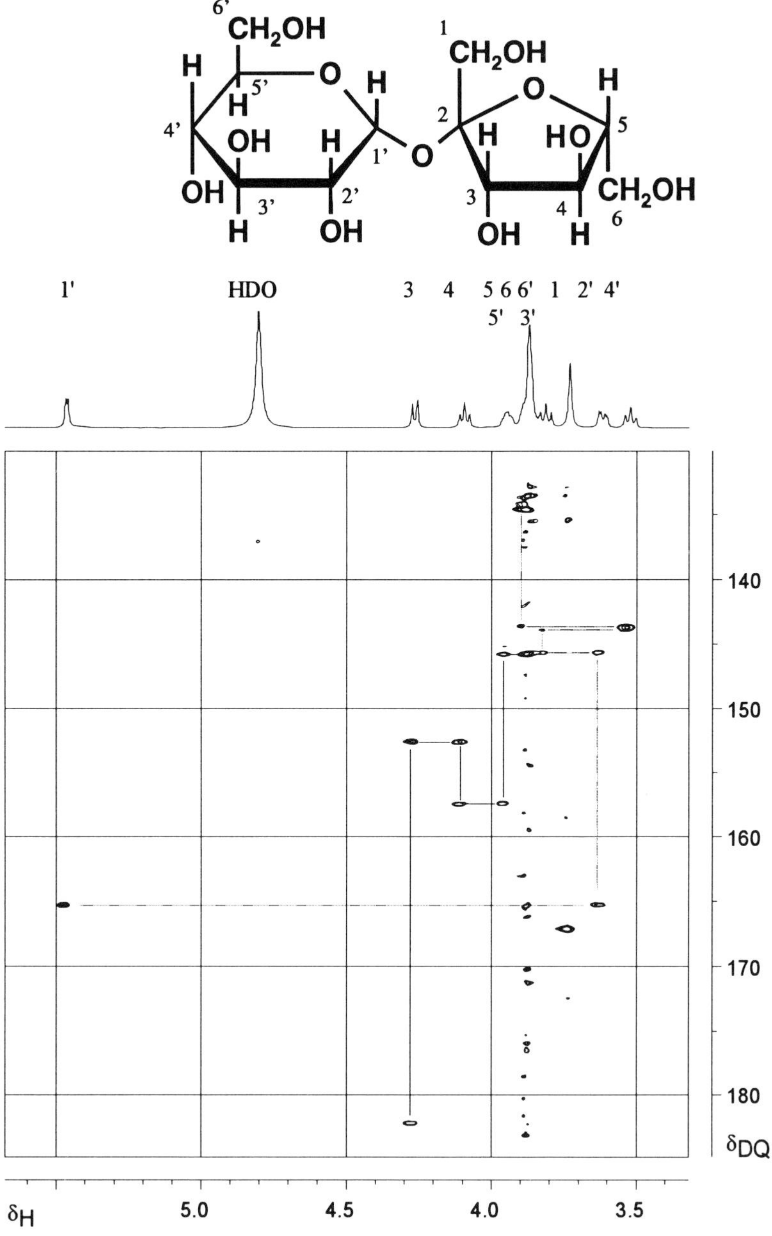

The figure on page 515 shows the ^{1}H detected INEPT-INADEQUATE spectrum obtained on an AMX-500 spectrometer using a BGU (10 A) gradient unit and a multinuclear *z*-gradient probe-head. H-3 displays a double quantum signal at δ_{DQ} =182.2 which corresponds to $\delta_{C\text{-}2}$ = 104.7 + $\delta_{C\text{-}3}$ = 77.5 and another DQ signal at δ_{DQ} = 152.6 ($\delta_{C\text{-}3}$ + $\delta_{C\text{-}4}$). This connectivity is also seen in F_2 for the signal of H-4 which shows the next DQ signal at δ_{DQ} = 157.5 ($\delta_{C\text{-}4}$ + $\delta_{C\text{-}5}$) leading to H-5. This displays at δ_{DQ} = 145.9 the connectivity C-5–C-6. Thus the solid line gives the carbon-carbon connectivities of the fructose ring. The DQ signal of H-1 appears at δ_{DQ}= 166.0 ($\delta_{C\text{-}2}$ + $\delta_{C\text{-}1}$) and stands alone since C-1 has no further connectivities and C-2 is a quaternary carbon atom. Similarly the glucose ring can be traced (dashed line). Note that the DQ frequencies for C-3' + C-4' and C-4' + C-5' fall together as a strong signal for H-4 at δ_{DQ} = 144.

7. Comments

The sequence starts with an INEPT transfer from protons to ^{13}C. A proton spin-lock purging pulse p3 removes unwanted coherences, which arise from imperfect pulses during the first INEPT transfer. The antiphase ^{13}C magnetization $2I_{Hz}I_{Cy}$ present after p9 develops C,C spin coupling to a second ^{13}C nucleus yielding a term $4I_{Hz}I_{Cx}I_{Cz}$. This is transformed into double quantum coherence by p11. During t_1 the double quantum chemical shifts of carbon develop and are transformed back in two stages into proton magnetization, and therefore a H–C–C fragment is detected.
The gradients are applied within [delay–180° pulse–gradient] sandwiches to avoid phase errors due to the finite duration of the gradients. The first two gradients act at a time when double quantum carbon coherences $4I_{Hz}I_{Cx}I_{Cy}$ are present. The final gradient therefore rephases only the desired coherences.

8. Own Observations

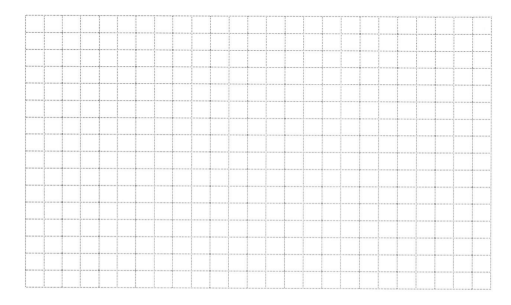

Experiment 12.11

gs-NOESY

1. Purpose

The standard NOESY experiment (see Exp. 10.19) needs at least 8 transients for each t_1 increment to suppress unwanted COSY type signals and axial peaks by the phase cycle employed. Compared with other techniques for structure elucidation (COSY, HMQC) it is therefore a rather lengthy procedure, which is inconvenient especially when several measurements with different mixing times are desired. The gs-NOESY method shown here replaces the phase cycling procedure by one pulsed field gradient during the entire mixing time. In practice, only two transients for each t_1 increment are needed. The technique is demonstrated for the same strychnine sample used throughout this book.

2. Literature

[1] J. Jeener, B. H. Meier, P. Bachmann, R. R. Ernst, *J. Chem. Phys.* **1983**, *71*, 4546–4553.
[2] R. Wagner, S. Berger, *J. Magn. Reson. Ser. A* **1996**, *123*, 119–121.
[3] T. Parella, F. Sánchez-Ferrando, A. Virgili, *J. Magn. Reson.* **1997**, *125*, 145–148.

3. Pulse Scheme and Phase Cycle

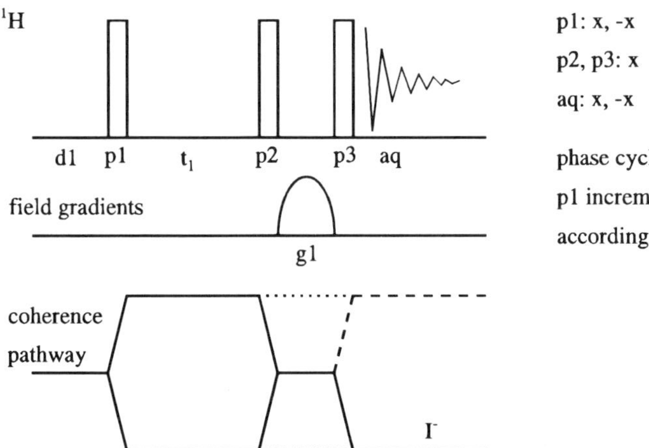

p1: x, -x
p2, p3: x
aq: x, -x

phase cycle for p1 incremented according to TPPI

4. Acquisition

Time requirement: 20 min

Sample: 10% strychnine in $CDCl_3$

Record a normal ^1H spectrum of the sample and optimize the spectral width. Change to the 2D mode of the spectrometer and load the phase-sensitive gs-NOESY pulse program. You have to set:

td2: 2 k data points in F_2
td1: 256 data points in F_1
sw2: 10 ppm
sw1: 10 ppm
o1: middle of ^1H NMR spectrum
p1, p2, p3: 90° ^1H transmitter pulse
d1: 2 s
initial value for t_1 evolution: 3 μs
increment for t_1 evolution: 1/[2·sw1]
g1: sinusoidal shaped field gradient with 5% truncation, 250 ms duration and ca. 0.005 T/m strength, with gradient loop counters, ring-down delays (100 μs), lock blanking and gradient coil blanking switches according to actual instrumentation used. Note that the length of the gradient pulse replaces the usual mixing time and its strength must therefore be adjusted appropriately to rather weak values.
ds: 2
ns: 2

5. Processing

Apply zero filling in F_1 to 512 words in order to have a matrix of 512·512 real data points. Use an exponential window in F_2 with lb = 2 Hz and a π/2 shifted squared sinebell in F_1. Phase correction is usually only necessary in the F_2 dimension.

6. Result

The figure on page 519 shows an expansion of the spectrum obtained on an AMX-500 spectrometer using a BGU (10 A) gradient unit and a multinuclear z-gradient probehead. Note that the spectrum looks almost identical to the result of Experiment 10.19, which was obtained with the full phase cycle requiring 16 transients.

7. Comments

The second r.f. pulse in the NOESY sequence creates $-z$ magnetization, which is frequency labelled with the proton chemical shift. This pathway is shown in the coherence diagram on page 517. In addition, however, this pulse can generate zero-, double quantum- and antiphase coherences, since during t_1 H,H spin coupling is also evolved. The gradient pulse which replaces the mixing time dephases all these components except the zero quantum coherences. Furthermore, it dephases axial signals of those protons which have relaxed during t_1 and are excited again by p2. Thus, instead of the phase cycle, in principle one transient for each t_1 increment is sufficient; in practice two transients yield better results, of course, also in terms of signal-to-noise ratio.

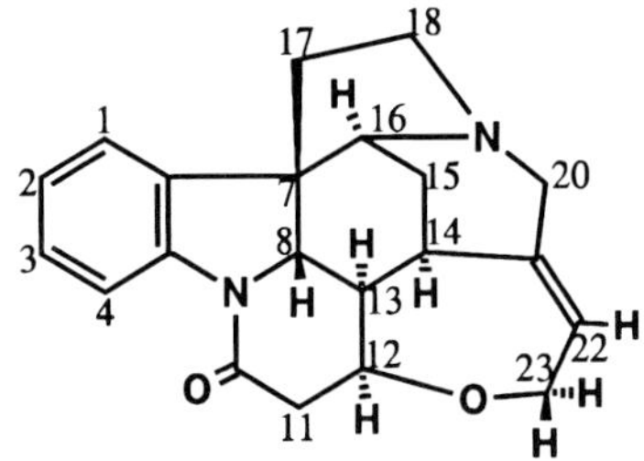

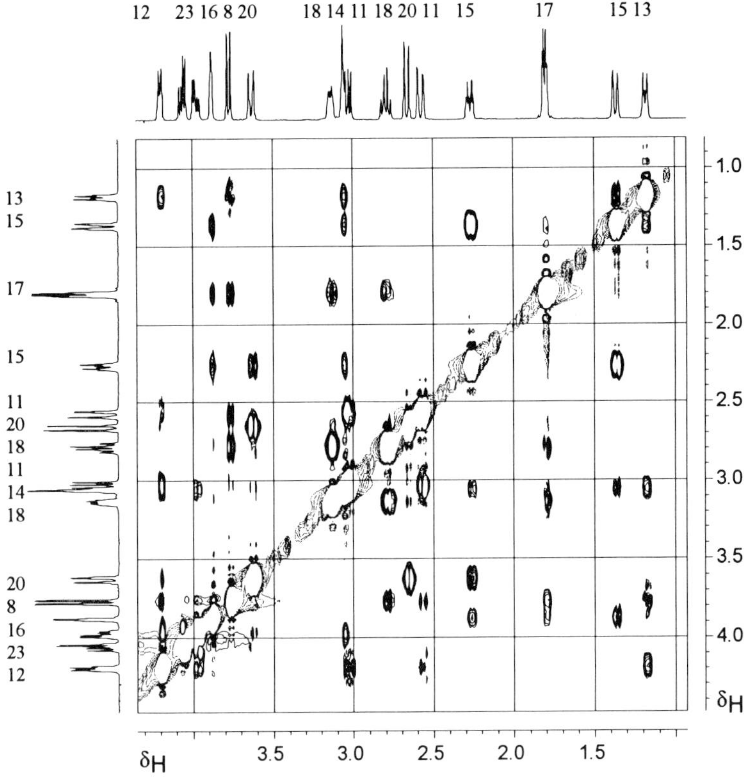

8. Own Observations

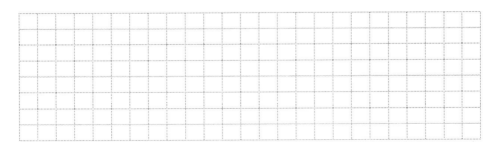

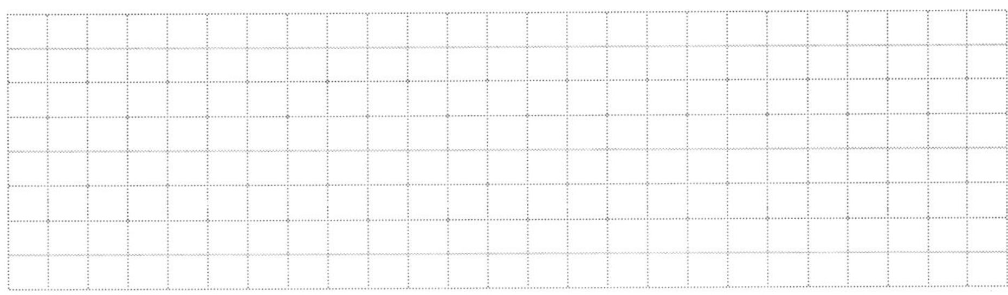

Experiment 12.12

gs-HSQC-NOESY

1. Purpose

It is very difficult to observe and evaluate NOESY cross peaks if the corresponding diagonal signals are very close together or overlap. For symmetric molecules it is even impossible to obtain distance information, with the standard NOESY technique, between protons related by symmetry. A remedy to these problems can be achieved by editing the NOESY spectra by the carbon chemical shift in a manner similar to that described for the HMQC-TOCSY technique in Experiment 12.8. The acquisition of the data is performed without ^{13}C decoupling, which allows one to observe an NOE effect between a proton bound to ^{13}C and a proton in the same molecule with the identical chemical shift but bound to ^{12}C. In the experiment described here we demonstrate the technique with a sample of phenanthrene.

2. Literature

[1] J. Kawabata, E. Fukushi, J. Mizutani, *J. Am. Chem. Soc.* **1992**, *114*, 1115–1117.
[2] R. Wagner, S. Berger, *Magn. Reson. Chem.* **1997**, *35*, 199–202.

3. Pulse Scheme and Phase Cycle

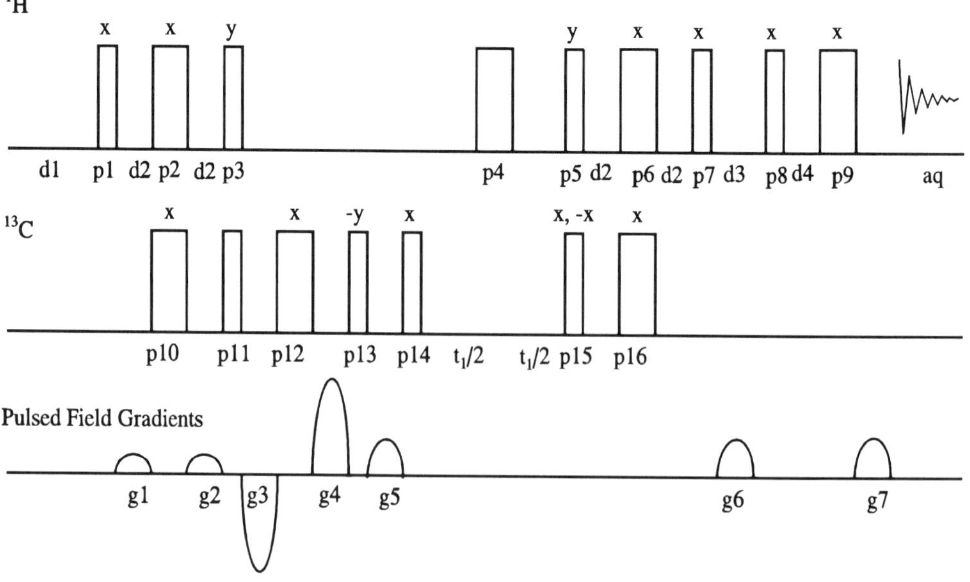

p4: x, x, -x, -x   p11: $(y)_4$, $(-y)_4$ aq: x, -x, -x, x, -x, x, x, -x
phase cycle for p14 incremented according to TPPI

4. Acquisition

Time requirement: 8 h

Sample: 25% phenanthrene in $CDCl_3$

Record normal ^1H and ^{13}C NMR spectra of the sample, note the required spectral widths, and note the offset of the middle of each spectrum. Change to the 2D mode of the spectrometer and load the gs-HSQC-NOESY pulse program. You have to set:

td2: 1 k data points in F_2
td1: 64 data points in F_1
sw2: 2.0 ppm
sw1: 12 ppm
o1: center of ^1H NMR spectrum
o2: center of ^{13}C NMR spectrum
p1, p3, p5, p7, p8: 90° ^1H transmitter pulse
p2, p4, p6, p9: 180° ^1H transmitter pulse
p11, p13, p14, p15: 90° ^{13}C decoupler pulse
p10, p12, p16: 180° ^{13}C decoupler pulse
d1: 1 s
d2: $1/[4J(C,H)] = 1.56$ ms, calculated from $^1J(C,H) \approx 160$ Hz
d3: NOE mixing time, 2 s
d4: set equal to gradient length; the time delays between the pulses p11–p14 are also set equal to the length of the gradient pulses
initial value for t_1 evolution: 3 µs
increment for t_1 evolution: $1/[2 \cdot sw1]$
g1–g7: shaped field gradients with a sinusoidal shape and 1% truncation, 1 ms duration and ca. 0.02 T/m strength, with gradient loop counters, ring-down delays [100 µs], lock blanking and gradient coil blanking switches according to actual instrumentation used. Gradient strength ratio: 5 : 5 : –40 : 40 : 15 : 25 : –20.1153
ns: 128

5. Processing

Apply zero filling in F_1 to 128 real data points to obtain a matrix of $512 \cdot 128$ real data points. Use exponential windows both in F_1 and in F_2 with lb = 10 and 3 Hz, respectively. Apply real Fourier transformation corresponding to the TPPI-type signal selection using the quadrature mode in F_1.

6. Result

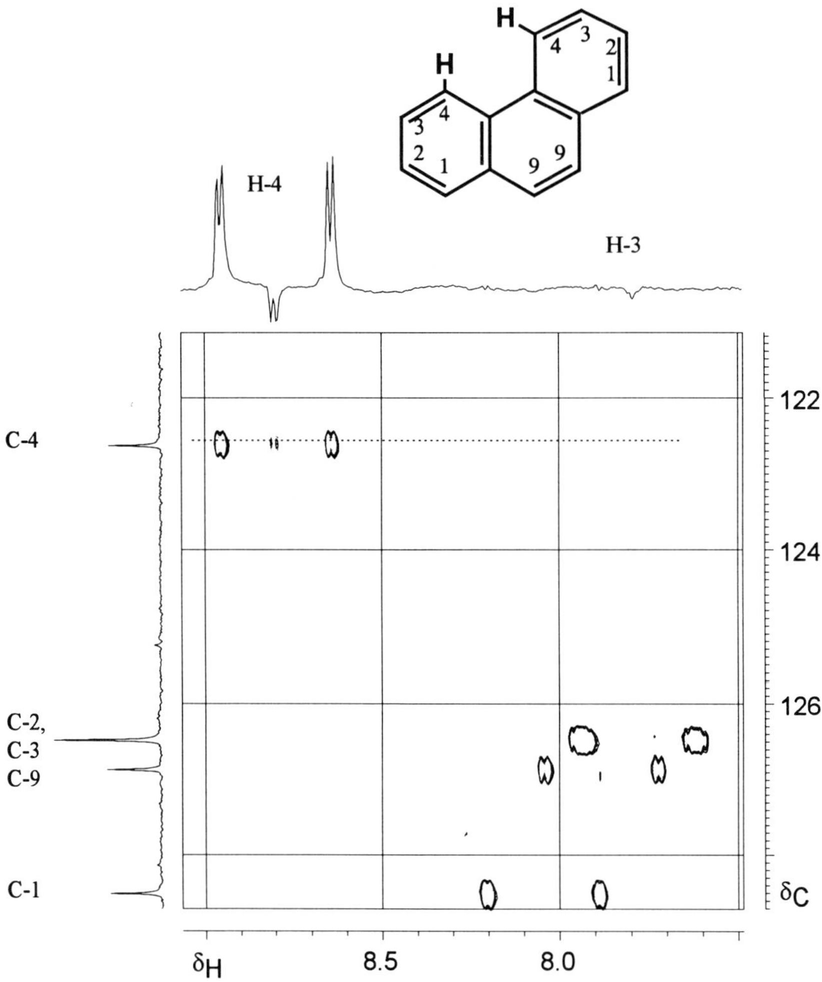

The figure shows an expansion of the ^1H-detected gs-HSQC-NOESY spectrum obtained on an AMX-500 spectrometer with a BGU (10 A) gradient unit using an inverse multinuclear *z*-gradient probe-head. Instead of the high-resolution ^1H NMR spectrum a row of the 2D matrix taken at the dotted line is plotted at the F_2 axis. This row shows clearly the negative NOE signal connecting the symmetrical protons H-4; in addition, a rather weak NOE signal is seen, indicating the interaction of H-4 and H-3.

7. Comments

The sequence starts with an INEPT transfer from protons to ^{13}C. A pair of weak gradients g1 and g2 removes signals which arise from imperfect 180° pulses during the first INEPT transfer. The antiphase ^{13}C magnetization $2I_{Hz}I_{Cy}$ present after p11 is dephased by the gradients g3 and g4, which are applied in the form of a [gradient, 180° pulse, gradient] bracket. The next step is a gradient zz-filter comprised of p13, g5 and p14 (see Exp. 11.7) to remove further unwanted signal contributions. During t_1, ^{13}C chemical shift develops, which is transferred back to protons via the back INEPT sandwich consisting of the pulses p5, p6, p15 and p16 (see Exp. 6.8). This back INEPT part serves at the same time as the start of the NOE part of the sequence. Pulse p7 transfers the magnetization into the *z*-direction, where cross-relaxation can occur during the mixing time d3. This situation is read by the reading pulse p8. The final gradient g7, which again is applied in a [delay, 180° pulse, gradient] bracket, rephases only the desired magnetization, whereas the gradient g6 removes any transverse magnetization build-up during the mixing time (see Exp. 11.10). Instead of the TPPI manner of sign determination in F_1, the echo/anti-echo technique as described in Experiment 12.6 could be used.

One should be aware that the method is very insensitive and comes close to the requirements of proton-detected INADEQUATE (Exp. 12.10). Since NOE signals are usually in the 5% range and only protons bound to ^{13}C are detected, the method reaches the limit of current instrumentation.

8. Own Observations

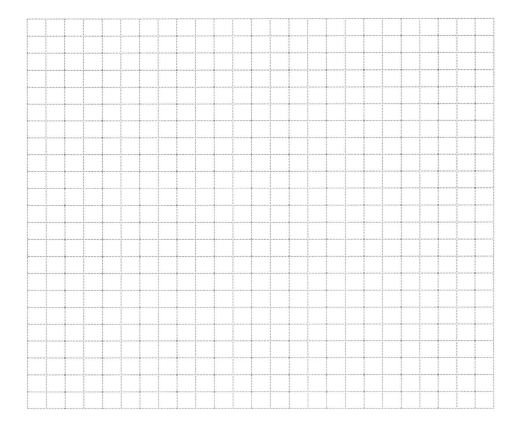

Experiment 12.13

gs-HOESY

1. Purpose

The gs-HOESY (**H**eteronuclear **O**verhauser **E**ffect **S**pectroscop**Y**) experiment is the gradient-selected inverse equivalent of Experiment 10.21, yielding information on the spatial relationship between spins in the heteronuclear case. It will be of main value in cases where information from spin–spin couplings is unhelpful or unavailable. Since it is proton-detected it has a much higher inherent sensitivity, and unwanted signals are effectively removed by the gradient-selection. It has been applied for the spin pairs $^{1}H,^{31}P$ and $^{1}H,^{7}Li$ [1]. The example shown here is taken from the field of organolithium chemistry with the same sample as used in Experiment 10.21 but using the ^{7}Li isotope. Selective 1D versions for the $^{1}H,^{13}C$ spin pair have also been reported [2].

2. Literature

[1] W. Bauer, *Magn. Reson. Chem.* **1996**, *34*, 532–537.
[2] K. Stott, J. Keeler, *Magn. Reson. Chem.* **1996**, *34*, 554–558.

3. Pulse Scheme and Phase Cycle

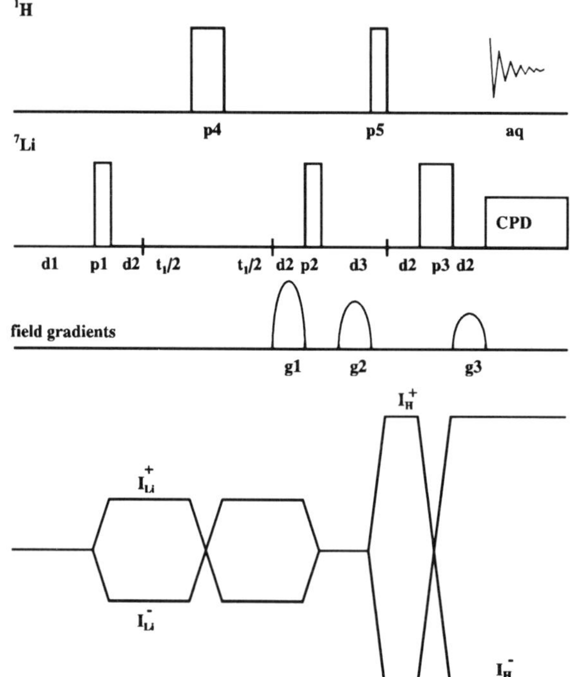

4. Acquisition

Time requirement: 10 min

Sample: Commercial 1.4 M *n*-butyllithium in hexane; add 10% *dry* [D_8]THF for locking purposes. Seal the sample with parafilm. The measurement can be done at room temperature.

Record normal ^{1}H and ^{7}Li NMR spectra, change to the 2D mode of the spectrometer, and load the gs-HOESY pulse program. You have to set:

td2: 2048 data points in F_2
td1: 32 data points in F_1
sw2: 5 ppm
sw1: 2.5 ppm
o1: middle of ^{1}H NMR spectrum
o2: middle of ^{7}Li NMR spectrum
p1, p2: 90° ^{7}Li decoupler pulse
p3: 180° ^{7}Li decoupler pulse
p4: 180° ^{1}H transmitter pulse
p5: 90° ^{1}H transmitter pulse
d1: 2 s
d2: 1 ms, set equal to gradient pulse-length
d3: 400 ms mixing time
g1, g2, g3: sinusoidal shaped field gradients 1 ms duration and ca. 0.2 T/m strength for the largest gradient, with gradient loop counters, ring-down delays (100 μs), lock blanking and gradient coil blanking switches according to actual instrumentation used. Gradient strength ratio: 36 : 50 : 14.
initial value for t_1 evolution: 3 μs
increment for t_1 evolution = 1/[2·sw1]
decoupler attenuation and 90° pulse for CPD
ds: 4
ns: 4

5. Processing

Apply zero filling in F_1 to 64 real data points. Use exponential windows with lb = 5 Hz in F_2 and a π/2 shifted squared sinusoidal window in F_1. Apply complex Fourier transformation corresponding to the N-type signal selection using the quadrature off mode in F_1. Phase correction is not necessary since the data are displayed in magnitude mode.

6. Result

The figure shows the result obtained on an AMX-500 spectrometer with a 5 mm inverse multinuclear *z*-gradient probe-head. The cross peaks with the α- and β-protons of butyllithium are clearly visible, although the signals of the β-protons are hidden under one of the resonances of hexane. Note the dramatic time savings compared to Experiment 10.21 where ^{6}Li is the detected nucleus.

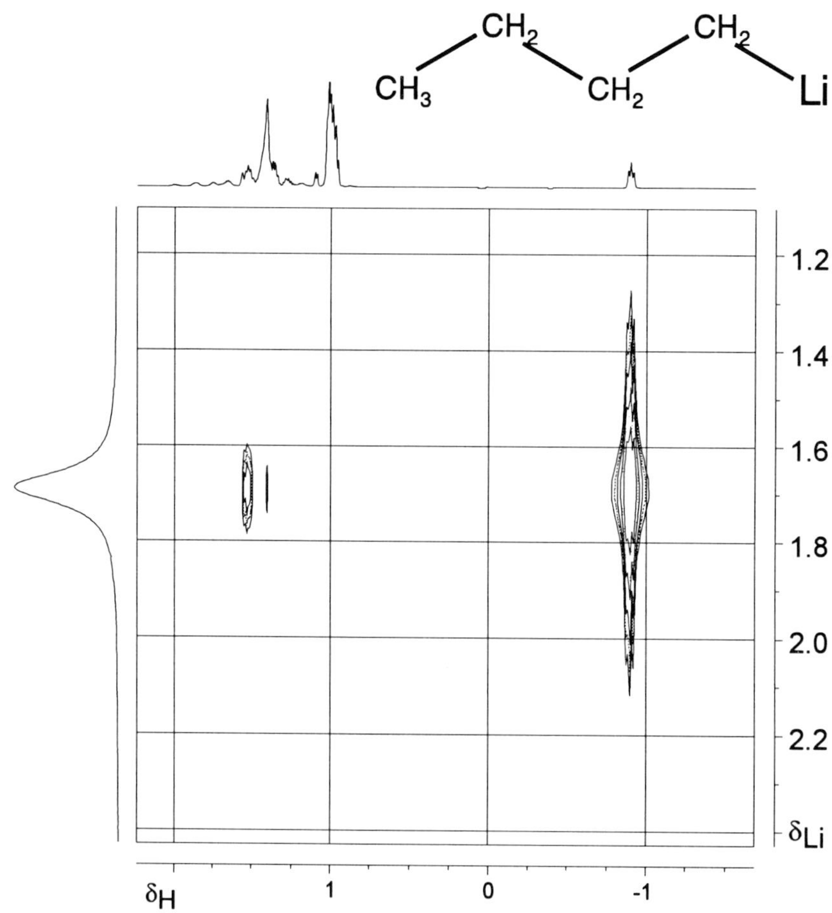

7. Comments

The product operator formalism for this experiment holds as given for the forward version of the HOESY method shown in Experiment 10.21. Here we comment only on the gradient selection scheme. As seen from the coherence pathway diagram, the first gradient g1 acts during the t_1 period, when transverse Li magnetization I^+_{Li} is present. Thus dephasing occurs with the factor g1·γ_{Li}. Directly before acqusition, when only I^-_H is present, the coherences are rephased with the factor g3·($-\gamma_H$). Therefore the gradients g1 and g3 must be in the ratio of the γ values of the two nuclei and they select the N-type signal pathway yielding a non phase-sensitive 2D NMR spectrum. The gradient pulse g2 dephases any left-over transverse magnetization present during the mixing time and does not take part in the actual signal selection.

In contrast to Experiment 10.21, here the F_2 dimension with the larger digital resolution is used for protons, whereas the less digitized F_1 dimension contains the Li signal. It should be noted that probably because of the long ^{6}Li relaxation time, the experiment was unsuccessful with this isotope.

8. Own Observations

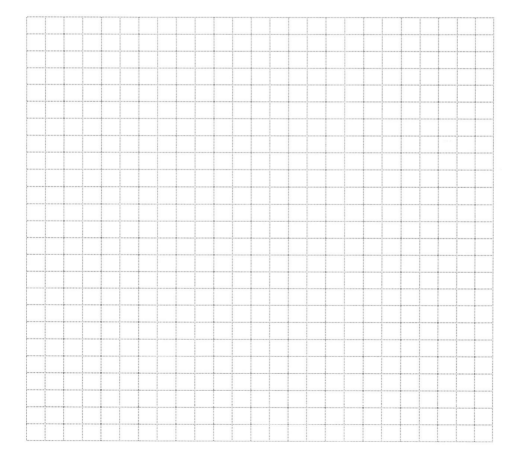

Experiment 12.14

$^1H,^{15}N$ Correlation with gs-HMQC

1. Purpose

Due to the low receptivity of ^{15}N it is very tedious to obtain ^{15}N NMR spectra of organic compounds if they are available only in milligrams. Even with the polarization transfer methods like DEPT (see Exp. 9.1) it will take hours to record such spectra, especially if the nitrogen atoms bear no directly attached proton. Inverse detection is therefore the method of choice, particularily if the unwanted signals can be effectively suppressed with pulsed field gradients (see Exp. 12.3). In the experiment described here we demonstrate the efficiency of this approach with the strychnine sample.

2. Literature

[1] R. E. Hurd, B. K. John, *J. Magn. Reson.* **1991**, *91*, 648–653.

[2] J. Ruiz-Cabello, G. W. Vuister, C. T. W. Moonen, P. van Gelderen, J. S. Cohen, P. C. M. van Zijl, *J. Magn. Reson.* **1992**, *100*, 282–303.

[3] W. Willker, D. Leibfritz, R. Kerssebaum, W. Bermel, *Magn. Reson. Chem.* **1993**, *31*, 287–292.

[4] G. Otting, B. A. Messerle, L. P. Soler, *J. Am. Chem. Soc.* **1996**, *118*, 5096–5102.

[5] K. A. Farley, G. S. Walker, G. E. Martin, *Magn. Reson. Chem.* **1997**, *35*, 671–679.

3. Pulse Scheme and Phase Cycle

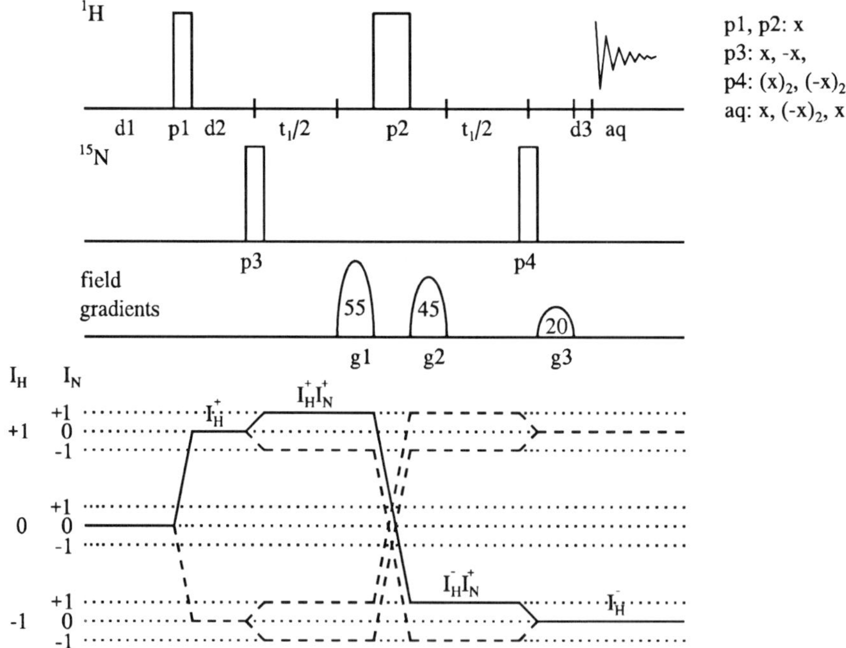

4. Acquisition

Time requirement: 20 min

Sample: 10% strychnine in $CDCl_3$

Tune the probe-head both for ^{15}N and ^{1}H. Record a normal ^{1}H spectrum of the sample and optimize the spectral width. Change to the 2D mode of the spectrometer and load the gs-HMQC pulse program. You have to set:

td2: 1 k data points in F_2
td1: 128 data points in F_1
sw2: 10 ppm
sw1: 400 ppm
offset of ^{1}H frequency: middle of ^{1}H NMR spectrum
offset of ^{15}N frequency: middle of ^{15}N NMR spectrum
p1: 90° ^{1}H transmitter pulse
p2: 180° ^{1}H transmitter pulse
p3, p4: 90° ^{15}N decoupler pulse
d1: 2 s
d2: $1/[2J(N,H)] = 50$ ms, calculated from $^{2,3}J(N,H) \approx 10$ Hz
d3: set equal to d2 minus gradient length
start increment for t_1 evolution: 3 μs
increment for t_1 evolution: $1/[2 \cdot sw1]$
g1, g2, g3: sinusoidal shaped field gradients with 5% truncation, 2 ms duration and ca. 0.01 T/m strength, with gradient loop counters, ring-down delays (100 μs), lock blanking and gradient coil blanking switches according to actual instrumentation used. Gradient strength ratio: 55 : 45 : 20.14
ds: 4
ns: 4

5. Processing

Apply zero filling in F_1 to 512 words in order to have a matrix of 512·512 real data points. Before Fourier transformation use $\pi/2$ shifted sinusoidal window functions both in F_2 and F_1. Phase correction is unnecessary, since the spectrum is displayed in the magnitude mode.

6. Result

The figure on page 530 shows an expansion of the 2D spectrum obtained on an AMX-500 spectrometer with an inverse multinuclear *z*-gradient probe-head and a BGU (10 A) gradient unit. Note that, contrary to all other examples in this book, the 1D spectrum on the F_1 axis is the internal projection of the 2D matrix since it would take exceedingly long to record a normal ^{15}N NMR spectrum from this sample.

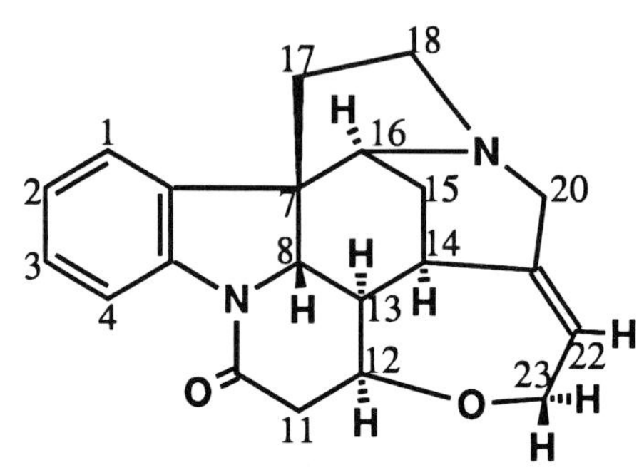

12 23 16 20 18 11 18 20 15 17 15 13
23 8 14 11

19

9

-350
-325
-300
-275
-250
-225
δ_N

δ_H 3 2 1

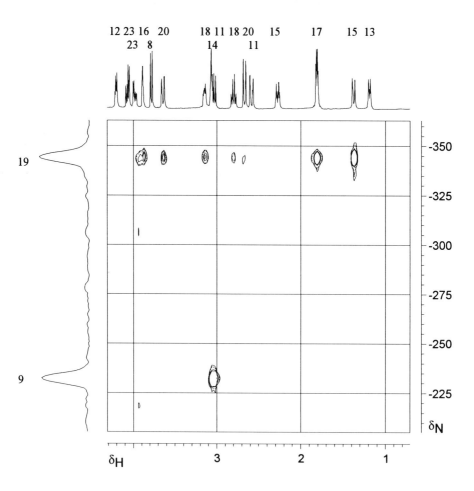

The amide nitrogen N-9 (δ_N = – 233) couples with one of the protons 11, whereas the other one, presumably because of the small coupling constant (Karplus relationship), does not give a correlation signal. The tertiary amine nitrogen (δ_N = – 345) does show correlation signals to all protons which are separated by two bonds (16, 18 and 20) and furthermore a correlation signal over three bonds to one of the protons 15.

7. Comments

The experiment uses nearly the same parameters as explained in Experiment 12.3. ^{15}N GARP decoupling was not applied for the same reasons outlined in Exp. 12.4. Note that the gradient ratios are quite different, since the gyromagnetic ratios of ^{15}N and ^{1}H are about 1 : 10. It is advantageous to use the exact gradient ratios as extracted from the carrier frequencies in both channels of the instrument. Double quantum magnetization, which in the coherence pathway diagram is labeled as $I_H^+ I_N^+$, is first dephased with the relative gradient strength g1 = 55, corresponding to $\gamma_H + \gamma_N = 11$. The 180° pulse in the proton channel transforms the coherence into $I_H^- I_N^+$. At this stage, the relative sum of γ values is –9. During acquisition, only I_H^- is present, with a relative γ value of –10. Thus, with the relative gradient strength used and applying the exact γ ratios, Eq. (1) yields 20.14 for the last gradient g3.

$$g1\,(\gamma_H + \gamma_N) + g2\,(-\gamma_H + \gamma_N) + g3\,(-\gamma_H) = 0 \qquad (1)$$

Gradients should not only select the desired coherences but also most efficiently deselect the undesired ones, and there are computer programs which perform this task. Another often used gradient ratio for the ^{15}N HMQC experiment is 70 : 30 : 50.

8. Own Observations

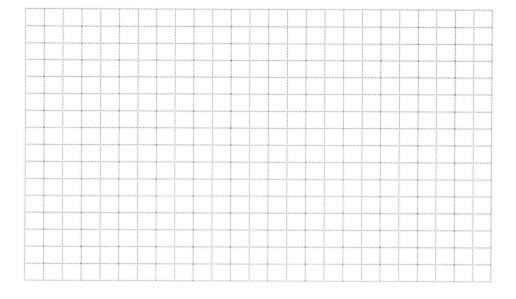

Chapter 13

The Third Dimension

Although 3D NMR experiments cannot be regarded as "basic experiments", we provide in this book four examples as a possible introduction to this exciting field of NMR spectroscopy.

3D experiments are typically constructed from two 2D sequences; thus one has a t_1 period from the first 2D experiment, a t_2 period from the second 2D experiment and the acquisition time, which is often designed as the t_3 period. The detection period of the first 2D sequence is replaced by the evolution part of the second 2D sequence, so that the first pulse of the second sequence is usually missing.

The remarks given in the introduction of Chapter 10 with regard to frequency discrimination and phase-sensitive detection in F_1 and F_2 similarly apply to the 3D case, except that now three time domains have to be considered.

Window functions, Fourier transformation, and phase correction are applied with respect to these three domains to finally yield a cuboid with three frequency axes, which are dependent on the particular experiment. They can be fully heteronuclear (e.g. ^{1}H, ^{13}C, ^{15}N) or mixed homo- and heteronuclear (e.g. ^{1}H, ^{1}H, ^{13}C). Up until now all reported 3D experiments have been proton-detected, which, for sensitivity reasons, has been the only viable option. Phasing of the indirect dimensions in 3D can be tedious, and is best performed by calculation as described for 2D on page 345.

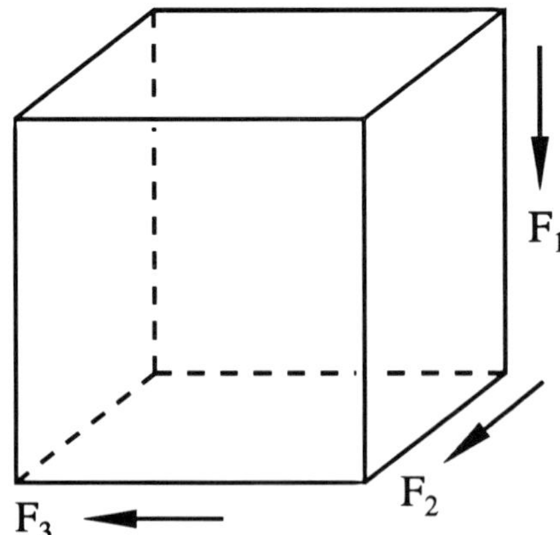

The processing of 3D NMR spectra involves very large data files and is therefore still limited to laboratories which are equipped with reasonably fast workstations and sufficient data storage capabilities. 3D spectra are inspected by choosing a horizontal or vertical plane through the cuboid which is then treated as a normal 2D spectrum.

Literature

[1] C. Griesinger, O. W. Sørensen, R. R. Ernst, *J. Magn. Reson.* **1989**, *84*, 14–63.
[2] C. Griesinger, H. Schwalbe, J. Schleucher, M. Sattler, in: W. R. Croasmun, R. M. K. Carlson (eds.), *Two-Dimensional NMR Spectroscopy*, VCH, Weinheim, **1994**, 457–580.

Experiment 13.1

3D HMQC-COSY

1. Purpose

HMQC (Exp. 10.13) and H,H-COSY (Exp. 10.3) are the most often used 2D pulse sequences for structure elucidation of organic compounds. However, severe signal overlap can occur in complicated molecules. One alternative is the 3D technique described here, in which the COSY spectra are "edited" via C,H correlation. The 3D spectrum leads to a cuboid in which one axis represents the carbon chemical shift and two axes the proton chemical shift. A C,H correlation signal can be found in a C,H plane of the cuboid for each protonated carbon atom. This signal also forms the diagonal peak of the corresponding COSY plane; thus by moving across between C,H and H,H planes unequivocal assignments are possible, even for very complicated cases. We show here a sequence which is phase-sensitive with respect to all three dimensions and uses the BIRD sandwich to suppress unwanted signals.

2. Literature

[1] C. Griesinger, O. W. Sørensen, R. R. Ernst, *J. Magn. Reson.* **1989**, *84*, 14–63
[2] S. W. Fesik, R. T. Gampe, E. R. P. Zuiderweg, *J. Am. Chem. Soc.* **1989**, *111*, 770–772.

3. Pulse Scheme and Phase Cycle

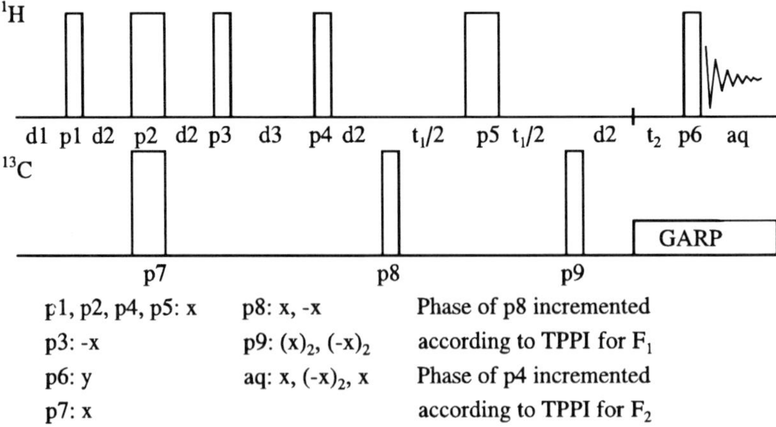

4. Acquisition

Time requirement: 24 h

Sample: 10% strychnine in $CDCl_3$

Record normal 1D ^1H and ^{13}C NMR spectra, optimize the spectral widths for the aliphatic region only and note the offsets. Narrow spectral windows should be used because of the the digitization problem in 3D NMR. Switch to the 3D mode of the spectrometer software, and load the 3D HMQC-COSY pulse program. You have to set:

td3: 256 data points in F_3 (^1H)
td2: 64 data points in F_2 (^1H)
td1: 128 data points in F_1 (^{13}C)
sw3: 3.3 ppm
sw2: 3.3 ppm
sw1: 42 ppm
o1: middle of selected ^1H NMR region
o2: middle of selected ^1H NMR region
o3: middle of selected ^{13}C NMR region
p1, p3, p4, p6: 90° ^1H transmitter pulse
p2, p5: 180° ^1H transmitter pulse
p8, p9: 90° ^{13}C decoupler pulse
p7: 180° ^{13}C decoupler pulse
d1: 2 s
d2: $1/[2J(C,H)]$ = 3.5 ms, calculated from $^1J(C,H)$ = 145 Hz
d3: BIRD delay to be optimized for minimum FID, ca. 0.4 s (see Exp. 6.11)
^{13}C decoupler attenuation and 90° pulse for GARP
initial values for t_1 and t_2 evolution: 3 μs
increment for t_1 evolution: $1/[4 \cdot sw1]$; increment for t_2 evolution: $1/[2 \cdot sw2]$
ds: 2
ns: 4

5. Processing

Apply zero filling to 128 real data points in F_2 and to 256 real data points in F_1 to obtain a matrix of 128·128·256 real data points. This will result in 4 MB of processed data. Use an exponential window with lb = 5 Hz line broadening in F_3 and $\pi/2$-shifted squared sinusoidal windows in the other two dimensions. Apply the correct acquisition order parameter (3–1–2) before FT in all three dimensions. Phase correction is best performed after the FT of each dimension. Further details are very dependent on the particular software you use to process such a 3D data file.

6. Result

The figure on page 536 shows a H,H-COSY plane through the 3D cuboid which was obtained on an AMX-500 spectrometer. The plane **a** was chosen at the nearly identical chemical shifts of C-11 and C-17 (δ_C = 42.5). Since protons 11 are attached to the same carbon atom, they reveal a normal COSY pattern; furthermore, the proton 11 at δ_H = 3.05 reveals a cross peak with H-12. The diagonal peak of H-17 leads to two cross-signals of both H-18 protons. In addition, a C,H plane **b** is shown on page 537, chosen at the chemical shift of H-14 (δ_H = 3.05). Nearby are the resonances of one H-18 and one H-11 proton. One observes a C,H correlation signal of C-14 with H-14. Furthermore H,H correlation signals H-14, H-15 and H-14, H-13 can be seen. The C,H correlation signal of H-11 leads to the signal of the other H-11 and of H-12, and the C,H correlation peak of H-18 leads to the other H-18 and to H-17. Note that these carbon selected COSY spectra are not symmetrical. Under the recording conditions the C,H correlation signal of C-12 (δ_C = 78) is folded and is seen in the lower left corner of the cuboid at about δ_C = 53.

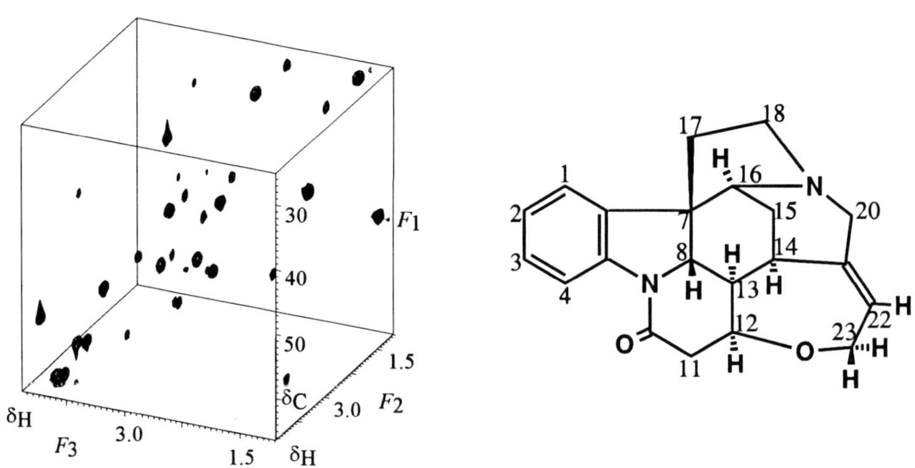

a: H,H plane at δ_C = 42.5

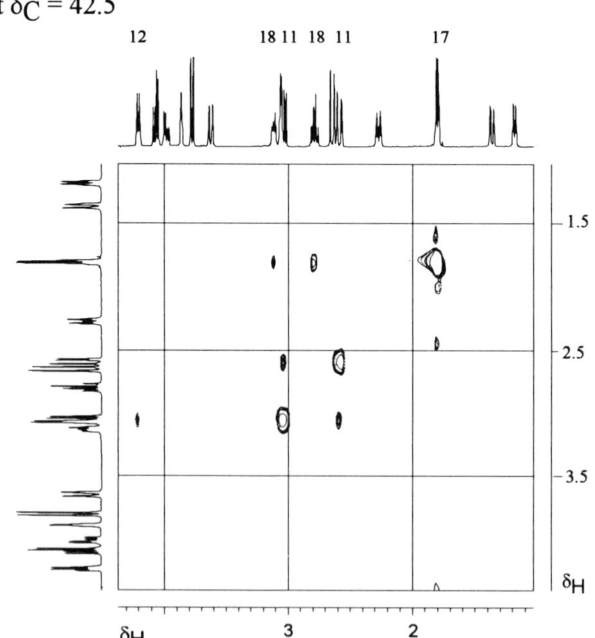

b: C,H plane at $\delta_H = 3.05$

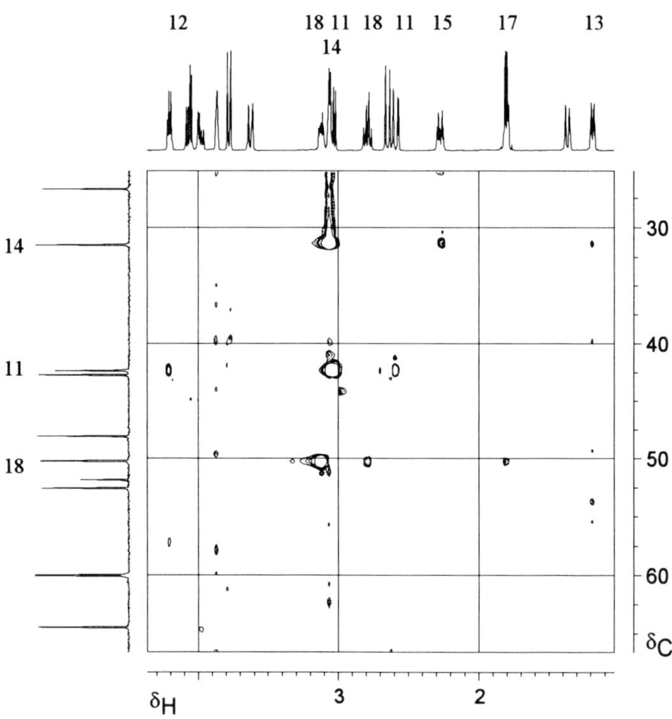

7. Comments

3D sequences are constructed by combining the two corresponding 2D sequences. The last pulse of the first sequence usually replaces the first pulse of the second pulse sequence. Thus, in the case described here, p9 transfers carbon magnetization back to proton magnetization and hence serves as the excitation pulse of the COSY part, so that only a second evolution time t_2 and the COSY read pulse p6 are required. GARP decoupling is switched on after the d2 delay in which the antiphase magnetization of protons with respect to ^{13}C nuclei has developed into in-phase magnetization.

8. Own Observations

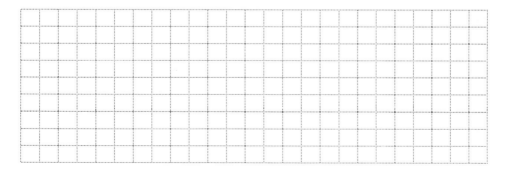

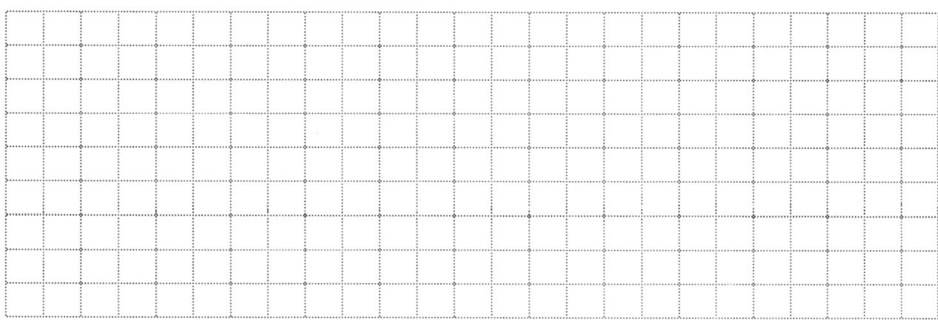

Experiment 13.2

3D gs-HSQC-TOCSY

1. Purpose

The 3D HMQC-COSY experiment (Exp. 13.1) is very time-consuming because of the required phase cycling. With gradient selection 3D experiments can be considerably shortened. Furthermore, the BIRD sandwich is no longer necessary at the beginning of the sequence and the gradient technique allows a far higher receiver gain. In the experiment described here we replace the HMQC part by the HSQC sequence, which has the advantage that no H,H spin coupling evolves during t_1. The COSY part is replaced by the TOCSY sequence so that only in-phase magnetization is transferred during t_2. This removes the problem of cancelling positive and negative signals due to poor digitization. The length of the spin-lock can be adjusted in order to observe several connectivities in a spin system. The pulse sequence given here is not phase-sensitive. Selective versions are also known [3].

2. Literature

[1] B. K. John, D. Plant, S. L. Heald, R. E. Hurd, *J. Magn. Reson.* **1991**, *94*, 664–669.
[2] W. Willker, D. Leibfritz, R. Kerssebaum, W. Bermel, *Magn. Reson. Chem.* **1993**, *31*, 287–292.
[3] T. Fäcke, S. Berger, *Tetrahedron* **1995**, *51*, 3521–3524.

3. Pulse Scheme with Phase Cycle

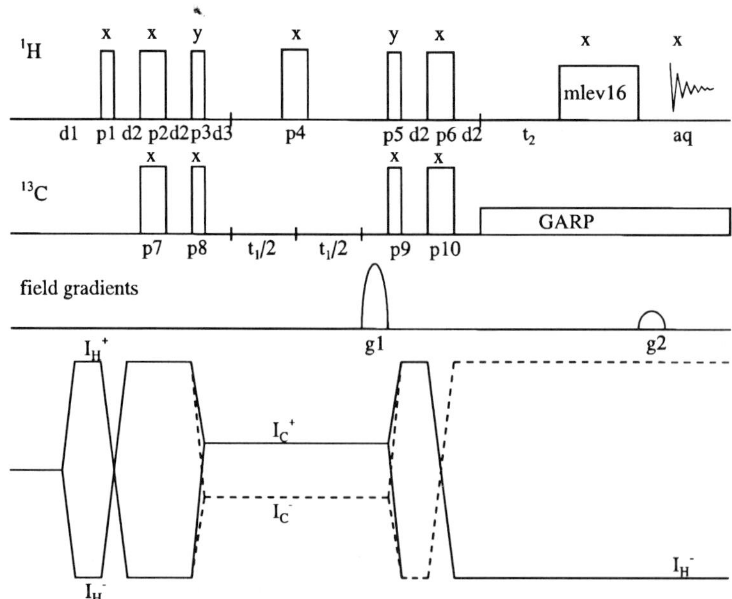

4. Acquisition

Time requirement: 5 h

Sample: 10% strychnine in $CDCl_3$

Record normal 1D ^1H and ^{13}C NMR spectra, optimize the spectral widths for the aliphatic region only, and note the offsets. A narrow spectral window should be used because of the the digitization problem in 3D NMR. Switch to the 3D mode of the spectrometer software and load the gs-HSQC-TOCSY pulse program. You have to set:

td3: 256 data points in F_3 (^1H)
td2: 64 data points in F_2 (^1H)
td1: 128 data points in F_1 (^{13}C)
sw3: 3.3 ppm
sw2: 3.3 ppm
sw1: 56 ppm
o1: middle of selected ^1H NMR region
o2: middle of selected ^1H NMR region
o3: middle of selected ^{13}C NMR region
p1, p3, p5: 90° ^1H transmitter pulse
p2, p4, p6: 180° ^1H transmitter pulse
p8, p9: 90° ^{13}C decoupler pulse
p7, p10: 180° ^{13}C decoupler pulse
90° and 180° ^1H pulses within MLEV-16 spin-lock = 40 and 80 μs at transmitter attenuation of 16 dB (see Exp. 11.9); loop counter for length of spin-lock = 40 (must be an even number), giving a spin-lock time of 100 ms
d1: 2 s
d2: $1/[4J(C,H)]$ = 1.75 ms, calculated from $^1J(C,H)$ = 145 Hz
d3: delay to compensate the length of field gradient = 1.5 ms
g1, g2: sinusoidal shaped field gradients with 5% truncation, 1.5 ms duration and ca. 0.01 T/m strength, with gradient loop counters, ring-down delays (100 μs), lock blanking and gradient coil blanking switches according to actual instrumentation used. Gradient strength ratio: 4:1
^{13}C decoupler attenuation and 90° pulse for GARP
initial values for t_1 and t_2 evolution: 3 μs
increment for $t1$ evolution: 1/[2·sw1]; increment for t_2 evolution: 1/sw2
ds: 4
ns: 1

5. Processing

Apply zero filling to 128 real data points in F_2 and to 256 real data points in F_1 to obtain a matrix of 128·128·256 real data points. This will result in 4 MB of processed data. Use an exponential window with lb = 5 Hz line broadening in F_3 and $\pi/2$ shifted squared sinusoidal windows in the other two dimensions. Apply the correct acquisition order parameter (3–1–2) before FT in all three dimensions. Phase correction is not necessary, since a magnitude calculation is performed after the last Fourier transformation in F_1. Further details are very dependent on the particular software you use to process such a 3D data file.

6. Result

The figure shows a H,H-TOCSY plane through the 3D cuboid which was obtained on an AMX-500 spectrometer using a BGU (10 A) gradient unit and a multinuclear *z*-gradient probe-head. The plane **a** was chosen at the ^{13}C chemical shifts of C-15 (δ_C = 26.7). Since two protons are bound to this carbon atom their C,H cross-peaks both appear, together with COSY-like H,H cross-peaks. Furthermore, the two cross-signals to neighbour protons H-16 and H-14 are observed.

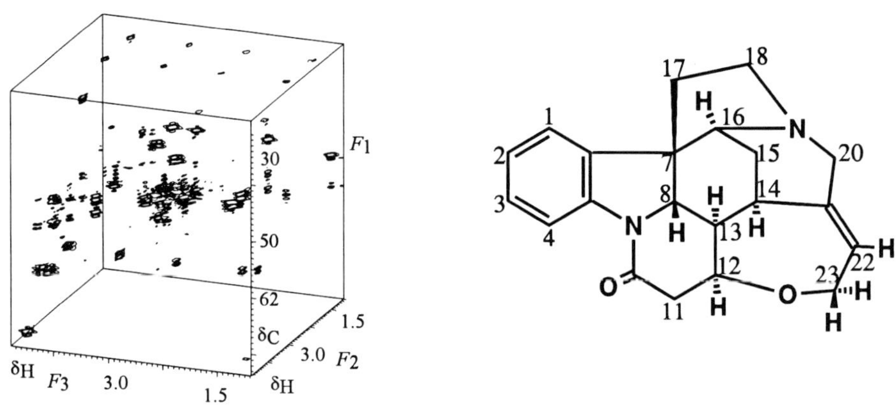

a: H,H plane at δ_C = 26.7

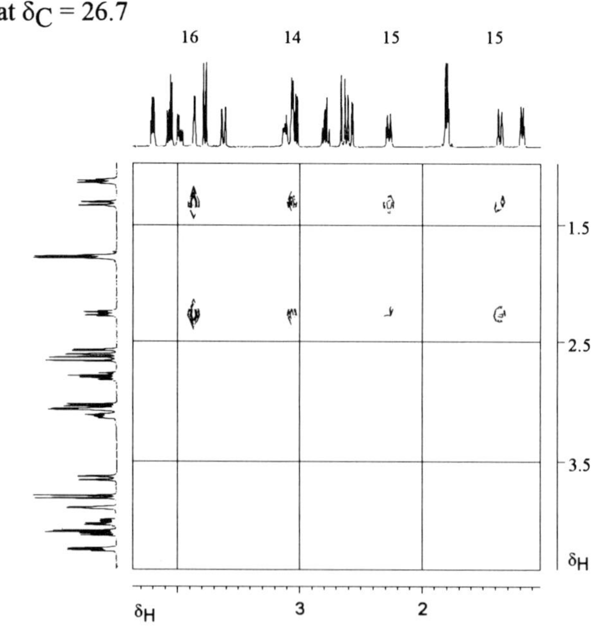

In addition, an expansion of a C,H plane **b** is shown, chosen at the chemical shift of proton 12. The corresponding C,H cross signal can be seen together with cross-peaks to the neighbour protons 11 and 12 and in addition to proton 8, which is a TOCSY result.

b: C,H plane at $\delta_H = 4.2$

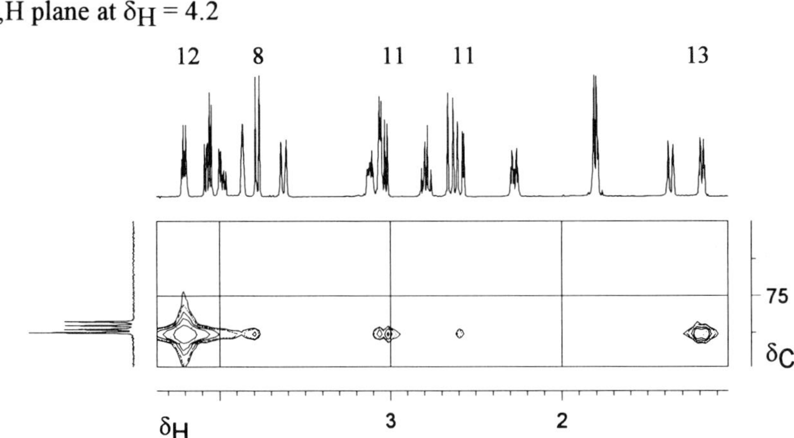

7. Comments

The initial part of the sequence uses the standard HSQC method (see Exp.10.17) while the second part uses the TOCSY technique (see Exps. 10.18 and 12.7). The MLEV-16 pulse train is used as a spin-lock, which consists of an even number of composite 180° pulses and therefore does not change the coherence order. Hence the two positive gradients select the N type pathway during t_1 and P type signals during t_2. The sequence is therefore not phase-sensitive and can be processed in magnitude mode. The overwhelming advantage of this experiment when compared with Experiment 13.1 is that the 3D measurement can be performed with only one transient and, because of the TOCSY part, gives more information in less than a quarter of time. Note, however, that for strychnine the same chemical information is provided by the 2D version of such an experiment, as shown in Experiment 12.8.

8. Own Observations

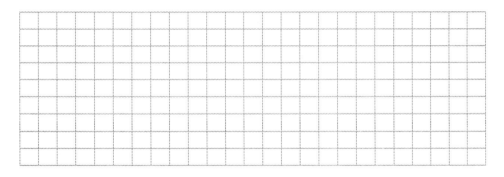

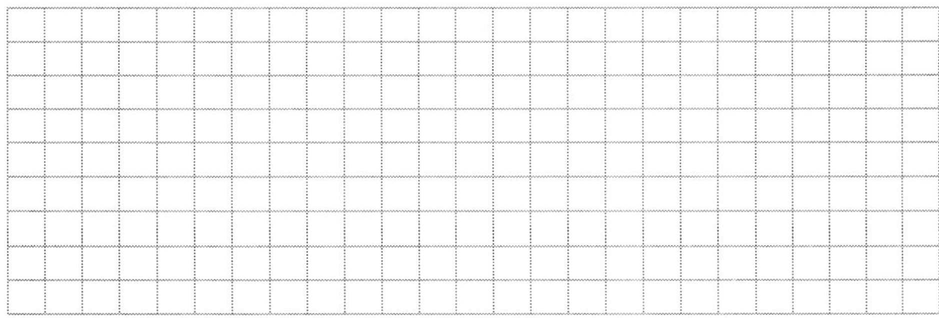

Experiment 13.3

3D H,C,P-Correlation

1. Purpose

In this 3D experiment three different nuclides are correlated with each other, so that one obtains a cuboid in which one axis represents the ^{1}H, the second axis the ^{13}C, and the third axis the ^{31}P chemical shift. Cross-signals appear at points where protons are coupled to ^{13}C nuclei which are in turn coupled to phosphorus. Experiments of this kind, using the spin trio ^{1}H, ^{13}C and ^{15}N, are very important in protein chemistry, whereas the type of experiment described here will find applications in the field of nucleic acids. The very simple example of triphenylphosphane is given here, from which a beginner in this field can gain considerable insight. The experiment is phase-sensitive with respect to all three dimensions.

2. Literature

[1] S. Berger, P. Bast, *Magn. Reson. Chem.* **1993**, *31*, 1021–1023.

[2] H. A. Heus, S. S. Wijmenga, F. J. M. van de Ven, C. W. Hilbers, *J. Am. Chem. Soc.* **1994**, *116*, 4983–4984.

[3] J. P. Marino, H. Schwalbe, C. Anklin, W. Bermel, D. M. Crothers, C. Griesinger, *J. Am. Chem. Soc.* **1994**, *116*, 6472–6473.

3. Pulse Scheme and Phase Cycle

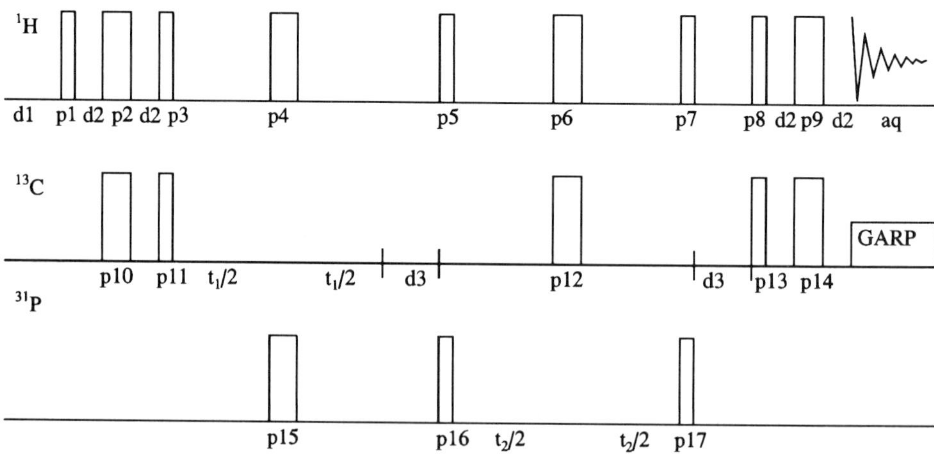

p1, p2, p4, p5, p6, p7, p8, p9, p13, p14. p15, p17: x

p3: y, -y

p10: x, -x

p16: $(x)_2$, $(-x)_2$, incremented according to TPPI during t_2

p12: $(x)_4$, $(y)_4$, $(-x)_4$, $(-y)_4$

p11: x, incremented according to TPPI during t_1

aq: x, $(-x)_2$, x, -x, $(x)_2$, -x

4. Acquisition

Time requirement: 6 h

Sample: 10% triphenylphosphane in $CDCl_3$

This experiment requires a three channel spectrometer and a probe-head, which is tunable to ^{1}H, ^{13}C, and ^{31}P. Use appropriate r.f. pass and stop filters in all three channels. Record normal 1D ^{1}H, ^{13}C and ^{31}P NMR spectra, optimize the spectral widths (^{1}H, ^{13}C) for the aromatic region only, and note the offsets. Switch to the 3D mode of the spectrometer software and load the H,C,P correlation pulse program. You have to set:

td3: 256 data points in F_3 (^{1}H)
td2: 16 data points in F_2 (^{31}P)
td1: 64 data points in F_1 (^{13}C)
sw3: 0.6 ppm
sw2: 0.5 ppm
sw1: 12 ppm
o1: middle of selected ^{1}H NMR region
o2: middle of selected ^{13}C NMR region
o3: middle of selected ^{31}P NMR region
p1, p3, p5, p7, p8: 90° ^{1}H transmitter pulse
p2, p4, p6, p9: 180° ^{1}H transmitter pulse
p11, p13: 90° ^{13}C decoupler pulse
p10, p12, p14: 180° ^{13}C decoupler pulse
^{13}C decoupler attenuation and 90° pulse for GARP
p16, p17: 90° ^{31}P pulse in third channel of the spectrometer
p15: 180° ^{31}P pulse in third channel of the spectrometer
d1: 2 s
d2: $1/[4J(C,H)]$ = 1.56 ms, calculated from $^{1}J(C,H)$ = 160 Hz
d3: $1/[2J(C,P)]$ = 38 ms, calculated from $^{n}J(C,P)$ = 13 Hz (average)
initial values for t_1 and t_2 evolution: 3 μs
increment for t_1 evolution: $1/[4\cdot sw1]$; increment for t_2 evolution: $1/[4\cdot sw2]$
preacquisition delay: as small as possible
ds: 2
ns: 8

5. Processing

Apply zero-filling to 32 real data points in F_2 and to 128 real data points in F_1 to obtain a matrix of 128·32·128 real points. This will result in 1 MB of processed data. Use an exponential window with lb = 3 Hz line broadening in F_3 and $\pi/2$ shifted squared sinusoidal windows in the other two dimensions. Apply the correct acquisition order parameter (3–1–2) before FT in all three dimensions. Phase correction is best performed after the FT of each dimension. Further details are very dependent on the particular software you use to process such a 3D data file.

6. Result

The figures **a** and **b** show planes through the 3D cuboid obtained on an AMX-500 spectrometer equipped with an inverse multinuclear probehead containing an additional r.f. channel fixed on the ^{13}C frequency. The plot **a** is a C,H plane chosen at the ^{31}P chemical shift position. Only the signals of the *ortho* and *meta* hydrogen nuclei are seen, since the *para* ^{13}C nucleus does not have a significant C,P spin coupling constant. On page 545 a C,P plane **b** is shown, chosen at the chemical shift of the *ortho* hydrogens.

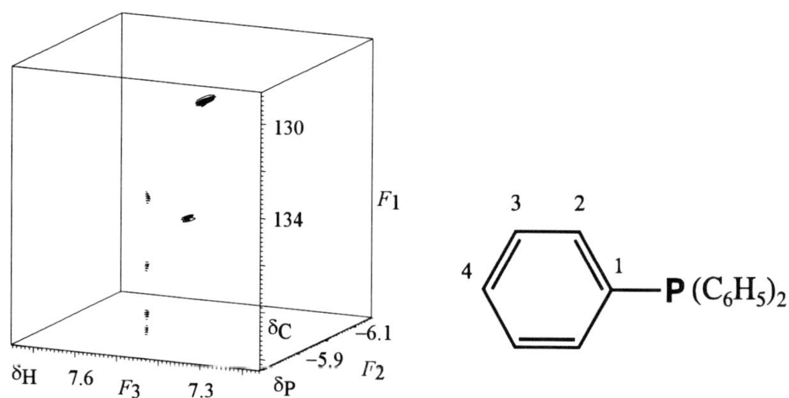

a: C,H plane at $\delta_P = -6$

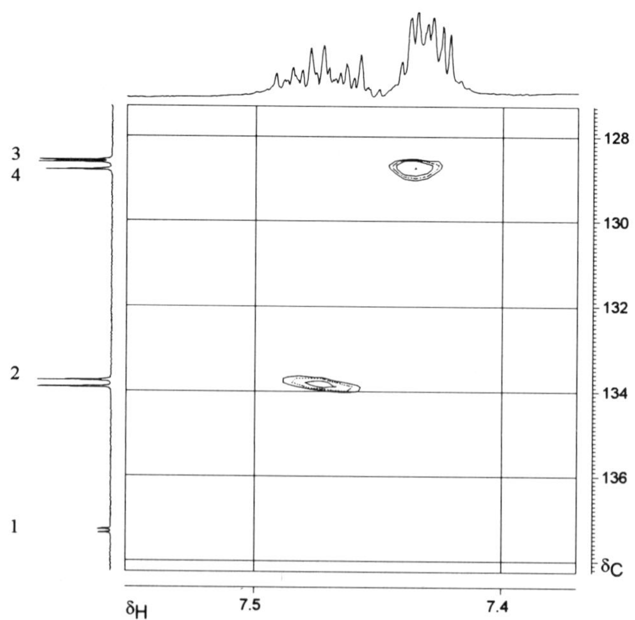

b: C,P plane at $\delta_H = 7.48$

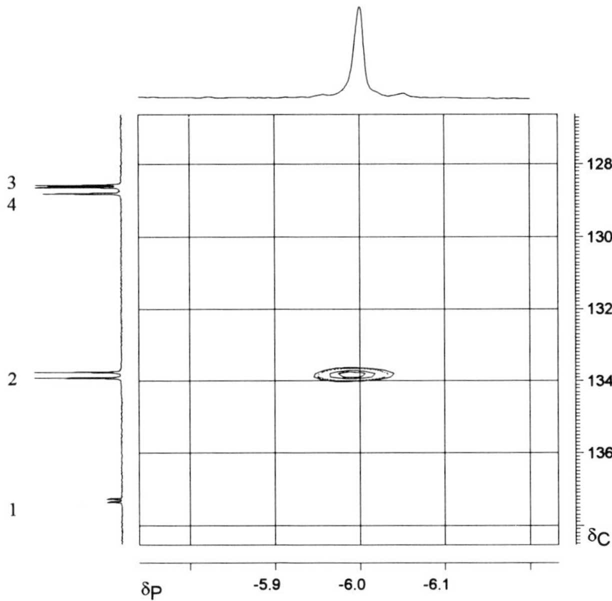

7. Comments

The pulse train begins as for the normal HSQC sequence (see Exp. 10.17). During t_1, when carbon chemical shift develops, a 180° ^{31}P pulse p15 removes any H,P and P,C spin couplings. Antiphase magnetization $4I_{Hz}I_{Cx}I_{Pz}$ of carbon with respect to phosphorus develops during the delay d3, and is subsequently transformed into three-spin coherence $4I_{Hy}I_{Cx}I_{Py}$ by the simultaneous proton and phosphorus 90° pulses p5 and p16. During t_2 phosphorus chemical shift develops. Two simultaneous pulses p7 and p17 create antiphase magnetization $4I_{Hz}I_{Cx}I_{Pz}$ which refocuses to give antiphase magnetization $2I_{Hz}I_{Cy}$ during the second d3 period. This is transferred by the reverse INEPT part of the sequence to protons for detection.

8. Own Observations

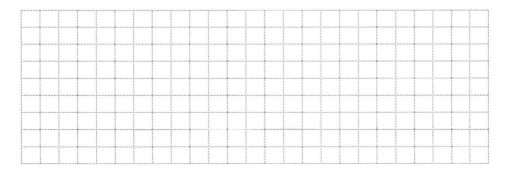

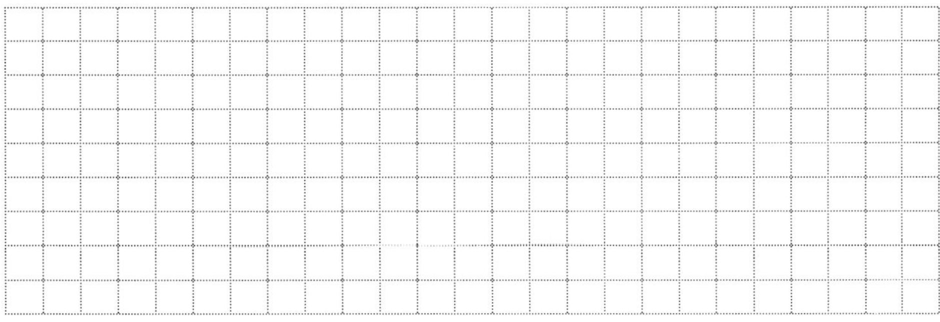

Experiment 13.4

3D-HMBC

1. Purpose

The gs-HMBC experiment (Exp. 12.4) is one of the most powerful methods for structural elucidation of organic compounds. One drawback of the method, however, is that with the typically chosen HMBC delay of 60 ms not all long-range interactions between carbons and protons are met in an optimal way because of the variation of ^{n}J(C,H). Instead of measuring several HMBC spectra with different delays, a 3D version was recently proposed in which the corresponding delay is incremented; thus the whole range of C,H long-range coupling constants are actually used for double quantum excitation. Here we demonstrate the non phase-sensitive pulse sequence with the sample of strychnine.

2. Literature

[1] K. Furihata, H. Seto, *Tetrahedron Lett.* **1996**, *37*, 8901–8902.

3. Pulse Scheme and Phase Cycle

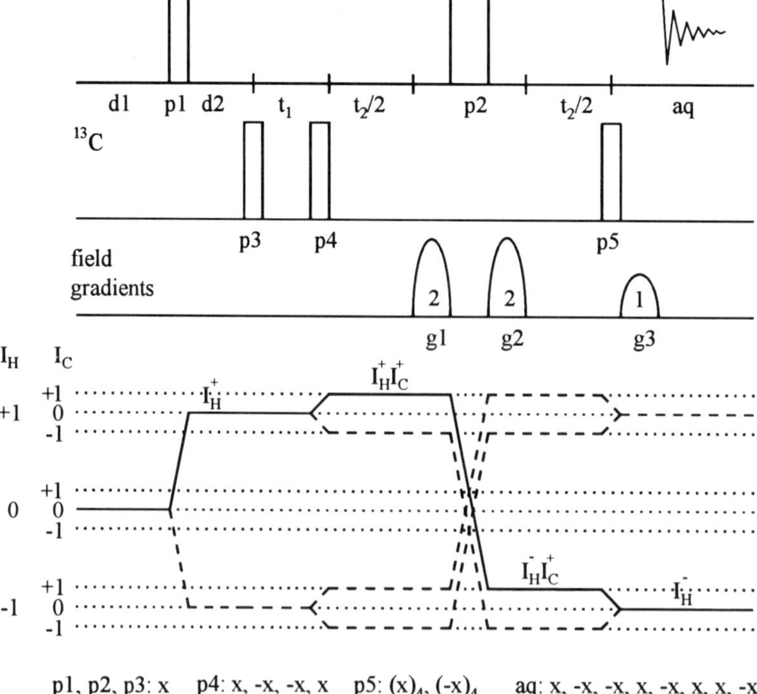

p1, p2, p3: x p4: x, -x, -x, x p5: $(x)_4$, $(-x)_4$ aq: x, -x, -x, x, -x, x, x, -x

4. Acquisition

Time requirement: 5 h

Sample: 10% strychnine in $CDCl_3$

Record normal 1D ^1H and ^{13}C NMR spectra, optimize the spectral widths and note the offsets. Switch to the 3D mode of the spectrometer software and load the 3D-HMBC pulse program. You have to set:

td3: 512 data points in F_3 (^1H)
td2: 256 data points in F_2 (^{13}C)
td1: 16 data points in F_1 (J(C,H))
sw3: 9.5 ppm
sw2: 187 ppm
sw1: 250 Hz
o1: middle of ^1H NMR spectrum
o2: middle of ^{13}C NMR spectrum
p1: 90° ^1H transmitter pulse
p2: 180° ^1H transmitter pulse
p3, p4, p5: 90° ^{13}C decoupler pulse
d1: 2 s
d2: $1/[2J(C,H)]$ = 3.5 ms, calculated from $^1J(C,H)$ = 145 Hz
g1, g2, g3: sinusoidal shaped field gradients with 5% truncation, 1 ms duration and ca. 0.01 T/m strength, with gradient loop counters, ring-down delays (100 μs), lock blanking and gradient coil blanking switches according to actual instrumentation used. Gradient strength ratio: 2 : 2 : 1
initial values for t_1 evolution: 20 ms
initial values for t_2 evolution: 3 μs
increment for t_1 evolution: 1/sw1; increment for t_2 evolution: 1/[2·sw2]
ds: 2
ns: 2

5. Processing

Apply zero filling to 512 real data points in F_2 and to 32 real data points in F_1 to obtain a matrix of 256·512·32 real data points. This will result in 4 MB of processed data. Use a $\pi/2$ shifted squared sinusoidal in all three dimensions. Apply the correct acquisition order parameter (3–1–2) before FT. Phase correction is not necessary, since a magnitude calculation is performed after the last Fourier transformation in F_1. Further details are very dependent on the particular software you use to process such a 3D data file. Note that the 3D cuboid obtained is not further inspected, instead; the F_2-F_3 *projection* is calculated and displayed.

6. Result

The figure shows the F_2-F_3 projection of the 3D-HMBC spectrum which was obtained on an Avance DRX-400 spectrometer using a multinuclear *z*-gradient probe-head. By a detailed comparison with the result of Experiment 12.4, which can be performed best on a computer screen, significantly more cross peaks especially in the aliphatic region have been detected. Note that the digital resolution used here is far less, however.

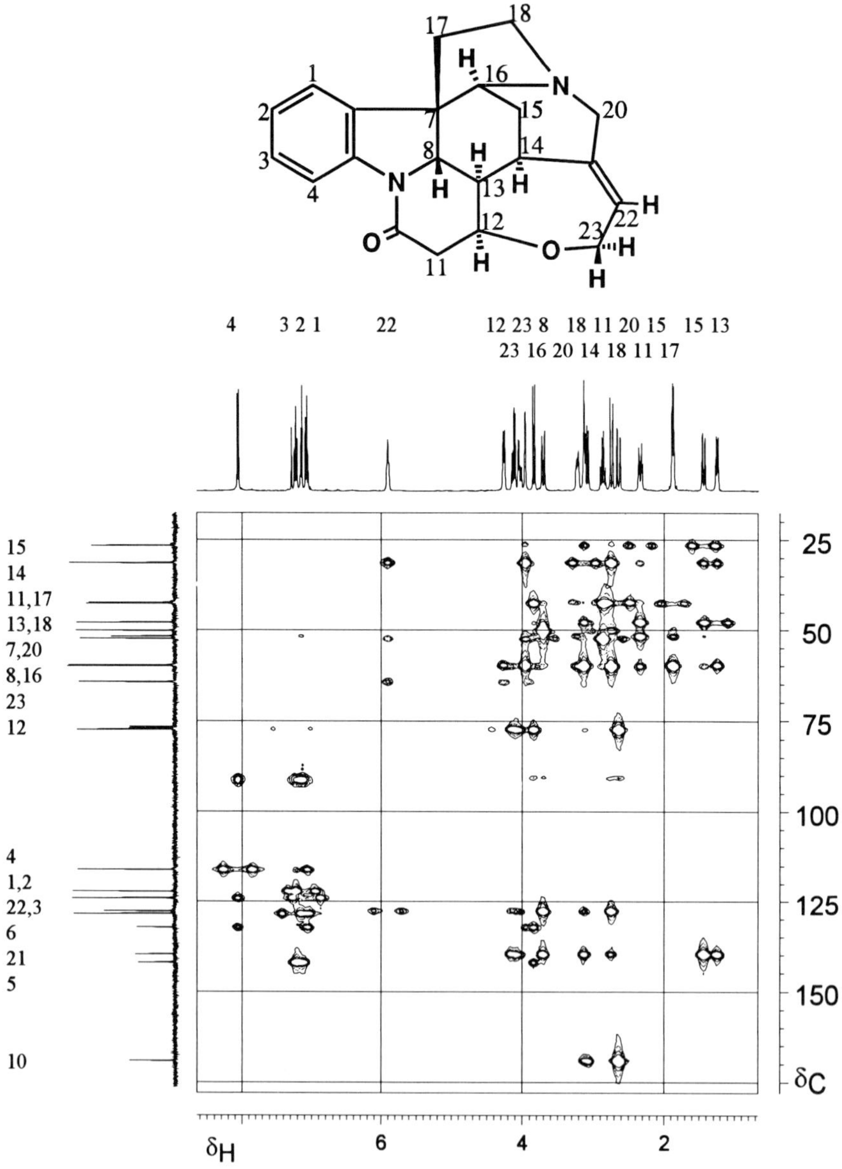

7. Comments

The sequence is identical to Experiment 12.4 with the only difference that the delay d3 of Exp. 12.4 is replaced by the t_1 evolution. When using an initial delay for t_1 evolution of 20 ms and a "spectral width" of 250 Hz in 16 steps, this correponds to a 4 ms increment in t_1 and the final t_1 delay will be 80 ms. Therefore in this experiment spin coupling constants ranging from 25 Hz to 6.25 Hz have been chosen to contribute to the HMBC transfer. Of coarse, this range may be extended if desired.

The gradient strength chosen here can be rationalized from an inspection of the coherence pathway diagram. The first gradient g1 acts when the term $I_H^+ I_C^+$ is present, thus the coherence will be dephased with 5·g1; g2 acts when $I_H^- I_C^+$ is present, therefore the result will be −3·g2. Finally, g3 acts when only I_H^- is present, therefore the rephasing will occur with −4·g3. Thus choosing the gradient ratio of 2 : 2 : 1 only the desired coherences are observed. GARP decoupling is not applied for the same reason as stated in Exp. 12.4; since the low pass filter of the HMBC sequence does not work perfectly decoupling would made it impossible to distinguish between direct and long-range correlations.

8. Own Observations

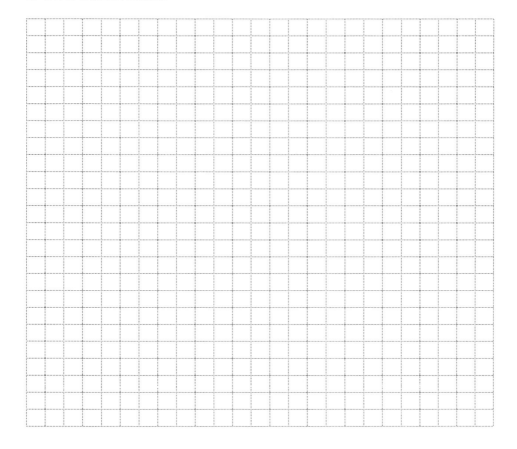

Chapter 14

Solid-State NMR Spectroscopy

Many problems in chemistry cannot be tackled with solution NMR either because the material cannot be dissolved or because special interactions only present in the solid state are to be investigated. These are especially the study of the anisotropy of the NMR parameters such as the chemical shift or the spin–spin coupling.

The scope of solid-state NMR is extremely wide, reaching from solid-state physics and material science to structural biology. Corresponding to this there exist a multitude of methods which are currently applied and still being developed. Similarly, there is a very large variety in the available instrumentation, reaching from hybrid instruments which are laid out to perform both high resolution in the liquid and the solid states to very specialized solid-state instruments which are specially designed to cover wide-line NMR applications of quadrupolar nuclei or even NMR microscopy of solid materials.

Whereas in high-resolution NMR of the liquid state as described in the first 13 chapters of this book one can hardly cause serious problems for the instrument by choosing wrong parameters, this can easily happen at the high power levels typically used in solid-state NMR. Thus, for instance, one uses in solid-state NMR high-power continous-wave decoupling instead of low-power composite-pulse decoupling. The novice in this field should therefore be extremely careful and always double check on all settings of both hardware connections and software parameters before starting up an experiment. Probe-head coils, preamplifiers and other parts of the instrument can easily be destroyed by wrong settings. For instance, it is advantageous to remove the proton preamplifier when using high-power decoupling.

In this chapter we provide descriptions for shimming a solid-state probe-head, setting the magic angle, finding the Hartmann-Hahn matching condition, and performing the basic CP/MAS experiment. We also introduce some side-band suppressing (TOSS and SELTICS) and editing experiments such as NQS.

[1] C. A. Fyfe, *Solid State NMR for Chemists*, C.F.C. Press, Guelph, **1983**.
[2] R. Völkel, *Angew. Chem. Int. Ed. Engl.* **1988**, *27*, 1468–1483.
[3] B. Blümich, H. W. Spiess, *Angew. Chem. Int. Ed. Engl.* **1988**, *27*, 1655–1672.
[4] C. P. Slichter, *Principles of Magnetic Resonance*, 3rd Edition, Springer, Berlin, **1989**
[5] B. Blümich (ed.), *NMR-Basic Principles and Progress* **1994**, *Volumes 30-34*.
[6] K. Schmidt-Rohr, H. W. Spiess, *Multidimensional Solid-State NMR and Polymers*, Academic Press, London, **1994**.
[7] E. O. Stejskal, J. D. Memory, *High-Resolution NMR in the Solid State,* Oxford University Press, New York, **1994**.

Experiment 14.1

Shimming Solid-State Probe-Heads

1. Purpose

In solid-state NMR there is usually no lock channel provided by the probe-head. Although the line-widths are considerably larger than those typical in solution NMR, a reasonable basic shim is necessary to provide a Lorenzian line-shape and to assure good results. This is usually achieved by first shimming on the FID of a water sample which is encapsulated in a solid-state rotor. In a second step, one optimizes the shim using adamantane as a sample. Here we show how to perform this two-step procedure. Using the water sample, the field position can be easily controlled.

2. Literature

[1] H. Förster, *Avance DSX Operators Manual, Bruker, Rheinstetten*, **1995**.

3. Pulse Scheme and Phase Cycle

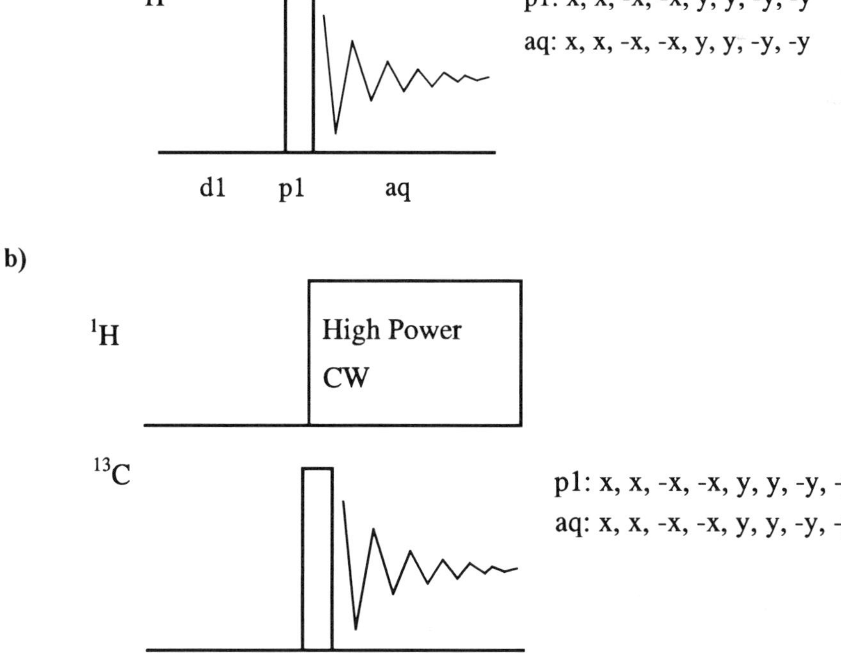

4. Acquisition

Time requirement: 20 min

Sample **a**: For step fill a solid-state rotor with normal water. Be sure that the rotor is completely filled and no air bubble render the sample inhomogeneous. It often helps to drill a tiny hole into the rotor cap to avoid this situation.
Sample **b**: For the second step **b** fill a solid-state rotor with finely powdered adamantane

Step **a**: Load standard proton parameters. Do not spin the sample. You have to set:

td: 4 k
sw: 125 kHz
o1: 2000 Hz to lower frequencies from water signal
p1: 1 μs ^{1}H transmitter pulse
d1: 1 s
transmitter power level (1 μs pulse should correspond to a pulse angle of apprimately 20°)
rg: receiver gain for correct ADC input
ns: 1

Using the set-up mode of the spectrometer, where the individual FIDs are not accumulated, display the FID of the water signal, turn the field sweep off, and optimize the various shims of your instrument by measuring the area of the FID. If the transformed signal is satisfactory proceed to step **b**.

Step **b**: Load the adamantane sample. Turn the spinner to 2500 Hz and load standard ^{13}C NMR parameters with high-power continous-wave decoupling during acquisition. You have to set:

td: 8 k
sw: 20 kHz
o1: middle of ^{13}C NMR spectrum
o2: middle of ^{1}H NMR spectrum
p1: 4 μs ^{13}C transmitter pulse
d1: 3 s
transmitter power level
decoupler attenuation for high power cw decoupling
rg: receiver gain for correct ADC input
ns: 1

Again, shim on the area of the FID in the set up mode. Record a spectrum with 1 transient.

5. Processing

For **a** use standard proton processing (see Exp. 3.1) with zero filling to 8 k. However, due to the cut-off of the FID after the short acquisition time, apodization artefacts may occur; a π/3-shifted squared sine window should then be performed. For **b** use standard ^{13}C NMR processing (see Exp. 3.2) with zero filling to 16 k and exponential weighting with lb = 5 Hz.

6. Result

The figures show in **a** the water signal obtained in a multinuclear solid-state probe-head with a 7 mm rotor on an AM-400 spectrometer with a wide-bore magnet. The line-width at half height in this case was measured to be 70 Hz; a value around 50 Hz is considered to be good. In **b** the spectrum of adamantane is shown, obtained on the same spectrometer and probe-head. The line-width of the signal for C-1 was 5 Hz.

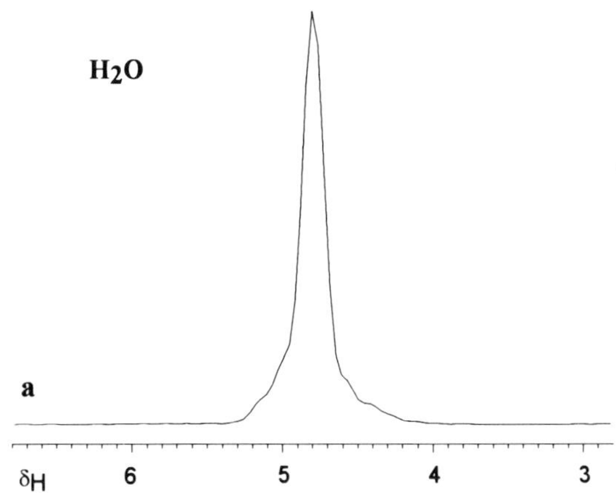

7. Comments

Although shimming for solid-state applications is far less critical and time demanding than for solution NMR (see Chapter 1.4), one should go through this procedure regularly and document the results in the log-book of the instrument. The signal-to-noise ratio will be severely affected if the line-shape is not satisfactory. The result of the adamantane spectrum is therefore also a suitable check of the sensitivity of the current solid-state set-up.

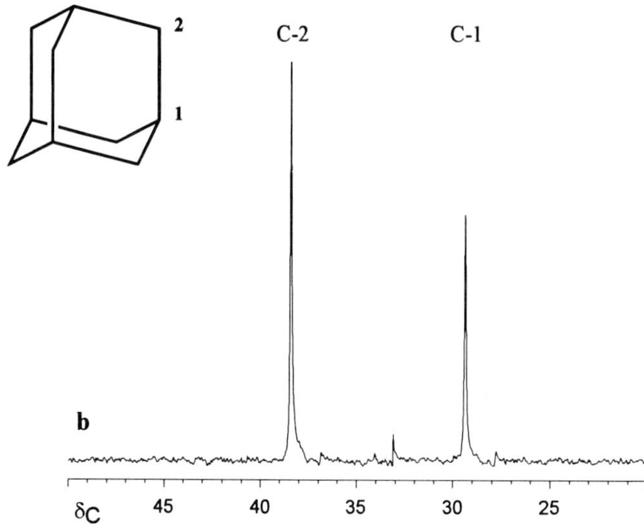

Currently, instrument manufacturers offer high-resolution MAS probe heads. These do provide a deuterium lock channel and thus are shimmed as usual in solution NMR. With these probe-heads chemists investigate problems which lie at the borderline between the real solid state and solution, such as food preparations or preparations obtained from syntheses in combinatorial chemistry. The MAS spinning assures high resolution for these samples.

8. Own Observations

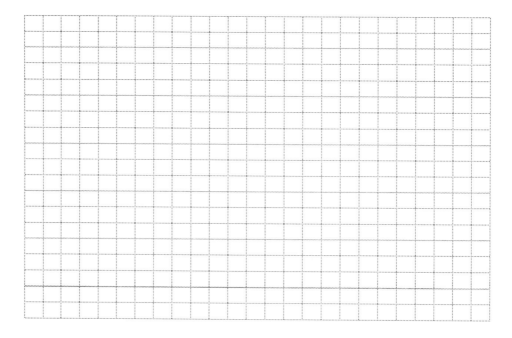

Experiment 14.2

Adjusting the Magic Angle

1. Purpose

Static solid-state NMR spectra are governed by chemical-shift anisotropy, dipolar spin coupling and quadrupolar interactions. If the information inherent in these physical effects is not wanted, one can significantly narrow the spectral response by rapidly spinning the sample in the magic angle. The time-averaged Hamiltonian of the above mentioned interactions contains a factor of $(1 - 3\cos^2\theta)$; thus, if the angle θ between the magnetic field direction and the spinning axis is adjusted to 54° 44', these interactions will vanish. Current MAS (**M**agic **A**ngle **S**pinning) probe-heads have a mechanical device for fine adjustment of the magic angle, which should be regularly performed to obtain optimum results. In this experiment we describe the procedure using a KBr sample which has the advantage that the result is independent of the shims and any decoupler adjustment [4].

2. Literature

[1] H. S. Gutowski, G. E. Pake, *J. Chem. Phys.* **1948**, *16*, 1164–1165; ibid. **1950**, *18*, 162–170.

[2] E. R. Andrew, A. Bradbury, R. G. Eades, *Nature*, **1958** *182*, 1659–1659.

[3] I. J. Lowe, *Phys. Rev. Lett.* **1959**, *2*, 285–287.

[4] J. S. Frye, G. E. Maciel, *J. Magn. Reson.* **1982**, *48*, 125–131.

[5] C. P. Slichter, *Principles of Magnetic Resonance*, 3rd Edition, Springer, Berlin, **1989**, 392–406.

[6] E. W. Wooten, K. T. Mueller, A. Pines, *Acc. Chem. Res.* **1992**, *25*, 209–215.

3. Pulse Scheme and Phase Cycle

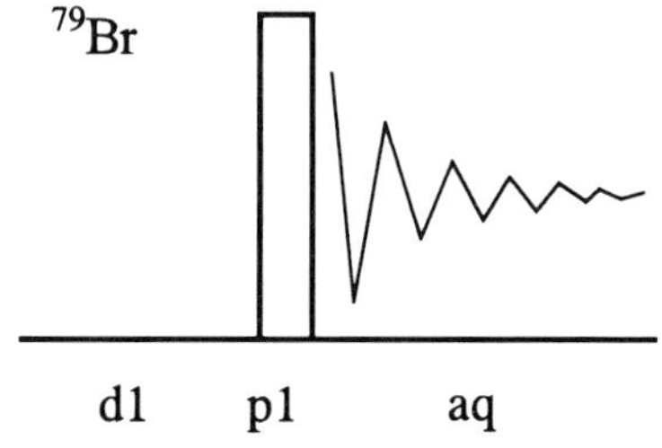

p1: x, x, -x, -x, y, y, -y, -y

aq: x, x, -x, -x, y, y, -y, -y

4. Acquisition

Time requirement: 5 min

Sample: Fill a solid-state rotor tightly with finely powdered KBr. Use spacers if the sample does not spin.

Load standard ^{13}C NMR parameters without proton decoupling. As can be seen from the table on page 306, the resonance frequency of ^{79}Br is very close to that of ^{13}C; therefore, you usually do not need to retune the probe-head if it was previously used for ^{13}C. Spin the sample at 4 kHz. You have to set:

td: 2 k
sw: 125 kHz
o1: on resonance of KBr signal
p1: 2 μs ^{79}Br transmitter pulse
d1: 50 ms
transmitter power level (choose value typically used for ^{13}C work)
rg: receiver gain for correct ADC input
ns: 128

Record 128 transients, perform the Fourier transformation and set the offset o1 on resonance of the sharp center signal. For better observation of the effect, adjust the transmitter phase relative to the receiver phase so that mainly the left quadrature channel receives the FID signal. Turn the magic angle adjustment in either direction and observe the sideband signals in the FID. Adjust for maximum side-bands at the end of the FID.

5. Processing

For the adjustment no processing is required, since the FID is directly observed.

6. Result

The figure on page 557 shows in **a** the FID of KBr with a slightly deadjusted magic angle obtained in a multinuclear solid-state probe-head with a 7 mm rotor on an AM-400 spectrometer with a wide-bore magnet. In **b** the FID is given after magic angle adjustment.

7. Comments

^{79}Br is a quadrupolar nucleus with $I = 3/2$. The crystal symmetry of KBr is cubic, and therefore a sharp central transition for $m_I = -1/2$ to $m_I = +1/2$ is observed. The other transitions, however, also contribute to the spectrum. The side-bands generated by the spinning frequency are easily seen and are very sensitive to the exact setting of the magic angle; misadjustment by 0.5° can clearly be observed.

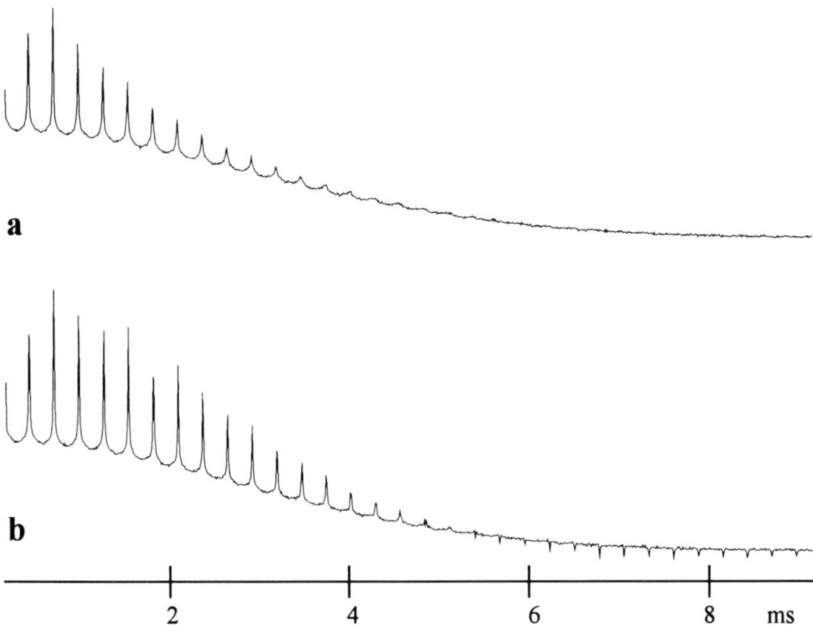

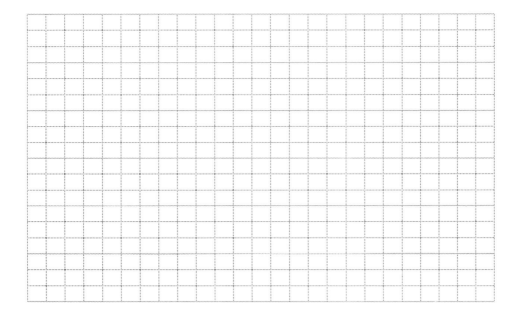

The magic angle can also be set using signals of nuclei with a high chemical-shift anisotropy, such as the ^{13}C signal of the carboxy C-atom of glycine. For HR-MAS ^{1}H NMR the use of $Ba(ClO_3)_2{\cdot}H_2O$ has been suggested.

8. Own Observations

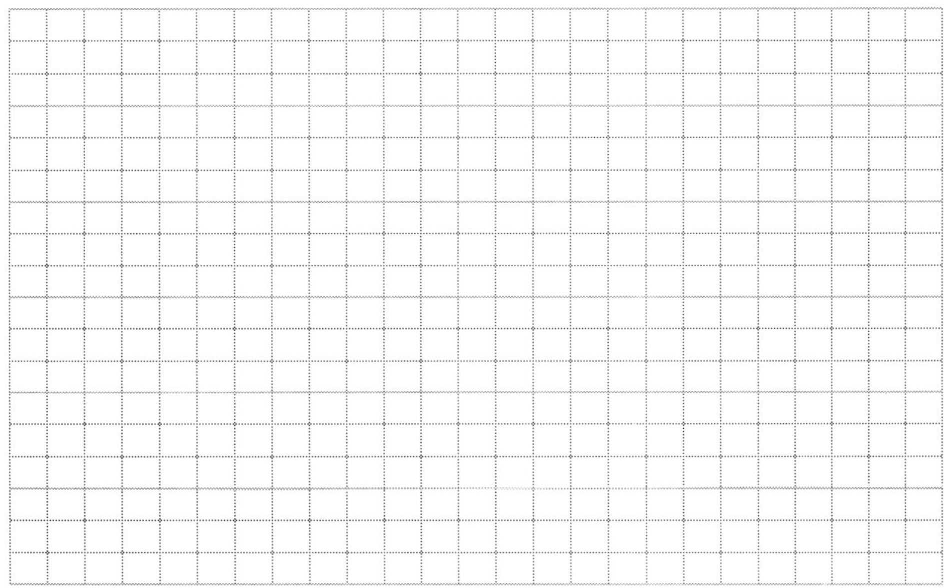

Experiment 14.3

Hartmann-Hahn Matching

1. Purpose

Standard CP/MAS spectra (see Exp. 14.4) are acquired with **C**ross **P**olarization from protons to carbon. This allows the adjustment of the pulse repetition time according to the relaxation time of the abundant proton spins and in addition enhances the carbon spectra by a factor of $\gamma_H/\gamma_C \approx 4$. Experimentally this requires the matching of the proton radiofrequency field strength B_{1H} with the radiofrequency field of carbon B_{1C} according to the Eq. (1). In the experiment described here we demonstrate the procedure to obtain this Hartmann Hahn match by using a sample of adamantane.

$$\gamma_H B_{1H} \cong \gamma_C B_{1C} \qquad (1)$$

2. Literature

[1] S. R. Hartmann, E. L. Hahn, *Phys. Rev.* **1962**, *128*, 2042–2053.
[2] A. Pines, M. G. Gibby, J. S. Waugh, *J. Chem. Phys.* **1972**, *56*, 1756–1777; *ibid.* **1973**, *59*, 569–590.

3. Pulse Scheme and Phase Cycle

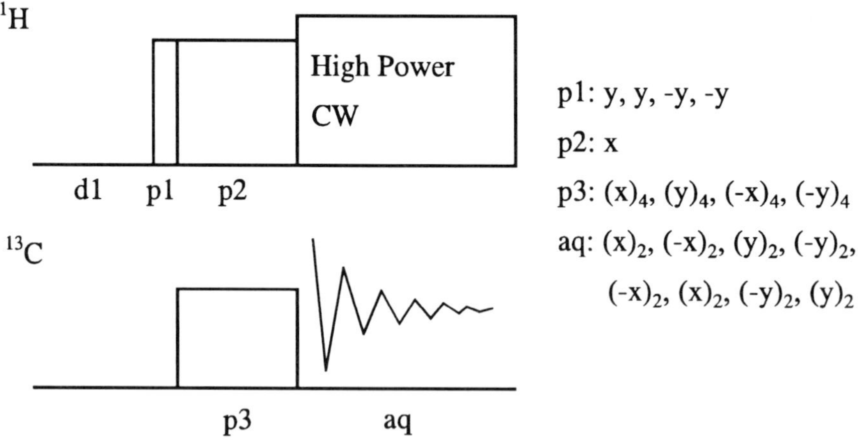

4. Acquisition

Time requirement: 20 min

Sample: Fill a solid-state rotor with finely powdered adamantane

Spin the sample with 4 kHz in the magic angle (see Exp. 14.2). Load standard ^{13}C NMR parameters. You have to set:

td: 4 k
sw: 20 kHz
o1: middle of ^{13}C signals of adamantane
o2: middle of ^{1}H NMR spectrum
p1: 4 μs ^{1}H decoupler pulse, to be varied
p2: 5 ms spin-lock decoupler pulse
p3: 5 ms spin-lock transmitter pulse
d1: 4 s
decoupler attenuation for cross polarization, to be varied
decoupler attenuation for high power cw decoupling, typically 2 dB less than for cross polarization; on instruments with no fast power switching use same attenuation as for cross polarization
transmitter attenuation
rg: receiver gain for correct ADC input
ns: 1

Go into the set-up mode of the instrument and observe the incoming FID on the screen. Start with a relatively high decoupler attenuation (90° pulse of 6 μs) and change the attenuation until a maximum FID is observed. This ensures the best Hartman-Hahn match. For these power settings then increase p1 until a minimum FID is observed. This will indicate the 180° proton excitation pulse under the given power settings. Divide this value by two and run the spectrum again. If the 90° proton pulse determined after this procedure is too long, repeat the sequence; however, start with higher transmitter power for ^{13}C. Note the signal-to-noise ratio obtained in the log-book of the instrument.

5. Processing

Use standard ^{13}C NMR processing with exponential multiplication (lb = 5 Hz) as described in Experiment 3.2.

6. Result

The figure on page 560 shows the CP/MAS spectrum of adamantane obtained with one scan in a multinuclear solid-state probe-head with a 7 mm rotor on an AM-400 spectrometer with a wide-bore magnet. The ^{13}C signals of adamantane have no significant chemical-shift anisotropy; therefore no spinning sidebands are observed in the spectrum.

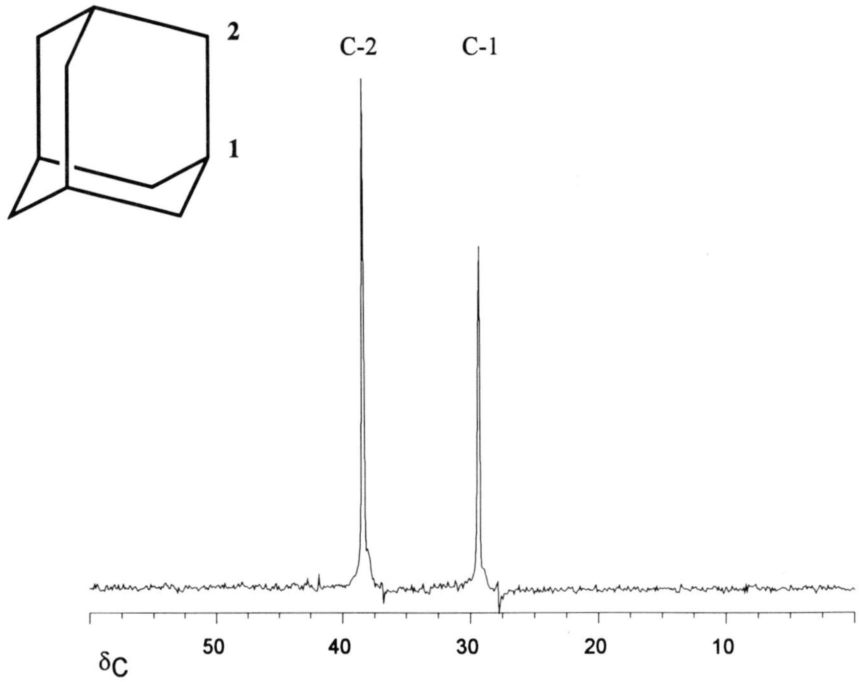

7. Comments

For the Hartmann-Hahn match Eq. (1) has to be satisfied. This can be performed by either changing the decoupler or the transmitter power. On older instruments it is easier to adjust the former and subsequently search for the 180° pulse under these conditions as described above. Another approach to find the Hartmann-Hahn match is to determine the 90° pulses on both channels and vary the the power of the frequency sources until both 90° pulses have the same length. Be sure not to shorten d1; for high-power decoupling the duty cycle should not exceed 5% of the pulse repetition time. Note that the Hartmann-Hahn matching condition is spinning rate dependent.

8. Own Observations

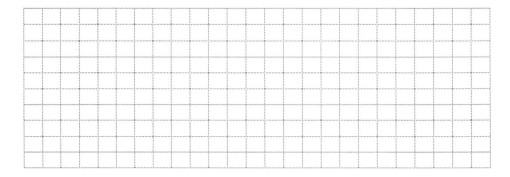

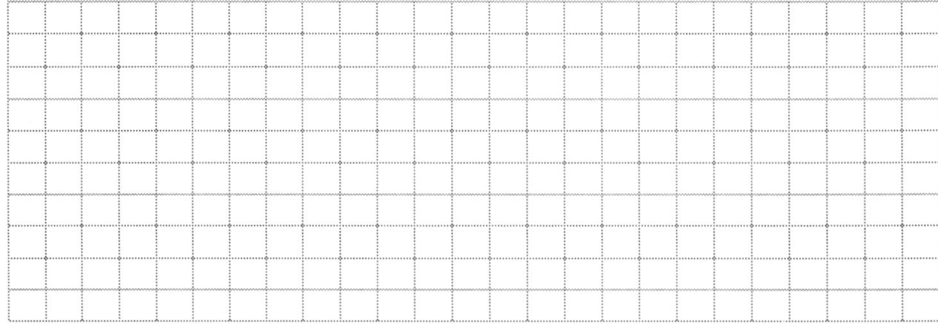

Experiment 14.4

The Basic CP/MAS Experiment

1. Purpose

The CP/MAS (**C**ross **P**olarization/**M**agic **A**ngle **S**pinning) method provides high resolution NMR spectra in the solid state and is mostly performed on ^{13}C with cross polarization from ^{1}H. The pulse repetition time is governed by the proton relaxation and the magic angle spinning removes effects of chemical-shift anisotropy. High-power proton decoupling during acquisition finally provides ^{13}C NMR spectra nearly as well resolved as solution spectra. Originally the cross polarization method was called proton-enhanced nuclear-induction spectroscopy [1]; however, the corresponding acronym was not accepted in the literature. In the experiment described here we demonstrate the CP/MAS technique using a sample of glycine and show the effects of different spinning rates.

2. Literature

[1] A. Pines, M. G. Gibby, J. S. Waugh, *J. Chem. Phys.* **1972**, *56*, 1756–1777; *ibid.* **1973**, *59*, 569–590.
[2] J. Schaefer, E. O. Stejskal, *J. Am. Chem. Soc* **1976**, *98*, 1031–1032.
[3] E. O. Stejskal, J. Schaefer, R. A. McKay, *J. Magn. Reson.* **1977**, *25*, 569–573.
[4] J. Herzfeld, A. E. Berger, *J. Chem. Phys.* **1980**, *73*, 6021–6030.

3. Pulse Scheme and Phase Cycle

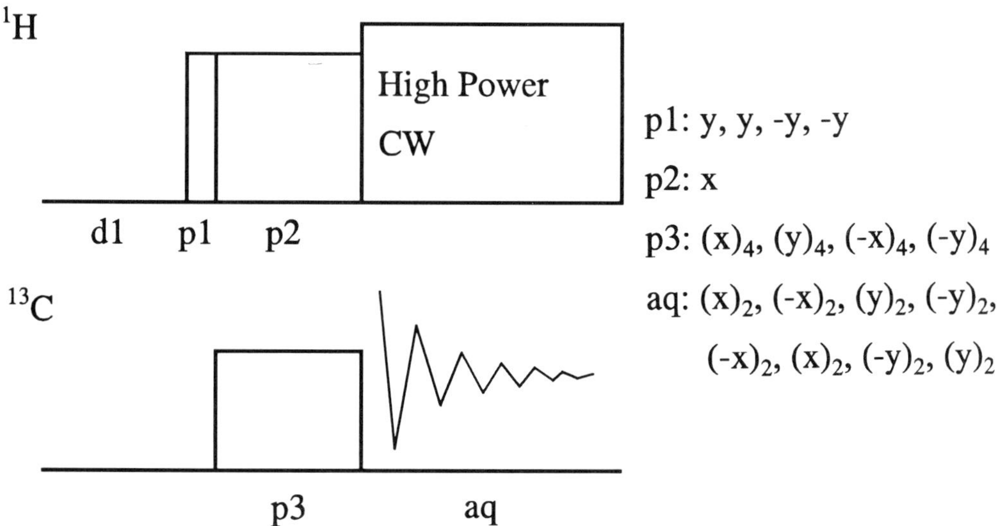

4. Acquisition

Time requirement: 30 min

Sample: Fill a solid-state rotor with finely powdered glycine.

Load standard ^{13}C NMR parameters. You have to set:

td: 4 k
sw: 500 ppm
o1: middle of ^{13}C NMR spectrum
o2: middle of ^{1}H NMR spectrum
p1: 90° ^{1}H decoupler pulse
p2: 5 ms spin-lock decoupler pulse
p3: 5 ms spin-lock transmitter pulse
d1: 3 s
decoupler attenuation for cross polarization
decoupler attenuation for high-power cw decoupling, typically 2 dB less than for cross polarization; on instruments with no fast power switching use same attenuation as for cross polarization
rg: receiver gain for correct ADC input
spinning rate ν_R **a**: 5000 Hz, **b**: 4000 Hz, **c**: 3000 Hz, **d**: 2000 Hz, **e**: 1000 Hz, **f**: 500 Hz, **g**: 0 Hz
ns: 16 in Experiment **a** to **c**, 64 in **d** to **f** and 256 in **g**

5. Processing

Use standard 1D processing for ^{13}C NMR as described in Experiment 3.2 with zero filling to 4 k and different exponential multiplication corresponding to the line-width ranging from lb = 25 Hz in **a** to lb = 100 Hz in **g**.

6. Result

The figure on page 563 shows the spectra of glycine obtained in a multinuclear solid-state probe-head with a 7 mm rotor on an AM-400 spectrometer with wide-bore magnet. As can be seen from the figure, signals of nuclei with high chemical-shift anisotropy such as the signal from the carboxyl carbon atom generate spinning side-bands; at low spinning speed the signal of the methylene carbon atom also yields side-bands. The spinning rate can be measured from the distance of the spinning side-bands and the side-band pattern can be analyzed to obtain the chemical-shift tensor [4].

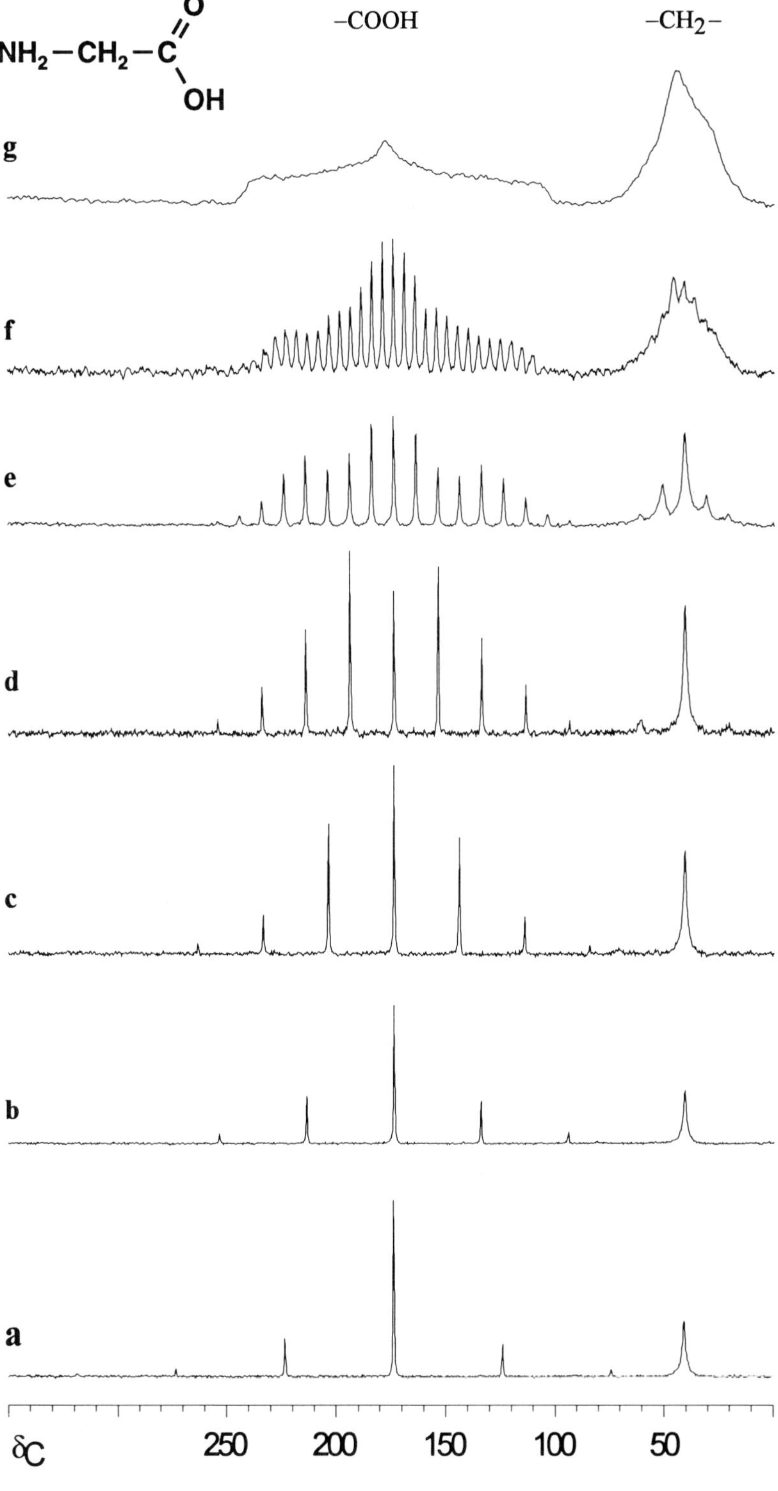

$NH_2-CH_2-C(=O)OH$
–COOH
–CH2–
g
f
e
d
c
b
a
δC
250
200
150
100
50

7. Comments

After the first proton pulse, which aligns the magnetization towards the $+x$ axis, a proton spin-lock pulse p2 with phase x locks this magnetization in the radiofrequency field. A simultaneous ^{13}C spin-lock pulse p3 assures for the ^{13}C spins the same precession frequency if the Hartmann-Hahn condition (see Experiment 14.3) is obeyed. In this situation the carbon and proton spins can exchange energy which leads to a polarization of the carbon spins in the large bath of the abundant protons. High-power proton decoupling removes all dipolar couplings to the protons and magic angle spinning removes the chemical-shift anisotropy, however, creates spinning side bands. These render the evaluation of the CP/MAS spectra often very difficult; methods to remove the spinning side-bands are described in Experiment 14.5 and 14.6. The repetition time of the experiment is determined by the proton relaxation.

Note that for the sample used, ^{14}N quadrupolar interaction can be observed at some field strengths. With the CP/MAS method an extremely large variety of chemical problems can be investigated, ranging from physical organic questions such as the structure of carbocations to applications in material science such as the composition of rubber used for car tires.

8. Own Observations

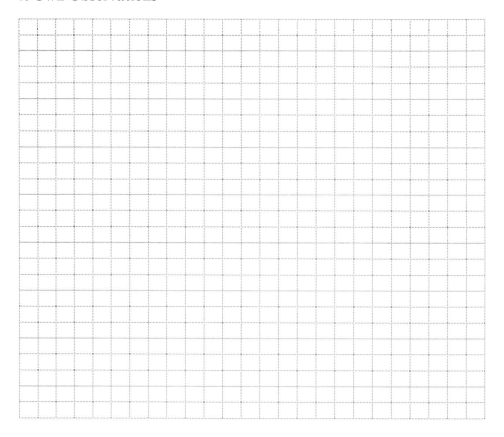

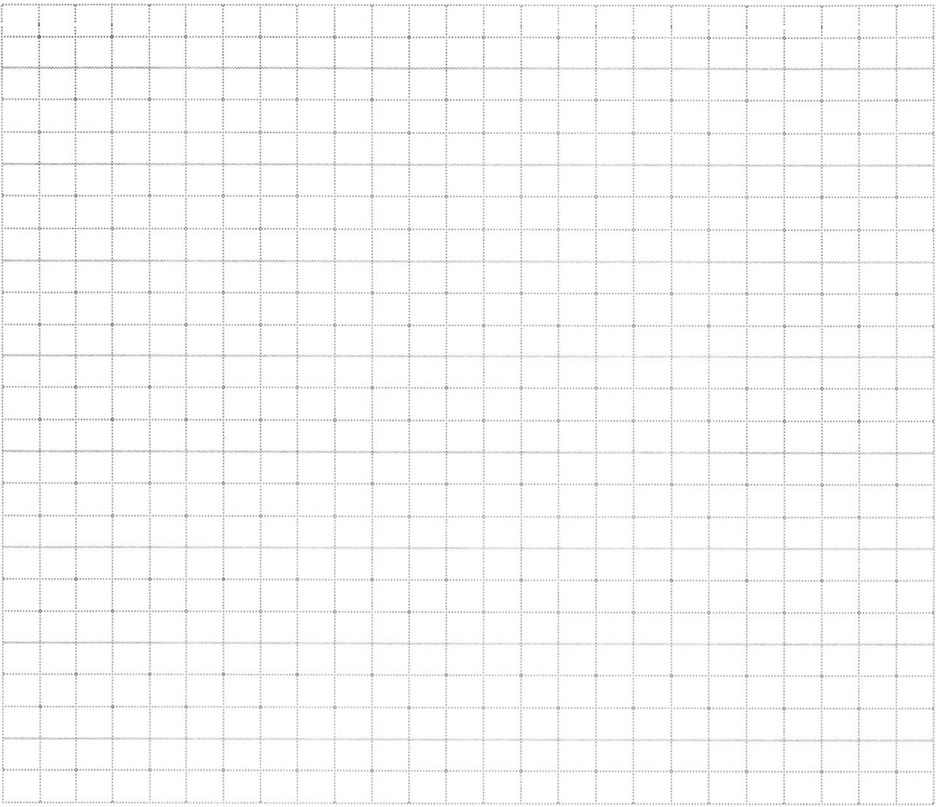

Experiment 14.5

TOSS

1. Purpose

The CP/MAS method as described in Experiment 14.4 produces side-bands, depending on the chemical-shift anisotropy of the signals and the spinning rate. This can lead to difficulties in the assignment of spectra with many ^{13}C signals. Although the center band can be identified by altering the spinning rate, a method which suppresses the spinning side-bands should be very useful. One of the first techniques to achieve this goal was christened TOSS (**TO**tal **S**uppression of **S**ide-bands). In the experiment described here we demonstrate the technique with a sample of glycine.

2. Literature

[1] W. T. Dixon, J. Schäfer, M. D. Sefcik, E. O. Stejskal, R. A. McKay, *J. Magn. Reson.* **1982**, *49*, 341–345.

[2] W. T. Dixon, *J. Chem. Phys.* **1982**, *77*, 1800–1809.

[3] D. P. Raleigh, E. T. Olejniczak, R. G. Griffin, *J. Magn. Reson.* **1991**, *93*, 472–484.

[4] S. H. Lang, *J. Magn. Reson. Ser. A* **1993**, *104*, 345–346.

[5] H. Geen, G. Bodenhausen, *J. Am. Chem. Soc.* **1993**, *115*, 1579–1580.

3. Pulse Scheme and Phase Cycle

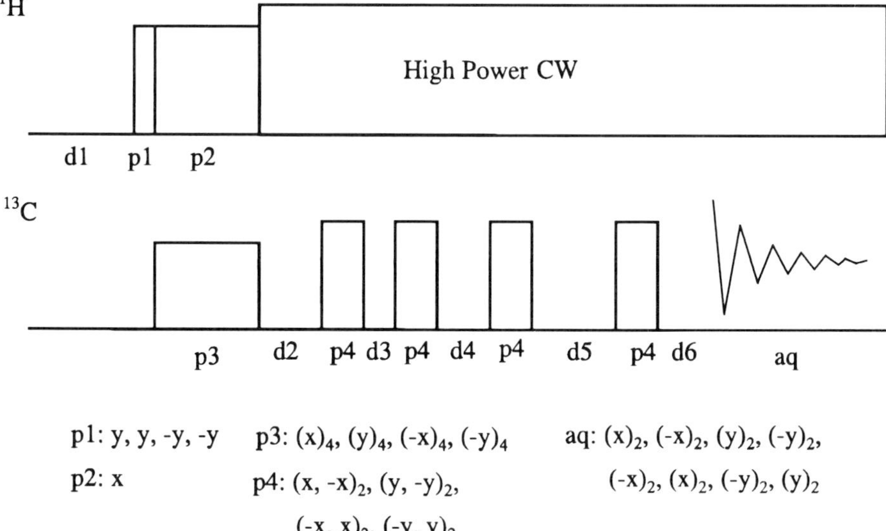

p1: y, y, -y, -y

p2: x

p3: $(x)_4$, $(y)_4$, $(-x)_4$, $(-y)_4$

p4: $(x, -x)_2$, $(y, -y)_2$, $(-x, x)_2$, $(-y, y)_2$

aq: $(x)_2$, $(-x)_2$, $(y)_2$, $(-y)_2$, $(-x)_2$, $(x)_2$, $(-y)_2$, $(y)_2$

4. Acquisition

Time requirement: 10 min

Sample: Fill a solid-state rotor with finely powdered glycine.

Load standard ^{13}C NMR parameters. Spin the sample at 4 kHz and run first a normal CP/MAS spectrum (see Exp. 14.4). Then load the TOSS pulse sequence. You have to set:

td: 4 k
sw: 500 ppm
o1: middle of ^{13}C NMR spectrum
o2: middle of ^{1}H NMR spectrum
p1: 90° ^{1}H decoupler pulse
p2: 5 ms spin-lock decoupler pulse
p3: 5 ms spin-lock transmitter pulse
p4: 180° ^{13}C transmitter pulse [11 μs]
d1: 3 s
d2: 25.2 μs, calculated from $(0.1226/\nu_R) - p4/2$
d3: 8.3 μs, calculated from $(0.0773/\nu_R) - p4$
d4: 44.9 μs, calculatcd from $(0.2236/\nu_R) - p4$
d5: 250 μs, calculated from $(1.0433/\nu_R) - p4$
d6: 183.2 μs, calculated from $(0.7744/\nu_R) - p4/2 - de$
decoupler attenuation for cross polarization
decoupler attenuation for high-power cw decoupling, typically 2 dB less than for cross polarization; on instruments with no fast power switching use same attenuation as for cross polarization
rg: receiver gain for correct ADC input
preacquisition delay: as short as possible
ns: 16

5. Processing

Use standard 1D processing for ^{13}C NMR as described in Experiment 3.2 with zero filling to 4 k and exponential multiplication corresponding to the line width using lb = 50 Hz.

6. Result

The figure on page 567 shows the spectra of glycine obtained in a multinuclear solid-state probe-head with a 7 mm rotor on an AM-400 spectrometer with wide-bore magnet. The spinning rate was 4000 Hz as seen from the normal CP/MAS spectrum in **a**, whereas in **b** the result of the TOSS sequence is given.

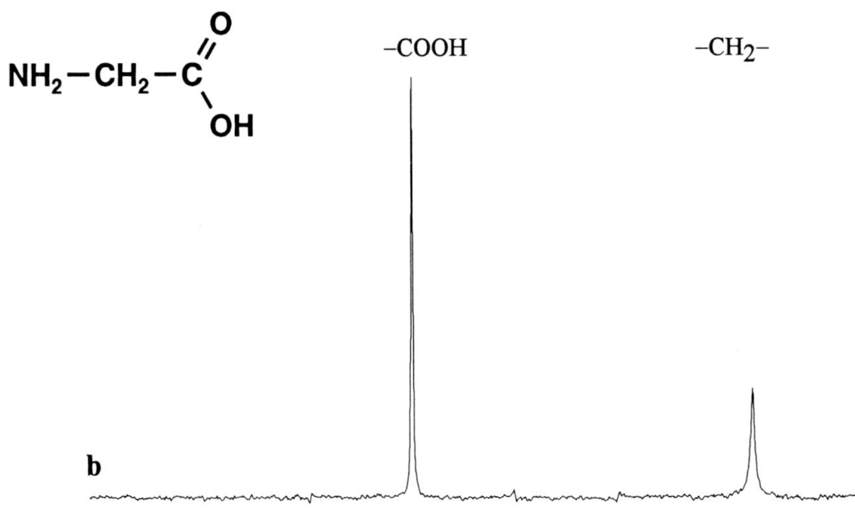
NH2−CH2−C(=O)OH
−COOH
−CH2−
b

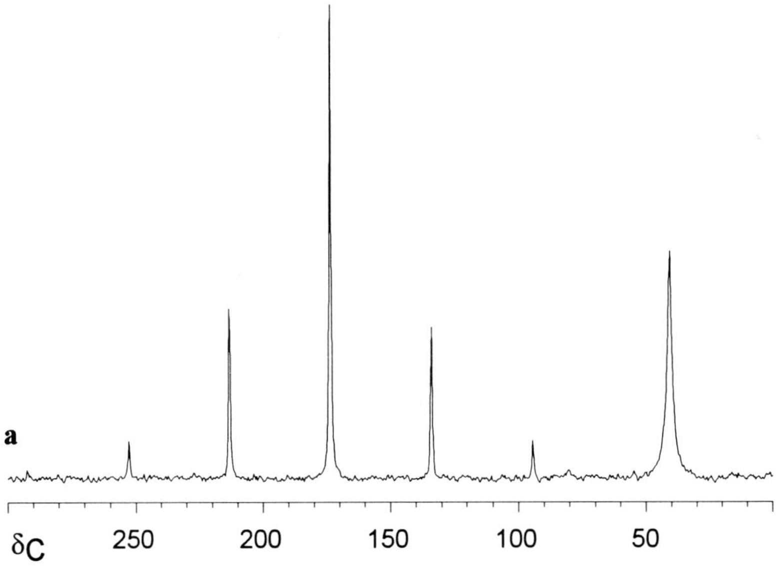
a
δC
250
200
150
100
50

7. Comments

The sequence is very similar to the standard CP/MAS procedure with the difference that before the acquisition four 180° ^{13}C pulses are applied. The delays between these pulses are derived from a graphical analysis of the side-band pattern and are a function of the spinning frequency [2]. In principle, the spinning frequency ν_R should be higher than the breadth of the static chemical-shift powder pattern $\Delta\sigma = \sigma 11 - \sigma 33$. In our example (400 MHz instrument, ^{13}C NMR frequency 100.6 MHz) $\Delta\sigma$ equals to about 15000 Hz; thus $\nu_R/\Delta\sigma$ amounts to $\approx$ 0.3. TOSS yields very satisfactory results; for $\nu_R/\Delta\sigma$ ratios less than 0.3, however, intensity losses or even disappearance of signals may occur. A drawback of the method is the long time between the Hartmann Hahn contact and the start of the acquisition, and for some samples the relaxation losses will be severe. Furthermore, the 180° pulses are not delta functions as assumed in the derivation of the spin echo delays d2 to d6, but have finite length. For high spinning frequencies this can lead to difficulties in obtaining the correct delays. Several variations of the original TOSS sequence have been proposed [3–5].

8. Own Observations

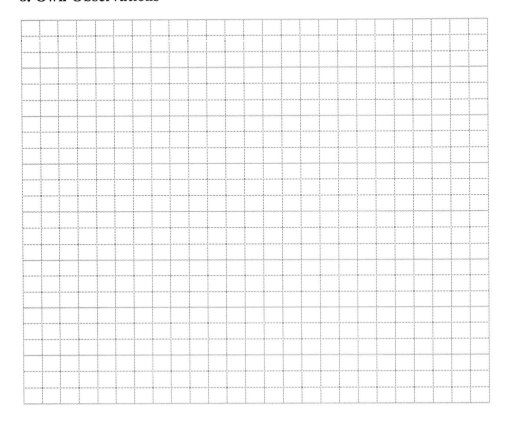

Experiment 14.6

SELTICS

1. Purpose

The CP/MAS method as described in Experiment 14.4 produces side-bands, depending on the chemical-shift anisotropy of the signals and the spinning rate. This can lead to difficulties in the assignment of spectra with many ^{13}C signals. There are several methods which provide a suppression of the side-bands such as TOSS (see Exp. 14.5) or methods which use multipulse narrowing of the chemical shift scale. A relatively recent method has the acronym SELTICS (**S**ideband **EL**imination by **T**emporary **I**nter-ruption of the **C**hemical **S**hift). Similar to TOSS, SELTICS causes a destructive interference between side bands but with a different working principle. In the experiment described here we demonstrate the technique with a sample of glycine.

2. Literature

[1] J. Hong, G. S. Harbison, *J. Magn. Reson. Ser. A* **1993**, *105*, 128–136.

3. Pulse Scheme and Phase Cycle

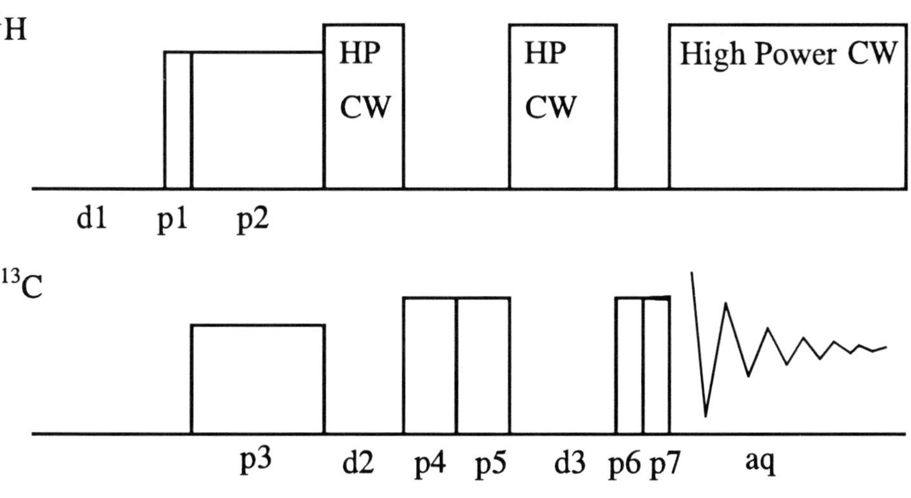

p1: y, y, -y, -y p3: $(x)_4$, $(y)_4$, $(-x)_4$, $(-y)_4$ aq: $(x)_2$, $(-x)_2$, $(y)_2$, $(-y)_2$, $(-x)_2$, $(x)_2$, $(-y)_2$, $(y)_2$

p2: x

4. Acquisition

Time requirement: 10 min

Sample: Fill a solid-state rotor with finely powdered glycine.

Load standard ^{13}C NMR parameters. Spin the sample at 4 kHz so that the duration τ_R of one rotor period is 250 μs and run first a normal CP/MAS spectrum (see Exp. 14.4). Then load the SELTICS pulse sequence. You have to set:

td: 4 k
sw: 500 ppm
o1: middle of ^{13}C NMR spectrum
o2: middle of ^{1}H NMR spectrum
p1: 90° ^{1}H decoupler pulse
p2: 5 ms spin-lock decoupler pulse
p3: 5 ms spin-lock transmitter pulse
p4: ^{13}C transmitter pulse, duration $\tau_R/12 = 20.8$ μs
p5: ^{13}C transmitter pulse, duration $\tau_R/12 = 20.8$ μs
p6: ^{13}C transmitter pulse, duration $\tau_R/24 = 10.4$ μs
p7: ^{13}C transmitter pulse, duration $\tau_R/24 = 10.4$ μs
d1: 3 s
d2: 20.8 μs, calculated from $\tau_R/12$
d3: 41.6 μs, calculated from $\tau_R/6$
decoupler attenuation for cross polarization
decoupler attenuation for high-power cw decoupling, typically 2 dB less than for cross polarization; on instruments with no fast power switching use same attenuation as for cross polarization
rg: receiver gain for correct ADC input
de: as short as possible
ns: 16

5. Processing

Use standard 1D processing for ^{13}C NMR as described in Experiment 3.2 with zero filling to 4 k and exponential multiplication corresponding to the line width using lb = 50 Hz.

6. Result

The figure on page 571 shows the spectra of glycine obtained in a dual solid-state probe-head with a 4 mm rotor on an MSL-300 spectrometer. The spinning rate was 4000 Hz as seen from the normal CP/MAS spectrum in **a**, whereas in **b** the result of the SELTICS sequence is given.

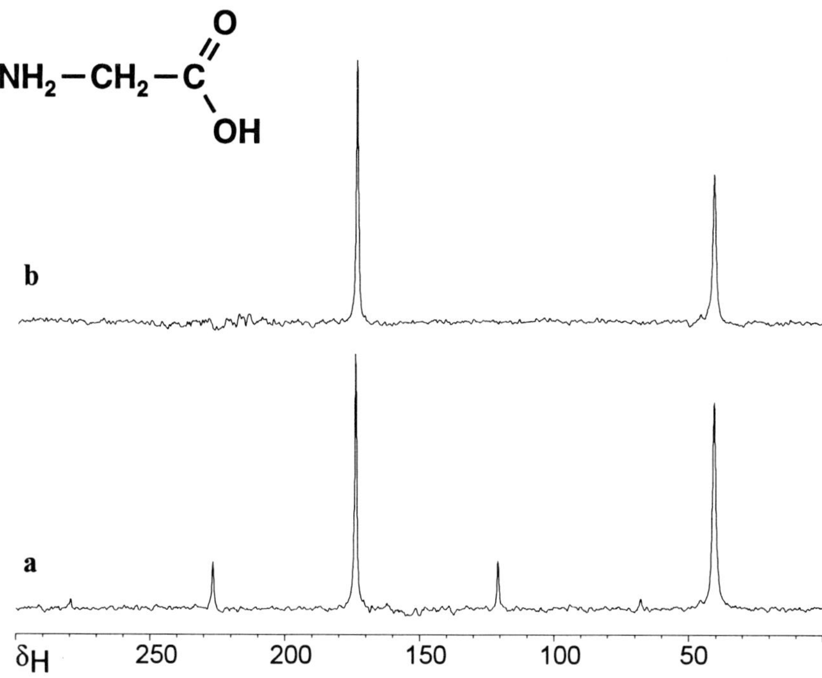

7. Comments

The sequence is similar to the basic CP/MAS procedure, with the difference that before the acquisition two pairs of ^{13}C pulses are applied with durations and delays between these pulses of integer divisions of one rotor period. The method is significantly shorter than TOSS with respect to the time between the end of the cross polarization step and the start of the acquisition. Note, however, that with this method a rather high first-order phase correction may be necessary.

8. Own Observations

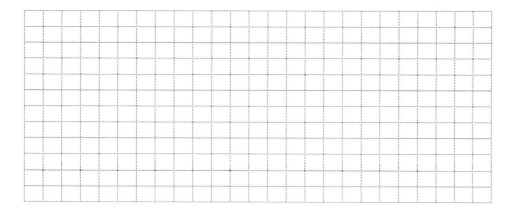

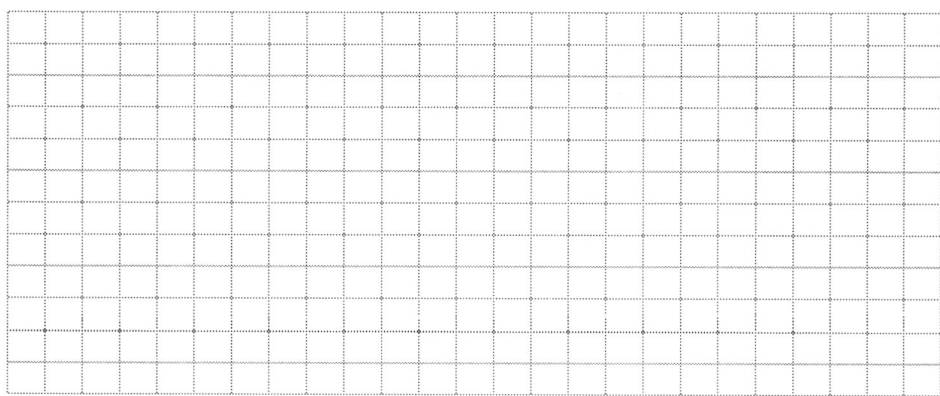

Experiment 14.7

Multiplicity Determination in the Solid-State

1. Purpose

There are many methods to the multiplicity determination of ^{13}C NMR spectra recorded in solution, and several of these techniques are described in Chap. 6. For the solid-state it would also be very helpful to be able to distinguish at least between signals which stem from quaternary carbon atoms and those from protonated ones. In contrast to solution spectra, the *J* coupling *J*(C,H) cannot be used for this purpose, since the solid-state spectra are dominated by the dipolar coupling between ^{13}C and ^{1}H. The first editing method was baptized NQS (**N**on **Q**uaternary **S**uppression), and we demonstrate this technique here with a sample of glycine using the variant which applies a single ^{13}C 180° pulse during the dephasing step [2,3].

2. Literature

[1] S. J. Opella, M. H. Frey, *J. Am. Chem. Soc.* **1979**, *101*, 5854–5856.

[2] P. D. Murphy, *J. Magn. Reson.* **1983**, *52*, 343–345; *ibid.* **1985**, *62*, 303–308.

[3] L. B. Alemany, D. M. Grant, T. D. Alger, R. J. Pugmire, *J. Am. Chem. Soc.* **1982**, *105*, 6697–6704.

[4] R. K. Harris, P. Jonsen, K. L. Packer, *Org. Magn. Reson.* **1984**, *22*, 269–271.

3. Pulse Scheme and Phase Cycle

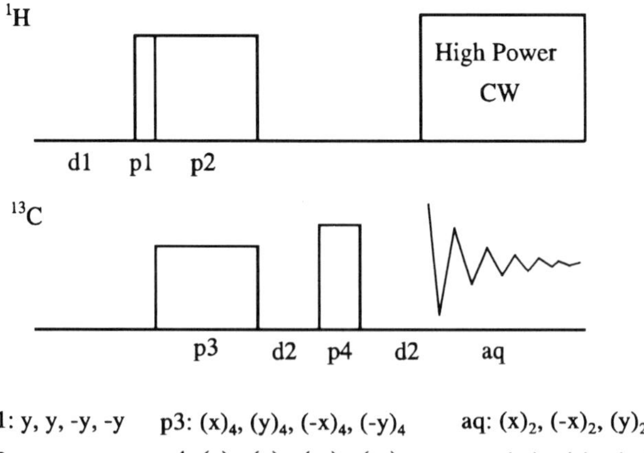

p1: y, y, -y, -y p3: $(x)_4$, $(y)_4$, $(-x)_4$, $(-y)_4$ aq: $(x)_2$, $(-x)_2$, $(y)_2$, $(-y)_2$,

p2: x p4: $(x)_4$, $(y)_4$, $(-x)_4$, $(-y)_4$ $(-x)_2$, $(x)_2$, $(-y)_2$, $(y)_2$

4. Acquisition

Time requirement: 5 min

Sample: Fill a solid-state rotor with finely powdered glycine.

Load standard ^{13}C NMR parameters. Spin the sample at 4 kHz and run first a normal CP/MAS spectrum (see Exp. 14.4). Then load the NQS pulse sequence. You have to set:

td: 4 k
sw: 500 ppm
o1: middle of ^{13}C NMR spectrum
o2: middle of ^{1}H NMR spectrum
p1: 90° ^{1}H decoupler pulse
p2: 5 ms spin-lock decoupler pulse
p3: 5 ms spin-lock transmitter pulse
p4: 180° ^{13}C transmitter pulse
d1: 3 s
d2: 25 μs
decoupler attenuation for cross polarization
decoupler attenuation for high-power cw decoupling, typically 2 dB less than for cross polarization; on instruments with no fast power switching use same attenuation as for cross polarization
rg: receiver gain for correct ADC input
de: as short as possible
ns: 16

5. Processing

Use standard 1D processing for ^{13}C NMR as described in Experiment 3.2 with zero filling to 4 k and exponential multiplication corresponding to the line width using lb = 50 Hz.

6. Result

The figure on page 574 shows the spectra of glycine obtained in a multinuclear solid-state probe-head with a 7 mm rotor on an AM-400 spectrometer with wide-bore magnet. The spinning rate was 4000 Hz as seen from the standard CP/MAS spectrum in **a**, whereas in **b** the result of the NQS sequence is given.

7. Comments

The sequence is similar to the standard CP/MAS procedure with the difference that before the acquisition the decoupler is switched off for a very short time, typically 50 μs. During this time protonated carbon atoms will feel the dipolar coupling and dephase rapidly. A 180° ^{13}C pulse refocuses chemical shift evolution during this time and thus removes phase errors. For rapidly rotating groups, e.g. methyl groups, the NQS method is not efficient. By taking the difference of a normal CP/MAS spectrum

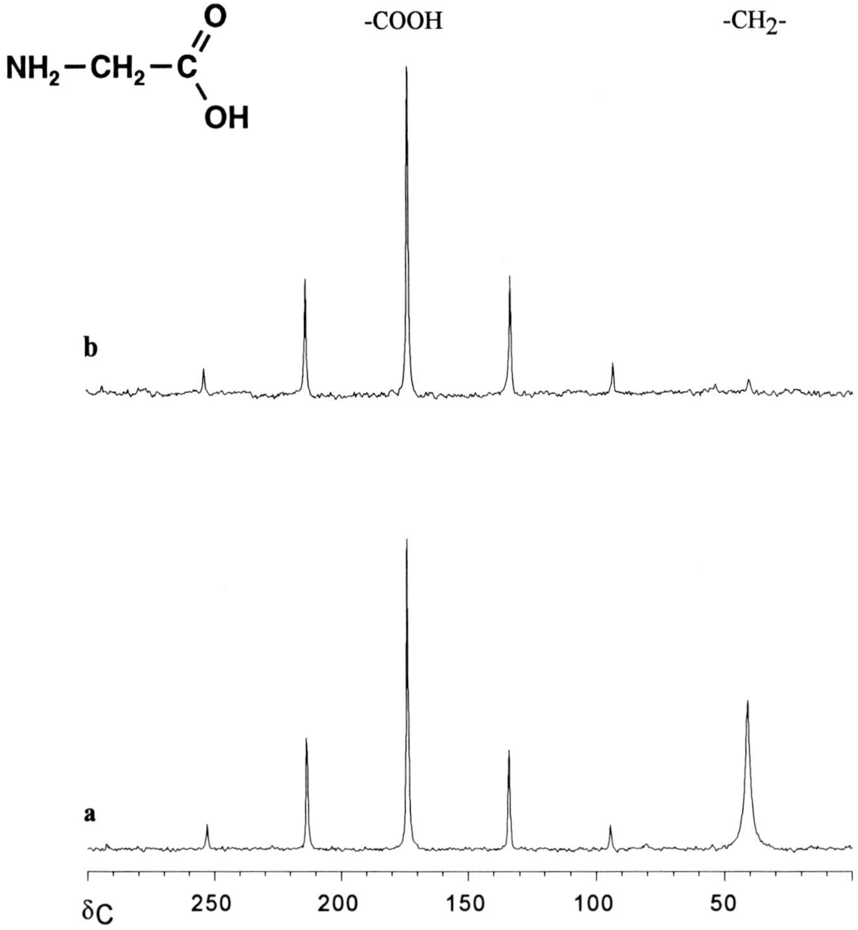

$NH_2-CH_2-C(=O)OH$

-COOH

$-CH_2-$

b

a

δ_C 250 200 150 100 50

and a NQS spectrum one can obtain spectra which have only signals of CH_n carbon atoms [4]. The NQS pulse technique can be combined with the TOSS procedure.

8. Own Observations

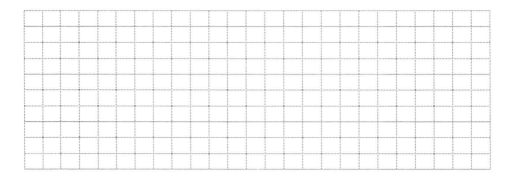

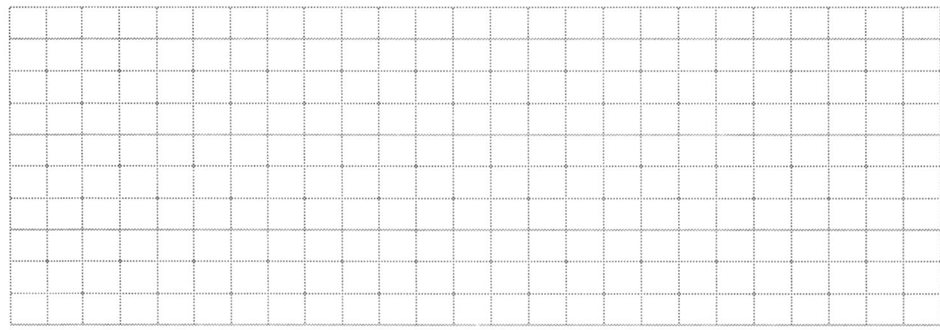

Appendix 1

Instrument Dialects

In recent years the manufacturers of NMR spectrometers have used a variety of instruments, different computers, operating systems, and software. The experiments described in this book are mainly given in the notation of Bruker AMX spectrometers using the UXNMR system. For comparison purposes we give here a glossary which should enable the users of other instruments to find the appropriate parameters to set up the experiments.

Manufacturer	Bruker	Bruker	Bruker	Varian	Varian	Jeol	Jeol
Instrument	AMX ARX	AM AC	DMX DRX DPX	Gemini H/C BB	Gemini-2000 UNITY-plus UNITY-Inova	GX, GSX EX, Alpha, Lambda	Eclipse Delta
Computer	X32, Aspect Station-1	Aspect-3000	SG-Indy	Motorola 68000	Sun	DEC-PDP11, VAX Alpha AXP	SG-Indy
Operating System	Unix	Adakos	Irix	VXR-4000	Solaris	RSX11M, VMS, Unix	Irix
time domain data size	td	TD	td	np	np	SAMPO, EX: SPO	x_points
processed data size	si	SI	si	fn	fn	POINT	
time domain in F_2 or F_1	td2 td1	TD2 TD1	td2 td1			F_2: FREQU F_1: CLFRQ	F_2: x_sweep F_1: y_sweep
processed data size in F_2 or F_1	si	SI2/SI1	si	fn fn1	fn fn1	F_2: POINT F_1: CLPNT	
transmitter offset	o1	O1	o1	to	tof	OBSET + OBFIN	x_offset
decoupler offset	o2	O2	o2	do	dof	IRSET + IRFIN	irr_ offset
spectral width [ppm]	sw	-	sw	sw	sw		x_sweep

Manufacturer	Bruker	Bruker	Bruker	Varian	Varian	Jeol	Jeol
pulse width	px, x = 0–n	PW, Px, x = 0–9	px, x = 0–n	pw	pw	PWx, x = 1–n	user defined
delay	dx, x = 0–n	Dx, x = 0–n	dx, x = 0–n	dx, x=1–n	dx, x=1–n	PIx, x = 1–n	user defined
receiver gain	rg	RG	rg	gain	gain	RGAIN	recvr_ gain
transmitter power level	thi, tlo, tlx, x=0–31	THI, TLO	plx, x = 0–31	tpwr	tpwr	OBATN	obs_ attenuator
decoupler power level	dhi, dlo,dlx, x=1–31		plx, x = 0–31	dhp,dlp	dpwr	IRATN	irr_ attenu-ator
proton trans-mitter power level	hlx, x = 1–4		plx, x = 0–31				module_ config(irr.a mp_full_p wr)
proton decoupler power level	hlx, x =1–4	Sx, x=1–4	plx, x = 0–31	pplvl	pplvl		normal irr oparation mode
power level for soft pulses on the trans-mitter channel	tpx, x=0–15		spx, x = 0–15		selpwr	OBATN	obs_atte-nuator
power level for soft pulses on the de-coupler	dpx, x=0–15		spx, x = 0–15			IRATN	irr_atte-nuator
number of scans	ns	NS	ns	nt	nt	SCANS	scans
homonuclear decoupling mode	hd	HD	hd	homo	homo	EXMOD =SGHOM IRMOD= HOM	module_ config(pulser.time _share)
dummy scans	ds	DS	ds	ss	ss	DUMMY INDMY	x_prescans
composite pulse decoupling	cpd	CPD	cpd	w	w	EXMOD =SGCOM IRMOD= COM	irr_noise
continous wave decoupling	cw	CW	cw	c	q	EXMOD =SGSEL IRMOD= SEL	on (irr.gate)
preacquisition delay	de	DE	de	rof2	rof2pad	PREDL	dead_time & delay

Manufacturer	Bruker	Bruker	Bruker	Varian	Varian	Jeol	Jeol
number of incremented periods during t1	nd0	ND0	ndo			CLPNT	
increment for t1 evolution	in0	IN0	in0			PI1	
phase diffe-rence bet-ween pulses	phcorx, x=0–31		phcorx, x= 0–31				

Appendix 2

Elementary Product Operator Formalism Rules

In the **Comments** section of the experiments the product operator formalism is often used to follow the course of the magnetization. This table provides a summary of the various interrelationships for quick reference.

1. 90° Pulses

$I_{1z} \xrightarrow{90°\ I_x} -I_{1y}$ $I_{1x} \xrightarrow{90°\ I_x} I_{1x}$ $I_{1y} \xrightarrow{90°\ I_x} I_{1z}$

$I_{1z} \xrightarrow{90°\ I_y} I_{1x}$ $I_{1x} \xrightarrow{90°\ I_y} -I_{1z}$ $I_{1y} \xrightarrow{90°\ I_y} I_{1y}$

$I_{1z} \xrightarrow{90°\ I_{-x}} I_{1y}$ $I_{1x} \xrightarrow{90°\ I_{-x}} I_{1x}$ $I_{1y} \xrightarrow{90°\ I_{-x}} -I_{1z}$

$I_{1z} \xrightarrow{90°\ I_{-y}} -I_{1x}$ $I_{1x} \xrightarrow{90°\ I_{-y}} I_{1z}$ $I_{1y} \xrightarrow{90°\ I_{-y}} I_{1y}$

$2I_{1x}I_{2z} \xrightarrow{90°\ I_y} -2I_{1z}I_{2x}$

$2I_{1x}I_{2z} \xrightarrow{90°\ I_x} -2I_{1x}I_{2y}$

2. Chemical Shift

$I_{1x} \xrightarrow{\Omega I_z t} I_{1x}\cos\Omega t + I_{1y}\sin\Omega t$ $-I_{1x} \xrightarrow{\Omega I_z t} -I_{1x}\cos\Omega t - I_{1y}\sin\Omega t$

$I_{1y} \xrightarrow{\Omega I_z t} I_{1y}\cos\Omega t - I_{1x}\sin\Omega t$ $-I_{1y} \xrightarrow{\Omega I_z t} -I_{1y}\cos\Omega t + I_{1x}\sin\Omega t$

$I_{1z} \xrightarrow{\Omega I_z t} I_{1z}$

3. Spin–Spin Coupling

$$I_{1x} \xrightarrow{\pi J\, 2I_{1z}I_{2z}t} I_{1x}\cos\pi J\,t + 2I_{1y}I_{2z}\sin\pi J\,t$$

$$-I_{1x} \xrightarrow{\pi J\, 2I_{1z}I_{2z}t} -I_{1x}\cos\pi J\,t - 2I_{1y}I_{2z}\sin\pi J\,t$$

$$I_{1y} \xrightarrow{\pi J\, 2I_{1z}I_{2z}t} I_{1y}\cos\pi J\,t - 2I_{1x}I_{2z}\sin\pi J\,t$$

$$-I_{1y} \xrightarrow{\pi J\, 2I_{1z}I_{2z}t} -I_{1y}\cos\pi J\,t + 2I_{1x}I_{2z}\sin\pi J\,t$$

$$2I_{1x}I_{2z} \xrightarrow{\pi J\, 2I_{1z}I_{2z}t} 2I_{1x}I_{2z}\cos\pi J\,t + I_{1y}\sin\pi J\,t$$

$$-2I_{1x}I_{2z} \xrightarrow{\pi J\, 2I_{1z}I_{2z}t} -2I_{1x}I_{2z}\cos\pi J\,t - I_{1y}\sin\pi J\,t$$

$$2I_{1y}I_{2z} \xrightarrow{\pi J\, 2I_{1z}I_{2z}t} 2I_{1y}I_{2z}\cos\pi J\,t - I_{1x}\sin\pi J\,t$$

$$-2I_{1y}I_{2z} \xrightarrow{\pi J\, 2I_{1z}I_{2z}t} -2I_{1y}I_{2z}\cos\pi J\,t + I_{1x}\sin\pi J\,t$$

$$I_{1z} \xrightarrow{\pi J\, 2I_{1z}I_{2z}t} I_{1z}$$

4. Shift Operators

$$I^+ = I_x + iI_y$$

$$I^- = I_x - iI_y$$

$$I_x = \frac{1}{2}(I^+ + I^-)$$

$$I_y = \frac{1}{2i}(I^+ - I^-)$$

$$2I_{1x}I_{2z} = I_{2z}(I^+ + I^-)$$

$$I^+ \xrightarrow{90^\circ\, I_x} \frac{1}{2}(I^+ + I^-) + iI_z$$

$$I^- \xrightarrow{90^\circ\, I_x} \frac{1}{2}(I^+ + I^-) - iI_z$$

$$I^+ \xrightarrow{\Omega I_z t} I^+ \mathrm{e}^{-i\Omega t}$$

$$I^- \xrightarrow{\Omega I_z t} I^- \mathrm{e}^{+i\Omega t}$$

Literature

[1] O. W. Sørensen, G. W. Eich, M. H. Levitt, G. Bodenhausen, R. R. Ernst, *Prog. NMR Spectrosc.* **1983**, *16*, 163–192.
[2] H. Kessler, M. Gehrke, C. Griesinger, *Angew. Chem. Int. Ed. Engl.* **1988**, *27*, 490–536.
[3] J. Cavanagh, W. J. Fairbrother, A. G. Palmer III, N. J. Skelton, *Protein NMR Spectroscopy*, Academic Press, San Diego, **1996**, Chapter 2.

Glossary and Index

(*italic numbers* refer to the most relevant experiments)